**An Introduction to
Molecular Biotechnology**

*Edited by
Michael Wink*

Related Titles

Klipp, E., Liebermeister, W., Wierling, C., Kowald, A., Lehrach, H., Herwig, R.

Systems Biology
A Textbook

2009
ISBN: 978-3-527-31874-2

Ninfa, A. J., Ballou, D. P., Benore, M.

Fundamental Laboratory Approaches for Biochemistry and Biotechnology

2009
ISBN: 978-0-470-08766-4

Borbye, L., Stocum, M., Woodall, A., Pearce, C., Sale, E.,
Barrett, W., Clontz, L., Peterson, A., Shaeffer, J.

Industry Immersion Learning
Real-Life Industry Case-Studies in Biotechnology and Business

2009
ISBN: 978-3-527-32408-8

Gruber, A. C.

Biotech Funding Trends
Insights from Entrepreneurs and Investors

2009
ISBN: 978-3-527-32435-4

Behme, S.

Manufacturing of Pharmaceutical Proteins
From Technology to Economy

2009
ISBN: 978-3-527-32444-6

Helms, V.

Principles of Computational Cell Biology
From Protein Complexes to Cellular Networks

2008
ISBN: 978-3-527-31555-0

An Introduction to Molecular Biotechnology

Fundamentals, Methods, and Applications

Edited by
Michael Wink

Second, Updated Edition

ⓦ WILEY-BLACKWELL

The Editor

Prof. Dr. Michael Wink
Institute of Pharmacy and
Molecular Biotechnology
University of Heidelberg
Im Neuenheimer Feld 364
69120 Heidelberg
Germany

Cover
Pictures courtesy of Michael Knop, EMBL,
Heidelberg (gel chromatography, pipet),
National Human Genome Research Institute,
Bethesda, USA (DNA), Fotolia/Franz Pfluegl
(cereals), PhotoDisc/Getty Images (pills),
Fotolia/SyB (stock exchange charts),
Fotolia/Aintschie (law code)

Library of Congress Card No.:
applied for

British Library Cataloguing-in-Publication Data
A catalogue record for this book is available from
the British Library.

**Bibliographic information published by
the Deutsche Nationalbibliothek**
The Deutsche Nationalbibliothek lists this
publication in the Deutsche Nationalbibliografie;
detailed bibliographic data are available on the
Internet at http://dnb.d-nb.de.

© 2011 WILEY-VCH Verlag GmbH & Co. KGaA,
Boschstraße 12, 69469 Weinheim

Wiley-Blackwell is an imprint of John Wiley
& Sons, formed by the merger of Wiley's global
Scientific, Technical, and Medical business with
Blackwell Publishing.

Composition   K+V Fotosatz GmbH, Beerfelden
Printing and Binding   betz-druck GmbH,
Darmstadt
Cover Design   Formgeber, Eppelheim

Printed in the Federal Republic of Germany

Printed on acid-free paper

ISBN   978-3-527-32637-2

Contents

An Introduction to Molecular Biotechnology, 2nd Edition.
Edited by Michael Wink
Copyright © 2011 WILEY-VCH Verlag GmbH & Co. KGaA, Weinheim
ISBN: 978-3-527-32637-2

Preface

The term biotechnology was only coined in 1919 by the Hungarian engineer Karl Ereky. He used it as an umbrella term for methods by which microorganisms helped to produce valuable products. Humankind has been using biotechnological methods for thousands of years – think of the use of yeast or bacteria in the production of beer, wine, vinegar, or cheese.

Biotechnology is one of the key technologies of the twenty-first century. It includes established traditional industries such as the production of milk and dairy products, beer, wine, and other alcoholic drinks, as well as the production and biotransformation of enzymes, amino acids, vitamins, antibiotics, and other fine chemicals. This area, including the associated process engineering, is referred to as **white or industrial biotechnology**. As it is well established, it will only be treated in passing in Chapter 34. Many good books have been written to cover the field.

Breathtaking progress has been made in molecular and cell biology in the past 50 years, particularly in the last 20–30 years. This opens up new exciting perspectives for industrial applications. This area of applied biology is clearly distinguished from the traditional biotechnological fields and is known as **molecular biotechnology**. In a few years' time, however, it may well be regarded as another established branch of traditional biotechnology.

Molecular biology and cell biology have revolutionized our knowledge about the function and structure of macromolecules in the cell and the role of the cell itself. Major progress has been made in genomics and proteomics. A historic milestone was the sequencing of the human genome in 2001. At present, more than 1200 genomes of diverse organismal groups (including more than 100 genomes of eukaryotes) have been completely sequenced (http://www.ebi.ac.uk/genomes). As a next milestone it has been proposed to sequence 10 000 genomes from species covering the tree of life (http://www.genome10k.org). With the new generation of DNA sequencers it is now possible to sequence the human genome in a matter of weeks. This new knowledge has had direct repercussions on medical science and therapy, as it is now possible for the first time to study the genetic causes of diseases. It should thus be possible in due course to treat the causes rather than the symptoms. High-throughput sequencing will probably become a routine diagnostic, which will allow personalized medical treatment. Opportunities open up for the biotech industry (**red biotechnology**) to develop new diagnostics and therapeutics such as recombinant hormones, enzymes, antigens, vaccines, and antibodies that were not available before the genetic revolution. In the field of **green biotechnology**, targeted modification of crop cultivars can improve their properties, such as resistance to pests or the synthesis of new products (including recombinant human proteins). In **microbial biotechnology**, production processes can be improved and new products can be created through combinatorial biosynthesis.

The term **molecular biotechnology** also covers state-of-the-art research in genomics, functional genomics, proteomics, transcriptomics, systems biology, gene therapy, or molecular diagnostics. The concepts and methods are derived from cell and molecular biology, structural biology, bioinformatics, and biophysics.

The success of molecular biotechnology has been considerable, if you look at the scientific and economic prowess of companies like Genentech, Biogen, and

An Introduction to Molecular Biotechnology, 2nd Edition.
Edited by Michael Wink
Copyright © 2011 WILEY-VCH Verlag GmbH & Co. KGaA, Weinheim
ISBN: 978-3-527-32637-2

others. Already today total annual revenues from recombinant drugs exceeds US $ 20 billion. Over 100 recombinant proteins have been approved by the US Food and Drug Administration and several hundred others are in the developmental pipeline.

As textbooks covering this extensive subject are few, a group of experts and university teachers decided to write an introductory textbook that looks at a wide variety of aspects. This is the English language version of the second edition of *An Introduction to Molecular Biotechnology*, which has been thoroughly updated, a new chapter on systems biology has been added (Chapter 23), and many illustrations are now in color.

The comprehensive introductory chapters (*Part I*) provide a brief compendium of the essential building blocks and processes in a cell, their structure, and functions. This information is crucial for the understanding of the following chapters, and while it cannot be a substitute for the profound study of more substantial and extensive textbooks on cell and molecular biology (Alberts *et al.*, 2008; Campbell and Reece, 2006), it gives a quick overview and recapitulation.

Part II contains short chapters discussing the most important methods used in biotechnology. Again, for a more thorough approach to the subject, consult the relevant textbooks.

Part III explores the different fields of molecular biotechnology, such as genome research, functional genomics, proteomics, transcriptomics, bioinformatics, systems biology, gene therapy, and molecular diagnostics. It not only gives a summary of current knowledge, but also highlights future applications and developments.

Part IV discusses the industrial environment of molecular biotechnology, including the business environment and difficulties young biotech firms have to cope with and their chances of success.

To give a snapshot of state-of-the-art research in an area where things move faster than anywhere else is next to impossible. Thus, it is inevitable that by the time this book goes into print, some developments will have superseded those described here. Although we have tried to include most relevant issues, the choice of topics must naturally limited in a such a textbook.

Forty-two coauthors worked on this project, and although we tried to find a more or less uniform style, the authors with their different views and values are still recognizable.

The publisher and editors would like to thank all authors for their constructive cooperation. Special thanks go to the team at Wiley-VCH (Dr. A. Sendtko, M. Petersen, H.-J. Schmitt) who gave their enthusiastic support to this project.

Heidelberg, Winter 2011   *Michael Wink*

List of Contributors

Michael Breuer
BASF SE
Fine Chemicals & Biocatalysis
Research
GVF/B – A030
67056 Ludwigshafen
Germany

Benedikt Brors
German Cancer Research Center
Computational Oncology
Im Neuenheimer Feld 580
69120 Heidelberg
Germany

Ulrich Deuschle
Phenex Pharmaceuticals AG
Waldhofer Str. 104
69123 Heidelberg
Germany

Stephan Diekmann
Leibniz Institute for Age Research
Fritz Lipmann Institute
Beutenbergstraße 11
07745 Jena
Germany

Stefan Dübel
Institute for Biochemistry
and Biotechnology
Technical University of Braunschweig
Spielmannstr. 7
38106 Braunschweig
Germany

Rainer Fink
Institute for Physiology
and Pathophysiology
University of Heidelberg
Im Neuenheimer Feld 326
69120 Heidelberg
Germany

Gert Fricker
Institute of Pharmacy
and Molecular Biotechnology
University of Heidelberg
Im Neuenheimer Feld 366
69120 Heidelberg
Germany

Marcus Frohme
Molecular Biology
and Functional Genomics
Technical University of Applied
Sciences
Bahnhofstraße
15745 Wildau
Germany

Reinhard Gessner
Visceral, Transplantation, Thorax
and Vascular Surgery
University Hospital Leipzig
Liebigstr. 20
04103 Leipzig
Germany

Ariane Groth
General, Visceral and Transplantation
Surgery
Molecular OncoSurgery
University Hospital Heidelberg
Im Neuenheimer Feld 365
69120 Heidelberg
Germany

Bernhard Hauer
Institute of Technical Biochemistry
University of Stuttgart
Allmandring 31
70569 Stuttgart
Germany

An Introduction to Molecular Biotechnology, 2nd Edition.
Edited by Michael Wink
Copyright © 2011 WILEY-VCH Verlag GmbH & Co. KGaA, Weinheim
ISBN: 978-3-527-32637-2

Rüdiger Hell
Heidelberg Institute of Plant Sciences
University of Heidelberg
Im Neuenheimer Feld 360
69120 Heidelberg
Germany

Ingrid Herr
General, Visceral and Transplantation
Surgery
Molecular OncoSurgery
University Hospital Heidelberg
Im Neuenheimer Feld 365
69120 Heidelberg
Germany

Helke Hillebrand
European Molecular Biology
Laboratory (EMBL)
Meyerhofstr. 1
69117 Heidelberg
Germany

Ana Kitanovic
Institute for Pharmacy & Molecular
Biotechnology
University of Heidelberg
Im Neuenheimer Feld 364
69120 Heidelberg
Germany

Manfred Koegl
Boehringer Ingelheim Vienna
Oncology Research
Dr. Boehringer Gasse 5–11
1121 Vienna
Austria

Rainer König
Institute of Pharmacy and
Molecular Biotechnology
University of Heidelberg Bioquant
Im Neuenheimer Feld 267
69120 Heidelberg
Germany

Robert Kraft
Carl Ludwig Institute of Physiology
University of Leipzig
Liebigstr. 27
04113 Leipzig
Germany

Claus Kremoser
PheneX Pharmaceuticals AG
Im Neuenheimer Feld 515
69120 Heidelberg
Germany

Stefan Legewie
Institute of Molecular Biology
Ackermannweg 4
55128 Mainz
Germany

Wolf-Dieter Lehmann
German Cancer Research Center
Molecular Structure Analysis
Mass Spectroscopy
Im Neuenheimer Feld 280
69120 Heidelberg
Germany

Susanne Lutz
Institute of Experimental and Clinical
Pharmacology and Toxicology
University of Heidelberg
Maybachstraße 14
68169 Mannheim
Germany

Nils Metzler-Nolte
Chair of Inorganic Chemistry I
Bioinorganic Chemistry
Ruhr-University of Bochum
Universitätsstr. 150
44801 Bochum
Germany

Andrea Mohr
National Center for Biomedical
Engineering Science
National University of Ireland
University Road
Galway
Ireland

Ehmke Pohl
Department of Chemistry & School
of Biological and
Biomedical Sciences
Durham University
Durham, DH1 3LE
Great Britain

David B. Resnik
National Institute of Environmental
Health Science
National Institutes of Health
111 T.W. Alexander Drive
Research Triangle Park, NC 27709
USA

Andreas Schlosser
Center for Biological Systems
Analysis (ZBSA)
University of Freiburg
Habsburgerstr. 49
79104 Freiburg
Germany

Hannah Schmidt-Glenewinkel
German Cancer Research Center
Theoretical Systems Biology
Im Neuenheimer Feld 280
69120 Heidelberg
Germany

Julia Schüler
BioMedServices
Hecker-Str. 20
68199 Mannheim
Germany

Anna Sosniak
Chair of Inorganic Chemistry I
Bioinorganic Chemistry
University of Bochum
Universitätsstr. 150
44801 Bochum
Germany

Rolf Sprengel
Max Planck Institute
for Medical Research
Jahnstrasse 29
69120 Heidelberg
Germany

Ralf Tolle
Center for Molecular Biology (ZMBH)
University of Heidelberg
Im Neuenheimer Feld 282
69120 Heidelberg
Germany

Peter Uetz
Delaware Biotechnology Institute
University of Delaware
15 Innovation Way
Newark, DE 19711-5449
USA

Martin Vogel
Max Planck Institute of Biophysics
Max-von-Laue-Str. 3
60438 Frankfurt
Germany

Gary Walsh
Department of Chemical &
Environmental Sciences
Plassey Park
University of Limerick
Limerick
Ireland

Hans Weiher
Bonn-Rhein-Sieg University
of Applied Sciences
Department of Natural Sciences
Von-Liebig-Str. 20
53359 Rheinbach
Germany

Thomas Wieland
Institute of Experimental and Clinical
Pharmacology and Toxicology
University of Heidelberg
Maybachstraße 14
68169 Mannheim
Germany

Stefan Wiemann
German Cancer Research Center
Molecular Genome Analysis
Im Neuenheimer Feld 580
69120 Heidelberg
Germany

Michael Wink
Institute of Pharmacy and Molecular
Biotechnology
University of Heidelberg
Im Neuenheimer Feld 364
69120 Heidelberg
Germany

Stefan Wölfl
Institute for Pharmacy & Molecular
Biotechnology
University of Heidelberg
Im Neuenheimer Feld 364
69120 Heidelberg
Germany

Ralf Zwacka
National Center for Biomedical
Engineering Science
National University of Ireland
University Road
Galway
Ireland

Abbreviations

1 Å	=0.1 nm
aa-tRNA	aminoacyl-tRNA
AAV	adeno-associated virus
ABC	ATP binding cassette
Acetyl-CoA	acetyl coenzyme A
AcNPV	*Autographa californica* nuclear polyhedrosis virus
ACRS	amplification-created restriction sites
ACTH	adrenocorticotropic hormone
ADA	adenosine deaminase
ADEPT	antibody-directed enzyme pro-drug therapy
ADME-T	absorption, distribution, metabolism, excretion and toxicity
ADP	adenosine diphosphate
ADRs	adverse drug reactions
AEC	aminoethylcysteine
AFLP	amplified fragment length polymorphism
AFM	atomic force microscope
AIDS	acquired immune deficiency syndrome
ALS	amyotrophic lateral sclerosis
AMP	adenosine monophosphate
AMPA	α-amino-3-hydroxyl-5-methyl-4-isoxazol-propionate
Amp^r	ampicillin resistance gene
AMV	avian myeloblastosis virus
ANN	artificial neural network
AO	acridine orange
AOX1	alcohol oxidase 1
APC	anaphase promoting complex
ApoB100	apolipoprotein B100
ApoE	apolipoprotein E
APP	amyloid precursor protein
ARMS	amplification refractory mutation system
ARS	autonomously replicating sequence
ATP	adenosine triphosphate
att	attachment site
BAC	bacterial artificial chromosome
bcl2	B-cell leukemia lymphoma 2 (protein protecting against apoptosis)
BfArM	German Bundesinstitut für Arzneimittel und Medizinprodukte
β-Gal	β-galactosidase
BHK-21	baby hamster kidney cells
BLA	biologics licence application
BLAST	basic local alignment search tool
BMP	bone morphogenetic proteins
bp	base pairs
BrdU	bromodeoxyuridine

An Introduction to Molecular Biotechnology, 2nd Edition.
Edited by Michael Wink
Copyright © 2011 WILEY-VCH Verlag GmbH & Co. KGaA, Weinheim
ISBN: 978-3-527-32637-2

CA	correspondence analysis
CAD	coronary artery disease
CaM-Kinase	Ca^{2+}/calmodulin-dependent protein kinase
cAMP	cyclic AMP
cap	AAV gene mediating encapsulation
CARS	coherent anti-Raman scattering
CAT	Committee for Advanced Therapies
CBER	Center for Biologics Evaluation and Research
CC	chromatin remodelling complex
CCD	charge-coupled device
CDER	Center for Drug Evaluation and Research
CDK	cyclin-dependent kinase
cDNA	copy DNA
CDR	complementary determining region
CDRH	Center for Devices and Radiological Health
CEO	chief executive officer
CFP	cyan fluorescent protein
CFTR	cystic fibrosis transmembrane regulator
CGAP	cancer genome anatomy project
CGH	comparative genome hybridization
CHMP	Committee for Medicinal Products for Human Use
CHO	Chinese hamster ovary
CIP	calf intestinal phosphatase
CML	chronic myeloic leukemia
CMN	*Corynebacterium-Mycobacterium-Nocardia* group
CMV	cauliflower mosaic virus
CMV	Cytomegalovirus
CNS	central nervous system
COMP	Committee on Orphan Medicinal Products
COS-1	simian cell line, CV-1, transformed by origin-defective mutant of SV40
cpDNA	chloroplast DNA
CPMV	cowpea mosaic virus
cPPT-sequence	central polypurine tract – regulatory element in lentiviral vectors that facilitates double strand synthesis and the nuclear import of the pre-integration complex
CSF	colony-stimulating factor
CSO	contract service organisation
CTAB	cetyltrimethylammonium bromide
CVM	Center for Veterinary Medicine
CVMP	Committee for Medicinal Products for Veterinary Use
2D	two-dimensional
Da	Dalton
DAG	diacylglycerol
DAPI	4,6-diamidino-2-phenylindole
dATP	deoxyadenosine triphosphate
DBD	DNA-binding domain
DAC	divide-and-conquer strategy
DD	differential display
DDBJ	DNA Data Bank of Japan
ddNTP	dideoxynucleotide triphosphate
DEAE	diethylaminoethyl
dHPLC	denaturing HPLC
DIC	differential interference contrast
DIP	Database of Interacting Proteins
DNA	deoxyribonucleic acid
DNAse	deoxyribonuclease

dNTP	deoxynucleoside triphosphate
Dox	doxycycline
ds diabodies	disulfide-stabilized diabodies
dsDNA	double-stranded DNA
dsFv-fragment	disulfide-stabilized Fv fragment
dsRNA	double-stranded RNA
DtxR	diphtheria toxin repressor
Ebola-Z	envelope protein of the Ebola-Zaire virus, which has a high affinity to lung epithelial cells
EC_{50}	effective concentration, the dose or concentration that produces a 50% effect in the test population within a specified time
ECD	electron capture dissociation
EDTA	ethylenediaminetetraacetic acid
ee	enantiomeric excess
EF2	elongation factor 2
EF-Tu	elongation factor Tu
EGF	epidermal growth factor
EGFP	enhanced green fluorescent protein
EGTA	ethyleneglycol-bis-(2-aminoethyl)-tetraacetic acid
EIAV	equine infectious anaemia virus
ELISA	enzyme-linked immunosorbent assay
EM	electron microscope
EMA	European Medicines Agency
EMBL	European Molecular Biology Laboratory
EMCV	Encephalomyocarditis virus
EMSA	electrophoretic mobility shift assay
EMEA	European Agency for the Evaluation of Medicinal Products
ENU	*N*-ethyl-*N*-nitrosourea
env	retroviral gene coding for viral envelope proteins
EPO	European Patent Office
EPR effect	enhanced permeability and retention effect
EPC	European Patent Convention
ER	endoplasmic reticulum
ESI	electrospray ionization
EST	expressed sequence tags
ES cells	embryonic stem cells
EtBr	ethidium bromide
Fab-fragment	antigen binding fragment
FACS	fluorescence-activated cell sorter
FAD	flavin adenine dinucleotide
FBA	flux balance analysis
FCS	fluorescence correlation spectroscopy
FDA	Food and Drug Administration
FFL	feed-forward loop
FGF	fibroblast growth factor
FISH	fluorescence in situ hybridization
FIV	feline immunodeficiency virus
FKBP	FK506-binding protein
FLIM	fluorescence lifetime imaging microscopy
FLIPR	fluorescent imaging plate reader
FMN	flavin mononucleotide
FPLC	fast performance liquid chromatography
FRAP	fluorescence recovery after photobleaching
FRET	fluorescence resonance energy transfer
FT-ICR	Fourier transformation cyclotron resonance, method in mass spectroscopy

FtsZ	prokaryotic cell division protein
Fur	ferric uptake regulator
Fv-fragment	variable fragment
FWHM	full width at half maximum
GABA	gamma aminobutyric acid
Gag	retroviral gene coding for structural proteins
Gal	galactose
GAP	GTPase-activating protein
GAPDH	glyceraldehyde 3-phosphate-dehydrogenase
Gb	Gigabases
GCC	German cDNA consortium
GCG	genetics computer group
GCP	good clinical practice
ΔG_d	free enthalpy
GDH	glutamate dehydrogenase
GDP	guanosine diphosphate
GEF	guanine exchange factor
GEO	gene expression omnibus
GFP	green fluorescence protein
GM-CSF	granulocyte/macrophage colony-stimulating factor
GO	gene ontology
GOI	gene of interest
GPCR	G-protein-coupled receptor
GPI anchor	glycosylphosphatidylinositol anchor
GRAS	generally regarded as safe
GST	glutathione-S-transferase
GTC	guanidinium isothiocyanate
GTP	guanosine triphosphate
GUS	glucuronidase
GMO	genetically modified organism
HA	hemagglutinin
HCM	hypertrophic cardiomyopathy
HCV	Hepatitis C virus
HEK	human embryonic kidney
HeLa cells	human cancer cell line (isolated from donor Helene Larsen)
HER 2	human epidermal growth factor 2
HGH	human growth hormone
HIC	hydrophobic interaction chromatography
His_6	hexahistidine tag
HIV	human immunodeficiency virus, a retrovirus
HIV 1	human immunodeficiency virus 1
HLA	human leukocyte antigen
hnRNA	heterogeneous nuclear RNA
HPLC	high performance liquid chromatography
HPT	hygromycin phosphotransferase
HPV	human papilloma virus
HSP	high-scoring segment pairs
HSP	heat shock protein
HSV-1	Herpes simplex virus
HTS	high-throughput analysis
HUGO	Human Genome Organisation
HV	Herpes virus
IAS	international accounting standard
ICDH	isocitric dehydrogenase

ICH	International Conference on Harmonization of Technical Requirements for the Registration of Pharmaceuticals for Human Use
ICL	isocitric lyase
ICP-MS	inductively coupled-plasma mass spectrometry
ICR-MS	ion cyclotron resonance mass spectrometer
IDA	iminodiacetic acid
IEF	isoelectric focusing
Ig	immunoglobulin
IHF	integration host factor
IMAC	immobilized metal affinity chromatography
IND-Status	investigational new drug status
IP_3	inositol-1,4,5-triphosphate
IPO	initial public offering
IPTG	isopropyl-b-D-thiogalactoside
IR	inverted repeats
IR	investor relations
IRES	internal ribosome entry site
ISAAA	International Service for the Acquisition of Agri-Biotech Applications
ISH	in situ hybridization
ISSR	inter simple sequence repeats
ITC	isothermal titration calorimetry
ITR	inverse terminal repeats – regulatory elements in adenoviruses and AAV
i.v.	intravenous
k_a	second order velocity constant in bimolecular association
Kan^r	kanamycin resistance gene
K_{av}	specific distribution coefficient
kb	Kilobases
k_d	first order velocity constant in unimolecular dissociation
$K_d = k_d/k_a$	velocity constant in dissociation/K_a in association
kDa	Kilodalton
KDEL	amino acid sequence for proteins remaining in the ER
KDR receptor	kinase insert domain containing receptor
KEGG	Kyoto Encyclopedia of Genes and Genomes
Lac	lactose
LASER	Light Amplification by Stimulated Emission of Radiation
LB	left border
LB	Luria-Bertani medium
LCR	ligation chain reaction
LDL	low-density-lipoprotein
LIMS	laboratory information management systems
LINE	long interspersed elements
LSC	Laser scanning-cytometer
LTQ	linear trap quadrupole
LTQ-FT-ICR	linear trap quadrupole-Fourrier transformation-ion cyclotron resonance
LTR	long terminal repeats; regulatory elements in retroviruses
LUMIER	LUMInescence-based mammalian intERactome
MAC	mammalian artificial chromosome
mAChR	muscarinic acetylcholine receptor
MAGE-ML	microarray gene expression markup language
MALDI	matrix-assisted laser desorption/ionization
6-MAM	6-monoacetylmorphine
MAP	microtubule-associated protein
MAP	mitosis-activating protein

Mb	Megabases
MBP	maltose-binding protein
MCS	multiple cloning site
M-CSF	macrophage colony-stimulating factor
MDR protein	multiple drug resistance protein
MDS	multidimensional scaling
MGC	mammalian gene collection
MHC	major histocompatibility complex
MIAME	minimum information about a microarray experiment
miRNA	microRNA
MIT	Massachusetts Institute of Technology
MoMLV	moloney murine leukemia virus
Mowse	molecular weight search
MPF	M-phase promotion factor
MPSS	massively parallel signature screening
Mreb/Mbl	proteins of prokaryotic cytoskeleton
mRNA	messenger RNA
MRSA	methicillin-resistant *S. aureus*
MS	mass spectrometry
MSG	monosodium glutamate
MS-PCR	mutationally separated PCR
MTA	material transfer agreement
mtDNA	mitochondrial DNA
MULVR	Moloney Murine Leukemia Virus
MW	molecular weight
μF	μFarad
nAChR	nicotinic acetylcholine receptor
NAD	nicotinamide adenine dinucleotide
NAPPA	nucleic acid programmable protein array
NCBI	National Center for Biotechnology Information
NDA	new drug application
NDP	nucleoside diphosphate
NDPK	nucleoside diphosphates kinase
NFjB	nuclear factor jB
NIH	National Institutes of Health
NK cell	natural killer cell
NMDA-receptor	*N*-methyl-D-aspartate-receptor
NMR	nuclear magnetic resonance
NPTII	neomycin phosphotransferase II
NSAID	non-steroidal anti-inflammatory drug
NTA	nitrilotriacetic acid
NTP	nucleoside triphosphate
OD	optical density
ODE	ordinary differential equation
ODHC	2-oxoglutarate dehydrogenase
OMIM	online Mendelian inheritance in man
ORF	open reading frame
ori	origin of replication
OXA complex	membrane translocator in mitochondria
PAC	P1-derived artificial chromosome
PAGE	polyacrylamide-gel electrophoresis
PAZ-domain	*PIWI Argonaute Zwille domain*
PCA	principal component analysis
PCR	polymerase chain reaction
PDB	protein data bank
PEG	polyethylene glycol
PFAM	protein families database of alignments and HMMs

PFG	pulsed-field gel electrophoresis
PI	propidium iodide
PIR	protein information resource
piRNA	piwi-interacting RNA
PKA	protein kinase A
PKC	protein kinase C
PK data	pharmacokinetic data
Plos	Public Library of Science
PMSF	phenylmethylsulfonyl fluoride
PNA	peptide nucleic acid
PNGaseF	peptide N-glycosidase F
PNK	T4-polynucleotide kinase
pol	retroviral gene coding for reverse transcriptase and integrase
P_{PH}	polyhedrin promoter
PR	Public Relations
psi	retroviral packaging signal
PTGS	posttranscriptional gene silencing
PTI	pancreatic trypsin inhibitor
Q-FT-ICR	Q-Fourier transform ion cyclotron resonance
Q-TOF	Quadrupole-Time-of-Flight
RACE	rapid amplification of cDNA ends
Ran	protein involved in nuclear import
RAPD	random amplification of polymorphic DNA
RAP-PCR	RNA arbitrary primed PCR
RB	right border
RBD	RNA-binding domain
Rb-gene	retinoblastoma gene
RBS	ribosome binding site
RDA	representative difference analysis
RdRp	RNA-dependent RNA polymerase
rep	AAV gene, mediating replication
RES	reticuloendothelial system
RFLP	restriction fragment length polymorphism
Rf-value	retention factor
RGS	regulator of G-protein signaling
RISC	RNA-induced silencing complex
RNA	ribonucleic acid
RNAi	RNA interference
RNP	ribonucleoprotein
rpm	revolutions per minute
RRE	regulatory element in a lentiviral vector, enhancing the nuclear export of viral RNA
rRNA	ribosomal RNA
RSV	respiratory syncytial virus
RSV	promoter of the Rous sarcoma virus
RT	reverse transcriptase
rtTA	tetracyclin-sensitive regulatory unit
SAGE	Serial Analysis of Gene Expression
SALM	spectrally assigned localization microscopy
SAM	S-adenosylmethionine
sc diabodies	single-chain diabodies
scFab	single-chain Fab-fragment
scFv/sFv fragment	single-chain Fv fragment
SCID	severe combined immunodeficiency
SCOP	structural classification of proteins
SDS	sodium dodecyl sulfate

SDS-PAGE	sodium dodecyl sulfate polyacrylamide gel electophoresis
SELEX	systematic evolution of ligand by exponential enrichment
SEM	scanning electron microscope
Sf cells	*Spodoptera frugiperda* cells
SFM	scanning force microscope
SFV	Semliki-Forest virus
SH1	Src-homology domain 1 = kinase domain
SH2	Src-homology domain 2
SH3	Src-homology domain 3
SHG	second harmonic generation
SIM	single input
SIN	self-inactivating lentiviral vectors, due to a 3′ LTR mutation
SINE	scattered or short interspersed elements
siRNA	small interfering RNA
SIV	simian immunodeficiency virus
SNARE proteins	SNAP-receptor proteins
SNP	single nucleotide polymorphism
snRNA	small nuclear RNA
snRNP	small nuclear ribonucleoprotein
SOP	stock option program
SP function	sum-of-pairs function
SPA	scintillation proximity assay
SPDM	spectral precision distance microscopy
SPF	S-phase promotion factor
SRP	signal recognition particle
SSB	single strand binding proteins
SSCP	single-strand comformation polymorphism
ssDNA	single-stranded DNA
SSH	suppressive subtractive hybridization
SssI methylase	methylase from *Spiroplasma*
ssRNA	single-stranded RNA
STED	stimulated emission depletion
STEM	scanning transmission electron microscope
stRNA	small temporal RNA
STS	sequence-tagged site
SV40	Simian-virus-type 40
TBP	TATA-binding protein
T_C	cytotoxic T-cells
Tc	tetracycline
T-DNA	transfer DNA
TEM	transmission electron-microscope
TEV	Tobacco Etch Virus
T_H	T helper cell
THG	third harmonic generation
TIGR	The Institute for Genome Research
TIM	translocase of inner membrane
T_m	melting temperature of dsDNA
TNF	tumor necrosis factor
TOF	time of flight
TOM	translocase of outer membrane
t-PA	tissue plasminogen activator
TRE	tetracycline-responsive element
TRIPs	Trade-Related Aspects of Intellectual Property Rights
tRNA	transfer RNA
Trp	tryptophan
t-SNARE	protein in target membrane to which vSNARE binds
TSS	transformation and storage solution

tTA	tetracycline-controlled transactivator
TY	transposon from yeast
UPOV	Union for the Protection of New Varieties of Plants
US-GAAP	US generally accepted accounting principle
UV	ultraviolet
V_0	empty volume
VC	venture capital
V_e	elution volume
VEGF	vascular endothelial growth factor
VIP	vasoactive peptide
VNTR	variable number tandem repeats
v-SNARE	protein in vesicular membrane, binding to t-SNARE
VSV-G	envelope protein of vesicular stomatitis virus, great affinity to a wide range of cells
V_t	total volume
wNAPPA	modified nucleic acid programmable protein array
WPRE	woodchuck hepatitis virus posttranscriptional regulatory element
X-Gal	5-bromo-4-chloro-3-indolyl-b-D-galactopyranoside
YAC	yeast artificial chromosome
YEp	yeast episomal plasmid
YFP	yellow fluorescence protein
YIp	yeast-integrating plasmid
YRp	yeast-replicating plasmid
Yth	yeast two-hybrid

Part I
Fundamentals of Cellular and Molecular Biology

1

The Cell as the Basic Unit of Life

Learning Objectives

This chapter offers a short introduction into the structure of prokaryotic and eukaryotic cells, as well as that of viruses.

The base unit of life is the **cell**. Cells constitute the base element of all **prokaryotic cells** (cells without a cell nucleus, e.g., **Bacteria** and **Archaea**) and **eukaryotic cells** (or **Eukarya**) (cells possessing a nucleus, e.g., protozoa, fungi, plants, and animals). Cells are small, membrane-bound units with a diameter of 1–20 μm and are filled with concentrated aqueous solutions. Cells are not created *de novo*, but possess the ability to copy themselves, meaning that they emerge from the division of a previous cell. This means that all cells, since the beginning of life (around 4 billion years ago), are connected with each other in a continuous lineage. In 1885, the famous cell biologist Virchow conceived the law of *omnis cellula e cellulae* (all cells arise from cells), which is still valid today.

The structure and composition of all cells are very similar due to their shared evolution and phylogeny (Fig. 1.1). Owing to this, it is possible to limit the discussion of the general characteristics of a cell to a few basic types (Fig. 1.2):

- Bacterial cells.
- Plant cells.
- Animal cells.

Fig. 1.1 Tree of life – phylogeny of life domains. Nucleotide sequences from 16S rRNA, amino acid sequences of cytoskeleton proteins, and characteristics of the cell structure were used to reconstruct this phylogenetic tree. Prokaryotes are divided into **Bacteria** and **Archaea**. Archaea form a sister group with eukaryotes; they share important characteristics (Tables 1.1 and 1.2). Many monophyletic groups can be recognized within the eukaryotes (diplomonads/trichomonads, Euglenozoa, Alveolata, Stramenopilata (heterokonts), red algae and green algae/plants, fungi and animals; see Tables 6.3–6.5 for details).

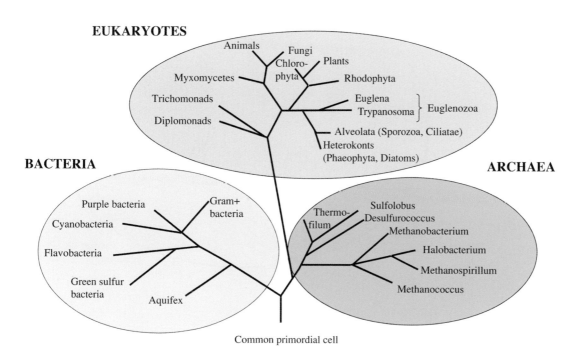

An Introduction to Molecular Biotechnology, 2nd Edition.
Edited by Michael Wink
Copyright © 2011 WILEY-VCH Verlag GmbH & Co. KGaA, Weinheim
ISBN: 978-3-527-32637-2

Fig. 1.2 Schematic structure of prokaryotic and eukaryotic cells. (A) Bacterial cell. (B) Plant mesophyll cell. (C) Animal cell.

(A)

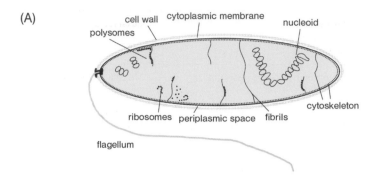

(B)

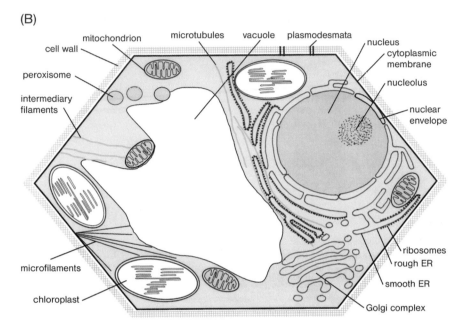

(C)

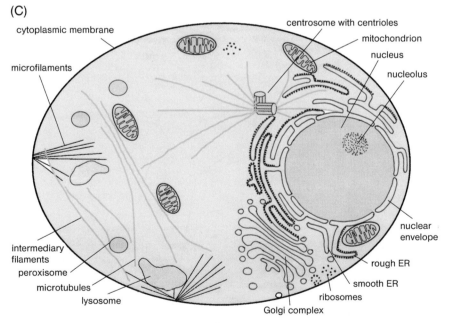

Character	Prokaryotes		Eukaryotes
	Archaea	Bacteria	
Organization	Unicellular	Unicellular	Unicellular or multicellular
Cytology			
Internal membranes	Rare	Rare	Always (Table 1.2)
Compartments	Only cytoplasm	Only cytoplasm	Several (Table 1.2)
Organelles	No	No	mitochondria; plastids
Ribosomes	70S	70S	80S (mt, cp: 70S)
Membrane lipids	Ether lipids	Ester lipids, hopanoids	Ester lipids, sterols
Cell wall	Pseudopeptidoglycan, polysaccharides, glycoproteins	Murein (peptidoglycan), polysaccharides, proteins	PL: polysaccharides, cellulose F: chitin A: no
Cytoskeleton	FtsZ and MreB protein	FtsZ and MreB protein	Tubulin, actin, intermediary filaments
Cell division	Binary fission	Binary fission	Mitosis
Genetics			
Nuclear structure	Nucleoid	Nucleoid	Membrane-enclosed nucleus
Recombination	Similar to conjugation	Conjugation	Meiosis, syngamy
Chromosome	Circular, single	Circular, single	Linear, several
Introns	Rare	Rare	Frequent
Noncoding DNA	Rare	Rare	Frequent
Operon	Yes	Yes	No
Extrachromosomal	DNA plasmids (linear)	Plasmids (circular)	mtDNA, cpDNA, plasmids in fungi
Transcription/ translation	Concomitantly	Concomitantly	Transcription in nucleus; translation in cytoplasm
Promotor structure	TATA box	−35 and −10 sequences	TATA box
RNA polymerases	Several (8–12 subunits)	1 (4 subunits)	3 (with 12–14 subunits)
Transcription factors	Yes	No (sigma factor)	Yes
Initiator tRNA	Methionyl-tRNA	*N*-formylmethionyl-tRNA	Methionyl-tRNA
Cap structure of mRNA polyadenylation	No	No	Yes

PL, plants; F, fungi; A, animals; mt, mitochondria; cp, plastid.

Table 1.1 Comparison of important biochemical and molecular characteristics of the three domains of life.

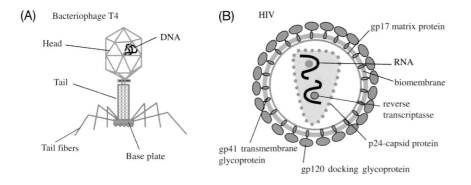

Fig. 1.3 Schematic structure of bacteriophages and viruses. (A) Bacteriophage T4. (B) Structure of a retrovirus (human immunodeficiency virus causing AIDS).

Table 1.2 Compartments of animal and plant cells and their main functions.

Compartment	Occurrence		Functions
	Animal	Plant	
Nucleus	A	P	Harbors chromosomes; site of replication, transcription, and assembly of ribosomal subunits
Endoplasmic reticulum			
rough ER	A	P	Posttranslational modification of proteins
smooth ER	A	P	Synthesis of lipids and lipophilic substances
Golgi apparatus	A	P	Posttranslational modification of proteins; modification of sugar chains
Lysosome	A		Harbors hydrolytic enzymes; degrades organelles and macromolecules, macrophages eat invading microbes
Vacuole		P	Sequestration of storage proteins, defense and signal molecules, contains hydrolytic enzymes, degrades organelles and macromolecules
Mitochondrium	A	P	Organelle derived from endosymbiotic bacteria; contains circular DNA, own ribosomes; enzymes of citric acid cycle, β-oxidation, and respiratory chain (ATP generation)
Chloroplast		P	Organelle derived from endosymbiotic bacteria; contains circular DNA, own ribosomes; chlorophyll and proteins of photosynthesis, enzymes of CO_2 fixation and glucose formation (Calvin cycle)
Peroxisome	A	P	Contains enzymes that generate and degrade H_2O_2
Cytoplasm	A	P	Harbors all compartments, organelles, and the cytoskeleton of a cell; many enzymatic pathways (e.g., glycolysis) occur in the cytoplasm

A, animal; P, plant.

The most important **biochemical and cell biological characters** of Archaea, Bacteria, and Eukarya are summarized in Table 1.1.

As **viruses** and **bacteriophages** (Fig. 1.3) do not have their own metabolism they therefore do not count as organisms in the true sense of the word. They share several macromolecules and structures with cells. Viruses and bacteriophages are dependent on the host cells for reproduction, and therefore their physiology and structure are closely linked to that of the host cell.

Eukaryotic cells are characterized by **compartments** that are enclosed by biomembranes (Table 1.2). As a result of these compartments, the multitude of metabolic reactions can run in a cell at the same time.

In the following discussion on the shared characteristics of all cells, the diverse differences that appear in **multicellular organisms** should not be forgotten. The human body has more than 200 different cell types, which show diverse structures and compositions. These differences must be understood in detail if cell-specific disorders, such as cancer, are to be understood and consequently treated.

Before a detailed discussion of cellular structures and their functions (see Chapters 3–5), a short summary of the biochemical basics of cellular and molecular biology is given in Chapter 2.

2

Structure and Function of Cellular Macromolecules

Learning Objectives
This chapter introduces the structure of polysaccharides, lipids, proteins, and nucleic acids, built from simple monomers (sugars, amino acids, and nucleotides), and illustrates how they are derived from simple monomers. Their most important functions are summarized.

In contrast to the diversity of life forms found in nature with several million species, the cells that make up all of these diverse organisms contain only a limited number of types of ions and molecules (Table 2.1). Among the most important **macromolecules** of prokaryotic and eukaryotic cells are **polysaccharides**, **lipids**, **proteins**, and **nucleic acids**, which are constructed from comparatively few **monomeric building blocks** (Table 2.2). The **membrane lipids** (phospholipids, cholesterol) will also be considered in this context because they spontaneously form supramolecular biomembrane structures in the aqueous environment of a cell.

Inorganic ions, sugars, amino acids, fatty acids, organic acids, nucleotides, and various metabolites are counted among the **low-molecular-weight components** and building blocks of the cell. The qualitative composition of cells is similar in prokaryotes and eukaryotes (Table 1.1), even though eukaryote cells generally have a higher protein content, and bacterial cells a higher RNA content. Animal cells have a volume that is 10^3 times larger than that of bacterial cells.

Owing to their shared evolution, the structure and function of the important cellular molecules is very similar in all organisms, often even identical. Apparently, reliable and functional biomolecules were developed and, if useful for the producer, were selected early in evolution (Table 2.2) and are therefore still used today.

Table 2.1 Molecular composition of cells.

Contents	Bacterium (% of cell mass)	Animal cell (% of cell mass)
Water	70	70
Inorganic ions	1	1
Small molecules (sugars, acids, amino acids)	3	3
Proteins	15	18
RNA	6	1.1
DNA	1	0.25
Phospholipids	2	3
Other lipids	—	2
Polysaccharides	2	2
Cell volume (ml)	2×10^{-12}	4×10^{-9}
Relative cell volume	1	2000

An Introduction to Molecular Biotechnology, 2nd Edition.
Edited by Michael Wink
Copyright © 2011 WILEY-VCH Verlag GmbH & Co. KGaA, Weinheim
ISBN: 978-3-527-32637-2

Table 2.2 Formation and function of the cellular macromolecules.

Basic building blocks	Macromolecule	Function
Simple sugar	Polysaccharide	**Structural substances:** composition of the cell walls (cellulose, chitin, peptidoglycan); constituents of connective tissues **Storage substances:** starch, glycogen
Amino acids	Proteins	**Enzymes:** important catalysts for anabolic and catabolic reaction processes **Hemoglobin:** O_2 and CO_2 transport **Receptors:** recognition of external and internal signals **Ion channels, ion pumps, transporters:** transport of charged molecules across biological membranes **Regulatory proteins:** signal transduction through protein–protein interactions **Transcription factors:** regulation of gene activity **Antibodies:** recognition of antigens **Structural proteins:** structural organization of supramolecular complexes **Cytoskeleton:** formation of molecular networks in the cell that are important for shape and function **Motor proteins:** muscle contraction
Phospholipids, cholesterol		Elements of biomembranes
Deoxynucleotide	DNA	Storage, replication, and safe transfer of genetic information; recombination
Nucleotide	RNA	**rRNA:** structural molecules for the construction of ribosomes **ribozymes and siRNA:** catalytic and regulatory processes **tRNA:** mediators in translation **mRNA:** messengers and mediators between genes and proteins **snRNA:** splicing of mRNA

2.1
Structure and Function of Sugars

Monosaccharides occur in cells either as **aldoses** or **ketoses** (Fig. 2.1 A). The most important monosaccharides have a chain length of three, five, and six carbon atoms, and are called **trioses**, **pentoses**, and **hexoses**. Under physiological conditions, pentoses and hexoses can form ring structures through hemiacetal and hemiketal formation (Fig. 2.1 B).

Many important nitrogen-containing derivatives of these monosaccharides (Fig. 2.1 C) use galactose and glucose as a base. Examples include **glucosamine**, **N-acetylglucosamine**, and **glucuronic acid**. These derivatives can be present either as glycosides or as part of a polysaccharide.

Condensation reactions between sugar molecules result in the formation of **glycosidic bonds** with the elimination of a water molecule. As hydroxyl groups can be present in either the α or β position, the stereochemistry of sugar molecules is of great importance. The condensation of two sugar molecules results in the formation of a **disaccharide** (Fig. 2.1 D); that of three sugar molecules, correspondingly, is a **trisaccharide**. **Oligosaccharides** are built from a few sugar monomers and **polysaccharides** (e.g., starch, glycogen, cellulose, chitin, etc.) are made up of many sugar monomers.

Sugar molecules can be easily activated through esterification with an acid; one important example being esterification with phosphoric acid. Sugar phosphates are important in glycolysis.

The most important polysaccharide in animal cells is **glycogen**, which is stored as an energy source in liver and muscle. Glycogen can be quickly transformed into glucose-1-phosphate and then channeled into glycolysis. Glycogen is a branched polysaccharide formed from glucose molecules linked by α-$(1\rightarrow4)$-glycosidic bonds or α-$(1\rightarrow6)$-glycosidic bonds (Fig. 2.1 D). This results in many free ends on which the enzyme glycogen phosphorylase can begin degradation simultaneously.

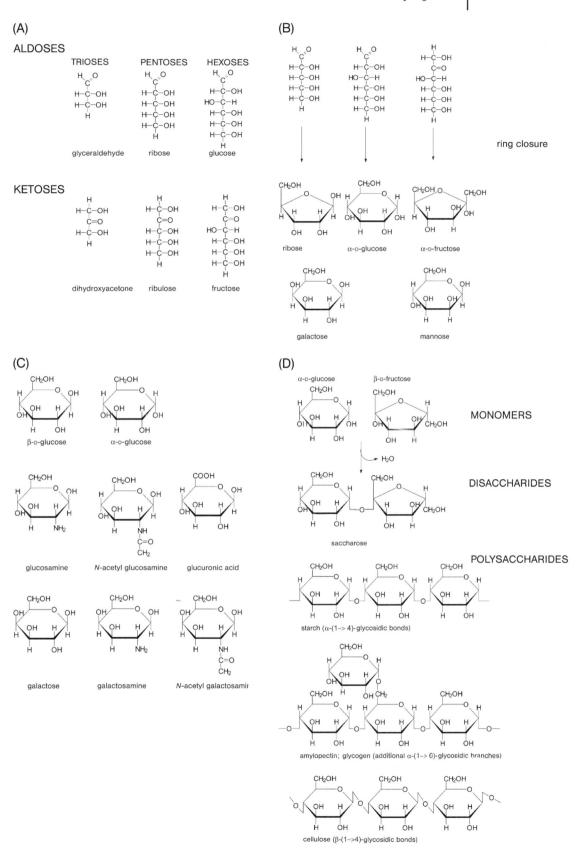

Fig. 2.1 Composition and structure of sugar molecules. (A) Structures of the most important aldoses and ketoses. (B) Ring structures of pentoses and hexoses (hemiacetal and hemiketal formations); important isomers of glucose. (C) Important derivatives of glucose and galactose. (D) Formation of disaccharides and polysaccharides (starch (amylose), amylopectin, glycogen, cellulose).

Starch or **amylose** (Fig. 2.1 D) consists of glucose residues linked by α-(1→4)-glycosidic bonds. In amylopectin, additional glucose residues linked by α-(1→6)-glycosidic bonds are built in. Amylopectin, therefore, has a similar structure to glycogen, but is less strongly branched. Starch is formed by **photosynthesis** in plant cells, where it is stored in amyloplasts. Starch can be broken down easily by animals and is therefore an important part of human nutrition.

Glucose is also used as a building block for **cellulose** (Fig. 2.1 D), which is necessary for formation of the plant cell wall. Cellulose is an unbranched polymer made from glucose molecules linked by β-(1→4)-glycosidic bonds. Cellulose cannot be broken down in the human digestive tract. Conversely, the rumen (first stomach) of ruminants (animals that chew the cud) contains microorganisms that produce **cellulase** – an enzyme that makes it is possible for cows, for example, to use cellulose as a nutrient. Additional polymers present in the plant cell wall include polysaccharides, so-called glycans made up of cellulose fibers linked together in a diagonal fashion, **pectin** (basic unit: **galacturonic** acid), and **lignin** (made from the coumaroyl, coniferoyl, and sinapoyl alcohols). Using cellulases, it is possible to digest the cell walls of plant cells. Cells without cell walls are called **protoplasts**. They are important in plant biotechnology because they are easily transformable by genetic engineering (see Chapter 32). In many plant species it is possible to regenerate intact plant cells from protoplasts. Cell walls of fungi and the exoskeletons of insects are composed of chitin, which has *N*-acetylglucosamine as a building block in β-(1→4)-glycosidic bonds.

Further important polysaccharides are found in animals. **Hyaluronic acid** is made up of many disaccharide building blocks, which themselves consist of glucuronic acid and *N*-acetylglucosamine. Hyaluronic acid has a very high viscosity, and is therefore found in synovial fluid in the joints and in the vitreous humor in the eye. Furthermore, polysaccharides made from disaccharides consisting of **sulfated glucuronic acid** and *N*-acetylglucosamine or *N*-acetylgalactosamine units, respectively, are found in the connective tissues. Examples include chondroitin-4-sulfate, chondroitin-6-sulfate, dermatan sulfate, and keratin sulfate. **Heparin**, involved in the control of blood coagulation, also falls into this structural group.

2.2
Structure of Membrane Lipids

Biological membranes consist of a **lipid bilayer** (Fig. 2.2). They are formed from **phospholipids**, **glycolipids**, and **sterols** (e.g., in animal membranes, cholesterol), which have **lipophilic** (fat loving, water repelling) and **hydrophilic** (water loving,

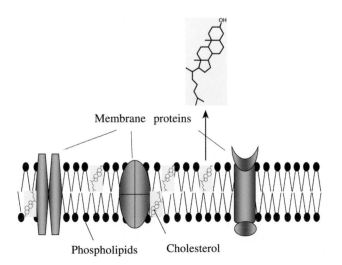

Fig. 2.2 Structure of the cytoplasmic membrane. Schematic diagram of the lipid bilayer containing phospholipids, cholesterol, and membrane proteins.

Membrane proteins

Phospholipids   Cholesterol

phosphatidylcholine phosphatidylethanolamine phosphatidylserine phosphatidylinositol sphingomyelin

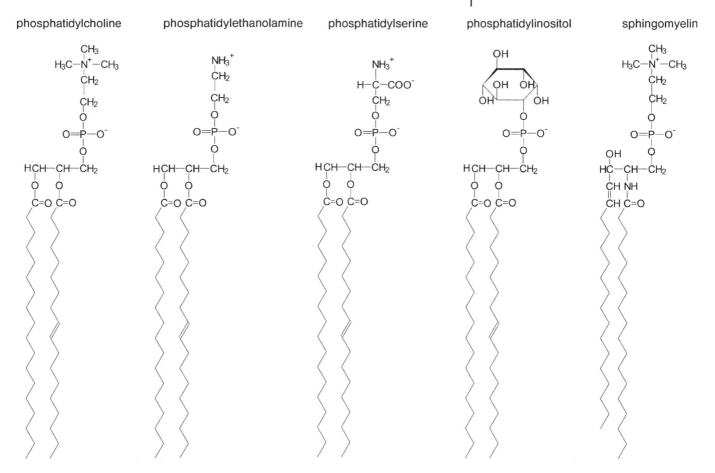

Fig. 2.3 Structures of important phospholipids. Phosphatidylcholine, phosphatidylethanolamine, phosphatidylserine, phosphatidylinositol, and sphingomyelin (a ceramide).

Table 2.3 Important fatty acids in membrane lipids.

Trivial name	Abbreviation	Melting temperature T_m (°C)	Structure
Saturated fatty acids			
Myristic acid	14:0	52.0	$CH_3(CH_2)_{12}COOH$
Palmitic acid	16:0	63.1	$CH_3(CH_2)_{14}COOH$
Stearic acid	18:0	69.1	$CH_3(CH_2)_{16}COOH$
Unsaturated fatty acids			
Palmitoleic acid	16:1	−0.5	$CH_3(CH_2)_5CH=CH(CH_2)_7COOH$
Oleic acid	18:1	13 .2	$CH_3(CH_2)_7CH=CH(CH_2)_7COOH$
Linoleic acid	18:2	−9.0	$CH_3(CH_2)_4(CH=CHCH_2)_2(CH_2)_6COOH$
γ-Linolenic acid	18:3	−17.0	$CH_3(CH_2)_4(CH=CHCH_2)_3(CH_2)_3COOH$
Arachidonic acid	20:4	49.5	$CH_3(CH_2)_4(CH=CHCH_2)_4(CH_2)_2COOH$

fat repelling) structural elements. Furthermore, biomembranes carry a diversity of membrane proteins (see Chapter 3). Biomembranes generate a diffusion barrier and enclose all cells, and in eukaryotes enclose all internal organelles (mitochondria, plastids) and compartments (see Chapter 3).

Figure 2.3 describes the structure of **phospholipids**. Of the three hydroxyl groups of the alcohol **glycerol**, two are linked to fatty acids (length usually 16 or 18 carbon atoms; Table 2.3) and the third is linked by an ester bond to a phosphate residue. An additional ester bond links the negatively charged phosphate residue to either an amino alcohol (choline or ethanolamine), the amino acid serine, or the sugar alcohol inositol. In the case of **phosphatidylcholine** (lecithin), the nitrogen atom is present as a quaternary amine and is therefore posi-

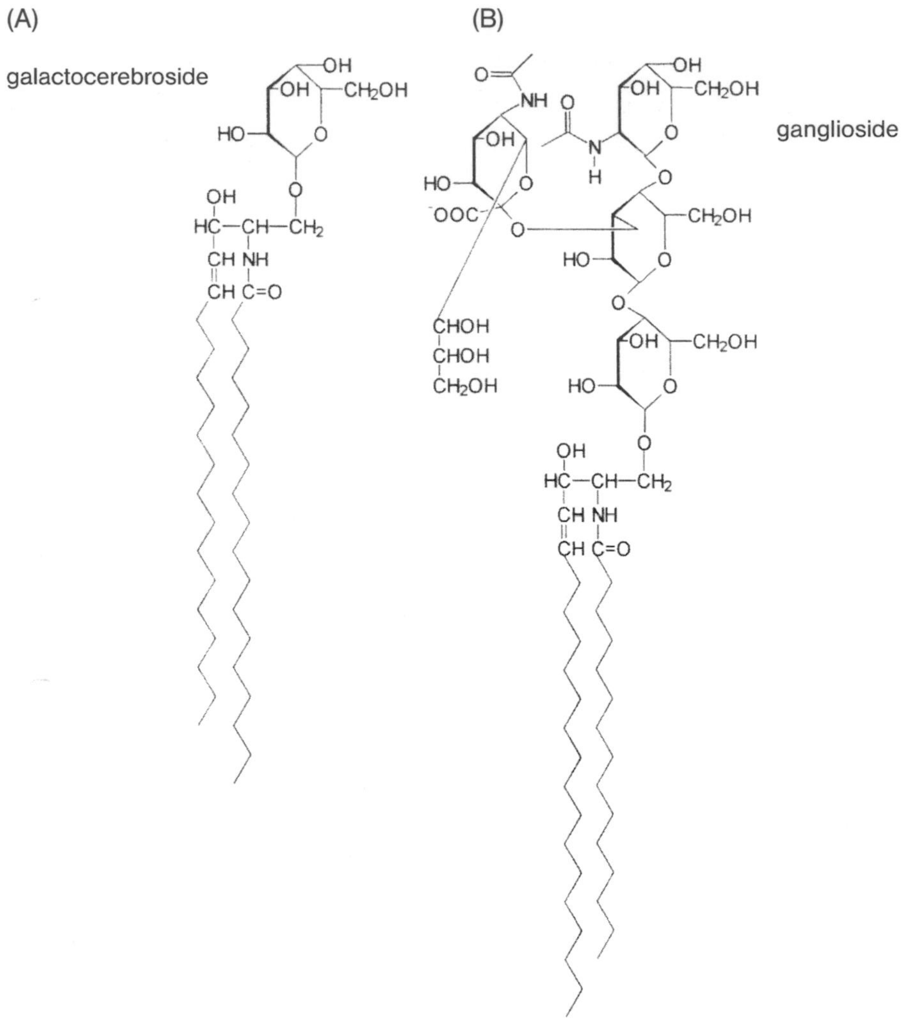

Fig. 2.4 Chemical structure of cerebrosides (glycolipids).
(A) Galactocerebroside. (B) Ganglioside (GM2).

(A)

galactocerebroside

(B)

ganglioside

tively charged. **Phosphatidylinositol** is a precursor for **inositol-1,4,5-triphosphate (IP₃)** – an important signaling molecule in signal transduction pathways of the cell (see Chapter 3.1.1.3).

Phospholipids are **amphiphilic molecules**; their fatty acid residues are strongly lipophilic while their charged head group is hydrophilic. Of the two fatty acids, one is generally **unsaturated** (i.e., one or more double bonds are present). As the single phospholipids constantly rotate, the fatty acid, which is kinked due to the inflexible double bond, has a significantly greater radius than that of two saturated fatty acids. This increases the fluidity of the biomembrane and the formation of **paracrystalline structures** is avoided. In bacterial or yeast cells that are exposed to different temperatures, the fluidity is constantly adjusted according to the surrounding temperatures by incorporation of phospholipids with different lengths of fatty acid residues, with or without double bonds.

In addition to the membrane lipids that are derivatives of glycerol, animal cells contain additional lipids and phospholipids. These have the amino alcohol **sphingosine** as a base and are referred to as **sphingolipids**. The *N*-acyl fatty acid derivatives of sphingosine are termed **ceramides**. **Sphingomyelin**, one of the most important of the sphingolipids, has a structure analogous to that of phosphatidylcholine (Fig. 2.3). Therefore, it is very common in the **myelin sheaths** found around the axons of neurons.

If the sphingomyelin head group is substituted with a sugar residue (e.g., galactose or glucose), a **cerebroside** results. These membrane lipids are missing the phosphate residue and are therefore uncharged. Cerebrosides are common in the brain, where they are oriented towards the cell exterior. **Gangliosides** are

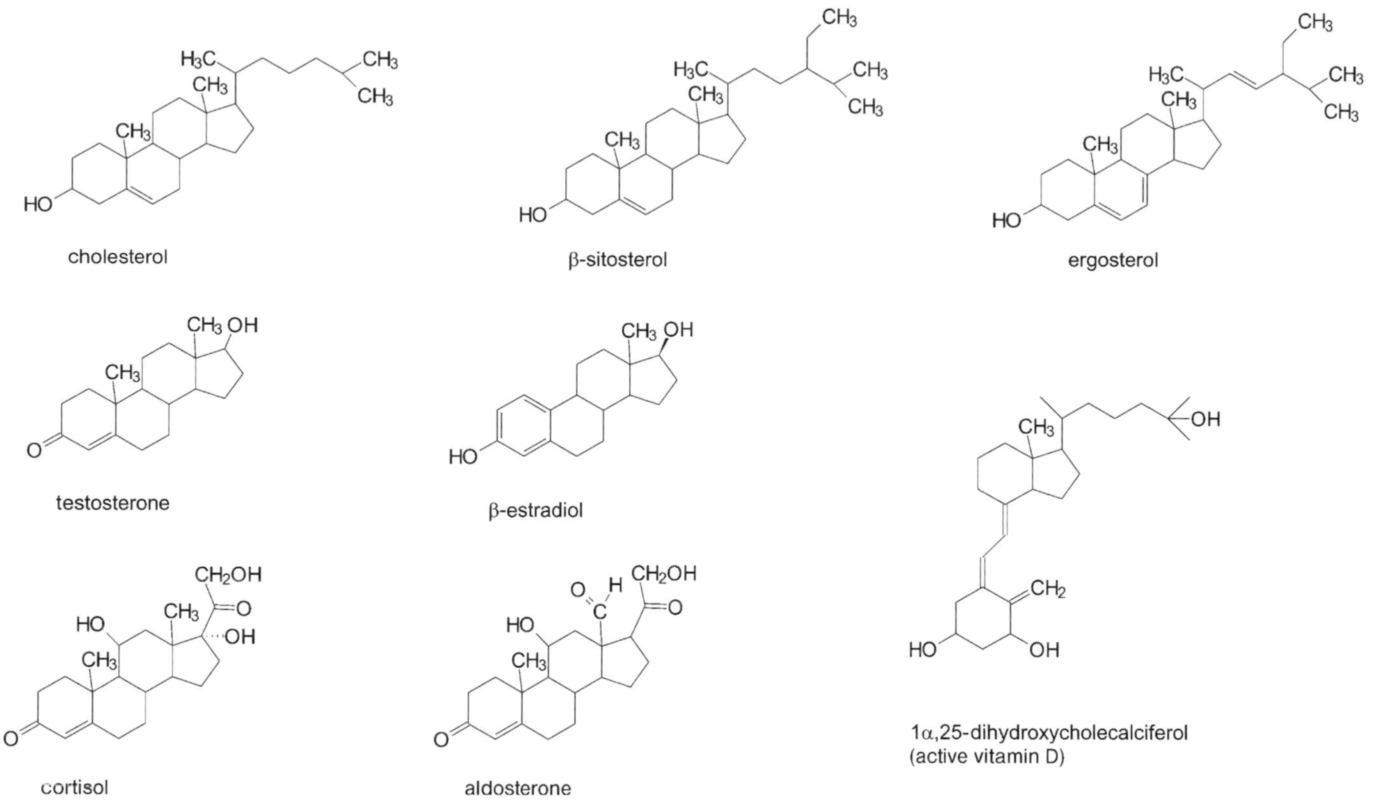

Fig. 2.5 Cholesterol and related sterols. Cholesterol; *β*-sitosterol replaces cholesterol in plants; ergosterol is present in the membranes of fungi; testosterone; *β*-estradiol; cortisol; aldosterone; active vitamin D.

sphingolipids with an especially complex structure. They contain oligosaccharides and at least one **sialic acid** unit (Fig. 2.4). In the brain, 6% of lipids are present in the form of gangliosides. Sphingolipid storage diseases (e.g., Tay-Sachs disease), which result in early neurological deterioration, are of great medical importance.

Phospholipids are cleaved by different **phospholipases**. **Phospholipase A_2** cleaves the central fatty acid at C2 of glycerol residues. The resulting lysophospholipid can lyse cell membranes; interestingly, many snake venoms contain high dosages of **phospholipase A_2**. **Phospholipase A_1** hydrolyzes the fatty acid at C1 of glycerol, while **phospholipase C** opens the phosphate ester bonds with glycerol.

A pharmacologically important lipid class, the **eicosanoids**, is only mentioned briefly here. To summarize, this class includes **prostaglandins, thromboxanes,** and **leukotrienes**. These play many roles and act as paracrine mediators (e.g., in pain, fever, inflammation, blood pressure, and blood coagulation). Phospholipase A2 releases **arachidonic acid** from phosphatidylcholine, which contains the 4-fold unsaturated arachidonic acid in its C2 position. Arachidonic acid is converted into prostaglandin (e.g., example by **cyclooxygenase**). This enzyme is an important target for many drugs (the so-called **nonsteroidal antiinflammatory drugs (NSAIDs)**), among which aspirin (acetylsalicylic acid) is the most famous. Inflammation can also be effectively suppressed by inhibition of phospholipase A_2 by corticoids (e.g., cortisone medications).

Triacylglycerides, not phospholipids, are present in the **storage tissue** of plants and animals. These are broken down by lipases.

The steroid **cholesterol** (Fig. 2.5) is an important and common building block of animal membranes (it is missing in the membranes of bacteria, fungi, and plants). It is stored in the membrane, parallel to the phospholipids (Fig. 2.2), with its polar hydroxyl group oriented towards the cell exterior. Cholesterol is a stiff molecule that stabilizes biological membranes, and lowers their fluidity and permeability. In biological membranes, local assemblies of membrane proteins usually rich in cholesterol, known as **rafts**, have been found. Cholesterol is transported as cholesteryl ester, such as **cholesterol-3-stearate** in lipoproteins (see Chapter 5.4).

Cholesterol can be synthesized in the body; the biggest portion, however, is obtained from food. It is important not only to build up membranes, but also as a precursor for the synthesis of important hormones and vitamins (Fig. 2.5):

- **Glucocorticoids:** for example, **cortisol** (from the adrenal gland) influences the metabolism of carbohydrates, proteins and lipids; cortisol inhibits phospholipase A2, induces several genes such as NF-κB, and thus suppresses inflammation processes.
- **Mineral corticoids:** for example, **aldosterone** (from the adrenal gland) regulates the secretion of salt and water through the kidneys.
- **Sexual hormones: androgens** (**testosterone**, formed in the testicles) and **estrogens** (**β-estradiol**, formed in the ovaries) are important male and female sexual hormones. They bind intracellular receptors that, as transcription factors, control the expression of sex-dependent genes (see Chapter 4.2).
- **Vitamin D:** increases the calcium concentration in the blood and assists in the formation of bones and teeth. Vitamin D deficiency is known as **rickets** in children and osteomalacia in adults.

2.3
Structure and Function of Proteins

Proteins represent the most important tools of the cell (Table 2.2). They catalyze chemical reactions, transport metabolites through membranes, recognize other molecules, and can regulate gene activity. If we consider genes as the legislative branch, proteins then function as the executive branch (i.e., as the executing organs). Proteins are built according to the same principles in both prokaryotes and eukaryotes.

Twenty amino acids serve as building blocks for peptides and proteins, linked to one another by **peptide bonds** (Fig. 2.6). **Polypeptides**, therefore, are polymers made from amino acids. Polypeptides are polar molecules, possessing a NH$_2$ group (**amino- or N-terminal**) on one end and a COOH group (**carboxyl- or C-terminal**) on the other. The diverse tasks and functions of proteins result from different arrangements (sequences) of amino acids.

The 20 amino acids differ in their side chains (Fig. 2.7). The functional groups of the side chains, which protrude from the a-C atom, dictate the conformation and later functionality of the protein by molecular recognition or biocatalysis. Amino acids exist in two optical isomers: the D- and L-forms. Polypeptides are composed exclusively of L-**amino acids**. D-**Amino acids** can be found in bacterial cell walls and in many **antibiotics** (gramicidin, valinomycin). Since proteases can only cleave peptides comprising of L-amino acids, the incorporation of D-amino acids results in a certain protection from untimely degradation.

The proteinogenic amino acids can be divided into different groups according to their functional groups and residues (Fig. 2.7 and Table 2.4):

- Amino acids with apolar, lipophilic residues.
- Amino acids with polar but uncharged residues (i.e., with hydroxyl or amide groups).
- Amino acids with acid groups that are negatively charged.
- Amino acids with basic groups that are positively charged.

The human body is capable of synthesizing some amino acids; others must be obtained through nutrition (essential amino acids). The amino acids phenylalanine, tryptophan, lysine, methionine, valine, leucine, isoleucine, histidine, and threonine belong to the **essential amino acids**.

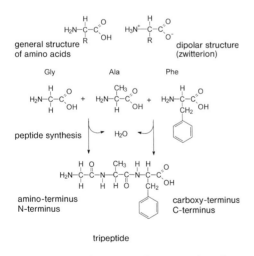

Fig. 2.6 General structure of amino acids and peptides.

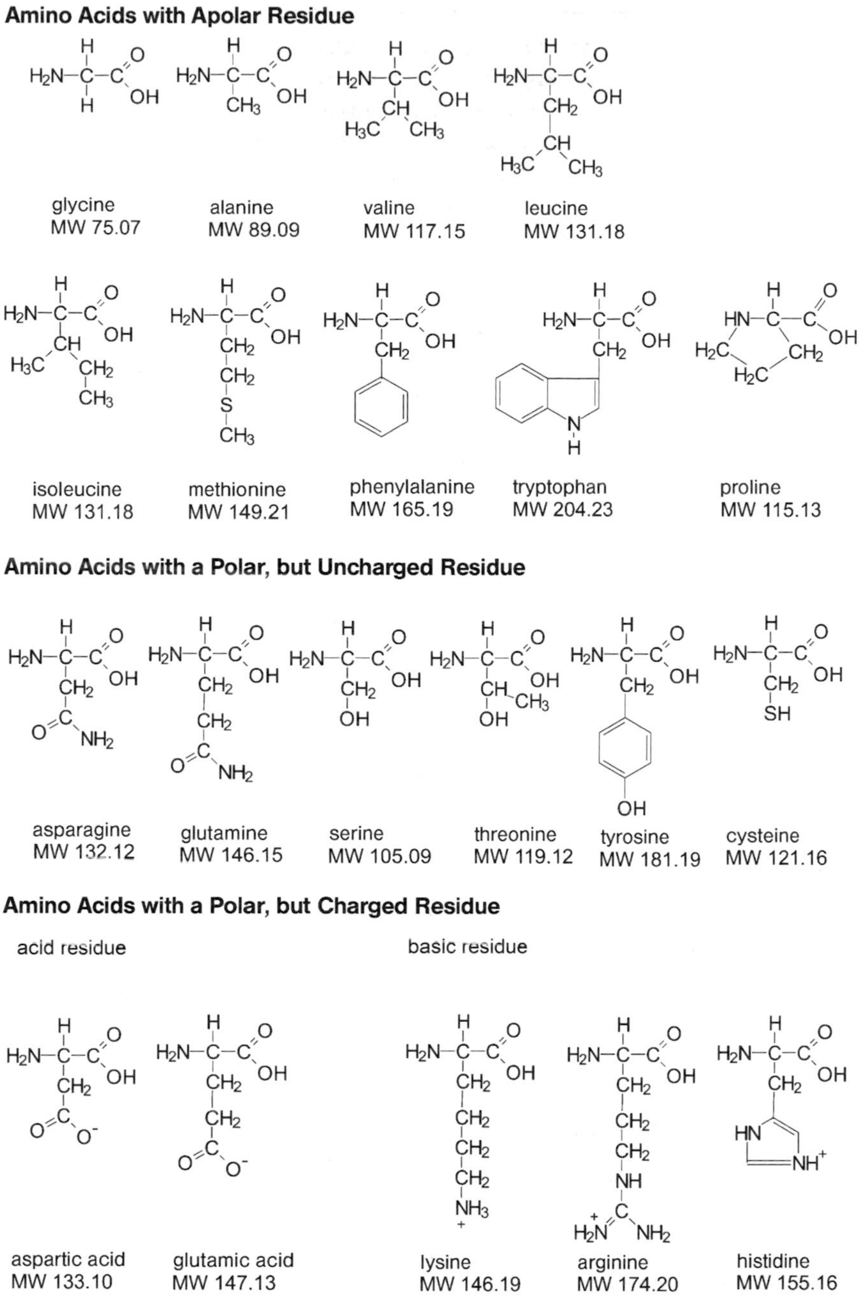

Amino Acids with Apolar Residue

Fig. 2.7 Structures of proteinogenic amino acids.

glycine
MW 75.07

alanine
MW 89.09

valine
MW 117.15

leucine
MW 131.18

isoleucine
MW 131.18

methionine
MW 149.21

phenylalanine
MW 165.19

tryptophan
MW 204.23

proline
MW 115.13

Amino Acids with a Polar, but Uncharged Residue

asparagine
MW 132.12

glutamine
MW 146.15

serine
MW 105.09

threonine
MW 119.12

tyrosine
MW 181.19

cysteine
MW 121.16

Amino Acids with a Polar, but Charged Residue

acid residue

basic residue

aspartic acid
MW 133.10

glutamic acid
MW 147.13

lysine
MW 146.19

arginine
MW 174.20

histidine
MW 155.16

Proteins often undergo **posttranslational modification**, by transferring oligo-saccharide residues to asparagine (*N*-glycosidic) or serine residues (*O*-glycosidic) (see Chapter 5.4). **Glycoproteins** are found on the outside of the cell, in cell walls, and in the extracellular matrix, especially in connective tissue. Glycosylation is important for the biological activity and antigenic properties.

While the peptide bond itself is inflexible, the substituents at the *a*-C atom of an amino acid can rotate freely. As a result, a polypeptide chain can engage in a number of spatial structures (**conformations**). Under aqueous conditions found in the cell the polypeptide chains are not present in a linear form, but form spontaneous **secondary** and **tertiary structures**, which are energetically more favorable. These structures rely on many noncovalent bonds and forces; those that are important include:

- **Hydrogen bonds** (bond strength of 4 kJ mol^{-1} under aqueous conditions).
- **Ionic bonds** (bond strength of 12.5 kJ mol^{-1}).

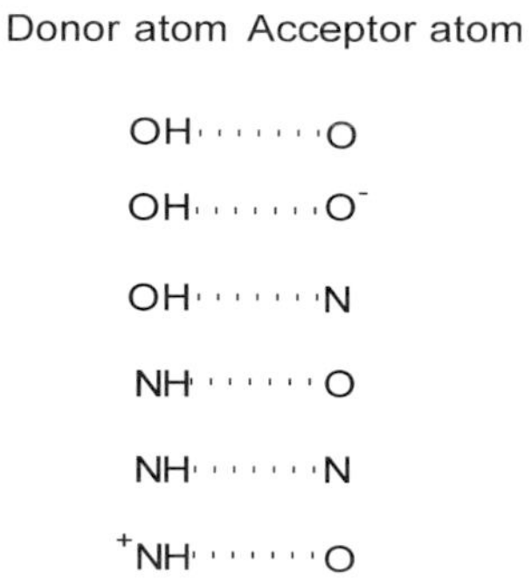

Donor atom Acceptor atom

OH·······O

OH·······O$^-$

OH·······N

NH·······O

NH·······N

$^+$NH·······O

Fig. 2.8 Important hydrogen bonds in biomolecules.

Table 2.4 Compilation and grouping of the proteogenic amino acids: two types of abbreviations are recognized internationally, which either consist of one or three letters; the codons that represent the amino acids in the genetic code are also given.

Classification	Symbols	Codons
Neutral and hydrophobic amino acids		
Glycine	Gly; G	GGA GGC GGG GGU
Alanine	Ala; A	GCA GCC GCG GCU
Valine	Val; V	GUA GUC GUG GUU
Leucine	Leu; L	UUA UUG CUA CUC CUG CUU
Isoleucine	Ile; I	AUA AUC AUU
Tryptophan	Trp; W	UGG
Phenylalanine	Phe; F	UUC UUU
Methionine	Met; M	AUG
Neutral and polar amino acids		
Cysteine	Cys; C	UGC UGU
Serine	Ser; S	AGC AGU UCA UCC UCG UCU
Threonine	Thr; T	ACA ACC ACG ACU
Tyrosine	Tyr; Y	UAC UAU
Aspargine	Asn; N	AAC AAU
Glutamine	Gln; Q	CAA CAG
Proline	Pro; P	CCU CCC CCA CCG
Basic amino acids		
Lysine	Lys; K	AAA AAG
Arginine	Arg; R	AGA AGG CGA CGC CGG CGU
Histidine	His; H	CAC CAU
Acidic amino acids		
Aspartate	Asp; D	GAC GAU
Glutamate	Glu; E	GAA GAG

Fig. 2.9 Noncovalent bonds and disulfide bridges lead to a spatial folding and stabilization of a peptide. Bond types: hydrogen bonds, ionic bonds, van der Waals forces, and disulfide bridges.

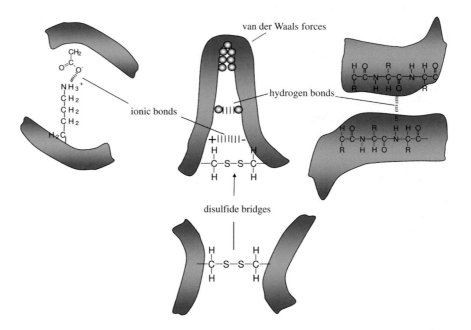

- **van der Waals forces** (bond strength of 0.5 kJ mol^{-1}).
- **Hydrophobic attractions**.

Figure 2.8 summarizes the most common hydrogen bonds present in a cell. **Electronegative atoms**, such as oxygen and nitrogen, try to withdraw electrons from neighboring atoms such as hydrogen. This results in oxygen and nitrogen having a slight negative charge, while hydrogen is slightly positively charged. Positive and negative charges attract one another. The resulting attractions are known either as hydrogen bonds or hydrogen bridges. The ability to form hydrogen bonds is espe-

Primary structure

Tertiary structure

polar residues

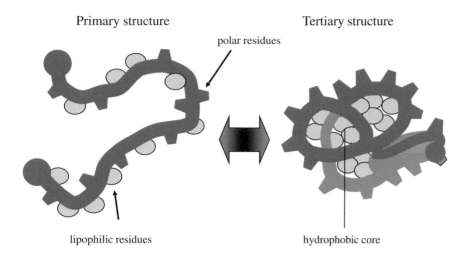

lipophilic residues

hydrophobic core

Fig. 2.10 Folding of peptide chains under aqueous conditions leads to a compact globular conformation.

cially present in water molecules (the hydrogens are positive, the oxygen atom is negatively charged) and water is therefore considered as the universal solvent of the cell. Biomolecules with polar groups easily take up water molecules (they are water soluble) while nonpolar residues repel water (**hydrophobic**) and group together with other apolar molecules (which are fat soluble). Figure 2.9 illustrates the importance of **noncovalent** and **covalent** bonds for the formation of protein folds. Through the formation of **disulfide bridges** between two cysteine residues, the conformation of a protein can also be covalently influenced (Fig. 2.9).

In comparison to **covalent bonds** (bond strength of 348–469 kJ mol^{-1}), **noncovalent bonds** are 5–100 times weaker. Many noncovalent bonds that are present simultaneously can work **cooperatively**, leading to the formation of stable and thermodynamically favored structure elements in polypeptides. Hydrophobic amino acid residues cluster together, in order to lock water out. In polypeptides this can lead to a globular tertiary structure, while the hydrophobic residues are oriented towards the inside, and the polar and charged residues are oriented towards the outside (Fig. 2.10). Under aqueous conditions proteins usually fold spontaneously into a stable conformation in which the free energy is at the lowest.

However, the conformation of proteins can easily change if they come into contact with other proteins or contents of the cell. Other examples of protein modifications are **phosphorylation** (hydroxyl groups of tyrosine, serine, and threonine) or **dephosphorylation** that lead to a change in conformation. It is experimentally simple to alter the conformation of a protein using detergents or urea. For example, when globular proteins are dissolved in a 4 M **urea** solution, the polypeptide chain unfolds (i.e., the protein is **denatured**). If the urea is removed the polypeptide chain refolds into the previous conformation (**renaturing**).

Even though each protein has an individual conformation, when the structures of many proteins are compared, two folding patterns that regularly appear are recognized. These structural elements are:
• *a*-Helix structures.
• *β*-Pleated sheet structures.

a-**Helix structures** and *β*-pleated sheet structures arise from hydrogen bonds between the N–H and C=O groups in the backbone of the polypeptide chain. Functional groups on the side chains do not take part in these structural elements. Figure 2.11 describes the structure of helices and pleated sheets more precisely. Other structures include loops and random coils.

A *β*-**sheet structure** element is often found at the inner core of many proteins. The *β*-pleated sheet can appear between neighboring polypeptide chains that have the same orientation (**parallel chain**). When a polypeptide chain folds back on itself and is aligned in parallel, the chains are termed **antiparallel chains**. In

(A)   (B)

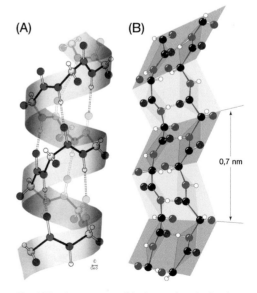

0,7 nm

Fig. 2.11 Importance of hydrogen bonds for the construction of *a*-helix and *β*-sheet structures. (A) The right twisting helix has 3.6 residues per turn. The dotted lines represent the hydrogen bonds between C=O and N=H groups. (From Voet et al., 2002, p. 129.) The zigzag-shaped representation of a *β*-pleated sheet. Dotted lines symbolize hydrogen bonds. The side chains alternate between being present below and above the folded plane. (From Voet et al., 2002, p. 131.)

both cases the chains are being held strongly together by hydrogen bonds (Fig. 2.11).

An *α*-helix forms when a single peptide chain winds around itself and forms a sturdy cylinder. In doing so, a hydrogen bond forms between each fourth peptide bond (i.e., between the C=O group of one peptide bond and the N=H group of the other peptide bond). This results in the formation of an ordered helix with a complete turn every 3.6 amino acids. Short *α*-helix structures can be found in membrane proteins that possess a **transmembrane region**. In this case, the *α*-helix contains only amino acids with nonpolar residues. The nonpolar residues are oriented towards the outside of the helix and shield the hydrophilic backbone of the peptide chain, and interact with the lipophilic components of the phospholipids.

In fibrous proteins (e.g., *α*-keratin), two or three longer helices can twist around each other (**coiled coil**) and form long rope-like structures.

The structure of proteins is very complex, because there are thousands of covalent and noncovalent bonding possibilities between the atoms of the peptide chains and the amino acid residues. Through **x-ray** and **nuclear magnetic resonance (NMR)** analysis, the spatial structures of many hundreds of proteins have been determined. Structure analysis is a challenge not only for basic research, but also for applied pharmaceutical research. If the structure or binding sites of a receptor or enzyme are known in detail, it should be possible to **design** new active substances that have the correct fit and act either as an agonist or antagonist. Successes in rational **drug design** so far concern active substances in the area of AIDS (HIV protease inhibitors; Viracept, Agenerase) and influenza (neuraminidase inhibitors: Relenza, Tamiflu).

There are four structural levels of protein structure:
- **Primary structure:** corresponds to the amino acid sequence.
- **Secondary structure:** corresponds to *α*-helix and *β*-pleated sheet formations.
- **Tertiary structure:** corresponds to the three-dimensional conformation of a polypeptide chain.
- **Quaternary structure:** if a protein complex consists of several subunits (i.e., hemoglobin), then the entire structure is referred to as the quaternary structure.

The proteins of a cell usually contain between 50 and 2000 amino acid residues. Theoretically, each of the 20 amino acids can appear at each location of a polypeptide chain. In an oligopeptide, with a length of four amino acids, there are $20 \times 20 \times 20 \times 20 = 160\,000$ different oligopeptides. The number of possible peptide molecules can be calculated as 20^n, where n denotes the chain length. For a protein with the average length of 300 amino acids (Fig. 2.12), $20^{300} = 10^{390}$ possible variations are derived. However, not even our universe has that many atoms. From the great number of variants only a comparatively small number were seemingly realized by nature. Through the course of evolution many more proteins have been

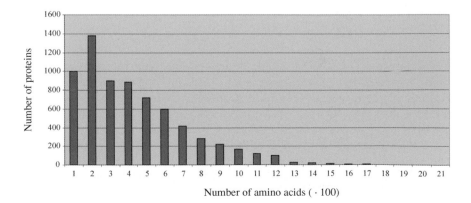

Fig. 2.12 Size of proteins in yeast (*Saccharomyces cerevisiae*). The yeast genome project allowed a first estimate of the size of yeast proteins.

created. However, following natural selection only those proteins that have proven to be of value remain. During the course of evolution **protein families** deriving from the first proteins with defined functions have developed through gene duplication. The original sequence has been changed in the *new* proteins.

During analysis of genome projects, individual **structural domains** of many proteins have been identified with the help of bioinformatics. Large proteins are usually made up of several functional domain or modules. Domains usually have defined structures and functions (Figs 2.13 and 2.14). They often correspond to the **exons in a eukaryotic gene** (see Chapter 4.2). They developed in early evolution, obviously independent of each other. In a later evolutionary phase the gene sections coding for a domain were newly combined. Through **domain shuffling**, proteins with new characteristics could thus be created. As a consequence, most proteins can be seen as variants of previously existing proteins or of their domains. Figure 2.13 shows as an example the structure of an Src protein that has four domains. Examples for domain shuffling are illustrated in Fig. 2.14. Domain shuffling is important for the explanation of evolutionary development. It is not only individual point mutations that bring evolutionary advancement, but mainly new combinations of functional modules (prefabricated building blocks).

Many proteins contain **binding sites** for **ligands**; ligands can be lower-molecular-weight substances, but also macromolecules such as nucleic acids or other proteins. The binding of a ligand to a binding site can be viewed as a **molecular recognition process**. Such molecular recognition processes are common in the cell, but these processes are only understood in detail in a few cases. However, these processes have an important relevance to cell function, metabolism, and "life" that should not be underestimated. Experiments in structural biology have already shown that the binding of a ligand in a binding site functions according to the **lock-and key principle**. The **binding site** has a specific spatial structure in which a ligand fits selectively. Binding of the ligand involves the formation of several noncovalent bonds (Fig. 2.15) between the functional groups of the ligand and those of the protein. **Binding generally brings about a change of the protein conformation (induced fit)**. The binding site is not formed by amino acid residues that lie beside each other on the peptide chain, but often consists of amino acids located in different parts of a peptide chain and spatially form a binding site by appropriate specific folding (Fig. 2.15).

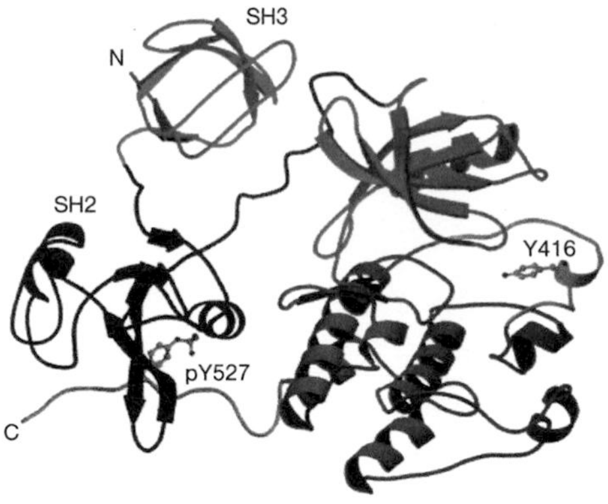

Fig. 2.13 Structure of Src protein with four domains. The four domains are the (a) small kinase domain, (b) large kinase domain, (c) SH2 domain, and (d) SH3 domain. This figure also appears with the color plates.

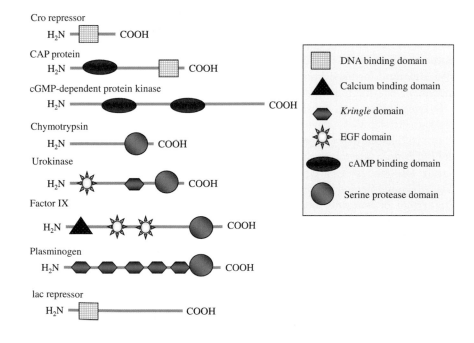

Fig. 2.14 Occurrence of domains in different proteins.

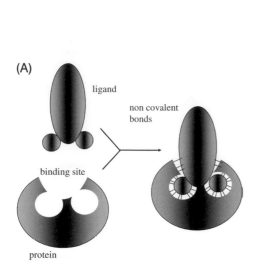

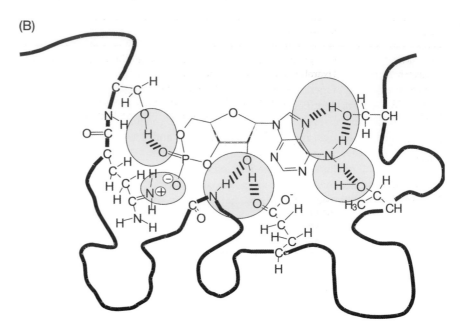

Fig. 2.15 Structure of binding sites within proteins. (A) Schematic illustration of the significance of noncovalent bonds in the lock and key principle. (B) cAMP is locked into a binding site via ionic and hydrogen bonds.

Interactions that occur between **antigens** and **antibodies** (see Chapter 28), between **ligands** and **hormone** receptors as well as between **enzymes** and their **substrates** are particularly intimate and selective. The topic of protein–protein interactions is discussed further in Chapter 24.

Most of the cellular building blocks are inert molecules that are not prone to react chemically. Significant **activation energy** has to be overcome in order to start an energy-consuming chemical reaction. In the laboratory this can be achieved by heating and adding acids or bases. In biological systems evolution has developed **enzymes as biological catalysts** that are able to catalyze all necessary reactions without higher temperatures being necessary. Enzymes do not change the **reaction equilibrium**, but usually alter the **reaction rate**. Enzymes contain an active center in which a substrate is bound. After the enzyme has catalyzed a reaction the product is released, but the enzyme remains unchanged and is ready for a new reaction. Noncovalent interactions (hydrogen bonds, ionic bonds) and transient covalent bonds between protein and substrate play a key role during the binding and catalysis. Detailed elucidation of such interactions at the atomic scale is the task of biophysics and biochemistry. This research is also important for biotechnology in relation to the synthesis of new enzyme inhibitors or enzyme modulators.

Enzymes show high **substrate specificity**. It is believed that for almost every biosynthetic step that happens in the cell, a specific enzyme is also present. This does not rule out that enzymes that catalyze chemically similar reactions can be derived from a common original enzyme. Such enzymes belong to one **protein family**. Most enzymes have particular **pH** and **temperature optima**. Enzymes are divided into different classes according to the processes catalyzed (Table 2.5). **Coenzymes** or **inorganic ions** often take part in the catalysis itself. Biochemists and biotechnologists are interested in the elucidation of the enzymatic reaction mechanisms because hints for new catalysts for organic synthesis can be obtained. Apart from this, scientists are attempting to create new biological catalysts through the production of artificial enzymes.

In addition to a **catalytic center**, many enzymes (especially those composed of several subunits) also have a **regulatory center** where **allosteric ligands** bind. For example, the second messenger cAMP binds to the tetrameric protein kinase A

Table 2.5 Important classes of enzymes.

Enzyme	Reaction catalyzed
Hydrolases	Catalyze hydrolytic cleavage (amylase, lipase, glucosidase, esterase)
Nucleases	Hydrolyze nucleic acids (DNase, RNase)
Proteases	Cleave peptides (pepsin, trypsin, chymotrypsin)
Isomerases	Catalyze the rearrangement of bonds within a molecule
Synthases	General name for an enzyme that catalyzes condensation reactions in anabolic processes
Polymerases	Catalyze the formation of RNA and DNA
Kinases	Transfer phosphate residues; the protein kinases (PKA, PKC) are particularly important
Phosphatases	Remove phosphate residues from a molecule
ATPases	Require ATP (e.g. H^+-ATPase, Na^+, K^+-ATPase, Ca^{2+}-ATPase); motor proteins, such as myosin
Oxidoreductases	Enzymes that catalyze redox reactions, in which one molecule is reduced and another is oxidized; they are grouped into oxidases, reductases, and dehydrogenases

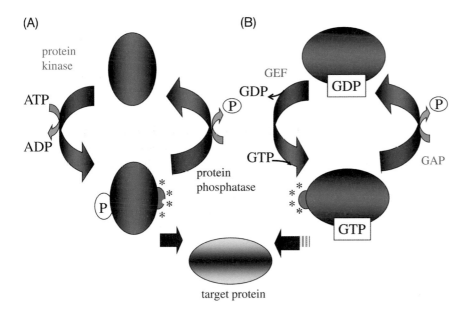

Fig. 2.16 Reversible activation and inactivation of enzymes and regulatory proteins. (A) Phosphorylation/dephosphorylation. (B) Binding of GTP/GDP. GEF = guanine nucleotide exchange factor; GAP = GTPase-activating protein.

complex; after binding both regulatory protein subunits dissociate from both catalytic subunits, which results in their activation (Fig. 3.9). Enzymes can be inhibited by **inhibitors**. We distinguish between reversible, irreversible, competitive, and noncompetitive inhibitors.

A further important way to regulate the activity of enzymes or regulatory proteins is that of reversible **conformational change**. This is achieved by phosphorylation/dephosphorylation with the help of **protein kinases** or **phosphatases**, respectively. It can also be achieved through the binding of **GTP** and **GDP** (Fig. 2.16). A reversible reduction of **disulfide bridges** (e.g., through thioredoxin) plays an important role during the regulation of light-dependent chloroplast enzymes. Biochemists and cell biologists are working extensively to define all cellular proteins that are regulated through phosphorylation and GTP/GDP to gain a better understanding of regulation processes inside the cell.

2.4
Structure of Nucleotides and Nucleic Acids (DNA and RNA)

Nucleotides play important roles in the cell: as **energy carriers** (ATP, ADP), as **coenzymes** (FAD, NAD^+, coenzyme A), during the transfer of sugar moieties (ADP-glucose), and as **building blocks for nucleic acids** (Fig. 2.17A). Nucleotides

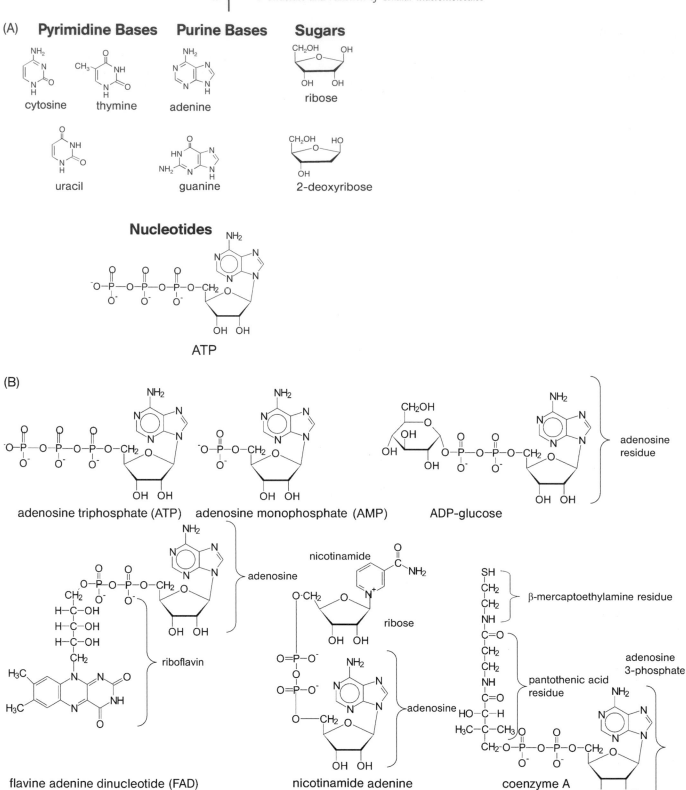

Fig. 2.17 Structure of nucleotides. (A) Structures of purine and pyrimidine bases, pentoses, and ATP (as an example of a nucleotide). (B) Structures of ATP, AMP, ADP, glucose, FAD+, and coenzyme A.

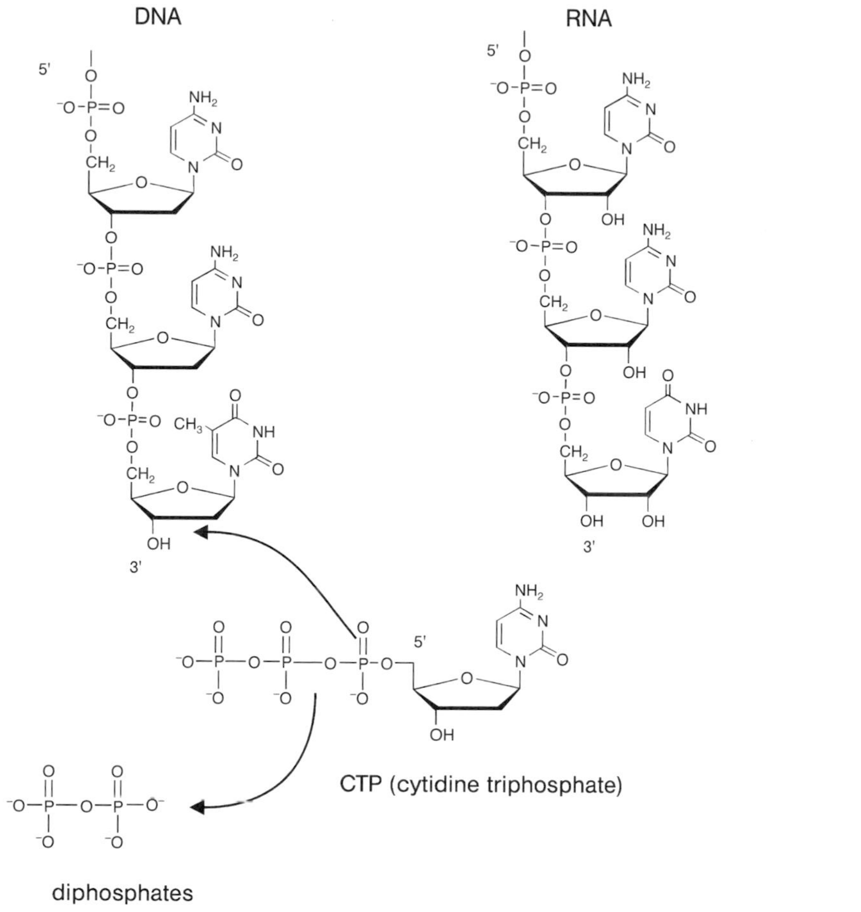

DNA

RNA

CTP (cytidine triphosphate)

diphosphates

Fig. 2.18 Linear structure of DNA and RNA. In nucleic acid biosynthesis the α-positioned phosphate group of a nucleotide triphosphate (NTPs in RNA; dNTPs in DNA) is linked to the free 3'-OH group of the available strand.

Table 2.6 Nomenclature of DNA and RNA building blocks.

Base	Nucleotide (abbreviation)	Nucleotide (number of phosphate groups)					
		RNA			DNA		
		1	2	3	1	2	3
Adenine	Adenosine (A)	AMP	ADP	ATP	dAMP	dADP	dATP
Guanine	Guanosine (G)	GMP	GDP	GTP	dGMP	dGDP	dGTP
Cytosine	Cytidine (C)	CMP	CDP	CTP	dCMP	dCDP	dCTP
Thymine	Thymidine (T)				dTMP	dTDP	dTTP
Uracil	Uridine (U)	UMP	UDP	UTP			

AMP, adenosine monophosphate; ADP, adenosine diphosphate; ATP, adenosine triphosphate; d, deoxy.

consist of the purine bases adenine and guanine and the pyrimidine bases cytosine and thymine or uracil, which form N-glycosidic bonds with ribose or deoxyribose. The 5'-hydroxyl group of the pentose is esterified with one, two, or three phosphate residues (Fig. 2.17B).

Our genetic information is stored in the form of **deoxyribonucleic acid (DNA)**. DNA is a macromolecule and is made up of nucleotide subunits bound together linearly (Fig. 2.18). DNA contains the bases A, T, G, and C; RNA contains the bases A, U, G, and C. The nomenclature of the bases, nucleosides, and nucleotides is explained in Table 2.6.

The nucleotides are the **building blocks for DNA and RNA**. Nucleotides are esterified into polynucleotide chains via a phosphate backbone. The 5'-hydroxyl group ("five prime hydroxyl group") of a pentose is linked via a phosphodiester bond to the 3'-hydroxyl group of a second pentose (Fig. 2.18). During the bio-

Fig. 2.19 Structure of the DNA double helix. The spatial orientation of the base pairs in the double helix and the principle of complementary base pairing between A and T and G and C, respectively, via the formation of hydrogen bonds. (A) Schematic structure of the double helix. (B) Structural formulae.

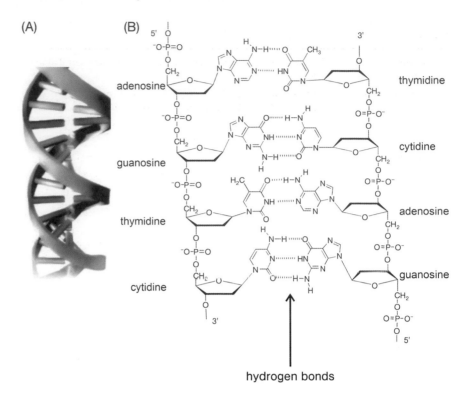

(A)　　**(B)**

hydrogen bonds

synthesis of the nucleic acids, the respective nucleotide triphosphates are needed whose phosphoric acid anhydride bonds are especially rich in energy. In the completed nucleic acid only nucleotide monophosphates are present. After cleavage of a diphosphate residue, the α-phosphate group attacks the free 3'-end of the already existing nucleic acid strand and forms a new ester bond. The synthesis is said to occur in the 5'→3' direction.

DNA exists as a **double helix** whereby the bases A and T and G and T, respectively, face each other in a **complimentary** manner (Fig. 2.19). Both DNA strands are arranged antiparallel to each other (i.e., within a helix one of the strands runs in the 5'→3' direction, while the complementary partner strand is oriented in the 3'→5' direction). The DNA double helix has a diameter of 2 nm.

Complementary base pairing is achieved through the specific formation of two or three **hydrogen bonds** between A–T and G–C pairs, respectively (Fig. 2.19). This is an important example of a molecular recognition reaction via noncovalent bonds. Base pairing occurs spontaneously should the two bases meet. This results in the ability to self organize and to form supramolecular structures without the requirement of energy or regulatory helpers. The selectivity of complementary base pairing is an important requirement for basic genetic processes (e.g., replication, transcription, and recombination) and diagnostic procedures (e.g., Southern hybridization, DNA fingerprinting with DNA probes, and DNA microchips; see Chapters 21, 22, and 27).

In eukaryotes the multiple negative charges on the backbone of the DNA double helix are complexed with basic, positively charged **histone proteins** (Fig. 4.6); in prokaryotes, positively charged **polyamines** take over this role. The bases are arranged inside of the helix and form planar stacks (Fig. 2.19). The inside of the helix is anhydrous – only lipophilic substances, especially if they are also planar, can be inserted in between the base stacks (so-called **DNA intercalators**). Such intercalation often leads to errors during replication which can initiate **frame shift mutations** (see Chapter 4.1.5).

Determined by the **cooperativity of many hydrogen bonds** and the lipophilic interactions between the base stacks, the DNA double helix is very stable and can only be separated into the single strands by high temperatures. This process

Enzyme	Reaction
Restriction endonuclease	Cuts DNA at specific palindromic recognition sequences that are 4–6 bp long
DNA polymerase I	Synthesis of the complementary DNA strand; requires a primer with a free 3'-end; important for DNA sequencing
DNA ligase	ligates (joins together) DNA strands; the enzyme forms phospho-diester bonds between neighboring phosphate residues
Telomerase	Synthesizes telomere sequences at the end of chromosomes
DNA topoisomerases	Cuts DNA strands, either single or double stranded
Taq polymerase	Heat-stable DNA polymerase from *Thermus aquaticus*; important for PCR
DNase	Hydrolase that cleaves double-stranded DNA
RNase	Hydrolase that degrades single- or double-stranded RNA
RNA polymerase	Copies DNA into mRNA and rRNA
Reverse transcriptase	Copies RNA into DNA

Table 2.7 Enzymes that use DNA as a substrate and are used in genetic engineering.

is also called **melting**; T_m **(melting temperature) indicates the temperature at which 50% of the DNA is already present as single strands.** T_m is dependent on the GC content of the DNA, which varies significantly between organisms. The higher the GC content, the higher the average T_m (caused by three hydrogen bonds in G–C pairs versus two hydrogen bonds in A–T pairs); this is practically important when primers or DNA probes are to be designed. If these primers/probes are to be hybridized under stringent conditions, primers with a higher GC content are preferred.

Important **enzymes that use DNA** as their substrate are summarized in Table 2.7. Many of these enzymes are important tools in molecular biology and bio-technology (see Chapter 12).

As apposed to DNA, the **RNA world** is much more complex. The basic structure of **RNA**, from the four **ribonucleotides** A, U, G, and C, is valid for all RNA species. RNA molecules initially occur as single strands. As partial sequences within an RNA molecule are often complementary, RNA double strands form spontaneously (so-called **stem structures**). Nonpaired regions form single-stranded **loop structures**. RNA can interact with several diverse molecules via the nonpaired bases and can be catalytically active (e.g., by formation of peptide bonds in ribosomal protein biosynthesis or the splicing of nucleic acids).

RNA often exhibits characteristic structures and functions (Fig. 2.20):

- **mRNA:** messenger RNA codes for proteins; in eukaryotes with a cap structure on the 5'-end and a poly(A) tail on the 3'-end.
- **tRNA:** transfer RNA, adaptor between mRNA and amino acids; with post-transcriptional base modifications in loop regions.
- **rRNA:** 5S, 23S, and 16S rRNA in prokaryotic ribosomes with characteristic secondary and tertiary structures.
- **rRNA:** 5S, 5,8S, 18S, and 28S rRNA in eukaryotic ribosomes with characteristic secondary and tertiary structures.
- **snRNA:** small nuclear RNA; catalyzes pre-mRNA splicing.
- **snoRNA:** small nucleolar RNA; chemically modified rRNA.
- **siRNA:** small interfering RNA; small double-stranded RNA molecules that can influence gene expression by directing degradation of selective mRNAs and the establishment of compact chromosome structures.
- **miRNA:** microRNA; small single-strand RNA molecules that can control gene activity, development and differentiation by specifically blocking translation of particular mRNA.
- **ribozymes:** RNA with catalytic activity.

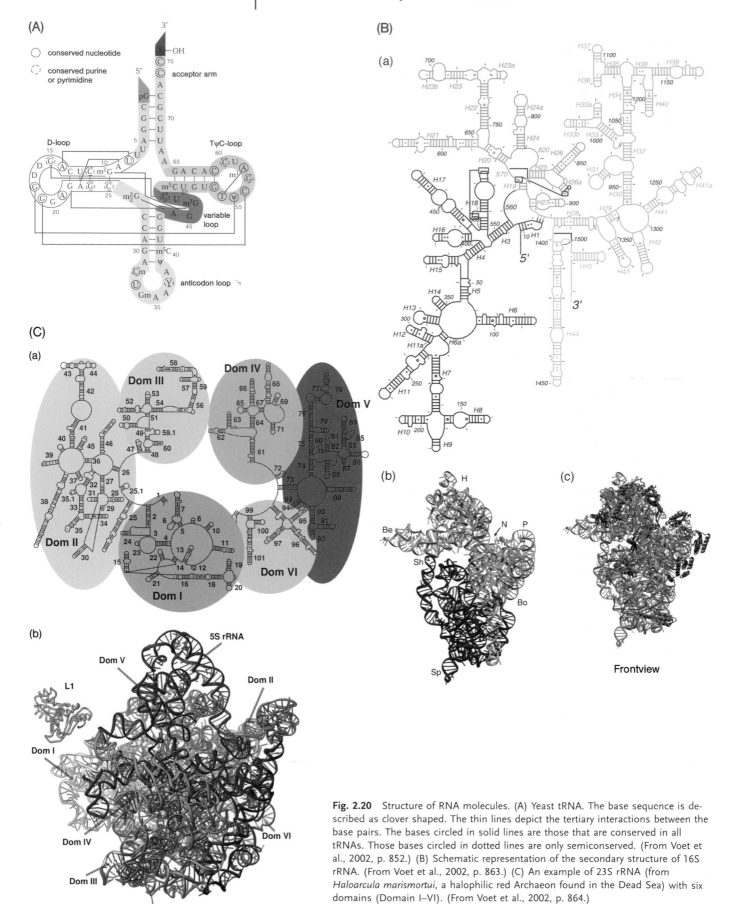

Fig. 2.20 Structure of RNA molecules. (A) Yeast tRNA. The base sequence is described as clover shaped. The thin lines depict the tertiary interactions between the base pairs. The bases circled in solid lines are those that are conserved in all tRNAs. Those bases circled in dotted lines are only semiconserved. (From Voet et al., 2002, p. 852.) (B) Schematic representation of the secondary structure of 16S rRNA. (From Voet et al., 2002, p. 863.) (C) An example of 23S rRNA (from *Haloarcula marismortui*, a halophilic red Archaeon found in the Dead Sea) with six domains (Domain I–VI). (From Voet et al., 2002, p. 864.)

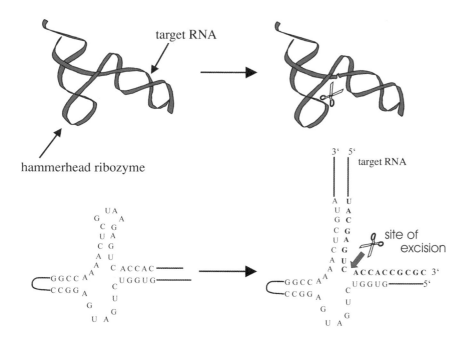

Fig. 2.21 Structure and function of a hammerhead ribozyme.

RNA interference (RNAi) describes a widely distributed phenomenon in which double-stranded RNA molecules lead to the breakdown of complementary mRNA. In the cell there is a ribonuclease (so-called **Dicer**), which can cleave the double-stranded RNA into short, 21- to 23-nucleotide **siRNA (short interfering RNA) molecules.** The siRNA assembles itself together with proteins and forms the **RISC (RNA-induced silencing complex) complex**, which binds to the mRNA that is complementary to siRNA (e.g., of viruses or transposons). By cleaving the mRNA, the associated gene activity is inhibited. SiRNA regulates gene expression and **rearrangements**, by switching off transposons.

A further group of small noncoding RNA molecules are the **miRNAs (microRNAs)**. An endogenous single-stranded RNA molecule is produced by RNA polymerase II, which is then trimmed to miRNA 21–23 nucleotides in length by **Dicer**. miRNAs have been found in plants and animals. miRNA binds and inactivates complementary mRNA molecules, and seems to play a very important role in gene regulation, differentiation, and tissue development

The **RNAi method** is an important tool for basic research in order to examine the function of genes. By introducing double-stranded siRNA through transfection or with the help of a particle gun, targeted inhibition of gene activity is possible. It is also possible to produce transgenic cells that produce siRNA themselves. siRNA is a further development of the **antisense RNAs**, and plays an important role as a tool for cellular/molecular biology and developmental biology, in order to silence all the genes of an organism in a specific way. Biotechnologists are also working on developing these molecules as therapeutics.

Catalytically active RNA molecules were supposedly present in early evolution. These RNAs were surrounded by a simple biological membrane. They contained the genetic information, and were also responsible for structure formation and catalysis. In addition to other tasks, they carried out protein synthesis. It is assumed that there was a division of labor further in the course of evolution, so that DNA took over the storage of genetic information and proteins took over the role as catalysts and structure carriers. Today, RNA has important roles both as a messenger between DNA and protein, as well as a catalytic and regulatory molecule.

Ribozymes are short RNA molecules that recognize and specifically cleave their target RNA via shared base sequences (Fig. 2.21). Through selection of new ribozymes, biotechnologists are attempting to develop new enzyme-like catalysts or therapeutics that can switch off unwanted gene activity.

3
Structure and Functions of a Cell

Learning Objectives

This chapter gives an introduction to the structure of eukaryotic and prokaryotic cells, including the compartments they contain and their function. All cells are totally surrounded by a semipermeable cytoplasmic membrane. In eukaryotes, there is a range of various inner compartments that form separate entities in which diverse reactions can take place. Biomembranes ensure that the entry and exit of polar or charged molecules or ions to cells or compartments is controlled. Biomembranes can easily merge into each other, incorporating or releasing vesicles. As cells must be able to take up compounds from their environment or to release compounds that cannot diffuse through the membrane, specific membrane proteins are needed that act as transport proteins (for polar or charged molecules) or ion channels (for Na^+, K^+, Ca^{2+}, and Cl^-). Many cells contain receptors that communicate with other cells, tissues, and organs. These recognize signaling substances and pass on information to the inner cell, using complex pathways. This chapter summarizes essential information about the endoplasmic reticulum, the Golgi complex, lysosomes, and vacuoles as well as mitochondria, chloroplasts (and their evolution), the cytoskeleton, and the cell walls. Bacteria and viruses are also briefly discussed. Bacteria are among the earliest organisms that emerged in the evolutionary process. Compared to eukaryotic cells, their structure is simple. This makes them suitable models for biochemistry and molecular biology, as basal processes can be easily studied. Furthermore, bacteria are significant infective agents as well as producer organisms in biotechnology. Viruses, on the other hand, do not possess an independent metabolism and rely on host cells for their proliferation. Apart from being major infective agents in plants and animals, they are important model systems in molecular biology and are used as vectors in gene technology.

3.1
Structure of a Eukaryotic Cell

3.1.1
Structure and Function of the Cytoplasmic Membrane

The hydrophilic or hydrophobic interactions of many lipid molecules in the aqueous cell environment give rise to the spontaneous formation of energetically favorable **membrane bilayers**. These are fluid, plastic, and mobile (Figs 2.2 and 3.1). Although the individual phospholipids spin around themselves and constantly move laterally, the resulting membrane is not easily permeable for ions and charged or polar molecules.

Under cellular conditions, **biomembranes** tend not to lie flat like a carpet, but assume a spherical shape (Fig. 3.2 A). Should holes and ruptures in the cytoplasmic membrane occur, they are only transient and immediately resealed. This remarkable self-organization and formation of supramolecular structures were pre-

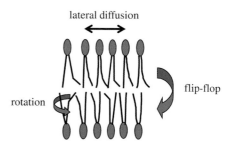

Fig. 3.1 Mobility of phospholipids in a biomembrane. Three types of movement are possible: rotation (spin), lateral diffusion, and flip-flop, which occurs rarely. A flip-flop can be brought about with the enzyme flippase.

An Introduction to Molecular Biotechnology, 2nd Edition.
Edited by Michael Wink
Copyright © 2011 WILEY-VCH Verlag GmbH & Co. KGaA, Weinheim
ISBN: 978-3-527-32637-2

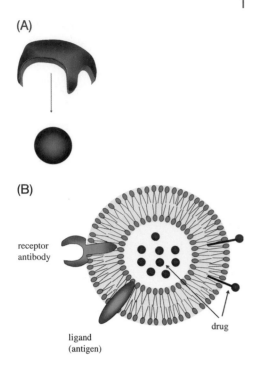

(A)

(B)

receptor
antibody

ligand
(antigen)

drug

Fig. 3.2 Vesicle and liposome formation. (A) In a watery environment, lipid bilayers spontaneously form spherical vesicles, which makes them energetically favorable. (B) Schematic view of a liposome. Receptors, antibodies, and ligands may be integrated into the outside, which enables the liposome to recognize its target. Active compounds may be stored inside the liposome or bound to the outside of the membrane using nanoparticles or carrier molecules.

requisites for the emergence of cells – and thus of life itself. Membranes can easily invert to form **vesicles** that, in turn, can merge with other membranes. When a vesicle is pinched off from a biomembrane, this is called **endocytosis**. When a vesicle is absorbed by a compartment membrane, it is called **exocytosis**.

Small closed vesicles consisting of synthetic phospholipids are also called **liposomes** (Fig. 3.2 B). These play an important role in medicine and biotechnology, as they serve as vehicles for pharmaceutical compounds. They can be loaded with aggressive toxins. Researchers are trying to modify liposomes so that they can direct them to their targets via receptors or antibodies that are embedded in the liposomal membrane (see Chapter 26). This could prevent chemotherapeutics, such as those used in cancer therapy, from attacking and damaging healthy cells.

Cellular membranes have an **asymmetric structure**. Their building blocks on the inside of the cell differ from those on the outside (Fig. 3.3). Due to the presence of negatively charged **phosphatidylserine**, the inside of a membrane is negatively charged. Biomembranes owe their specificity to the integration of certain membrane proteins and lipids. In the **endoplasmic reticulum**, new membrane sections are synthesized, allowing for their asymmetric structure. The enzyme **flippase** has an additional role to play in this context – facilitating a change of orientation in individual phospholipids.

3.1.1.1 Membrane Permeability
Biomembranes serve primarily as **permeability barriers**. The lipophilic inside of the membrane is an effective barrier against the diffusion of polar and charged substances, while **membrane proteins** enable the controlled import and export of ions and metabolites. The effectiveness of the membrane as a permeability barrier becomes apparent when looking at the difference in ion concentrations inside and outside a cell (Table 3.1). The differences in ion concentration may be as large as several powers of 10.

Figure 3.4 shows a schematic view of the barrier function, using the example of an artificial lipid bilayer. Given sufficient time, any substance will diffuse through a membrane. The diffusion rate, however, varies considerably, depending on size, charge, and lipophilic properties of a molecule. The smaller and more hydrophobic a molecule is, the faster it will diffuse across a cell membrane. The following rules apply:

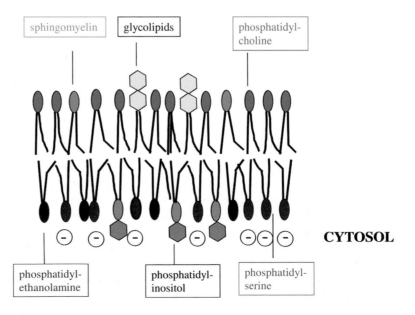

sphingomyelin

glycolipids

phosphatidyl-choline

CYTOSOL

phosphatidyl-ethanolamine

phosphatidyl-inositol

phosphatidyl-serine

Fig. 3.3 Asymmetric structure of biomembranes.

Ion	Intracellular concentration	Extracellular concentration
Cations		
Na^+	5–15 mM	145 mM
K^+	140 mM	5 mM
Mg^{++}	0.5 mM[a]	1–2 mM
Ca^{++}	100 nM[a]	1–2 mM
H^+	$10^{-7.2}$ M (= pH 7.2)	$10^{-7.4}$ M (= pH 7.4)
Anions		
Cl^-	5–15 mM	110 mM

Table 3.1 Ion concentrations inside mammalian cells and in the extracellular space.

a) Ca^{++} and Mg^{++} also occur bound to proteins within the cell (1–2 mM or 20 mM, respectively).

- **Smaller, nonpolar molecules** such as O_2, CO_2, and N_2 are lipid-soluble, diffusing rapidly through biomembranes. This is also true for lipophilic organic molecules such as benzene or chloroform. Many therapeutics are strongly lipophilic and can thus diffuse freely into the body.
- **Small uncharged polar molecules** are slightly slower to diffuse through the membrane. This category includes molecules such as H_2O, ethanol, urea, or glycerol.
- The biomembrane forms an effective barrier to **larger charged molecules** (sugar, amino acids, and nucleotides).
- **Small charged inorganic ions** such as Na^+, K^+, Ca^{2+}, or Cl^- are unable to permeate the lipid bilayer through free diffusion.

Membrane permeability can be modified by certain agents. For example, bacteria are vulnerable to certain antibiotics, such as the peptide antibiotics tyrothricin, polymyxin B, gramicidin, and valinomycin or the polyene antibiotic amphotericin B. These substances act on the biomembrane and disrupt the ion balance specifically or nonspecifically; some are ionophores. Many plants produce saponins, which unselectively disturb membrane permeability. Furthermore, the effect of some inhaled anesthetics can be put down to a disturbance of the biomembrane and of the ion channels.

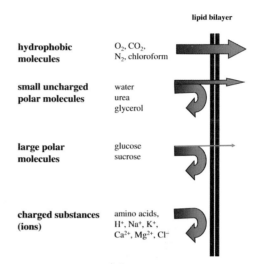

Fig. 3.4 Permeability of artificial lipid membranes for biologically relevant substances.

3.1.1.2 Transport Processes across Biomembranes

The properties of artificial lipid bilayers (Fig. 3.4) also apply to biomembranes. Water and other small nonpolar molecules enter the cell by free diffusion. While cells possess additional specific water absorption mechanisms (aquaporins), they also need to take up polar and charged nutrients, and to release waste products. Polar charged components include inorganic ions, sugars, amino acids, organic acids, nucleotides, and various other metabolites. As normal diffusion through the membrane would be too slow, the cell uses specific membrane proteins to speed up the process (Fig. 3.5):

- **Ion channels** or **ion pumps** for inorganic ions, above all Na^+ channels, K^+ channels, Ca^{2+} channels, and Cl^- channels.
- **Transporters** or **carriers** for organic molecules.

The concentration of the transported substance on either side of the membrane plays a crucial part (Fig. 3.5 B). **Free and facilitated diffusion** (if at all possible) happens spontaneously, from a compartment containing a high concentration of the compound in question towards another compartment containing only very few of these molecules. The net diffusion comes to an end when a concentration equilibrium is reached. For energetic reasons, this process cannot be reversed.

These rules also apply within the cell. Ion channels and passive transporters can only rebalance concentration levels. Where ions or metabolites are to be

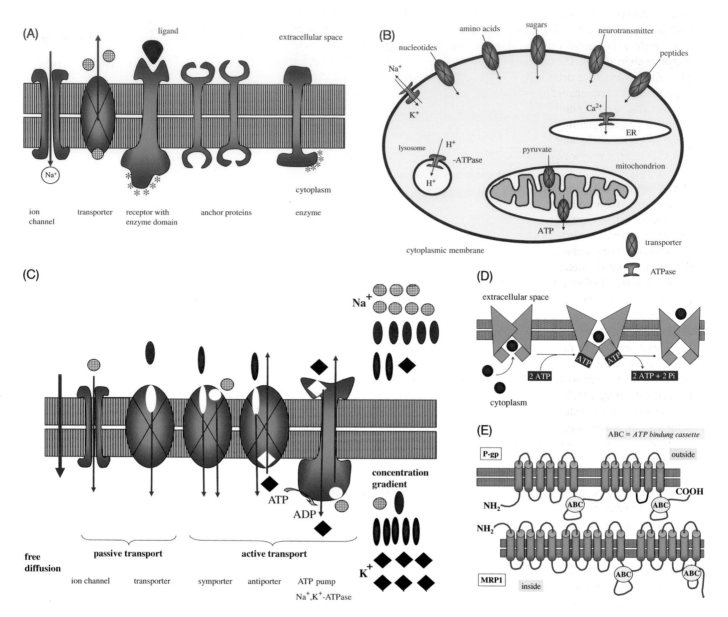

Fig. 3.5 Important membrane proteins and transport processes. (A) Schematic view of ion channels, transporters, receptors, enzymes, and protein anchors. (B) Comparison of simple diffusion, and active and passive transport. (C) Examples of transporters and ion pumps in an animal cell. (D) Mechanism of ABC transporters. (E) Structures of the MDR proteins P-gp and MRP1.

transported against a concentration gradient, additional energy is required. Specific **ion pumps** in the biomembrane are in place to build up the ion gradient described in Table 3.1, which is needed for many intracellular processes (particularly secondary active transport, action potential, and signal transduction):

- **Na⁺,K⁺-ATPase** uses ATP to pump Na⁺ ions out of the cell and K⁺ ions into the cell.
- **Ca²⁺-ATPase** pumps Ca²⁺ into the endoplasmic reticulum.

Several strategies of transporting organic compounds against a concentration gradient are available to a cell (Fig. 3.5 B–E).

- For **active transport** through **ABC (ATP-binding cassette) transporters**. ATP can be used as energy source. ABC transporters are widespread among organisms (prokaryotes as well as eukaryotes) and are encoded by many genes. In humans, more than 50 ABC transporter genes are known. Particularly important are the MDR genes coding for **multidrug resistance proteins** such as **P-glycoprotein (P-gp)**. They are often very strongly expressed in diseased tissue. If, for example, a drug has penetrated a tumor cell through diffusion, it is pumped out immediately into the extracellular space, thus losing its effectiveness (Fig. 3.5 D). The best known is P-gp (MW 170 kDa; two ATP-binding sites and 12 transmembrane regions), encoded by the MDR1 gene. P-gp is active in gut epithelia and in many other tissues. Furthermore, MRP1 and 2 (*multiple resistance-associated protein*; 190 kDa; two ATP-binding sites and 17 transmembrane regions) are important, which are encoded by MRP1 and MRP2 genes (Fig. 3.5E). Overexpression of MDR proteins is also responsible for drug resistance in some malaria-causing parasites.

- Apart from active transport with direct use of ATP, there are many other active transport mechanisms in a cell. These are referred to as **secondary active transporters**. They make use of an ion gradient that has been built up to transport a specific metabolite against the concentration slope, using up ATP. Depending on whether the ions that share the pathway are concentrated on the same or the opposite side of the biomembrane, the transport is called **symport** or **antiport**. The transport mechanism resembles a revolving door, which can be operated from the inside as well as the outside. Within an individual cell, more than one transporter may be needed for one specific substance, depending on the concentration within the cell and in the extracellular space. This has been well researched by studying **glucose transporters** in intestinal cells (Fig. 3.6). There is a Na^+ symporter on the luminal side, pumping glucose into the intestinal cell against a gradient. As the glucose concentration in the blood is lower, all that is needed is a simple uniporter to carry the glucose along the concentration gradient. The sodium ions that have been enriched inside the cell are pumped out of the cell using Na^+,K^+-ATPase.

Research on genome projects of diverse organisms showed that genomes contain many transporter genes, although their specificity and function is still partly unknown. Finding answers to these questions is not only relevant to the understanding of cellular transport processes, but is also extremely important for pharmaceutical research. The **pharmacokinetics** of a compound is a crucial aspect. Although we often know that an active agent is taken up (i.e., it is bioavailable), we still do not know whether the uptake is the result of diffusion, the use of a transporter, endocytosis, or receptor-mediated endocytosis.

3.1.1.3 Receptors and Signal Transduction at Biomembranes

Apart from ion channels and transporters, there are many other membrane proteins contained in the cytoplasmic membrane, such as **receptors**, **enzymes**, and **anchor proteins**. Some of these are schematically shown in Fig. 3.5(A).

In a multicellular organism, the cells must be able to recognize and process signals from outside, coming from other cells or tissues. There are several cellular communication options (Fig. 3.7):

- **Endocrine signals (hormones)** are produced by endocrinal gland cells (Table 3.2) and are released into the bloodstream. They circulate through the body and are picked up by receptors in the **target cells** – sometimes in a very distant part of the body – where they spring into action. In other words, hormones have a systemic effect. **Hydrophilic and polar hormones** (adrenaline and growth factors) bind to **membrane receptors**, whereas **lipophilic hormones** (e.g., steroidal hormones, thyroxine, retinoic acid, vitamin D_3) diffuse into the target cells to bind to **intracellular receptors**. These act as **transcription factors**, controlling the expression of hormone-regulated genes.

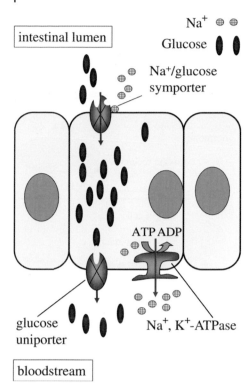

Fig. 3.6 Glucose transporters in an intestinal cell. Glucose is pumped from the intestine into the cell by a Na^+/glucose symporter and leaves it again using a uniporter, following the concentration gradient.

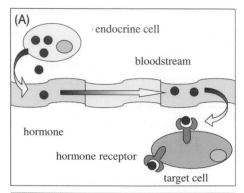

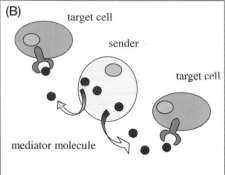

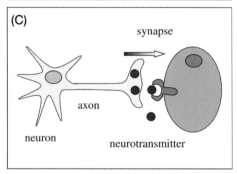

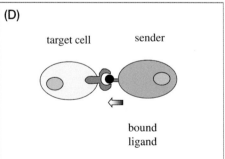

Fig. 3.7 Schematic view of communication pathways between cells (Alberts *et al.*, 2008) (A) Endocrine hormones, (B) Paracrine mediators. (C) Synaptic signal transduction. (D) Cell-to-cell communication.

- **Paracrine signals** have an effect on their immediate surroundings. Released from a tissue cell, they are recognized and processed by neighboring cells. Their effect is local (e.g., prostaglandins).
- In direct **cell-to-cell interaction**, a cell presents a membrane-bound signaling molecule to another cell carrying a membrane receptor that recognizes the molecule. Examples are found in the **immune system** (e.g., MHC complex and T cell receptors).
- In **neuronal signal transduction**, an electric signal (**action potential**) is transformed into a chemical signal at the **synapse**. **Neurotransmitters** are released that are recognized and processed by the receptors of a postsynaptic target cell.

Polar signaling molecules that are unable to pass the biomembrane through diffusion are recognized by receptors on the cell surface. There are three categories of such receptors (Fig. 3.8):
- **Ion channel-linked receptors** are activated by specific ligands. As a reaction, the conformation of the channel protein is modified, leading to the opening or closure of the channel in question. Ions are let in or out accordingly. The changes in ion concentration produce a change in the membrane potential. In this way, the tension in ion channels can be regulated or new action potentials released. **Ion channel-linked receptors** are mainly found in the neuronal system, such as the **nicotinic acetylcholine receptor (nAChR)**, the **GABA receptor**, the **NMDA receptor**, and the **glycine receptor**.
- **G-protein-linked receptors (GPCR)** communicate with a **G-protein** that is bound either to **GTP** or **GDP**. The activation of this type of receptor by a ligand causes a **conformation change**, which is recognized by the G-protein. The G-protein (or, to be more precise, its α-subunit) is activated and can, in turn, interact with a membrane-bound effector protein. The effector protein is often an enzyme (**adenylate cyclase** or **phospholipase**), which produces second messengers. This mechanism whereby a single signaling molecule activates a multitude of effector proteins, which, in turn, release a host of second messengers, results in an effective amplification of the signal. **Adenylate cyclase** turns ATP into **cAMP**, which acts as second messenger, regulating **protein kinase A** allosterically. Once protein kinase A has been activated, it may phosphorylate other enzymes or proteins (e.g., transcription factors), which then spring into action (Fig. 3.9). After the dissociation of the α-subunit, the $\beta\gamma$-complexes of the activated G-protein can also be biologically active. In the cardiac muscle, acetylcholine binds to a **muscarinergic receptor (mAChR)**, thus activating the $\beta\gamma$-complex. The $\beta\gamma$-complex binds to K^+ channels and opens them. cAMP is degraded by **phosphodiesterase** – an enzyme that is considered a target structure for several pharmaceutical products. Table 3.3 gives an overview of some essential hormones that are amplified by cAMP.
- **Phospholipase C** is another important effector protein, cleaving **phosphatidylinositol** into **inositol-1,4,5-triphosphate (IP$_3$)** and **diacylglycerol (DAG)** after activation (Fig. 3.10). IP$_3$ acts as second messenger, binding to ryanodine receptors in the endoplasmic reticulum and thus activating a calcium channel. Calcium can also act as a signaling substance, activating, for example, **protein kinase C (PKC)**, various **calmodulin-dependent (CaM) kinases**, and many other proteins (Fig. 3.11). PKC, a modulator of many target proteins (such as transcription factors), can also be activated by DAG (Fig. 3.11). Table 3.3 summarizes the major signaling processes involving phospholipase C. For medical research, G-protein-linked signaling pathways are of major interest, as they are used by many currently available pharmaceuticals. There are still many unknown steps in the process, which could prove interesting targets for new drugs to be developed.
- **Enzyme-linked receptors** can be activated by a signaling molecule (e.g., various **growth factors** that stimulate cell division) (Figs 3.8 and 3.11). In dimeric

Table 3.2 Most important hormones in humans.

Hormone	Hormone gland	Target	Activity/function
Releasing hormones (P)	Hypothalamus	Adenohypophysis	Regulate release of hormones from adenohypophysis
Inhibitory hormones (P)	Hypothalamus	Adenohypophysis	Regulate release of hormones from adenohypophysis
Oxytocin (P)	Hypothalamus	Uterus, mammary gland	Stored and released from neurohypophysis; stimulates uterus contractions and milk secretion
Thyreotropin (GP)	Adenohypophysis	Thyroid	Stimulates synthesis and secretion of thyroxin
Adrenocortico-tropic hormone (ACTH) (P)	Adenohypophysis	Adrenal cortex	Stimulates secretion of hormones of adrenal cortex
Luteinizing hormone (LH) (GP)	Adenohypophysis	Gonads	Stimulates secretion of sex hormones from ovary and testes
Follicle-stimulating hormone (FSH) (GP)	Adenohypophysis	Gonads	Stimulates development of egg and sperm cells
Somatotropin (hGH) (P)	Adenohypophysis	Bones, liver, muscles	Stimulates protein synthesis and growth
Prolactin (P)	Adenohypophysis	Mammary	Stimulates milk production
Melanocyte stimulating hormone (MSH) (GP)	Adenohypophysis	Melanocytes	Regulates pigmentation of skin
Endorphins, enkephalins (P)	Adenohypophysis	Neurons of spinal cord	Analgesic properties
Adiuretin (ADH, vasopressin) (P)	Neurohypophysis	Kidneys	Stimulates water reabsorption and increases blood pressure
Melatonin (AA)	Epiphysis	Hypothalamus	Regulates biological rhythms (e.g., day/night rhythm)
Thyroxin (AA)	Thyroid	Many tissues	General stimulant of metabolism
Calcitonin (P)	Thyroid	Bones	Stimulates bone formation, lowers Ca^{2+} levels in blood
Parathormone (P)	Parathyroid	Bones	Stimulates bone absorption, increases Ca^{2+} levels in blood
Thymosins (P)	Thymus	Leukocytes	Activates T cell activity
Glucagon (P)	Panreas	Liver	Stimulates glycogen breakdown; increases blood sugar levels
Somatostatin (P)	Pancreas	Pancreas	Inhibits release of glucagon, insulin and digestive enzymes
Insulin (P)	Pancreas	Liver, muscles	Stimulates uptake of glucose and glycogen formation
Gastrin (P)	Stomach	Stomach	Stimulates release of digestive juices; enhances motility of stomach
Secretin (P)	Duodenum	Pancreas, stomach Gall bladder	Regulates digestion processes Induces contraction of gall bladder
Adrenaline/noradrenaline (AA)	Adrenal medulla	Heart, liver, blood vessels	Stimulates glycogen breakdown; stimulate heart, circulation and blood pressure
Cortisol (glucocorticoid) (S)	Adrenal cortex	Muscles, many tissues	Regulates stress reactions; stimulates metabolism of proteins and lipids; gluconeogenesis, inhibits inflammatory reactions
Aldosterone (mineral corticoid) (S)	Adrenal cortex	Kidneys	Stimulates excretion of K^+ and ammonium ions, and Na^+ reabsorption
Estrogen (S)	Ovary	Mammary, uterus	Regulates development and function of female sexual characters and sexual behavior
Progesterone (gestagen) (S)	Ovary (corpus luteum)	Uterus	Important for pregnancy and embryonic development
Testosterone (S)	Testes	Diverse tissues	Regulates formation of sperm cells and development and function of male sexual characters and sexual behavior
Atrial natriuretic peptide (ANP) (P)	Heart	Kidneys	Stimulates Na^+ excretion
Vitamin D (S)	Skin	Bones, kidneys	Enhances blood Ca^{2+} level

P, peptide or protein; S steroid, GP, glycoprotein; AA, amino acid derivative.

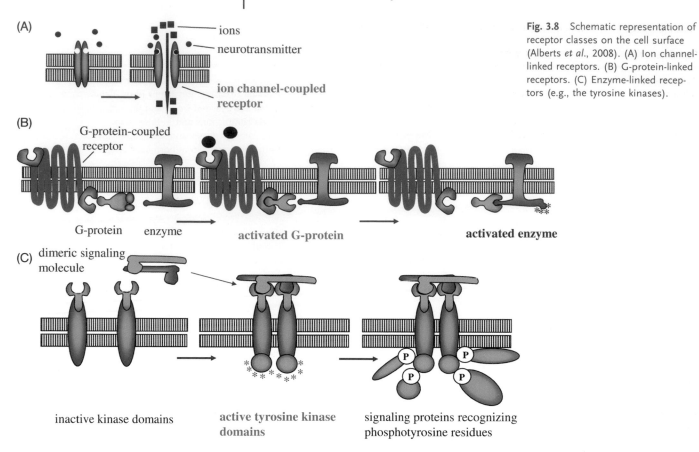

(A)

ions

neurotransmitter

ion channel-coupled receptor

(B)

G-protein-coupled receptor

G-protein   enzyme   **activated G-protein**   **activated enzyme**

(C) dimeric signaling molecule

inactive kinase domains   **active tyrosine kinase domains**   signaling proteins recognizing phosphotyrosine residues

Fig. 3.8 *Schematic representation of receptor classes on the cell surface (Alberts et al., 2008). (A) Ion channel-linked receptors. (B) G-protein-linked receptors. (C) Enzyme-linked receptors (e.g., the tyrosine kinases).*

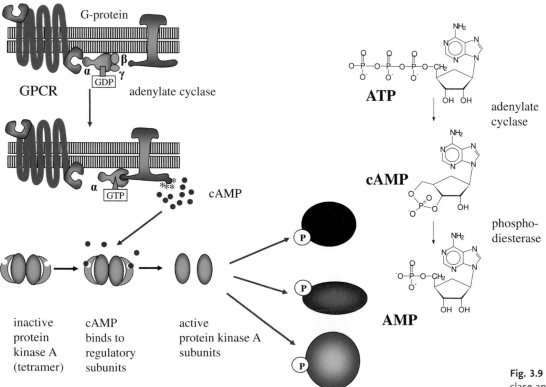

G-protein

GPCR   adenylate cyclase

cAMP

inactive protein kinase A (tetramer)   cAMP binds to regulatory subunits   active protein kinase A subunits

target proteins

ATP

adenylate cyclase

cAMP

phospho-diesterase

AMP

Fig. 3.9 *Activation of adenylate cyclase and formation from cAMP as second messenger (Alberts et al., 2008).*

Table 3.3 The role of adenylate cyclase and phospholipase C in signal transduction.

Signaling molecule	Target tissue	Main reaction
Adenylate cyclase		
Adrenaline	Heart	Raising heart frequency and enhancing contraction, muscles glycogen degradation
ACTH	Adrenal gland	Secretion of cortisone
ACTH, adrenaline,	Fat tissue	Fat digestion
glucagon	Liver	Glycogen degradation, incease of blood glucose levels
Phospholipase C		
Vasopressin	Liver	Glycogen degradation
Acetylcholine	Pancreas	Secretion of amylase
	smooth muscles	muscle contraction
Thrombin	Platelets	Platelet aggregation

Table 3.4 Signal proteins that act via receptor tyrosine kinases.

Signal protein	Receptor	Activity
Epidermal growth factor	EGF-R	Stimulates cell growth and differentiation
Insulin	Insulin-R	Enhances glucose consumption and protein synthesis
Insulin-like growth factor	IGF-1-R	Stimulates cell growth
Nerve growth factor (NGF)	TrkA	Stimulates cell growth and survival of neurons
Platelet-derived growth factor	PDGF-R	Stimulates cell growth; differentiation and cell migration
Macrophage colony-stimulating factor	MCSF-R	Stimulates cell growth and differentiation of macrophages and monocytes
Fibroblast growth factors (FGF1–FGF24)	FGF-R	Stimulates cell growth and differentiation
Vascular endothelial growth factor	VEGF-R	Stimulates angiogenesis
Ephrins	Eph-R	Stimulates angiogenesis and axon orientation

R, receptor.

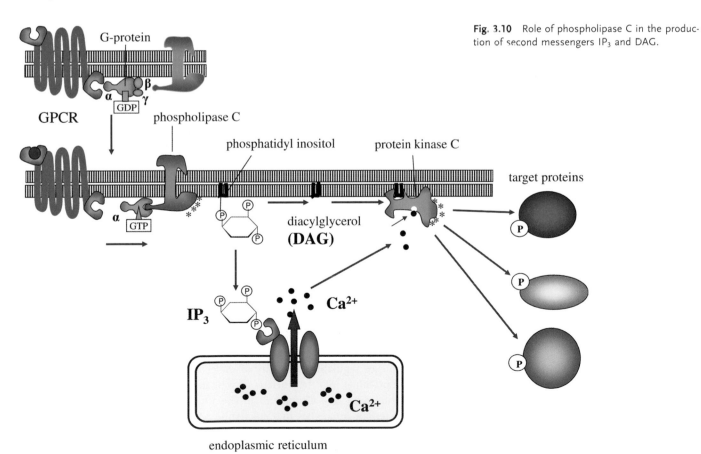

Fig. 3.10 Role of phospholipase C in the production of second messengers IP$_3$ and DAG.

Fig. 3.11 Signal transduction after activation of G-protein and enzyme-linked receptors (Alberts *et al.*, 2008). GPCR=G-protein coupled receptor; GEF=guanine exchange factor.

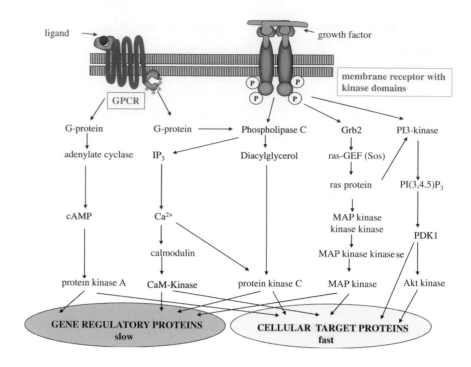

Fig. 3.11 Signal transduction after activation of G-protein and enzyme-linked receptors (Alberts *et al.*, 2008). GPCR=G-protein coupled receptor; GEF=guanine exchange factor.

receptors, two units form an active receptor with enzyme domains on the cytosolic side. The dimerization process activates **tyrosine kinases** (Table 3.4) that begin to phosphorylate each other. The **phosphotyrosine** residues are recognized by specific adapter proteins that are activated by them and then cause the release of other signaling proteins. Since such enzyme-linked receptors are often found in tumor cells where they are overexpressed or permanently activated, their inhibition, especially the inhibition of tyrosine kinase, is a major strategy in the **treatment of cancer**. The drug Gleevec (STI-571) binds to the ATP-binding site and thus inhibits tyrosine kinases effectively.

3.1.2
Endomembrane System in a Eukaryotic Cell

Most eukaryotes have an extensive **endomembrane system** covering the entire intracellular space. The most striking parts are the **endoplasmic reticulum (ER)** and the **Golgi complex**. Other compartments are also contained in biomembranes and form separate reaction entities within the cell. The characteristics of internal biomembranes may vary, depending on the membrane proteins and lipids they contain.

The ER (Fig. 3.12) is an extensive labyrinth of tubules and sacs pervading the entire eukaryotic cell. It is here that the components of biomembranes are put together and it is here that the **posttranslational modification** of proteins takes place. The rough ER contains **ribosomes**, which produce (translate) proteins ready for export (see Chapter 5.3), while the **smooth ER** is ribosome-free and contains enzymes requiring a lipophilic environment, such as **cytochrome oxidases**.

The ER envelops the **eukaryotic nucleus**, which is thus surrounded by two biomembranes (Fig. 3.12). The nuclear membrane has two characteristic nuclear pore complexes regulating the entry of molecules (e.g., transcription factors and ribosomal proteins) into the nucleus as well as the export (e.g., of mRNA and ribosomal subunits) (see Chapter 5.1). The nucleus is one of the most striking and characteristic organelles in a eukaryotic cell, containing genetic information, stored as DNA. DNA is usually found in several linear double strands or chromosomes, as seen under the light microscope. It is usually surrounded by pro-

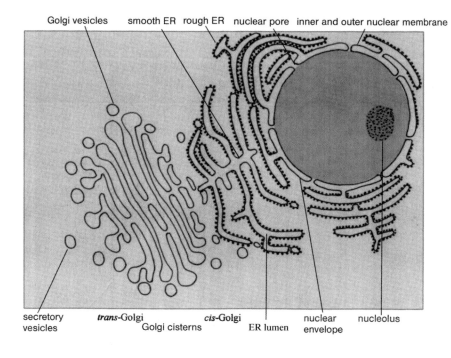

Golgi vesicles   smooth ER rough ER   nuclear pore   inner and outer nuclear membrane

secretory   *trans*-Golgi   *cis*-Golgi   nuclear   nucleolus
vesicles   Golgi cisterns   ER lumen   envelope

Fig. 3.12 Schematic representation of the endo-membrane system of the cell: nuclear envelope, rough and smooth endoplasmic reticulum (ER), and Golgi complex.

teins (e.g., **histones**) that form specific structures known as **nucleosomes** (see Chapter 4.1.2). The **nucleolus**, a distinct structure visible under the electron microscope, contains rRNAs serving as scaffolding for the assembly of ribosomal subunits.

Piles of membranous tubules form the **Golgi complex** (Fig. 3.12). The Golgi complex receives vesicles containing proteins from the ER on the *cis* side and passes them on for transport on the *trans* side to **lysosomes** or to the cytoplasmic membrane for export. The proteins are modified in the Golgi complex – sugar residues are cleaved off or added (see Chapters 5.3 and 5.4). The Golgi complex is particularly developed in glandular cells.

Lysosomes (Fig. 3.13) are small, membrane-enclosed organelles with an irregular structure. They contain a range of hydrolytic enzymes (**nucleases, proteases, glycosidases, lipases, phosphatases, sulfatases,** and **phospholipases**) that degrade lipids, polysaccharides, proteins, and nucleic acids. Liposomes also degrade and **recycle** defective macromolecules or organelles. Monomers released from proteins, polysaccharides, and lipids are often recyclable. Lysosomes evolve from vesicles cut off from the Golgi complex, also known as **endosomes**. Their pH value is acidic, due to membrane-bound H^+-ATPases pumping protons into the lysosomes. Hydrolytic enzymes have a pH optimum of 4–5 and become inactive at pH 7. Thus, should any of the hydrolytic enzymes escape into the cytoplasm, which has a pH around 7.4, they cannot cause any harm. **Lysosomes** fuse with **endosomes** or **phagosomes** that are pinched off from the cytoplasmic membrane by endocytosis and filled with protein complexes or microorganisms (see Chapter 5.4).

Plant cells do not contain lysosomes, but **vacuoles**. These can make up by far the largest compartments in adult plant cells (Figs 1.2 and 3.13). Vacuoles store inorganic ions and low-molecular-weight metabolites (e.g., sugar, organic acids, and amino acids). All plants produce **secondary compounds** such as flavonoids, phenylpropanes, tannins, terpenes, iridoid glycosides, alkaloids, glucosinolates, and cyanogenic glycosides, which are not needed for their primary metabolism. Contrary to earlier beliefs, they are not waste products, but ensure the survival of the plant, defending it against herbivores and microorganisms. As signal compounds, they can also help communication with other organisms by attracting insects for pollination or animals for seed propagation. **Polar secondary**

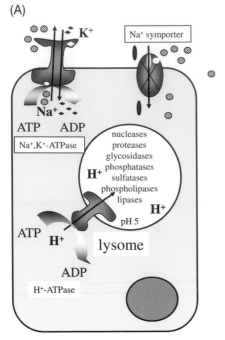

animal cell

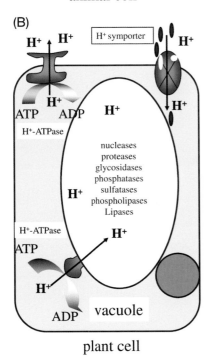

plant cell

Fig. 3.13 Similarities of lysosomes and plant vacuoles. (A) Schematic structure of lysosomes. (B) Schematic structure of plant cells with vacuoles.

compounds are frequently stored in vacuoles, whereas **lipophilic compounds** are kept in oil vessels, resin channels, or glandular cells. Often, secondary compounds are stored as prodrugs and only when the plant is wounded or an infection occurs will they be activated – mostly by **β-glucosidase** cleaving off a glucose residue. Seeds contain vacuoles storing a protein reserve. Vacuoles can have various functions in a plant cell – providing additional **storage space** or acting as a **defense or signaling compartment**. The storage function in particular requires high osmotic pressure inside the vacuole, which is crucial for the stabilization and growth of the plant (**turgor regulation**). Active transport in plants mostly functions via **proton gradients**, as opposed to animal cells, which rely more on Na^+/K^+ gradients, using Na^+,K^+-ATPase (see Section 3.1.1.2).

Peroxisomes are small membrane-enclosed, mostly rounded vesicles in which H_2O_2 is produced or degraded (e.g. by the enzyme catalase).

3.1.3
Mitochondria and Chloroplasts

Mitochondria are very striking organelles that are found in all eukaryotic cells (Fig. 3.14). They look like worms or sausages and are between 1 μm and several micrometers long and 0.5 μm thick. Mitochondria have two separate membrane systems. The inner membrane forms a series of infoldings (cristae) that extend the surface considerably. The large surface area is needed because proteins and enzymes of the **respiratory chain** are found in or on the inner mitochondrial membrane (Fig. 3.15).

The **respiratory chain** produces ATP from the **reduction equivalents NADH** and **FADH$_2$**. During this process, electrons are transported through several intermediate stages and a **proton gradient** is built up to provide energy for **ATP synthase**. This is also referred to as cellular respiration because the respiratory chain uses up oxygen. Without mitochondria, aerobic organisms such as animals, fungi, and plants would not be able to use oxygen from the air for the oxidation of organic matter – in other words, to **produce energy**. There are, however, some bacteria and a few eukaryotes that are anaerobic (i.e., they do not need oxygen). These organisms do not contain mitochondria.

During the **citric acid cycle (Krebs cycle)**, which takes place in the mitochondria, acetyl CoA is introduced, and in each run of the cycle, CO_2 and reduction equivalents are generated. The acetyl CoA is derived from pyruvate, a product of **glycolysis**, which has been taken up by the mitochondria through a **pyruvate transporter**. It is then transformed into acetyl CoA by a pyruvate decarboxylase complex. Another way of generating acetyl CoA is by the **β-oxidation** of fatty acids – a process that also takes place in mitochondria (Fig. 3.15).

Mitochondria contain their own ring-shaped DNA (Fig. 3.16). In animals, the mitochondrial genome (mtDNA) is significantly smaller (16–19 kb) than in plants. It contains 13 genes coding for enzymes or other proteins involved in electron transport, 22 genes for tRNAs, and two genes for rRNAs. As every animal cell contains several hundred or even thousand mitochondria, each of which contains 5–10 mtDNA copies, the total of mtDNA copies amounts to several thousand per cell. mtDNA makes up about 1% of the total amount of DNA contained in a cell. Plant mitochondria, by contrast, have large genomes (150–2500 kb). Some of their genes even have an intron/exon structure.

Mitochondria contain **functional ribosomes** equivalent to the prokaryotic 70S type, and the nucleotide sequences in mitochondrial genes and the amino acid sequences of the respective proteins are more closely related to the corresponding prokaryotic genes than to equivalents coded in the nucleus. The genetic code of mitochondria shows a few differences to the universal code: UGA (stop codon) codes in animals and fungi for tryptophan, AUA (for isoleucine) codes in animals and fungi for methionine, and AGG (arginine) codes in mammals for stop and in invertebrates for serine.

(A)

(B)

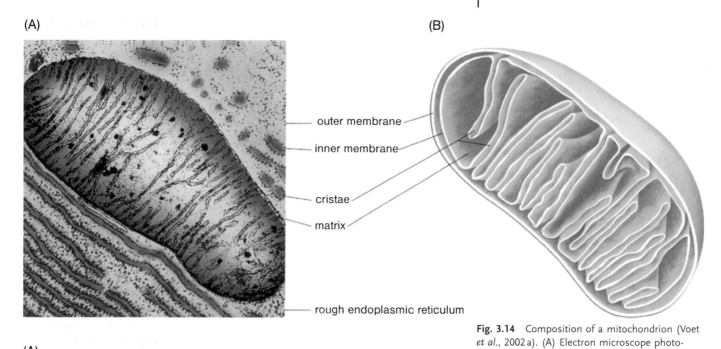

outer membrane

inner membrane

cristae

matrix

rough endoplasmic reticulum

Fig. 3.14 Composition of a mitochondrion (Voet *et al.*, 2002a). (A) Electron microscope photograph. (B) Schematic representation.

(A)

(B)

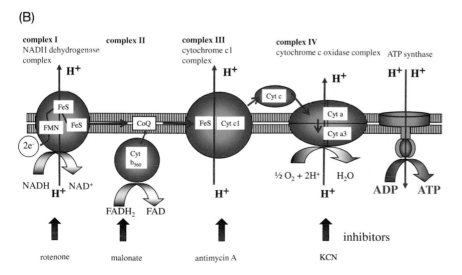

Fig. 3.15 Function of mitochondrion: metabolism and respiratory chain. (A) Metabolism and respiration in mitochondrion. (B) Schematic representation of the respiratory chain with the complexes I–IV; the proton gradient is used by ATP synthase to produce ATP. Rotenone, malonate, antimycin A, and KCN are the inhibitors of complexes I–IV. FeS, iron-sulfur cluster; Cyt, cytochrome; CoQ, ubiquinone; FMN, flavin mononucleotide.

Fig. 3.16 Schematic overview of the arrangement of genes in the mtDNA of mammals.

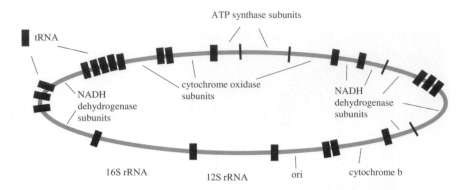

Fig. 3.17 Development of an early eucyte and origin of mitochondria. α-Purple bacteria were ingested by the early eucyte in a kind of phagocytosis. Hence, the outer mitochondrial membrane is derived from the host cell, whereas the inner mitochondrial membrane is the original bacterial cytoplasmic membrane.

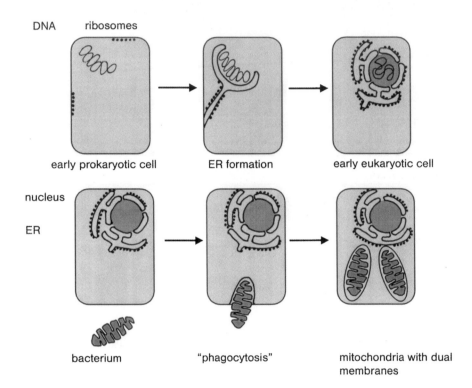

These findings as well as other mitochondrial characteristics led to the **endosymbiont hypothesis**, which states that mitochondria are derived from *α-purple bacteria* that were ingested by an ancestral eucyte 1.2 billion years ago and lived on as endosymbionts. The cell provides nutrients for the endosymbionts and receives ATP in return. Figure 3.17 shows a likely ingestion path for the α-purple bacteria into the ancestral eucyte. It is assumed that the ancestral eucyte came into being by the infolding of a bacterial cytoplasmic membrane to form an ER. The membrane then began to surround the chromosome, thus forming a nucleus.

Green plants and **algae** contain an additional organelle, the conspicuous **chloroplasts**, which are significantly larger and structurally more complex than mitochondria (Fig. 3.18). Apart from the surrounding inner and outer biomembranes, the chloroplast contains an extensively folded membrane system, known as **thylakoids**. These contain **chlorophyll**, as well as the proteins and enzymes required for photosynthesis, to enable the plants to turn sunlight into energy in the form of ATP and NADPH (Fig. 3.19). The electron transport between photosystem II and I and the production of NADPH are explained in Fig. 3.19(B). The light reaction leads to the build-up of a proton gradient, which is then used by **ATP synthase** to **produce ATP**. During the subsequent CO_2 fixation process (Calvin cycle), CO_2 is first bound to ribulose-1,5-biphosphate, which is then

(A)

(B)

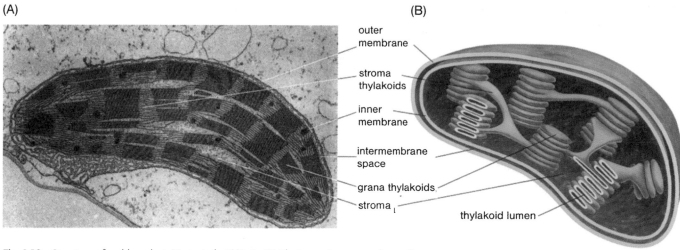

outer
membrane

stroma
thylakoids

inner
membrane

intermembrane
space

grana thylakoids

stroma

thylakoid lumen

Fig. 3.18 Structure of a chloroplast (Voet *et al.*, 2002a). (A) Electron microscope photo of a chloroplast. (B) Schematic representation.

(A)

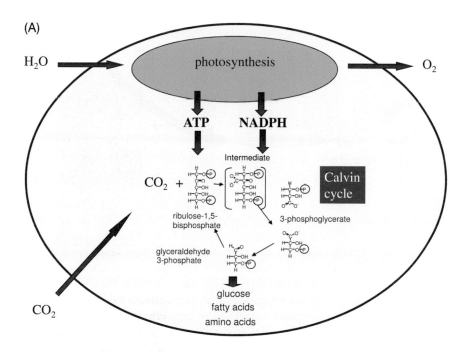

(B)

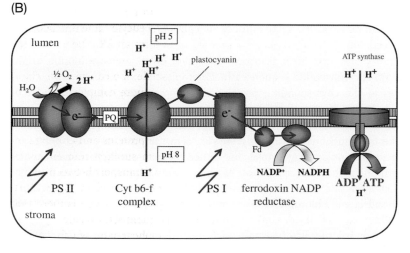

Fig. 3.19 Essential steps in photosynthesis. (A) Overview of photosynthetic reactions in chloroplasts. (B) Electron transport between photosystem II and I across the thylakoid membrane, resulting in NADPH production. ATP synthase uses the proton gradient for the production of ATP. Q, plastoquinone; FD, ferredoxin; PS I, photosystem I.

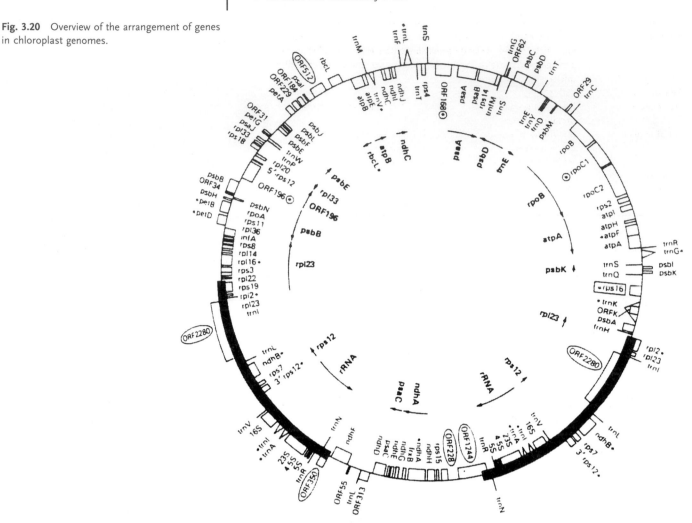

cleaved into two C3 units (3-phosphoglycerate). 3-Phosphoglycerate is transformed into glycerol aldehyde-3-phosphate, which is used for the regeneration of ribulose-1,5-biphosphate, and for building glucose, fatty acids, and amino acids. A plant cell can generate additional ATP from glucose for the energy supply of the cell. This makes plants autotrophic and a suitable basic nutrient for heterotrophic animals that live on organic matter.

Like mitochondria, chloroplasts contain their own **ring-shaped DNA** (cpDNA) as well as independent replication, transcription, and protein biosynthesis. The chloroplast genome has a size of 120–200 kb (Fig. 3.20); it encodes 120 genes and is present at 20–80 copies in a single chloroplast. As a plant cell contains up to 40 chloroplasts, the total number of cpDNA copies is between 800 and 3200 per cell.

For chloroplasts, too, it is assumed that there is an **endosymbiontic origin** (Table 3.5). The nucleotide sequences in chloroplast genes and the amino acid sequences of the corresponding proteins are more closely related to those of **cyanobacteria** than to the respective genes in the plant cell nucleus. Figure 3.21 gives a schematic overview of the presumptive origin of chloroplasts. Similar to the intake of mitochondria, an early eucyte seems to have taken up photosynthetic bacteria through phagocytosis and tamed them to develop an **endosymbiosis**. It is thought that the acquisition of chloroplasts happened **several times** in the phylogeny of photosynthetically active algae and plants.

Mitochondria and **chloroplasts** never emerge *de novo*, but replicate through **division**. When the cell divides, mitochondria are distributed over the daughter cells. Mitochondria can also fuse with each other. Both mitochondria and mostly

Genome	Mostly circular DNA adhesive to biomembrane without histones and nucleosomes, several copies concentrated in nucleoids; gene arrangement more or less prokaryotic (operon structure); repetitive sequences rare or nonexistent
Ribosomes	70S-type
Translation	No *Cap* structure at the 5′ end of mRNAs; prokaryotic complement of initiation factors
Tubulin, actin	Not found in organelles; FtsZ, a bacterial, tubulin-homologous cell division protein is involved in the division of plastids
Plastid fatty acid synthesis	As in bacteria, using acyl carrier proteins
Cardiolipin	Membrane lipid found in many bacteria. Not present in eucaryotic membranes except the inner mitochondrial membrane

Table 3.5 Prokaryotic properties of plastids and mitochondria.

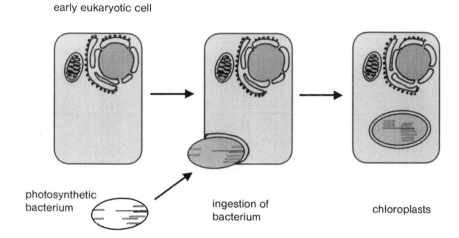

early eukaryotic cell

photosynthetic bacterium

ingestion of bacterium

chloroplasts

Fig. 3.21 Development of chloroplasts through phagocytosis of cyanobacteria.

also chloroplasts are inherited **maternally**. Mitochondria in sperm cells are not incorporated into the fertilized egg. Although replication, transcription, and protein biosynthesis still happen in the same way in mitochondria and chloroplasts, they have become organelles and are no longer autonomous. They import most of their proteins from the cytoplasm. These proteins carry **signaling sequences** that bind to receptors on the organelles (see Chapter 5), and through complex transport mechanisms, they finally reach their **working place** inside the mitochondria and chloroplasts. The corresponding genes used to be part of the endosymbionts, but have increasingly been moved into the nucleus. Only a relatively small set of genes has remained in mitochondria and chloroplasts. While this applies mostly to protein-coding genes, tRNA and rRNA genes have remained in the organelles.

3.1.4
Cytoplasm

The **cytoplasm** or **cytosol** of a eukaryotic cell is what is left when all membrane systems and organelles have been removed. In most cells, this is the largest compartment. In bacteria, it is the only existing compartment. It contains a multitude of low-molecular-weight compounds and proteins, including hundreds of **regulatory proteins** that are interlinked and communicate through complex interaction, such as phosphorylation and dephosphorylation of proteins, modulation by the binding of GTP or GDP, and conformational changes (cell biologists coined the term **cross-talk** for protein interaction). They can pick up signals and pass them on (**signal transduction**), and it will require extensive research to understand these processes in detail.

Fig. 3.22 Synopsis of the breakdown pathways and energy-producing pathways in heterotrophic organisms (e.g., in humans).

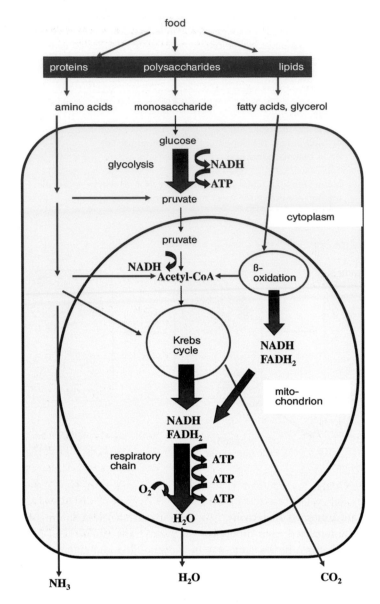

At the level of the cellular metabolism, there is a fundamental distinction between **catabolism** and **anabolism**. Catabolism is the **degradation of organic matter** (mostly polysaccharides, protein, and lipids) in order to provide chemical bonding energy through oxidation, which can be transferred to ATP. Polysaccharides are degraded to simple sugars such as glucose. Anabolism is the biosynthesis of monomers (e.g., amino acids, organic acids, and fatty acids) required for macromolecules, of macromolecules and other cell building blocks. Many catabolic and anabolic pathways involve the cytosol, but other compartments may also be involved (Fig. 3.22).

The **degradation of glucose to pyruvate** is an important energy-producing process. On balance, glycolysis produces 8 mol ATP per 1 mol glucose (2 mol NADH and 2 mol ATP). Pyruvate is transported into the mitochondria where it is transformed into acetyl CoA while producing NADH. In the mitochondria, acetyl CoA is further processed in the citric acid cycle, using up O_2, and releasing CO_2 and H_2O (Fig. 3.15). What matters for the energy balance is the provision of 4 mol NADH, 1 mol $FADH_2$, and 1 mol GTP from 1 mol pyruvate. In the respiratory chain, they produce 12 mol ATP per mol acetyl CoA and 15 mol per 1 mol pyruvate. One mole of glucose, when completely oxidized, produces 38 mol ATP. **Lipids** are hydrolyzed into fatty acids by lipases. Fatty acids are particularly rich in energy. During β-oxidation, they are broken down into acetyl

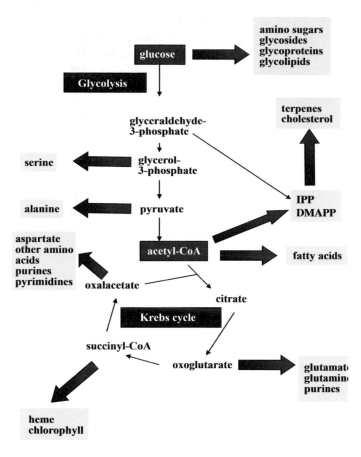

Fig. 3.23 Importance of glycolysis and the citric acid cycle as a point of departure for diverse biosynthetic pathways. IPP, isopentenyl pyrophosphate; DMAP, dimethylallyl pyrophosphate.

CoA in the mitochondria to provide NADH and $FADH_2$. One mole of stearic acid yields 9 mol NADH, $FADH_2$, and acetyl CoA, which is then further oxidized in the citric acid cycle. The total balance amounts to $9 \times 5 + 9 \times 12 = 153$ mol ATP. **Proteins** are broken down by proteases (pepsin, trypsin, and chymotrypsin) into amino acids. These can be entered into the degradation pathways at various stages, thus also producing ATP.

The synthetic pathways of the various low-molecular-weight building blocks are complex. They can often be derived from precursors in glycolysis or the citric acid cycle (Fig. 3.23). In cell biology, physiology, medicine, and biotechnology, it is important to have a good understanding of these various pathways. This introduction can only scratch the surface and readers should deepen their knowledge in the relevant textbooks.

3.1.5
Cytoskeleton

The cytoplasm is by no means an unstructured, soup-like fluid. It contains a complex network of thread-shaped proteins, which are part of the cytoskeleton. These networks that can be made visible using fluorescent dye or electron microscopy are often connected to the cytoplasmic membrane or cellular organelles:
- Actin filaments.
- Intermediary filaments.
- Microtubules.

The most subtle filaments are the **actin filaments** (Fig. 3.24), consisting of G actin monomers. Actin filaments are interconnected by a multitude of connecting and anchoring proteins. They are also in close contact with various membranes. The interaction of cytoskeletal proteins is particularly complex in muscle cells.

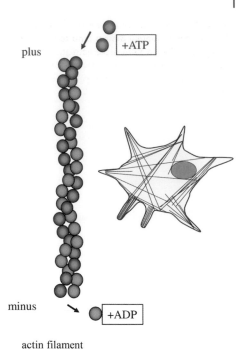

Fig. 3.24 Schematic composition of actin filaments (microfilaments).

The best studied cells are striated muscle cells. A single muscle fiber that has merged from several cells, thus containing several nuclei, contains many myofibrils. In these myofibrils, actin and myosin filaments cooperate, forming a highly organized nanomachine. Any contraction of a muscle is the result of highly coordinated interaction between actin filaments and myosin (Fig. 3.25).

The thickness of **intermediary filaments** lies in the middle between actin filaments and microtubules. Their main task is to stabilize the cell. The filaments are interconnected with many other proteins to create complex networks that are firmly anchored to the cytoplasmic membrane.

The thickest filament in the cytoskeleton are the **microtubules**, which form hollow tube systems and may be looked at as polymers of tubulin dimers (α- and β-tubulin) (Fig. 3.26). Microtubules play a special part in the intracellular transport of vesicles. During cell division, they form the spindle apparatus that transfers chromosomes into the daughter cells. During the metaphase, the condensed chromosomes line up along the equatorial plate of the cell. The microtubules bind to the centromeres of the chromatids and pull them into the new daughter cells. The microtubules extend from polar-bound centrioles.

Flagella and cilia contain microtubules as supramolecular complexes (9+2 structure, Fig. 3.26). Contact between two neighboring microtubules is mediated by dynein. The movement of microtubules against each other causes the cilia to bend, which, in turn makes them move.

In **cancer treatment**, microtubules are important target structures for chemotherapeutics. The vinca alkaloids vinblastine and vincristine or colchicine inhibit the polymerization of tubulin dimers, which form microtubules. By contrast, taxol or paclitaxel derived from the yew tree stabilize microtubules and

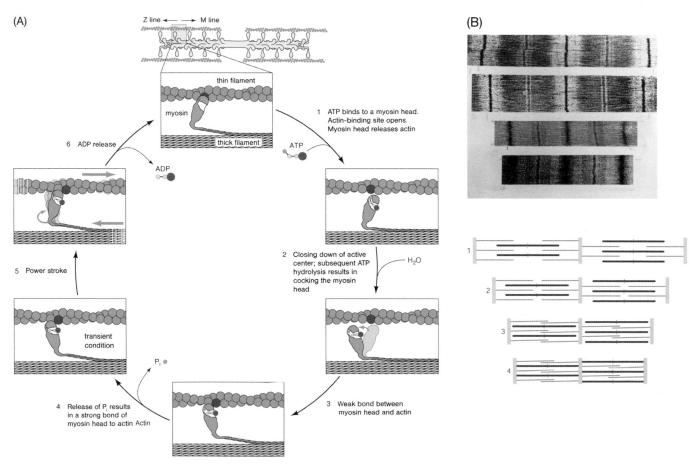

Fig. 3.25 Mechanism of muscle contraction. (A) Molecular mechanism of muscle contraction (see Voet *et al.*, 2002b, p. 185). (B) Contraction of myofibrils; the thin filaments are actin filaments, the thick filaments consist of myosin.

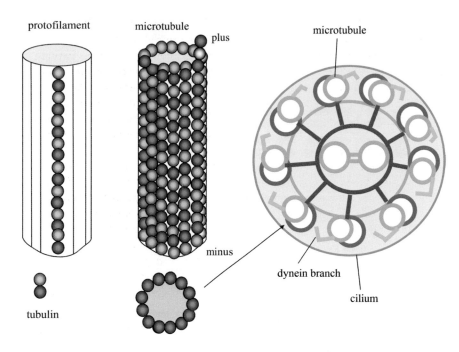

protofilament   microtubule

plus

microtubule

minus

tubulin

dynein branch

cilium

Fig. 3.26 Schematic view of microtubules and cilia structures. Tubulin dimers bind to GTP and then polymerize, to form protofilaments. Thirteen protofilaments are required to form a microtubule. Cilia and flagella are composed of 9+2 microtubules.

prevent their depolymerization. Also, actin and actin filaments serve as targets for some toxins. Phalloidin (one of the toxins from *Amanita phalloides*) binds to actin filaments and stabilizes them. Cytochalasin B (a mycotoxin) caps the plus site of actin filaments, swinholide (from a sponge) severs actin filaments while latrunculin (also from a marine sponge) inhibits polymerization G actin into actin filaments.

Cytoskeletal filaments form complex networks, while also providing a matrix that organizes the other organelles and multienzyme complexes within the cell. Complex regulation mechanisms control the build-up and breakdown of cytoskeletal elements. ATP-consuming reactions (i.e., phosphorylation and dephosphorylation) and microtubule-binding proteins play an important part in the process (e.g., microtubule-associated proteins).

3.1.6
Cell Walls

Some cell types are enclosed by a cell wall:
- **Bacterial cells** are surrounded by a peptidoglycan layer (Fig. 3.27). **Gram-positive bacteria** (e.g., members of the *Bacillus* genus) have a thick cell wall, which borders immediately to the outside milieu, whereas in **Gram-negative bacteria** (e.g., *Escherichia coli*) a thin cell wall is surrounded by a second lipopolysaccharide membrane as an outer shell. The outer membrane has porin proteins that allow the entry of food molecules. The cell wall is an important target structure for **antibiotics** – penicillins and cephalosporins inhibit the cross-linking of linear glycopeptide strands. Bacitracin inhibits the synthesis of polyprenol, which is a prerequisite for the formation of a murein sacculus.
- **Fungal cells** are surrounded by a chitin wall.
- **Plant cells** have cell walls consisting of cellulose, hemicellulose, and pectin. They can be enzymatically digested by cellulases, producing protoplasts, which are useful for plant biotechnology.

Cell walls serve mostly to protect and stabilize cells. They ensure that in difficult circumstances the cells will not take up too much water through osmosis and burst.

Fig. 3.27 Schematic view of bacterial cell walls. (A) Gram-positive bacteria. (B) Gram-negative bacteria.

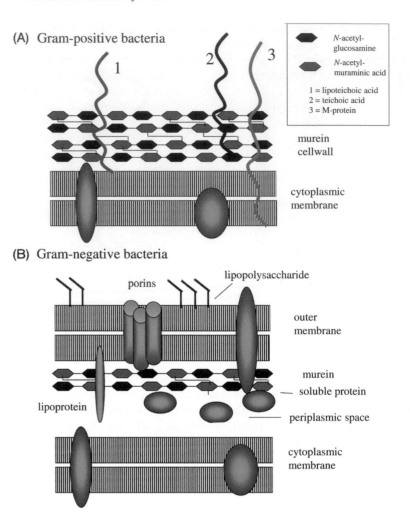

3.2
Structure of Bacteria

Compared to eukaryotic cells, **bacteria** have a fairly simple structure (Fig. 1.2 A). The outside of Gram-positive bacteria is shielded by a **peptidoglycan cell wall**, known as the **murein sacculus**. In Gram-negative bacteria the **periplasmic space**, where some of the metabolic processes take place, lies between the cell wall and the cytoplasmic membrane (Fig. 3.27). The cytoplasmic membrane contains many membrane proteins, including transporters, ABC transporters, receptors, and enzymes. There is no compartmentalization in bacterial cells (i.e., they do not contain organelles). However, the cytoplasmic membrane is sometimes in-folded, which makes it resemble an eukaryotic endomembrane system.

Contrary to earlier views, bacteria also contain various types of **cytoskeletal structures**, either based on FtsZ, Mreb/Mbl, or EF-Tu proteins. FtsZ seems to be related to tubulin and Mreb/Mbl to actin, as found in eukaryotes. There seems to be no eukaryotic equivalent to EF-Tu. All three forms may coexist in the same cell. Underneath their cytoplasmic membrane, bacteria have a cytoskeleton consisting of monomer proteins. Monomer EF-Tu proteins, for example, can form protofilaments. There are also cytoskeleton-like interconnections or fibers.

The proteins are synthesized in **ribosomes** that lie freely in the cytoplasm or are associated with the inside of the cytoplasmic membrane (Fig. 1.2A).

Bacteria carry their genetic information on one single chromosome. This ring-shaped DNA strand is also known as a **nucleoid**. There are additional ring-shaped molecules called **plasmids**, which also carry genetic information and may include **antibiotic resistance genes**. Modified plasmids play an important role as cloning vectors in molecular biology and biotechnology (see Chapter 15).

Bacteria continue to be the favorite "pets" of molecular biologists and biotechnologists. Basic research on genetics, molecular biology, and biochemistry is often first carried out in bacteria such as *E. coli*. Some bacteria are even indispensable for the cloning and expression of DNA (see Chapters 15 and 16).

Bacterial infections are the cause of many **diseases** in humans, animals, and plants. Some bacteria damage the host through sophisticated toxins that interfere with the signaling structure. **Tetanus toxins** from *Clostridium tetani* act as proteases. In the synapses, they specifically hydrolyze SNARE proteins, thus blocking neuronal signal transduction. **Cholera-causing *Vibrio cholerae*** produces an enzyme that redirects the transfer of ADP ribose from NAD^+ to the a-subunit G_s of a G-protein. This inhibits GTPase, and the once activated adenylate cyclases remain permanently active, producing cAMP. As a consequence, intestinal cells secrete excessive amounts of Cl^+ ions and water, resulting in diarrhea. *Bordetella pertussis*, the cause of whooping cough, produces enzymes that activate the a-subunit G_i of the G-protein, preventing G_i from regulating its target proteins.

The discovery and the development of new antibiotics from various *Streptomyces* and fungal species in the second half of the twentieth century was a milestone in medical history, saving millions of lives. However, **pathogenic bacteria** (e.g., *Pseudomonas aeruginosa* and *Staphylococcus aureus*, such as methicillin-resistant *S. aureus* (MRSA)) are becoming increasingly resistant to effective antibiotics. This is why the development and production of new antibiotics remains a high priority for the biotechnological industry.

Organic low-molecular-weight compounds such as amino acids or recombinant proteins are often produced in bacteria (see Chapter 16). Sometimes, genetic manipulation can give a substantial boost to the yield (Chapter 33).

3.3
Structure of Viruses

Viruses (or **phages** when found in bacteria) are not autonomous organisms. Although they have some cell elements in common with bacteria (DNA or RNA as genetic information) (Table 3.6), they depend on host cells for their propagation.

Table 3.6 Classification of major animal and human pathogenic viruses.

Class	Example/Disease
I. dsDNA (double-stranded DNA)	
Papovavirus	Papilloma (cervical cancer)
Adenovirus	Infections of the respiratory tract, tumors in animals
Herpes virus	HV I (blisters on skin), HV II (blisters on genitals), Varicella zoster (chicken pox, shingles), Epstein-Barr virus (mononucleosis, Burkitt lymphoma)
Pox viruses	Smallpox, vaccinia, cowpox
II. ssDNA (single-stranded DNA)	
Parvovirus	Phlebotomus fever
III. dsRNA (double-stranded RNA)	
Reovirus	Diarrhea viruses, diseases of the respiratory tract
IV. ssRNA (working as mRNA)	
Picornavirus	Polio virus, common cold viruses, enteroviruses
Togavirus	Rubella, yellow fever, encephalitis
V. ssRNA (used as matrix for mRNA synthesis)	
Rhabdovirus	Rabies
Paramyxovirus	Measles, mumps
Orthomyxovirus	Influenza viruses
VI. ssRNA (used as matrix for DNA synthesis)	
Retrovirus	RNA tumor viruses, HIV (AIDS)

Fig. 3.28 Infection cycle and genome of retroviruses. (A) Genome composition of retroviruses: *gag*, genes coding for capsid proteins, which will be further processed by a protease; *pol*, codes for reverse transcriptase; *env*, codes for envelope proteins, which are also cleaved through proteolysis; *onc*, oncogene. (B) Infection cycle of a retrovirus. *RVT* = reverse transcriptase.

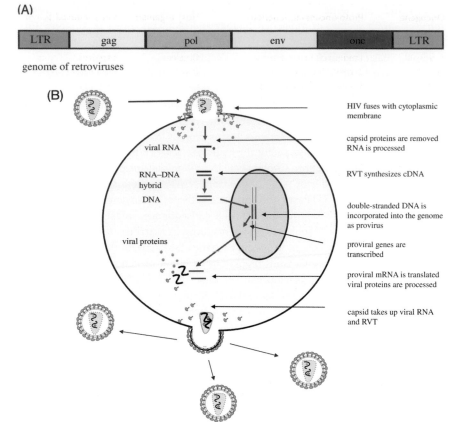

(A)

| LTR | gag | pol | env | onc | LTR |

genome of retroviruses

(B)

viral RNA

RNA–DNA hybrid

DNA

viral proteins

HIV fuses with cytoplasmic membrane

capsid proteins are removed RNA is processed

RVT synthesizes cDNA

double-stranded DNA is incorporated into the genome as provirus

proviral genes are transcribed

proviral mRNA is translated viral proteins are processed

capsid takes up viral RNA and RVT

They invade bacterial, plant, or animal host cells to live as parasites. Excessive viral multiplication causes the death of host cells and thus disease in the host.

Viral nucleic acid (Table 3.6) is enclosed by a protein envelope or capsid. Many viruses carry a biomembrane on the outside, which is derived from the host cell. It contains viral proteins (envelope proteins) that act as antigens. Viral proteins are often very variable. By **modifying their surface antigen** whenever they multiply, they are able to circumvent the immune system, which cannot keep up the speed to produce the latest specific antibodies. Viral proteins are tailor-made for each other. This enables them to spontaneously form supramolecular complexes and infectious viral particles.

Retroviruses such as the HIV pathogen are medically very significant (Table 3.6). They carry genetic information as RNA (Figs 1.3 B and 3.28). The retroviral genome codes for a relatively low number of gene products, amongst others for **reverse transcriptase**, which translates viral RNA into DNA (cDNA). **Oncogenes** have the ability to transform cells into tumor cells. The discovery of viral oncogenes was essential for the understanding of regulatory mechanisms that are involved in cell division, cell differentiation, and the development of cancer (Table 3.7). Phages and viruses are not only important as pathogens, but also as vectors for cloning and gene therapy (see Chapter 30).

3.4
Differentiation of Cells

Although many characteristics discussed in the earlier sections apply to all cells, we must remember that there are differences between monocellular organisms and that a multicellular organism contains a variety of cells that are differentiated in many ways according to the division of labor among them.

Oncogene	Proto-oncogenic function	Host organism	Virus-induced tumor
Abl	Tyrosine kinase	Mouse, cat	Pre-B cell leukemia
Erb-B	Epidermal growth factor	Chicken	Sarcoma
Fes	Tyrosine kinase	Cat, chicken	Fibrosarcoma
Fms	Receptor of macrophage colony-stimulating factor (M-CSF)	Cat	Sarcoma
Fos, jun	Join to produce gene regulatory protein	Cat, chicken	Osteosarcoma, fibrosarcoma
Myc	Gene regulatory protein	Chicken	Sarcoma
Raf	Serine/threonine kinase	Chicken, mouse	Sarcoma
H-ras	GTP-binding protein	Rat	Sarcoma
Rel	Gene regulatory protein	Turkey	Reticuloendotheliosis
Sis	Growth factor from platelets	Monkey	Sarcoma
Src	Tyrosine kinase	Chicken	Sarcoma

Table 3.7 Viral oncogenes that may play a part in the emergence of tumors. The cellular equivalent of a viral oncogene is known as proto-oncogene.

Table 3.8 Overview of important cell types in plants and animals.

Cell and tissue type	Function
A. Plant cells and tissues	
All plant organs consist of three basic kinds of tissue (epidermal, ground and vascular tissue)	
Epidermal tissue	
Epidermis	Epidermal cells form one or two layers of epidermis with a thick cuticle
Guard cells	Gas exchange
Trichomes	Epidermal hair cells: storage of terpenes; protection against evaporation
Root hairs	Uptake of water and ions
Endodermis	Innermost layer of the cortex
Protoderm	Primary meristem (growth of the epidermal tissues)
Ground tissue	
Parenchyma	Not very specialized; flexible primary walls
Mesophyll cells	Photosynthesis
Storage parenchyma	Storage tissue
Xylem parenchyma	Exchange of substances with xylem vessel elements
Collenchyma cells	Living cells with thick primary walls (support), no secondary walls and no lignin
Sclerenchyma	Dead cells with a support function
Fiber cells	Long extended lignified sclerenchyma cells
Sclereids (stone cells)	Irregularly shaped sclerenchyma cells with thick lignified secondary walls
Ground meristem	Primary meristem (growth of ground tissue)
Vascular tissue	
Phloem	Transport of synthesized nutrients (sucrose, amino acids) to roots, stems and fruits
Sieve tube element	Living cell without nucleus and ribosomes, sieve plates between neighboring sieve cells
Companion cell	Exchange of substances with sieve tube elements.
Xylem	Transport of water and inorganic ions
Tracheids	Long tubular system consisting of dead cells (sclerenchyma)
Vessel elements	Lignified secondary walls with pits, surrounded by living xylem parenchyma
Procambium	Primary meristem (growth of vascular tissue)
B. Animal cells	
The human body contains more than 200 cell types and 4 types of tissues (epithelia, connective tissue, nerves and muscles)	
Embryonic stem cell	Omnipotent cell that can differentiate into all other cell types
Epithelia	
Intestinal cells	Prismatic epithelial cells, secretion of digestive juices and absorption of nutrients
Ciliated epithelium	Prismatic epithelial cells, secretion and absorption; transport of mucus (bronchial epithelium)
Glandular cells	Cubic epithelial cells in glands and kidney tubules; main function secretion
Endothelial cells	Simple squamous epithelium inside blood vessels

Table 3.8 (continued)

Cell and tissue type	Function
Connective tissue	
Fibroblast	Production of proteins for the extracellular network, including collagen and elastin
Osteoblast	Bone-producing cell
Chondrocyte	Cartilage production; secretion of collagen and chondroitin sulfate
Adipocyte	Production and storage of fat in fat tissue
Mast cells	Storage and release of histamine
Blood	
Haematopoietic stem cell	Precursor cell of all other blood cells
Erythrocyte	Oxygen and CO_2 transport through hemoglobin
Platelet	Blood coagulation
Lymphocyte	Specificity and diversity of immune response
T cells	T helper cells (Th) recognize antigenes and activate B cells
Cytotoxis T cells	Tc cells recognize antigens and attack infected cells
B cells	Form antibody-secreting plasma cells
Monocytes	Migrate to infection foci and mature into macrophages, devouring bacteria and debris
Granulocytes (leucocytes)	
Neutrophilic granulocytes	Phagocytize bacteria
Eosinophilic granulocytes	Destroy parasites, important in allergies
Basophilic granulocytes	Release histamines in some immune reactions
Natural killer cells	Destroy infected body cells and tumor cells
Nerve tissue	
Neuron	Reception, storage and transport of information
Glial cell	Supporting the structure and metabolism of neurons
Schwann cell	Forming a myelin sheath around the axons of the peripheral nervous system
Oligodendrocyte	Forming a myelin sheath around the axons of the CNS
Astrocyte	Large glial cells that give structural and metabolic support to neurons are crucial for the blood-brain barrier
Sensory cells	
Mechanoreceptor cells	Cells containing mechanoreceptors that are sensitive to pressure, touch, stretching, movement and sound
Hair cells	Cells (in the ear of vertebrates, in the side lines of fish) with mechanoreceptors that pick up movement in relation to their surroundings and sounds
Pain receptor cells	Cells containing nocireceptors; free nerve endings (dendrites), e.g. in the epidermis. Nocireceptors react to heat, pressure and irritants and are sensitized by prostaglandins
Temperature receptor cells	Cells containing thermoreceptors that gauge the temperature
Taste receptor cells	Cells containing chemical and taste receptors. They can distinguish the categories sweet, sour, salty and bitter
Smell receptor cells	Cells containing smell receptors
Light receptor cells	In the retina of vertebrae, cones and rods serve as photoreceptors
Muscles	
Striated muscle cell	Rapid and forceful contractions (skeletal muscle), controlled via the somatic nervous system
Smooth muscle cell	Slow and sustained contractions (in the intestine tract, in bladder, arteries and veins); no striation, controlled via the autonomous nervous system
Heart muscle cell	Striated; heart contraction
Gametes	
Sperm cells	Male gamete (haploid)
Egg cells	Female gamete (haploid)

Many simple organisms (bacteria, but also eukaryotes such as yeast, algae, or protozoa) consist of a **single cell**, whereas more highly developed organisms are **multicellular**. The level of bacteria and monocellular eukaryotes already shows a fascinating degree of differentiation and variety of shapes that are genetically controlled.

In multicellular organisms, an increasing **specialization** and **division of labor** can be observed in the cells. Through differentiation, huge differences occur in aspect, size, and function of the cells. The differentiated cells form specific tissues and organs that communicate with each other. In humans, more than 10^{14} cells of 200 different types (Table 3.8) are found in various tissues and organ systems. The human genome contains about 25 000 genes, of which less than 1000 are needed to provide proteins for a cell. What makes the division of cells and tissue possible is the differential expression of the genome. During the differentiation process, further genes are activated, while the majority of genes in a cell remain switched off. The specific selection and combination of expressed genes makes a wide range of functions and structures possible.

To find out which genes are active in which cell type is one of the major tasks of cell and molecular biology. Variations during the development of an organism or as a result of environmental changes further complicate the analysis. This area is known as functional genomics and proteomics. Developmental biologists try to find out which differentiation factors are necessary to change a totipotent **stem cell** (as present in the early embryonic stages) into a differentiated cell. This knowledge is essential for the use of stem cells in gene therapy or tissue engineering. Table 3.8 shows the major types of plant and animal cells and their main functions.

4
Biosynthesis and Function of Macromolecules (DNA, RNA, and Proteins)

Learning Objectives

This chapter introduces the composition of the genome, as well as the structure and function of chromosomes. Important processes that occur at chromosomes are DNA replication, DNA repair, recombination, and transcription. The mRNA is translated into proteins by the ribosomes. Basic principles of gene regulation are discussed. In the synthesis of nucleic acids, proteins, and polysaccharides, simple building blocks (nucleotides, amino acids, and sugar monomers) are joined together following condensation reactions in which a molecule of water is eliminated. As the condensation reaction is an energy-consuming process, these reactions do not take place spontaneously. Instead, they require energy and versatile multienzyme complexes. Hydrolysis of macromolecules is thermodynamically favored and is catalyzed by simple enzymes. Proteases (trypsin, chemotrypsin, and pepsin) break down proteins and peptides, DNases and RNases break down DNA and RNA, respectively, and glucosidases (e.g., amylase) break down polysaccharides.

4.1
Genomes, Chromosomes, and Replication

In the past few decades, **genomics** has developed into a new specialized area of genetics and biotechnology. The aim is the complete molecular and functional characterization of genomes of all important organisms. It is divided into **structural or functional genomics** (see Chapter 21). As part of the human genome project HUGO (Human Genome Organization), the nucleotide sequence of a human haploid chromosome has been almost completely determined. More than 1150 other genomes are already completely sequenced (as of 2010); including 100 genomes of Eukarya, 970 of Bacteria and 70 of Archaea (Table 4.1). By comparing nucleotide sequences obtained from various organ- and tissue-specific cDNA and **expressed sequence tag (EST)** banks, or through the construction of **knockout RNAi**, or antisense mutants, assigning the genomic sequences to functional units or genes is being attempted. Finally, **functional genomics** (see Chapter 21) will supply an exact answer to the question of which regions of the genome have a function (today it is estimated that the information necessary for survival constitutes 85–95% of bacteria and only 10% of the whole DNA for vertebrates) and which parts can be regarded as apparently functionless evolutionary remnants.

4.1.1
Genome Size

The total DNA of a cell is referred to as a **genome**. Genome sizes of major organism groups are shown schematically in Fig. 4.1. When the minimal genome size of organisms is examined (i.e., only the left side of the bar), an increase in size can be seen that mainly runs parallel to the organizational level. Bacteria

An Introduction to Molecular Biotechnology, 2nd Edition.
Edited by Michael Wink
Copyright © 2011 WILEY-VCH Verlag GmbH & Co. KGaA, Weinheim
ISBN: 978-3-527-32637-2

Table 4.1 Overview of a few of the genomes that are already sequenced and published (Mb = one million bases).

Organism	Size (Mb)
Archaebacteria	
Archaeoglobus fulgidus	2.18
Methanobacterium thermoautotrophicum	1.75
Methanococcus jannaschii	1.66
Pyrococcus horikoshii	1.80
Eubacteria	
Bacillus subtilis (gram-positive bacterium)	4.21
Borrelia burgdorferi (borreliosis pathogen)	1.44
Chlamydia trachomatis (pathogen of urogenital tract)	1.05
Escherichia coli (intestinal bacterium)	4.64
Haemophilus influenzae (pathogen of purulent throat infections)	1.83
Helicobacter pylori (stomach ulcer pathogen)	1.67
Mycobacterium tuberculosis (tuberculosis pathogen)	4.45
Mycoplasma pneumoniae (pneumonia pathogen)	0.81
Rickettsia prowazekii (typhus fever pathogen)	1.10
Treponema pallidum (syphilis pathogen)	1.14
Eukaryotes	
Plasmodium falciparum (Malaria pathogen)	1.00
Saccharomyces cerevisiae (Brewer's yeast)	12.069
Arabidopsis thaliana (Arabidopsis)	142
Caenorhabditis elegans (nematode)	97
Drosophila melanogaster (fruit fly)	137
Mus musculus (house mouse)	3000
Homo sapiens (human)	3200

Fig. 4.1 Number of nucleotides in the haploid genomes of important groups of organisms.

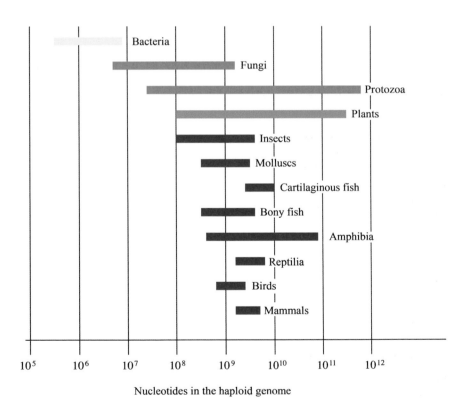

Nucleotides in the haploid genome

and fungi with simple structures have smaller genomes than structurally complicated multicellular organisms. It is presumed that the genome was enlarged particularly through **genome duplications**. Protostomia and the Deuterostomia ancestors of the vertebrates (see Chapter 6) contain generally only one copy of a gene, while several copies of a gene are often found in the genomes of chordates. As a result, it is supposed that the chordate genomes have doubled at least 2 or 3 times (**1-2-4 rule**). The first genome duplication during the **evolution of chordates** had already taken place before the **Cambrian explosion**, whereas the second and next doubling occurred in the early **Devonian period**. In the evolution of fish a further doubling of the genome occurred with up to eight copies of the original Deuterostomia (1-2-4-8 hypothesis) in the late Devonian period. This took place after the Actinopterygii and Sarcopterygii had already divided. Among the Sarcopterygii are the famous *Coelacanthus* and lung fishes. All land vertebrates (amphibians, reptiles, birds, and mammals) have apparently descended from them. Within the eukaryotes, the maximum genome size has only a small relationship to the developmental level. This is because many plants and amphibians have genomes with up to 10^{11} bases and the genomes are therefore one to two orders of magnitude higher than the genome of humans – it is obvious that many genome duplications must have taken place in these groups.

When the human genome (which has already been sequenced) is considered, it is obvious that a massive amount of information is present. If the DNA in an individual human cell was stretched out it would be 2 m long. With around 10^{13} cells in our body, the total length of DNA in all cells is 2×10^{10} km. It is therefore theoretically possible to stretch a DNA strand many times from the Earth to the sun and back again!

Of the 3.2 million bases that are present in human haploid chromosomes, about 25% of the DNA defines genes, but only 1.5% of the DNA codes directly for proteins (Table 4.2 and Fig. 4.2). The rest of the DNA is made up of RNA genes and noncoding sequences, which often either serve no function or their function is still unknown. In recent years microRNAs have been detected encoded in the "functionless" DNA, which are important for gene regulation (see Chapters 3 and 31).

Table 4.2 Relation between genome size and the number of genes of a few selected species whose genomes have been sequenced (see www.ebi.ac.uk/genomes).

Organisms	Genome size (bp)[a]	Number of genes
Archaea		
Archaeoglobus fulgidus	2.18×10^6	2405
Methanothermobacter thermoautotrophicus	1.75×10^6	1866
Pyrococcus furiosus (Archaea)	1.91×10^6	2057
Sulfolobus acidocaldarius (Archaea)	2.99×10^6	2221
Bacteria		
Clostridium tetani	2.8×10^6	2373
Escherichia coli	4.67×10^6	4288
Haemophilus influenzae	1.83×10^6	1702
Mycoplasma genitalium	0.58×10^6	476
Rhodospirillum rubrum	4.35×10^6	3791
Fungi		
Aspergillus fumigatus	2.9×10^7	9920
Saccharomyces cerevisiae	1.3×10^7	6275
Candida glabrata	1.4×10^7	5180
Sporozoa		
Plasmodium falciparum (causes malaria)	2.3×10^7	5300
Plants		
Arabidopsis thaliana	1.4×10^8	26000
Animals		
Caenorhabditis elegans (nematode)	1.0×10^8	20000
Drosophila melanogaster (fruit fly)	1.6×10^8	14000
Danio rerio (zebra fish)	1.0×10^9	25000
Mus musculus (mouse)	3.0×10^9	25000
Homo sapiens (human)	3.2×10^9	25000

a) Haploid genome.

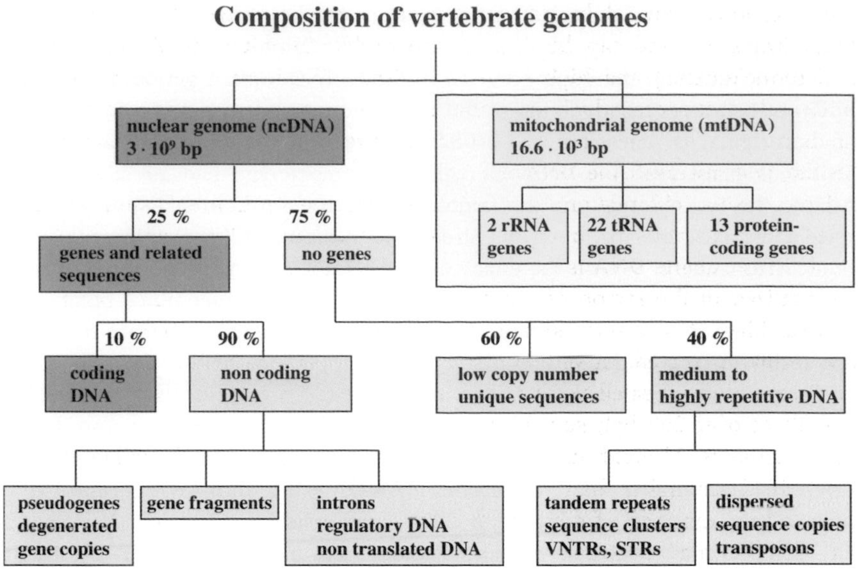

Fig. 4.2 Composition of eukaryotic genomes and a fraction of a few DNA elements of the entire human genome.

Composition of vertebrate genomes

Table 4.3 A few characteristics of the human genome.

Parameter	Human genome
Genome size	3.2×10^9 bp
Number of genes	25 000
Longest gene	3.4×10^6 bp
Medium gene size	27 000 bp
Smallest number of exons/gene	1
Highest number of exons/gene	178
Mean number of exons/gene	10.4
Largest exon	17 106 bp
Mean exon size	145 bp
Number of pseudogenes	> 20 000
Percentage of protein-coding sequences	1.5%
Percentage DNA in rRNA, functional DNA	3.5%
Percentage in repetitive DNA elements	∼ 50%

Possibly the largest part of the genome (over 50% with higher eukaryotes) is not transcribed and is partially functionless. Important elements are **pseudogenes** and **repetitive DNA** sequences (Table 4.3 and Fig. 4.2).

It is usually (but not always) the case that new functional genes develop through the doubling or duplication of genes in the progress of evolution. New genes can also be generated by **combining domains** or partial gene sequences. **Horizontal gene transfer** (which happened when bacteria became mitochondria) also helped to enlarge the eukaryotic genomes. In contrast, pseudogenes, which are nontranslatable copies of genes, show **frameshifts**, **nonsense** mutations, deletions, and insertions (see Section 4.1.4). Pseudogenes do not have any further function today. **Pseudogenes** can be divided into two groups: the first arose from **gene duplication**, the second from **retroposons**. In the second case, the genes were transcribed and processed, and following reverse translation in DNA, were inserted into a location in the genome. It is usually the case that these **retropseudogenes** have no introns, but frequently poly(A) tails, and unlike the pseudogenes they are not present in the vicinity of the original gene from which they arose. Surprisingly, nature can afford to reproduce this junk DNA in every generation, even though replication is an energy-consuming process. Perhaps these DNA sections that today appear to be useless will become functional in a later evolutionary phase as molecular **replacement parts**.

When the duplicated DNA sequence lies beside the original gene it is termed a **tandem repeat**. These tandem repeats are the starting point for further DNA amplifications, induced by **uneven crossing over.** **Repetitive DNA** is quantitatively important and can be divided into **middle repetitive DNA** (transposons and retroelements) and **highly repetitive DNA.** The latter class includes short nucleotide sequences, which are present in great numbers in chromosomes in a tandem-type style. There are also further divisions into **telomere, satellite, minisatellite,** and **microsatellite DNA.**

Upon cesium chloride gradient centrifugation, the DNA of eukaryotes is separated and two bands are often observed, the smaller of which contains satellite DNA. This **satellite DNA** is especially rich in repetitive sequences and prefers to be localized in the region of the centromeres. In insects and other arthropods this satellite DNA is very homogenous, meaning that their sequence elements are highly conserved. In vertebrates the repeated sequence units contain up to 1000 repetitions of satellite DNA, and it is significantly longer and more variable (length of over 200 bp); subelements such as GA_5TGA can often be found in these elements. Through uneven crossing over, the variability of satellite DNA is about 10 times higher than with genes that only have a low copy number. Division and organization of the repetitive DNA elements in the centromere region are chromosome- and type-specific. It is assumed that the repetitive DNA at the centromere region is responsible for homologous chromosome recognition and the fact that they arrange themselves next to each other during meiosis.

In the actual satellite DNA of both plants and animals, elements are found that are repeated 5–50 times; each being 15–100 bp. The sequence elements can be attributed to the original sequence that was varied through point mutations. This repetitive DNA, each about 500–5000 nucleotides in length, is significantly shorter than the satellite DNA and is termed **minisatellite** or **VNTRs (variable number tandem repeats).** It exhibits a large variability in length in every locus and a very high mutation rate is present as a result of uneven crossing over (as the number and length of repeats is changed), which can amount to 5% of the gamete. **Minisatellite DNA** is therefore termed the **hot spot** of **meiotic recombination.** Minisatellite DNA is especially suitable for the identification of individuals, and also for clarification of paternity and homozygosity in a population. Many VNTR loci each have dozens of alleles, which are codominantly inherited. This characteristic is used in **DNA fingerprinting.** The possibility that two unrelated individuals have the same DNA fingerprints is less than 1 in 10 million.

In addition, there are still shorter repeats that arise in animal and plant genomes. These consist of a basic unit of two (sometimes as many as five) nucleotides, such as $(GC)_n$ or $(CA)_n$, which are repeated up to 100 times. Of these elements, termed **microsatellites** or **STRs (short tandem repeats),** about 30 000 loci are found in humans, which are of great importance for the recognition of tissues and individuals, paternity and populations studies, and genome mapping. STR analysis is the method of choice for the determination of sexual crimes or murder in forensic medicine or criminal studies. The alleles allow amplification through **polymerase chain reaction (PCR)** (see Chapter 13). **Microsatellite PCR** is currently the method of choice for many forensic, biotechnological, and biological investigations due to the fact that it requires only the smallest amounts of DNA. The variability of microsatellite DNA is strongly increased during meiosis via uneven crossing over and slippage of the DNA polymerase, so that the short sequence elements can be mutated, duplicated, and deleted.

Additional 500-base long DNA sections are found in animal and plant genomes. These so-called scattered or **short interspersed elements (SINEs),** or 1000- to 5000-nucleotide **long interspersed elements (LINEs),** appear in high copy numbers (although not in tandem style repeats) (Fig. 4.2). The DNA elements **Alu** (which is recognized by the restriction enzyme *Alu*I), **Kpn,** and **poly(CA)** are also counted among the SINEs. The percentage of these elements in the human genome is about 20% of the entire genome. It is presumed that these elements, which are also called **mobile genetic elements** or **retrotransposons,** arise through reverse transcription. From an evolutionary point of view, transposons (with **long terminal repeats (LTRs)** or **inverted repeats (IRs)**), retro-transposons, and retroposons (transposons without LTRs) could be considered as examples of active **egoistic** genes (**selfish DNA**), which only have their own replication in mind. On the other hand, these mobile elements lead to genetic variability (an increased exon shuffling or enhancer shuffling) that in the long run can also have positive effects. In areas of Alu sequences chromosomes exhibit increased rates of new orientation. When Alu elements jump into active genes, most of them are inactivated; conversely, sleeping genes can be activated, in that the skipped elements can function as enhancers. Finally, the selection of new characteristics is made available. Sexual isolation and type formation can be increased through this mechanism.

The relative percentage of nonrepetitive DNA in bacteria is 100% and decreases in the higher developed eukaryotes: 70% in *Drosophila*, around 55% in mammals, and 33% in plants.

The percentage of repetitive DNA increases correspondingly. Assisted by uneven crossing over, the percentage of repetitive DNA in the genome of eukaryotes in future evolutions will probably increase further. As explained above, the function of about 50% of the genome remains unknown. Whether or not repetitive DNA is really functionless or egoistic DNA, as is often speculated, will be determined by future research.

4.1.2
Composition and Function of Chromosomes

With eukaryotes, the DNA in chromosomes is present as a **linear double helix**. In humans there are 22 paired **autosomes** (a copy from both the father and mother) and two **sex chromosomes** (XY in males and XX in females), giving a total of 46 chromosomes in the cell nucleus (Fig. 4.3). In some cases genes causing specific diseases have already been assigned to specific chromosomes. A selection is presented in Fig. 4.3. Through specific hybridization procedures (e.g. **fluorescence *in situ* hybridization (FISH)**) it is possible to locate genes on individual chromosomes and make them visible. Such location is an important assignment in human genetics and the diverse genome projects.

Chromosomes consist of a **centromere**, to which the microtubules attach during cell division, diverse replication starting points (origins of replication), and **telomere** sequences on the ends (Fig. 4.4). These telomeres are made up of over 1000 short repetitive sequence elements (e.g., GGGTTA in humans) and are then attached by a **telomerase** to the chromosomes (Fig. 4.5). The telomeres prevent the exonucleases from cleaving the chromosomes from the ends. Telomerase is only active in embryonic cells and synthesizes long telomere residues on the ends of the chromosomes. The telomerase is later inactivated (except in tumor cells, in which it is usually permanently activated), so that the telomeres with later replication cycles cannot be lengthened further. After 70–80 cell divisions, cell division usually succumbs to the fact that the exonucleases have been able to nibble away the telomere and have disrupted an important part of functional genes. It is speculated that the **aging process** and **death** are controlled by this internal clock.

DNA is not present in the chromosomes as free strands, but is wrapped around **histone proteins** of a basic nature. DNA is wound around four histone proteins (H2A, H2B, H3, and H4), which exhibit many positively charged lysine residues and form octameric cylinders (Fig. 4.6). This is how **nucleosomes**, which each contain about 145 bp of DNA, are constructed. Here, ionic bonds between the positively charged lysine residue and the negatively charged phosphate groups of the DNA play an important role. A linear DNA section of about 80 bp of DNA, which binds sequence-specific proteins, is usually present between the two neighboring nucleosomes.

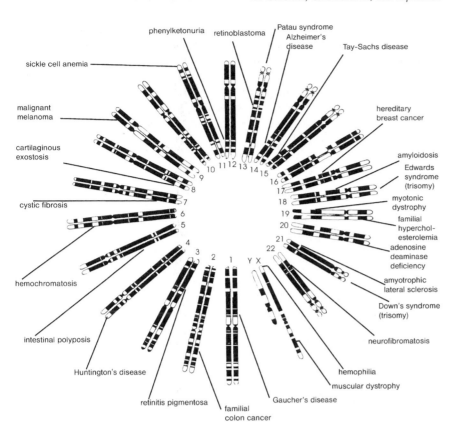

Fig. 4.3 Schematic illustration of human chromosomes. The indentations indicate centromeres. Staining the chromosomes results in a typical band pattern. A number of genes involved in diseases have already been located.

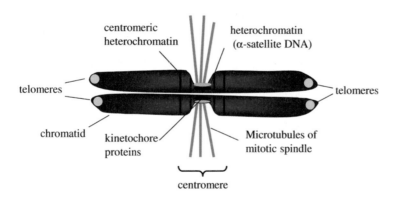

Fig. 4.4 Important structural elements of chromosomes necessary for the replication and separation of chromatids. The centromere region consists of repetitive α-satellite DNA, which is rich in A-T base pairs. It is flanked by centric heterochromatin. Kinetochore proteins that form an inner and an outer kinetochore plate bind to the area of α-satellite DNA. The kinetochore proteins bind the microtubules of the spindle apparatus, which pulls the chromatid halves apart.

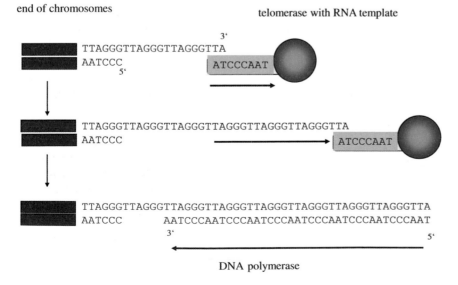

Fig. 4.5 Principle of telomere replication. The telomerase exhibits an RNA template, through which it binds a TA residue of DNA. The telomerase lengthens the DNA strands complementary to the RNA template. When a repeat region is synthesized, the telomerase jumps to the next TA residue and begins the synthesis. This can be repeated over 1000 times. The opposite strand will be synthesized by DNA polymerase, which through the help of its own primase uses a complementary RNA primer on the 5′-end and links this to the next nucleotide.

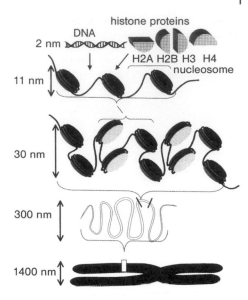

Fig. 4.6 From the nucleosome to the condensed metaphase chromosome. The DNA winds itself around the octamer histone protein complex forming nucleosomes. The nucleosomes are organized into 30-nm thick chromatin fibrils that in interphase are arranged together forming 300-nm thick bundles. In metaphase chromosomes the chromatin is densely packed through repeated bundling.

When a gene is transcribed, the tight complex between DNA and histones must be loosened. This is achieved by diverse protein modifications, such as acetylation, methylation, phosphorylation, and ATP-dependent chromatin modeling complexes (together with histone H1).

Chromosomes are present in their more extended form in the interphase of the cell cycle. Only in metaphase do we see the well-known metaphase chromosomes, in which the DNA is highly condensed (1000 times shorter than in the extended form) (Fig. 4.6).

4.1.3
Mitosis and Meiosis

In fertilization, an **oocyte** fuses with a **spermatozoon**; both cells are **haploids** resulting from meiosis (Fig. 4.7), therefore making the **zygote** a **diploid**. All other somatic cells undergoing mitotic division are also diploid. Only the gametes are haploid.

During **mitotic cell division**, the entire DNA is duplicated (replication); this results in two identical sister chromatids, which are held tightly together by a shared centromere (Fig. 4.4). The condensed chromatids are pulled apart by the **spindle apparatus**, so that each new cell has a complete set of diploid chromosomes. In many eukaryotes the phases of mitosis can be divided into distinct phases: prophase, metaphase, anaphase, and telophase.

Following the duplication of the chromosomes in the **interphase nucleus** (during S phase), the chromatin condenses to form discrete chromosomes in **prophase** (Fig. 4.6 and Fig. 4.7), which consist of **two identical chromatids**. At the end of prophase and before the beginning of **prometaphase** the nuclear membranes and nucleoli disappear; the nuclear spindle (consisting of **polar microtubules** and **kinetochore microtubules**) is formed. In **prometaphase** the microtubules attach themselves to the **centromeres** of the chromosomes with the help of special protein complexes (**kinetochores**) (Fig. 4.4). The chromosomes are pulled by the microtubules (Fig. 3.26) to the cell equator. In **metaphase** the chromosomes line up at the equator and microtubules bind the centromeres with both spindle poles. In **anaphase** the kinetochore microtubules shorten and therefore the chromatids are pulled towards each spindle pole. In the following **anaphase II** the polar microtubules lengthen, so that the cell begins to stretch. In **telophase** all of the daughter chromosomes are found at their corresponding spindle poles, the nuclear membrane reforms, and the nucleoli become visible again. At the same time, the polar microtubules push the cells further apart. Finally, in **cytokinesis** both daughter cells fully pinch off from one another; this results in two independent cells, each with an identical set of chromosomes.

In **meiosis** the sets of chromosomes are divided in half (**reduction division**), therefore resulting in the haploid genome. Meiosis is required for sexual reproduction in diploid organisms. If the gametes were diploid, every new zygote formed would contain double the number of chromosomes. This dilemma can only be solved through haploid gametes. Meiosis also serves to mix the paternal genes and increases genetic variability, which is an important prerequisite for natural selection.

The most important difference between meiosis and mitosis lies in the pairing of homologous chromosomes and the consequent reduction of the chromosome set. **Meiosis** is divided into two divisions: the first and second reduction division (or **meiosis I** and **meiosis II**). In **prophase** of the first reduction division there are five stages: **leptotene**, **zygotene**, **pachytene**, **diplotene**, and **diakinesis**. After doubling of the chromosomes, two sister chromatids become apparent in the leptotene. In **zygotene** the pairing of the homologous maternal and paternal chromosomes begins (**synapsis**). The chromosomes paired in each case correspond to each other at the sequence level, enabling **crossing over** and **recombination** to take place. These processes enhance the exchange of genetic informa-

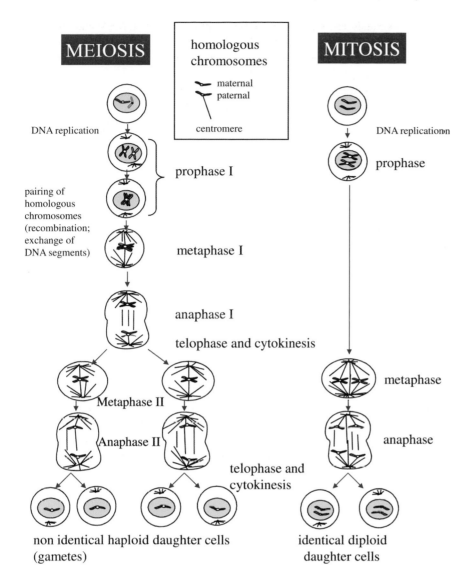

Fig. 4.7 Schematic overview of mitosis and meiosis.

tion from both parents. In **pachytene** the pairing of homologous chromosomes is completed. In the following **diplotene** the chromosome pairs separate from one another, but cling together at the sites on which crossing over took place (so-called **chiasmata**). In this phase, the chromosomes unwind and are transcriptionally active. In **diakinesis** the transcription ends and the chromosomes condense again.

In the following **metaphase I** the nuclear membrane and nucleolus break down, and the spindle apparatus is formed. The chromosome pairs assemble themselves on the equator, so that the centromeres are orientated to the spindle poles. However, the kinetochore microtubules do not attach to the centromeres of individual chromatids (as in mitosis), but to a shared centromere of each chromatid pair. In meiotic **anaphase I**, the chromosome pairs are pulled apart over the shortening kinetochore microtubules to the cell poles. The recombinant chromosome areas separate at the chiasmata. The sister chromatids remain joined together via the centromere.

After a short interphase the second reduction division occurs. This division contains the mechanics of mitosis. The chromosomes are ordered again as in metaphase: **metaphase II**. The chromatids are pulled apart by the kinetochore microtubules to the cell poles. This process is completed with **anaphase II**. After the telophase II and **cytokinesis**, haploid cells remain (so-called **meiospores** or **meiogametes**), each with a haploid chromosome set available. During the forma-

Replication

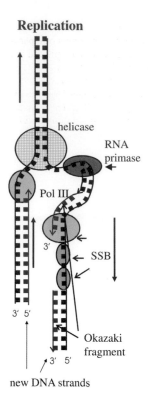

Fig. 4.8 Schematic summary of DNA replication. SSB, single-strand binding proteins. Pol III = DNA polymerase III.

tion of haploid gametes the chromosomes from the parents are sorted randomly, which again increases genetic variability.

4.1.4
Replication

With every **cell division** (mitosis; Fig. 4.7), the entire genome of the cell is duplicated. This means that two identical **chromatids** result from every chromosome. These chromatids are identical daughter chromosomes that will be distributed to the daughter cells following cell division. The duplication of the DNA, which is referred to as **DNA replication**, occurs in a **semiconservative manner**.

During semiconservative replication, the DNA double strand is locally separated into single strands and a **replication fork** is formed. The single strands serve as a matrix for the synthesis of each complementary new strand. The DNA replication is a complex process, in which many proteins and enzymes are involved (Fig. 4.8). To open the double strands a **helicase** is needed. The **leading strand**, orientated in the 3′→5′ direction, can be copied directly by the DNA polymerase, as synthesis occurs in a 5′→3′ direction. The opposing strand, termed **lagging strand**, cannot be copied in the same way as it is orientated in a 5′→3′ direction. As soon as the DNA is present as single strands, specific proteins bind (**single-strand binding proteins**) and prevent the reformation of the double helix. **DNA primase** places short RNA primers, which are complimentary to the DNA sequence, at regular intervals on the lagging strand. These RNA primers can be lengthened by DNA polymerase until the next RNA primer is reached (referred to as **Okazaki fragments**). The RNA is then removed and replaced by dNTPs. The Okazaki fragments are linked through the **DNA ligase**. Enzymes involved in replication vary between prokaryotes and eukaryotes. However, the general composition of the multienzyme complex is similar. During the unwinding of the double helix DNA topoisomerases cut the DNA at regular intervals to prevent the rotation of the double helix. **DNA topoisomerase I** catalyzes single-strand breaking, while **DNA topoisomerase II** can cut both DNA strands at the same time.

Replication begins at specific DNA sequences termed **origins**. Here the **replication bubble** opens, and replication occurs in parallel on both the right and left replication forks (Fig. 4.9). Whereas in circular bacterial genomes only one **origin of replication** is present, a replication start site is positioned on the linear chromosomes every 1000 bp. In this way, even long chromosomes can be replicated in a short time.

DNA polymerases copy the original nucleotide sequence flawlessly (the error rate during synthesis is one incorrect nucleotide per 10000 nucleotides). However, special **repair enzymes** play a large role. Incorrectly paired nucleotides are removed by specific **exonucleases** and then replaced through DNA polymerase; finally, the phosphoester bond is covalently linked through **DNA ligase**.

4.1.5
Mutations and Repair Mechanisms

The structure of DNA must be relatively stable and replicated almost flawlessly, in order to serve as an information and inheritance carrier. DNA is a relatively stable macromolecule; however it is liable to have constant **mutations** in the body, due to internal or external causes. Internal mechanisms are due to spontaneous **depurination** and **deamination** of the DNA bases; external factors include **energy-rich radiation** (UV, x-rays, and radioactivity) and **mutagens**. Natural mutation rates in bacteria are estimated to be 10^{-5} to 10^{-6} mutations per gene locus and generation. With eukaryotes these rates are difficult to determine, but should also be in the same range.

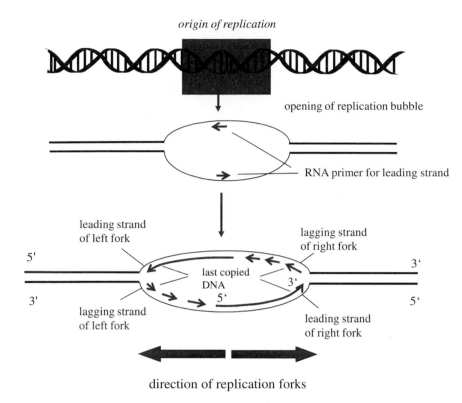

origin of replication

opening of replication bubble

RNA primer for leading strand

leading strand
of left fork

lagging strand
of right fork

5'

last copied
DNA

3'

3'

lagging strand
of left fork

leading strand
of right fork

direction of replication forks

Fig. 4.9 Asymmetric composition of replication bubbles. DNA is unwound at the origin of replication, and a replication bubble with a right and left replication fork is formed. Replication proceeds in parallel within the replication bubble. DNA primase introduces a complimentary RNA primer on each leading strand, so that DNA polymerase III can carry out replication. The individual lagging stands are synthesized as shown in Fig. 4.8.

Mutations where only one or a few nucleotides are exchanged are termed point mutations; other types of mutations include **chromosome mutations** or **rearrangements** when larger sequence sections are cut out (**deletion**) or put in (**insertion** or **translocation**), doubled (**duplication**), or oriented inversely (**inversion**). If such mutations occur within a transcription unit, they are referred to as **gene mutations**.

In the human body, nucleotide **deamination** of nucleotides also spontaneously arises with a rate of 100 deaminations per day and per cell (Fig. 4.10). Cytidine is converted to uracil by deamination. If, following replication, U pairs with A instead of with G, as the original C had done, then the resulting CG pair is completely replaced with a TA pair (Fig. 4.11). The **purine** residues guanine and adenine can be removed spontaneously from DNA by hydrolysis (Fig. 4.10). **Depurination** is considered as one of the most common spontaneous mutations and usually leads to **transversions**, but also to the deletion of individual bases; over 5000–10000 purine bases are depurinated daily in every human cell. Under **UV radiation** (e.g., from extensive sunbathing) **neighboring thymine** or **cytosine** residues can be activated, which then form dimers (**dimerization**). The **oxidation** of guanosine to 8-oxoguanosine by **reactive oxygen species (ROS)** can also induce point mutations (Fig. 4.10).

In rare cases the bases can be present in a **tautomeric form** (Fig. 4.12). If they are selected during replication they can lead to point mutations. G and T are normally present in the **keto form** and very rarely take the **enol form**. The amino group of A and C can in rare cases convert into an imino function. Tautomeric adenine pairs with cytosine instead of with thymine, tautomeric thymine pairs with guanine instead of with adenine, and vice versa. The resulting nucleotide substitutions fall into the class of the **transitions**. This includes substitutions of pyrimidine bases with another pyrimidine (T to C or *vice versa*) and the substitution of a purine base by another (A→G or *vice versa*). **Transversions** include the exchange of a purine base with a pyrimidine (A→C or T/G→C or T and *vice versa*).

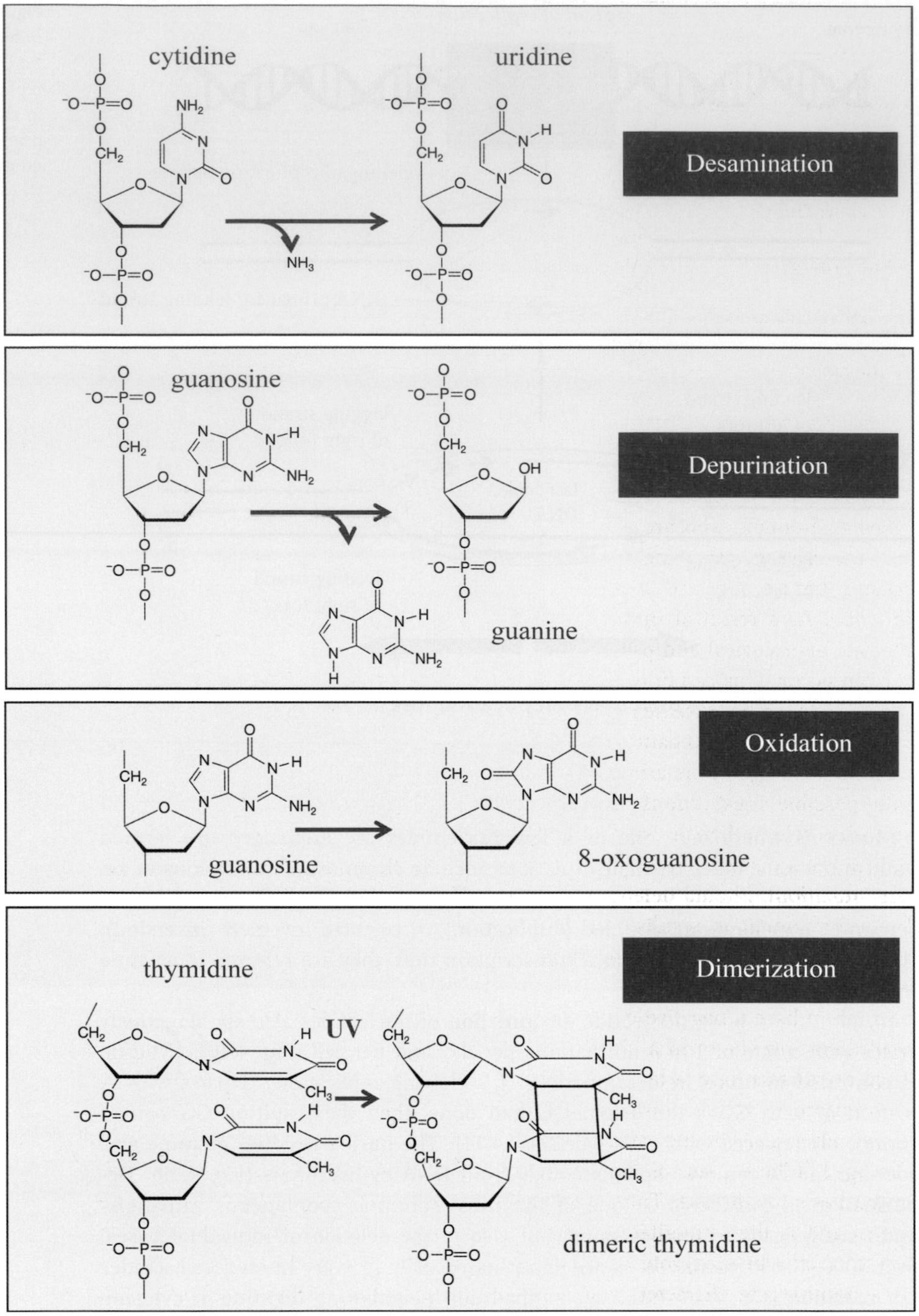

Fig. 4.10 Depurination, deamination, oxidation, and dimerization as examples of major mutation mechanisms.

Most of the primary gene changes (deamination, depurination, dimerization, and oxidation) are recognized by **repair enzymes** (such as AP endonuclease and DNA glycosylases), cut out (as long as the second DNA strand is not also damaged), and repaired by DNA polymerase and DNA ligase. In addition, **alkyltransferases**, **photolyases**, and incorrect pairing repair and recombination repair systems are also available, which are active after replication. The double helix is also advantageous for any repair process, as genetic information is complementarily saved. Even if the information on one strand is lost, the complementary strand is still available and can be used as a template for the needed correction. In **gametes**, such as those of humans, thanks to the effectiveness of the repair systems there are only 10–20 nucleotide substitutions per year in relation to the

Syndrome	Phenotype	Damage
MH2,3,6; MLH1; PMS2	Colon cancer	Mismatch repair
Xeroderma pigmentosum	Skin cancer, neurological disorders	Excision repair
BRCA-2	Breast and ovarian cancer	Repair through homologous recombination
Werner syndrome	Premature aging, many tumors	3-Exonuclease, DNA helicase
Bloom syndrome	Many tumors, stunted growth, genome instability	DNA helicase
Fanconi anemia groups A–G	Malformations, leukemia, genome instability	DNA cross link repair
46 BR-patient	Hypersensitivity for mutagenic substances	DNA ligase I

Table 4.4 Genetic diseases which are associated with defective DNA repair system.

available 3.2×10^9 bases. The significance of the repair system is easily recognized in humans who are affected by **xeroderma pigmentosum**, a rare autosomal recessive neurocutaneous disease, in which certain elements of the repair system that are required to repair DNA damage caused by UV radiation do not function. As a result of mutagenic UV radiation from sunlight, aside from numerous neurological and psychiatric symptoms, skin discoloration and skin cancer can occur. This can only be prevented by complete avoidance of sunlight. Table 4.4 lists further diseases caused by defective repair enzymes.

Owing to the redundant genetic code, not every point mutation in a gene leads to a change in the amino acid sequence. Twenty-five percent of all theoretically possible substitutions are **synonymous**, 4% lead to **stop codons**, and 71% to amino acid exchange. Nucleotide substitutions in the third codon position do not lead to a change in the amino acid in about 69% of cases (referred to as a **silent mutation**). Should deletions or insertions occur within a coded sequence, a frameshift mutation results, which almost always leads to severe damage of the corresponding proteins (Fig. 4.13).

Point mutations that cause amino acid exchange (nonsynonymous substitution) often have a negative effect on the corresponding proteins. If the mutation is in the active site or in a binding site, a total loss of function can result. As diploid organisms have at least two copies of every gene, positioned on autologous chromosomes, such a point mutation usually does not lead to physical damage if the other copy of the gene is still intact. Only after both copies have been damaged is there a loss of the corresponding protein (Fig. 4.14). Such disruption is the basis for disease. This is especially true when the mutation arises in gametes and is then inherited. If the disorder is only in one allele, then it is referred to as **heterozygote character**. If both alleles are identical, this is known as a **homozygote character** (Fig. 4.14). In certain genes, such point mutations and the consequences on the health of the individual are widely known. They are referred to as **SNPs (single nucleotide polymorphisms)**. One of the most important tasks for molecular biotechnologists is the development of diagnostic systems to quickly and reproducibly detect such SNPs. For this purpose mass spectrometry, DNA sequencing, PCR methods, and DNA chip strategies can be used (see Chapters 13 and 14). This information can help to rationally treat diseases and can lead to a better understanding of their causes.

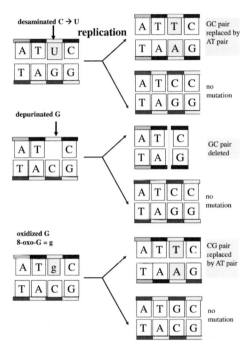

Fig. 4.11 Consequences of deamination, depurination, and oxidation. Cytidine is deaminated, resulting in uracil that then pairs with adenine during replication. If a depurinated nucleotide is not repaired DNA polymerase passes over the depurinated position during replication. A point deletion that can lead to frame shift mutations results.

Fig. 4.12 Base pairing of tautomeric DNA bases. The correct base pairings for A–T and G–C pairs are illustrated in I. Base pairs between tautomers A, G, C, and T are shown in II and III.

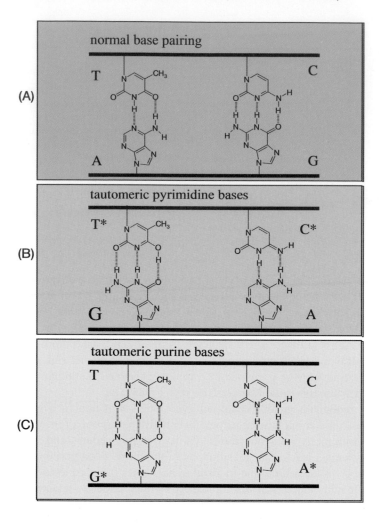

Fig. 4.13 Consequences of gene mutations.

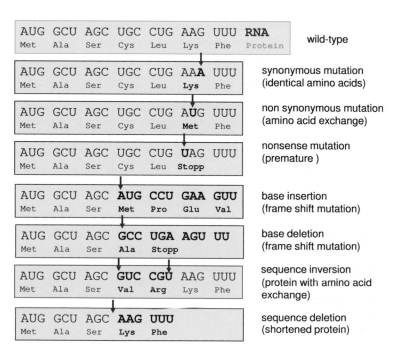

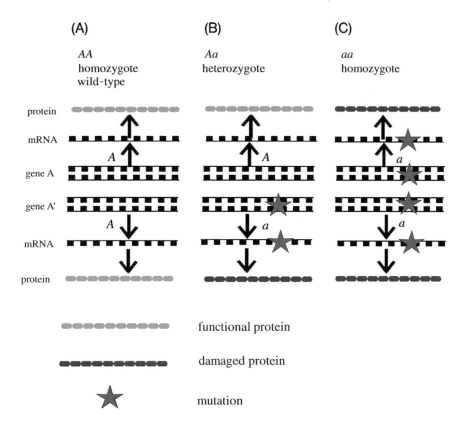

(A)

AA
homozygote
wild-type

(B)

Aa
heterozygote

(C)

aa
homozygote

protein

mRNA

gene A

gene A'

mRNA

protein

functional protein

damaged protein

★ mutation

Fig. 4.14 Inheritance of mutations leading to the loss of protein function. Gene A codes for a functional protein; Gene a codes for a protein rendered functionless by a mutation. (A) Wild-type genotype *AA*. (B) Heterozygote genotype *Aa*. (C) Homozygote recessive genotype *aa*.

4.2
Transcription: From Gene to Protein

Originally, mutation and recombination units were regarded as genes; in the 1950s the **"one gene, one protein"** hypothesis was developed (**DNA makes RNA, which makes proteins**). Today, the gene is defined as a **transcription unit**. In the meantime the intron/exon structure and the noncoding regulatory sequences, which also belong to the gene, have been recognized. Since mRNAs can be alternatively spliced, the statement "one gene, one protein" is no longer true in the strictest sense. The genetic information flows in all organisms from the gene to the mRNA and to the protein (Fig. 4.15). Only retroviruses can reversely translate RNA into DNA using a **reverse transcriptase**, but in no case has a translation of the amino acid sequence of a protein in a nucleotide sequence been shown.

In eukaryotes, three different **RNA polymerases** exist, which transcribe DNA into mRNA (**RNA polymerase II**), rRNA (**RNA polymerase I**), or into other functional RNAs (e.g., tRNAs, 5S rRNA, snRNA; **RNA polymerase III**). In prokaryotes, only one RNA polymerase is present. The translation of DNA into RNA is termed **transcription**.

As with replication, in transcription the DNA double helix is locally unwound, so that the RNA polymerase can synthesize the RNA (mRNA, rRNA, or tRNA) complementary to the **template DNA strand** (Fig. 4.16). The DNA strand bearing an identical sequence to the mRNA (except that the T has been replaced with U) is referred to (in a confusing manner) as the **coding strand**. In addition the sequence of the coding strand is written in the 5′→3′ orientation and is also stored in this format in sequence data banks.

coding strand	5′-GGC TCC CTA TTA GCA GTC TGC CTC ATG ACC-3′
template strand	3′-CCG AGG GAT AAT CGT CAG ACG GAG TAC TGG-5′
mRNA	5′-GGC **UCC CUA UUA** GCA GUC UGC CUC AUG ACC-3′

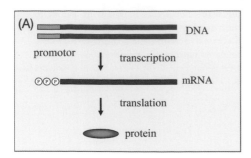

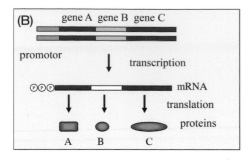

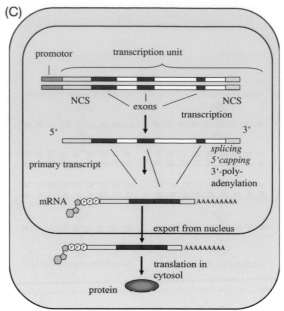

Fig. 4.15 From gene to protein: comparison of prokaryotes and eukaryotes. (A) Simple prokaryotic gene: the mRNA is translated to a protein. (B) Bacterial operon: the primary transcript holds the genetic information for many genes (polycistronic mRNA). In protein biosynthesis the protein units are synthesized separately. (C) Eukaryotic system: in the nucleus a primary transcript from RNA polymerase II is synthesized from which the intron regions are removed in preceding steps. At the 5'-end a 7-methylguanosine cap is added and a poly(A) tail is added to the 3'-end. The completed mRNA is transported through the nuclear pore complex into the cytosol where it is translated into proteins by the ribosomes. NCS = noncoding sequence.

Transcription

The bacterial **RNA polymerase** is a multienzyme complex containing a removable sigma factor. The sigma factor recognizes promoter regions of genes and assists the RNA polymerase in finding the transcription start. In *Escherichia coli* the promoter is made up of two hexamer sequence motives, which are positioned 10 or 35 bases in front of a gene. The consensus sequences are ...TTGACA...TATAAT... Prokaryotic genes are usually organized in the form of **operons** (Fig. 4.15 B): genes that belong together, such as those that code for enzymes of a biosynthesis pathway, lie beside one another and are controlled by a common promoter, which consists of an **operator** as the control element (Fig. 4.17 A).

Control of gene expression in eukaryotes is very complex. In eukaryotic genomes there are considerably more genes present than proteins required for a single cell. Therefore, it is necessary to express genes in a cell-, tissue- and development-specific fashion. This means that out of the estimated 25 000 genes encoding proteins in humans, only a small number of the genes are activated in individual differentiated cells. Research and documentation of differential gene expression patterns is part of the enormous task for current molecular biology.

The **transcription** of eukaryotic genes (Fig. 4.17 B) is controlled by neighboring **regulatory DNA** regions (**promoter regions**) that are themselves controlled by transcription factors, which are responsible for the activation or inactivation of a gene. As well as the promoter region which is in close proximity to the coding sequences, further regulatory elements (**enhancer**, **silencer**) can also be positioned further away (Fig. 4.17 B). The eukaryotic RNA polymerase II is only activated when diverse transcription factors have bound to the promoter (Fig. 4.17 B). Table 4.5 reviews the most important control elements and the associated **consensus sequences**. As most genes in eukaryotic cells are expressed in a cell-, tissue- and development-specific manner, additional specific transcription factors play a decisive role. Very many of these factors have not yet been discovered.

Fig. 4.16 Schematic overview of the function of RNA polymerase and transcription.

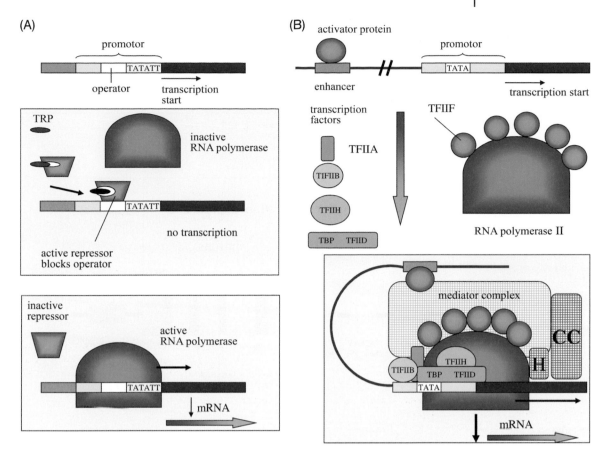

Fig. 4.17 Simplified schematic illustration of the control of gene expression in prokaryotes and eukaryotes. (A) Bacteria: example tryptophan operon. When the amino acid tryptophan (TRP) is available in excess, the transcription of tryptophan biosynthesis enzymes is then inhibited by a repressor that is activated through the tryptophan, blocking the operator in the promoter. If no tryptophan is available, then the repressor dissociates from operator, and RNA polymerase can begin with transcription (bottom illustration). (B) Eukaryotes: transcription can only begin when an activated protein has bound to the enhancer and the complete transcription factors (Table 4.5) form a transcription complex together with the RNA polymerase II. The connections between the activator protein and the transcription complex are established through a mediator protein, which collaborates with a chromatin remodeling complex (CC) and a histone-modifying enzyme (H). In addition, proteins are present that dissolve nucleosome complexes so that the DNA is accessible to the RNA polymerase.

As opposed to bacteria, eukaryotic protein-coding genes usually consist of **exons** and **introns** (Fig. 4.18), and are therefore referred to as **mosaic genes**. The primary transcript deriving from the transcription is completely processed in the nucleus. It is spliced so that each noncoding intron region, which is flanked by GU and AG sequences, is removed. **snRNAs (small nuclear RNAs)** are catalytically involved in splicing. The snRNA can be seen as a type of **ribozyme** (see Chapter 2.4).

The assignment of template or coding strand does not apply for a complete chromosome; the orientation within chromosomes can change from gene to gene, meaning that gene A can be read from the template strand and the neighboring gene B from the strand lying opposite. In eukaryotes, the genes are arranged in a linear manner, one after the other, on chromosomes. In prokaryotes overlapping genes are found, which are coded for either by the same DNA strand or the complementary DNA strand lying opposite. This results in more dense information, but prevents the independent evolution of the DNA sequences. In eukaryotes, **differential** or **alternative splicing** of the genes is ob-

Table 4.5 Consensus sequences in eukaryotic promoter regions.

Box	Consensus sequence	Transcription factor
BRE	G/C G/C G/A C G C C	TFIIB
TATA	T A T A A/T A /A/T	TBP
INR	C/T C/T A N T/A C/T C/T	TFIID
DPE	A/G G A/T C G T G	TFIID

For the position of the consensus boxes see Fig. 4.17

Fig. 4.18 Structure of a eukaryotic gene. NCS, noncoding sequence.

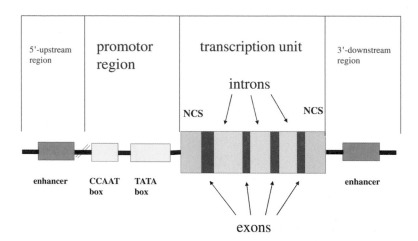

Fig. 4.19 Schematic representation of alternative splicing processes. The letters A, B, C, and so on indicate exons. After the complete primary transcript is produced, further selection occurs in the splicing process, in which not all exons remain but a few are removed with the introns. In this way many different proteins are synthesized from one gene, which differ in domain composition. NCS=noncoding sequence.

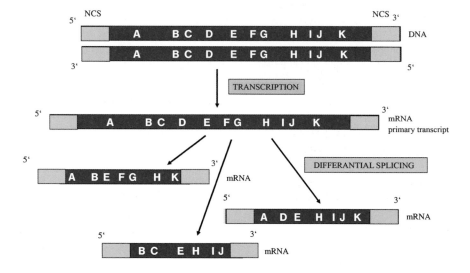

served (Fig. 4.19). That is, not all exons will be present in the final mRNA. Due to alternative splicing a single gene can lead to more proteins depending on the tissue in which they are expressed (this is the reason why the number of proteins in humans is higher than the number of genes).

In **gene regulation**, the **methylation of cytosine (5-methylcytosine** in plants and animals) and **adenine** (N^6**-methyladenine** in prokaryotes) also plays an important role. As a rule, genes that are transcribed are less methylated than genes that are turned off (**silent**). After each replication the methylation of the newly replicated DNA strands must take place; an inhibition of the corresponding methyl transferases strongly influences gene expression and cell differentiation. DNA methylation is also important for **DNA repair**, being that the repair enzymes can recognize a newly constructed and defective DNA strand by the absence of methylation. Methylation and changes in chromatin structure changes

Genetic inheritance   Epigenetic inheritance

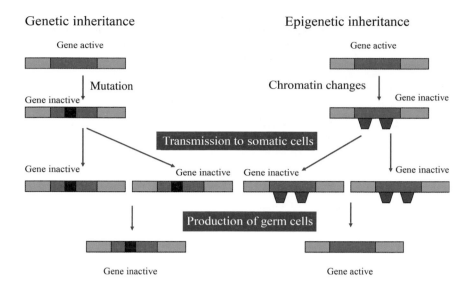

Fig. 4.20   Differences between genetic and epigenetic inheritance

the expression patterns of genes; these changes are inherited to daughter cells (so-called **genetic imprinting** or **epigenetic inheritance**) (Fig. 4.20). Usually epigenetic changes are not transferred via the germ line to the next generation, whereas mutations in gametes are inherited.

The nucleotide sequence of mRNA is translated using the **genetic code** into amino acid sequences. **tRNA**, with its specific **anticodon**, serves as a mediator between the mRNA and the protein. A central event in the progress in molecular biology was the discovery of the unit-less, comma-less, nonoverlapping code in all living organisms. In each case three nucleotides code for a specific amino acid in each protein (Table 2.4). Using a triplet code with four bases there are $4^3 = 64$ available combinations. As there are only 20 amino acids that are used to synthesize proteins (Table 2.4), there are more codons than are actually necessary. This problem was solved by evolution in such a way that most of the amino acids are not be coded from only one, but from two to at the most six different **synonymous codons** (Table 2.4).

The widely **universal triplet code has a specific start signal**. Since **methionine** (in eukaryotes) and **N-formylmethionine** (in bacteria and chloroplasts) is the first amino acid to be built into polypeptides, the **universal start codon is AUG** (far more seldom, GUG is present). In most cases, however, methionine is removed by specific proteases following translation. When the start of the translation shifts only one or two nucleotides, resulting in a shift of the **reading frame (frameshift)**, a totally new protein results. This means that the start codon must be strictly preserved in order to produce reproducible proteins. In **animal** (but not in plant) **mitochondria**, there is a deviation from the universal genetic code (e.g., AUA is used for translation initiation and codes for methionine). However, in eukaryotic ribosomes this codon codes for isoleucine; AGG/A is used as a termination codon by vertebrate mitochondria, while it usually codes for arginine. UGA, which is usually a stop codon, codes for tryptophan in animal mtDNA.

Usually the codons that code for the same amino acid differ in the **third codon position**. Every codon is recognized by tRNA via the anticodon sequence. Within the so-called **degenerate codons** that all code for the same amino acid, usually only one tRNA exists, one which tolerates a **mismatching** in the third codon position. Overall, about 31 tRNAs have been discovered in the eukaryotic system and 22 tRNAs in mitochondria.

Fig. 4.21 Structure of RNA cassettes and synthesis of rRNA. ITS, internal transcribed spacers; IGS, intergenic spacers; 5S rRNA genes are transcribed separately.

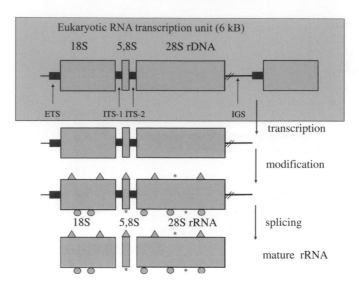

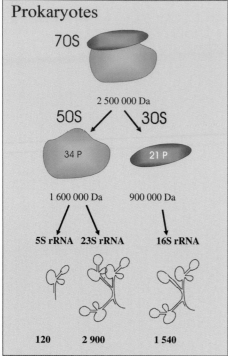

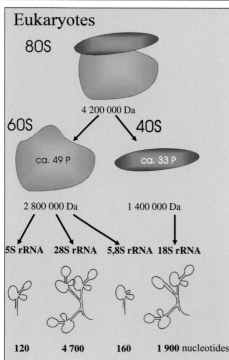

Fig. 4.22 Structure of prokaryotic and eukaryotic ribosomes. For the structure of rRNA, see Fig. 2.20.

4.3
Protein Biosynthesis (Translation)

Protein biosynthesis takes place in **ribosomes** – intricately constructed multi-enzyme complexes in which different rRNAs play an important role (Fig. 4.22). **Ribosomal RNAs (rRNAs)** belong to the most prevalent macromolecules of the cell; just for *E. coli* alone, the number of rRNA molecules is estimated to be 38 000. The numerous copies of the rDNA cassettes in the genome (Fig. 4.21) indicate that this gene must be transcribed very often, in order to produce the large number of rRNA molecules that every cell requires. The rRNA genes for **18S**, **5.8S**, and **26S rRNA** are transcribed together, and the individual rRNAs are produced afterwards by splicing. Nucleotides of the precursor RNA are chemically modified by **small nucleolar RNAs (snoRNAs)** before splicing (Fig. 2.20).

Figure 4.22 shows the assembled building blocks of **prokaryotic and eukaryotic ribosomes**. As **mitochondria** and **chloroplasts** contain their own ribosomes, which originated from bacteria (see Chapter 3.1.3), the expected type of rRNAs corresponding to bacteria is found in mtDNA and cpDNA (note that in mitochondria a 12S rRNA is present instead of the 23S rRNA).

16/18S rRNA and 23/28S rRNAs exhibit complex spatial structures, which are conserved over a wide range of organisms (Fig. 2.20). Even though the RNAs are present as single strands, they form **complementary double strands** (so-called **stem structures**) at many sites in aqueous environments. The nucleotide sequence of stem structures is very strongly preserved in evolution. The situation is different for the **loops**, in which the nucleotides have been modified post-transcriptionally. This phenomena of **base modification** is especially observed with tRNAs (but also in rRNAs), in which more than 50 modified nucleotides have been discovered. **Substituted bases** are thiouracil, 5-methylcytosine, dihydrouracil, thiothymine, thiocytosine, N^4-acetylcytosine, 1-methylhypoxanthine, 1-methylguanine, and N^6-methyladenine. There are comparatively many substitutions, deletions, insertions, and inversions present in the loops. Genetic trees of all organisms can be reconstructed from the nucleotide sequences of the rRNAs, giving them a special role in molecular evolution. The **tree of life** and the classification of species are largely based on the analysis of conserved rDNA genes (see Chapter 1).

The ribosomal proteins are arranged around the rRNA, together constituting a complex nanomachine known as the **ribosome** (Fig. 4.22 and 4.23). Both ribosomal subunits are assembled in the cell nucleus and are transported into the cytosol through the nuclear pores. Free mRNA molecules are recognized by the small subunits, which are first loaded with **methionine tRNA** and GTP activated **initiation factors (eIF-2)**. The small subunit slides along the mRNA until the first start codon **AUG** is reached, where methionine tRNA is bound via its anticodon UTC. Following the dissociation of the initiation factor eIF-2 the large ribosomal subunit is able to bind and the ribosome is positioned ready to begin translation. There are three formally distinguished binding sites: the arriving **aminoacyl-tRNAs** bind to the **A-site**, the tRNA with the peptide chain sits in the **P-site**, and the **E-site** releases the free tRNA after peptide transfer (Fig. 4.23).

In the A-site the arriving aminoacyl-tRNAs (loaded with amino acids) are hybridized via their anticodon to the corresponding triplet codon on the mRNA (Fig. 4.24). In the next step, the peptide residue on the tRNA in the P-site is transferred to the aminoacyl-tRNA in the A-site (peptidyl transfer is catalyzed by the rRNA; Fig. 4.25). Next the ribosome moves along three nucleotides on the mRNA and releases the free tRNA from the P-site, which now carries the tRNA with the growing peptidyl residue. These steps are repeated until a **stop codon** is reached. A specific **release factor** then binds and blocks access for further aminoacyl-tRNAs to the A-site. As a consequence, the peptide chain is released. After protein synthesis, the newly synthesized proteins fold themselves into the correct conformation; aided in many cases by **chaperones** (e.g., diverse **heat-shock proteins**, HSP70 and others) acting as auxiliary enzymes. Incorrectly folded or incorrectly synthesized proteins (e.g., protein fragments resulting from strand breaking) are coupled with the protein **ubiquitin** and are broken down in a cellular "shredder" – the **proteasomes**.

Protein biosynthesis can occur on free ribosomes in the cytoplasm or on ribosomes which bind to the rough ER (see Chapter 5).

Prokaryotic and eukaryotic ribosomes are constructed according to a very similar pattern (Fig. 4.22), and protein biosynthesis is conducted according to very similar principles. However, the particular rRNAs and ribosomal enzymes exhibit important differences. The importance of many antibiotics depend on these differences to specifically inhibit prokaryotic ribosomes. Many antibiotics intervene in bacterial protein biosynthesis (Table 4.6).

Owing to their selectivity, antibiotics are generally substances with few side effects in humans. The search for new and more effective antibiotics is still one of the most important challenges of biotechnology because many pathogens have become resistant (overexpression of ABC transporters, target site muta-

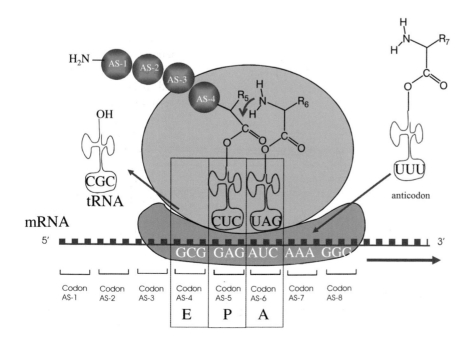

Fig. 4.23 Schematic illustration of protein biosynthesis in ribosomes. Three binding sites are distinguished in ribosomes: E, P, and A.

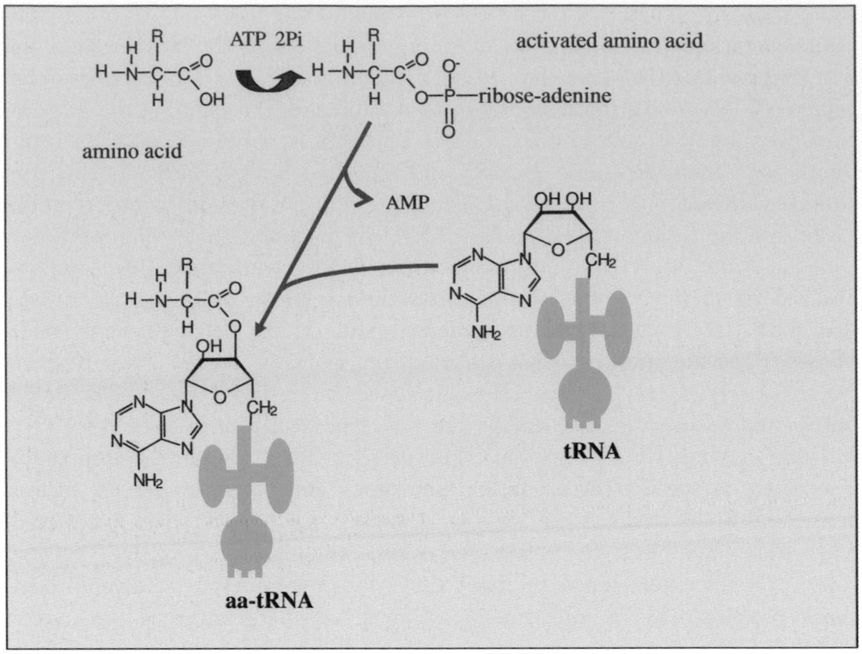

Fig. 4.24 Loading tRNA with an amino acid. First the amino acid is activated through the binding of ATP. The activated amino acid is transferred to the 3′-OH group of the terminal adenine residue of the tRNA and an AMP residue is set free. This reaction is catalyzed by aminoacyl-tRNA synthetase that is specific for every amino acid. aa-tRNA, aminoacyl-tRNA (i.e., a tRNA loaded with an amino acid).

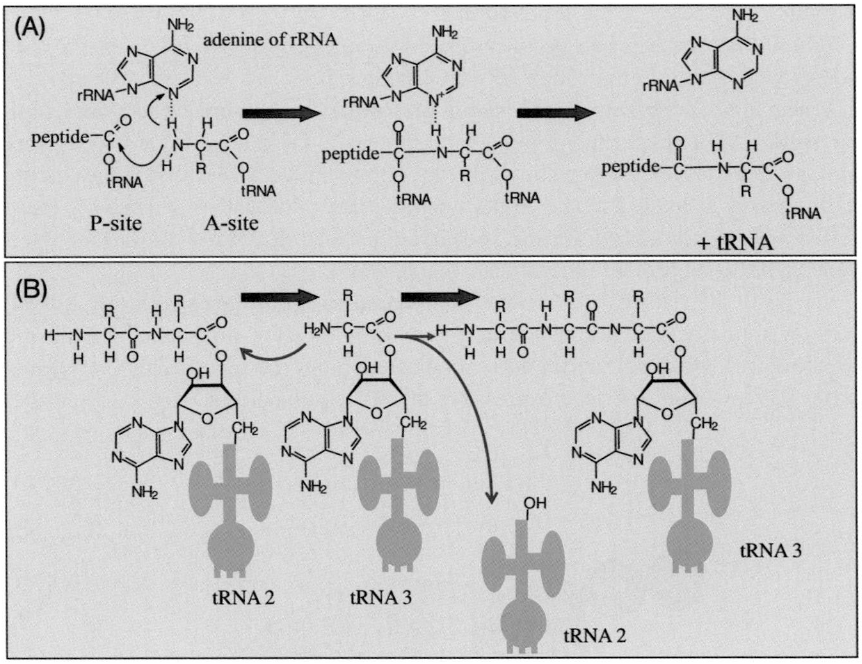

Fig. 4.25 rRNA-catalyzed peptide transfer in ribosomes. (A) Possible reaction mechanism with an adenine residue of the rRNA participating in catalysis. (B) Reaction pathway of peptidyl transfer.

Table 4.6 Protein biosynthesis in ribosomes as a target for antibiotics.

Antibiotic	Mode of action
Tetracycline	Inhibits A-site in ribosomes
Aminoglycosides (streptomycin)	Disturbs anticodon–codon recognition and chain elongation
Erythromycin	Binds to 50S subunit, blocks exit site (E), and inhibits chain elongation
Chloramphenicol	Binds to 50S subunit and inhibits peptidyl transfer
Puromycin	Induces a premature chain termination

tions) to existing antibiotics. A number of pathogenic strains of *Staphylococcus aureus* that have become resistant to most antibiotics (so-called methicillin-resistant *S. aureus* (MRSA)) are particularly dangerous.

5

Distributing Proteins in the Cell (Protein Sorting)

Learning Objectives
In this chapter, the principles of protein sorting to the individual cellular compartments will be discussed. The import and export of proteins into and out of the cell nucleus via the nuclear pore complex are described. The uptake of proteins by mitochondria, chloroplasts, and peroxisomes occurs through specific protein transporters. When protein synthesis occurs at the rough endoplasmic reticulum (ER), the proteins are first transported to the ER; from there they arrive in the Golgi apparatus. They exit the Golgi apparatus via vesicles, to lysosomes and endosomes or through exocytosis into the extracellular space. Membrane vesicles are formed via endocytosis; they later fuse with the endosomes.

The **cellular compartments** were introduced in Chapter 3. All compartments are enclosed by a biomembrane and contain a multitude of proteins. In many cases, the separation of proteins in a cell is compartment specific, meaning that every compartment harbors its own set of proteins. Every animal cell contains about 10^{10} single protein molecules, whose synthesis begins on the **ribosomes** in the cytoplasm. Every protein must finally arrive in the part of the cell where it is to be functional. One of the central questions in molecular biology concerns the mechanism of **protein sorting**. The understanding of this issue is important for biotechnology, especially when it comes to direct recombinant proteins into the correct compartments.

Three important pathways of protein sorting (Fig. 5.1) are known:

- Transport via the **nuclear pore complex** into the **cell nucleus**. The nuclear pores exhibit selective channeling, allowing entry only for certain macromolecules. The export out of the nucleus also proceeds selectively via nuclear pores.
- Uptake of a protein produced in the **cytosol** by an **organelle** via specific **protein translocators**. This is the pathway for proteins taken up by the **mitochondria, plastids**, and **peroxisomes**.
- Proteins secreted in the **endoplasmic reticulum (ER)** undergo a series of **posttranslational modifications** in the ER and in the **Golgi apparatus**. The finished proteins are packed into vesicles and sent to the **lysosomes, endosomes**, or **cytoplasmic membrane**. There the vesicle fuses with the membranes of the organelles or the cell and the content of the vesicle is released through **exocytosis**.

The **selectivity of protein transport** is based on **recognition signals** that proteins must carry. If a protein does not have a signal, it remains in the cytoplasm. All other proteins contain address labels that determine the designated location. They are either coherent signal sequences, with 15–60 amino acids, or recognition spots, which are only recognizable in a three-dimensional state and are made up of signal sequences from many protein domains. The signal sequences are very conservative in their structure. Important examples are shown in Table 5.1. Signal sequences are usually found on the N- or C-terminal of a protein.

An Introduction to Molecular Biotechnology, 2nd Edition.
Edited by Michael Wink
Copyright © 2011 WILEY-VCH Verlag GmbH & Co. KGaA, Weinheim
ISBN: 978-3-527-32637-2

Fig. 5.1 Schematic overview of protein transport inside a cell.

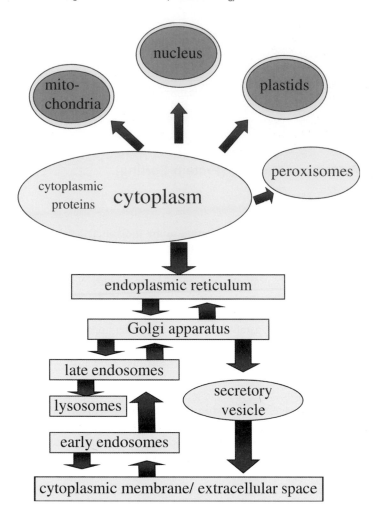

Table 5.1 Examples of typical recognition sequences. Amino acids printed in bold are especially important for the signal sequence.

Targeted compartment	Sequence
Nuclear import	-Pro-Pro-**Lys-Lys-Lys-Arg-Lys**-Val-
Nuclear export	-**Leu**-Ala-**Leu**-Lys-**Leu**-Ala-Gly-**Leu**-Asp-**Ile**-
Mitochondria	H_3N^+-Met-Leu-Ser-Leu-**Arg**-Gln-Ser-Ile-**Arg**-Phe-Phe-**Lys**-Pro-Ala-Thr-**Arg**-Thr-Leu-Cys-Ser-Ser-**Arg**-Tyr-Leu-Leu-
Plastids	H_3N^+-Met-Val-Ala-Met-Ala-Met-Ala-**Ser**-Leu-Gln-**Ser-Ser**-Met-**Ser-Ser**-Leu-**Ser**-Leu-**Ser-Ser**-Asn-**Ser**-Phe-Leu-Gly-Gln-Pro-Leu-**Ser**-Pro-Ile-**Thr**-Leu-**Ser**-Pro-Phe-Leu-Gln-Gly-
Peroxisomes	-**Ser-Lys-Leu**-COO⁻
ER import	H_3N^+-Met-Met-Ser-Phe-Val-Ser-**Leu-Leu-Leu-Val-Gly-Ile-Leu-Phe-Trp-Ala**-Thr-**Glu**-Ala-**Glu**-Gln-Leu-Leu-Thr-**Lys**-Cys-**Glu**-Val-Phe-Gln-
ER retention	-**Lys-Asp-Glu-Leu**-COO⁻-

They are usually removed by **signal peptidases** as soon as a protein has reached its destination.

5.1
Import and Export of Proteins via the Nuclear Pore

Every cell nucleus contains over 3000–4000 **nuclear pore complexes**. The nuclear pore complex in animals has a molecular weight of 125 million Da and is made up of 50–100 proteins, which are termed **nucleoporins**. Nuclear pore complexes are able to import (e.g., histone proteins) or export (e.g., the subunits of the ri-

Fig. 5.2 Structure of a nuclear pore (reconstructed from electron microscopy images). The nuclear pore complex contains between 50 and 100 different proteins. Inner diameter=9 nm. The upper side is oriented towards the cytosol. (From Voet *et al.*, 2002 b, p. 833.)

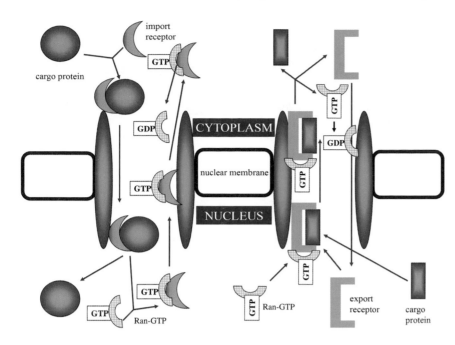

Fig. 5.3 Simplified model of the import and export of proteins via the nuclear pore. The left side represents protein import via a nuclear pore; the right side represents the export of cargo proteins.

bosomes that are assembled in the nucleolus) a large number of proteins in a short time. The nuclear pores are filled with water and allow substances smaller than 5000 Da to pass through unhindered. For larger molecules, they are highly selective. **Cargo proteins** must bear the correct signal sequence (Table 5.1). The structure of a nuclear pore is schematically represented in Fig. 5.2.

For the import or export, **mobile nuclear import receptors** and **export receptors** are required. These receptors must, on the one hand, recognize the recognition signal of the protein to be transported (**cargo protein**) (Table 5.1) and, on the other hand, interact with the nucleoporins of the nuclear pores. Nuclear import and export is shown schematically in Fig. 5.3. First, a **cargo protein** and **nuclear import receptor complex** is formed. As soon as a cargo protein/import receptor complex has arrived at the inner side of the nuclear membrane, a **GTP-binding protein** (**Ran-GTP**) binds to the import receptor. A conformational change occurs and a cargo protein is released. The complex of Ran-GTP and the import receptor binds to nucleoporin and transports it through the pore in the direction of cytosol. Once it arrives, Ran-GTP is dephosphorylated and dissociated from the import receptor as **Ran-GDP**, whereby the receptor is reactivated. Export out of the nucleus occurs with a similar principle (Fig. 5.3). The change from Ran-GTP to Ran-GDP is catalyzed by a **GTPase activating protein** (**GAP**); the exchange of GDP to GTP in the nucleus is assisted by a **guanine exchange factor (GEF)**.

5.2
Import of Proteins in Mitochondria and Chloroplasts

Proteins that should function inside the **mitochondria** or **chloroplasts** are synthesized as **precursor proteins** on the cytosolic ribosomes and carry a **recognition sequence on the N-terminal** (Table 5.1). After uptake by the organelle this signal sequence is removed by a **signal peptidase**. The import progresses via a multienzyme complex: the **TOM complex** binds a precursor protein and transports it over the outer mitochondrial membrane. Further transport over the inner mitochondrial membrane is taken over by **TIM22** and **TIM23** complexes (Fig. 5.4). When membrane proteins are imported, they contain an additional signal sequence, which is then recognized by the **OXA complex**. The OXA com-

Fig. 5.4 Schematic overview of the uptake of a precursor protein by the mitochondria and the assembly of membrane proteins in the inner mitochondrial membrane. Eukarya: translocase of outer membranes; TIM: translocase of inner membranes. (A) Setup of transport systems. (B) Cooperation between Eukarya and TIM complexes. (C) Function of the OXA complex.

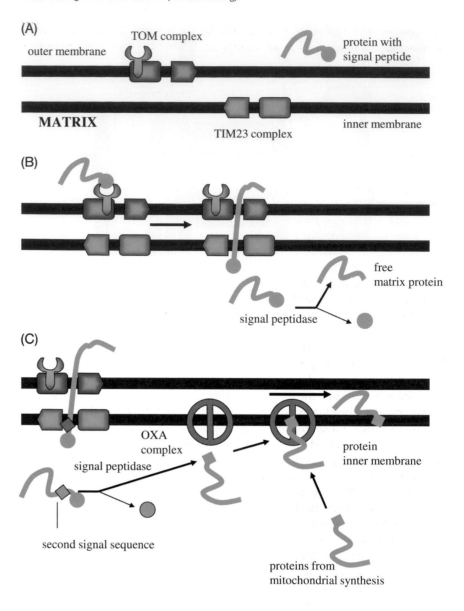

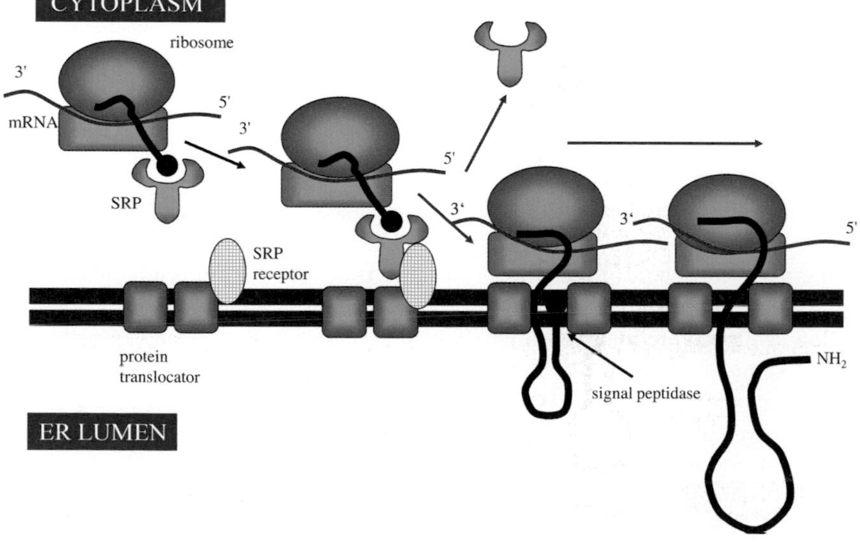

Fig. 5.5 Simplified scheme of the import of a protein into the ER lumen.

plex ensures that membrane proteins, whether synthesized by the mitochondria or imported out of the cytosol, are incorporated correctly in the inner mitochondrial membrane.

Transporting precursor proteins into **chloroplasts** follows a similar scheme. A second signal is necessary for transport into the **thylakoid**.

5.3
Protein Transport into the Endoplasmic Reticulum

In electron microscope photographs the **rough ER** is recognized by its large number of **ribosomes**, which look as if they are tightly bound to the ER membrane (Fig. 1.2). These **ribosomes** are in the process of synthesizing proteins, which are then secreted into the **ER lumen**. These proteins are characterized by a **specific signal peptide on the N-terminal** (Table 5.1).

In principle, protein biosynthesis begins on the **free ribosomes in the cytoplasm**. When a protein exhibiting an **ER import signal peptide** is synthesized, a **signal recognition particle (SRP)** will bind to the signal sequence. In the next step, **SRP binds an SRP receptor** present at the ER membrane and therefore brings the translating ribosome into the vicinity of a **protein translocator**. Figure 5.5 schematically shows the import of a protein into the ER lumen. As soon as a protein is completely synthesized and the C-terminal of the protein has arrived in the ER lumen, a **signal peptidase** cleaves the signal recognition sequence and the protein is freed into the ER lumen.

The import of **membrane proteins** is similar. The growing polypeptide chain is internalized until a **second signal sequence**, which corresponds to a transmembrane domain, is reached (Fig. 5.6). The cleavage of the first signal sequence results in a transmembrane protein with a transmembrane region. The C-terminal lies in the cytosol and the N-terminal in the ER lumen. The formation of membrane proteins with many transmembrane regions occurs in a similar way.

Proteins that remain in the ER and are not channeled out through the Golgi apparatus have a retention signal at the C-terminal. Such ER proteins serve among others as **chaperones**.

Upon entry into the ER, most proteins that are to be exported are coupled with an **oligosaccharide residue**. An oligosaccharide is linked to an **asparagine residue** via an **N-glycosidic bond**. Oligosaccharides with 14 sugar residues (above all those containing N-acetylglucosamine, mannose, and glucose) are present as

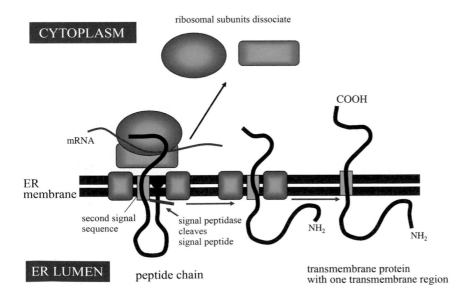

Fig. 5.6 Simplified scheme of the integration of a membrane protein into the ER membrane.

Fig. 5.7 Assembly of glycoproteins in the ER. The oligosaccharide exists as a dolichol diphosphate ester in its activated form and can be transferred onto an asparagine residue of the growing peptide chain.

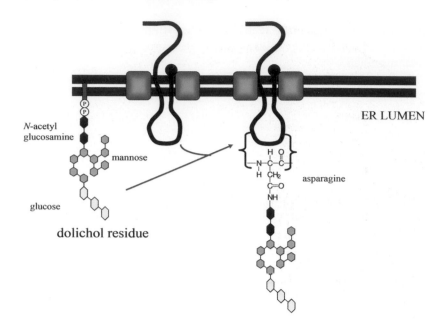

dolichol residue

dolichol diphosphate esters in the activated form, in which the lipophilic dolichol residue is anchored in the biomembrane (Fig. 5.7). Also present in the cell are glycoproteins whose sugar residues are linked to threonine or serine with an O-glycosidic bond. Their synthesis occurs in the **Golgi apparatus** and not in the ER. The sugar residues are altered again in the Golgi apparatus, where they obtain their final specificity.

A few proteins are associated with the cell membrane. This usually occurs through a **glycosylphosphatidylinositol (GPI) anchor**, which can be attached to the C-terminal of a protein.

5.4
Vesicle Transport from the ER via the Golgi Apparatus to the Cytoplasmic Membrane

The **endomembrane system** of the cell shows a high degree of dynamics through the uptake and secretion of vesicles. Proteins from the ER are also transported in this way to the **Golgi apparatus**, and from the Golgi apparatus to the **lysosomes** and **endosomes** as well as the **cytoplasmic membrane** (Fig. 5.8).

The **pinching off of vesicles** and their uptake is a complex process that involves a large number of internal and external proteins (many of them not yet known). The **budding of vesicles** only occurs when a specific protein coat is formed on the vesicle surface:

- Vesicles that bud from the ER carry **COPII proteins**.
- Vesicles that migrate between the *cis* and *trans* sides of the Golgi apparatus carry **COPI proteins**.
- Vesicles that are sent from the *cis*-Golgi to the endosomes or endocytotic vesicles are covered with a coat of **clathrin molecules** (Fig. 5.9)

These surface proteins are connected to membrane-bound cargo receptors via **adapter proteins**, which recognize cargo proteins that are present in the vesicle.

Vesicles must be able to recognize a target compartment and to bring the content to the correct location. Further receptor molecules termed **SNARE proteins** serve this purpose. Every vesicle carries specific **v-SNARE** proteins on the surface, which can be recognized by the target compartment with specific **t-SNARE receptors**. In this context **Rab proteins** are important: Rab proteins are monomeric GTPases that ensure that the vesicle finds the right partner. The

Fig. 5.8 Vesicle transport pathways in the cell.

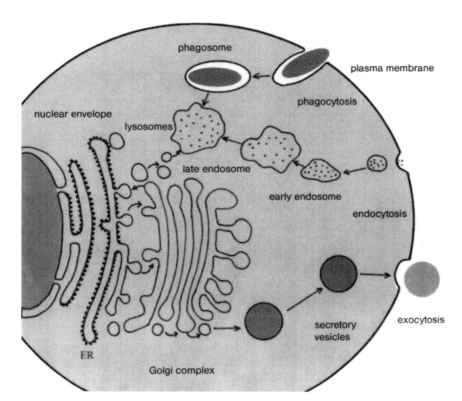

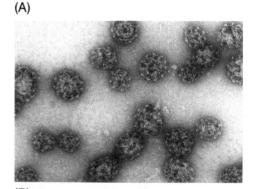

(A)

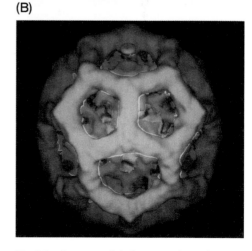

(B)

Fig. 5.9 Structure of clathrin-coated vesicles: (A) electron micrograph and (B) three-dimensional representation of a clathrin coat, derived from an electron microscope photo. (From Voet *et al.*, 2002 b, p. 259.)

most researched SNARE proteins are those associated with the **neurovesicles** in the **presynapse**. Neurovesicles can only carry out exocytosis when **synaptobrevin** (v-SNARE) on the vesicle membrane interacts with **syntaxin** (t-SNARE) on the inside of the presynapse. Additionally a further peripheral membrane protein, **SNAP25** (t-SNARE), must enter the complex. The exocytosis is initiated via a calcium signal: when an **action potential** occurs in the synapse the voltage-gated calcium channels open and Ca^{2+} flows into the synapse for a short time.

In the different compartments of the **Golgi apparatus** the sugar residues of the proteins are altered in different ways. For example, the **mannose residues** of the lysosomal proteins are phosphorylated and therefore recognized by their **mannose-6-phosphate residues**. In other proteins the mannose residues are removed and replaced by *N*-acetylglucosamine, galactose, or *N*-acetylneuraminic acid (NANA).

In the *trans*-Golgi, proteins with **mannose-6-phosphate residues** are recognized by a specific transmembrane receptor. The loading of these receptors results in a conformation change in the proteins, which is then recognized by **clathrin molecules** (Fig. 5.9). This leads to the budding of the vesicle, which is loaded with lysosomal enzymes. These vesicles fuse with vesicles of the **late endosomes**, finally resulting in the formation of the endosomes.

Proteins that are sent to the cytoplasmic membrane, where they bud into the extracellular space via **exocytosis**, are also processed in the Golgi apparatus. The fusion of the Golgi vesicle with the cytoplasmic membrane is termed **exocytosis**. In this process water-soluble proteins, such as **peptide hormones** or **antibodies**, are released into the extracellular space (e.g., the blood). Membrane-associated proteins remain as membrane proteins in the cytoplasmic membrane and are orientated with their sugar residues into the extracellular space. Exocytosis can be both continuous and signal controlled. An example of the latter is the release of **insulin** or **histamine** from their respective **storage vesicles**.

The opposite process, **endocytosis**, also occurs continuously at the cytoplasmic membrane. In this process vesicles bud off and migrate **from the early endo-**

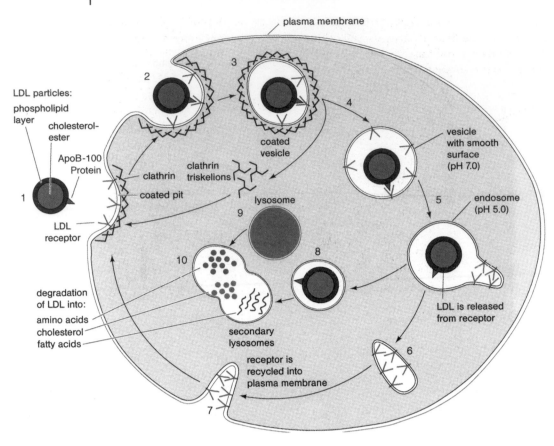

Fig. 5.10 Schematic progression of receptor-mediated endocytosis of LDL. (From Voet *et al.*, 2002b, p. 263.)

somes to the late endosomes, and finally deliver their contents to the lysosomes or the Golgi apparatus (Fig. 5.8).

The processes of endocytosis are subdivided:
- **Phagocytosis** (uptake of microorganisms or dead cells).
- **Pinocytosis** (uptake of liquids and smaller molecules).

Phagocytosis (Fig. 5.8) is the function of the phagocytes (macrophages, neutrophils, and dendritic cells) of the cellular immune system. The phagocytosed cells are degraded in the **lysosomes** (for details about the immune system, see Further Reading at the end of Chapter 6).

Pinocytosis is a continuous process: macrophages take up about 25% of their cell volume per hour via pinocytosis; in relation to the cytoplasmic membrane, this corresponds to a budding rate of the membrane in vesicles of 3% per minute. The surface of the membrane, which is taken up via endocytosis, corresponds to the cell surface, which is released via **exocytosis (endocytosis–exocytosis cycle)**. During endocytosis the vesicles are filled with liquids and molecules that are present in the extracellular space. Through **fluid-phase endocytosis** polar molecules can also enter the cell, which otherwise would not be taken up via diffusion or carriers.

An important variation of endocytosis is **receptor-mediated endocytosis**. This is how **lipoproteins** such as **LDL (low-density lipoprotein) particles**, which are loaded with cholesterol ester in blood, are recognized and bound by **LDL receptors** of the target cell (Fig. 5.10). After binding, the clathrin-coated endocytosis vesicle buds off and migrates via the endosomes to the lysosomes. There the receptors with exocytotic vesicles are returned to the cytoplasmic membrane, while the lipoproteins are degraded in the lysosome. The **cholesterol ester** is also cleaved by an **esterase**. Cholesterol is then available to the cell for synthesis or as a membrane lipid. Patients with **defective LDL receptor genes** (prevalence of 1 in 500) have an increased risk of myocardial infarction, as higher levels of cholesterol lead to **arteriosclerosis**.

6
Evolution and Diversity of Organisms

Learning Objectives
Modern biology is mostly concerned with a few model organisms. For biotechnologists it is nevertheless important to have an overview of the diversity of living organisms. This chapter gives a short overview of the evolution and systematics of Bacteria, Archaea and Eukaryota.

Molecular cell biology focuses more and more on **model organisms**. Examples are, among others, the Gram-negative bacteria *Escherichia coli*, the brewer's yeast (*Saccharomyces cerevisiae*), the nematode worm (*Caenorhabditis elegans*), the fruit fly (*Drosophila melanogaster*), the common mouse (*Mus musculus*), the human (*Homo sapiens*), and wall cress (*Arabidopsis thaliana*) as a representative of higher plants. Extensive knowledge about molecular and cellular biology was gained from these model organisms. As these organisms share a common evolutionary history, it is assumed that basic characteristics found in one model organism are also valid for all other organisms. This can, but must not always be true. In some cases nature has also found different solutions to the same problem (**convergent evolution**).

In the future of biotechnology we must not ignore the diversity of organisms, with perhaps over 10 million species and complex adaptations. Many of them offer evolutionarily derived solutions for problems that are of great interest for biotechnological purposes or applications.

6.1
Prokaryotes

Figure 6.1 shows a **tree of life** reconstructed over nucleotide sequences of 31 genes from 191 species whose genomes were completely sequenced (Ciccarelli *et al.*, 2006). This tree illustrates the lines of development for the different kingdoms. Within **prokaryotes**, two large domains are recognizable: the **Bacteria** (or simply bacteria) and the **Archaea** (or archaebacteria). Important biochemical differences were already summarized in Table 1.1.

6.2
Eukaryotes

The evolution of the ancestor of a **eukaryotic cell** and the uptake of bacteria (endosymbiotic origin of mitochondria and chloroplasts) was a key innovation of early evolution. While the incorporation of mitochondria only occurred once in evolution, there is reasonable evidence for the assumption that the incorporation of cyanobacteria (leading to chloroplasts) occurred many times (especially within the different groups of algae).

An Introduction to Molecular Biotechnology, 2nd Edition.
Edited by Michael Wink
Copyright © 2011 WILEY-VCH Verlag GmbH & Co. KGaA, Weinheim
ISBN: 978-3-527-32637-2

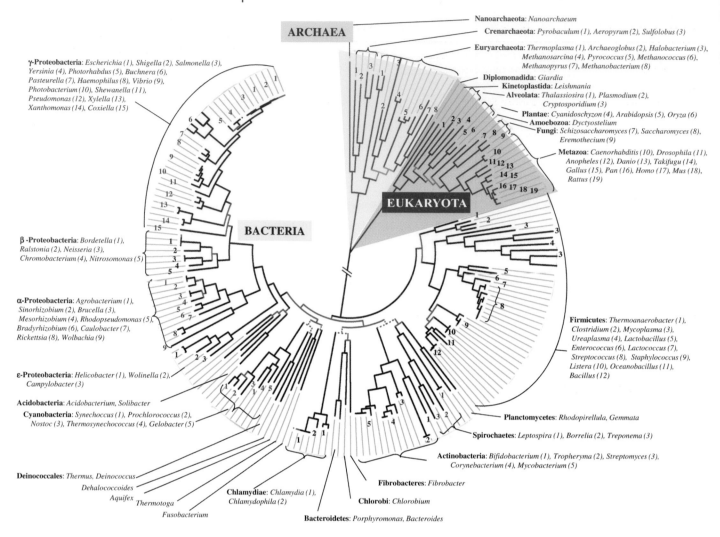

Fig. 6.1 Tree of life: molecular phylogeny of Bacteria, Archaea and Eukaryota.

There are large differences in cellular structure and function between prokaryotes and eukaryotes. Table 1.1 summarized the important characteristics. The eukaryotic cell is distinctly further developed (Fig. 1.2) and is able to carry out different processes at the same time in a single cell. This required the development of separated reaction spaces – **cellular compartments** (Table 1.2) – in the early stages of evolution.

A simplified overview of the origin of organisms is shown in Figs 1.1 and 6.1. Due to lack of space it is not possible to go into more detail for the different organisms in the specific individual domains of the living kingdoms. To give biotechnologists a quick orientation about which organism they are focusing on, and where these organisms stand in the tree of life, a short systematic synopsis of the organisms is put together in the following. For simplicity, only the large groups of protists (Table 6.1; Fig. 6.2), plants (Table 6.2; Fig. 6.3), and animals (Table 6.3, Figs 6.4 and 6.5) will be more closely characterized (a good short overview can be found in Campbell and Reece, 2006). Apparently, the protozoa do not form a monophyletic clade, but several independent evolutionary lineages. Traditionally, algae, and sometimes even fungi and bacteria, have been included in botany. As can be seen from Figs 6.1 and 6.2 only the metabionta with red algae, green algae, and land plants form a monophyletic unit. Fungi cluster with opisthokonta and thus much closer to animals then to plants. Among animals, the protostomia have now been separated in ecdysozoa and lo-

Major protist clades	Characteristics	Example
Tetramastigota	Second ary loss of mitochondria	
Diplomonadida	Two separate cell nuclei	*Giardia*
Parabasalia		
Trichomonadida	Undulating membrane	*Trichomonas*
Euglenozoa	Flagellates with or without photo synthesis	
Euglenophyta	Paramylon as storage polysaccharide	*Euglena*
Kinetoplastida	With kinetoplast	*Trypanosoma* (sleeping sickness)
Chromalveolata	With chloroplasts from secondary endosymbiosis	
Alveolata	Alveoli under the cell surface	
Dinoflagellata	Shell from cellulose plates	*Pfiesteria*
Apicomplexa (Sporozoa)	Apical complex for penetration of hosts	*Plasmodium* (malaria), *Toxoplasma*
Ciliata (ciliates)	Cilium for movement and nutrient uptake	*Paramecium*
Stramenopilata or Heterokonts	With trailing and flimmer flagellum	
Oomyceta	Hypha; cell walls from cellulose	
Bacillariophyceae (diatoms)	Glassy; walls separated into two	*Pinnularia*
Chrysophyceae (golden algae)	Two flagellate cells	*Dinobryon*
Phaeophyceae (brown algae)	Brown accessory pigments	*Laminar ia*
Metabionta	With chloroplasts from primary endosymbiosis	
Rhodobionta (red algae)	Without flagellate stage; phycoerythrin	*Porphyra*
Chlorobionta (green algae)	With chloroplasts (similar to land plants)	*Chlamydomonas*
Charophyceae → Land plants		
Unikonta		
Amoebozoa	With sheet-like form pseudopods	*Amoeba*
Mycetozoa (slime mold)	Saprophyte; amoeboid stages forms colonies	*Physarum*, *Dictyostelium*
Opisthokonta	protruding flagellum	
Fungi (Ascomycetes, Basidomycetes)	Cell walls from chitin, saprophytic	*Saccharomyces cerevisiae* (yeast) *Amanita phalloides* (deadly agaric)
Choanoflagellata → Metazoa (animals)	With microvilli	

The red, brown, and green algae were previously grouped with the plants; due to new molecular systematics a new order has been proposed.

Table 6.1 Important groups of protists (model organisms or diseases caused by pathogens) (important model organisms are given in bold).

Fig. 6.2 Phylogenetic relationships between protists and transition to plants and animals.

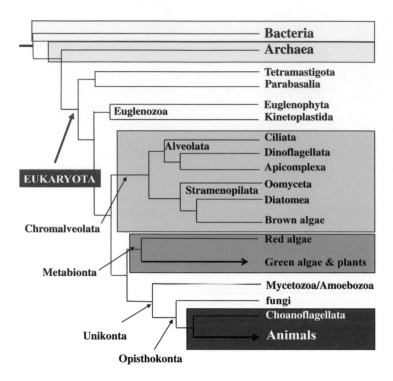

Table 6.2 Systematic classification of the land plants (important model organisms are given in bold).

Subdivision	Class
Sporophyta (spore-bearings plants)	
Moss plants	Marchantiophyta (Marchantiopsida, liverwort)
	Anthoceratophyta (Anthoceratopsida, hornwort)
	Bryophyta (Bryopsida, moss)
Lycophytes (club mosses)	Lycopodiophyta (Lycopodiopsida, lycopod)
Pteridophyta (Euphyllopytes; fern and other seedless vascular plants)	Psilotophyta (Psilotopsida, wisk fern)
	Sphenophyta (Equisetopsida, horsetail)
	Filicophyta (Filicopsida, fern)
Spermatophyta (seed-bearing plants)	
Gymnospermae (naked seed plants)	Ginkgophyta (Ginkgopsida, Ginkgo plant)
	Cycadophyta (Cycadopsida, palm fern)
	Gnetophyta (Gnetopsida, joint-fir family)
	Pinophyta (Pinopsida, conifers)
Angiospermae (flowering plants)	Magnoliophyta (Magnoliopsida) (*Arabidopsis thaliana, Nicotiana tabacum*)

Fig. 6.3 Phylogeny of land plants.

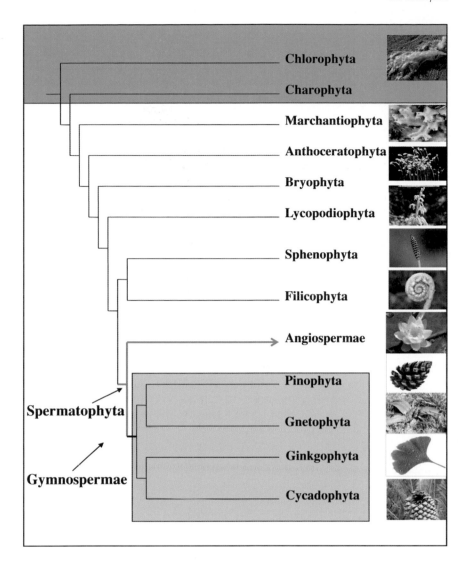

photrochozoa on account of molecular and anatomical data (Fig. 6.4). According to the rules of cladistics only monophyletic groups should be accepted. This requires a restructuring of some of the groups of organisms that had been grouped together, such as protists, mosses, fishes, and reptiles (Lecointre and Le Guyader, 2001).

Table 6.3 Systematic classification of multicellular animals (important phyla) (important model organisms are given in bold).

Category	Phylum	Characteristics
Parazoa	Porifera (sponges)	Simple multicellular animals with choanocytes, that can take up bacteria by phagocytosis; cells are mostly totipotent
Radiata	Cnidaria (anemones and jelly fish) (*Hydra*)	Stingi ng cells (cnidocytes) with nematocysts; developed gastrovascular system (gastric space with mouth, without anus)
	Ctenophora (comb jellies)	Adhesive cells (colloblasts) to catch prey; eight rows of rows of fused cilia; gastrovascular system
Bilateria		
Protostomia		
Lophotrochozoa (150 000 species)		With lophophor and trochophora larvae
	Plathelminthes (flatworms)	Dorsoventrally flattened; unsegmented; no coelom
	Rotifera (rotifers)	Pseudo coele with digestive tract; rotary organ; without circulatory system
	Ectoprocta/Bryozoa (moss animals)	With coelom; with ciliated tentacles (lophophor) for uptake of nutrients; colonial
	Nemertea (ribbon worms)	Co elom-like structure for storing proboscis; closed circulatory system with blood vessels; digestive tract with mouth and anus
	Mollusca (mollusks)	With small coelom; three body parts: foot, visceral mass, mantle; head often reduced
	Annelida (segmented worms)	With small coelom and epitheliomuscular tube; segmented body and segment specialization
Ecdysozoa (>1 million species)		
	Nematoda (round worms) (*Caenorhabditis elegans*)	Cylindrical, unsegmented pseudocoelomates; complete digestive tract without circulatory system
	Arthropoda	With coelom and segmented body, jointed appendages; ectodermal exoskeleton
	Chelicerata (Arachnida)	
	Myriapoda (millipedes and centipedes)	
	Hexapoda (insects) (*Drosophila melanogaster*)	
	Crustaceae (crustaceans)	
Deuterostomia (60000 species)		
	Echinodermata (echinoderm) (star fish, sea urchin, sea cucumber)	With coelom; larvae with bilateral symmetry; adult animals with radial symmetry; ambulacral system; mesodermal endoskeleton
	Hemichordata	With coelom and trimeric abdominal cavity; reduced chorda; branchial gut (pharyngial gill)
Chordata (chordates)		With coelom; chorda dorsalis; dorsal tubular nerve cord branchial gut (pharyngial gill)
	Urochordata (Tunicata, tunicates)	
	Cephalochordata (Acrania, skull-less) (*Brachiostoma*)	
	Vertebrata (vertebrates)	Neural crest; cephalization; spinal column; closed circulatory system
	Agnatha (lamprey)	
	Chondrichthyes (cartilaginous fish)	
	Osteichthyes (bony fish) (*Danio rerio*)	
	Lisamphibia (amphibians) (*Xenopus laevis*)	
	Reptilia (reptiles) (turtle, lizard, crocodile)	
	Aves (birds) (*Gallus gallus*)	
	Mammalia (mammals) (*Mus musculus, Homo sapiens*)	

Fig. 6.4 Phylogeny of Deuterostomia and vertebrates.

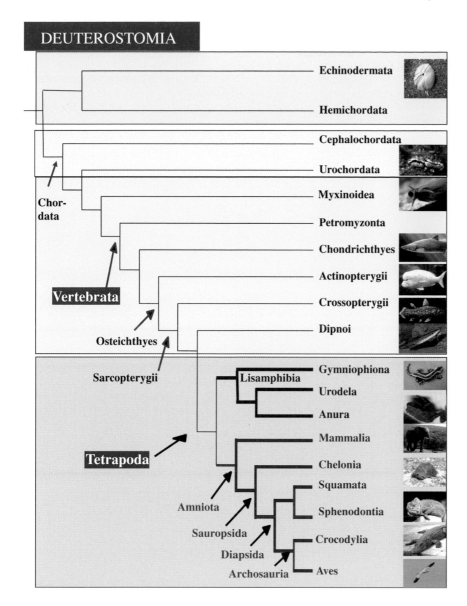

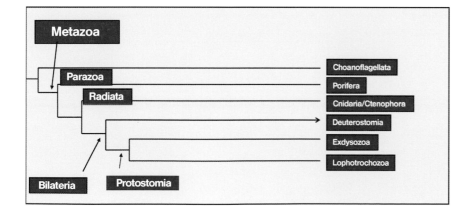

Fig. 6.5 Evolutionary trends in animal phylogeny.

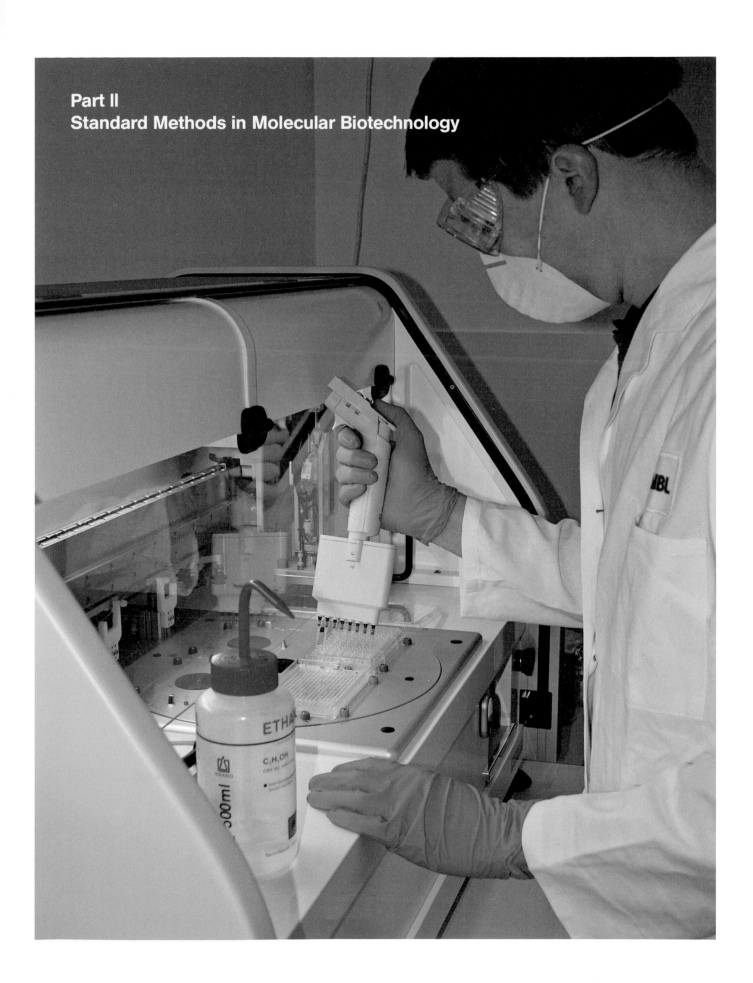

Part II
Standard Methods in Molecular Biotechnology

7
Isolation and Purification of Proteins

Learning Objectives
Isolated and pure proteins are required for many purposes in modern biotechnology. This chapter introduces the most important techniques used for the isolation and analysis of proteins, especially several methods of electrophoresis and column chromatography.

7.1
Introduction

Many experiments that involve the **characterization of proteins** (e.g., exploring the mode of action of a given enzyme) can only be successful if the relevant protein can be isolated and separated from the many other proteins in the cell. Homogeneous solutions of proteins or protein complexes are also a prerequisite for other analytical procedures, such as sequence analysis, x-ray structure analysis of protein crystals, and mass spectrometric examinations. The foundations of protein purification were laid by Otto Warburg in the 1930s. Over recent decades, the focus has been on miniaturization, automation, and optimization of existing principles. Today, the **(over)expression of recombinant proteins** in *Escherichia coli*, yeast, insect, plant, or even mammalian cells strongly facilitates the extraction of large amounts of purified protein (see Chapter 16). This purification strategy was again greatly simplified with the introduction of **suitable peptide sequences** (usually N- or C-terminal tags such as the hexahistidine tag, see Section 7.6.1).

When isolating **enzymatically active proteins**, the purification process can be monitored by measuring the specific activity, defined as the ratio of enzyme activity to the amount of total protein. This ratio should increase with every successful purification step, resulting in the preparation of a homogeneous enzyme. However, purification often goes hand in hand with an increasing instability of the protein, diminishing the theoretically expected increase in enrichment. Once enzymes are removed from the protective environment of the cell interior, the purification process might bring them into contact with metals, oxygen, high ion concentrations, and further potentially harmful influences. These influences often lead to an irreversible denaturation of the protein's fragile spatial structure (its **conformation**). Proteins should thus be treated and looked upon as unstable biomolecules that require special attention in handling.

Some general rules can help avoid the loss in activity:
1. Work swiftly and in the cold. To avoid protease digestion, protease inhibitors should be added as soon as possible (Table 7.1). Avoid storing protein solutions at room temperature unless there is a reason for it. Always store them on ice or in the refrigerator.
2. Avoid contact with metal. Metallic surfaces can contaminate the solution with heavy metal ions.
3. Minimize oxygen exposure (i.e., avoid excessive shaking or stirring). It often helps to add reducing agents such as 2-mercaptoethanol.

An Introduction to Molecular Biotechnology, 2nd Edition.
Edited by Michael Wink
Copyright © 2011 WILEY-VCH Verlag GmbH & Co. KGaA, Weinheim
ISBN: 978-3-527-32637-2

Table 7.1 Commonly used protease inhibitors in protein purification.

Inhibitor	Effective against	Effective concentrations	Special properties
Pefabloc® SC [a, c]	Serine proteases	0.4–4 mmol L^{-1}	Mixture of inhibitor and stabilizing agent
EDTA [a]	Metalloproteases	0.1–1 mmol L^{-1}	
Aprotinin [b]	Serine proteases	0.01–0.3 µmol L^{-1}	Inactivation at pH > 10
Pepstatin [b]	Acid proteases	1 mmol L^{-1}	Stock solution in methanol, 1 g L^{-1}
Leupeptin [b]	Serine and cysteine proteases	1–10 µmol L^{-1}	

a) When isolating recombinant proteins from bacteria, adding these protease inhibitors has often proven sufficient.

b) The threat posed by proteases is much greater when isolating, for example, from animal tissues. Use of a mixture of inhibitors is thus recommended. A combination of the five inhibitors listed here covers a wide spectrum of proteases and is available as a premixed cocktail.

c) The less expensive phenylmethylsulfonyl fluoride (PMSF) can be used instead of Pefabloc. However, PMSF is barely soluble in water, unstable in aqueous solution, and of considerably higher toxicity.

4. The protein solution should not be too strongly diluted, as activity could be reduced through adsorption to surfaces.

Before setting up the experiment, try to obtain the following information from the relevant literature in order to develop an effective purification strategy:

1. What is the protein's stable pH range (choosing the correct buffer pH value)?
2. Does the enzyme require certain cations like Ca^{2+} or Mg^{2+} as stabilizing cofactors?
3. Under what conditions is the protein soluble? Pay attention to the following:
   – Some proteins will not dissolve at low ionic strength.
   – All proteins precipitate at high ionic strength.
   – Membrane-anchored proteins can only be solubilized with detergents, which may affect the purification behavior.

7.2
Producing a Protein Extract

The **material** from which proteins are obtained can be very diverse. In a straightforward case, the desired protein already exists in aqueous solution (e.g., in blood, milk, or cell culture supernatant) and purification can begin right away. In most cases, however, a protein-containing extract needs to be produced from the starting materials. In the past, **animal and plant tissues** were often the only accessible protein source. Nowadays, they have been replaced by **cell cultures**, **yeasts**, or **bacteria** that express recombinant proteins.

Most materials like animal tissues are often broken up mechanically by shredding or exposure to high shearing forces (e.g., ULTRA-TURRAX® or Potter-Elvehjem homogenizers). The disruption is usually performed in the presence of an extraction buffer, which is customized for the relevant protein (in terms of pH, detergents, ion concentrations, and stabilizing agents). The tissue should be cut into small pieces beforehand.

The **disintegration** of yeasts and especially bacteria requires some special effort; the available methods are basically divided into **enzymatic and mechanical lysis**. Enzymatic lysis means the digestion of the **bacterial cell wall** components (peptidoglycans) with reagents like **lysozyme**. The exposed protoplasts are very sensitive and can be disintegrated with detergents (Triton X-100), osmotic shock, or mechanically exerted shearing forces (homogenizing through a narrow needle). Enzymatic cell lysis minimizes protein denaturing, succeeds independently of the sample volume, and sometimes leads to a prepurification by a selective release of the cellular components. As a disadvantage, added substances (lysozyme, detergents) may interfere with the following purification procedures.

Mechanical lysis is performed with, for example, a swing mill or a so-called "French press". The swing mill is a closed chamber containing fine glass beads; the chamber is filled with a cell suspension and then shaken at high frequencies. The impact and co-occurring shearing forces then fragment the cells. The **French press** puts a cell suspension under high pressure and releases it through a tiny opening; the cells are fragmented by the dramatic decrease in pressure and strong shearing forces.

Ultrasonic treatment is another common method of disintegration: an ultrasonic tip is immersed into the cell suspension, which is then sonicated repeatedly for 30–45 s.

All mentioned, homogenization methods involve considerable amounts of heat development. It is therefore recommended to cool the samples with ice (from the outside), to keep the homogenization steps short (10–30 s), and to allow for cooling breaks.

All methods of **cell disintegration** result in a homogenate, which is subsequently processed into a **protein extract**. This requires removing all insoluble components, usually by sedimentation in a cooled centrifuge (e.g., 30 min at 1000 g). Remaining fragments of the endoplasmic reticulum and the Golgi apparatus (the so-called microsomes) may cause a light opalescence in the solution. Should they interfere with the purification, they can be removed by ultracentrifugation (60 min at 100 000 g). If the samples are separated by automated high-pressure column systems **(fast protein liquid chromatography (FPLC), high-performance liquid chromatography (HPLC))**, this step is highly recommended to avoid clogging of the columns.

We will now discuss the major **separation principles in protein purification**, illustrated by two simple examples.

7.3
Gel Electrophoretic Separation Methods

7.3.1
Principles of Electrophoresis

Proteins in aqueous solution carry a **defined electrical charge** at all pH values except their **isoelectric point**. Hence, they migrate in an electric field. The specific mobility v of this migration is proportional to the number of charges per molecule z, and inversely proportional to the Stokes radius r and the medium's viscosity η of the particle:

$$v = z/6\pi\eta r$$

Gel electrophoresis separates proteins much better than electrophoresis in free solution. It separates based on a net-like matrix with pores of varying diameters. The sizes of the pores and of the migrating molecules determine the effective viscosities of the medium, and proteins are thus separated based on both **charge** and **size**. The separation range can be optimized by altering the gel's **degree of cross-linking**. In most applications, gels are run with neutral or weakly alkaline pH values, at which most proteins migrate towards the anode. Gel systems minimize protein convection and diffusion, so protein bands on the gel are sharply separated. A decisive drawback to gel electrophoresis is the **low amount of protein** that can be separated in a single gel. Therefore, gels are mainly used for analytic purposes, like sequence analysis of a cut-out band to identify unknown proteins (see Section 7.3.5).

7.3.2
Native Gel Electrophoresis

This method separates proteins in their unchanged **active (native) conformation**. To this end, the sample and running buffers contain neither sodium dodecyl sulfate (SDS) nor urea. The weakly alkaline running buffer used (pH 8–9) causes most proteins to carry a negative charge and thus to migrate towards the **anode**. However, all proteins that carry a positive charge under these conditions do not enter the gel, but diffuse towards the **cathode** instead. An advantage of this method is that proteins from native gels are not denatured, and can be identified after **excision** and **elution** (e.g., by their **enzymatic activity**). Oligomeric proteins that consist of several noncovalently linked peptide chains also remain intact.

7.3.3
Discontinuous Sodium Dodecyl SulfatePolyacrylamide Gel Electrophoresis (SDS-PAGE)

This standard method of protein analysis uses the detergent **SDS** to denature proteins before electrophoresis. Oligomeric proteins with noncovalently linked subunits are split into individual subunits after the existing disulfide bonds are reduced by added **2-mercaptoethanol** or **dithiothreitol**. SDS binds to proteins very strongly (Fig. 7.1 A) and in proportion to the number of amino acids; resulting polypeptide chains contain one SDS molecule per two amino acids. Each SDS molecule carries a negative charge, so a peptide chain of 180 amino acids (about 20 000 Da) will carry 90 SDS molecules. Likewise, a polypeptide chain of 900 amino acids will carry 450 SDS molecules (and the same amount of negative charges). The large number of negative charges outweighs the protein's net charge at the buffer's pH value by far, so that the **charge/size ratio** is virtually the same for all proteins. The separation of unmodified peptide chains in the gel thus results from the pore's **molecular sieving** effect exclusively and in proportion to chain size. Compared to native gel electrophoresis, SDS-PAGE offers the following advantages:
1. Aggregates and insoluble particles are dissolved in SDS and converted to single peptide chains.
2. Protein separation is almost exclusively dependent on peptide chain size.
3. The entire gel can be calibrated by loading a size standard (a protein mixture of defined composition).
4. A half-logarithmic plot of the molecular weights against their mobility in the gel shows a linear dependency over a certain range (Fig. 7.1 C).
5. The gel's degree of cross-linking, and thus the pore size, can be adjusted over a wide range of protein size to be separated (changing the content of **acrylamide** (320%) or the cross-linker **bisacrylamide** (0.11%)).

The disadvantage of the method lies in protein **denaturation** (i.e., the loss of enzymatic activity), which is irreversible in the most cases. The method therefore does not lend itself to preparative purifications. However, it is well suited for monitoring the efficiency of a purification process (see examples in Section 7.6), especially when combined with the staining methods mentioned in Section 7.3.5.

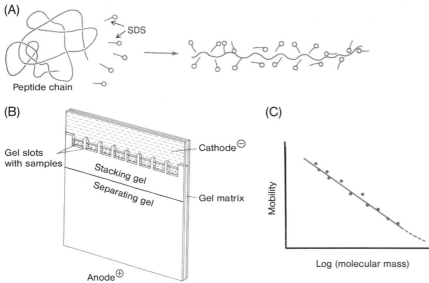

Fig. 7.1 SDS gel electrophoresis. (A) Denaturing effect of SDS. (B) Setup of a gel sandwich for SDS gel electrophoresis. (C) Separation of a standard protein mixture. Plot of migration length versus molecular weight.

7.3.4
Two-Dimensional (2D) Gel Electrophoresis, Isoelectric Focusing (IEF)

Complex protein mixtures can be separated with this combination of methods that is frequently used in proteomics. **IEF** provides the first dimension for the 2D separation. It is performed with prefabricated gel strips with a linear pH gradient formed by several **ampholytes**. When an electrical current is applied, loaded proteins migrate along the gradient until they reach the pH level equal to their **isoelectric point** (i.e., they carry a neutral net charge). The gel has no molecular sieve effect, it simply serves to stabilize the pH gradient; the proteins should not be hindered in their migration to their isoelectric points.

Some features of proteins may complicate IEF, however:
1. The protein precipitates at the isoelectric point (e.g., hydrophobic membrane proteins).
2. The protein is unstable at the isoelectric point (e.g., disintegration to several protein chains).
3. The protein forms complexes with the ampholytes (resulting in the appearance of several bands).

The second dimension is added with SDS-PAGE (the IEF gel strip is simply laid onto a normal SDS-PAGE gel); by combining these two methods, it is possible to separate proteins with very high resolution.

7.3.5
Detecting Proteins in Gels

After electrophoresis, the separated proteins can be **visualized** in the gel with **dyes** that firmly bind to proteins. Individual dyes differ in sensitivity and the ability to stain all types of proteins equally. The most frequently used dye is **Coomassie Blue R-250** (with a detection limit of about 1 mg of pure protein), although Coomassie Blue G-250, **Amido black**, and **Nigrosine** are also used. Staining with dyes follows a similar basic protocol:
1. Denature the proteins immediately after removal from the electrophoresis chamber by fixation, which will prevent further diffusion through the gel. This is commonly done with a mixture of methanol, acetic acid, and water in a 3:1:6 ratio.

2. Immerse the gel in staining solution and shake it until it is completely soaked. Moderate heating (e.g., in a microwave) can help accelerate this step.

Remove excess dye by shaking in **destaining solution**. This step can also be sped up by heating the gel or by adding paper towels, to which the dye can adsorb.

To avoid the potentially hazardous use of organic solvents such as methanol, **aqueous** solutions containing **Coomassie-based staining reagents** are commercially available.

The **silver stain method** is even more sensitive than staining with Coomassie Blue (about 10- or 20-fold). Soaking the gel in a silver nitrate solution leads to a nonstoichiometric binding of silver ions to proteins. After reduction, these complexes become visible as black to brownish bands. Unfortunately, silver stains are inconsistent as some proteins are hardly stained by silver at all. Still, silver staining is the **most sensitive method** of protein detection in gels.

7.4
Methods of Protein Precipitation

These techniques date back to the early days of protein purification. In the past, precipitating proteins by **altering their solubility** was often the only enrichment method available. Nowadays, that enrichment is often achieved with **hydrophobic interaction chromatography** (see Section 7.5.1.2). A protein's solubility is mainly determined by the distribution of hydrophilic and hydrophobic areas over the surface. While hydrophobic elements are preferably found within a protein, a characteristic amount is also located on the surface and interacts with the solvent. There are multiple ways to alter protein solubility (changing ionic strength, pH value, or temperature, adding soluble organic solvents or organic polymers, or any combination thereof). Most commonly used is **salt-induced precipitation** with high concentrations of neutral salts. This approach takes advantage of the presence of hydrophobic and hydrophilic structures on the protein surface. Water molecules aggregate around hydrophilic structures that neighbor hydrophobic surfaces. These ordered hydration cages around the protein molecule prevent the convergence and aggregation of two or more hydrophobic surfaces. The more salt is added to this system, the more water molecules are required to hydrate the introduced ions. These water molecules are increasingly drawn from the protein's hydrate cage, which exposes hydrophobic surfaces. Their surface areas are now able to interact and the protein is precipitated. Proteins with more **hydrophobic surface** elements precipitate at lower salt concentrations than those with mainly **hydrophilic surfaces**. The ion nature is of great importance when salt precipitating; univalent cations (NH_4^+, K^+, Na^+) and polyvalent anions (SO_4^{2-}, PO_4^{3-}) are preferred. **Ammonium sulfate** is the most frequently used ionic precipitant for several reasons: it has a very high solubility in water (up to $4 \, mol \, L^{-1}$), it dissolves endothermically (no risk of denaturing the protein by heat), its density in water is advantageous, and microbial growth is prevented in concentrated solutions. Another advantage of precipitation with ammonium sulfate is a frequently observed stabilization of the precipitated proteins. Enzymes can be returned to full activity even after long storage in ammonium sulfate. If an **enrichment procedure** needs to be interrupted for longer periods of time, storage as an ammonium sulfate precipitate at $4 \, °C$ is recommended. Adding small amounts of EDTA in order to complex traces of possibly present heavy metal ions can be recommended. Ammonium sulfate precipitation is often performed as fractionated precipitation at the beginning of a purification procedure (see Section 7.6.2).

It is also possible to precipitate proteins with **water-soluble organic solvents** (e.g., acetone or primary alcohols), but this protocol is less frequently followed. Since this approach lowers the solubility of charged molecules, it can be performed in addition to a previously performed salt precipitation.

7.5
Column Chromatography Methods

7.5.1
General Principles of Separation

The separation principles mentioned in this section can be generally used for purifying all sorts of proteins. The majority are based on the **specific adsorption** of proteins to a gel matrix and subsequent elution with specialized reagents, which usually contain linear or step-graded concentration gradients. Use of adsorptive techniques is highly popular and usually yields highest enrichments. Greatest success (and reproducibility) is achieved with industrially prefabricated columns and **automated pump systems** (FPLC, HPLC).

7.5.1.1 **Size Exclusion Chromatography (Gel Filtration)**
The frequently used name of **gel filtration** when referring to this procedure is misleading, because, unlike regular filtration, no components of the applied sample are retained. As protein fractionation occurs due to size differences, a more fitting name is **size exclusion chromatography**. **Adsorptive interactions** between the protein sample and the gel matrix are undesired with this method, which makes it different from the other chromatographic techniques used for protein purification. The absence of adsorptive effects has advantages and drawbacks. On the one hand, sensitive proteins are not affected by binding to a matrix. On the other hand, the lack of specific binding worsens the chromatographic resolution the volume in which a protein is eluted from the column is always greater than the volume of the applied sample. Thus, a protein sample should be applied to the column in the smallest volume possible. The sample size should never exceed 5% of the matrix volume, and best resolutions are achieved with sample volumes between 0.1 and 1%. The gel matrix consists of **porous materials** (e.g., cross-linked dextrans and cross-linked agarose) with a pore size defined as closely as possible. Columns packed with such globular gel particles possess two differently defined fluid volumes: (1) the **void volume (exclusion volume)**, which corresponds to the volume outside of and between the gel particles, and (2) the **inclusion volume**, which mainly consists of the volume contained within the gel particles. As proteins in the sample migrate through the gel matrix with the elution buffer, they are separated according to differing running behavior. Very large proteins (such as **Dextran Blue 2000**, molecular weight above 2 MDa) cannot diffuse into the particle pores and are the first to migrate through the column within the void volume. Smaller molecules partially diffuse into the porous gel particles and elute correspondingly later (Fig. 7.2 A and B). Both pore size and the diameter of the molecules (as defined by the Stokes radius) determine the separation. Assuming that all proteins of a sample have a similar globular structure, the order of elution is inversely proportional to the respective molecular weights.

Figure 7.2 (C and D) shows a gel filtration elution profile. A column's void or **exclusion volume** V_0 is determined by eluting a compound with very high molecular weight (e.g., Dextran Blue 2000). The **total volume** V_t that a completely enclosed molecule elutes corresponds to the sum of void volume and the gel matrix inclusion volume. Each protein has a characteristic **elution volume** (V_e,

Fig. 7.2 Size exclusion chromatography.
(A) Time course of size exclusion chromatography: large molecules are excluded from entering the greatest part of the available bed volume
and migrate through the matrix nearly unham-
pered. (B) Schematic illustration of the separation
of different sized proteins by the gel matrix pores.
(C) Separation of three substances in a gel matrix,
the first of which is totally excluded from the
matrix and elutes with the void volume (V_0). The
second substance is partially, the third substance
completely included. The latter elutes with the
total volume of the column (V_t). (D) Separation
of a complex substance mixture by a commercially
available size exclusion chromatography column.

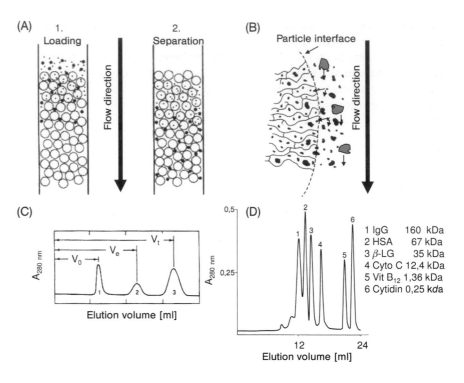

see Fig. 7.2 C). The protein's **partition coefficient** K_{av} can be calculated from the protein's elution volume and the two column volumes, V_0 and V_t:

$$K_{av} = V_e - V_0/V_c - V_0$$

Half-logarithmic plotting of the molecular weight against K_{av} yields a sigmoidal curve. Gel matrices separate best in the linear range ($K_{av} = 0.2–0.8$); this range is referred to as the fractionation range of a gel matrix. When separating proteins with a great difference in molecular weights, a matrix with a correspondingly large fractionation range should be used. Similarly, when separating proteins with only slightly differing molecular weights, use column materials with a frac-tionation range as small as possible. Separation of the latter type of protein can also be improved by separating over longer distances (i.e., column length). How-ever, increasing the separation distance also broadens the protein peaks due to diffusion. Good separation by a gel filtration column can only be expected from samples that contain less than 10 different proteins. The procedure's enrich-ment is moderate, so it is mostly performed at a later point during the purifica-tion process, when there is only a small number of contaminants. However, **gel filtration** is often used as an in-between step to gently switch buffers or desalt proteins due to its ability to remove small molecules (i.e., salts and other buffer components). Prefabricated single-use columns can be obtained commercially.

7.5.1.2 **Hydrophobic Interaction Chromatography**

Section 7.4 has already demonstrated the great importance of **hydrophobic inter-actions** for a protein's characteristic biochemical features. They are integrally in-volved in the stabilization of tertiary structures, as well as in proteinprotein in-teractions and enzymesubstrate binding reactions. The term hydrophobic inter-action describes the phenomenon that hydrophobic molecules spontaneously aggregate in **polar environments** (e.g., water). Dissolving salts and raising a medium's ionic strength increases its polarity. Since proteins possess smaller or larger numbers of hydrophobic surface structures, they attach to the hydropho-bic surfaces of a matrix at correspondingly high ionic strengths. The strength of that interaction can be adjusted by altering the buffer's ion strength, its **lipophilicity** (e.g., through the ratio of glycerol, ethylene glycol, or detergents), or the choice of adsorbent. Materials with a **low hydrophobicity** (e.g., butyl residues covalently linked to the matrix) are preferably used for strongly hydrophobic proteins; materials with **high hydrophobicities** (e.g., octyl residues) are correspondingly used for more hydrophilic proteins. Matrices with covalently bound phenyl residues rank between butyl and octyl residues for hydrophobicity, so they are suited for most proteins. Since the adsorption occurs in the presence of high salt concentrations, elution occurs by lowering the salt content in a linear gradient. When purifying lipophilic proteins that bind very strongly (like membrane proteins), the elution can be improved by adding a rising, similarly linear gradient of detergents (e.g., $500 \rightarrow 0$ mmol L^{-1} NaCl; $0.4 \rightarrow 4.0\%$ Na cholate). The same criteria that were already discussed in relation to precipitation techniques apply for the choice of ions.

7.5.1.3 Ion Exchange Chromatography

Protein surfaces carry electrical charges due to the side chains of certain **amino acids** (aspartate, glutamate, histidine, lysine, and arginine; see Chapter 2) and can thus bind to the surfaces of corresponding ion exchangers, displacing a corresponding number of counterions in the process. Due to its potentially **high protein binding capacity**, ion exchange chromatography suggests itself as an introductory step to the purification of unmodified native proteins. As a general rule, the ion exchanger's capacity to bind a protein decreases with the increasing size of the protein. The criteria for choosing a particular matrix are as follows:
1. The charge of the protein (i.e., positive or negative at a given pH value).
2. The chemical nature of the ion exchanger's charged group.
3. The nature of the matrix (particle shape and size, binding capacity).

The predominant number of proteins carries a negative net charge at pH values between 7 and 8, and thus binds to an **anion exchanger material** under these conditions (Fig. 7.3). A common anion exchanger for an initiating step of a purification is **diethylaminoethyl (DEAE)-Sepharose**. Additionally, matrices with diethyl hydroxypropyl amino ethyl groups are also used as anion exchangers; matrices with carbonic or sulfonic acid groups are used as cation exchangers. The choice of exchanger and the optimal elution conditions become much easier, when the **isoelectric point** of the protein in question is known (Table 7.2).

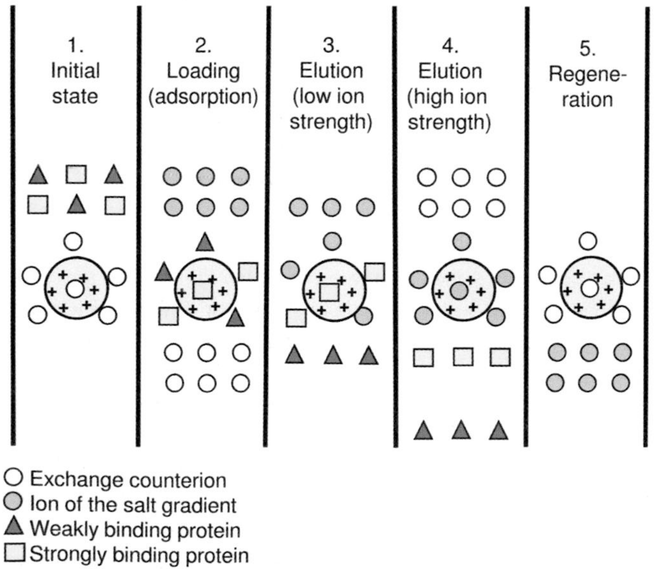

○ Exchange counterion
◒ Ion of the salt gradient
▲ Weakly binding protein
□ Strongly binding protein

Fig. 7.3 Anion exchange chromatography. Illustration of the time course of an anion exchange chromatography procedure. Negatively charged proteins bind to the matrix and displace the counterions of the covalently matrix-bound exchanger (2). Uncharged and positively charged proteins migrate with the flow-through. Weakly negative proteins elute at lower ion strength in the column (3), while high salt concentrations are needed to elute strongly negative proteins (4). Very high concentrations (usually 1–2 mol L^{-1}) of salts that contain the exchanger's original counterion elute all other bound anions. The matrix is thus regenerated for reuse (5).

Table 7.2 Suggestions for choosing ion exchangers when enriching proteins of known isoelectric point.

Isoelectric point	Ion exchanger	pH value of loading buffer
8.5	Cationic	7.0
7.0	Cationic	8.0
7.0	Anionic	6.0
5.5	Anionic	6.5

Note, however, that the pH value within the ion exchanger is not equal to that within the elution buffer. This difference is caused by the **Donnan effect**, which describes the adsorption and liberation of protons to and from the matrix. Generally, the pH value within an **anion exchanger** is roughly one unit higher than in the buffer. Conversely, the pH value within a **cation exchanger** is about one unit lower than in the buffer. The Donnan effect should always be considered in relation to the pH stability optimum of a protein, if such is known. Generally, two methods of protein elution are possible:

1. Changing the eluent's pH level (lowering the pH for anion exchangers, raising it for cation exchangers).
2. Raising the eluent's ionic strength.

Since the pH method is often faced with difficulties (e.g., when generating homogeneous pH gradients) and is problematic with regard to protein stability, the **standard elution method** is the use of increasing salt concentrations. Salts commonly used for elution are sodium chloride or potassium chloride. **Protein desorption** is caused by two effects. On the one hand, salt ions displace the charged amino acid side chains as counterions (ion exchanger effect). On the other hand, the increasing ion strength weakens the electrostatic interactions required for binding (compare salt-induced precipitation in Section 7.4).

7.5.1.4 Hydroxyapatite Chromatography

Hydroxyapatite $[Ca_5(PO_4)_3(OH)_2]$ is a crystalline special form of calcium phosphate that can be used for the purification of proteins, nucleic acids, and other macromolecules. Calcium cations and phosphate anions are involved in the electrostatic interaction with proteins. In general, it is quite difficult to predict how a given protein will interact with hydroxyapatite; still, a good interaction of acidic proteins (such as **phosphoproteins**) with the matrix is considered certain, so the matrix can be employed to enrich such proteins. The protein elution is performed with an increasing phosphate concentration in the elution buffer.

7.5.2
Group-specific Separation Techniques

Covalently linking defined molecules or reactive groups to, for example, cyanobromide-activated agarose generally allows for a great spectrum of purification strategies. In some cases, a given protein can be purified to homogeneity with a single purification step (from cell lysate), such as by use of a **specific antibody** that recognizes the native protein. Since the great number of specific purification techniques cannot be covered adequately in a textbook chapter, a chosen number of frequently used techniques will be presented.

7.5.2.1 Chromatography on Protein A or Protein G

Protein A from *Staphylococcus aureus* and **protein G** from *Streptococcus* sp. bind immunoglobulins, especially IgGs, with high capacity. Thus, matrices carrying covalently bound protein A or G are used to purify **monoclonal antibodies** from cell culture supernatants. The proteins are eluted by lowering the pH value of the buffer (e.g., with 0.1 mol L^{-1} citric acid, pH 4 for protein A or 0.1 mol L^{-1} glycine-HCl, pH 2.7 for protein G, respectively).

7.5.2.2 Chromatography on Cibacron Blue (Blue Gel)

Cibacron F3G-A is a synthetic, polycyclic dye and an aromatic anion, which binds several proteins (albumin, interferon). Due to structural similarities to adenylyl or guanylyl residues, **purine nucleotide-binding proteins** (e.g., kinases, GTP-binding proteins, and NAD$^+$-dependent enzymes) are bound as well. Elution is performed with sodium chloride or potassium chloride, which lower the electrostatic interactions necessary for binding. Nucleotide-binding proteins can also be eluted by adding the respective nucleotides in excess to the elution buffer. If an enzyme has a high specificity for its nucleotide substrate (see Section 7.6.1), use of said substrate as eluting agent is advantageous over unspecific elution, because the protein to be purified can thus be eluted with some selectivity.

7.5.2.3 Chromatography on Lectins

Lectins are proteins that interact with certain sugar residues selectively and reversibly. **Matrix-bound lectins** are thus very well suited to enrich **glycoproteins** such as cell membrane surface proteins. The choice of lectin depends on the known or expected sugar modification of the protein. Theoretically, the elution of lectin matrices could be performed by raising the elution buffer's ion strength. However, since lectins are charged proteins and can thus function as ion exchangers, chromatography is often performed at high ion strengths to counter this ion exchanger effect. Rising concentrations of **interacting sugars** (such as α-methylmannoside for a concanavalin A matrix) are used as eluents instead (Table 7.3).

7.5.2.4 Chromatography on Heparin

Heparin is a highly sulfated **glycosaminoglycan** (see Chapter 2), which interacts with a multitude of biomolecules. Heparin that has been covalently bound to a matrix can be used to purify a number of proteins. Good enrichments are achieved for DNA-binding proteins (initiation and elongation factors, restriction enzymes, DNA ligase, etc.), coagulation factors (antithrombin III), growth factors (epidermal growth factor, fibroblast growth factor), extracellular matrix proteins (fibronectin, vitronectin, laminin), corticoid hormone receptors, and lipoproteins. Heparin interacts with proteins in two ways: (1) it can imitate the DNA's polyanion structure (e.g., when interacting with DNA-binding proteins) and (2) it can also serve as a specific high-affinity interaction partner (e.g., when

Lectin	Specificity	Eluent	Special properties
Concanavalin A	α-D-mannosyl-, α-D-glucosyl residues in presence of Mn^{2+} or Ca^{2+}	Methyl α-D-mannoside (0.1–0.2 mol L^{-1})	No EDTA in buffer
Wheat germ agglutinin	N-acetyl-β-D-glucosaminyl residues	N-acetyl-β-D-glucosamine (0.02–0.2 mol L^{-1})	Stable in 0.07% SDS and 1% deoxycholate
Lentil lectin	α-D-mannosyl-, α-D-glucosyl residues in presence of Mn^{2+} or Ca^{2+}	Methyl-α-D-mannoside (0.1–0.2 mol L^{-1})	No EDTA in buffer, stable in 1% deoxycholate
Soybean lectin	N-acetyl-D-glucosaminyl residues	N-acetyl-D-glucosamine	

Table 7.3 Commonly used lectins for the enrichment of glycoproteins.

binding coagulation factors). In both cases, the interaction can be weakened by increasing the elution buffer's ionic strength. Thus, elution from heparin matrices is often performed with high salt gradients of NaCl or KCl.

7.5.3
Purification of Recombinant Fusion Proteins

Large amounts of purified protein are no longer extracted from their naturally occurring sources, but from suitable organisms that have been genetically modified to (over)express the **recombinant protein** instead (see Chapter 16). Separating the recombinant protein from the host proteins can be greatly facilitated by adding a so-called tag (a peptide sequence of defined size and with known characteristics) to the protein sequence. The tag can also be used to detect the protein within the host organism (e.g., with a **tag-specific antibody**). Tags most frequently added to proteins to aid in their purification are the **GST tag** of GST fusion proteins (containing glutathione-S-transferase from *Schistosoma japonicum*) and the **polyhistidine tag** (usually hexahistidines (His$_6$)). Using molecular biological methods, tags are often added to the respective protein N- or C-terminally. Many constructs also contain cleavage sites for endoproteases (thrombin, factor Xa) that allow proteolytic cleavage of the tag after purification. Meanwhile, novel systems are available that utilize the maltose binding protein of *E. coli* in addition to the His$_6$ tag. This additional tag enhances the solubility of the recombinant protein in *E. coli* and offers also the possibility for an additional purification step via affinity chromatography on a matrix with covalently linked α-amylose. These constructs encode fusion proteins with a specific cleavage site for the protease of the **tobacco etch virus (TEV)**. Apart from using this highly effective protease for removing the tags from the purified recombinant, the protease can be coexpressed in *E. coli* and is able to selectively cleave the recombinant protein in the living protein. This offers an advantage for purification procedures in which the cleavage of fusion protein *in situ* is essential for obtaining functional active protein.

7.5.3.1 **Chromatography on Chelating Agents**

Polyhistidine-containing proteins are often purified with matrices that covalently bind **chelating agents**, such as iminodiacetic acid or nitrilotriacetic acid (NTA), which in turn are loaded with Ni^{2+} ions. Alternatively, systems can also be loaded with Co^{2+}. The polyhistidine sequence binds the complexed metal ions via its imidazole side chains. Since polyhistidine sequences are extremely rare in naturally occurring proteins, host proteins do not bind to the matrix with exceptional strength and can often be removed with a single washing step (using a buffer that contains **imidazole** in the range of 20–50 mmol L^{-1}). The His-tagged proteins are subsequently eluted with buffers that contain imidazole in the range of 200–500 mmol L^{-1} (see Section 7.6.2). Alternatively, chelating agents such as EDTA can be used for elution. If the column is to be reused, however, it needs to be reloaded with the respective cations after elution with chelating agents. An advantage of purification via **His tag** is that this procedure can also be performed under **denaturing conditions** (68 mol L^{-1} urea or 3–4 M guanidinium hydrochloride).

7.5.3.2 **Chromatography on Glutathione Matrices**

Matrices that carry covalently bound glutathione (glutathione-agarose, glutathione-Sepharose) are used to purify **GST fusion proteins**. Since the fusion protein's GST part binds glutathione with high affinity (unlike the host proteins), this technique allows for high enrichment rates. The employed elution buffer contains glutathione in a concentration of 10 mmol L^{-1}. One advantage of this technique is GST's considerable hydrophilicity. GST fusion proteins are often more soluble (e.g., in the cytosol of *E. coli*) than their unmodified counterparts, which helps achieve greater yields. A drawback lies in the size of the GST tag (around 24 kDa), which can limit the functionality of the protein by altering its spatial structure. In extreme cases, the protein function can be blocked entirely.

7.6
Examples

7.6.1
Example 1: Purification of Nucleoside Diphosphate Kinase from the Cytosol of Bovine Retina Rod Cells

Nucleoside diphosphate kinases (NDPKs) are ubiquitous, mainly cytosolic proteins that enable the transfer of high-energy tertiary phosphate residues from 5′-nucleoside triphosphates (NTPs) to nucleoside diphosphates (NDPs). They are thus essential for the synthesis of other NTPs from ATP and NDPs in cells. To characterize NDPK's enzymatic activity, the enzyme needs to be purified from a cell extract and separated from other proteins of the nucleotide metabolism. At least 100 isolated bovine retinae are required to provide sufficient amounts of protein; they are resuspended in 170 mL of NDPK isolation buffer (10 mmol L^{-1} Na_2PO_4; 10 mmol L^{-1} K_2PO_4; 10 mmol L^{-1} H_2PO_4; 0.2 mmol L^{-1} $MgCl_2$; 0.2 mmol L^{-1} EGTA; 0.2 mmol L^{-1} Pefabloc; 0.02% NaN_3, pH 7.4). The suspension is stirred in a glass beaker at 4 °C for 30 min in the cold storage room. The outer segments of the rod cells break off during this treatment. As the next step, raise the concentrations of sodium chloride and magnesium chloride in the isolation buffer to 150 and 4 mM, respectively. Stir the suspension again at 4 °C for 30 min, then centrifuge in a cooled centrifuge (4 °C) at 30 000 g for 1 h to remove insoluble material. Centrifuge the supernatant once more in order to quantitatively remove remaining membranes from the soluble components (in the cold for 1 h at 100 000 g). Transfer the supernatant to a glass beaker and add an equal amount of cold saturated **ammonium sulfate solution**. Stir in the cold for 2 h; the resulting precipitate is pelleted by centrifugation for 40 min at 40 000 g. Since cytosolic NDPK is an extremely hydrophilic protein, it does not precipitate at 50% ammonium sulfate. The supernatant (containing the NDPK) is carefully transferred to a new beaker and mixed with ammonium sulfate solution until reaching 75% saturation. Place the beaker in the cold and stir overnight. The next day, centrifuge for 40 min at 40 000 g; discard the supernatant and resuspend the second precipitate (that, among others, contains NDPK) in 40 mL of TMES buffer (10 mM TrisHCl, pH 7.4; 2 mM $MgCl_2$; 0.1 mM EDTA;

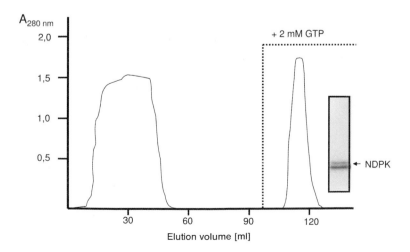

Fig. 7.4 Purification of NDPK with a Cibacron Blue-Sepharose column. Plot of UV light absorption (280 nm wavelength) by proteins in the flow-through against the flow-through volume. The broad, first peak contains the proteins that do not bind to the matrix. After loading a buffer that contains 2 mmol L^{-1} GTP, NDPK elutes from the column in a single, sharp peak. The enzyme's purity is demonstrated with a Coomassie Blue R-250 staining after SDSgel electrophoresis. The gel shows only the protein double band (in a molecular weight range of around 20 000 Da) that is characteristic for NDPK.

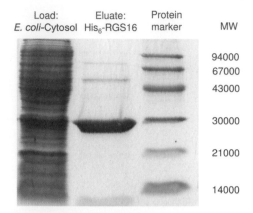

Load: Eluate: Protein
E. coli-Cytosol   His$_6$-RGS16   marker   MW

94000
67000
43000

30000

21000

14000

Fig. 7.5 Purification of His$_6$-RGS16: Coomassie Blue R-250 stain of a 15% SDS gel. Lane 1 has been loaded with *E. coli* cytosol. Lane 2 shows the eluate of the Ni-NTA matrix with a buffer containing 400 mmol L^{-1} imidazole. Lane 3 has been loaded with a molecular weight standard.

1 mM dithiothreitol; 300 mM NaCl). Thirty minutes of centrifugation at 100000 g separates insoluble materials; the clear supernatant contains the NDPK in solution. Press this solution through a sterile filter (diameter = 0.2 mm) and load it onto a **FPLC system**. FPLC is performed on a **Cibacron Blue-Sepharose CL-6B column** (volume = 20 mL) that has been equilibrated with TMED buffer (pump rate = 1 mL min^{-1}). NDPK, as a purine nucleotide-binding enzyme, binds to the dye (see Section 7.5.2.2). After washing the column with two column volumes of TMED buffer, elute the enzyme with TMED buffer that contains 2 mM GTP. The elution's specificity is based on NDPK's relatively high affinity for GTP, which cannot serve as a substrate to many other ATP-utilizing enzymes. The eluate is collected in a fraction collector (fraction volume = 1 mL; pump rate = 1 mL min^{-1}; original chromatogram, see Fig. 7.4). Validate the content and purity of the enriched NDPK by SDS-PAGE and subsequent protein staining (Fig. 7.5).

7.6.2
Example 2: Purification of Recombinant His$_6$-RGS16 after Expression in *E. coli*

RGS16 is a **GTPase-activating protein** that interacts specifically with the α-subunits of signal-transducing heterotrimeric G-proteins. For *in vitro* analysis of this interaction, both proteins (i.e., RGS16 as well as G-protein α-subunits) need to be available in sufficient amounts and purity. A **purification procedure** of recombinant RGS16 with an **N-terminal His$_6$ tag** from *E. coli* is described below: protein expression is induced in *E. coli* cells of the BL21(DE3) strain that were transformed with the **prokaryotic expression vector pET15b-RGS16**. First, a preculture is grown by inoculating 40 mL of bacterial growth (LB) medium (containing 100 µg mL^{-1} ampicillin) with bacteria from a single colony and incubation in a shaking incubator overnight. Thereafter, inoculate 1 L of LB medium with 100 µg mL^{-1} ampicillin with the preculture and incubate in the shaking incubator (with 37 °C and 150 rpm) until reaching an optical density of 0.5–0.7 at 600 nm. Protein expression is selectively induced by addition of 0.1 mmol L^{-1} isopropylthiogalactoside (IPTG). The bacteria synthesize the desired protein during the following 2.5 h of incubation in the shaking incubator at 30 °C. Subsequently, the bacteria are pelleted by centrifugation for 10 min at 10000 g and at 4 °C, and then resuspended in 40 mL of buffer A (50 mmol L^{-1} TrisHCl, pH 8.0; 100 mmol L^{-1} NaCl; 2 mmol L^{-1} MgCl$_2$; 6 mmol L^{-1} β-mercaptoethanol; 5% (v/v) glycerol). The cells are lysed with an ultrasonic homogenizer using five pulse intervals of 30 s each, followed by 2 min of cooling (perform the entire procedure on ice). Centrifuge for 15 min at 25000 g at 4 °C to pellet cell debris and particles. Then add the protein-containing supernatant to 1 mL of Ni-NTA-Sepharose matrix that has been equilibrated before in buffer A for 10 min. The protein solution and Ni-NTA matrix are stirred for 20 min at 4 °C and subsequently loaded onto a column. After the flow-through has dripped off, wash the matrix with 60 mL of buffer A with 25 mM imidazole. This step removes unspecifically bound protein. Elute the RGS16 protein with 5 mL 400 mmol L^{-1} imidazole in buffer A. Validate success with SDS-gel electrophoresis and subsequent Coomassie Blue R-250 staining (Fig. 7.5).

8

Peptide and Protein Analysis with Electrospray Tandem Mass Spectrometry

Learning Objectives

This chapter introduces a modern technique important to protein analytics: electrospray tandem mass spectrometry, which allows for the detection of amino acid sequences of peptides and posttranslational modifications. Electrospray ionization mass spectrometry has hence become an important tool in proteomics and bioanalytics.

8.1
Introduction

Mass spectrometry (MS) is a highly sensitive technique of instrumental molecular analysis that was invented about 90 years ago. The physicists Thomson and Aston developed the technique in Cambridge, England, and initially used it for elemental analysis. In the 1950s, mass spectrometry expanded into organic chemistry. Today, a wide range of mass spectrometry types exists that are specialized for the analysis of elements, small gaseous molecules, or biomolecules and biopolymers. Here, we will introduce **electrospray ionization (ESI) mass spectrometry** as an important ionization method in biological mass spectrometry, and describe its applications in peptide and protein analysis.

8.2
Principles of Mass Spectrometry

Mass spectrometers are made up of three functional units: **ion source**, **mass analyzer**, and **detector**. For mass spectrometric analyses, free gaseous ions are generated from the sample in the ion source and then focused into an ion beam in high vacuum. The mass analyzer separates ions in this beam according to their **mass/charge (m/z) ratio**; these ions are then registered by the detector. Individual measurements are plotted in a mass spectrum with m/z (x-axis) and intensity (y-axis), as shown in Fig. 8.1. The MS raw data (Fig. 8.1, bottom spectrum) show the experimental width of the peak, which largely depends on the mass spectrometer used. To reduce the amount of data and more precisely localize the peak position, the raw data can be converted to centroid data. This transforms each peak into a single line positioned on the original peak center. When analyzing biomolecules, the m/z ratio represents the "hard" and easily reproducible information of MS. The intensity is considered "soft" information, because it is subject to much greater fluctuation.

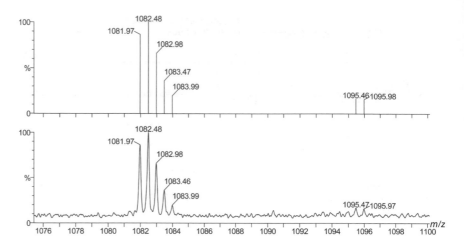

Fig. 8.1 Partial ESI mass spectrum with two signals from doubly charged peptides. The registered ion *m/z* values (mass-to-charge ratio) are plotted on the *x*-axis, their intensity on the *y*-axis. The same spectrum is presented in two different ways: upper spectrum, raw data; lower spectrum, centroid data after background subtraction.

8.3
Mass Precision, Resolution, and Isotope Distribution

Every mass spectrometer needs to be calibrated with a **reference substance** in order to ensure exact mass determination (**external calibration**). Even more precise data can be gathered with so-called **internal calibration**, which involves a peak of known *m/z* ratio occurring within the recorded spectrum. Signals from analytes already present in the sample (e.g., known contaminants) can be used as reference signals or, alternatively, a calibration compound can be added. Of course, the significance of any measured molecular mass depends on its error. The true error can be determined experimentally by repeated measurements of known analytes; it is specified in absolute (Da or mDa) or relative terms (usually ppm). Errors of present-day mass determinations range from about ±1000 ppm (±1 Da at an *m/z* ratio of 1000) to ±1 ppm (±1 mDa at an *m/z* ratio of 1000).

The mass spectrometer resolution has a direct influence on the achievable mass precision. Exact definitions of resolution depend on the type of spectrometer used. A simple definition commonly applied for **time-of-flight (TOF)** analyzers is the **FWHM (full width at half maximum)** definition: it defines the resolution as the quotient of the *m/z* value and the peak width at half maximum value. The better the resolution, the more exact the experimental determination of the peak width.

As shown in Fig. 8.1, any biomolecule's mass spectrometry signal consists of a group of signals with differing *m/z* values (**isotopomers**: the same substance consisting of different isotopes). This occurs due to the **natural isotope distribution** of the bioelements C, H, N, O, and S, which all feature a light main isotope and one or two rare heavy isotopes. An element's isotope distribution can be calculated from the molecular ion gross formula and the natural isotope frequencies of the incorporated elements. Comparison between experimental and calculated isotope patterns can yield additional information in biological mass spectrometry.

8.4
Principles of ESI

The two decades between 1970 and 1990 were marked by a rapid development of new ionization techniques. Two of these techniques have since established themselves in mass spectrometry of biomolecules, in particular: **matrix-assisted laser desorption/ionization (MALDI)** and **electrospray ionization (ESI)**. These soft methods of ionization can transfer even large biomolecules like proteins to the gaseous phase without disintegration. Known peptides can quickly be identi-

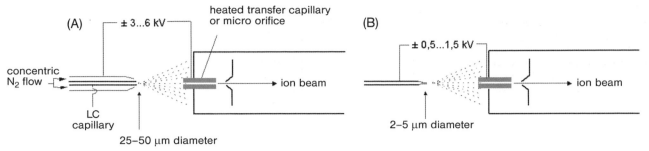

Fig. 8.2 Schematic view of an electrospray ion source. Analyte solution is sprayed at atmospheric pressure, droplets enter the evacuated analyzer area through a micro-orifice, and an ion beam is formed. (A) Classic ESI with gas-supported spray and flow rates of 2–5 µL min⁻¹. (B) NanoESI with flow rates of 10–40 nL min1. Due to the smaller spray and flow rate, the required potential difference is smaller than with classic ESI (around 0.5–1.5 compared to 3–6 kV).

fied with MALDI-MS. The technique is therefore routinely used for analysis of tryptic fragments from 2D gels. ESI was developed by Fenn, who received the Nobel Prize in Chemistry for this achievement in 2002. This method involves spraying the analyte solution from a microcapillary that carries a high (negative or positive) potential in reference to the mass spectrometer. When the electrostatic force of the applied current exceeds the surface tension of the analyte solution, a **Taylor cone** forms at the tip of the microcapillary. Highly charged droplets form and solvent evaporation disintegrates them further to a fine spray. This analyte spray is then sucked into the evacuated mass analyzer through a micro-orifice. In the interface area, the droplets are dried and ion formation occurs. The working schematic of an ESI ion source is shown in Fig. 8.2.

All the substance classes relevant to bioanalytics (i.e., lipids, proteins, nucleic acids, and carbohydrates) can be analyzed with **ESI-MS**. ESI is most sensitive for compounds with basic or acidic groups, because ionization largely occurs through proton addition (formation of **positive ions**) or proton abstraction (formation of **negative ions**). This fact makes compounds like triacylglycerides somewhat difficult to analyze. Detecting them as anion adducts (e.g., with Cl⁻) or cation adducts (e.g., with Na⁺) is possible, but the sensitivity is often much lower than for protonatable or deprotonatable analytes. The occurrence of ion series, which consist of several charge states of the analyte, is typical for ESI (see Section 8.8).

8.5
Tandem Mass Spectrometers

8.5.1
Mass Analyzers

The introduction of effective and reliable mass analyzers was a basic principle to establish mass spectrometry in bioanalytics. Key technologies in this respect are **quadrupole analyzers, ion traps,** and **TOF analyzers**. The working principles of TOF-MS were already developed in the 1940s; the concepts of ion traps and quadrupole analyzers were designed in the 1950s by Wolfgang Paul, who was later awarded the Nobel Prize in Physics. Quadrupole analyzers are composed of four parallel rods in symmetrical arrangement, through which a central ion beam is directed. Two opposing rods are impressed with a potential of:

$$U_1 = U + V \cos(\omega t)$$

the other two with a potential of:

$$U_2 = U - V\cos(\omega t)$$

At given ratios of U/V, only ions with a corresponding m/z ratio reach the detector on stable trajectories; ions with other m/z values have unstable trajectories and therefore are not detected. Mass spectra are recorded by scanning the quadrupole. During this scan, the values U and V are varied while their ratio U/V remains constant. Further development of the quadrupole analyzer led to the **quadrupole ion trap**, which traps ions in a 3D quadrupole field.

A combination of two analyzers with a collision cell in between is known as a **tandem mass spectrometer**. Exceptional accomplishments in structure analysis were possible with this configuration, so today most of the devices used in modern biological mass spectrometry are tandem mass spectrometers. The most important types of mass analyzers are described below.

8.5.2
Triple Quadrupole

A **triple quadrupole mass spectrometer** consists of three quadrupole analyzers in a row, of which Q1 and Q3 are used as mass analyzers (Fig. 8.3). The intermediate system Q2 functions as a collision cell that can be filled with a collision gas (usually argon); collisions of ions and gas atoms fragment the molecular ions. Triple quadrupole spectrometers can be run in four different scanning modes: product ion scan, neutral loss scan, precursor ion scan, and selective reaction monitoring. Product ion scan mode is frequently used for peptide sequencing, for example. In this mode, a given molecular ion is selected in Q1 to be fragmented in the collision chamber; the resulting fragment ions are analyzed in Q3. The other scanning modes are particularly suited for selectively detecting single analytes or compound classes from complex mixtures. For instance, both neutral loss scan and precursor ion scan can selectively detect phosphopeptides by measuring phosphopeptide-specific fragmentation reactions.

On top of the numerous applications offered by the different scanning modes, triple quadrupole spectrometers are very well suited for quantitative analysis due to their high dynamic range. Applications are, however, limited by the fact that quadrupole analyzers normally produce spectra of low or moderate resolution and moderate mass precision.

8.5.3
Linear Trap Quadrupole (LTQ) and LTQ Orbitrap

Linear ion traps are related to quadrupole analyzers. These are currently replacing the originally developed more compact ion traps (3D traps) due to their substantially higher sensitivity. Multistage MS/MS (MS^n) experiments are possible using ion traps, (i.e., fragment ions can be isolated and selectively fragmented). Ion traps can be used for highly sensitive and fast acquisition of MS/MS spectra; however, these spectra are recorded at low resolution. To overcome this limitation, linear ion traps are combined with a second high-resolution mass analyzer, such as an Orbitrap or an FT-ICR analyzer (see Section 8.5.5). Orbitrap analyzers operate with electric fields only and require very high vacuum conditions. Currently, they reach resolution data between 30 000 and 100 000.

Fig. 8.3 Schematic presentation of a triple quadrupole spectrometer. Q1, first quadrupole; Q2, collision cell for the formation of fragment ions; Q3, second quadrupole. This variant of tandem mass spectrometry is also referred to as tandem in space because MS and MS/MS analyses are spatially separated.

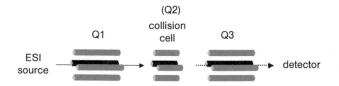

8.5.4
Q-TOF

Spectrometers with two different mass analyzers are known as **hybrid tandem spectrometers**. The most successful hybrid analyzers at present consist of a combination of a quadrupole (Q) and a time-of-flight (TOF) analyzer. This system has a high sensitivity in product ion scanning, because detection of all fragment ions occurs almost simultaneously. Compared to a quadrupole spectrometer, the sensitivity is increased by almost two magnitudes, since the quadrupole system needs to scan the entire mass range to record a fragment ion spectrum. Furthermore, the reflector TOF analyzer ensures high resolution (between 5000 and 20 000). These features have established **Q-TOF systems** as the standard high-performance MS/MS systems in bioanalytics. Routine operations with these tandem spectrometers achieve mass precisions between 5 and 100 ppm, depending on the application and applied definition.

8.5.5
Q-FT-ICR

In principle, the mass analyzer of an ion cyclotron resonance mass spectrometer (ICR-MS) is a cell containing a superconducting magnet (1.0–9.4 T). Within the ICR cell, the Lorentz force guides ions onto circular orbits, whose circulation frequency (the so-called cyclotron frequency) is inversely proportional to the ion m/z values. The circulating ions induce periodical currents that can be measured and used to calculate the cyclotron frequencies with **Fourier transformations**. The working principle of Fourier transform (FT)-ICR analysis produces mass spectra with very high resolution (50 000 to around 106). Hybrid tandem MS systems (e.g., **Q-FT-ICR**) using this separation principle are currently being introduced to bioanalytic research, so the following years will see an increase of precision mass measurements with errors of a few parts per million.

8.6
Peptide Sequencing with MS/MS

Peptides contain the 20 proteinogenic amino acids linearly linked to each other by peptide (acid amide) bonds. This means that each peptide has exactly one free N-terminus and one free C-terminus (see Chapter 2). Conventionally, peptide sequences (whether in three- or one-letter code) are presented with the **N-terminus** to the left and the **C-terminus** to the right, e.g.:

V-S-I-N-E-K   or   Val-Ser-Ile-Asn-Glu-Lys

ESI spectra of peptides show practically nothing but molecular ions. The fragmentation of positively charged peptide molecular ions by collision activation occurs almost exclusively at the peptide bond. The resulting fragments contain either the N-terminus (b-ions) or the C-terminus (y-ions). The mass distance between ions of ion series corresponds to the mass of the amino acid units (amino acid-H_2O), so the sequence can often be read right off the spectrum.

The appearance of b- or y-series is determined by the distribution of basic amino acids throughout the peptide. Figure 8.4 shows a tryptic peptide that features an arginine (R) or lysine (K) residue at its C-terminus. This is typical for tryptic peptides and is due to trypsin's cleavage specificity. Tryptic peptides usually have pronounced y-ion series; the b-ions tend to be small (b_2, b_3, or smaller internal b fragments) and of low intensity (with the exception of the b_2-ion). The b-type ions often lose CO and can then be clearly identified (and distinguished from y-ions) by satellite peaks at 28 Da.

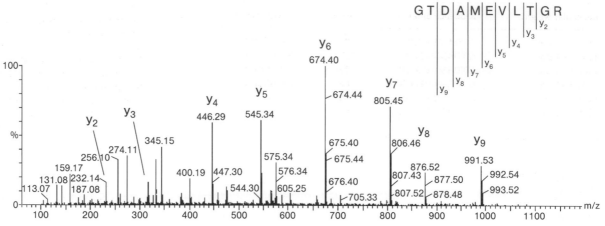

Fig. 8.4 Electrospray MS/MS spectrum of an [M + 2H]²⁺ ion (*m/z* is 575.3) derived from a peptide with 11 amino acids. Almost the entire sequence can be read from the y-series of this product ion spectrum.

This technique can be used to sequence peptides of up to about 20 amino acids in length. If the peptides are hitherto unknown, the procedure is referred to as *de novo* sequencing. Success depends largely on the peptide sequence: regions with many basic amino acids (R, K, H) do not fragment well and are thus difficult to sequence with MS/MS, whereas regions with largely neutral (L, V, N, Q, F, M) or acidic (D, E) amino acids are generally quite accessible to MS/MS sequencing.

8.7
Identifying Proteins with MS/MS Data and Protein Databases

Modern **protein databases** contain sequences of several million proteins from diverse species. Since life has developed in an evolutionary and conservative manner, only a miniscule fraction of the statistically possible protein sequences has actually been realized. Due to this, only partial sequences are needed to safely identify a protein. This circumstance is taken advantage of when unidentified proteins are proteolytically degraded and the peptides are then analyzed with tandem mass spectrometry. The alignment between protein database information and **MS/MS data** can be performed in two ways: either on the level of amino acid sequence or on the level of MS/MS fragment ion masses. Both approaches are discussed in the following sections.

8.7.1
Database Search with MS/MS Raw Data

Faster results are achieved when uninterpreted **MS/MS spectra** are used for a database search instead of a manually read sequence. To identify a given protein, the protease digest and MS/MS fragments are simulated computationally (*in silico* digest and *in silico* MS/MS of database protein sequences); the simulation data is then aligned with the MS/MS raw data. In practice, every protein analysis with **ESI-MS/MS** creates a large number of MS/MS spectra (up to several thousand), which are then summarized in a single data file. Since a single MS/MS spectrum may be enough to identify a protein with certainty (as shown in Table 8.1), this approach has a very high specificity. Figure 8.5 shows the result of a raw data analysis of a batch of MS/MS spectra that were created with nanoESI-MS/MS of a tryptic digest of dynamin A. The protein is identified with a very high score of about 1500 (significance threshold at 80), because the identification is based on an entire batch of MS/MS spectra (about 30). Only one other protein is identified a dynamin precursor protein, from which only one peptide was detected.

Molecular weight	Sequence entered[a]	Score	Interpretation
1148.5 ± 1 Da	GTDAMEVLTGR	120	Significant
1148.5 ± 1 Da	GTDAMEVLTGR	107	Significant
1148.5 ± 1 Da	GTDAMEVLTGR	96	Significant
1148.5 ± 1 Da	GTDAMEVLTGR	80	Significant
1148.5 ± 1 Da	GTDAMEVLTGR	68	Not significant
1148.5 ± 1 Da	GTDAMEVLTGR	56	Not significant

a) The sequence entered for the search is in bold print.

Table 8.1 Sequence-based protein database search with partial sequences of different lengths; all sequences were read from the MS/MS spectrum of the peptide GTDAMEVLTGR shown in Fig. 8.4. Dynamin A (SwissProt No. Q94464) was identified in all significant cases; the significance threshold (p < 0.05) is at a score of 72. A sequence length of at least 6 amino acids (plus information on peptide mass and trypsin specificity) scored a significant search result.

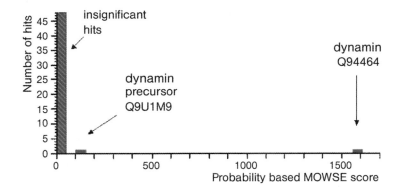

Fig. 8.5 Search result for a set of nanoESI tandem spectra performed with the search engine MASCOT. The spectra were produced from a digest of the protein dynamin A with trypsin. The x-axis shows the probability score for the hit's significance (MOWSE = molecular weight search). The significance threshold (P < 0.05) was 80. The protein is identified with high certainty.

Sequence coverage in this example was about 60%. Usually, it ranges from 5 to 50% according to the protease used and the amount, sequence, and purity of the protein to be analyzed. Owing to its high specificity, **protein identification with ESI-MS/MS** is well suited for the analysis of protein mixtures.

8.8
Determining Protein Molecular Mass

ESI allows proteins with molecular weights of up to several hundred kDa to be transferred to the gaseous state intact. A spectrum of such an ionized protein contains several peaks (Fig. 8.6A), each of which represents a different charge state of the protein. The nanoESI spectrum for the catalytic subunit of protein kinase A (PKA, shown in Fig. 8.6A) displays a distribution that ranges from 23- to 53-fold protonated PKA. By applying the correct algorithms, the molecular weight can be calculated from the peak series. Figure 8.6 (B) shows this transformed (deconvoluted) spectrum with the molecular mass on the x-axis (instead of the m/z value). It shows the presence of two different isoforms of PKA (Cα and Cβ), differing in molecular mass by 26 Da. The measured mass data diverges from the mass calculated from the amino acid sequence by less than 1 Da.

Salts should be removed as completely as possible before measuring to avoid them interfering with the correct mass determination. Remaining salts may form adducts that lead to incorrect results. They are often removed via microscale solid-phase extraction (e.g., ZipTip® C₄).

Molecular mass is an important piece of information for the analysis of recombinant proteins. If it matches the molecular mass calculated from the amino acid sequence within the measuring accuracy (usually a few Da), it is highly probable that the protein possesses the correct sequence. Molecular mass information is equally important for the analysis of covalent protein modifications (see Section 8.9), as experimental data can provide clues for the presence and sometimes even the nature of modifications.

ESI even allows for the analysis of **noncovalent protein complexes** with sizes in the MDa range. This requires special optimization of the experimental setup, however, both with regard to instrumental parameters (e.g., pressures during

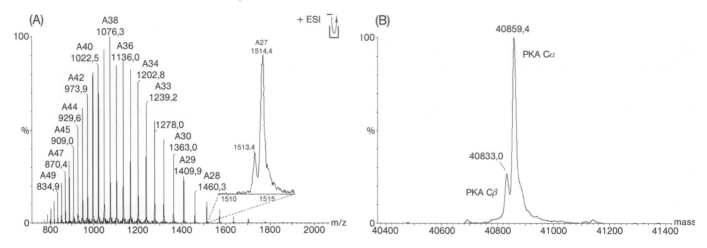

Fig. 8.6 Mass spectrum of the catalytic subunit of PKA. (A) NanoESI spectrum with a zoom of the 27-fold protonated ion. (B) Deconvoluted spectrum.

desolation) as well as the solving conditions. Only a few buffers are compatible with both ESI and the formation of the protein complex, which requires all components involved in their native conformation.

When performed with proteins with molecular masses up to 100 kDa, FT-ICR mass spectrometry can resolve isotope peaks. This can be used to identify the oxidation state of metalloproteins (e.g., Fe^{2+} and Fe^{3+}).

8.9
Analysis of Covalent Protein Modification

The large majority of a cell's proteins are modified on a co- or posttranslational level. **Covalent modification** most frequently occurs on the N-terminal amino group or on the side chains of the amino acids Cys, Ser, Thr, Tyr, Lys, Arg, and Asn. The possible modifications range from chemically simple (e.g. methyl groups) to highly complex structures (such as branched oligosaccharides). One type of **posttranslational modification** that is involved in the regulation of numerous cellular functions (especially in eukaryotes) is reversible phosphorylation of Ser, Thr, and Tyr. Further frequently found modifications include, but are not limited to, acetylation (N-terminal or Lys), myristoylation (N-terminal), palmitoylation (Cys), farnesylation (Cys), methylation (Arg, Lys), nitration (Tyr), sulfation (Tyr), glycosylation (Ser, Thr, Asn), and deamidation (Asn).

Any **covalent modification** (with the exception of isomerization and racemization) alters a protein's molecular mass and is thus detectable with mass spectrometry. Thus, a protein's molecular mass can provide the first hints on the nature and number of modifications as described in Section 8.8. The exact localization of modified amino acids then usually requires cleaving the protein and analyzing it on the peptide level. Modified proteins often display a characteristic fragmentation behavior that distinguishes them from unmodified proteins: Ser/Thr phosphorylated peptides, for example, show a neutral loss of phosphoric acid; similarly, Tyr sulfated proteins show a neutral loss of sulfur trioxide. Myristoylated peptides (as well as Tyr nitrated and Tyr phosphorylated peptides) form particular fragment ions during collision-induced fragmentation, whose appearance is specific for the respectively modified proteins. By applying specialized scanning techniques (such as neutral loss and precursor ion scan), modified peptides can be selectively detected even from complex peptide mixtures.

In the case of **protein phosphorylation**, another method of separating modified and unmodified peptides exists beyond selective detection with neutral loss and precursor ion scan: **immobilized metal ion affinity chromatography (IMAC)** (Fig. 8.7). This method employs a complex of trivalent metal ions, usually Fe(III) or Ga(III), with immobilized iminodiacetic acid (IDA) or nitrilotriacetic acid (NTA). The metal ions in this complex still possess free coordination sites, to

phosphoprotein **peptide mixture** **phosphopeptides**

Fig. 8.7 IMAC for the selective enrichment of phosphopeptides.

which phosphopeptides can coordinate via their phosphate groups. After a washing step to remove nonphosphorylated proteins, bound phosphopeptides can be eluted with phosphate buffer, for example. In addition to IMAC, phosphopeptide enrichment via TiO_2 has been established as additional powerful method.

Protein glycosylation is an especially complex mode of **posttranslational modification** that occurs on nitrogen or oxygen atoms. *N*-Glycosidically (Asn) and *O*-glycosidically (Ser/Thr) **bound oligosaccharides** (see Section 5.3) can be differentiated with the protein PNGase F, which selectively cleaves only *N*-glycosidically bound oligosaccharides. Carbohydrate-specific fragment ions can be used to selectively detect glycopeptides. Also, due to their high oxygen content and oxygen's relatively low atomic mass of 15.9949, peptides with glycosidic modifications have a molecular mass significantly lower than unmodified proteins of the same nominal mass. An unmodified protein with a nominal molecular mass of 1000 Da, for example, has an average molecular mass of 1000.50 Da; the molecular mass of a highly glycosylated peptide would be expected at below 1000.38 Da. Thus, a peptide's precise mass can provide important clues for the presence of glycosylation. Another way to differentiate glycosylated and unglycosylated peptides is using unspecific proteases (e.g., pronase a mixture of different endo- and exoproteases). Proteolysis of a glycoprotein with unspecific proteases produces small peptides with low mass throughout; only peptides with a high content of carbohydrates are also found in the high mass range, as carbohydrate units cannot be cleaved by the proteases.

Covalent modifications often impart peptides with physicochemical characteristics that complicate their mass spectrometric detection. As such, glycosylated and phosphorylated peptides often have a lower ionization efficiency compared to unmodified peptides. To avoid such problems of the analysis of peptide mixtures, the mass spectrometry analysis can be performed without prior proteolysis. **Electron capture dissociation (ECD)** and **electron transfer dissociation (ETD)** are newly developed fragmentation techniques that allow for this kind of analysis. They are currently mainly used in combination with FT-ICR mass spectrometry.

The analysis of reversible covalent modifications that have regulatory functions (e.g., phosphorylation) raises a second question beyond that of the modified position: what fraction of the protein is modified? To answer this question, special methods of mass spectrometry have been developed that allow for the quantitative analysis of posttranslational modifications. In most cases such methods involve **stable isotopes** (^{2}H, ^{13}C, or ^{15}N). The isotope-labeled peptides serve as an internal standard and allow for direct quantification from the mass spectrum's peak intensities. If the modified peptides contain elements like phosphorus (phosphopeptides), selenium (seleno-cysteine), or iodine (iodo-tyrosine), **inductively coupled plasma (ICP)-MS** can also be used for quantification.

8.10
Relative and Absolute Quantification

Quantitative proteomic analyses open access to studies of biological functions of proteins and their covalent modifications. The simple comparison of absolute signal intensities delivers quantitative data of moderate accuracy. More accurate

Fig. 8.8 Schematic overview of quantitative proteomic methods. (a) Levels of methods for relative quantification. (b) Levels for the addition of labeled standards for absolute quantification. Literature on these methods can be easily accessed using the given acronyms.

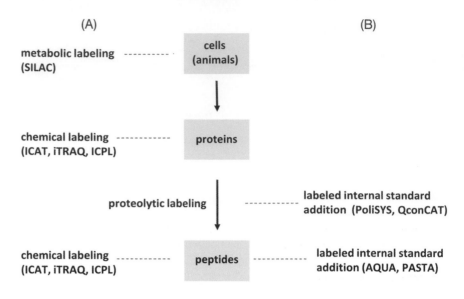

Fig. 8.9 Mass spectrum of the phosphopeptide SNTHPpSPDDAFR. The unlabeled peptide was mixed 1:1 with the peptide that bears a completely ^{13}C- and ^{15}N-labeled arginine.

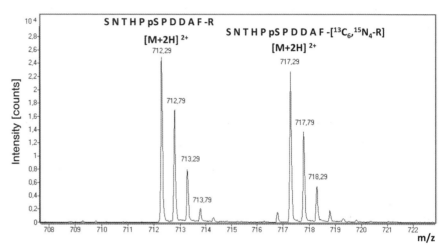

data require the use of stable isotope labeling. In this context, methods for relative quantification describe the changes in concentrations between two or more samples and absolute quantification methods deliver peptide or protein concentrations. For relative quantifications, the peptide or protein samples to be compared are in general differently labeled, then combined and analyzed as a mixture. Absolute quantifications require the addition of an absolutely quantified and stable-isotope-labeled standard. Figure 8.8 gives an overview of methods of relative and absolute quantification, and indicates on which level in the analytical work flow they are applied or added, respectively.

One way for absolute quantification is the application of AQUA peptides. Within these peptides one amino acid (e.g. a C-terminal Arg or Lys residue) is exchanged against a fully ^{13}C- and/or ^{15}N-labeled amino acid. The labeled and unlabeled peptides have the same physicochemical properties (e.g., retention time in HPLC) but different masses (e.g. 6 or 10 Da). Thus, these peptides can be used as internal standards for relative or absolute quantification. Figure 8.9 shows a 1:1 mixture of such an AQUA peptide.

9
Isolation of DNA and RNA

Learning Objectives

Isolation of DNA and RNA is one of the basic tasks of modern biotechnology. Any biological material can provide DNA as well as RNA for analytic or preparative purposes. When isolating DNA, one needs to pay special attention to its physical vulnerability to shearing forces; RNA isolation, on the other hand, is always vulnerable to ubiquitously present RNases. The classic methods of nucleic acid isolation are increasingly being replaced with commercially developed techniques based on specific affinities to column materials.

9.1
Introduction

Deoxyribonucleic acid (DNA) and **ribonucleic acid (RNA)** can be isolated from living or conserved tissues, cells, virus particles, or other samples for both analytic and preparative purposes. Since the intended use and origin of the genetic material can vary greatly, no general method of isolation exists. Still, the great chemical universality of RNA and DNA throughout nature enables us to exploit general properties when extracting them from organisms: both types of nucleic acids share a **high solubility in water**, and precipitate as macromolecules in suitable **mixtures of alcohol and water**. Furthermore, both are weakly soluble in organic solvents such as chloroform or phenol. These solvents can hence be used to extract proteins and hydrophobic components from nucleic acid solutions with relative ease.

There are many important differences between DNA and RNA (see Chapter 2). DNA, for instance, usually occurs as a double-stranded, extremely inflexible molecule that is highly viscous in solution. By contrast, single-stranded nucleic acids (such as many RNA species) are often coiled or adopt inter- and intramolecular secondary structures. Such interactions also account for the spatial structures of **ribosomal RNA (rRNA)** and **transfer RNA (tRNA)**. Many regulatory and structural RNA molecules assume their spatial structure through complexing with proteins. Moreover, RNA molecules are unstable at high pH.

Due to these physicochemical differences and varying sensitivities to different nucleases (see below), different methods for the isolation of RNA and DNA are necessary and available. Individual descriptions are given in the following.

9.2
DNA Isolation

DNA isolation from prokaryotes, eukaryotes, and viruses alike begins with **disintegration of the cells or virus shells**. This is achieved through enzymatic digest, detergents, or mechanical force. Isolation can begin as soon as the cells have been disintegrated. When extracting from eukaryotes, an isolation of the nuclei can be performed in between to deplete organelle DNA (e.g., mitochondrial DNA).

An Introduction to Molecular Biotechnology, 2nd Edition.
Edited by Michael Wink
Copyright © 2011 WILEY-VCH Verlag GmbH & Co. KGaA, Weinheim
ISBN: 978-3-527-32637-2

Depending on the required purity, proteins and hydrophobic components are removed through single or multiple **extractions** of the aqueous phase with phenol/chloroform or another hydrophobic extraction agent. DNA can then be precipitated from the aqueous phase (or the supernatant) with addition of alcohol (2.5 volumes of ethanol or 1 volume of isopropanol) and high salt concentrations (e.g., 0.3 M sodium acetate). After desalting in 70% ethanol, the sample is dried and then resuspended in any desired volume of aqueous buffer. Alcohol precipitation removes hydrophilic components of low molecular weight. For a quantitative removal of such components, a gel filtration column can be used to purify the sample (see Chapter 7). The yield can be determined **photometrically** due to the specific absorption of the bases at a wavelength of 260 nm. As a rule of thumb, one OD 260 unit roughly corresponds to 50 µg mL^{-1} of double-stranded nucleic acids (37 µg mL^{-1} of single-stranded DNA or 40 µg mL^{-1} of RNA). The ratio of OD 260 to OD 280 should be roughly 2; significantly lower ratios suggest protein contamination.

Chromosomal DNA from eukaryotic cells is very long. When precipitating with alcohol, these molecules precipitate as large, thread-like aggregates that can be fished from the liquid with a glass rod or a pipette tip (Fig. 9.1). DNA's inflexibility and length account for very high viscosities in aqueous solutions and the molecule's extraordinary vulnerability to shearing. As a normal precaution during isolation of genomic DNA, small diameter pipettes or vigorous shaking during the extraction should be avoided. Still, the shattering of DNA to fragments of 50–100 kb in length is hard to avoid. If greater gene spans are to be analyzed (e.g., by use of appropriate electrophoresis methods), the nuclear lysis can to be performed directly in the gel pocket. In this way even entire chromosomes (consisting of a single DNA molecule) can be extracted and separated without breaking apart.

Many applications (such as PCR in gene diagnostics) are largely unaffected by shearing of high-molecular-weight DNA; the gene sequences to be amplified are usually rather short and fragments of smaller length (less than 10 kb) are relatively insensitive to shearing forces.

Smaller **circular species of DNA** like plasmid, virus, or organelle DNA can be enriched by centrifugation in an ethidium bromide (EtBr)-cesium chloride (CsCl) gradient after alcohol precipitation and resuspension. This method is based on the **intercalation of EtBr** into the double-stranded DNA, which alters the buoyant density of the molecule in highly concentrated CsCl solutions. Owing to intramolecular tension, covalently closed circular molecules (i.e., plasmids) can incorporate less EtBr per base pair compared to open (i.e., nicked circles) or linear molecules. Thus, they accumulate at higher densities in the CsCl gradient (Fig. 9.2). After extracting the respective band, the hydrophobic EtBr is removed with appropriate hydrophobic solvents; the DNA is reprecipitated with alcohol.

Plasmid DNA from bacteria can also be concentrated by precipitating high-molecular-weight aggregates from the cell lysate in high salt concentrations. This procedure is also followed by alcohol precipitation of the plasmid DNA.

When isolating **DNA from viruses or bacteriophages**, it may be necessary to prepare or enrich the particles (e.g., by centrifugation) before proteolytic digest or chemical disintegration.

DNA isolation from plant cells can also be performed with a nonionic detergent (**cetyltrimethylammonium bromide (CTAB)**). This established method also uses organic solvents and alcohol precipitation in later steps.

Several manufacturers offer simple DNA extractions kits for standardized routine isolations. The kit protocols release nucleic acids with proteases. The extract is then purified over **affinity chromatography** columns that mainly consist of treated glass beads; the beads selectively bind the negatively charged nucleic acids depending on salt and alcohol concentrations. Potentially copurified RNA molecules can be digested with RNases and the fragments removed afterwards. An example of a simple affinity chromatography extraction is shown in Fig. 9.3.

Fig. 9.1 Mammalian chromosomal DNA in solution (right) and precipitated after the addition of 2.5 volumes of ethanol (left).

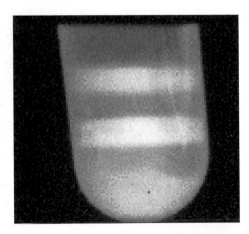

Fig. 9.2 Separated plasmid DNA after ultracentrifugation in a CsCl-EtBr gradient. DNA is visible under UV light due to the intercalation of EtBr. Upper band: relaxed DNA (plasmid with single- or double-strand breaks, chromosomal DNA). Lower band: intact circular plasmid DNA. (Image kindly provided by Andreas Lössl.)

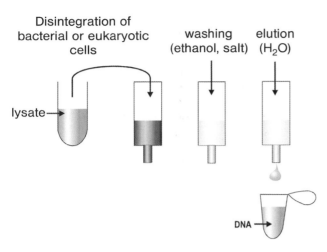

Disintegration of
bacterial or eukaryotic
cells

washing
(ethanol, salt)

elution
(H₂O)

lysate

DNA →

Fig. 9.3 Scheme of DNA purification for prokaryotes or eukaryotes using a common commercial affinity column.

These standardized methods have become routine procedures in both applied and basic research in **gene diagnostics**.

9.3
RNA Isolation

Just like DNA extraction, **RNA isolation** includes removing other molecule classes such as proteins or lipids; as with DNA isolation, organic solvents are used to separate RNA from such contaminants. However, RNA needs to be treated differently from DNA in several respects. Most cell types contain a large number of **RNases** in their cytoplasm along with the mRNA, which is usually the focus of the analytic extraction. In order to prevent the action of these ubiquitous RNases, the isolation must be performed swiftly and the cell lysate must be brought into an RNase denaturing environment as fast as possible. Strongly denaturing **guanidinium isothiocyanate (GTC)** solutions can be used for this purpose. Freshly sampled or harvested tissues can be homogenized and largely dissolved in 4 M GTC. The RNA can then be pelleted with CsCl density gradient centrifugation and thereby separated from DNA, which does not precipitate under appropriate conditions. A diverse number of methods can then be used to purify the RNA pellet further.

Alternatively, RNA can also be extracted from tissues that have been homogenized in other denaturing salt solutions (e.g., 4 M **lithium chloride**). Subsequently, a phenol extraction is performed to rid the homogenate of proteins and the RNA is then precipitated with alcohol.

9.3.1
Enrichment of mRNA

More than 90% of mammalian RNA is rRNA. This excess of noncoding sequences sometimes complicates the analysis of gene expression. However, the polyadenylation at the 3′-end, common to most coding eukaryotic mRNA species, can be utilized for extraction (see Chapters 2 and 4). If a mixture of **polyadenylated** and **nonpolyadenylated RNA** is run over column material that is covalently linked with **oligo(dT)**, selective base pairing causes the polyadenylated RNA to be retained by the column.

Column kits for extracting complete RNA as well as mRNA are commercially available. Detailed protocols for the extraction of DNA and RNA can be found in regularly updated publications (Ausubel *et al.*, 2009).

The systematic analysis of large numbers of nucleic acid samples is now facilitated by the development of instrumental equipment allowing the extraction and purification of many different samples of RNA or DNA from biological material in parallel.

10
Chromatography and Electrophoresis of Nucleic Acids

Learning Objectives
Separating nucleic acids or fragments of nucleic acids is one of the standard tasks in molecular biology and biotechnology. This chapter gives a description of the chromatography and electrophoresis techniques used for this purpose.

10.1
Introduction

Several methods are used to separate nucleic acids from other substances. **Chromatography** (Greek *chroma*: color; *graphein*: to write) is a physicochemical method, separating compounds by exploiting the fact that the components to be analyzed vary in their affinity to two different phases. In **electrophoresis**, the substances are separated using an electric field. The speed at which the molecules reach the poles depends on the voltage, the properties of the carrier, and the charge and shape of the molecule.

10.2
Chromatographic Separation of Nucleic Acids

Our current chromatography methods are derived from a procedure developed by Tswett in 1903 that enabled him to separate dissolved plant pigments using solid adsorbents. A major prerequisite for the application of chromatography is that the substances contained in the mixture to be analyzed do not undergo any chemical changes when dissolved or vaporized. In most chromatographic methods, a **liquid or gaseous mobile phase (eluent) migrates with the analyte over a solid or liquid stationary phase** (adsorbent). The analyte can be separated either by distribution of the components between mobile and stationary phases (**partition chromatography**), through differences in adsorption by the stationary phase (**adsorption chromatography**), through effects of ion exchange (**ion exchange chromatography**), or through selective binding to the stationary phase (**affinity chromatography**) (see also Chapter 7).

For the **separation of nucleic acids**, partition, adsorption and affinity chromatography are the methods of choice. In **partition chromatography**, the difference in polarity of the components is used for separation. If the components have a high affinity to the stationary phase, the migration speed is slow. If the affinity to the stationary phase is low, the substances move faster. The composition of the solvent used has a major impact on the affinity of the analyte. A hydrophobic substance that is difficult to dissolve in water migrates a long distance if an organic solvent is used. The migration speed of a substance is defined by the R_f value, which is calculated by dividing the migration distance of the substance by the migration distance of the eluent. A contemporary form of partition chromatography is known as **high-performance liquid chromatography (HPLC)**, which is used for the separation and purification of oligonucleotides, for example.

An Introduction to Molecular Biotechnology, 2nd Edition.
Edited by Michael Wink
Copyright © 2011 WILEY-VCH Verlag GmbH & Co. KGaA, Weinheim
ISBN: 978-3-527-32637-2

Many of the chromatographic techniques used for the separation of proteins can also be used for the separation of nucleic acids. Often, **hydroxyapatite** is used as an adsorbent because double-stranded DNA binds to it more tightly than most other molecules. Thus, DNA can be isolated fairly quickly by adding a cell lysate to a hydroxyapatite column and washing it with a low-concentration phosphate buffer in order to elute proteins and RNA. Then, the DNA is eluted using a concentrated phosphate solution.

Affinity chromatography can be used to purify mRNA. Most eukaryotic mRNAs have a poly(A) sequence at the 3'-end (see Section 4.2). **Poly(dT)** sequences are therefore used as adsorbent material (e.g., bonded to cellulose). At high salt concentration and low temperatures, the poly(A) sequences bind specifically to the complementary poly(dT) residues and can be released later by dissociating conditions.

Exclusion chromatography or **gel filtration** is a special type of chromatography. It is a method by which dissolved macromolecules can be separated (see Section 7.5.1.1). The stationary phase consists of expanded gel particles with a defined pore size. The separation process is determined by the size of the particles as large molecules cannot pass through the pores and thus migrate faster than smaller molecules. The latter are retained in the pores, and therefore can be separated and isolated as the last molecules eluted from the column. This method permits the separation of nucleic acids from low-molecular substances (e.g., from nucleotides after a labeling reaction).

10.3
Electrophoresis

Nucleic acids are (poly) acids and thus negatively charged, which makes them migrate to the positive pole in the electric field. The separation process depends on the voltage used, the properties of the gel, as well as the charge and shape of the molecule in question.

The method used for the separation of nucleic acids is chosen according to the size of molecules to be separated and the desired resolution capacity. The most frequently used methods use the following gel systems: **agarose gel** for submarine electrophoresis and pulsed-field methods, and **polyacrylamide gel** for high resolution.

10.3.1
Agarose Gel Electrophoresis: Submarine Electrophoresis

Agarose gel electrophoresis is a standard procedure to separate DNA fragments that vary in size. Agarose is a polysaccharide obtained from marine red algae. It is added to an electrophoresis buffer and is then dissolved by heating it. The presence of many hydroxyl groups (R-OH) enables hydrogen bonds to form, which lends firmness to the large-pored gel matrix. In submarine electrophoresis, the agarose gel is kept in a horizontal position. It is completely covered by the buffer that prevents it from drying out.

The speed at which DNA fragments migrate through agarose gels in the **electric field** depends above all on the **size of the DNA fragments**. The migration speed of linear double-stranded DNA molecules is inversely proportional to the logarithm of their size. Apart from DNA molecule size, factors such as the properties of the buffer used, the concentration of agarose within the gel, the strength of the current used, as well as the conformation of DNA molecules affect the traveling speed. DNA can be made visible through staining, such as with **ethidium bromide (EtBr)**, which intercalates and the DNA then shows up as pink bands under UV light. Figure 10.1 demonstrates the separation of different forms of the same plasmid DNA in the presence of this dye. The velocity of

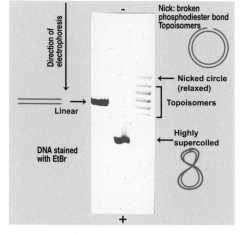

Fig. 10.1 Agarose gel electrophoresis of plasmid DNA in the presence of EtBr. All forms of the DNA molecule have the same molecular mass. Left: plasmid, linearized by restriction enzyme. Middle: covalently closed, highly superhelical plasmid DNA from a plasmid preparation. Right: different, experimentally produced topoisomers of the plasmid, showing different degrees of superhelicity. Negative image.

migration of these different forms depends upon the amount of intercalated, positively charged EtBr. The naturally occurring, covalently closed, highly supercoiled form can, due to intramolecular tension, bind only a limited amount of dye and runs the fastest. Less superhelical, covalently closed forms take up more EtBr and therefore run slower. Linearized DNA binds the most dye and therefore runs the slowest.

10.3.2
Pulsed Field Agarose Gel Electrophoresis

Very long DNA molecules (over 20 kb) cannot be sufficiently separated in standard agarose gel electrophoresis because they stretch out in the electric field to an extent inhibiting the passage through the gel matrix. To accomplish this, **pulsed field gel electrophoresis (PFG)** is employed. In this method the direction of the direct current field is periodically changed. This keeps the molecules in a compact configuration, which makes their charge-dependent migration in the main running direction possible. Using this modified submarine electrophoresis technique, **very large nucleic acid molecules, even whole chromosomes**, can be separated.

10.3.3
Polyacrylamide Gel Electrophoresis (PAGE)

This method is primarily used to resolve small differences in the size of nucleic acid molecules. The high-resolution polyacrylamide gel is generally arranged vertically. Depending on the running conditions, such as the presence or absence of urea in the gel, double-stranded, single-stranded, or even heteroduplexes consisting of double- or single-stranded portions can be separated. Moreover, due to their high-resolution capacity, polyacrylamide gels allow us to separate molecules differing by only one nucleotide in size. Therefore, they can be used for DNA sequencing purposes as well as to identify point mutations, for example.

For DNA sequencing, the gel contains a high urea concentration and is run at a high voltage. This heats up the gel and, in conjunction with the urea, contributes to the denaturation of DNA. In automated sequencing devices (capillary electrophoresis sequencer) separation of the DNA molecules generally occurs in gel matrix containing capillaries (see Chapter 14).

In order to detect DNA fragments in a polyacrylamide gel after separation, nucleic acids are usually labeled. While initially incorporated nucleotides were marked using **radioactive labels** (^{32}P, ^{33}P, ^{35}S), which could be detected through **autoradiography**, more recent procedures are based on the incorporation of **fluorescent dyes** (e.g., **Cy5**). These can be selectively excited by laser light and detected by photodiodes. It is also possible to transfer DNA fragments from the gel to a nylon or nitrocellulose membrane (**Southern blotting**). Detection on the membrane is then accomplished through hybridization (see Chapter 11) using specific gene probes, through autoradiographic, immunological, or fluorescence-based methods.

11
Hybridization of Nucleic Acids

Learning Objectives
The ability of nucleic acids to form complementary hybrid complexes in the form of double-stranded molecules is one of their most important properties – for experimental diagnostics as well as preparative purposes. Selective molecular recognition forms the basis of our classical diagnostic techniques – such as Southern and Northern blotting used in gene identification and gene expression analysis – as well as the polymerase chain reaction and all its applications. The formation of hybrids is also the key principle in systematic expression and gene analysis using gene chip technology, as well as in a range of techniques to localize and study genes in chromosomes and tissues.

11.1
Significance of Base Pairing

The formation of **double strands of complementary nucleic acid sequences**, as first described by Watson and Crick in 1953, is the basis of gene replication and expression in the entire living world. DNA base pairing between guanine and cytosine (G-C pairs) involves three hydrogen bonds, whereas pairing between adenine and thymine (A-T pairs) involves two hydrogen bonds (Fig. 2.19). In DNA-RNA, as well as RNA-RNA complexes, A-T pairs are replaced by A-U pairs (see Chapter 2.4). The stability of base pairs, given the same sequence, is highest in RNA-RNA hybrids and these RNA-DNA hybrids are more stable than DNA-DNA hybrids. G-C pairs are more stable than A-T or A-U pairs, as they are able to form three instead of two hydrogen bonds (Fig. 2.19). The process of complementary single strands coming together to form double strands is called **hybridization**.

The stability of a hybrid in solution at a given ion strength is defined by its **melting temperature (T_m)**, which is the temperature at which 50% of a given hybrid denature into single strands. This parameter mainly depends on the proportion of G-C base pairs – stability increases with a higher proportion of G-C pairs. There is a wide variety of naturally occurring base combinations, which define the physical properties of the genetic material, including its melting point. Comparatively heat-resistant organisms, such as *Thermus aquaticus* – a bacterium found in geysers at temperatures above 90 °C, have a high G-C content (65%).

11.2
Experimental Hybridization: Kinetic and Thermodynamic Control

The ability of nucleic acids to hybridize opens up a wide range of **diagnostic and preparative possibilities**. The important feature is that in an experiment, nucleic acids can be kept as single strands using high temperatures, and when cooling down, they will recognize and find their complementary partner molecules in solution and bind to them. The **binding of complementary nucleic acid**

An Introduction to Molecular Biotechnology, 2nd Edition.
Edited by Michael Wink
Copyright © 2011 WILEY-VCH Verlag GmbH & Co. KGaA, Weinheim
ISBN: 978-3-527-32637-2

Fig. 11.1 Classical setup of a Southern blot after Southern (1975).

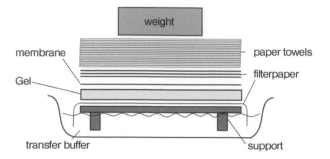

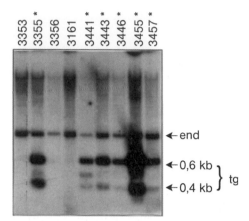

Fig. 11.2 Genetic analysis of transgenic mice by Southern blotting. Genomic DNA obtained from biopsy samples of mice was cleaved with a restriction endonuclease and separated in an agarose gel. A radioactive probe, also recognizing homologous endogenous sequences ("end"), was used to identify the introduced transgene, seen here as two additional fragments ("tg"). Different animals appear to carry different numbers of transgene copies, as can be concluded from the variation in signal strength in comparison with the endogenous band (end). (From Jäger, 1996.)

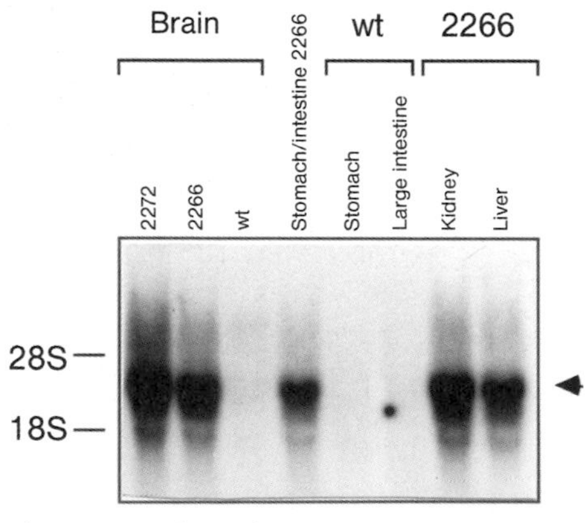

strands is a bimolecular process that depends primarily on the concentration of the reactants. However, as explained above, the binding stability depends on the G-C content, and thus on the temperature and the length of the hybridizing strands. The hybridization reactions can be controlled in various ways by choosing the appropriate conditions. If one or both complementary strands are highly concentrated, if there is short homology or a low temperature, the reaction is primarily controlled by kinetics. Within a short time short hybrids with relatively low stability form preferentially. This is exploited in the **polymerase chain reaction (PCR)**, as discussed in Chapter 13, but also in preparative enrichment and depletion of repetitive sequences in mammalian DNA.

In order to perform a quantitative hybridization of specific sequences present at low concentrations, the reaction should be thermodynamically controlled. This is accomplished by keeping the temperature as high as possible in order to avoid the formation of less-specific hybrids. The reaction times should thereby be chosen as long as possible to ensure that the hybridization process comes to completion. These are the conditions under which **Southern blotting** and **Northern blotting** are performed in diagnosis (see Section 11.3.1), and for preparative processes, such as enriching differentially expressed genes when cross-hybridizing several expression libraries.

11.3
Analytical Techniques

11.3.1
Clone Detection, Southern Blotting, Northern Blotting, and Gene Diagnosis

Once cloned or amplified DNA fragments are available, it is possible to attach **radioactive** or **fluorescent labels** to specific nucleic probes. Such probes can be used as **hybridization probes** for the analysis of immobilized nucleic acids (e.g., on **nylon** or **nitrocellulose membranes**). An early example of this is the detection of specific cloned DNA fragments in bacteriophages or plasmids, on the basis of a technique used for the first time by Grunstein and Hogness in 1975. Bacterial colonies or phage plasmids on plates are blotted onto nitrocellulose or nylon carrier membranes in order to immobilize their DNA. These filters are then hybridized using a labeled probe. The base-paired sequences can be visualized using an appropriate detection system.

This method can also be used for nucleic acids that have been separated in a gel. Figure 11.1 shows the principle setup by which DNA molecules that have been cut by restriction enzymes can be transfered to a membrane and characterized. This technique has been named after its inventor Ed Southern **Southern blotting**. Figure 11.2 gives an example of the application of this method: genomic mouse DNA that has been cleaved by restriction enzymes is used to find out if a transgene is present. Using hybridization probes recognizing so-called polymorphic loci in the genome to be analyzed, this method can be employed in forensic genetic analyses (**DNA fingerprinting**).

Fig. 11.3 Analysis of gene expression in two strains of transgenic mice (2272, 2266) using Northern blotting. Total RNA from several tissue samples was electrophoresed in an RNA gel and transferred to a membrane. The hybridization was carried out with a radioactive probe derived from the transgene. wt, wild-type. (From Jäger, 1996.)

A variation of this technique is called **Northern blotting**. Instead of DNA fragments, **RNA fragments** are seperated in a gel and transferred to a membrane. After hybridization, the length and amount (i.e., expression intensity) of different RNA species from different samples can be determined (Fig. 11.3).

In Southern or Northern blotting, both RNA and DNA molecules – the latter usually molecularly cloned DNA fragments – make suitable labeled probes. Alternatively, synthetic oligonucleotides can serve as probes. Thus, for example, employing selective hybridization conditions, allele-specific oligonucleotides can be used to identify and distinguish different alleles of a particular gene. This technique is also suited to type different DNA samples on the basis of genetic polymorphisms.

11.3.2
Systematic Gene Diagnosis and Expression Screening based on Gene Arrays

The decoding of whole genome sequences and the development of **gene chips** or **gene arrays** has made the systematic analysis of whole genomes possible. For this purpose libraries of specific oligonucleotides are immobilized in an ordered array on a glass matrix; such matrices (chips, arrays) can then be hybridized with fluorescently labeled DNA from the cell or tissue to be analyzed. This technique allows the characterization of whole genomes in one hybridization reaction. This method is used, for instance, to characterize individuals with respect to many different gene variants at a time. Potential applications include tumor diagnostics, mapping of genetic diseases, determination of **genetic risks**, and **pharmacogenetics**.

Analogously, systematic expression screening can be carried out. To this end, fluorescently labeled cDNA molecules made from the expressed RNA from the sample to be analyzed are hybridized to an appropriate oligonucleotide array. From the intensity and the location of the fluorescence after hybridization, expression profiles can be established. Figure 11.4 shows the original data obtained in such a hybridization project using an Affymetrix GeneChip® with about 200 000 oligonucleotides from about 20 000 genes (www.affymetrix.com). Such analyses allow comprehensive investigations, for instance of the action of active substances on cells or within experimental animals, in order to identify new target molecules for therapeutic treatment.

11.3.3
In Situ Hybridization

In situ **hybridization (ISH)** with radiolabeled probes is a long-established method in **cytogenetics**. A probe is hybridized onto a chromosome preparation that has been immobilized on a slide. The hybridization site can thus be visualized, localizing the gene in question on the chromosome. If a probe is labeled with a fluorophore, this is called **fluorescence *in situ* hybridization (FISH)**. The localization technique is shown in Fig. 11.5. By using several fluorescently labeled probes, a number of localizations can be carried out simultaneously in one

Fig. 11.4 Result of the expression screening of thousands of genes using labeled cDNA on a single GeneChip array supplied by Affymetrix. Every spot represents the hybridization of an individual oligonucleotide. The color indicates the quantity of the hybridized gene sequence to each particular spot. Special software is used for evaluation. (Reproduced with kind permission from Affymetrix.).

(A)

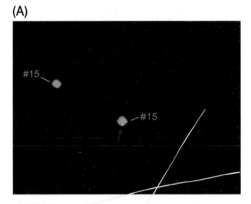

(B)

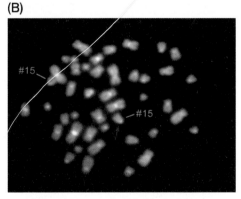

Fig. 11.5 FISH in chromosome preparations. (A) Detection of a deletion in what is known as the PraderWilli region in an allele of chromosome 15 on a metaphase chromosome preparation. The gene-specific probe (arrow) can only be seen in one of the two chromosomes 15 that have been marked by chromosome-specific probes (green). (B) Control staining with a nucleic acid-specific dye, 4,6-diamidino-2-phenylindole (DAPI). (Reproduced with kind permission from K. Teller, I. Solovei, and T. Cremer, Ludwig-Maximilians-Universität München.)

Fig. 11.6 ISH of two developmental genes (*even skipped* (blue) and *fushi tarazu* (brown)) in the cellular blastoderm stage of *D. mela-* *nogaster* larvae. (Reproduced with kind permission from P. Gergen, State University of New York, Stony Brook, NY.)

experiment, as shown in Fig. 11.5. The FISH technique is widely used in the analysis of chromosome structures, but is also applicable on whole cells (e.g., to karyotype tumor cells or to test for fetal trisomy). The hallmark of this technology is its potential to genetically analyze single cells.

Another application of ISH is the localization of RNA transcripts in tissue, in analogy to the immunohistological analysis of proteins. Probes that have been labeled either radioactively or optically are hybridized with the RNA of a histological tissue section or preparation. Figure 11.6 shows the expression of two genes responsible for the development of *Drosophila melanogaster* larvae as an example.

12

Use of Enzymes in the Modification of Nucleic Acids

Learning Objectives

It was not until the 1970s that tools suitable for modern gene technology were developed. These included the discovery and utilization of restriction endonucleases. Meanwhile, a wide range of enzymes that can modify nucleic acids has been added to the arsenal, such as the frequently used ligases, methyltransferases, polymerases, nucleases, kinases, and phosphatases.

12.1

Restriction Enzymes (Restriction Endonucleases)

Restriction enzymes are **endonucleases** that cleave DNA sequence-specifically (Fig. 12.1). They were discovered when researchers wanted to find out how bacteria protect themselves from viral intruders. When foreign DNA is introduced into a bacterium (e.g., via a phage infection) it is dissected into dysfunctional segments by restriction enzymes. The bacterial DNA is protected by **methylation** of the recognition sites of restriction enzymes (see Section 4.2). Bacteria possess a DNA-modifying enzyme, which methylates DNA in those places where their own restriction enzyme would cut the DNA.

To date, we know a range of nearly 1000 individual restriction endonucleases, which makes it necessary to agree on a standard nomenclature. The enzyme is named after a letter code derived from the name of the bacterial species the enzyme was isolated from. EcoRI, for example, stands for an enzyme that has been isolated from *Escherichia coli*. If more than one restriction enzyme has been isolated from the same bacterial species, they carry additional Roman numerals. The enzymes HaeI and HaeII have both been isolated from *Haemophilus aegypticus*.

Restriction enzymes recognize specific sequences of 4–8 bp in double-stranded DNA, cleaving their phosphodiester bonds. These restriction sites usually have a central symmetrical structure and are called **palindromes**. Palindromes (Greek *palindromos*: running backwards) give an identical reading whether you look at them from the left or right: you will end up with the same information (Fig. 12.2).

There are **three types of restriction endonucleases (types I, II, and III)**. Types I and III are generally large enzyme complexes, consisting of many subunits that perform the tasks of endonucleases as well as **methyltransferases**. The specific recognition sequence of type I restriction enzymes lies about 1000 bp in the 3'-direction from the recognition sequence where DNA is cleaved nonspecifically. Owing to the nonspecific occurrence of cleavage, this group of enzymes is only useful for a small number of applications. In type III enzymes, the cutting site lies at a known distance of up to 14 nucleotides from the DNA-binding site. What is striking in these enzymes is that the recognition sequence does not necessarily need to be a palindrome.

An Introduction to Molecular Biotechnology, 2nd Edition.
Edited by Michael Wink
Copyright © 2011 WILEY-VCH Verlag GmbH & Co. KGaA, Weinheim
ISBN: 978-3-527-32637-2

Fig. 12.1 The discovery of restriction endonucleases such as HindIII was a milestone in the history of genetic engineering. It won Arber and Smith the Nobel Prize in 1978. This type of enzyme recognizes certain sequences at which they cut DNA. In bacteria, restriction enzymes act as a protection from viruses. The effect of these enzymes is shown here using the example of EcoRI. The recognition site GAATTC is enclosed by restriction enzymes. One DNA strand is cut at one place, the other at another between G and A. The separated fragments have sticky ends and so another DNA fragment featuring a sticky end can latch onto the complementary end. The newly paired DNA fragments are ligated by ligase.

Restriction enzymes

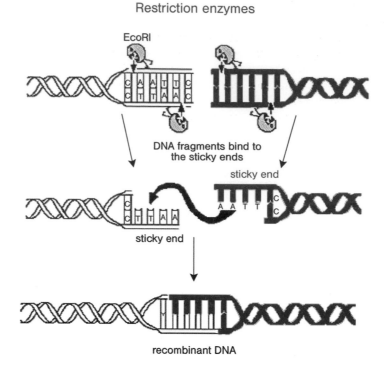

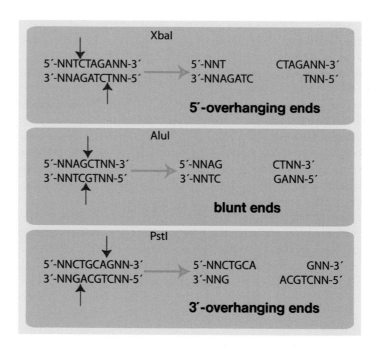

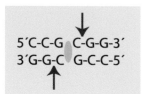

Fig. 12.2 Palindromic sequence recognized by a restriction enzyme. The symmetry axis is marked by an ellipse and the restrictions sites by arrows.

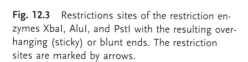

Fig. 12.3 Restrictions sites of the restriction enzymes XbaI, AluI, and PstI with the resulting overhanging (sticky) or blunt ends. The restriction sites are marked by arrows.

The cleavage of DNA either results in **blunt ends** or complementary 3'- or 5'-overlapping sticky ends. Figure 12.3 shows the three possible forms of DNA ends after digestion by a restriction enzyme.

Many companies now offer a wide range of restriction and other modification enzymes. These enzymes are derived from microorganisms and are delivered ready to use with the appropriate reaction buffer.

Sometimes it is necessary to run two simultaneous digestion processes with two different enzymes. In such cases, extra attention should be given to the compatibility of the two buffer systems. Due to the stringent requirements of their reactive conditions, some enzymes cannot be used in dual digestion and the enzymes must be applied one after the other to the same restriction preparation. The enzyme that has the buffer with the lower salt concentration should be used first. For the second enzyme, adding the second buffer can raise the salt concentration. Sometimes it may be necessary to inhibit the activity of an enzyme after a digestion process. This can be done by applying heat (usually over 60 °C). Phenol extraction is required to remove the enzyme completely.

12.2
Ligases

Ligases are enzymes that connect DNA molecules through phosphodiester bonds between a 5′-phosphate and a 3′-hydroxyl end. Along with restriction enzymes, they are basic tools in genetic engineering. In contrast to restriction enzymes, ligases need either ATP or NAD^+ as cofactors. Two compatible sticky or blunt ends can be coupled by ligases.

If no suitable restriction sites can be found for two DNA fragments to be ligated, linkers can be used. These are short stretches of double-stranded DNA of length 8–14 bp and have recognition sites for three to eight restriction enzymes. These linkers are ligated to blunt-end DNA by ligases. Linkers are synthesized as oligonucleotides and are commercially available. New restriction sites can also be introduced using approaches employing PCR.

The following is an overview of various types of ligases and their characteristics:

- **T4 DNA ligase** is isolated from cells infected with bacteriophage T4. It ligates the ends of double strands of DNA or RNA. This enzyme brings blunt as well as complementary sticky ends together. This enzyme repairs single-strand ruptures (nicks) in double-stranded DNA, RNA, or DNARNA hybrids. The cofactor involved is ATP.
- **Taq DNA ligase** catalyzes the phosphodiester bond formation between two oligonucleotides that are hybridized to a complementary target DNA. The enzyme is only effective at comparatively high temperatures (45–65 °C) and requires NAD^+ as a cofactor.
- **T4 RNA ligase** catalyzes a phosphodiester bond between RNARNA, RNADNA, or DNADNA oligonucleotides. ATP is needed as a cofactor, but no template strand is required.
- **DNA ligase** (*E. coli*) catalyzes phosphodiester bonding between double-stranded DNA with sticky ends, whereas fragments with blunt ends are not ligated efficiently. NAD^+ serves as a cofactor.

12.3
Methyltransferases

Many organisms have enzymes that **methylate DNA**. Most restriction enzymes are unable to cut a methylated recognition sequence. However, there are restriction enzymes that only cut a recognition sequence if the DNA is methylated at that site (e.g., DpnI). Furthermore, there are restriction enzymes that can digest both methylated and nonmethylated recognition sequences (e.g., BamHI).

Methyltransferases and their corresponding restriction endonucleases recognize identical restriction sequences. All methyltransferases transfer the methyl group from **S-adenosylmethionine (SAM)** to a specific base of the recognition sequence – SAM itself also takes part in the methylation reaction. Normally, methylation protects DNA from the corresponding restriction endonucleases. However, there are also low-specificity methyltransferases, such as SssI methy-

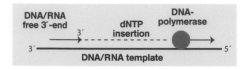

Fig. 12.4 In order to incorporate nucleotides, a polymerase requires a DNA or RNA template and a free 3′-end of DNA or RNA that can serve as a primer.

lase, which methylate cytosine residues in the sequence 5′-CG-3′. In this case, the DNA is protected from digestion by a whole set of restriction endonucleases.

Plasmid DNA generated in *E. coli* contains certain methylated sequences. As there is great variation between the methylation patterns of the numerous *E. coli* strains, the success of a restriction digestion depends on the *E. coli* strain from which the plasmid DNA was obtained.

12.4
DNA Polymerases

To date, a large number of various polymerases have been characterized and are commercially available. A common feature is the addition of nucleotides to a free 3′-end of a DNA strand. The precise sequence in which the nucleotides are inserted is determined by a template (Fig. 12.4).

In addition to their **5′-polymerase activity**, they can also act as **exonucleases**, working either in the 5′→3′ direction or in the 3′→5′ direction. In the **3′→5′ exonuclease** activity, a process called "proofreading"allows the enzyme to check each nucleotide during DNA synthesis and excise mismatched nucleotides in the 3′→5′ direction. These exonucleases can also help to slowly degrade overhanging 3′-ends to create blunt ends.

The **5′→3′ exonuclease** action degrades all hybridized primers present. They are absolutely necessary to get rid of blocking primers.

This variation of action in polymerases makes them suitable for a wide range of different applications. For example, polymerases cannot only amplify or repair DNA; they can fill up sticky ends that have been produced by the use of restriction endonucleases. If the 5′-end overhangs, it can be filled in using 5′→3′ active polymerase. Conversely, if the 3′-end sticks out, T4 DNA polymerase is used to cut off the superfluous nucleotides to produce blunt ends.

With nick translation, radioactively marked single-stranded fragments of DNA are manufactured. These are then inserted with the help of the exonuclease activity of some polymerases, such as *E. coli* DNA polymerase. DNase I is used to produce a nick in double-stranded DNA. In the next step, DNA polymerase I is added together with radioactive nucleotides. The 5′→3′ exonuclease activity degrades the 5′-end on the nicked strand, while the polymerase inserts the radioactively marked nucleotides. The resulting polynucleotide carries a distinctive radioactive label and can be hybridized with a corresponding DNA sequence.

Thermostable polymerases retain their stability even at temperatures high enough to melt the DNA double helix and separate it into single strands. This is exploited in the **polymerase chain reaction (PCR)** (see Chapter 13).

The error rates for inserting nucleotides vary within individual polymerases and there is also variation in the length of polymerizations produced (Table 12.1).

Table 12.1 Features of different polymerases.

	E. coli DNA polymerase I	E. coli DNA polymerase I (Klenow fragment)	T4 DNA polymerase	T7 DNA polymerase	Taq DNA polymerase	M-MuLV reverse transcriptase
5′→3′ Exonuclease activity	Yes				Yes	
3′→5′ Exonuclease activity	Yes					
Error rate ($\times 10^{-6}$)	9	40	<1	15	285	
Displacement of DNA strand		Yes				
Inactivation through heat	Yes	Yes	Yes	Yes		Yes

12.5
RNA Polymerases and Reverse Transcriptase

In certain applications such as the generation of RNA from DNA or the **quantitative analysis of RNA (real-time PCR (RT-PCR))** different specialized polymerases are needed. **T7 RNA polymerase** or **SP6 RNA polymerase** from *Salmonella typhimurium* generate RNA from a DNA sequence. **M-MuLV reverse transcriptase** from Moloney murine leukemia virus and **AMV reverse transcriptase** from avian myeloblastosis virus are able to synthesize a complementary DNA strand initiated from a primer using either RNA (cDNA synthesis) or single-stranded DNA as a template.

12.6
Nucleases

Several types of nucleases are used in genetic engineering. Their applications include the removal of 3′-overhangs, filling in or cleavage of 5′-overhangs (mung bean nuclease), removal of DNA in RNA preparations (DNase I), removal of oligonucleotides post-PCR (exonuclease I), and the generation of singe-stranded DNA from linear double-stranded DNA (exonuclease III).

- **Desoxyribonuclease I (DNase I)**, from bovine pancreatic cells, is an endonuclease that nonspecifically cleaves DNA to release di-, tri-, and oligonucleotide products with 5′-phosphorylated and 3′-hydroxylated ends. DNase I acts on single- and double-stranded DNA, chromatin, and RNADNA hybrids. In the presence of Mg^{2+} ions, DNase will attack each strand separately, producing random nicks, which are needed in nick translation. The function of DNase I depends specifically on its buffer composition. In the presence of Mn^{2+} ions, the enzyme will cleave both DNA strands at roughly the same site, leaving ragged ends.
- **Nuclease BAL-31** is an exonuclease that degrades 3′- as well as 5′-ends of double-stranded DNA. It does not create nicks, but it functions as a single-strand endonuclease at existing internal nicks and single-stranded gaps. However, the degradation process is incomplete, producing ragged rather than blunt ends. These can then be filled using a polymerase such as T4 polymerase.
- **Exonuclease III** attacks 3′-hydroxyl groups from the blunt DNA ends that occur at the end of a DNA double helix or from the internal nicks within it. As it relies on duplex DNA, exonuclease III is unable to degrade overhanging 3′-end. This function can be fulfilled by **exonuclease I**.
- **Mung bean nuclease** is isolated from mung bean sprouts. It is a specific DNA and RNA endonuclease that degrades overhangs of DNA or RNA ends, leaving blunt ends in both 5′- as well as 3′-end direction.

12.7
T4 Polynucleotide Kinase

T4 polynucleotide kinase (PNK) catalyzes the transfer and exchange of phosphate groups from the ATP γ-position to the 5′-hydroxyl terminal of double- or single-stranded DNA or RNA, and of nucleoside 3′-monophosphates. The enzyme also removes 3′-phosphate groups. PNK can be used to phosphorylate the 5′-ends of polynucleotides. This could be necessary, for instance, in automatically produced oligonucleotides, which do not contain 5′-phosphate groups and could thus not be ligated to other unmodified polynucleotides.

12.8
Phosphatases

Phosphatases catalyze the removal of 5′-phosphate groups. **Shrimp alkaline phosphatase (SAP)** and **calf intestinal alkaline phosphatase (CIP)** remove 5′-phosphate groups from RNA, DNA, and desoxyribonucleoside triphosphates (e.g., NTP, dNTP). Cleaved and CIP-treated double-stranded DNA can thus not religate with itself and prevents recirculation of plasmids. The 5′-end can then labeled differently.

13
Polymerase Chain Reaction

Learning Objectives

The polymerase chain reaction has revolutionized diagnosis as well as molecular biology and medical research. It is a technique that is used universally, ranging from the analysis of hereditary human disease to the diagnosis of viral infection; from paternity tests to the investigation of evolutionary links. Used in combination with reverse transcription, it is also one of the most powerful techniques used to study gene expression.

13.1
Introduction

At the end of the 1970s, **molecular cloning technology** provided DNA for practically any kind of genetic analysis. However, those early cloning methods were technically very complicated and time-consuming, and sometimes the problem was that there was not enough genetic material available (see Section 13.3.1). In 1983, so we are told, biochemist Kary Mullis was on his way to his woodland cabin in Mendocino County when he hit on the principle of the **polymerase chain reaction (PCR)**. At that time, Mullis was working on oligonucleotides for Cetus, a biotech company near San Francisco – oligonucleotides being the crucial starting point for amplifying DNA by polymerization. He was jointly awarded the Nobel Prize in Chemistry in 1993 for his development of the PCR.

In the following, we introduce the most important, but by no means all, of the methods that are based on the PCR principle, and briefly discuss some of their areas of application in research, medicine, and forensic science.

13.2
Techniques

13.2.1
Standard PCR

With PCR, gene sections that are in limited supply can be targeted and replicated. This can only be done if the parts of the sequence in question are known. These are used to produce oligonucleotides, usually 20–25 bases in length, known as **primers**. They mark the starting points of DNA synthesis once DNA polymerase and deoxynucleoside triphosphates have been added. As Fig. 13.1 shows, two primers are chosen that ensure through their position on the DNA fragment that synthesis starts at opposite ends of the fragment, bearing in mind that enzymatic polymerization runs $5' \rightarrow 3'$. Usually, the DNA to be replicated is double stranded, so in a first step it must be **denatured** at a temperature of 94–96 °C in order to produce **single-stranded templates**. In a second step, the primers are **annealed to their complementary sites** on the templates. This is done at a temperature between 40 and 60 °C, close to the **melting temperature (T_m)** of

An Introduction to Molecular Biotechnology, 2nd Edition.
Edited by Michael Wink
Copyright © 2011 WILEY-VCH Verlag GmbH & Co. KGaA, Weinheim
ISBN: 978-3-527-32637-2

Fig. 13.1 Schematic outline of PCR. Double-
stranded DNA is first denatured. Then Primer-1
(forward primer) and Primer-2 can anneal to the
corresponding DNA strands. During an extension
reaction, these are elongated by Taq polymerase
along their DNA templates, resulting in two double-stranded DNA fragments that enter the next
PCR cycle to be duplicated, etc.

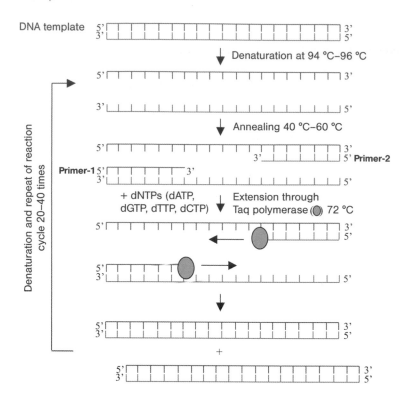

the relevant DNAoligonucleotide hybrid. The melting temperature is usually calculated using one of the commonly used DNA analysis programs or the approximation formula:

$$T_m = 4 \times (G + C) + 2 \times (A + T) \tag{13.1}$$

The following step – **elongation of the oligonucleotide primers** – reveals the
strength of PCR technology: the DNA polymerase (**Taq polymerase**) used is derived from a heat-stable bacterium (*Thermus aquaticus*) and works at temperatures of 72 °C. Unlike other proteins, it is not destroyed in the denaturation process. The reaction can thus be repeated without having to add new polymerase
for each cycle. Once the elongation reaction is completed, the resulting DNA
hybrids are again denatured to start a new cycle. In current PCR machines, the
repetition of cycles has been automated (Fig. 13.2), which, at least in theory,
opens up the possibility of exponential replication of the original template. In
reality, the productivity of the cycles declines due to a dwindling amount of
primers and nucleotides, and reduced Taq polymerase activity in later cycles.

13.2.2
RT-PCR

PCR is used not only for the amplification of genomic DNA, but also for detecting and analyzing **RNA expression** as well as **cloning expressed genes**. In these
cases, the amplification process described above is preceded by a **reverse transcription (RT)** reaction. From the RNA material to be analyzed, single-stranded
cDNA is synthesized. This is done using specific DNA primers and the enzyme
reverse transcriptase, which is able to synthesize DNA from RNA templates.
The use of **oligo(dT) primers** that specifically bind to **3′-polyadenylated mRNA**
ensures that only mRNA undergoes reverse transcription (Fig. 13.3). Reverse
transcriptase was originally isolated from retroviruses. Its biological function is
to copy the viral RNA genome after infection and thus integrating it into the
host genome (Fig. 3.28). Nowadays, cloned reverse transcription enzymes pro-

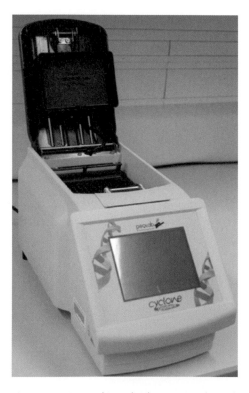

Fig. 13.2 PCR machine, also known as a thermal
cycler.

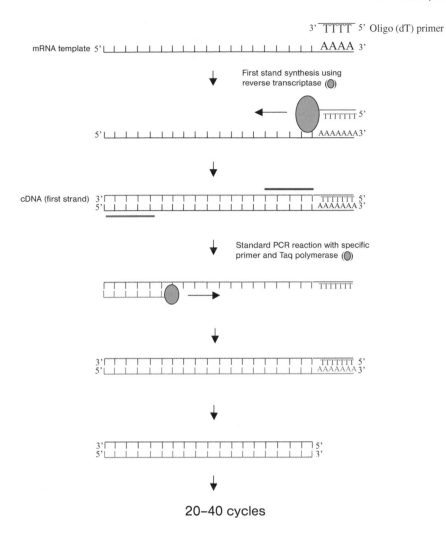

3' ‾TTTT‾ 5' Oligo (dT) primer

mRNA template 5' |_|_|_|_|_|_|_|_|_|_|_|_|_|_|_|_|_|_|_| AAAA 3'

First stand synthesis using
reverse transcriptase (●)

←● TTTTTT 5'
5' |_|_|_|_|_|_|_|_|_|_|_|_|_|_|_|_|_| AAAAAAA 3'

cDNA (first strand) 3' |_|_|_|_|_|_|_|_|_|_|_|_|_|_| TTTTTT 5'
5' |_|_|_|_|_|_|_|_|_|_|_|_|_|_|_|_| AAAAAAA 3'

Standard PCR reaction with specific
primer and Taq polymerase (●)

|_|_|_|_|●→|_|_|_|_|_|_|_|_|_| TTTTTT

3' |_|_|_|_|_|_|_|_|_|_|_|_|_|_| TTTTTT 5'
5' |_|_|_|_|_|_|_|_|_|_|_|_|_|_| AAAAAAA 3'

3' |_|_|_|_|_|_|_|_|_|_|_|_|_|_|_| 5'
5' |_|_|_|_|_|_|_|_|_|_|_|_|_|_|_| 3'

20–40 cycles

Fig. 13.3 RT-PCR reaction process. In a first step, oligo(dT) primers anneal to the poly(A) tail – a feature shared by all eukaryotic mRNAs. Alternatively, the first strand can also be synthesized using a random primer or a primer specific to the mRNA sequence in question. mRNA is transcribed into single-stranded DNA (cDNA). Sequence-specific primers then initiate the standard PCR.

duced within bacteria are used. The most widely used enzymes are those derived from **avian myeloblastosis virus (AMV)** and **Moloney murine leukemia virus (M-MuLV)**.

13.2.3
Quantitative/Real-Time PCR

The number of cycles in a standard PCR reaction is normally chosen to ensure that the reaction can run its course as completely as possible. If the number of cycles is very high, the reaction is slowed down by a lack of primers, a surplus of templates, or eventually by a loss of enzyme activity in the late cycles. In the early cycles, the reaction yields approximately exponential amplification, which makes it possible to compare the results of several reactions. Thus, standard amplification can also be used in quantitative analysis. In order to make this approach work, the number of cycles required for the amplification of the gene in question and its control must be carefully chosen (i.e., the sample as well as the standard reaction must yield exponential amplification and produce detectable amounts of amplified material). It is also important that the amplified fragments are of roughly the same size. As they undergo gel electrophoresis, the intensity of the signals can be compared and the number of molecules in the original RNA sample determined.

Alternatively, the PCR products can be analyzed *in situ*. This is called **real-time PCR**. It involves the **detection of intercalated fluorescent molecules** in double-

Fig. 13.4 Typical output of a real-time PCR analysis. In this case, several cell lines were tested for the expression of TRAIL receptor 1. RT-PCR was carried out first and then followed by real-time PCR. A DNA-intercalating dye – in this case SYBR Green – gives off a fluorescent signal that is detected by the real-time PCR machine after each cycle. The fluorescent intensity on the y-axis is plotted against the number of cycles on the x-axis. The different curves represent various expression strengths. It becomes apparent that in some cells in cycle 25, for example, the fluorescence measured is low. In other words, the transcript of the gene in question is only available in small amounts in these cells. In other cells, the result may look very different indeed.

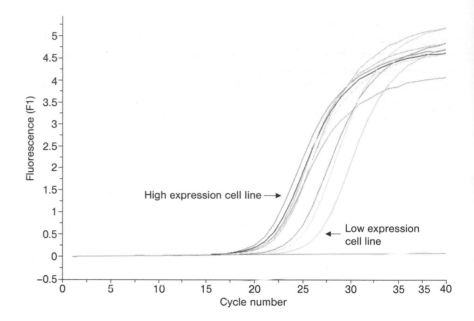

stranded DNA, such as SYBR® Green (Molecular Probes). Figure 13.4 shows the result of real-time PCR from a computer connected to a PCR machine, translated into a graph by suitable evaluation software. The TaqMan® (Applied Biosystems) system uses yet another technique. A detection oligonucleotide is added to the oligonucleotides, its binding site lying between the primers. It carries a fluorophore at its 5′-end, which is removed due to the 5′→3′ exonuclease activity of Taq polymerase during the elongation process. At the same time, the fluorophore is also removed from its quencher at the 3′-end of the detection oligonucleotide and can be picked up by the real-time PCR. There are a variety of real-time PCR methods currently in use and, in all of these, the fluorescence intensity provides a criterion for the strength of expression of the gene concerned. It is currently the most sensitive way of measuring expression. For all expression studies, a reverse transcription reaction must be carried out as a first step.

13.2.4
Rapid Amplification of cDNA Ends (RACE)

This is a procedure used for **amplifying 5′-cDNA ends** whose sequence is unknown. After the first cDNA strand has been synthesized from an RNA template using **reverse transcriptase**, an oligonucleotide (e.g., oligo(dT)) can be ligated to the 3′-end of the newly synthesized strand. In the subsequent PCR reaction, a gene-specific primer is used as well as an oligonucleotide homologous to that ligated 3′-end. A slightly modified procedure can be used to reproduce unknown 3′-ends.

13.3
Areas of Application

13.3.1
Genome Analysis

Due to its high sensitivity and throughput, PCR is widely used in genetic analysis. In research, it is used for **genotyping** the entire plant and animal kingdom (e.g., to analyze transgenic animals, to search for mutants, or in evolutionary biology). In a **medical context**, it has become the method of choice in tumor characterization, in paternity tests, and in forensic science (DNA fingerprinting). With the help of PCR, forensic scientists are able to amplify minute quantities of genetic material found at the site of the crime and thus identify the perpetrator. Here, polymorphic microsatellite alleles are studied that have been amplified using PCR (see Section 4.1). Other applications of PCR include the detection of genetic contamination (recombinant DNA in the environment), such as genetically modified corn in food or in fields of conventionally grown corn.

13.3.2
Cloning Techniques

The PCR ensures that minute amounts of genetic material can be used not only for analysis, but also for cloning. Often the cDNA of expressed genes needs to be cloned in order to investigate their functions and expression plasmids can be employed here. If fragments obtained through PCR are to be used in this way, it is particularly important that the polymerases used in the process make as few errors as possible. This can be ensured by using **proofreading polymerases** such as Pwo, Pfu, or Vent polymerases that can be obtained from various suppliers. What is often used is a mixture of proofreading enzymes and ordinary Taq polymerase, which provides a balance between exact reading and processivity (i.e., long fragments are amplified before the enzyme falls off the DNA template). PCR fragments can generally be molecularly cloned in the same way as restriction fragments (i.e., by attaching **adaptor sequences** containing restriction sites to the primers). Once the PCR has been completed, the fragments are digested by the relevant restriction enzymes. This results in compatible ends, ready to be ligated into specific plasmid vectors. Alternatively, there are cloning strategies in use that are based on the finding that Taq polymerase tends to attach an adenosine nucleoside to the 3'-end of each amplified sequence. These ends can easily be ligated to open plasmid vectors featuring specific 5'-T overhangs. Flanking restriction sites in these **TA cloning vectors** help to transfer the fragments into expression vectors, for example.

13.3.3
Expression Studies

RT-PCR has become the most important technique used in gene expression analysis. This is an area where basic research still takes center stage. The techniques described above are applied, for example, in comparative studies of expression patterns in normal and pathological tissue (e.g., tumor tissue). The effect of active pharmaceutical agents on gene expression can thus be tested in cell cultures and animal models. The aim of such studies is to achieve a better understanding of pathomechanisms and the action of medications. This will enable us to develop new treatments and/or customize treatments for individual patients.

14
DNA Sequencing

Learning Objectives

DNA sequencing has revolutionized biomedical research. Without this technology, the progress that led to the biotechnological production of enzymes, therapeutic agents as well as antigens for vaccines would not have been possible. Over recent years, interest has been focused on the sequence analysis of entire genomes with the objective of providing a comprehensive database of sequenced genomes for functional studies. It is also expected that new genes associated with diseases or polymorphic variants will be found and studied. The sequencing of nucleic acids has its place in disease diagnosis, and the analysis of genetic changes in the plant and animal kingdom.

14.1
Introduction

The primary structure of proteins and nucleic acids is arranged in linear order. This is the **primary sequence** that contains most of the information about the functions of genes and their products (see Chapters 2 and 4). Hence, from the 1950s onward, research has been focused on finding methods for the sequencing of these macromolecular building blocks. The first pioneering achievement in this field was the determination of the **amino acid sequence of insulin** by Frederick Sanger, which earned him the Nobel Prize in 1957. Spurred on by the discovery of the DNA double helix structure by Watson and Crick, the sequencing of nucleic acids could begin. Soon afterwards, it was discovered that the analysis of nucleic acid sequences held the key not only to the primary structure of proteins, but also to the regulatory areas of DNA (e.g., signals for the expression and processing of genetic information). The first nucleic acid sequencing techniques were based on **sequence-specific RNases**. By combining various cleavage products, it was possible to create overlapping sequences of significant sizes. Sequences of parts of RNA tumor viruses, of entire RNA viruses (phages) as well as structural rRNA molecules were identified using this method. For analysis, DNA sequences had to be transcribed into RNA, which slowed down the method considerably, and the need for higher performing DNA sequencing methods became apparent.

14.2
DNA Sequencing Methods

During the 1970s, several breakthroughs supported the development of principles of DNA sequencing methods that are still in use today. **Molecular cloning** was an initial step that made it possible to provide DNA for analysis in the desired quantities. This was complemented by **gel electrophoretic separation methods** that separated DNA fragments that differed in length by just one nucleotide.

An Introduction to Molecular Biotechnology, 2nd Edition.
Edited by Michael Wink
Copyright © 2011 WILEY-VCH Verlag GmbH & Co. KGaA, Weinheim
ISBN: 978-3-527-32637-2

In the laboratories of Walter Gilbert and Frederick Sanger, two different sequencing methods were developed independently, which won them both the Nobel Prize in 1980. In the 1990s, a faster and more efficient sequencing method, called **pyrosequencing**, was developed.

14.2.1
Chemical Sequencing Method (MaxamGilbert Method)

Chemical sequencing according to Maxam and Gilbert is based on the **base-specific chemical cleavage** of a DNA molecule that has been labeled at one end. The first step consists of labeling the DNA sample to be analyzed. This is achieved by labeling a double-stranded DNA molecule at one end (5′-end) with radioactive phosphorus (^{32}P) using the enzyme polynucleotide kinase. The strands are then denatured and undergo base-specific chemical treatment, partially cleaving them into fragments that, on average, are cleaved only once. This process is carried out separately for all four bases and yields a mixture of single DNA strands of various lengths (between one and several hundred nucleotides). The labeled subfragments created by all four reactions have the ^{32}P nucleotide label at one end and the chemical cleavage point at the other. The samples are then transferred to a high-resolution polyacrylamide gel (**sequencing gel**) in parallel lanes and separated electrophoretically. This creates a specific band pattern that can be directly read after it has undergone autoradiography. As chemical sequencing remains quite labor- and time-intensive, it has now been largely replaced by enzymatic sequencing.

14.2.2
Enzymatic Sequencing (SangerCoulson Method)

For this method the DNA fragment has to be located close to a known sequence and has to be available in single strands (Fig. 14.1). To achieve the first, the fragment is cloned into a plasmid vector that contains known sequences further along from the cloning sites. Some of such sections are used very frequently, but generally, any known sequence will serve the purpose. The single DNA strands are usually obtained by a gentle denaturation process involving sodium hydroxide (NaOH). DNA synthesis is initiated from a nucleotide primer binding to the known sequence and then continued along the unknown sequence by DNA polymerase. During the synthesis, radioactively or fluorescence-labeled nucleoside triphosphates are introduced, while specific competitive inhibitors are also added. These are **dideoxyribonucleoside triphosphates (ddNTPs)** that bind to the single-stranded DNA in the same way as nucleoside triphosphates would. However, lacking the essential hydroxyl group in the 3′-position, they are unable to bind to the next nucleotide to be incorporated. Four separate reactions are run for each of the nucleotides (ddATP, ddGTP, ddCTP, or ddTTP) that compete with their normal counterparts.

In each of the four reactions, the chain elongation stops where a ddNTP is introduced. If the ddNTP/dNTP ratio is right (usually, about 1% added ddNTP is sufficient), the chain termination sites should be evenly distributed over the unknown sequence stretch. The resulting fragments of varying length then undergo high-resolution **polyacrylamide gel electrophoresis (PAGE)**, which is run in parallel for the four reactions. As schematically shown in Fig. 14.1, the sequence can be directly read from the band pattern.

Over the past 25 years, sequencing procedures have become faster through increasing automation. The enzymatic sequencing technique, in particular, has undergone modification. After DNA synthesis and base-specific fluorescent marking, all four reaction products can be simultaneously separated by **PAGE** or **capillary electrophoresis** in the machines currently used. A fluorescence detector reads the nucleotide sequences at the end of the separation process. Fig-

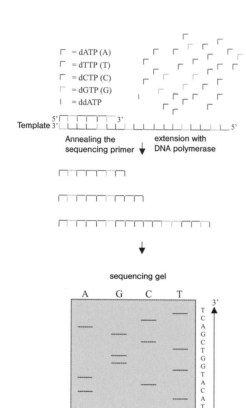

Fig. 14.1 Schematic representation of a Sanger sequencing reaction, independent of the labeling technique used (previously ^{35}S, ^{33}P, and ^{32}P radiolabeling; nowadays various fluorescent dyes). The sequencing primer, the four dNTPs and a representative ddATP are shown. DNA polymerase puts the dNTPs in place, as the template requires, until the ddNTP is introduced, breaking off the chain. A parallel process would take place with each of the other ddNTPs, resulting in labeled fragments of varying lengths, depending on the position of the complementary base on the template strand. The fragments resulting from the ddATP reaction are shown and can be seen in lane A of the sequencing gel. The gel shown here is typical for a result obtained from radiolabeling. Although this method has been largely superseded, it shows very clearly how sequencing works. The sequence from 5′→3′ is read from bottom to top. In real life, no readable sequence can be obtained right next to the sequencing oligonucleotide, as assumed in this simplified diagram, but only at a distance of about 30 bp.

ure 14.2 shows a section of such a chromatogram. As a consequence of the high demand for DNA sequencing arising from genome projects, the performance of automatic DNA sequencers has improved dramatically. There are now sequencing robots available that can carry out several hundred sequencing reactions simultaneously.

14.2.3
Pyrosequencing

Ever since large sequencing projects started, one goal was to reduce costs and sequence faster. To this end, **pyrosequencing** was developed by Pal Nyren and Mostafa Ronaghi at the Royal Institute of Technology in Stockholm in the mid-1990s. It is a novel DNA sequencing method that is based on the "sequencing by synthesis" principle. **"Sequencing by synthesis"** uses an immobilized single-strand DNA molecule and synthesizes the complementary strand from a sequencing primer using DNA polymerase. During this process A (dATPαS is used, which is not a substrate for the enzyme luciferase), C (dCTP), G (dGTP) and T (dTTP) nucleotides are sequentially added and removed. Upon addition of the correct nucleotide solution to the newly synthesized strand, pyrophosphate (PP$_i$) is stoichiometrically released. A second enzyme in the mix, ATP sulfurylase, converts the generated PP$_i$ to ATP, which in turn is utilized by a third enzyme, luciferase, to generate oxyluciferin from luciferin. This reaction generates visible light that is detected by a camera while a computer program produces the final sequencing results. Finally, a fourth enzyme in the reaction, apyrase, degrades unincorporated nucleotides and ATP, and the reaction can restart with another nucleotide. In summary, a chemiluminescent signal is generated only when the respective nucleotide solution complements the first unpaired base on the template. The order of solutions that produce signals determines the sequence of the DNA stretch. Currently, a limitation of the method is that the lengths of individual reads of DNA are around 300–500 nucleotides, which is somewhat shorter than the 800–1000 obtainable with conventional Sanger sequencing.

Pyrosequencing was first commercialized by a Swedish company called Pyrosequencing AB (later renamed Biotage and acquired by Qiagen in 2008). Furthermore, the technology was out-licensed to the US company **454 Life Sciences**. 454, now owned by Roche Diagnostics, developed a high-throughput, array-based technology that has emerged as a platform for large-scale DNA sequencing. The latest platform can generate around 400 million sets of nucleotide data in a 12-h run on a single machine costing around US $ 5000. This means an entire mammalian genome can now be sequenced for under US $ 1 million. Further developments are expected that will push these costs down into the US $ 100000 or even US $ 10000 range.

Besides pyrosequencing, another method of high throughput sequencing has been developed. The Illumina sequencing system is a groundbreaking platform for genetic analysis and functional genomics. It dramatically improves speed and reduces costs, as it produces up to two billion reads per run, generating the industry's highest sequencing output and fastest data generation rate.

14.3
Strategies for Sequencing the Human Genome

During the mid-1980s, a proposal was made to analyze the sequence of the entire human genome. Although technology had been constantly improving, such a project called for the development of specific strategies. This led to two complementary approaches. The first one is based on detailed mapping prior to sequencing. This ensures that the sequence information obtained can be easily lo-

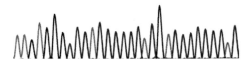

AAGAGGGCCTATTTCCCATGATTCCTTCAT.

Fig. 14.2 Automatic sequencing with capillary electrophoresis.

calized. The downside of this method is the labor intensity of the mapping process. As an alternative, randomly cloned fragments were sequenced using high-throughput machines. Special software then puts the sequences in place. This is known as the **shotgun approach**. The disadvantage of this method is the amount of sequencing involved. By combining the two approaches and involving various companies and institutions, it was possible to reveal the first almost complete sequence of the human genome, comprising 3.2 billion base pairs, to the public. Meanwhile, the genomes of other organisms, including mice (*Mus musculus*), baker's yeast (*Saccharomyces cerevisiae*), thale cress (*Arabidopsis thaliana*), and a nematode worm (*Caenorhabditis elegans*) have been sequenced (see Sections 4.1, 21).

14.4
Practical Significance of DNA

The availability of the sequences of the entire **human genome** (i.e., all human genes) does not necessarily mean that we understand their functions, but it means a great acceleration in the growth of medical knowledge. Through comparative studies of individuals, variants of a gene could be identified that are either associated with a functional change or could help in the mapping of disease (**single nucleotide polymorphisms (SNP)**). In routine diagnosis, sequence analysis has become a tool for **recognizing congenital disease** and **assessing the risk** of certain diseases, such as screening for the BRCA1 or BRCA2 genes that transmit susceptibility for breast cancer. Another important application area is the discovery and analysis of somatic mutations in certain tissue during malignant processes.

Nucleotide sequences play an important role in many fields of biological research. For example, the comparison of sequences is essential where the evolutionary history of the plant and animal kingdoms is concerned (**molecular systematics and phylogeny**) (see Fig. 6.1). DNA sequencing is also used to assess the genetic impact of environmental factors.

15

Cloning Procedures

Learning Objectives

This chapter gives an introduction to one of the main topics of molecular biology and biotechnology – cloning of genetic information. The construction of vectors for cloning and expressing foreign DNA in bacteria, yeast, viruses, and animal cells is described, as are methods of transformation or transfection of cells through vectors.

15.1
Introduction

In molecular biology, cloning means the amplification of any DNA fragment through **recombinant DNA technology**. If the DNA fragment is a well-defined, complete gene, the process is called **gene cloning**. This must under no circumstances be confused with the cloning of a whole organism, which involves producing genetically identical copies of organisms through nonsexual reproduction. Both cloning procedures are, strictly speaking, methods of replication, but they differ in their end products, which are unfortunately both called **clones**.

Human interference with the DNA of organisms has been taking place for a long time (e.g., **targeted selection** and **cross-breeding** in order to obtain higher yields in agriculture). Direct intervention through **DNA manipulation**, however, has only been possible in the last 40 years. Important milestones in molecular biology include the deciphering of the genetic code, discoveries of restriction endonucleases and ligases, antibiotic resistance plasmids, and methods of introducing and multiplying heterologous DNA in bacteria. The first cloning process was described by Cohen in 1973. It involved two plasmids that were linearized by the restriction enzyme EcoRI and then ligated to form one plasmid using ligase. While the two original plasmids each contained one antibiotic resistance gene, the newly created recombinant plasmid contained both and the colony of cloned bacteria that developed from the bacterium that had been modified by the plasmid turned out to be resistant to both antibiotics.

About 5 years later, the first **recombinant plasmid vector** was created that fulfils all of the necessary conditions for DNA fragment cloning. It consists of three segments: a **tetracycline resistance gene** that occurs naturally in the *Salmonella* plasmid pSC101, the ampicillin resistance gene of the transposon *Tn3*, and the **replication region** (*ori*), as well as neighboring sequences of the *Escherichia coli* plasmid pMB1. This vector is known as **pBR322**, named after the two researchers who first described it, Bolivar and Rodriguez (1977).

15.2
Construction of Recombinant Vectors

Cloning of a DNA fragment involves several methodical steps, such as the amplification and purification of the DNA fragment to be cloned, also called an

Fig. 15.1 Cloning, amplification, and selection of heterologous DNA in host organisms.
(1) Heterologous DNA (PCR product, restriction digest product, genomic DNA, etc.) is introduced into a suitable vector (2), which has the ability to replicate (*ori*) in the chosen host organism and carries at least one selection marker (resistance gene). This is done by ligation (cohesive or blunt-end ligation, TOPO TA, UA cloning) or by recombination (e.g., Gateway system) (3). After the transformation (4) of the host organism (e.g., *E. coli* or *S. cerevisiae*), the recombinant vector is amplified through replication. The replication of the transformed host organism in a selective medium (5) gives rise to colonies of clones from which the amplified DNA can be obtained and used to express a recombinant protein.

insert. The **vector DNA** must be linearized, and the insert and vector must be ligated. The DNA is transformed into bacteria and the transformed bacteria are selected. The recombinant bacterial DNA is then purified and the cloning process verified. For each of these steps, there is a choice of largely standardized possibilities and methods available (see overview in Fig. 15.1). A wide range of **restriction enzymes**, **vectors**, **specialized cloning bacteria**, and commercially available kits are contributing to a high success rate of the procedure. Nevertheless, it cannot be taken for granted that the cloning of a cDNA fragment in an appropriate vector will succeed at the first attempt. Problems during the cloning process usually stem from flaws in the approach.

15.2.1
Insert

As a general rule, any **double-stranded DNA** can be cloned and amplified in bacteria, whether they are **cDNAs** or **genomic DNA** fragments originating from various donor organisms. However, the **length** of the DNA fragments and the method through which the fragment has been obtained or amplified affect the cloning process in various ways. Depending on the length of the DNA fragment to be cloned, various types of vectors with different integration capacities are to be considered. The most frequently used vectors are derived from bacterial plasmids and can integrate fragments from a few base pairs up to about 10 kb of **heterologous DNA**. This is sufficient for the cDNA of most genes, but if, for example, a **cDNA library** is to be created, larger genes (more than 10 kb) must be included. This is often done with the help of **phage vectors**. Most of these are derived from the **lambda (λ) phage** (see Section 3.3). It has a genome of 49 kb, of which 40% can be replaced by foreign DNA. For the transfer of even larger DNA fragments, such as genomic DNA fragments, **cosmids** (vectors including the *cos* sites of bacteriophage lambda), **P1-derived artificial chromosomes (PACs)**, **bacterial artificial chromosomes (BACs)**, **yeast artificial chromosomes (YACs)**, and **mammalian artificial chromosomes (MACs)** are available (Table 15.1).

When cloning a DNA fragment, **restriction sites** for the restriction enzymes are chosen that are unique in the vector DNA. In modern vectors, these are found close together in a vector region known as a **multiple cloning site (MCS)** or **polylinker** (see Section 15.2.3). Restriction enzymes that cut in these sites are called rare cutters because the sequence they recognize does not occur often, from a statistical point of view. Given the limited number of rare cutter enzymes, the cloning of large inserts can be problematic. The longer the fragment to be cloned is, the more likely it is that it carries one or two of the restriction sites from the MCS. If no suitable enzyme can be found for the insertion of large genes, there is always the option of **sequential cloning** of several fragments or simultaneously ligating several fragments. This requires meticulous verification of the orientation of the individual DNA fragments in respect of each other as well as the vector (through a **restriction digest** or **sequencing**). In classical cloning of restriction fragments, the ligation of compatible ends must not be ne-

Table 15.1 Vectors, heterologous DNA uptake capacity, and host organisms.

Vector	Host organism	Uptake capacity
Plasmid	*E. coli*	Max. 10 kb
Lambda phage	*E. coli*	Max. 25 kb
Cosmids	*E. coli*	35–45 kb
P1 phages (PAC, P1-derived artificial chromosomes)	*E. coli*	100–300 kb
BAC (bacterial artificial chromosome)	*E. coli*	Max. 300 kb
YAC (yeast artificial chromosome)	*S. cerevisiae*	100–2000 kb
MAC (mammalian artificial chromosome)	Mammalian cells	Max. 500 Mb

glected. Restriction enzymes that differ in their recognition sequences may produce **compatible sticky ends**. Thus, a fragment cut with BamHI can be ligated to a fragment cut by BglII, as both enzymes produce a 5'-overhang containing the sequence GATC. Once the fragments have been ligated, often neither of the enzymes can cut the resulting sequence. This applies also to the ligation of two **blunt ends** resulting from the use of different restriction enzymes.

If it is neither possible to use an appropriate restriction site nor to ligate compatible ends, a filling or digestion reaction can be performed to produce blunt ends out of sticky ends. The resulting fragment can then be cloned into a blunt end restriction site in the vector. Table 15.2 shows a list of enzymes suitable for these reactions.

Another decisive factor in the choice of **cloning strategy** is the method by which the fragment to be cloned has been created. This applies particularly to the cloning of **polymerase chain reaction (PCR)** products. Currently, many suppliers offer quick and very efficient cloning systems for PCR fragments. Most of them build on the fact that during the amplification process with **Taq polymerase** (see Chapter 13), deoxyadenosine is attached to the 3'-end of the newly synthesized strand, no matter what the actual sequence is. Once the cohesive DNA ends have been ligated through a T4 ligase (e.g., in the UA cloning system), the PCR product is cloned into a linearized vector containing complementary deoxyuracil residues. As an alternative, the **TOPO TA Cloning* system** (Invitrogen) could be used, where the overhangs of the PCR products are attached to overhanging deoxythymidine residues of the linearized vector and then supercoiled through **topoisomerase**. Any remaining breaks are closed by bacterial ligase once the fragment has been transformed into bacteria. However, these cloning processes, known as **TA or UA cloning**, only work with PCR products synthesized with Taq polymerase or a polymerase mix containing Taq.

Proofreading polymerases with high correction rates produce **blunt DNA ends**. This makes a special form of ligation possible, known as a **cut ligation**. The vector is linearized with the help of a restriction enzyme that does not cut the insert DNA and also leaves blunt ends. The insert and the vector are then ligated with a ligase in the presence of the restriction enzyme. The restriction enzyme very efficiently prevents a religation of the vector. The drawback of ligating DNA fragments with blunt ends is the low efficiency rate and nondirectional insertion. The insert can be ligated in 5' → 3' orientation as well as in the opposite direction, which makes an orientation check indispensable.

The greatest disadvantage in all PCR-based cloning lies in the high error rate of the polymerases during the synthesis of the fragments. The risk of producing mutations is significantly lower in cloning fragments obtained from restriction digestion.

	Klenow	T4 DNA polymerase	T7 DNA polymerase	Mung bean nuclease
Filling in 5'-overhangs	👍	☞	☞	👎
Digesting 5'-overhangs	👎	👎	👎	👍
Filling in 3'-overhangs	👎	👎	👎	👎
Digesting 3'-overhangs	👎	☞	☞	☞

Table 15.2 Properties of enzymes for turning sticky into blunt DNA ends.

15.2.2
Vector

The **choice of vector** is determined by the size of the insert and by the purpose of the cloning process. The vector should meet all requirements relevant to subsequent processes (e.g., sequence analysis, DNA amplification, and expression of a recombinant protein). There is a wide range of state-of-the-art vectors commercially available, combining various functional elements. Given the wide range of available molecular building blocks, it should nearly always be possible to find the best-suited vector for any cloning application. If the problem at hand is too specific for a commercially available vector, there is always the possibility of creating one that is tailor-made. The required elements can be isolated from one vector through restriction digest and then introduced into others.

There has been a traditional distinction between various categories of vectors. **Plasmids**, for example, are vectors derived from extrachromosomal bacterial ring-shaped DNA molecules. **Phagemids** are vectors containing partly bacterial plasmid sequences and partly sequences from bacteriophages. However, depending on their function, many vectors nowadays contain sequences from a variety of origins, including lower and higher eukaryotes, various nonprokaryote-specific viruses, and tailor-made artificial sequences. This renders the traditional categories obsolete.

They have been replaced by functional categories, leading to the following **vector groups**:

- Cloning vectors.
- Shuttle vectors.
- Prokaryotic and eukaryotic expression vectors.
- Viral vectors, etc.

This categorization is fraught with its own problems, as many vectors combine several functions. Most vectors share what will be described as essential components in the following sections.

15.2.3
Essential Components of Vectors

15.2.3.1 Bacterial Origin of Replication (*ori*)

A **bacterial replication origin** is essential in a vector, as it is needed to ensure its amplification in bacteria (see Section 4.1.4). Many vectors carry an origin that is derived from the ColE1 plasmid in *E. coli*, such as **pBR322-*ori*** or **pUC-*ori***. These "high-copy vectors" ensure a high amplification rate because their replication does not depend on the bacterial chromosome. Vectors containing a p15A origin can also be replicated independently. Their copy rate is below that of ColE1-derived vectors. Vectors with a p15A origin are relatively rare, but they have their place in the cotransformation of two different plasmids into one bacterium in order to express two different proteins. If vectors share a replication origin, they are incompatible because a bacterium can only replicate one type of vector in one origin. Therefore, if two vectors are to be replicated in a bacterium, they must have two different replication origins (e.g., the ColE1 and p15A origins). There are so-called **low copy vectors** that yield a defined low number of copies (one to five) in a bacterium. Their replication origin is derived from **F plasmids** in *E. coli*. These plasmids are duplicated during the replication of the bacterial chromosome.

15.2.3.2 Antibiotic Resistance

Antibiotic resistance is used to select the **transformed bacteria**. The most frequently used resistance-conveying genes are *Amp*^r and *Kan*^r. *Amp*^r codes for the enzyme *β*-lactam amidohydrolase (*β*-lactamase), which cleaves ampicillin and penicillin into ineffective degraded components. Using ampicillin for the selection of transformed bacteria is usually an efficient method. Alternatively, the more stable antibiotic carbenicillin can be used. The gene *Kan^r codes for a phosphorylating enzyme* that inactivates kanamycin – an antibiotic far more stable at high temperatures than ampicillin. It even survives autoclaving. Note that excessive kanamycin concentrations have a strongly negative selection effect that can impair or even prevent the growth of transformed bacteria.

15.2.3.3 Polylinkers

Every vector contains a defined number of recognition sites for restriction enzymes that cut the vector sequence only once (single cutters) to enable the cloning of a DNA fragment. These sites usually lie closely together and these sections are referred to as polylinkers (or multiple cloning sites, MCSs). These regions comprise 50–100 bp on average and may contain up to 25 restriction sites for single-cutting restriction enzymes.

There are also many vectors that carry several sequence elements from bacteriophagal genomes right next to the polylinkers. These phage components can act as promoters that enable the *in vitro* transcription of the cloned genes (e.g., the SP6 or T7 promoters derived from the eponymous phages). Genes transcribed *in vitro* are used, for example, as probes for the hybridization of RNA or as starting material for *in vitro* translation (cell-free expression of proteins, see Chapter 16).

15.2.4
Cloning Using Recombination Systems

In contrast to traditional cloning based on restriction enzymes and ligases, more recent cloning systems rely on a sequence-specific recombination of DNA molecules. In the following section, the Gateway® system will be described in more detail as an example.

The Gateway system uses recombination elements of the bacteriophage lambda. Depending on the surrounding circumstance, it can be integrated (lysogenic cycle) or disintegrated (lytic cycle). These processes are not homologous, but sequence-specific recombinations. The lambda phage has specific attachment sites attP/attP' that recombine via the recombination site attB/attB' of the bacterial genome. Two enzymes act as catalysts during the process – the phage protein integrase and the bacterial integration host factor (IHF). After integration, the phage genome is present in the bacterial genome as a prophage, flanked by the newly formed recombination sites attR (attP/attB') and attL (attB/attP'). Under certain circumstances (e.g., if the survival chances of the bacterium are deteriorating) another sequence-specific recombination between attR and attL takes place, releasing the phage genome from the bacterial genome. Integrase and IHF are again needed in this process, plus an additional enzyme called excisionase, which is coded for by the phage.

In order to increase its recombination efficiency, the Gateway system has introduced some modifications that also enable directed cloning. The Gateway system comprises eight different recombination sites, each of which only reacts specifically with its corresponding site. This makes it possible to retain the orientation of a gene while it is recombined into another vector. Successful cloning using the Gateway system involves the following successive steps.

The DNA to be cloned is amplified through PCR, using primers that also carry the recombination sites attB1 (forward primer) and attB2 (reverse primer) in addition to the gene-specific sequence. The PCR product is then recombined *in vitro*, using a donor vector carrying the recombination sites attP1 and attP2. The recombination process requires a BP Clonase^TM enzyme mix, containing, amongst others, integrase and IHF. Alternatively, a restriction fragment can be cloned into an **entry vector**. Both methods result in a recombinant entry vector that is transformed into bacteria. The resulting entry clone contains the recombination sites attL1 and attL2, flanking the cloned gene. The Gateway system offers a wide choice of **target vectors**, including those that enable the **heterologous expression** of proteins in bacteria, yeasts, insect, and mammal cells. All vectors carry the recombination sites attR1 and attR2, which enables them to recombine with the entry vector. This reaction takes place *in vitro* in the presence of the LR Clonase enzyme mix, which contains integrase, IHF, and excisionase.

Two selection criteria are in place to select the desired recombinant target clone. The entry vector carries the **kanamycin resistance gene**, while the target vector carries the **ampicillin resistance gene**. The second criterion is the ability to replicate. Bacteria transformed by nonrecombinant target vectors are unable to divide because of the ***ccdB* gene**, which is inserted between the two recombination sites of the target vector. When expressed, the ccdB protein interacts with bacterial **gyrase**, thus preventing bacterial growth.

15.2.5
Further Components of Vectors for Prokaryotic Expression Systems

Presently, *E. coli* plays an almost exclusive role as the **host organism** in heterologous gene expression (see Chapter 16). *E. coli* expression vectors feature more functional units than cloning vectors. These enable and regulate the transcription of a cloned DNA fragment and its subsequent translation of the mRNA into a **recombinant protein**.

15.2.5.1 Promoter

In order to transcribe a DNA fragment in bacteria, a promoter is needed that ensures reliable and strong mRNA synthesis with **RNA polymerase** (see Chapter 4.2). The strongest promoters are found in bacteriophages, such as the **T5** or **T7 promoter** in the eponymous phages. There are also **hybrid promoters** composed from various bacterial promoters. The **ptac promoter**, for example, is a hybrid of the promoter of the *lacZ* gene, which can be induced by the addition of isopropyl-b-d-thiogalactoside (IPTG) and the promoter of the **tryptophan operon** (Fig. 4.17). The transcriptional activity of the hybrid is stronger than that of each of its parents and can be induced through IPTG. Depending on the promoter, the transcription is carried out using *E. coli* RNA polymerase (T5) or a bacteriophagal polymerase (T7) inserted into the bacterial genome. As the **constitutive expression** of a recombinant protein in bacteria can be problematic (see Chapter 16), the activity of the promoter must be closely regulated. Most promoters are under strict repression, which is mediated by a regulatory sequence in the **Lac (lactose) operon** in *E. coli*. Only by the addition of IPTG – which binds to the **Lac1 repressor**, thus reducing its affinity to the **Lac operator** – is protein synthesis initiated. In some strains of bacteria that have been specifically developed for the expression of recombinant proteins, IPTG also induces the transcription of a bacteriophage RNA polymerase, which in turn mediates very efficiently the transcription of the DNA fragment that was going to be expressed. Depending on the expression system chosen, the Lac1 repressor may also be coded into the bacterial expression vector (*in cis*). Alternatively, it can also be introduced into the bacteria via cotransformation through an **auxiliary vector** (*in trans*). Care must be taken, however, that the expression vector and the auxiliary vector coding for the Lac1 repressor carry compatible replication origins (see above). Some recently developed *E. coli* strains carry a mutation in their own Lac1 repressor gene, which initiates a sufficiently strong endogenous synthesis of the repressor in the bacterium. This makes an auxiliary vector redundant.

15.2.5.2 Ribosome-Binding Site

When it comes to the **translation of mRNA** that has been synthesized in bacteria, a specific sequence is needed that directs the ribosomes to the translation starting point. This is why the promoter and its regulatory components are followed by a **ribosome-binding site** (RBS, also known as a **Shine-Dalgarno sequence**) in front of the cloned cDNA. The RBS is often derived from viruses, which helps to enhance the efficiency of translation significantly. There are also artificial RBS sites featuring optimized sequences. If the RBS site is not followed by a start codon determined by the vector, the start codon of the gene to be expressed must follow the RBS at a distance of seven to nine nucleotides in order to ensure efficient translation of the mRNA.

15.2.5.3 Termination Sequence

Regulation of the termination of transcription is just as important as the regulation of its initiation, which is why all bacterial expression vectors carry specific sequences that enable them to form stable mRNA secondary structures after transcription. These prevent RNA polymerase from continuing the synthesizing process beyond this site. Without **transcription terminators**, a whole vector sequence would be transcribed into one long mRNA in a runaway transcription. A termination of the transcript also enhances the stability of mRNA. Some transcription terminators consist of partly viral (e.g., phage lambda) and partly bacterial termination sites, while others are derived from exclusively viral sequences (e.g., **T7 bacteriophage**).

Apart from transcription termination sites, many bacterial expression vectors also carry **translation termination sites**. Often, only fragments of genes are cloned into vectors without their sequence-specific stop codons. A short TG-rich sequence before the transcription terminator acts as a stop codon in each of the possible reading frames.

15.2.5.4 Fusion Sequence

Often, DNA is cloned into a bacterial expression vector to purify greater quantities of recombinant proteins in order to achieve homogeneity (see Chapter 16). The purification process of the recombinant protein through affinity chromatography can be facilitated by the **expression of a fusion protein** (see Chapter 7). This is why many vectors already contain sequences leading to the expression of **N- or C-terminal peptide sequences (tags)**. There are several factors that should be taken into account when developing a cloning strategy that does not depend on the size of the fusion component. Points to consider are: would N-terminal, C-terminal, or even internal positioning of the fusion component be more suitable for further applications? Care must be taken that no 5′-nontranslating gene regions are cloned into N-terminal fusion components. An **open reading frame** must be provided for the protein expression of the fusion component. If the bacterial expression vector does not carry a translation termination site, the gene-specific stop codon should still be present. By contrast, for producing a C-terminal fusion protein, the cloned gene must not contain its own stop codon, but feature a **start codon** at a distance of seven to nine nucleotides from the RBS, as mentioned above. Here, too, retaining the reading frame between the protein of interest and the fusion component is crucial.

15.2.6
Further Components of Eukaryotic Expression Vectors

It is often more advantageous to express **recombinant proteins** in eukaryotes rather than in prokaryotes (see Chapter 16). As a first step, a suitable expression system, including an appropriate vector, must be chosen. The most frequently used **eukaryotic expression systems** are yeast, cultured insect cells, and cultured mammalian cells (see Chapter 16).

15.2.6.1 Eukaryotic Expression Vectors: Yeast

Like prokaryotic expression vectors, yeast expression vectors have their idiosyncrasies. Apart from the sequences that ensure the propagation and selection of vectors in *E. coli*, yeast-specific **promoters**, **termination**, and **replication sequences** plus a **selection marker** are required.

Commercially available yeast expression vectors can contain **constitutively active or inducible promoters**. Constitutive promoters include, for example, the GAP promoter of the gene for glyceraldehyde-3-phosphate dehydrogenase. Examples for inducible promoters are: (1) the AOX1 promoter of the **alcohol oxidase gene**, which is induced by methanol and is suitable for protein expression in *Pichia pastoris*, (2) the **galactose-inducible promoters** Gal1 and Gal10 for protein expression in *Saccharomyces cerevisiae*, and (3) **thiamine-inducible promoters** nmt1, nmt42, and nmt81 for protein expression in *Schizosaccharomyces pombe*.

In eukaryotic cells, the termination of transcription is as important as in prokaryotes. In yeast expression vectors, sequences found in **auxotrophy genes** are often chosen as terminators (e.g., ura4TT).

A vector can only be persistent in its host if it has either the ability to integrate into the yeast or, if episomal, can replicate autonomously. Accordingly, **yeast vectors** are classified as **yeast-integrating plasmids (YIps)** or as **yeast episomal plasmids (YEps)**, also known as **yeast-replicating plasmids (YRps)**.

YIps integrate into the yeast genome through homologous **recombination** at a low frequency rate. By linearizing the vector, the efficiency of integration can be significantly raised by a factor of 1050. Recombination is mediated by **auxotrophy markers** coded (1) on the vector and (2) found in the genome in a mutated version. Usually, only one copy of the vector integrates into the yeast genome during the process. Alternatively, homologous recombination can take place, through repetitive sequences, such as **TY (transposon yeast)** sequences that are found throughout the yeast genome. Yeast clones produced by homologous cloning are very stable and thus suitable for the industrial production of heterologous proteins.

By contrast, YEps contain a sequence that ensures autonomous replication within the yeast. This is the 2 μm replication origin, obtained from a 6.3-kb long plasmid that is found episomally in the nuclei of most *S. cerevisiae* strains. Between 50 and 100 copies of the plasmid, known as the **2 μm circle plasmid**, are found per haploid genome. The cell replication mechanism replicates them once in every cell cycle. The 2 μm circle plasmid codes for the three genes *REP1*, *REP2*, and *REP3*, which are essential for the replication of the plasmid. YEps can contain either the entire 2 μm plasmid or just the replication origin with the *REP3* gene lying *in cis*. In this case, however, the yeast to be transformed must contain *REP1* and *REP2 in trans* to ensure the replication of the YEp vector. YEp transformants often lack stability, as between two to 10 copies are lost per replication generation. The problem can be circumvented by adding a mutated auxotrophy marker to the vectors, which is expressed far more rarely than the wild-type protein. This, therefore, requires the presence of a higher number of YEp copies in order to maintain growth in the selection medium.

The replication origins of YRps are often called **autonomously replicating sequences (ARSs)**. These are part of the yeast genome and are probably equivalent to the natural replication origins. The yeast genome features about 500 replicons of an average length of 40 kb. This is reflected in the number of ARSs that have so far been successfully cloned. They contain a short consensus region that almost exclusively consists of AT pairs. Only a few vectors with ARSs are found episomally in a yeast cell nucleus.

Often, **auxotrophy markers** such as *leu2, ura3, trp1,* or *his4* are used for the **selection of transformed yeasts**. These genes code for enzymes that catalyze partial reactions in essential metabolic pathways. They are recessive genes. The yeast to be transformed must be free of these auxotrophy markers and, in diploid strains, the mutation must only occur homozygously. After a successful transformation, positive clones can be propagated in **minimal media** and need not be fed the essential metabolite.

However, many industrially used yeast strains are polyploid and feature multiple copies of the genes in question. Thus, transformed yeasts can no longer be selected through an auxotrophy marker. They need a **dominant marker** (e.g., genes conveying resistance to cytostatic or cytotoxic substances). The **blasticidin resistance gene** is a popular choice because of its wide range of action. It specifically inhibits the formation of peptide bonds during translation, in prokaryotes such as *E. coli* as well as in eukaryotes (yeasts, insect, and mammal cells).

15.2.6.2 Eukaryotic Expression Vectors for Mammal Cells

Protein expression in mammalian cells is particularly cost- and labor-intensive (see Chapter 16). The greatest difficulty often lies in the introduction of recombinant DNA into the cells. The best method to achieve this depends on the type of cell. In some widely used cell lines (see Chapter 16), the DNA can be **transfected directly**, whereas in other cases, the heterologous DNA is introduced into the cells with the help of **recombinant viruses**.

For established cell lines, it is possible to increase **transfection efficiency** up to 100% by using commercially available transfection reagents that are tailor-made for the cell type in question. Over the last few years, various **viral expression systems** have also been developed that enable protein expression in cells that are difficult to transfect. The following criteria must be considered when choosing a suitable viral system: it is important to know whether the cells in question still have the ability to divide. A decision should be taken whether a **transient or stable expression** of the recombinant protein is intended. The **viral expression system** also depends on the species of the host (Table 15.1). Cloning genes into recombinant viral systems is often very time- and labor-intensive, and the subsequent propagation and purification of the viruses may cost a considerable amount of money.

No matter what expression system, state-of-the-art vectors offer a whole host of possibilities for the regulation of protein expression, and the modification and localization of recombinant proteins. In the following, functional components of eukaryotic expression vectors will be discussed.

Promoters in Eukaryotic Expression Vectors for Mammalian Cells   In order to express proteins in eukaryotes, a promoter must be located in front of the cloned cDNA to enable its transcription in the cellular system. Viral promoters are frequently used, as these ensure **strong constitutive expression**. The most often used promoters are the CMV promoter derived from the **cytomegalovirus** and the **SV40 promoter** of the simian virus 40. There are also nonviral promoters, such as the promoter of the **eukaryotic elongation factor EF2a**, which also ensures the strong expression of a recombinant protein. The constitutive expression of proteins, however, can become a problem where a high expression rate of some proteins may have cytotoxic effects. Expression systems for mammalian cells have been developed that can be regulated, as has been the case with bacterial expression systems.

The most widely used system is the **Tet system**. Its underlying principle is the tetracycline-dependent regulation of the **tetracycline resistance operon of *E. coli*.** In the absence of **tetracycline (Tc)**, the transcription of the operon is inhibited by the **negative regulatory protein Tet repressor (TetR)**. Transcription can only be activated through the binding of tetracycline to TetR. In the Tet system, the Tet repressor forms a fusion protein with the VP16 domain of the Herpes simplex virus. This turns the Tet repressor into a transcription activator. The hybrid protein is known as **tetracycline-controlled transactivator (tTA)** and is coded on one of the two vectors of the Tet expression system. The other vector, also known as the **response plasmid**, contains an MCS for the gene that is to be cloned and under the control of the **tetracycline-responsive elements (TREs)**. The cloned gene is transcribed and translated, as long as no Tc or **doxycycline (Dox)** is added. If one of these is added to the cell medium, no further transcription takes place. This system variant is called **Tet-off**, as the expression process is switched off by adding Tc or Dox. Through the introduction of mutations to the Tet-R/VP16AD fusion protein, it has been possible to create a **reverse tetracycline-controlled transactivator (rtTA)** that enables transcription only after the addition of Tc or Dox. Accordingly, this variant is called **Tet-on**. Various versions of the Tet system are now available, including viral expression systems. A problem that mainly concerns the culture of transgenic cells should be mentioned. Fetal calf serum, which is added to most culture media, can contain considerable amounts of tetracycline, residues from calf rearing. This may interfere with the intended repression or expression of recombinant proteins.

Termination Sequences in Eukaryotic Expression Vectors for Mammal Cells How the transcription of eukaryotic genes through **RNA polymerase II** is terminated has not been fully understood. In most genes, however, **polyadenylation** of the primary transcript seems to be a prerequisite for the formation of translatable mRNA. The process involves two steps: cleaving off the end of the transcript and attaching the poly(A) sequence. Several components are needed: a nucleolytic enzyme complex and **poly(A) polymerase**. Indispensable in the process is the polyadenylation signal AAUAAA, which, in all eukaryotic mRNAs except yeast, is found 11–30 nucleotides upstream from the polyadenylation site. Some termination sites, however, have been well characterized. One of them is the **SV40 termination site**. The sequence is comparable to the Rho-independent bacterial termination site where after a hairpin-forming sequence, a series of U bases occurs. In eukaryotic expression vectors, termination sites such as SV40 are always found after the MCS.

Sequences for the Replication of Eukaryotic Expression Vectors in Mammal Cells Expression vectors for the synthesis of heterologous proteins in mammalian cells do not normally carry replication sequences. In order to ensure their persistence, the vectors must **integrate** into the genome, which usually only happens infrequently and at random. There are, however, some exceptions. Some vectors carry the **SV40 replication origin**. Although the vectors are only episomally present after transfection, they are replicated in certain **cell lines**, including the cell lines COS1 and COS7 (CV1 transformed with an origin defective mutant of SV40). These express the large T-antigen of SV40, thus ensuring that the replication process starts from the SV40 origin.

Genes for the Selection of Stably Transfected Cell Clones Vectors that ensure the **heterologous expression** of genes in mammalian cells often contain – alongside the already mentioned antibiotic resistance genes for bacteria – resistance genes against certain cytostatic or cytotoxic substances. These enable the selection of **stably transfected cell clones** (Table 15.3). The selection genes are flanked by their own promoter and termination sequence in order to ensure correct transcription and translation. The exceptions are vectors carrying an **internal ribosome entry site (IRES)** – a 600-bp long sequence that has been isolated from the genome of the encephalomyocarditis virus (EMCV). It enables the translation of an mRNA, independent of the 5′-cap. **IRES vectors** carry one single promoter, which is followed by the MCS for the cloning of the desired DNA fragment. Next is the IRES sequence, followed by the resistance gene, and finally

Cytostatic	Effect	Concentration
G418 Geneticin	Blocks polypeptide synthesis, prevents chain elongation during translation	100–800 µg mL^{-1}
Bleomycin	Forms DNA complexes, causes strand-breakage	10–100 µg mL^{-1}
Hygromycin B	Blocks polypeptide synthesis, prevents chain elongation during translation	25–1000 µg mL^{-1}
Puromycin	Inhibits protein synthesis	10–100 µg mL^{-1}

Table 15.3 Commonly used cytostatic or cytotoxic selection markers.

Tag	Sequence	Localization	Maximum repeat
C-myc	EQKLISEEDL	N/C/internal	2×
Flag	DYKDHD	N/C	3×
HA	YPYDVPDYA	N/C	3×

N = N-terminal, C = C-terminal

Table 15.4 Commonly used antigenic fusion components (tags).

the termination site. The whole construct is read as a single bicistronic mRNA. During the translation process, the ribosomes bind (1) to the start codon of the cloned DNA and (2) to the internal ribosome entry site. The result is two different proteins, translated from a single mRNA.

Fusion Sequences in Eukaryotic Expression Vectors for Mammalian Cells The purpose of the expression of **heterologous genes** in mammalian cells is usually not to obtain large amounts of purified protein, but to assess their functionality. Such studies include research into the intracellular localization of a protein, its interaction with other proteins, and the regulation of enzymatic activity. For these purposes, the specific identification of a recombinant protein through immunological methods is usually indispensable. However, **specific antibodies** for every protein are not always commercially available, and the custom synthesis of such antibodies may prove too time-consuming and expensive. This is why many **expression vectors** offer the possibility of expressing **tagged proteins**. In contrast to the fusion components of prokaryotic expression vectors, which are very cost-effective in affinity purification, for the short peptide tags in many eukaryotic expression vectors, the most important criterion is their antigenicity. The most **frequently used tags** are the c-Myc tag, the hemagglutinin (HA) tag, and the FLAG tag. Their most important properties are listed in Table 15.4. The exceptions are those fusion components that are known as **living-color proteins**. When these proteins are stimulated with shortwave light, they emit a lower energy light that can be visualized by using specific filters for defined wavelengths. The best known example is the **Green Fluorescent Protein (GFP)** derived from the jellyfish *Aequorea victoria*. This protein has a length of 238 amino acids and a molecular mass of about 30 kDa – a heavy weight among the tags used in eukaryotic expression vectors. While the main advantage of using them lies in the easy detection of the fusion proteins within the cell through **fluorescence microscopy** (see Chapter 19), their large size may interfere with the localization, interaction, and function of the protein.

15.2.6.3 Viral Expression Systems for Mammalian Cells

Various viral vector systems (Table 15.5) offer a **viable alternative** for the transfection of mammal cells, particularly for those cell types that are difficult to transfect.

Table 15.5 Viral expression systems for mammal cells.

Virus	Advantages	Disadvantages	Commercially available systems
Adenoviruses	• High infection rate in various cell types. Particularly suitable for nondividing cells • Strength of expression is controllable via the virus-cell ratio • Ability to code for additional marker proteins (e.g., EGFP = Enhanced Green fluorescent protein)	• Cloning is labor-intensive • Amplification and purification is cost-intensive • Genes can only be cloned up to a certain size (about 7–9 kb) • Subject to regulations of safety level 2	• AdenoX • ADEasy system
Retroviruses	• Cloning straight-forward • Easy generation of stable cell clones	• Nondividing cells cannot be infected • Depending on their tropicity, they are subject to biological safety regulations level 1 or 2	ViraPort
Lentiviruses	• Cloning straightforward • Stable integration to the genome • Infection of both dividing and non-dividing cell types • Wide range of hosts	• Subject to safety level 2 regulations	ViraPower Lentiviral expression system
Semliki-Forest/ Sindbis viruses	• Wide range of hosts • High expression of recombinant DNA or protein	• Cotransfection of vector and auxiliary vector expressed *in vitro* are subject to biological safety regulations level 2 • Uptake capacity of the expression vector limited	

Adenoviral Expression Systems   Recombinant adenoviral systems are derived from the **Ad5 virus**. Independent of their ability to divide, they have the **ability to infect** many mammalian cells. Wild-type adenoviruses contain a double-stranded linear DNA genome of 32–36 kb length. In **recombinant adenoviruses**, their genome is deleted at least in the E1 gene, not only to make space for recombinant DNA, but also in order to produce viruses that cannot replicate. Normally, the gene to be expressed is cloned into a **shuttle vector** that has to be recombined with the deleted adenoviral genome in *E. coli*. This recombinant adenoviral genome is then linearized and transfected into a **packaging cell line** (e.g., HEK-293) that codes for the deleted regions of the adenoviral genome *in trans*. The packaging cell line is thus able to produce adenoviruses that are unable to replicate. One of the main advantages of using adenoviruses is that the expression level of the heterologous protein can be regulated. Vectors carrying inducible promoters for the expression of genes are commercially available and since several viruses can be introduced into a cell simultaneously, the ratio of viruses to a cell also determines the level of expression. The ability of cells to take up several viruses at a time makes it also possible to infect a cell with a variety of recombinant adenoviruses in order to express several proteins simultaneously.

The adenoviral genome is episomally present, which is a major drawback in proliferating cells, as information for the heterologous expression of the gene gets lost during the cell cycle.

Retroviral Expression Systems **Retroviruses** are RNA viruses that replicate via a DNA intermediate (provirus) with the ability to **integrate stably into the genome** of the infected cell. The genome of replication-competent retroviruses consists of two identical single-stranded RNA molecules, 710 kb long. Recombinant retrovirus are mostly derived from murine variants, such as the **Moloney murine leukemia virus (M-MuLV)**. The range of hosts they are able to infect depends on the envelope protein expressed and includes several categories. The most frequently used retroviruses are **ecotropic retroviruses**, which can only infect cells of mice and other rodents, and **amphotropic retroviruses** that have very large range of potential hosts, including human cells.

The **retroviral expression system** consists of two components – the **retroviral vector** and a **packaging cell line**. Apart from the essential components of vectors amplified and selected in *E. coli*, retroviral vectors carry an MCS for the heterologous gene, the retroviral packaging signal Y, are flanked by retroviral **long terminal repeats (LTRs)**. The packaging cell line provides retroviral proteins. Once it has been transfected with the recombinant retroviral vector, it forms replication-deficient virions that can infect various cell types, depending on the host range. The ease with which retroviruses integrate into the genome of the infected cells helps the formation of stable cell clones. Recombinant retroviruses derived from M-MuLV, however, can only infect cells with an ability to divide.

There is one exception in the retrovirus family – these are the lentiviruses. **Lentiviruses** are most often based on the **human immunodeficiency virus (HIV)-1**. To ensure a safety handling of this viral system, the necessary genetic information has been split up and is given by normally three distinct vectors. One vector is used to clone the gene of interest and is the only one holding the packaging signal, the second vector encodes for the essential lentiviral proteins, and the third encodes for the envelope protein. This envelope protein determines the host specificity of the replication-deficient viruses and is in the most case the **glycoprotein G from vesicular stomatitis virus (VSV-G)** as this protein allows the infection of many different cell types from a variety of species.

15.2.7
Nonviral Introduction of Heterologous DNA to Host Organisms (Transformation, Transfection)

15.2.7.1 **Transformation of Prokaryotes**
Independent of the cloning system used, recombinant vectors must normally be transformed into bacteria to be amplified. Several approaches can be used to achieve this. The most frequently used methods include **chemical transformation** with or without incubation at a raised temperature (42 °C), known as **heat shock**, and transformation through **electroporation**.

Before a transformation reaction can take place, **competent bacteria** must be produced. They come from bacterial cultures that are harvested during their logarithmic growth phase and then washed with an ice-cold water/glycerol mix (20%). These washed bacteria can be used immediately for electroporation. For other transformation methods, specific reagents (see below) must be added. Competent bacteria can be stored for later use at −80 °C without losing their competence.

Electroporation **Electroporation** is the most efficient method of transforming bacteria. A strong electrical impulse (2.5 kV, 25 µF, 200 Ω, about 5 ms) renders bacterial cell walls transiently permeable. It has an efficiency of 10^7–10^{10} colonies/µg DNA, which exceeds the efficiency of chemical transformation by a factor of 10–100. The method, however, has its drawbacks. An electroporator with suitable cuvettes must be available and salts used in vector preparation may interfere with the electroporation process. High salt content can be expected in ligation reactions, for example. Before transformation, a ligation reaction can be purified through phenol/chloroform extraction, alcohol precipitation, or by the use of a commercial purification kit, at the price of losing DNA.

Chemical Transformation   Depending on the method used, a transformation efficiency of 10^6–10^8/µg DNA can be achieved in **chemical transformation**. Buffers containing $CaCl_2$ and TSS (transformation and storage solution) are the most popular choice. Preincubation of bacteria in $CaCl_2$ damages the bacterial walls, thus facilitating the uptake of heterologous DNA during heat shock treatment. Transformation with TSS is based on a similar principle. TSS contains the reagent **dimethyl sulfoxide**, which damages the bacterial cell walls. A heat shock is not needed for the transformation with TSS.

15.2.7.2   Transformation of Yeast Cells

There are several widely used methods for the transformation of yeasts. These include electroporation, the fairly labor-intensive preparation of **spheroblasts**, and **lithium acetate-mediated transformation**, to name the most popular. For the latter, competent yeast cells are obtained by washing them in lithium acetate solution. The vector DNA is mixed with a surplus of carrier DNA (e.g., herring sperm DNA), and added to the cells with a mixture of polyethylene glycol and lithium acetate. Through addition of **dimethyl sulfoxide** and heat treatment at 42 °C (**heat shock**), the polyglycan shell and the plasma membrane of the yeast become permeable for the heterologous DNA.

15.2.7.3   Transfection of Mammal Cells

The introduction of heterologous DNA to mammalian cells is called transfection. In contrast to the transformation of bacteria, the DNA is not usually introduced as **naked DNA**, but actively taken up as precipitates, complexes with polymers, or packaged in lipid vesicles.

Calcium Phosphate-Mediated Transfection   Calcium ions bind to the phosphate groups of the backbone of the DNA helix, thus forming insoluble complexes (precipitates). When these are added to the cells, they are actively taken up through **endocytosis**. The advantage of the method is that it can be applied to nearly all kinds of cells, although its efficiency varies considerably, depending on the type of cells involved.

Liposomal Transfection   Optimized liposomal transfection reagents are available for commonly used cell lines. Depending on their charge, liposomes are classified as **cationic or anionic**. Due to their difference in charge, cationic liposomes form a stable complex with DNA. In anionic liposomes, the DNA is enclosed in the vesicles. Liposomes are also taken up by endocytosis.

Electroporation   There are two major points that distinguish electroporation from the biological transfection methods described so far: (1) the DNA that is transfected is not packaged or bound into complexes and (2) it is not actively taken up through physiological cell processes, but introduced through a physical impulse.

The electroporation of mammalian cells follows principles similar to those in prokaryotic electroporation. In a first step, adherent cells are suspended and incubated in a physiological phosphate buffer containing the heterologous DNA. A short **electric impulse** opens the cell membranes to let in the DNA. This method permits the **transfection of a vast range of cells** and often has a higher transfection efficiency than biological methods, as long as the experimental conditions are redefined for each cell type.

16
Expression of Recombinant Proteins

Learning Objectives

The possibility of expressing genes in various systems (bacteria, yeasts, animal, or plant cells) to produce recombinant proteins is particularly interesting for biotechnological research and industry. This chapter describes the most important of these systems that have been successful and tested.

16.1
Introduction

The completely sequenced human genome provides new challenges for scientific and medical research. A huge number of genomic sequences are currently being analyzed with the help of **bioinformatics** (see Chapter 24) in order to make predictions about the expression of proteins. A commonly used method for obtaining data about the function and structure of unknown proteins is to express the target gene – to make a **recombinant protein**. In many cases, the protein must be subsequently isolated to obtain it in a highly purified and concentrated form that is biologically active. Once the protein has been successfully enriched, further procedures such as crystallization, X-ray analysis, nuclear magnetic resonance (NMR), and proteinprotein interaction studies (see Chapter 23) can take place. The production of pure proteins (e.g., **monoclonal antibodies**) and their derivatives for pharmaceutical use has dramatically increased over recent years. As it can be expected that there will be an increasing demand for therapeutically effective proteins, the need for the development of new simple and cost-effective expression and purification systems is evident. Currently, the period needed for the development of therapeutically effective proteins – from preclinical experiments to the finished product – is 7–12 years. The financial demands until such a product is marketable are very high, compared to low-molecular-weight active compounds. However, with a potential turnover of over US$ 1 billion per year, this is an investment in future markets.

As shown in Chapter 7, enrichment of proteins expressed in organisms, tissues, and cells at their normal level is very difficult and labor-intensive. Two methods have proved very helpful: (1) **heterologous expression** of the target protein in a host organism with the help of a special expression system and (2) cell-free *in vitro* **translation**, using cellular lysates (e.g., reticulocyte lysates or *Escherichia coli* lysates). Among the many expression systems, most of which are commercially available, the most suitable has to be chosen. Apart from cost and labor intensity of the project, known or presumed properties of the protein must be taken into account when choosing an appropriate expression system (Fig. 16.1).

An Introduction to Molecular Biotechnology, 2nd Edition.
Edited by Michael Wink
Copyright © 2011 WILEY-VCH Verlag GmbH & Co. KGaA, Weinheim
ISBN: 978-3-527-32637-2

Fig. 16.1 Which organism for recombinant protein expression?

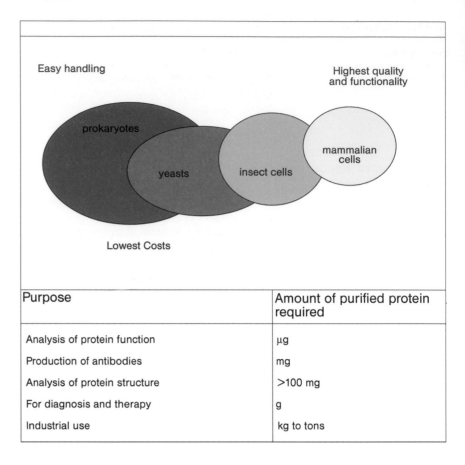

Purpose	Amount of purified protein required
Analysis of protein function	µg
Production of antibodies	mg
Analysis of protein structure	>100 mg
For diagnosis and therapy	g
Industrial use	kg to tons

16.2
Expression of Recombinant Proteins in Host Organisms

The most popular **expression systems are bacterial systems**, as they are cost- and time-effective, compared to other expression systems, and the yield in recombinant proteins is high. Overexpression in *E. coli* is perhaps the best known example. Commercially available *E. coli* strains have been optimized to cover all aspects of protein expression. There is also a wide variety of *E. coli* expression vectors with differentially regulated promotors to choose from (see Chapter 15). Other bacteria (e.g., *Staphylococcus, Bacillus, Caulobacter, Pseudomonas,* or *Streptomyces*) are also used for the expression of recombinant proteins. The main drawback of bacterial expression systems, however, lies in the fact that recombinant proteins cannot be **posttranslationally modified** and such modification is often needed in eukaryotic proteins. There are alternatives, such as various yeast expression systems (e.g., *Saccharomyces cerevisiae* or *Pichia pastoris*), which allow some of the modifications needed. Other processes, such as glycosylation, cannot be carried out correctly in yeast expression systems, so some proteins can only be enriched in **insect** (e.g., Sf9 cells) or **mammalian cells** (e.g., CHO cells). These cell culturing procedures are more difficult to carry out, and are cost- and labor-intensive. An overview of the most frequently used expression systems, including their pros and cons, is given in Table 16.1.

If the DNA sequence of the protein to be analyzed is known, a peptide sequence of a defined size with known properties can be introduced. This is known as a tag and is often used for the purification of the recombinant proteins by **affinity chromatography** (see Chapter 7). Many tags have been developed in recent years. They can be generally classified as **short tags** up to a length of 15 amino acids and longer tags. **Long tags**, such as glutathione-S-transferase, maltose-binding protein (MBP), chitin-binding domain, or calmodu-

Table 16.1 Comparison of prokaryotic and eukaryotic host organisms for the expression of recombinant proteins.

Host organism	Advantages	Disadvantages
Bacteria, e.g., *E. coli*	Many references, wide experience available	No modification possible after translation
	Wide choice of cloning vectors Protein expression easily controllable	Biological activity and immunogenicity can deviate from original protein
	Easy to cultivate, high yields (protein product can make up half the entire protein content) System can be modified that the protein is secreted into growth medium	High content of endotoxins in gram-negative bacteria
Bacteria, e.g., *Staphylococcus aureus*	Secretes fusion proteins into growth medium	Yield not as high as in *E. coli* Pathogenicity
Yeasts	No detectable endotoxins	Protein expression more difficult to control than in bacteria
	Generally considered biologically safe (GRAS)	
	Fermentation comparably cheap	Glycosylation not equivalent to glycosylation in mammalian cells
	Glycosylation and disulphide bonds possible	
	Only 0.5% of endogenous proteins are secreted, which makes isolation of secreted products easy	
	Well-established methods of mass production and down-stream processing	
Cultured insect cells with baculoviral vector	Posttranslational modifications similar to those in mammalian cells	The glycosylation mechanism has not been sufficiently researched
	Biologically safe, as only some arthropods are suitable host organisms for baculoviruses	Recombinant proteins are not always fully functional
	Products generated with baculoviral vectors have FDA approval for clinical studies	There are minor differences in function and antigenicity between recombinant and wild type proteins
	The virus stops the protein production of host cells	
	High yield of recombinant proteins	
Mammalian cells	Biologic activity similar to original proteins	Cells are difficult and expensive to grow
	Wide range of mammalian expression vectors available	Cell growth is slow
	Can be cultured in large quantities	Manipulated cells may be genetically unstable
		Small yields compared to microorganisms
Fungi, e.g., *Aspergillus sp.*	Well-established fermentation methods for molds	Expression rates achieved so far not very high
	Cost-effective culturing	Genetic characterization still insufficient
	Aspergillus niger is considered biologically safe (GRAS)	Cloning vectors not available
	Can secrete product into culture medium in large quantities, many industrially produced enzymes have been obtained from molds	
Plants	Easy to produce in large quantities	Low transformation efficiency
		Slow production rates

lin-binding peptide, can be a problem because they might interact with the protein to be analyzed, thus interfering with its functionality. They may also interface with protein crystallization and are strongly immunogenic. For these reasons, commercially available expression vectors often introduce a cleavage site for endoproteases into the peptide chain, separating the tag from the protein to be analyzed. After purification, the endoprotease can be used to cut off the tag, which is a tedious additional step, and there is also a risk that the enzyme may also cut within the actual protein sequence. This often leads to losses in protein yield.

Small tags, by contrast, are less immunogenic when used in recombinant proteins in an organism and need not be cut out using a protease before they can used, for example, for the production of specific antibodies. The oldest known tag is the His tag which consists of a series of six to 12 histidines. Fusion proteins containing this tag can be purified through **immobilized metal ion affinity chromatography (IMAC)** (see Chapter 7). Strep tag II (WSHPQFEK) can be used as an alternative to the His tag where bioactive proteins are to be enriched under physiological conditions. Fusion proteins carrying Strep tag II bind to the biotin-binding pocket of a modified streptavidin (Strep-Tactin*) with an affinity constant of 10^{-6} mol L^{-1}. An overview of pros and cons of the most frequently used tags (GST and His$_6$ tags), which can be used in nearly all expression systems, is given in Table 16.2.

Table 16.2 Comparison of the characteristics of the two most common modifications to the expression and purification of recombinant proteins from cellular systems.

GST-tag	His$_6$-tag
Can be used in any expression system	Can be used in any expression system
Purification results in high yields	Purification results in high yields
Wide range of purification products is available for any scale	Wide range of purification products is available for any scale
GST-tags are easily detectable through enzymatic or immunological systems	A small tag does not always need to be removed; if the immunogenicity of the gene is negligible, the fusion partner can be used as an antigen in antibody production
Easy purification, gentle elution, thus reducing the risk of damaging the functional or antigenic properties of the target protein	Specific proteases enable the removal of the tag if necessary
The GST-tag may help to stabilize the folding of the recombinant protein	It is preferable to use enterokinase sites which enable the excision of a tag without leaving amino acids behind
Improves solubility of hydrophobic proteins	His$_6$-tags are easily immunochemically detectable Easy purification, but elution is not as gentle as with the use of GST fusion proteins; if needed, purification can be combined with denaturation
Fusion proteins form dimers	High concentrations of imidazol can lead to precipitation; it may be necessary to remove imidazol through dialysis
	A His$_6$-hydrofolate reductase tag stabilizes smaller peptides during expression
	A small tag does not interact much with the structure and function of the fusion partner Mass determination through mass spectrometry for His$_6$-fusion proteins is not always reliable

16.2.1
Expression in *E. coli*

There is a whole range of expression vectors and *E. coli* strains commercially available for the expression of foreign proteins. Most vectors contain the following elements (their function has been discussed in detail in Chapter 15):

1. A **regulatory promoter**, such as the T7 polymerase promoter. An ideal promoter for the expression of a recombinant protein in bacterial systems enables a high synthesis rate while retaining its good regulatory properties. These are needed in order to keep metabolic stress for the cultured organisms at the lowest possible level and to minimize the often toxic effects of the overexpressed proteins. Another aspect to be considered when choosing an expression system is that it should not be too complex and easily inducible as well as cost-effective. The most frequently used promoters include classical examples like the tryptophan repressor (*trp*), lactose repressor (*lac*), or lambda CI repressor (PL).
2. A synthetic **ribosome-binding site (RBS)** for the initiation of the translation process.
3. A **multiple cloning site (MCS)** for the introduction of the cDNA encoding the target protein. Many vectors also contain DNA sequences encoding N- or C-terminal tags in the correct reading frame.
4. **Translation stop codons** in all three reading frames.
5. A gene for a selectable marker, mostly conveying **antibiotic resistance**, and the **replication origin** (*ori*) that are needed for the selection of transformed bacteria as well as for the replication of the plasmid.

At least as important as the choice of the expression vector is the choice of the *E. coli* strain in which the recombinant protein is to be expressed. As many expression vectors use the **T7 promoter**, an *E. coli* strain must be chosen that uses the T7 RNA polymerase for transcription (e.g., BL21). Table 16.3 gives an overview of frequent problems and suggestions for their solution. Once a strain has been chosen, it is then transformed using an **appropriate vector** (see Chapter 15) and selected on the grounds of its newly acquired antibiotic resistance. In a next step, the expression conditions must be optimized. Usually, the bacteria are cultured in the presence of the **selection marker** until the suspension culture

Table 16.3 Overview of problems for protein expression in *E. coli*.

Symptoms	Possible causes	Solution
No protein, truncated protein	Limited availability of tRNA for certain codons in *E. coli*	*E. coli* strain able to express rare tRNAs
Insoluble protein	Reduction of disulphide bonds	Minimize reduction in cytoplasm by using an *E. coli* strain with mutated thioredoxin reductase and glutathione reductase
	Expression too high	Reduce expression (reduce inductor)
No activity	Misfolded protein	Minimize reduction in cytoplasm (see above) Reduce expression (see above)
Cell death	Toxic protein	Stricter control of basic expression, e.g., using a strain of *E. coli* expressing T7 lysozyme
No colonies	High expression without inductor	Stricter control of basic expression (see above)

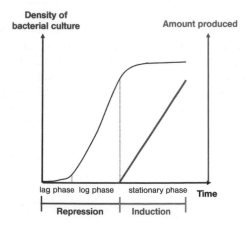

Density of bacterial culture

Amount produced

lag phase log phase stationary phase **Time**

Repression **Induction**

Fig. 16.2 Growth and protein induction in an *E. coli* culture using an inducible promoter. These enable the expression of recombinant proteins in *E. coli* with the lowest possible impact on the growth of bacteria. During the first growth period (lag phase) and the exponential proliferation period (log phase) of the bacteria, the promoter regulating the expression of the recombinant protein is kept under the control of a repressor. The expression of the protein is induced by adding an inductor that releases the activity of the promoter. If the inductor is added just before the stationary phase, protein production increases constantly while the density of the bacterial culture remains practically unchanged.

reaches the upper part of the growth phase curve shown in Fig. 16.2. This is usually defined by the optical density (OD) of the culture at 600 nm. When the extinction rate reaches a figure of 0.7–0.8, an inductor is added. As shown in Fig. 16.2, this leads to the production of the recombinant protein during the stationary phase. No general statements about the duration of the induction phase and the optimum concentration of the inductor can be made. It is recommended that before the expression of a protein is undertaken on a large scale, the production process should be tested and optimized on a small scale for every construct. This applies not only to quantitative production, but also to solubility. It often happens with large amounts of recombinant proteins in *E. coli* that they have not been folded correctly, forming what is known as **inclusion bodies**. Such proteins cannot function. If the quantity of soluble protein is insufficient, inclusion bodies may well hold the key to obtain a functional protein after all. Straightforward centrifugation is all it needs to enrich the inclusion bodies, and the proteins they contain are solubilized under **denaturing conditions** (6–8 mol L^{-1} urea or 3–4 mol L^{-1} guanidium hydrochloride). Proteins carrying a His tag can also be purified under denaturing conditions (see Chapter 7). Washing out the denaturing agent in the column or dialysis may induce a large proportion of proteins to fold correctly and be available for further analysis.

An alternative is the use of fusion proteins with tags that are known to increase the solubility of the recombinant proteins in *E. coli*. In addition to the classical GST tag, MBP fusion proteins have become more and more popular. They show an increased solubility in *E. coli* and additionally offer the possibility to be purified by matrices with covalently linked α-amylose. Modern construct also contain a cleavage site for a protease of the **tobacco etch virus (TEV)**. Apart from *in vitro* cleavage, this protease can be coexpressed in *E. coli* and thus allows the cleavage of the recombinant protein already in the living cell. This can offer advantages for the biological activity of certain proteins.

16.2.2
Expression in Yeasts

Yeasts (*S. cerevisiae*, *Pichia pastoris*, and *Schizosaccharomyces pombe*) provide an interesting alternative to other expression systems. As eukaryotic microorganisms, they combine two important characteristics – they possess an **eukaryotic protein secretion mechanism** and are able to perform posttranslational modifications, such as proteolytic processing, *N*- and *O*-glycosylation, formation of disulfide bridges, etc., and they also exhibit the fast growth rate typical of microorganisms when cultured in an inexpensive medium. At a dry weight of up to 120 g L^{-1} culture medium, the cell density in *P. pastoris* is extremely high and its wide commercial availability has certainly helped its breakthrough as a very efficient system for the expression of large amounts of recombinant proteins, which can be purified at high speed and low cost. It is suitable for a variety of applications, ranging from radioactive isotope *in vivo* labeling of proteins for NMR analyses, to the fast and cost-effective production of purified proteins for crystallization, and to **industrial-scale production** of recombinant proteins for commercial and pharmaceutical purposes.

The cDNA of the protein to be expressed in the yeast is cloned into **commercially available vectors** (see Chapter 15), which mostly work on a **shuttle principle**. This means that they carry sequences for the replication (*ori*) and selection (antibiotic resistance) in *E. coli* as well as sequences needed for the expression of proteins in yeasts (yeast-specific promoter, termination sequences, etc.). In older systems, the transformed yeasts express so-called **auxotrophy markers** encoded by the shuttle vector (see Chapter 15); they convey the ability to grow on minimal media. More recent systems use **antibiotic resistances** that permit a selection in *E. coli* as well as in yeast. There are two groups of expression systems – those in which the vector DNA is retained **episomally** in the yeast and those

Yeast	Promoter	Inductor	Selection	Fusion parts	Other
S. cerevisiae	GAL1	Galactose	URA3, blasticidin	His$_6$-tag, V5-tag, a factor (secretion)	Episomal (high and low copy)
P. pastoris	GAP (const) AOX1	Methanol	HIS4, blasticidin	His$_6$-tag, c-myc-tag, a factor (secretion)	Inserted into genome
S. pombe	Nmt1 Nmt41 Nmt81	Thiamine	LEU2	His$_6$-tag, V5-tag	Low, medium or strong expression, autosomal replication

Table 16.4 Typical properties of some yeast expression systems.

in which an expression cassette is inserted into the genome of the host cell. Apart from **constitutively active promoters** such as the GAP promoter, there is an increasing use of **inducible promoters** (Table 16.4), which are specific to the selected yeast system. Vectors for the production of suitable **fusion proteins** (GST tag, His tag, etc.) are available for all existing yeast systems. There is a wide variety of genetically defined *S. cerevisiae* strains available, which have all been classified as **generally recognized as safe (GRAS)**. However, a major drawback of protein expression in *S. cerevisiae* lies in the **hyperglycosylation** of the products, which is a fairly frequent complication. If a glycosylation of the target proteins similar (identical) to mammalian glycosylation is necessary to exhibit their biological activity (e.g., in erythropoietin), *S. cerevisiae* is an unsuitable host organism.

The tendency towards hyperglycosylation is less pronounced in *P. pastoris* than in *S. cerevisiae*, but the glycosylation patterns are not identical to those of mammalian cells, and a precise characterization of the **pharmacokinetic properties** and **potential immunogenic reactions** of the foreign protein is indispensable, at least for pharmaceutical applications. For the expression of foreign proteins in *P. pastoris*, an expression cassette, which is partly homologous to the DNA sequences of the yeast genome, is integrated into the yeast chromosome. This makes recombinant *P. pastoris* clones genetically very stable. *P. pastoris* is the most suitable of all yeasts for the expression of proteins on a large scale. The production of proteins is controlled effectively by a promoter that regulates the expression of **alcohol oxidase 1 (AOX1)** in wild-type yeast cells. It can be stringently induced by **methanol**. The foreign protein can either be produced intracellularly or it can be secreted into the medium. While intracellular protein productions often result in higher yields, the secretion of the protein into the medium makes subsequent purification steps easier, especially as *P. pastoris*, like most yeasts, only secretes a small number of **endogenous proteins**. The secretion of foreign proteins is often induced by inserted **signal peptides**, such as the factor from *S. cerevisiae*. Secreted recombinant proteins are often exposed to proteolytic degradation, however. This is a problem that occurs mainly during the fermentation of *P. pastoris*. Adding metabolically accessible sources of amino acids, such as total **casein hydrolysate** or **tryptone**, to the culture medium, modifying the pH level between 3.0 and 7.0, or using protease-deficient yeast strains can all help to diminish the risk.

The properties and modifications of recombinant proteins that have been expressed in *S. pombe* come closest to those of native proteins in higher eukaryotes. This is why this yeast type is a popular and reliable model organism for the expression and functional characterization of hitherto unknown proteins obtained from mammals. **Thiamine-induced promoters** for low, medium, and high expression levels of recombinant proteins in *S. pombe* have been developed (Table 16.4).

(A)

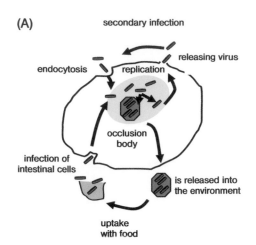

(B)

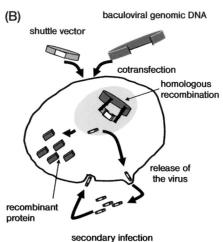

Fig. 16.3 Life cycle of wild-type and recombinant baculoviruses. (A) After an infection with wild-type baculoviruses, two types of viral population form within the infected cells. One population is made up of the viruses that are released when expelled from the cell and invading neighboring cells through secondary infection. The other population is to be found in the polyhedrin occlusion bodies. They are released into the environment, sometimes only with the death of the host. The occlusion bodies protect the virus from dehydration and other damaging environmental effects. If a host caterpillar takes up the occlusion bodies with its food, the viruses are released into the intestine during the digestion process and thus infect the host. (B) *In vitro* production and amplification of recombinant baculoviruses. After the cotransfection of baculoviral genomic DNA in a shuttle vector containing cDNA for a recombinant protein, a recombinant virus is produced through homologous recombination. The virus is then replicated in insect cells, while large amounts of the recombinant protein are produced. After the recombinant viruses have left the cell they infect others in the same culture. The number of viruses in the culture medium increases considerably. After the lysis of all cells grown in the culture, the supernatant is used for the infection of a new cell culture.

16.2.3
Expression in Insect Cells

The major advantages of protein expression in an **insect cell system** lie in the easy handling of the host cells and the relatively low cultivation costs. The resulting proteins are correctly folded, many modifications take place in the same way as in mammalian cells (formation of disulfide bridges, acylation, prenylation, etc.), and the synthesized proteins can form quaternary structures. In principle, two expression systems are available – systems based on the infection of the cultivated cells with **recombinant baculoviruses**, and systems that are, in analogy to mammalian expression systems, based on the **transfection** and **selection** of a stably transfected cell clone.

16.2.3.1 **Expression Based on Recombinant Baculoviruses**
The commercially available systems for the production of recombinant baculoviruses use viruses that derive from *Autographa californica* **nuclear polyhedrosis virus (AcNPV)**. All systems include a **shuttle vector** that can be replicated and selected in *E. coli* in the same way as yeast vectors. The vector also contains a promoter for viral proteins that are not essential for the replication for the viruses, but are produced in large numbers after the infection of the cells. The most commonly used promoters are the **polyhedrin (P_{PH})** and the **p10 promoters**, behind which the cDNA for the recombinant protein is cloned. Some vectors carry both promoters and can therefore be used for the expression of protein complexes consisting of two subunits. There are also vectors available that permit the generation of **fusion proteins** (GST, His, MBP, etc.). The shuttle vectors contain additional sequences that enable the insertion of heterologous cDNA into the genome of the virus through **homologous recombination**. The shuttle vector containing commercially available genomic baculovirus DNA is introduced into cultured ovary cells of *Spodoptera frugiperda*. This is done in a **liposome-mediated transfection** process. Several different cell lines of *S. frugiperda* are available, such as Sf9, Sf21, or Sf158H. These cells then produce **baculoviruses**. If a shuttle vector carrying a polyhedrin promoter is used, the cells producing **recombinant viruses** can be identified. Although they have been proven to be infected, they do not produce **polyhedrin-rich occlusion bodies** (Fig. 16.3). However, identifying these recombinant viruses and purifying them in a plaque assay is a very difficult lab procedure. As a first improvement, systems have been developed in which the gene for bacterial galactosidase (*β-gal*) was inserted into the bacuviral DNA. After a successful homologous recombination, the *β-gal* gene is replaced by fragments from the shuttle vector. Positive recombinant clones can be identified in a plaque assay because they do not exhibit the blue X-Gal staining. With this system, however, there remains a risk that wild-type viruses are isolated alongside the clones and in a subsequent amplification process these would have a significant replication advantage over the recombinant viruses. Safer systems are now available. One system permits the homologous recombination of baculoviral DNA with subsequent blue-white screening in bacteria (Bac-to-Bac™). Another system (BacPAKT™) uses baculoviral DNA that contains a deletion. Thus, no functional viruses can be produced without homologous recombination. However, **plaque purification** is highly recommended before the amplification of the viruses, whatever system is used.

Once the viruses have been amplified to a high titer, large amounts of cells can be used for the expression of the protein. As each cell is capable of hosting more than one virus, coinfection with several viruses is possible, which opens up the possibility of retaining complexes of several recombinant proteins within the same insect cell. Another advantage for large-scale protein production is the possibility to grow adherent Sf cells in a suspension culture. While, in general, protein expression can be expected to reach its maximum 30–60 h after infection, the optimal expression conditions must be worked out individually for each type of virus. A major advantage of baculoviral systems lies in their ability to clone and express large cDNAs. Genes containing genomic exonintron structures are correctly processed and expressed in baculoviruses. They are also GRAS, as they are only pathogenic to some arthropods. Furthermore, baculoviral systems allow for a high expression level of recombinant proteins. In some cases, the recombinant protein can make up 50% of the total protein content of an infected cell. If the proteins produced in insect cells are to be secreted into the medium, there is an alternative to Sf cells. These cells are derived from ovary cells of *Trichoplusia ni* (High Five™ cells) and may produce higher yields. A drawback of baculoviral systems is a possible aggregation of recombinant proteins within the insect cells due to their high expression rate. As the infected cell will be inevitably lysed by the virus, it is also necessary to initiate protein synthesis through new infection. **Glycosylation** in insect cells differs in many cases from the way glycosylation takes place in mammalian cells.

16.2.3.2 Expression of Proteins in Stably Transfected Insect Cells
Vector systems similar to those in mammalian cells have been developed that enable the selection of stably transfected insect cell cultures. There are vectors for Sf and High Five cells as well as *Drosophila* Schneider S2 cells, which are derived from a *Drosophila melanogaster* cell line. The vectors used for the SF cell system carry an **ampicillin resistance** and **pUC-*ori*** for the selection and replication in *E. coli*. The cDNA for the recombinant protein is cloned behind a **constitutively active promoter** (P_{OpIE2}). The vector conveys **blasticidin resistance** for the selection of stable transfectants. Vectors for generating the usual fusion proteins are also available. For the S2 system, vectors carrying either a constitutively active promoter (P_{Ac5}) or a P_{MT} promoter, which is inducible through $CuSO_4$, are on offer. In this system, the selection of stable transfectants is achieved through cotransfection of the expression vector with a vector conveying a **resistance to blasticidin** or **hygromycin**. The advantage of these systems over baculoviral systems lies in the easier management and proliferation of stably transfected cells. Due to a high consumption of selection markers, however, this can become very expensive in large-scale cultures. Another disadvantage is the sometimes significantly lower yield of the expressed recombinant protein. In comparison with expression systems in mammalian cells, insect cell cultures are a simpler and cheaper alternative.

16.2.4
Expression of Proteins in Mammalian Cells

A wide range of vectors and viral systems is available for the expression of proteins in mammalian cells (see Chapter 15). Often, virally immortalized cells or tumor cells are used as host cells with a high transfection efficiency. Some commonly used cell lines are listed in Table 16.5. For the production of large amounts of recombinant proteins, cells that can be grown in **suspension cultures** (such as CHO cells) are the most suitable. Thus, recombinant erythropoietin, factor VIII, or follicle-stimulating hormone for therapeutic purposes are obtained from CHO suspension cultures. Although mammalian cells have the enormous advantage of processing the recombinant proteins correctly, their culture is extremely cost- and labor-intensive. Therefore, they are often used on a laboratory scale for the functional characterization of unknown proteins. The procedure for the expression of proteins in mammalian cells largely follows the scheme described for insect cells (Section 16.2.3). However, if proteins are expressed in order to carry out a functional characterization, a **transient expression of proteins** is often the preferred method. As there is no need for the selection of stable transfectants, no artifacts will occur that were created by the undirected

Table 16.5 Typical properties of some important mammalian cell lines.

Cell line	Species	Specific properties	Requirements for high expression
CHO (Chinese hamster ovary cells)	Hamster	Suspension culture possible	Gene amplification
BHK-21 (baby hamster kidney cells)	Hamster	Efficient transient and stable expression Suspension culture possible	Gene amplification not needed
HEK-293 (human embryonic kidney cells)	Human	Constitutive expression of the E1a gene of adenoviruses Extremely high transient expression, high expression in stable transfectants Suspension culture possible	Promoters that can be efficiently activated by E1a
COS1/7 (CV-1 transformed by origin-defective mutant of SV40)	Monkey	Constitutive expression of the SV40 antigen; only efficient in transient expression	Circular plasmids containing an SV40 replication origin

insertion of a foreign gene into the genome of the host cells. In addition, adaptation processes can be avoided that result from a persistent high-level expression of the recombinant protein in the host cell. These may be very different from the function of the protein when expressed at a physiological level. As an alternative to transient protein expression, a stably transformed cell clone can be selected and the recombinant protein can be expressed under the control of an inducible **promoter** (Tet-on/Tet-off systems, see Chapter 15). This could make a *de novo* transformation of mammalian cells superfluous, which would be otherwise required for a repetition of the experiment.

Viral expression systems that can often infect efficiently a wide range of cell types (see Chapter 15) are frequently used where traditional transfection methods (liposomes, electroporation, calcium phosphate precipitates) are not efficient enough, as is often the case with **primary cell cultures**. Most viral systems are not considered as biologically safe and are therefore subject to strict safety regulations. Nevertheless, we would like to draw attention to several viral systems that are still in an experimental stage such as recombinant adenoviruses, adeno-associated viruses, and retroviruses, and are intended for therapeutic use in humans. Meanwhile, a more realistic appraisal of the possibilities of such systems has been achieved, and thus they are mainly used for the extracorporeal transformation and reapplication of specific patient cells.

16.3
Expression in Cell-Free Systems

An alternative to the expression in cells or organisms has been developed in recent years – cell-free expression systems. They produce sufficient amounts of recombinant proteins for experiments on a laboratory scale, at a still very high price/performance ratio. All systems are based on the *in vitro* translation of proteins. This is a clear advantage where proteins are concerned that would be toxic to the host organism or those that would be degraded rapidly by intracellular proteases. The systems also make it possible to carry out mutation studies quickly and efficiently, to define translation starting sequences, and to label proteins. An important criterion that has to be considered when choosing a method is whether it is suitable for **high-throughput screening**. *In vitro* translation is probably the best-suited method for the **automated protein expression** of many different genes.

Currently, three different systems are available for cell-free **protein expression**. These are based on **reticulocyte lysates** of rabbits, **wheat germ extracts**, or *E. coli* **extracts**. No matter what donor organism has been used, all extracts contain all macromolecular components (ribosomes, tRNAs, initiation, elongation, and termination factors) needed for a translation *in vitro*. In order to ensure an efficient translation process, amino acids, energy sources such as ATP and GTP as well as energy-restoring systems and cofactors must be added to the extracts.

Only RNA can be used as genetic material for *in vitro* translation. If the source matrix is DNA, an *in vitro* **transcription** must be carried out first. More recent commercial systems now include both *in vitro* transcription and translation in a single preparation.

16.3.1
Expression of Proteins in Reticulocyte Lysates

Reticulocytes are highly specialized cells that do not possess a nucleus. Their task is to translate cytoplasmic **hemoglobin mRNA**. Ninety percent of the total protein in an erythrocyte may be hemoglobin. In order to achieve an effective *in vitro* translation, the lysates must be treated with a calcium-dependent **micrococcal nuclease** that digests the cell's own mRNAs. The translation efficiency of lysates is comparable to that of intact reticulocytes. However, having no nucleus, these cells do not have a transcription mechanism, which makes it impossible to combine transcription and translation in this system. In comparison to cellular systems, the protein yield is extremely low, but reticulocyte lysates are suitable for radiolabeling proteins by adding, for example, [^{35}S]methionine. Due to the highly specific activity of the radioactive labeling, even small amounts of the radioactive protein can be used for **proteinprotein interaction** studies. Interactive surfaces on proteins, for example, can thus be identified by targeted mutagenesis, leading to an exchange of single amino acids.

16.3.2
Protein Expression Using *E. coli* Extracts

By contrast, state-of-the-art *in vitro* translation systems based on *E. coli* extracts are capable of producing up to 5 mg of protein within 24 h. In these systems it is possible to combine transcription and translation efficiently in one preparation, and substrates for *in vitro* translation are linear PCR products, linearized, or circular vectors. However, all useable sequences must contain a **T7 promoter** at the 5'-end, a RBS, a start codon, and a termination sequence at the 3'-end.

In the **Rapid Translation System**, a combined transcription/translation process is carried out in a special reaction unit. It consists of two chambers connected by a semipermeable membrane. A mixture of *E. coli* extract, amino acids, and DNA is fed into the actual reaction chamber. The other chamber or storage unit holds a nutrient solution containing all amino acids needed, various energy substrates, and nucleotides. During the reaction, the nutrients diffuse from the storage to the reaction chamber, while unwanted byproducts, such as nucleoside diphosphates and monophosphates, pyrophosphate, and DNA and RNA fragments, diffuse from the reaction into the storage chamber. The two-chamber system greatly enhances the efficiency of *in vitro* translation. The reaction is carried out in an instrument where the temperature can be precisely regulated. A shaking mechanism ensures the homogenous distribution of the reaction solutions, thus speeding up diffusion through the semipermeable membrane.

17
Patch Clamp Method

Learning Objectives

This chapter describes an essential method in membrane physiology. The patch clamp method enables scientists to follow the activity of individual ion channels in biomembranes and their modulation through active substances.

17.1
Biological Membranes and Ion Channels

All prokaryotic and eukaryotic cells are separated from their environment by a lipid bilayer that forms the **cytoplasmic membrane** (Section 3.1). Important cell functions such as reception and transmission of signals, and transport and conservation of energy depend on membranes, and are mediated through specific integral membrane proteins. **Ion channels** are an important group of such membrane proteins. They run through the entire lipid bilayer, forming water-filled pores. They can be mostly open or closed, depending on several factors. Switching between these two states ("gating") is mostly controlled by changes in the membrane potential or by the binding of signaling molecules. When the channels are open, an ion current, the magnitude of which is determined by the **equilibrium potential** of the ions in question and the electric potential across the membrane, flows through them. The equilibrium potential of ion type A (E_A) depends on the ion activity a_A in the aqueous phases I and II on both sides of the membrane. The activity of the ions is proportional to their concentration. The equilibrium potential can be worked out with the help of the Nernst equation:

$$E_A = \frac{RT}{zF} \ln \frac{a_A^{II}}{a_A^{I}}$$

The constants R and F are the molar gas constant and the Faraday constant, T is the absolute temperature, and z the valency of the ion type.

The majority of ion channels are impermeable for large organic molecules, but are permeable for single cations or a group of **cations such as Na^+, K^+, Ca^{2+}, H^+, and Mg^{2+}** or for **anions such as Cl^- and HCO_3^-**. The permeability for certain ions (selectivity) is defined by the specific pore structure (**selectivity filter**) of the channel protein. **Voltage-gated cation channels**, including voltage-gated potassium, sodium, and calcium channels as well as many nonselective cation channels, build the biggest group of signaling proteins besides the families of **G-protein-coupled receptors** and protein kinases. With the help of the patch clamp technique the current through single ion channels and, therefore, their functional properties can be determined.

The first **patch clamp measurements** were carried out by Sakmann and Neher, and published in 1976 in a study about the activity of single ion channels in frog muscles (Neher and Sakmann, 1976). Until then, it had only been possible to insert glass pipettes into large cells in order to measure the ion current

An Introduction to Molecular Biotechnology, 2nd Edition.
Edited by Michael Wink
Copyright © 2011 WILEY-VCH Verlag GmbH & Co. KGaA, Weinheim
ISBN: 978-3-527-32637-2

through various channels. Sakmann and Neher were awarded the Nobel Prize for Physiology and Medicine in 1991.

17.2
Physical Foundations of the Patch Clamp Method

The **patch clamp method** is derived from the **voltage clamp technique**, which is characterized by inserting two glass electrodes into large cells. One electrode sets the command potential. The second electrode enables the registration of membrane currents. The patch clamp method, however, unifies control of command potential and measurement of currents onto one electrode. In addition, the patch electrode is not inserted into the cell and thus prevents the development of leak conductance. Using micromanipulators, glass pipettes with an opening of about 1 μm are put on the surface of the lipid membrane. The membrane patch enclosed by the pipette opening can be sealed, so that the contact between glass and membrane is characterized by a high electrical resistance of several gigaohms ($10^9\,\Omega$). The patch pipette is filled with a salt solution in which an electrode is placed. A second electrode (signal ground) makes contact with the bath solution that surrounds the cell. In order to eliminate interfering electrode potentials triggered by chemical reactions on the electrodes, silver wires covered in a silver chloride layer are used. A command voltage can be imposed to the cell via the patch electrode. Once the gigaohm contact has been established, random fluctuations of the electric current (electric noise) are reduced. Generally, the noise of a resistance is in inverse proportion to the magnitude of the resistance. It is necessary to suppress the noise because the currents in the individual ion channels have very low amplitudes of only a few picoamperes ($10^{-12}\,A$). In order to measure these while building up a defined voltage across the membrane, special patch clamp amplifiers are used. These are based on the parallel connection of an operational amplifier and a feedback resistor (Fig. 17.1). The operational amplifier is an electronic module with two inputs – in this case, one for the pipette electrode and one for a regulatable voltage source, which is earthed via the bath electrode. The difference in voltage between the pipette electrode, the actual membrane voltage, and the command voltage is transformed into an amplified signal at the output of the operational amplifier. A feedback resistor is interposed between pipette electrode and output, through which the current runs, as long as there is a difference between membrane and command voltage. The current running through the feedback resistor is equivalent to the membrane current, but runs in the opposite direction. As the membrane current is carried by positive as well as negative ions and runs into two directions (from intracellular to extracellular and vice versa), a flux of cations into the cell or a flux of anions out of the cell has been defined as a negative current. A reversal of the direction leads to a positive sign of the current. Since the patch electrode – as described in Section 17.3 – can be turned to different sides of the plasma membrane, the membrane current has to be adapted to this convention.

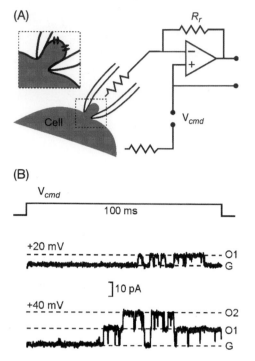

Fig. 17.1 Working principle of a patch clamp amplifier and the effect of the potential on the current in individual ion channels. (A) Diagram of a patch clamp measurement on an intact cell. A potential is directed to the electrode in the patch pipette. The command potential is specified at the amplifier. The feedback resistor R_r compensates for the current through the patch, which is measured at the output. (B) When +20 mV is directed to the pipette electrode, the activity of a channel can be seen alternating between a closed (C) and an open state (O1). At a voltage of +40 mV, current flow can sometimes be observed in two channels (O2). At the same time the amplitude (i.e., the difference between C and O1 or between O1 and O2) in the individual channel increases. Furthermore, the residence time in the O1 and O2 stage increases with increasing voltage. These channels (potassium channels of the BK type) are activated through voltage.

17.3
Patch Clamp Configurations

When the patch pipette is put on an intact cell and thereby establishes a gigaohm contact, the resulting measurement setup is called **cell-attached configuration** (Fig. 17.2). This arrangement makes it possible to recognize the gating of individual ion channels as abrupt changes in current amplitude. Ligand-gated ions cannot always be sufficiently characterized in the cell-attached configuration, as the composition of the cell interior is unknown and it is impossible to modify the composition of the pipette solution during the experiment. By contrast, the bath solution can be modified relatively easily, by simply letting it run off and replacing it. In order to influence the behavior of ion channels from the cytosolic side, it is possible to produce a patch that is detached from the cell. By cautiously retracting the pipette from a cell-attached configuration, a fragment is torn out of the membrane of a cell. The inside of the fragment is then brought into contact with the bath solution. This is called an **inside-out configuration** and requires cells that are attached to the bottom of the measuring chamber. Such a detached patch makes it also possible to study the individual ion channels it contains. For a better analysis of channel currents, giant patches can be separated from large cells by using pipettes with greater tip openings. This special form of the inside-out configuration enables the measurement of currents through multiple ion channels. In contrast to the cell-attached configuration, exposing the inside of the cell membrane makes it possible to define the ion composition on both sides of the patch. This facilitates the identification of ion channels according to their selectivity. This is done by calculating the reverse potential of the current running through the channels (i.e., the voltage at which the current is zero). According to Ohm's law:

$$V = IR$$

The voltage V driving the channel current is also zero at the reverse potential. The voltage consists of the command voltage V_{cmd} and the equilibrium potential E_A of the ions involved:

$$V = V_{cmd} - E_A$$

At the reverse potential, the formula looks like:

$$V_{cmd} = E_A$$

If, as in the case of an inside-out configuration, the ion concentrations on both sides of the membrane are known, the E_A value for the various types of ions (e.g., K^+, Na^+, or Cl^-) can be calculated using the **Nernst equation**. From the equilibrium potential, which is identical to the command voltage, conclusions can be drawn about the permeating ions. These can only be unequivocally identified if the ion distributions result in differing values for the individual equilibrium potentials. This is why the concentration of an ion in the bath solution is often changed during the experiment. If the reverse potential is shifted as a result, the ions in question are involved in the channel current.

In order to gain an insight into the activity of ion channels in connection with the whole cell, another variation of the patch clamp method, known as the **whole-cell configuration**, can be used. After a cell-attached configuration has been established, strong negative pressure is applied to the patch pipette, leading to a perforation of the membrane within the patch. If the tight contact between cell and pipette is retained during the process, this results in an electric resistance of several megaohms between the patch pipette and the interior of the cell, making it possible to measure membrane currents across the entire cytoplasmic membrane, including all ion channels contained in it. The whole-cell configuration has become the most frequently used of the patch clamp methods, because currents through multiple ion channels significantly improve the analysis of most channel properties. As in an inside-out configuration, the ional conditions on both sides of the membrane have been defined and the ion channel selectivity can be derived from the reversal potential.

Many **ligand-gated ion channels**, such as nicotinic acetylcholine receptors, GABA$_A$ receptors, or the ionotropic glutamate receptors, can be activated by bath application of neurotransmitters (Fig. 17.3). The whole-cell configuration is

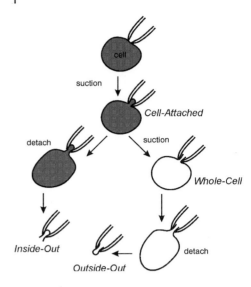

Fig. 17.2 Patch clamp configurations. When the patch pipette touches the cell membrane, a tight contact is formed that, through light suction, is turned into a gigaohm seal (cell-attached configuration). When the suction is increased, the membrane is ruptured (whole-cell configuration). If the membrane is pulled back, a membrane fragment is torn away (inside-out). The outside-out configuration is obtained by pulling back the pipette.

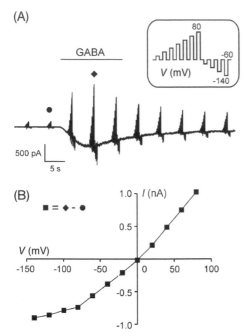

Fig. 17.3 Example of a whole-cell current through ligand-gated ion channels (GABA$_A$ receptors). (A) Starting from a potential of 60 mV, the voltage directed to the cell was modified to depolarizing (up to +80 mV) and hyperpolarizing (to −140 mV) values at regular intervals. During the measurement, the neurotransmitter γ-aminobutyric acid (GABA) was added to the bath solution to reach a final concentration of 100 μmol L^{-1}. This caused an increase in whole-cell currents at positive as well as negative potentials. (B) The difference in current amplitudes (■) during (▲) and before (●) GABA stimulation (see A) is shown in dependence of the command voltage. The reverse potential of the GABA-induced current is about zero. Since the distribution of chloride ions in the bath and pipette solution was equal, this suggests that there has been activity of a chloride channel.

particularly suited to study the blockage of different ion channel types by active substances (endogenous molecules, toxins, and pharmacological agents). Therefore, this technique belongs to the standard methods of pharmacological research.

Another important configuration, alongside the three patch clamp configurations mentioned thus far, is the **outside-out configuration**. It is obtained by pulling the pipette out of a whole-cell configuration from an adherent cell. The membrane fragments clinging to the pipette opening reseal spontaneously. Thus, the outside continues to be in contact with the bath solution. This configuration allows individual channel experiments to be carried out under conditions similar to the whole-cell current measurements. The risk of vesicle formation is lower, compared to the inside-out configuration. A special form of the whole-cell configuration is the nucleated patch. Suction of the nucleus of a neuron (in acute brain slices) and slow retraction of the pipette leads to a detachment of cellular processes and thus enables the measurement of somatic ion channels.

17.4
Applications of the Patch Clamp Method

Heterologous expression is the expression of proteins in cells that normally do not contain the protein to be analyzed or, if so, only at extremely low levels. Through the cloning of ion channels and their heterologous expression, functional properties can be assigned to certain molecularly defined channel proteins. The human embryonic kidney cell line HEK-293 or nonfertilized oocytes of the frog *Xenopus laevis* are frequently used as expression systems. In mammalian cells, the cDNA of an ion channel is inserted through transfection. This can either be done using liposomes or through injection or poration of the cell membrane using high electrical voltage. Oocytes are transfected through injection of DNA or RNA. Due to inconsistent expression efficiency a reporter gene is often transfected alongside the ion channel DNA. The **reporter** is usually some variant of **Green Fluorescent Protein (GFP)**. The combination of patch clamp methods with heterologous expression permits not only research into properties of known recombinant channels, but also of newly cloned ion channels and channel proteins modified through mutagenesis in ion channel DNA. Ion channels endogenously expressed in certain cell types can be identified by patch clamp and the previous application of **RNA interference (RNAi)**, which induces suppression of channel genes of interest. A further approach towards the molecular identification of ion channels is the **reverse transcription polymerase chain reaction (RT-PCR)** analysis of channel RNA extracted from cell material that was harvested via the patch pipette (**single-cell PCR**).

When ion channels are to be characterized in nerve cells, it is often necessary to keep them within the network of their natural environment. Neurons and glial cells located at the upper surface of slices prepared from mouse or rat brain can be investigated in the appropriate patch clamp configuration. The acute brain slices used are 100–400 μm thick. They remain vital for a few hours when kept in an oxygen- and glucose-enriched salt solution. In addition to the electrophysiological measurement, the whole-cell configuration permits the filling of single nerve cells with ion-sensitive fluorescent dyes, such as calcium-sensitive ionophores like fura or fluo dyes. Since these dyes diffuse into the whole cell, calcium signals can be measured (calcium imaging) in different cell compartments, including the axon as well as dendritic processes and spines.

The application range of the patch clamp method not only involves the analysis of currents through ion channels expressed in virtually every cell type (also plant and yeast cells), but also includes the determination of further electrical parameters, particularly the membrane potential and the cell capacitance. Thus, fundamental physiological properties of the cell, such as excitability (generation of action potentials) and secretion (vesicle budding), can be measured directly.

The traditional patch clamp technique requires a high work load and is connected with a low throughput. For high-throughput drug screening, therefore, further methods that enable automatic patch clamp have been developed. In principle, these methods are based on industrial patch clamp arrays (i.e. glass or polymer plates containing a microstructured aperture) that replace conventional patch pipettes. A single cell from a cell suspension can be positioned on the hole by suction. After establishing the whole-cell configuration, the planar patch clamp method enables, in comparison with the traditional technique, an automatic compound application and an additional perfusion of the intracellular side.

18
Cell Cycle Analysis

Learning Objectives

This chapter gives an introduction to the important topic of cell cycle control in eukaryotes, explained using the model organism *Saccharomyces cerevisiae*, and describes experimental methods used for the analysis of the cell cycle. Flow cytometry and laser scanning cytometry are discussed in more detail.

18.1
Analyzing the Cell Cycle

The cell cycle spans the time between the division of a mother cell and the subsequent division of its daughter cells. The coordinated processes of cell growth, DNA replication, and cell division are affected by various factors such as availability of nutrients, environmental conditions (including the cell environment within the organism and cell differentiation), or damage to the DNA.

The mitotic cell division in eukaryotes follows a predictable schedule, which is shared by all higher organisms (see Section 4.1.3). *Saccharomyces cerevisiae*, more commonly known as baker's yeast, has been established as a simple model for studies of the processes involved in the cell cycle and their control.

Two idiosyncrasies in the cell cycle of *S. cerevisiae* make it particularly suitable for the study of the genes controlling the cell cycle:

- Haploid as well as diploid cells undergo mitotic division, which means that recessive mutations in haploid cells can be isolated and complemented in diploid cells.
- Daughter cells can be recognized very early as buds sprouting on the surface of the mother cell. The changing size ratio between buds and mother cell serves as an indicator of the current status of the cell cycle.

The periodically repeated events can be divided into two or four main phases: interphase, consisting of G_1, S and G_2 Phases, and mitosis (M phase) (Fig. 18.1):

- **Interphase.** This is the phase between two cell divisions during which the newly formed cells grow until they reach a critical size and during which all critical steps for the preparation for cell division are performed. In baker's yeast, but also in mammalian cells that have not been transformed, such as fibroblasts, interphase is subdivided into several steps required for the preparation of the M phase. The transition between the different phases is regulated by the presence of critical control factors.
  i. **G_1 phase.** Cells increase in size, produce RNA, and synthesize proteins. Increasing levels of G_1 cyclins bind to their **cyclin-dependent kinases (CDKs)** and signal the cell to prepare the chromosomes for replication. The G_1 checkpoint controls the entrance into S phase, ensuring that progression through the cell cycle is avoided in cells carrying DNA damage until the damage is repaired.

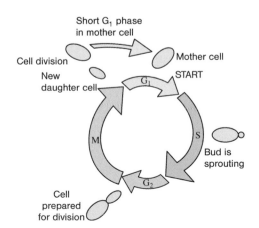

Fig. 18.1 The cell cycle and its phases in *S. cerevisiae*.

An Introduction to Molecular Biotechnology, 2nd Edition.
Edited by Michael Wink
Copyright © 2011 WILEY-VCH Verlag GmbH & Co. KGaA, Weinheim
ISBN: 978-3-527-32637-2

ii. **S phase.** Increasing levels of S-phase promoting factor (SPF) in the nucleus initiate DNA replication and duplication of the centrosomes (formation of a new bud in *S. cerevisiae*). Although the rate of protein synthesis is very low, most of the histone production occurs during the S phase. The intra-S-phase checkpoint network functions to avoid the duplication of damaged or broken DNA, which would be further propagated in mitosis, eventually leading to genomic instability.

iii. **G_2 phase.** An intermediate phase after DNA synthesis in which cyclin E is destroyed and the level of mitotic cyclins begins to rise (in *S. cerevisiae* this applies nearly exclusively to the bud). Significant protein synthesis occurs during this phase, mainly involving microtubule production required for the mitosis. The second G_2 checkpoint checks on the successful replication of DNA during the S phase.

- **Mitosis, M phase:** The cell division (Mitosis) consists of nuclear division (**karyokinesis**) and cytoplasmic division (**cytokinesis**). An increasing level of M-phase promoting factor initiates the serial of events in which duplicated chromosomes are divided between mother and daughter cells (in *S. cerevisiae* a septum is introduced underneath the bud to separate it from the mother cell). Karyokinesis consists of several distinct phases: prophase, prometaphase, metaphase, anaphase, and telophase. At the end of mitosis the anaphase promoting complex (APC) destroys B cyclins and ensures completion of mitosis

- **G_0 phase.** An additional critical phase for the cellular live cycle is the G_0 phase, which can be entered by cells alternatively to the G_1 phase. The G_0 phase represents a resting state in which cells, temporary or permanently, cease the cell cycle and stop dividing. The cells that permanently exit the cell cycle are also called quiescent or senescent cells, although the entrance to G_0 phase is not only characteristic of old, reproductively exhausted cells. In many somatic cells, G_0 arrest is often followed by terminal differentiation of cells and further specialization to carry out important everyday functions in different organs of multicellular organisms. Cancer cells, however, cannot enter or remain in G_0 and are able to divide indefinitely.

In order to divide, a yeast cell must first reach a **critical size** before DNA synthesis is initiated via a key control point in the cell cycle, called START. Cells that have developed beyond the START point are irreversibly geared to DNA replication and must go through the cell cycle. A lack of nutrients or the presence of mating signals may block the passage through START. There are further control points along the line in the cell cycle, preventing DNA damage or cell death through uncoordinated processes. These control points mark the transition between the four main cell cycle phases, G_1 and S and between G_2 and M, acting as internal control systems by blocking the cycle if certain vital prerequisites are not fulfilled.

The progression of the cell cycle is regulated by **cyclins**, which act in combination with **cyclin dependent kinases (CDKs)** the actual regulatory enzymes. Together they form cyclin CDK complexes, which are active protein kinases that control important phases in the cell cycle (chromosome condensation) through protein phosphorylation. Specific combinations of cyclin and CDKs act as regulatory cyclinCDK complexes at different stages of the cell cycle.

A cell cycle can only be completed if after an increase of cyclinCDK activity, the cyclinCDK complex is deactivated. This is done by a ubiquitin-mediated degradation of cyclin once a phase of the cell cycle has been successfully concluded (Fig. 18.2).

Positive CDK regulation through cyclins is complemented by negative regulation, such as small CDK-binding proteins such as **INKs (inhibitors of kinase)**, **CIPs (CDK-inhibiting proteins)**, and **KIPs (kinase-inhibiting proteins)**.

The **transcription of cyclins** is regulated by negative as well as positive controls. Some early CDK complexes stimulate the transcription of later cyclins

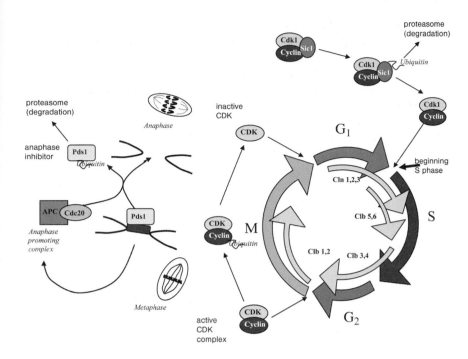

Fig. 18.2 Regulation of the cell cycle in the yeast *S. cerevisiae*. Addition of Ubiquitin (small tag) to proteins marks them for degradation through the proteasome, required for further progression of the cell cycle. APC, anaphase promoting complex; Pds1, anaphase inhibitor; Sic1, stoichiometric inhibitor1; Cln, cyclins for budding; Clb, B-type cyclins.

through the **activation of transcription factors** while suppressing their own expression. In mammalian cells, members of the E2F transcription factor family stimulate the transcription of cyclins E and A.

Following ubiquitinylation cyclins are degraded via the **proteasomal pathway**, thus ensuring the accurate termination of cyclin activity. In some cases, such as in rapidly dividing cells during early embryogenesis, cyclins are transcribed at a constantly high level and the degradation of cyclins is the only mechanism controlling the timing of the cell cycle.

18.2
Experimental Analysis of the Cell Cycle

The precision that underlies the progression of the cell cycle ensures the near-perfect accuracy of the copying process and thus the survival of living organisms. A loss of precision would lead to an increasing instability of the genome, which is a major factor in the emergence of cancer. The cell cycle is the essential process that ensures the survival and development of all life. Experimental studies of the cell cycle are therefore crucial in the context of many biological questions. The **status of the cell cycle** is commonly analyzed by measuring the **DNA content** of a cell. In the G_1 phase (i.e., before DNA replication) the DNA content in a cell is the exact DNA amount of the genome ($1N$ for haploid organisms or $2N$ for diploids). During the S phase, the amount is doubled to provide the genomes of the two daughter cells that are found during the G_2/M phase before mitotic division (haploid $2N$; diploid $4N$). Thus, the DNA content is an indicator for the progression of the cell cycle and can be used to monitor changes brought about by genetic modification or added compounds.

Cell cycle monitoring can be carried out in unsynchronized and synchronized cultures. In unsynchronized cells, all of the different stages of the cell cycle can be detected (cell cycle distribution), whereas in synchronized cells progression through the cell cycle is monitored (cell cycle progression).

This is important for many experiments, because the response to various signals and substances may depend on the phase of the cell cycle. In unsynchronized cell cultures, the individual cells are in different stages of the cell cycle and may thus vary in their reactions. Thus, results may be unclear and not reproducible, sometimes not even detectable. This can be avoided by using synchronized cell cultures in the experiments.

18.2.1
Preparing Synchronized Cell Cultures of *S. cerevisiae*

In synchronized cell cultures, all cells undergo the same processes in the cell cycle and pass the various points simultaneously. Synchronized yeast cultures can be maintained through induction or selection. However, the **synchronization** achieved can only be maintained for two or three cell cycles. This is because of the asymmetric division process in *S. cerevisiae* (mother cells start their next cycle before the daughter cells have reached the critical size) and the natural variation in the duplication time of individual cells. Numerous induction and selection synchronization methods can be applied to promote synchrony of cell division. Feeding/starving of key nutrients in the culture is the easiest, noninvasive method for cell cycle synchronization. Although this method is simple in principal to obtain yeast cells that are balanced with respect to growth rate and synchronous with respect to cell division, it requires a technically quite demanding precise control of nutrient supply, which is only achieved in tightly controlled "chemostat" cultures. Induction methods, in contrast, cause perturbation of normal growth events and cell division, and are often not very reproducible. The induction of cell synchronization can be achieved by chemical administration, heat shock, or DNA synthesis inhibitors, or cell division cycle (*cdc*) mutants. Alternatively, synchronization can be obtained with cell selection methods like centrifugal elutriation, which can be used to fractionate a population of asynchronous cells according to cell size.

18.2.1.1 **Centrifugal Elutriation**
Centrifugal elutriation is a method in which cells are continuously pumped into the bottom of a spinning centrifuge chamber, and cells and fluid are eluted from the top of the chamber (Fig. 18.3).Under the assumption that cell size reflects the stage of the cell cycle, it is possible to obtain synchronized cultures of clearly defined phases of the cell cycle, in particular of young daughter cells in the G_1 phase. The unsynchronized population is loaded into an **elutriation rotor**. The elutriation chamber is continuously flushed. The flow rate defines the equilibrium position for the cell size. The cells are thus separated in the elutriation chamber according to their size as a result of the balanced effects of centrifugation, inertia, and flow. When the flow rate increases, the smallest cells are accelerated fastest towards the outlet. The fractionated collection at the outlet provides synchronized cells for further cultivation. A stepwise increase of flow rates or the lowering of centrifugal velocity makes it possible to fractionate the whole cell population according to size. If the cells are elutriated in growth medium at the growth temperature a sample of cells that is uniform in size, morphology, and cell cycle position can be obtained for further experiments. This method re-

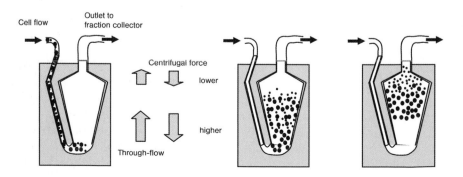

Fig. 18.3 Elutriation – schematic view (based on an instruction manual from Beckman Instruments, Palo Alto, CA, USA).

sembles the block-and-release method, except that no chemical induction that perturbs cell physiology must be used. The disadvantage of this method is that only a small proportion of the original culture, maximum 5%, is obtained. The second possibility is to perform elutriation in a chilled rotor at 4°C. In this case, about 15 to 20 different fractions are obtained, each with cells of increasing size, representing later and later points in the cell cycle. However, these fractions cannot be cultivated further and have to be examined immediately.

18.2.1.2 Cell Cycle Arrest Using α-Factor

Synchronized cells can also be obtained through induced cell cycle arrest, such as through **alpha-factor** in yeast cultures. The alpha-factor is a 13-amino-acid peptide (Trp-His-Trp-Leu-Gln-Leu-Lys-Pro-Gly-Gln-Pro-Met-Tyr) made by **mating type α (MATα)** cells that bind to a receptor found on **mating type a (MATa)** cells. Binding of α-factor leads to inactivation of G_1 cyclin/Cdc28 kinase complex in MATa cells and G_1 phase arrest. By adding the mating pheromone a-factor (3–5 μM for 2 h) to haploid yeast cells of the mating type MATa, the cells are kept in a "standby" position at the transition stage between the G_1 and S phase until they can fuse with cells of the other mating type (MATα) (Fig. 18.4). After one doubling, nearly all cells have been arrested as unbudded cells, showing a comma-like protrusion, which is known as the *shmoo* phenotype. The synchronized cells are then released from α-factor arrest by washing and resuspending them in fresh medium for further culture.

Difficulties can be faced when BAR1-expressing cells (wild-type cells) are used. BAR1, is a protease secreted by yeast and degrades α-factor. Thus, the rate and time of degradation of α-factor will depend on the number of cells in the yeast culture, meaning that the arrest will only be transient. Yeast cells can also be arrested either in S phase using hydroxyurea (0.1 M for 2 h) or in G_2 using nocodazole (15 μg/ml for 2 h).

18.2.2
Identification of Cell Cycle Stages

Several methods can be used to identify the cell cycle stages in cell culture and to measure their distribution.

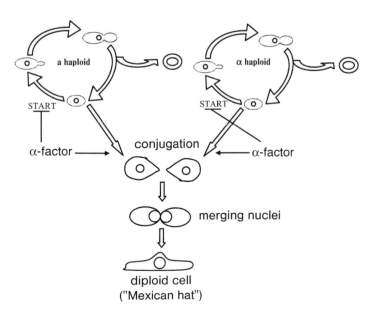

Fig. 18.4 Mating cycle of *S. cerevisiae*. The presence of a- and α-factors arrests the cells of the respective opposite mating type at the start control point of the cell cycle, inducing the *shmoo* phenotype. The fusion of both haploid cells produces a diploid cell. This can be avoided by a-factor synchronization in cultures containing one single mating type (MATa). (Based on Murray and Hunt, 1993.)

18.2.2.1 Budding Index

The budding index is defined as the percentage of the whole population carrying a bud. It helps to distinguish cells in the G_1 phase from those in the $S/G_2/M$ phases in experiments with the budding yeast *S. cerevisiae*. The budding factor is obtained by counting a sufficiently high number of cells (more than 200) under the microscope. It is a fast-track method of determining the distribution of cells at the various stages during an experiment. The samples should always be sonicated before buds are counted and coded samples should be used to prevent observer bias from skewing the results.

18.2.2.2 Fluorescent Staining of the Nucleus

High-throughput analysis of the cell cycle is widely used in studies of cell growth, defects in cell cycle regulation, oncology research, and DNA ploidy determinations. These applications require a fluorescent dye that binds to DNA in a stoichiometric manner. The fluorescent dye is added to a suspension of permeabilized single cells or nuclei. The principle is that the stained material incorporates an amount of dye proportional to the amount of DNA. The stained material is then measured in a flow cytometer or with laser scanning microscopy and the emitted fluorescent signal yields an electronic pulse with an amplitude proportional to the total fluorescence emission from the cell. Several dyes are used for the staining of DNA in cells:

- Propidium iodide (PI)
- Ethidium bromide
- Hoechst dyes (mainly Hoechst 33342 and 33258)
- Acridine orange (AO)
- Mithramycin
- DAPI (4,6-diamidino-2-phenylindole)
- SYBR Green I
- 7-Aminoactinomycin D
- TO-PRO-3
- Chromomycin
- Vybrant® DyeCycle™
- DRAQ5™
- SYBR® Green I
- TO-PRO®-3
- DRAQ5™

The most frequently used DNA dye in cell cycle analysis is **PI**. When attached to the DNA, it produces a strong red fluorescence (maximum emission 637 nm). The fluorescence is excited by light of 488 nm, which is within the range of most flow cytometers. In order to be stained with PI, the cells must first be fixed and rendered permeable. As PI binds to the base pair of guanine and cytosine, PI stains not only DNA, but also RNA. Therefore, digestion of RNA with ribonuclease is required before PI staining.

Hoechst 33342 and **33258** are bisbenzimide derivatives that bind AT-rich regions of the DNA and are ingested by living cells. As no fixation and permeabilization are needed, the cells can be retrieved for further cultivation. The drawback of this method is that Hoechst 33342 is excited only in the UV range (351–364 nm), which precludes its use in many flow cytometers. It is, however, possible to combine it with other dyes and fluorescent proteins, which often cannot be done with PI.

AO is used for staining DNA and RNA. AO fluorescent color varies, depending on its binding to DNA or RNA. Both are excited by 488 nm and become visible as green (526 nm, DNA) or red (630 nm, RNA) fluorescence. This property has been exploited in methods for simultaneously analyzing the DNA and RNA content of a cell culture.

DAPI is an AT-binding dye with properties similar to those of the Hoechst dyes. When stained with DAPI, the DNA appears as blue-white fluorescence under UV illumination. Most laser scanning microscopes and flow cytometers do not have a UV laser illumination system, so the use of DAPI staining is very much restricted.

SYBR Green I and **TO-PRO-3** preferentially stain the nuclear DNA, with negligible RNA staining. Thus, pretreatment with ribonuclease to remove RNA can be omitted. SYBR Green I intensively stains the DNA, appearing green upon excitation with blue light (488 nm), while TO-PRO-3 stains the DNA with far-red fluorescence under red light (647 nm) excitation and is therefore a convenient dye for multilabel staining. A disadvantage of SYBR Green I and TO-PRO-3 fluorescence labeling of DNA is the relative rapid fading of the fluorescence, making the staining unsuitable for comparative analysis.

Vybrant DyeCycle and **DRAQ5** are DNA-selective, cell membrane-permeant dyes that can be used in the presence of media components, including serum and divalent cations. The Vybrant DyeCycle dyes can be excited by 405 (Vybrant DyeCycle Violet; excitation/emission maxima around 396/437 nm), 488 (Vybrant DyeCycleTM Green; excitation/emission maxima around 506/534 nm), or 532 nm (Vybrant DyeCycle Orange; excitation/emission maxima around 519/563 nm) conventional laser lines, depending on the dye. Excitation of DRAQ5 is possible using a wide range of convenient laser light wavelengths (e.g., 488, 514, 568, 633, or 647 nm). Emission spectra extend from 670 nm into the low infrared, providing minimal overlap with the emission from visible range dyes including **Green Fluorescent protein (GFP)**. DRAQ5 shows no photobleaching, but it is ultimately toxic to the cells due to its persistence on the DNA and is therefore not useful for long-term tracking or viable sorting/cloning. The Vybrant DyeCycle dyes are not toxic, but cause some retardation of cell division.

Once the nuclei have been stained, the cells can be identified according to their current stage in the cell cycle. This can be done by counting cells under the microscope, flow cytometry, or laser scanning calorimetry.

Counting cells under the microscope (Fig. 18.5). As the form and position of the nucleus in the cell varies during each stage of the cycle, this can be used to distinguish between cells in G_1, S/G_2, and various mitotic phases. The cells are fixed onto a slide, meaning the cells cannot be grown in culture afterwards. The method is time-consuming and prone to researcher bias.

Flow cytometry is a method that measures various physical properties of cells or particles that flow past a measuring point. In some respects, flow cytometers can be regarded as highly specialized microscopes. A state-of-the-art flow cytometer consists of a light source, lenses, photodiodes or photomultipliers with amplifiers, and a computer for the transformation of signals into data.

The light source is usually a laser that provides coherent light of a specific wavelength. The emitted and reflected light is collected by various lenses, sepa-

DAPI   DIC

S. cerevisiae cells arrested in M phase

S. cerevisiae cells arrested in G_1 phase

S. cerevisiae cells with fragmented nucleus upon DNA damage

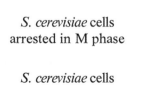
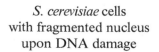
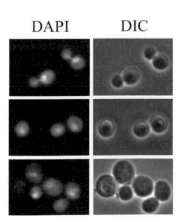

Fig. 18.5 DAPI staining and differential interference contrast (DIC) microscopy of yeast cells in different cell cycle phases: distribution of DNA between mother cell and bud in M phase; staining of nucleus in G_1 phase; and fragmented DNA after DNA damage.

(A) Mock-treated cells

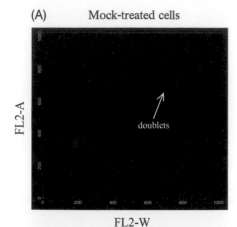

FL2-A

FL2-W

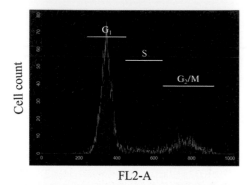

Cell count

FL2-A

(B) Colchicine-treated cells

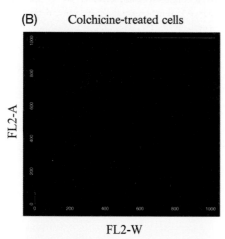

FL2-A

FL2-W

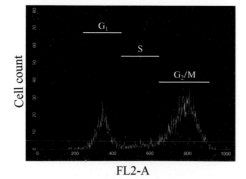

Cell count

FL2-A

Fig. 18.6 Cell cycle profiles after DNA staining and fluorescence-activated cell sorting (FACS) analysis. (A) Analysis of unsynchronised mammalian cell culture; top panel: fluorescence distribution, FL2-A: pulse-area, pulse intensity; FL2-W: pulse-width, lenght of pulse, this increases with cell diameter; lower panel: x-axis fluorescence intensity, y-axis number of cells. (B) Cell cycle distribution after inhibition of cell division by colchicine.

rated by filters and dichroic mirrors into fluorescent and excitation light, and then measured. Apart from fluorescence, physical properties such as cell size, shape, and internal complexity can also be assessed, which makes flow cytometers very versatile analytical tools for cell biology.

Measuring the DNA content of cells in a **flow cytometer** gives a very detailed picture of the **distribution of the cell population** in G_1, S and G_2/M phases. By binding to the DNA, the dyes make it possible to measure the DNA content (Fig. 18.6). G_2 and M phases, which both have an identical DNA content, cannot be discriminated based on their DNA content. To obtain a good discrimination between single cells (singlets) and cells sticking together (doublets) the cells need to pass the measuring points one by one, which is better obtained at a low flow rate below 1000 cells per second. In order to avoid the formation of clusters, the cells can undergo a special (e.g., ultrasonic) treatment before cytometry. Modern flow cytometers are also equipped with a "doublet discrimination module" that selects single cells on the basis of pulse processed data. The discrimination can also be obtained plotting pulse area (FL2-A) against pulse width (FL2-W) of the samples. This method is based on the fact that FL2-W increases with the diameter of the doublet particle, while both the G_1 doublet and the G_2/M singlet produce a comparable FL2-A signal. Cell density also plays an important role in efficient and accurate acquisition for DNA content with intercalating dyes. A standard acquisition is 20000 single events, but when staining is not homogenous and/or aggregates are present a much larger number of events may have to be acquired until a stable G_1 peak can be obtained. A poorly prepared single-cell suspension, incomplete fixation, or nonhomogenous staining may present false peaks or wide coefficient variations that the analytical software will interpret as an aneuploid population. By analyzing the cell cycle distribution within the cell population it is possible to calculate the mitotic index. The mitotic index is a measure for the proliferation status of a cell population. It is defined as the ratio between the number of cells in mitosis and the total number of cells, and is a good indicator of a disturbed cell cycle progression.

Using a single fluorochrome has its **limitations**. It will only give static, but not kinetic information. We do not know if cells containing amounts of DNA typical for the S phase are indeed undergoing a cell cycle, synthesizing DNA. This can be verified, for example, by using **bromodeoxyuridine (BrdU)**. BrdU is a thymidine analog inserted into the DNA during DNA synthesis. The inserted BrdU can be detected by denaturing (unwinding) the DNA and then adding an antibody against BrdU. Thus, cells in G_1, S, and G_2 phases can be clearly distinguished.

In **laser scanning cytometry** cells fixed to a slide can be analyzed using a laser scanning cytometer (LSC). This is based on the same optical principles as a flow cytometer, but the flow channel has been replaced by a microscope slide and the LSC scans the fixed cells during cytometric analysis. This new method of exposing cells to optical cytometry offers a range of options not available in flow systems, such as identifying the exact position of each cell on the slide, which allows the cytometer to find the cell again after analysis. The fixed cells can also be washed and stained with another dye, and the results can be compared on a cell-by-cell basis. As in flow cytometers, it is possible to carry out measurements using multiple staining.

This technology is particularly useful when small numbers of cells and subpopulations are analyzed, and additional information about morphology and phenotype of the cells are required. Small numbers of cells (a few hundred) that

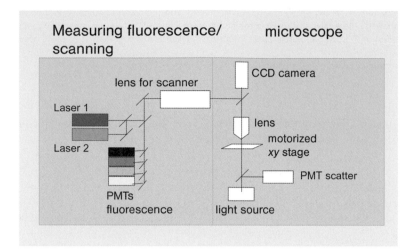

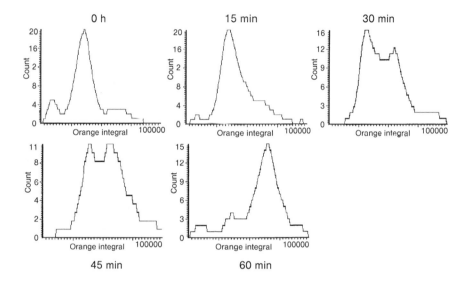

Fig. 18.8 Plotting the progression throughout the cell cycle in yeast cells, using a laser scanning microscope. The cells were taken from the culture in the middle of the log phase, washed and synchronized by MAT. After 2 h of synchronization, the cells were washed again to eliminate the mating factor and then re-established in fresh medium. Every 15 min, 250-μm samples were taken for cell cycle analysis. Cells were fixed in 70% ethanol treated with RNases and ultrasound, then stained with PI. Similar to FACS analysis, the varying fluorescence intensity reflects the number of cells in the various stages of the cell cycle.

are virtually impossible to analyze using traditional flow cytometers can be extensively evaluated using laser scanning cytometry.

One or two laser beams are directed through an oscillating lens (mirror) to a spot of only a few micrometers (e.g., 5 μm) in diameter and the slide on a motorized *xy* stage undergoes gradual computer-controlled scanning. The complete result is translated into a bitmap image. As in flow cytometry, the user defines threshold values to distinguish between signal and background noise (Fig. 18.7). The size (occupied area) and the specific fluorescence of each cell, including the highest fluorescence within the cell, is calculated. In the example in Fig. 18.8, the cell cycle analysis of *S. cerevisiae* (Y486), synchronized using MATa, is shown. The DNA has been stained with PI and analyzed using laser scanning cytometry.

Both analytical methods, flow cytometry and laser scanning microscopy can be adapted for high-throughput analysis using microtiter plate handling and automation.

Although in many cases cell behavior depends on the **stage in the cell cycle**, most experiments are currently carried out in unsynchronized cell populations. The synchronization of cell cultures (e.g., mammalian cells) is not always as straightforward as it has been described here for yeast cells. The analysis of the distribution of cell cycle phases, however, is indeed very easy and can be carried out using various methods. Changes in the distribution of cell cycle stages are simple indicators of the impact of the experimental conditions on the cell cycle.

19
Microscopic Techniques

Learning Objectives

Microscopes are used to view samples or objects. Three branches of microscopy can be distinguished: optical, electron, and scanning probe microscopy. Optical and electron microscopy involve the diffraction, reflection, or refraction of electromagnetic radiation, or an electron beam interacting with the studied subject, and the subsequent collection of this scattered radiation in order to build up an image. This process may be carried out by wide-field irradiation of the sample (e.g., standard light microscopy and transmission electron microscopy) or by scanning of a fine beam over the sample (e.g., confocal laser scanning microscopy and scanning electron microscopy). Scanning probe microscopy involves the interaction of a scanning element with the surface or object of interest. The resolution of an optical or electron microscope is determined by the wavelength of the radiation used. When using visible light, the resolution remains above 200 nm; the classical microscope is therefore not suited to study molecular structures. New techniques, however, allow us to circumvent the classical limit of microscope resolution deduced by Abbe, so that today molecular structures can be approached by light microscopes. According to de Broglie, electrons have wave properties. The wavelength of electrons is much smaller and allows for a resolution in the nanometer range, so that electron microscopes are ideally suited for cellular research on the molecular level. The electron microscope, however, suffers from other limitations. In this chapter, the physical origins of these limitations are described and modern techniques are discussed.

19.1
Electron Microscopy

In the **transmission electron microscope (TEM)**, developed by E. Ruska and M. Knoll in 1931, a beam of electrons is transmitted through a thin specimen, interacting with the specimen as it passes through (Fig. 19.1). An image is formed from the interaction of the electrons with the specimen; the image is magnified and focused onto an imaging device (such as a fluorescent screen, on a layer of photographic film, or to be detected by a sensor such as a CCD camera). TEMs are capable of imaging at a significantly higher resolution than light microscopes, owing to the small de Broglie wavelength of electrons. In a conventional TEM, the electron beam is produced by a hairpin-shaped cathode, and attracted and accelerated to the anode by applying a high voltage. The electron beam passes through a hole in the anode and then follows a path similar to a light beam in a light microscope. The higher the applied voltage, the shorter the electron wavelength and the higher the resolution obtained. Modern high-voltage electron microscopes for material sciences work at 700–3000 kV (i.e., a wavelength of about 1 pm) with a very high resolution. This allows objects to be thicker than in common electron microscopes. Electron microscopes for the analysis of biological probes work at only 80–200 kV acceleration voltage in order

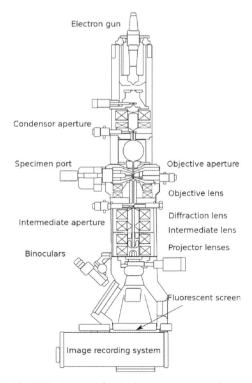

Fig. 19.1 Layout of optical components in a basic TEM. (Source: Wikipedia).

An Introduction to Molecular Biotechnology, 2nd Edition.
Edited by Michael Wink
Copyright © 2011 WILEY-VCH Verlag GmbH & Co. KGaA, Weinheim
ISBN: 978-3-527-32637-2

not to harm the biological samples. Additionally, errors in the lens systems and the methods of sample preparation have a strong influence on the resolution. For physical materials, modern electron microscopes reach a resolution of 0.2 nm. This high resolution is not obtained for biological samples due to the limited possibilities to vary the preparation protocols and the low contrast (see below) of biological objects.

Manipulation of the electron beam is performed using two physical effects. The interaction of electrons with a magnetic field will cause electrons to move according to the "right-hand rule", thus allowing electromagnets to manipulate the electron beam. Magnetic fields form a magnetic lens of variable focusing power; the lens shape is determined by the magnetic flux. Additionally, electrostatic fields can cause the electrons to be deflected by a constant angle. Coupling of two deflections in opposing directions with a small intermediate gap enables the formation of a shift in the beam path (used in the TEM for beam shifting). From these two effects, as well as the use of an electron imaging system, sufficient control over the beam path is possible for TEM operation. The optical configuration of a TEM can be rapidly changed, unlike that for an optical microscope, as lenses in the beam path can be manipulated, have their strength changed, or be disabled entirely simply via rapid electrical switching. The dynamic lenses of a TEM allow for beam convergence, with the angle of convergence as a variable parameter, giving the TEM the ability to change magnification simply by modifying the amount of current that flows through the coil, quadrupole, or hexapole lenses. Typically, a TEM consists of three stages of lensing. The stages are the condenser, objective, and projector lenses. The condenser lenses are responsible for primary beam formation, whereas the objective lenses focus the beam down onto the sample itself. The objective lenses are used to expand the beam onto the imaging screen. The magnification of the TEM is due to the ratio of the distances between the specimen and the objective lens' image plane.

The electron beam passes through the object where it is partially diffracted. The degree of diffraction depends on the electron density of the atoms in the sample – the higher the atomic mass of the sample, the stronger the diffraction (the elastic scattering increases with the atom order number Z more than proportionally according to $Z^{4/3}$). After passing through the sample, the scattered electrons are collected by an objective. The resulting image is visualized on a fluorescent screen and fixed on a photographic film or stored digitally. The resulting images are always black and white, and the degree of black density corresponds to the electron density of the analyzed sample. The figure contrast originates from the differences in atomic masses. As biological samples mainly contain atoms of low atom order number (C, H, N, and O), the observed contrast is low. In order to improve visualization of cellular structures, these samples are treated with special contrasting agents (heavy metals, such as uranyl acetate or lead citrate). Samples should not be thicker than 100 nm in order to avoid electron absorption (electron absorption increases the temperature, which might result in sample destruction).

With the **scanning transmission electron microscope (STEM)**, the electrons pass through the specimen, but the electron optics focus the beam into a narrow spot that is scanned over the sample. By using a STEM and a high-angle detector, it is possible to obtain atomic-resolution images where the contrast is directly related to the atomic number. The motivation for STEM imaging of biological samples in particular is to make use of dark-field microscopy; here, the STEM is more efficient than a conventional TEM, allowing high contrast imaging of biological samples without staining. Using this technique, large molecules and molecular complexes can be studied better and more carefully as compared with the TEM. Applying the scanning electron microscope (SEM), conductant surfaces are displayed. For this method, biological samples have to be made conductive by a thin metal film (mainly gold). In general, the SEM resolution is smaller as compared to the TEM; however, the depth of focus is much higher. The sample surface is scanned point by point by the electron beam, creating secondary electrons. The intensity of the secondary radiation depends on the topography of the sample surface. The secondary electrons are collected by a detector placed inclined behind the sample and the signal is amplified electronically.

19.1.1
Cryo-electron Microscopy

When using the **cryo-electron microscope (cryo-EM)**, the sample in solution is quick-frozen (in a few milliseconds) at very low temperatures (mainly in liquid ethane or propane) followed by electron microscope analysis. This method allows for the analysis of native samples in aqueous solution. The freezing process must be so fast that the water in the sample buffer cannot crystallize; thus, in the sample the water keeps it's glass-like structure. If the water would be allowed to form crystals, this would detract water molecules from the hydration shell of the biological molecules. Water crystal formation might influence or even destroy the structure of the biomolecules and their complexes.

The samples are frozen in thin slides (typically much less than 500 nm) since electrons cannot normally penetrate thick probes without multiple scattering. The thin frozen slides are then analyzed using a cryo raster electron transmission microscope. Due to the low sample temperature of liquid nitrogen (about −196 °C) or even lower, they can be analyzed in the high vacuum of the electron microscope. Most biological samples are sensitive to radiation. Therefore, they have to be studied at low radiation doses. The very low temperature is an additional protection against radiation damage.

The electron microscope images obtained have a very low contrast, since biological samples rarely contain heavy atoms. The images also have a low signal-to-noise ratio. In order to increase the signal-to-noise ratio and to obtain high-resolution information on the sample, it is necessary to superimpose many single images. Here, it is essential that the superimposed images indeed are the same views of the specimen. For biological samples with high internal symmetry (e.g., for some phages), this is easy to perform. Most biological samples, however, have no or only a very small degree of internal symmetry. In these cases, the obtained low-contrast electron microscope images must be grouped so that only images of the same sample orientation are superimposed. This treatment of the low-contrast images requires intense image processing by sophisticated software. In order to support this data analysis, the sample should not contain any contamination. This requires elaborate sample preparation before the structure is studied by the cryo-EM.

When successful, this method delivers a 3D structure of the unlabeled sample with a resolution in the range of 1 nm, which correctly displays the topological architecture of the protein or complex. The atomic structure, however, is normally not obtained. Single protein subdomains can be localized by obtaining a second cryo-EM structure, although this time with a label marking a single subdomain by antibody binding or fusion of a marker protein (e.g., **Green Fluorescent Protein (GFP)**) to the N- or C-terminus of a protein.

The various interactions within a complex can then be verified independently by applying other techniques to determine protein-protein interactions (e.g., the yeast two-hybrid method).

19.1.2
Electron Tomography

Normal TEM images are always 2D projections of 3D structures (even ultrathin samples have a 3D expansion). Electron tomography tries to reconstruct this third dimension. In order to do this, images are taken from the same sample area at a series of different inclination angles (e.g., in steps of 1° between +70° and −70°). Then, the images in the series are aligned and a 3D reconstruction of the common volume is obtained by filtered reprojection. The tomogram obtained can be used to segment different structures in the volume; these structures can be displayed in 3D models. As its ultimate goal, cryo electron tomography aims to obtain a 3D structure of an entire cell in its native state at molecular resolution.

19.2
Atomic or Scanning Force Microscopy

The **atomic (AFM) or scanning force microscope (SFM)** has a resolution of fractions of a nanometer – more than 1000 times better than the optical diffraction limit. The precursor to the AFM, the scanning tunneling microscope, was developed by Binnig and Rohrer in the early 1980s (earning them the Nobel Prize for Physics in 1986). Binnig, Quate, and Gerber invented the first AFM in 1986. The AFM is a modern tool for imaging, measuring, and manipulating matter at the nanoscale. The information is gathered by scanning the surface with a mechanical probe. Piezoelectric elements, which facilitate tiny but accurate and precise movements on (electronic) command, enable very precise scanning.

The AFM consists of a microscale cantilever with a sharp tip at its end that is used to scan the specimen surface (Fig. 19.2). The cantilever is typically silicon or silicon nitride with a tip radius of curvature of the order of nanometers. When the tip is brought near to a sample surface, forces between the tip and the sample lead to a deflection of the cantilever. Depending on the situation, forces that are measured in the AFM include mechanical and chemical, electromagnetic, and capillary forces. Typically, the deflection is measured using a laser spot reflected from the top surface of the cantilever into an array of photodiodes. If the tip was scanned at a constant height, the tip might collide with the surface, causing damage. Hence, in most cases a feedback mechanism is employed to adjust the tip-sample distance to maintain a constant force between the tip and the sample. Traditionally, the sample is mounted on a piezoelectric tube that can move the sample in the z direction for maintaining a constant force, and in the x and y directions for scanning the sample. In newer designs, the tip is mounted on a vertical piezo scanner while the sample is being scanned in the x and y directions using another piezo block. The resulting map represents the topography of the sample.

Depending on the application, the AFM can be operated in a number of modes. In general, possible imaging modes are divided into static (contact) modes and a variety of dynamic (noncontact) modes where the cantilever is vibrated. In static mode operation, the static tip deflection is used as a feedback signal. As the measurement of a static signal is prone to noise and drift, low stiffness cantilevers are used to amplify the deflection signal. However, close to the surface of the sample, attractive forces can be quite strong, causing the tip to snap-in to the surface. Thus, static mode AFM is almost always used in contact where the overall force is repulsive. In contact mode, the force between the tip and the surface is kept constant during scanning. In dynamic mode, the cantilever is externally oscillated at or close to its fundamental resonance frequency. The oscillation amplitude, phase, and resonance frequency are modified by tip-sample interaction forces; these changes in oscillation with respect to the exter-

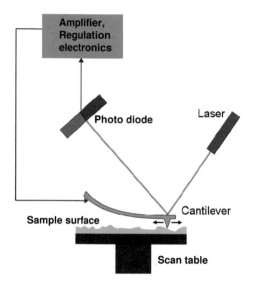

Fig. 19.2 Functional principle of the AFM. The scan table moves the sample under the sharp tip of the cantilever (see arrows). The different sample heights induce a stronger or weaker cantilever bending, thus changing the reflection of the laser beam that, as a consequence, will hit the photo diode at a different location. (Source: Wikipedia).

nal reference oscillation provide information about the sample characteristics. Schemes for dynamic mode operation include frequency modulation and the more common amplitude modulation. In frequency modulation, changes in the oscillation frequency provide information about tip-sample interactions. Frequency can be measured with very high precision and thus the frequency modulation mode allows for the use of very stiff cantilevers. Stiff cantilevers provide stability very close to the surface. In amplitude modulation, changes in the oscillation amplitude or phase provide the feedback signal for imaging. In amplitude modulation, changes in the phase of oscillation can be used to discriminate between different types of materials on the surface. Amplitude modulation can be operated either in the noncontact or in the intermittent contact regime. In the dynamic contact (tapping) mode, the cantilever is oscillated such that the separation distance between the cantilever tip and the sample surface is modulated. The tapping mode is gentle enough even for the visualization of supported lipid bilayers or adsorbed single polymer molecules in the low nanometer thickness range under liquid medium. Amplitude modulation has also been used in the noncontact regime to image with atomic resolution by using very stiff cantilevers and small amplitudes in an ultra-high vacuum environment.

19.2.1
Force Spectroscopy

Another major application of the AFM (besides imaging) is force spectroscopy – the measurement of force-distance curves. For this method, the AFM tip is extended towards and retracted from the surface as the static deflection of the cantilever is monitored as a function of piezoelectric displacement. These measurements have been used to measure nanoscale contacts, atomic bonding, van der Waals forces, dissolution forces in liquids, and single molecule stretching and rupture forces. Using this approach, folding forces of biomolecules can be measured when the biomolecule is linked to the tip by one end and to the surface by the other end, followed by slow tip retraction from the surface. Forces of the order of a few piconewtons can be routinely measured with a vertical distance resolution of better than 0.1 nm.

19.2.2
Advantages and Disadvantages

The AFM has several advantages over the SEM. Unlike the electron microscope, which provides a 2D image of a sample, the AFM provides a true 3D surface profile. Additionally, samples viewed by the AFM do not require any special treatments (such as metal or carbon coatings) that might change or damage the sample. While a classical electron microscope needs an expensive vacuum environment for proper operation, most AFM modes can work perfectly well in ambient air or a liquid environment. This allows the study of biological macromolecules and even living organisms. In principle, the AFM can provide higher resolution than the SEM. It has been shown to give true atomic resolution in ultrahigh vacuum and in liquid environments. High-resolution AFM is comparable in resolution to the scanning tunneling microscope and TEM. Compared to the SEM, a disadvantage of the AFM is the image size. The SEM can image an area of the order of millimeters by millimeters with a depth of field of the order of millimeters. The AFM can only image a maximum height of the order of micrometers with a maximum scanning area of around 150×150 µm. Furthermore, an incorrect choice of tip for the required resolution can lead to image artifacts. Traditionally, the AFM could not scan images as fast as an SEM, requiring several minutes for a typical scan, while a SEM is capable of scanning at near real-time (although at relatively low quality) after the chamber is evacuated. The relatively slow rate of scanning during AFM imaging often leads to thermal drift in the image. However, several fast-acting designs were suggested to increase microscope scanning productivity. Due to the nature of AFM probes, they cannot normally measure steep walls or overhangs. Specially made cantilevers can be modulated sideways as well as up and down (as with dynamic contact and non-contact modes) to measure sidewalls, at the cost of more expensive cantilevers and additional artifacts.

19.3
Light Microscopy

Light microscopes are well suited to visualize organelles, intact cells, or whole tissue. In recent years, the observation of single molecules became possible. In classical bright-field microscopy, the sample is illuminated via transmitted white light from below and observed from above. Limitations include the low contrast of most biological samples and the low apparent resolution due to the blur of out-of-focus material. Advantages are the simplicity of the technique and the minimal sample preparation required. The halogen lamps used produce light of high intensity in the whole visible spectrum. This light is focused onto the object by condenser lenses. The aperture of the illuminating light, determining the lit area of the object, can be varied by a variable shutter under the condenser. The condenser lens systems are corrected for chromatic and spherical aberration. By selecting a suitable objective, amplifications of 2- to 100-fold can be obtained. The eyepieces are constructed such that they produce a virtual amplified image. The amplification factor is in the order of 4- to 10-fold. Parallel light beams leaving the eyepiece are focused in the eye of the observer or in a camera, resulting in a sharpened image. Due to scattering effects, the resolution of the microscope is limited to about half the wavelength of the used light, as Abbe found out, and is larger than 200 nm in the xy plane and larger than 500 nm in the axial z plane – too large to observe single molecules or their complexes.

Limitations of **standard optical (bright-field) microscopy** are that it can only effectively image dark or strongly refracting objects, that diffraction limits resolution to approximately 200 nm, and that out-of-focus light from points outside the focal plane reduces image clarity. Since the internal structures of the cell are colorless and transparent, live cells generally lack sufficient contrast to be studied well without treatment. These limitations can be overcome to some extent by dark-field and phase contrast microscopy, which increase the contrast of the image by noninvasive methods. This technique makes use of differences in the refractive index of cell structures and changes this difference in phase into a difference in amplitude (light intensity).

Dark-field microscopy improves the contrast of an unstained transparent specimen. Here, illumination uses a carefully aligned light source to minimize the quantity of unscattered light entering the image plane, collecting only the light scattered by the sample. Dark-field can dramatically improve image contrast, especially of transparent objects, while requiring little equipment setup or sample preparation. However, the technique suffers from low light intensity in the final image of many biological samples and is affected by low apparent resolution.

Phase contrast microscopy, developed by Zernike in the 1930s (for which he was awarded the Nobel Prize in 1953), displays differences in refractive index as a difference in contrast (e.g., the nucleus in a cell will show up darkly against the surrounding cytoplasm). The contrast is excellent; however, it is not suitable for thick objects. Frequently, a halo is formed even around small objects, which obscures details. The system consists of a circular annulus in the condenser that produces a cone of light. This cone is superimposed on a similar sized ring within the phase objective. Every objective has a ring of different size, so that for every objective another condenser setting has to be chosen. The ring in the objective has special optical properties: it reduces the direct light in intensity, but more importantly, it creates an artificial phase difference of about a quarter wavelength. As the physical properties of the direct light have changed, interference with the diffracted light occurs, resulting in the phase contrast image.

The use of **interference contrast** is superior. Here, differences in optical density will show up as differences in relief. In the most often used differential interference contrast system according to Nomarski, a nucleus within a cell will actually show up as a globule. However, this is purely an optical effect and the relief does not necessarily resemble the true shape of the object. The contrast is very good and the condenser aperture can be used fully open, thereby reducing the depth of field and maximizing resolution. The system consists of a special (Wollaston) prism in the condenser that splits the light into an ordinary and an extraordinary beam. The spatial difference between the two beams is minimal – less than the maximum resolution of the objective. After passage through the specimen, the beams are reunited by a similar prism in the objective. In a homogeneous specimen, there is no difference between the two beams and no contrast is generated. However, near a refractive boundary, the difference between the ordinary and the extraordinary beam will generate a relief in the image. Differential interference contrast requires a polarized light source to function; two polarizing filters have to be fitted in the light path – one below the condenser (the **polarizer**) and the other above the objective (the **analyzer**).

19.3.1
Deconvolution

In a wide-field microscope, all parts of the specimen in the optical path are excited and the resulting fluorescence is fully detected by the microscope photodetector or camera as background signal; not only light of the focal plane of the objective but also nonfocused light from regions outside the focal plane of the object reaches the camera. Due to the superposition of the focused as well as unfocussed light, the fluorescence microscope image is blurred by the contribution of light from out-of-focus structures. This phenomenon becomes apparent as a loss of contrast especially when using objectives with a high resolving power – typically oil immersion objectives with a high numerical aperture. This phenomenon is defined by the optical properties of the image formation in the microscope. Light coming from a small fluorescent light source (a bright spot) spreads out in the axial dimension more, the further out-of-focus one is. It is possible to reverse this process to a certain extent by computer-based methods known as **deconvolution**. This can be an advantage over confocal microscopy, since no light is filtered out but all light is used for image construction.

19.3.2
Confocal Microscopy

Confocal microscopy is an optical imaging technique used to increase contrast and/or to reconstruct 3D images by using a spatial pinhole to eliminate out-of-focus light in specimens that are thicker than the focal plane. The principle of confocal imaging aims to overcome some limitations of conventional wide-field fluorescence microscopes. Confocal microscopes generate an image in a different way to normal wide-field microscopes. Using a scanning point of laser light instead of full sample illumination, confocal microscopy gives slightly higher resolution and significant improvements in optical sectioning by blocking the influence of out-of-focus light that would otherwise degrade the image. A confocal microscope uses point illumination and a pinhole in an optically conjugate plane in front of the detector to eliminate out-of-focus light (Fig. 19.3). As only fluorescence from the focal plane is detected, the image resolution, particularly in the sample depth direction, is much better than that of wide-field microscopes. However, as much of the light from sample fluorescence is blocked at

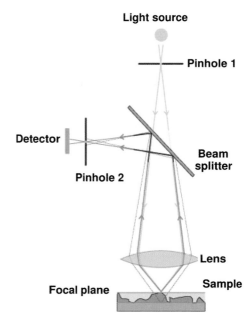

Fig. 19.3 Functional principle of the confocal microscope. Through the beam splitter and the lens, the light source illuminates a small part of the sample. All focal planes of the sample emit light that is reflected by the beam splitter. However, only the light of a single focal plane can pass pinhole 2 and reach the detector. (Source: modified from FRT GmbH (www.frt-gmbh.com/topo).)

the pinhole, this increased resolution comes at the cost of decreased signal intensity. As only one point in the sample is illuminated at a time, 2D or 3D imaging requires scanning over a regular raster (i.e. a rectangular pattern of parallel scanning lines) in the specimen. The thickness of the focal plane is defined mostly by the inverse of the square of the numerical aperture of the objective lens, and also by the optical properties of the specimen and the ambient index of refraction. The thin optical sectioning makes confocal microscopes particularly suitable for 3D imaging of samples.

Three types of confocal microscopes can be distinguished (**confocal laser scanning microscopes, spinning (Nipkow)-disk confocal microscopes, and programmable array microscopes (PAMs)**), which have their own particular advantages and disadvantages. Most systems are either optimized for resolution or high sensitivity for video capture. Confocal laser scanning microscopes generally yield better image quality than Nipkow and PAMs, but imaging frame rates are typically very slow (less than 3 frames/s). Spinning-disk confocal microscopes can achieve video rate imaging – a desirable feature for dynamic observations such as live cell imaging – but at lower resolution.

19.3.3
Why Fluorescence?

A dye molecule that absorbs light is excited into a higher energy state and re-emits a certain percentage (for good fluorescent dyes up to 80%) of the absorbed photons after typically a few nanoseconds. The emitted light is red-shifted towards longer wavelengths. When the illuminating light intensity is high enough, a single atom or molecule can in principle absorb and re-emit up to 100 million photons/s. However, the number of absorption and re-emission cycles that a single molecule can go through is limited by photochemical disruption of the molecule. Some dye molecules (such as coumarin) can go through only a few thousand such cycles emitting only a few thousand photons. Others emit up to a million photons before they are photo-bleached and thus are no longer available for observation. Since fluorescence emission differs in wavelength from the excitation light, the excitation light can be suppressed so that the fluorescent image ideally only shows an image of the molecule labeled with the fluorescent dye with an extremely high signal-to-noise ratio. This high sensitivity and specificity led to the widespread use of fluorescence light microscopy in biomedical research.

19.3.4
Nanoscopy

Two point-like objects lying next to each other cannot be separated by a conventional microscope when their distance is smaller than about 200 nm in the xy plane. Recently, however, a series of methods have been developed that circumvent Abbe's formula ruling the resolution of light microscopes. S. Hell succeeded in realizing the first laser microscopy technique that reduces the light optical resolution of two macromolecules of the same kind next to one another in a cell below the Abbe limit. His **stimulated emission depletion microscopy (STED)** uses two laser pulses. The first pulse is a diffraction-limited spot that is tuned to the absorption wavelength and excites any fluorophore in that region; an immediate second pulse is red-shifted to the emission wavelength and stimulates emission back to the ground state, thus depleting the excited state of any fluorophores in this depletion pulse. The depletion pulse illuminates the sample in the shape of a donut, so the outer part of the diffraction limited spot is depleted and the small center can still fluoresce. By saturating the depletion pulse, the center of the donut becomes smaller and smaller until resolution is in the nanometer range. A further approach to "nanoscopy" of cellular structures is **spectrally assigned localization microscopy (SALM)**. The basis for this rapidly developing technique is **spectral precision distance microscopy (SPDM)** or **spectral localization microscopy**. These methods enable the analysis of cellular nanostructures by optical detection of fluorescence. When several molecules in the sample, lying close to another, emit light, the optics of the microscope produce a diffraction (airy) disk with a diameter of about half the wavelength of the light used (above 200 nm) for each of these molecules. This cannot be avoided since light is a wave. These airy disks superimpose. The intensity profile through the center of these diffraction images results in a brightness curve that closely resembles that of a single molecule; it is hardly possible to identify a single molecule out of a collection of several close molecules, so that localization and relative distances of the single molecules cannot be deduced. However, if each of the light-emitting molecules, lying next to one another, could be identified specifically, determination of localization and distances becomes possible. Any selective property can be used which by optical methods allow us to identify the location of one light-emitting molecule and to separate it from other molecules in its proximity. One such property relates to the fluorescence spectral colors. In this case, the intensity maxima of the differently colored airy disks can be determined with high precision independent of one another and independent of the degree of superposition. Then, for every single light-emitting molecule, only the intensity maximum of the airy disk of a few nanometers diameter is registered and used for localization. In this way, the location of single molecules can be determined by optical methods even when the distance is as small as 50 nm or less. The precision, however, by which this technique can locate the center depends (amongst other parameters) on the number of photons collected. The spectral color is only one way to distinguish proximal molecules by optical methods. Alternatively, the molecules could be identified when having a specific blinking frequency. In a further procedure, the molecules are identified when they emit light at a time when most of the other molecules are dark and when they are dark while others emit light. When a sufficient amount of photons can be detected, a single light burst would be sufficient to measure the localization of a molecule with nanometer resolution. Which signal is measured first and which second is not relevant since the positions determined independently are composed to the same image that is built from many thousand contributions. This requires the use of sophisticated software for automated image acquisition and analysis. By combining thousands of images from the same cell, images with a strongly improved effective resolution are obtained. Thus, the resolution of the light microscope is driven below Abbe's limit by combining a high number of discrete experiments into a single image. Abbe's laws are still valid, of course, but their resolution limit is circumvented by experimental design.

Under specific photophysical conditions, such a blinking of chromophores can be realized for known fluorophorescent proteins like GFP and fluorescein dyes (as shown by C. Cremer). Already today, SPDM and STED techniques have improved the light optical resolution to values below 50 nm. Theoretical considerations and recent experiments showed that an even much better resolution is possible and that for some methods (like STED) no lower limit might exist.

The wide-field **structured illumination (SI)** approach also breaks the classical diffraction limit of light. As SI is a wide-field technique, it is usually able to capture images at a higher rate than confocal-based schemes like STED. The main concept of SI is to illuminate a sample with patterned light and increase the resolution by measuring the fringes in the Moiré pattern from the interference of the illumination pattern and the sample. SI enhances spatial resolution by collecting information from the frequency space outside the observable region. This process is done in reciprocal space: the **Fourier transform (FT)** of an SI image contains superimposed additional information from different areas of the reciprocal space; with several frames with the illumination shifted by some phase, it is possible to computationally separate and reconstruct the FT image, which has much more resolution information. The reverse FT returns the reconstructed image to a super-resolution image. However, this only enhances the resolution by a factor of 2 (because the SI pattern cannot be focused to anything smaller than half the wavelength of the excitation light). To further increase the resolution, nonlinear effects, represented in the FT as higher-order harmonics, can be considered. Each higher-order harmonic in the FT allows another set of images that can be used to reconstruct a larger area in reciprocal space and thus an increasingly higher resolution (down to less than 50 nm resolution).

19.4
Microscopy in the Living Cell

Several important cell biological results can best be obtained by experiments in the living cell, among them information on dynamic processes in the cell. *In vivo* experiments are of central importance for several reasons.

For **enzymes**, the situation is rather simple – these proteins have a well-defined biochemical function that can be verified not only *in vivo*, but also *in vitro*. When expressed heterologously, the correct folding of enzymes can be estimated by determining their specific activity. The situation is much more complicated for nonenzymatic proteins that function in the cellular context or as members in a protein complex. When these proteins are expressed heterologously, it remains unclear if these proteins are folded correctly, since correct folding can only be verified when the functionally relevant context within the cell is included in the analysis. The situation becomes even more complex when the proteins not only have one but more functions, potentially at different locations in the cell and/or at different time points during the cell cycle. Additionally, many proteins are chemically modified in order to activate, inhibit, or degrade them. In general, this modification is not or not correctly obtained in the heterologous system, in particular not in *Escherichia coli*. Many biochemical procedures respond to this situation by isolating the native active proteins in their cellular context (or as a member of a protein complex) and analyzing these isolated structures. Information on the time dependence of function during the cell cycle can be obtained by synchronizing the cells and isolating the proteins at particular time points. These experiments with synchronized cells are not always optimal: synchronization of the cells is lost over time and, for example, addition of nocodazol (mitotic arrest) modifies the protein composition of particular protein complexes and the modification of specific proteins. Fast changes in protein composition of protein complexes are thus difficult to measure *ex vivo*.

Endogenous proteins can be marked by antibodies and well studied when highly specific **antibodies** are available. Without support, antibodies are not able to pass the cell membrane. To get antibodies into the cell, the cell membrane must be treated or the antibodies must be microinjected – both processes affecting the cell. Thus, in general, when antibody labeling is used inside the cell, the cells are fixed – they have the correct internal structure, but they are no longer alive and the experiment is *in situ*, not *in vivo*. In most experiments, the epitope is recognized by a "primary" antibody. The Fc domain of this antibody is recognized by a "secondary" antibody that is responsible for detection (it is labeled by a fluorophore, a gold particle, or by other means). When multilabeling by different (differently colored) secondary antibodies is applied, the different primary antibodies must have different Fc domains – they thus must originate from different animals. A recent new alternative is to use the gene of a specific antibody, isolate it, and clone it into an expression vector that then is transfected into the cell where it expresses the (hopefully active) specific antibody. For this approach it is desirable to work with **single-chain (cameloid) antibodies** and to express the antibody gene under inducible external control so that the binding can be controlled since antibody binding might interfere with the cellular function of the labeled protein (see Chapter 28).

19.4.1
Analysis of Fluorescently Labeled Proteins *In Vivo*

The current improvements in imaging and labeling technologies, in particular fusions to fluorescent proteins, make fluorescence microscopy an important tool for quantitative biology. A broad variety of fluorescence spectroscopy techniques have been applied to monitor biomolecular function in cells via changes in spatial proximity, quenching and brightness, spectral shifts, or mobility.

A number of cellular processes can be studied *in vivo* without labeling (e.g., in phase contrast microscope or by autofluoresence). In general, however, it is necessary to label a specific protein in order to study its properties and function. The proteins labeled with fluorescent dyes can be placed in the cell (e.g., by microinjection). The fluorescent dyes, however, might influence the structure and function of the proteins, and the properties of the dyes might be modified by the cellular context; in general, this context is not known in detail. Furthermore, microinjection might influence cellular processes. For several years it has therefore become popular to clone proteins to be studied *in vivo* into a fusion with a fluorescent protein (e.g., with GFP or its mutants) and transfect the vectors into the cells. There, the genes are expressed so that they can be studied *in vivo*. This approach has several advantages but also drawbacks. If no information on the function of the protein studied is available, both termini of the protein, its N- as well as C-terminus, should be fluorescently labeled, both fusions should be studied, and the experimental results should be compared. The most used GFP (and its mutants) forms a barrel-like structure in the center of which three amino acids form a fluorophore. The β-sheets around the fluorophore protect it from the influence of the molecular environment so that its photophysical properties are very stable. This advantage is paid for by the rather large size of the fluorescent protein (27 kDa). A (much) smaller fluorophore would be desirable and might potentially be realized; this smaller fluorophore would, however, suffer from uncontrolled influences from the surrounding cellular medium. In some cases (in our hands below 20%), the fluorescent markers prevent the protein from remaining functional or further binding its complex. In these cases it sometimes helps to increase the amino acid linker length between protein and tag from about six to 30 or 50 amino acids.

In general, the gene of the fusion protein is not expressed under the control of the natural promoter. This might have the consequence that the fusion protein is not expressed in the same amount and/or at the same time during the cell cycle as the endogenous protein. Therefore, results obtained with fluorescent fusion proteins must be checked by control experiments with (antibody-labeled) endogenous proteins. By inspection in the microscope, those cells should be selected that show low expression levels, although still allowing for data with good quality. If a sensitive microscope is available, the selected cells can have an expression level of the fusion protein considerably lower than that of the endogenous protein. In any case, the expression level of the endogenous and the fusion protein must be determined by a Western blot. This experiment also tells us if the fusion protein is synthesized in full length. Stable cell lines that have integrated the fusion gene into their genome and thus always express the fusion protein in a rather constant lower level, are a big help for many experiments, in particular when the native promoter is introduced. If, furthermore, the protein level of the endogenous protein is specifically reduced (e.g., by **RNA interference (RNAi) knock-down**) to a few percent, the cell nearly exclusively expresses the fusion protein. Then, only the fusion protein is available for cellular function and the cell is forced to use it. This is an important point for the interpretation of the data. However, nearly completely replacing the endogenous by the fusion protein is not always advantageous. Some protein complexes in the cell can tolerate a large marker at one or the other protein and still retain their function; however, there might not be sufficient space available to sterically tolerate that *every* protein of a kind carries a large marker (like a GFP); the protein complex might lose its function. In this case, a mixture of endogenous and fusion proteins would be of advantage. On the other hand, in such sterically hindered cases it might be helpful to increase the linker length in the fusion between protein and fluorescent marker.

19.4.2
Fluorescence Recovery after Photobleaching

Fluorescence recovery after photobleaching (FRAP) is a method to measure the dynamics of biomolecules (mainly proteins and their complexes) in cells and liquid films. For FRAP the molecules of interest are labeled by fluorescent markers (e.g., proteins on the surface of cells with fluorescently labeled antibodies or proteins fused with GFP). The samples are analyzed in a fluorescence microscope. First, the fluorescence intensity is determined at the location of interest. At this location, the fluorescence is bleached by a laser pulse, the fluorescent molecules loose their fluorescence mostly irreversibly, and the selected location remains as a black spot or line in the cell. Then other fusion proteins with intact fluorophores can diffuse into the bleached region. Thus, by measuring the fluorescence intensity in the bleached region, the time of diffusion can be measured by following the fluorescence intensity over time. The slower the fluorescence intensity increases, the slower the diffusion of the fluorescent components. If the fusion proteins are bound in a protein complex, the fusion proteins with a bleached fluorophore must leave the complex before unbleached fusion proteins can bind. When the fusion protein is bound to a large extent and tightly to the complex, and the amount of binding sites remain constant, no fluorescence recovery will be observed. Alternative experimental FRAP variations allow us to measure dissociation rates (k_{off}).

19.4.3
Fluorescence Correlation Spectroscopy

Fluorescence correlation spectroscopy (FCS) is a highly sensitive optical detection that provides information from fluctuations in the fluorescence intensity. In general, using FCS diffusion values, the local concentrations and binding relations between diffusing molecules can be measured. In a confocal microscope, the exciting light is focused in the sample in a very small volume. When fluorescently active molecules (e.g., fluorescently labeled proteins) diffuse into the excited focal volume, they absorb light and fluoresce. The emitted photons are collected by the photodetector. The detectors must register if and when exactly a photon is detected (FCS measures the fluorescence intensity over time). The intensity profile shows peaks for every fluorescent particle diffusing through the focal volume. Each particle needs a particular time to pass through the volume; thus for a particular time photons are detected from this particle. After some time, another fluorescent particle passes through the confocal detection volume and the measuring process is repeated. This repeat is reflected in the time dependence of the measured intensities. For analysis, the measured intensities are correlated with themselves. The autocorrelation functions yield information on the concentrations of the fluorescent particles and their dynamics. FCS can be carried out in living cells. When two independent fluorescent molecules are measured at the same time and correlated by cross-correlation, the analysis might show if both molecules move together (in a complex) or alone and independently.

19.4.4
Förster Resonance Energy Transfer and Fluorescence Lifetime Imaging Microscopy

Förster resonance energy transfer (FRET) detects the radiationless energy transfer from a (fluorescent) donor to a (fluorescent) acceptor molecule; absorption and emission of a photon does not take place. By FRET, the emission of the acceptor is increased while the emission of the donor is decreased. The efficiency of energy transfer very strongly (by the sixth power) depends on the distance between the donor and the acceptor. FRET can be detected and measured by fluorescence microscopy. The measurement of FRET yields quantitative information in space and time *in vitro* and *in vivo* on the proximity of fluorescently labeled biomolecules if the donor and acceptor are closer than about 10 nm. Thus, the distance information is well below the resolution of the light microscope; it is smaller at least by a factor of 20. The development of a number of fluorescent proteins offers a number of FRET pairs to be studied (e.g., EGFP-mCherry). They allow us to measure the direct neighborhood and potentially also the interaction of biomolecules in living cells. Using FRET, not only distances but also changes in the spatial arrangement can be determined. Most often, the fluorescence intensity of the donor and, if possible, also of the acceptor is detected. Compared to detecting the fluorescence intensity, it is more elaborate but also more informative and conclusive to measure the **fluorescence decay or lifetime of the donor (fluorescence lifetime imaging microscopy (FLIM))**. In FLIM, the fluorescence decay time is recorded that is fitted by one or more exponentials. If the decay shows more than a single decay time, the energy transfer originates from more than one molecular situation. FRET offers the possibility to obtain detailed spatial information on biomolecules and their neighborhood relations to other biomolecules *in situ* and *in vivo*.

19.4.5
Single-Molecule Fluorescence

One of the best detectors of single-molecule fluorescence is the (human) eye, which needs around 40 photons to send a signal to the brain. Thus, for single-molecule fluorescence detection, it is not sensitivity that is the major problem, but distinguishing scattered stray light from the informative photons originating from the single molecule; in general, much more scattered light is seen than informative fluorescence light. Luckily, the scattered light has a different wavelength from the fluorescence light and thus can be partly removed by optical filtering. Unfortunately, such optical filters are not infinitely sharp – suppression by 10^4 is possible with simple filters and by 10^6 with high quality equipment. An additional technique for eliminating polluting photons uses the fact that scattered light reaches the detector usually after picoseconds while fluorescence light, owing to the corresponding lifetime of the states involved, arrives after nanoseconds. Thus, the ideal detector for single-molecule experiments is a sensitive (black and white) camera with a nanosecond gate and the highest quality optical filters.

tion of radiation energy by black bodies and its release also occurs in energy quanta or photons. However, these photons still have properties that could be attributed to waves rather than particles a phenomenon known as **waveparticle dualism**.

This result led to the theoretical and experimental development of quantum mechanics. It was shown, among other things, that the electron shell of atoms and molecules has a well-defined basic energy level that can only be raised or excited to higher levels in discrete steps. The change from one energy level to another is induced by the **absorption** or **emission** of a photon, such that the energy of the photon is equal to the difference between the initial and the final energy levels. Another result of Einstein's theoretical work in this area was the prediction of **stimulated emission** (1917). In **spontaneous emission**, an excited atom or molecule drops spontaneously back to its original energy level at a random moment in time by emitting a photon, whereas in stimulated emission, it is possible to induce the de-excitation process through the simultaneous presence of a second photon. However, a condition for this is that the second photon must have the same energy as the emitted photon. Thus, if there are a number of excited molecules in a given volume and one of them is spontaneously de-excited or loses energy by emitting a photon, then this photon can stimulate the de-excitement of further molecules and generate a cascade of photons. The interesting concept is that these generated photons are coherent or, in wave terminology, they oscillate in phase with one another.

Once the first experiments had confirmed the possibility of stimulated emission with gas discharge in 1928, it was discussed how light could be amplified using stimulated emission. Such amplification (**LASER**, **l**ight **a**mplification by **s**timulated **e**mission of **r**adiation) is only possible if there is population inversion between the two energy states involved in the stimulated emission process (i.e., in the relevant volume, more molecules are in an excited state than in the basic energy level state). It is, however, difficult to maintain the population inversion, unless the excited state is metastable and has a long lifetime.

The solution could be to produce an inversion between two excited states. A sufficient number of molecules is pumped into an excited state. From this excited level, the ground state is reached in several de-excitation steps, one of which is achieved through stimulated emission, producing the desired photon cascade. The first laser, a **ruby laser**, was produced by Maiman in 1960. Figure 20.1 shows the energy states of a Cr^{3+} ion in an Al_2O_3 crystal that are involved in the laser process. The chromium ion is excited into one of the excitation bands $^4F_1$ or $^4F_2$ through pumping with light at a wavelength of 404 or 554 nm, respectively. They decay quickly (50 ns) without emitting radiation into the state ^2E (which actually consists of two states of similar energy). Both ^2E states are metastable and are de-excited into the ground state through stimulated emission, producing laser radiation at wavelengths of 692.8 and 694.3 nm.

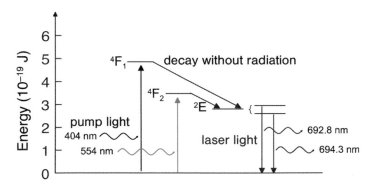

Fig. 20.1 Energy diagram of a ruby laser.

20.2
Properties of Laser Radiation

Laser radiation produced by stimulated emission differs from other light sources (e.g., light bulbs or gas discharge lamps) in several important ways. As mentioned in Section 20.1, the stimulating and the stimulated photons oscillate at the same frequency – they are coherent to one another. Although spatial coherence (i.e., the maximum distance between two photons oscillating in phase in a laser beam) and temporal coherence (i.e., the time span during which all photons passing one point are coherent to each other) varies between different types of lasers, the coherence phenomenon itself makes it possible to split a laser beam and superimpose it on itself (**interference**). This self-interference capacity can be harnessed, for example, to measure distances with utmost precision.

Another property of laser beams is their low **divergence**, which means that the diameter of a laser beam hardly increases even over many kilometers. This is due to its good focusability, which makes it possible to produce a very small spot with the help of a lens (e.g., a microscope objective). This property is not only harnessed in **laser scanning confocal microscopes**, but also used to produce high energy density. **CD and DVD players** are another application area.

Finally, lasers can be produced in such a way that the emitted radiation is **quasi-monochromatic**, (i.e., it only consists of light in a very narrow wavelength range). Line widths of a few nanometers or less are achievable in the visible range, which makes it possible to combine several laser lines without disruption. This is an advantage, for example, where several fluorescent dyes are to be excited.

20.3
Types of Lasers and Setups

Although we now know many different types of lasers, there is little difference in their basic setup. A laser-active medium is needed that is suitable for stimulated emission at the desired wavelength. This is often diluted with a host medium, as in the case of a **ruby laser** mentioned in Section 20.1. where the chromium ions are the laser-active medium, diluted in an Al_2O_3 crystal. The laser material is excited by pump light, gas discharges, electric currents, etc., and begins to emit light at the desired wavelength. For the light to be amplified and to form a laser beam, partial feedback of the photons into the laser medium is needed in order to stimulate further emissions of photons. This feedback can be realized, for example, through two mirrors. In the resulting resonator, consisting of the laser medium and the two mirrors, the laser photons oscillate back and forth. To decouple the laser beam, a mirror that is only partially reflecting is used on one side.

Figure 20.2 shows the diagram of the setup of a ruby laser. A flash tube is shown, which excites the Cr^{3+} ions, and the reflector with the two mirrors at both ends of the ruby rod.

To date, a wide range of laser types has been created. Their construction has been optimized in order to achieve the desired properties. There is a basic distinction between **continuous wave (cw) lasers** and **pulsed lasers**, where through the special configuration of the laser resonator almost all molecules are stimulated and emit light quasi-simultaneously. Pulsed lasers are mainly used where focused intensity has priority over purity of color, as pulsed lasers often show a broader color spectrum than their cw counterparts. Laser medium can be gases (He-Ne laser, argon ion laser, nitrogen laser, CO_2 laser), solids (ruby laser, neodymium laser, titanium-sapphire laser), dyes (**dye lasers**; a dye as a laser-active medium is diluted in a liquid such as methanol), or semiconductors (diode lasers).

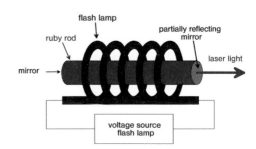

Fig. 20.2 General layout of a ruby laser.

20.4
Applications

Three representative examples of laser applications in biotechnology are given and will be discussed in more detail. These include laser scanning microscopy, optical tweezers, and laser microdissection. These three areas illustrate the most generally important properties of laser radiation – the specific excitation of molecules for light emission, the creation of a field of force for the transfer of impulses, and the targeted modification or destruction of cells and tissue through the high-performance density of a focused laser beam (laser microdissection).

20.4.1
Laser Scanning Microscopy

With laser scanning microscopy – as with conventional light microscopy – a sample is brought into the focus of a microscope objective lens. Instead of complete sample illumination, however, a quasi point-like focused laser spot probes the sample point by point. The light generated in that spot is measured with photodetectors such as photomultiplier tubes or an avalanche photodiode and, with a computer, is reassembled to a 2D image.

In confocal microscopes, the typical, clearly defined laser lines produced by beams such as argon, argon-krypton, or helium-neon lasers are used to visualize molecular structures or reactions in cells, organelles, membranes, or even molecular assays and single biomolecules that have been specifically marked by fluorescent dyes. These can then be recorded quantitatively with high temporal and spatial resolution (see Section 19.6). One example is measuring the intracellular release of **calcium ions** through calcium ion channels into **intracellular organelles** such as the endoplasmic reticulum in neurons or the sarcoplasmic reticulum in heart or skeletal muscle cells. Through intracellular release of calcium ions, the free calcium concentration of the release channels is raised from a nanomolar to a micromolar level. The local change in concentration is measured via the binding of calcium ions to fluorescent indicator molecules, such as Fluo-4, which are excited by a laser line (e.g., 488 nm of an argon laser). These changes in calcium concentration are measured quantitatively as light emissions in the region above the laser line (e.g., around 510 nm), producing microscopic flashes or sparks with a lateral localization precision of about 300 nm, while kinetics are measured in milliseconds during a simultaneous scanning of the deflection of the focused laser beam.

Such measurements can also be carried out by ultrashort pulsed lasers, such as picosecond or femtosecond lasers, which are used in multiphoton microscopy. As infrared radiation penetrates deeply into tissue and the excitation volume is extremely low because the emission depends on the square of the excitation intensity, changes in calcium concentration in neurons can be detected *in vivo*, even in the cortex of laboratory rats. Furthermore, active molecules such as **ATP** or calcium ions can be released from biologically inert **caged molecules**. Through combined laser applications, it is possible to control cellular reaction processes while taking measurements with high temporal and spatial resolution. It should be mentioned that there is already a wide range of fluorescent indicators for confocal and multiphoton microscopy, many of which are highly specific in their laser excitability, thus enabling the simultaneous observation of highly complex reaction processes. In a recent development of laser scanning microscopy, even self-resonant properties of biological molecules are harnessed for multiphoton excitation, resulting in intrinsic photon emission; next to multiphoton excited autofluorescence, these methods include **second harmonic generation (SHG)**, **third harmonic generation (THG)** and **coherent anti-raman scattering (CARS)**.

Additional laser scanning microscopy technologies with high application potentials in biotechnology are multifocal systems that allow for a significant increase in image acquisition rate or systems like **4Pi** microscopy or **stimulated emission depletion (STED)** microscopy that touch or even break the classic spatial Abbe resolution limit. The latter allow far-field imaging at an optical resolution of just a few tens of nanometers.

20.4.2
Optical Tweezers

Around 1990, lasers were used for the first time to exert low-level force (in the piconewton range) on single molecules, especially on motor proteins such as kinesin and myosin, and to measure their interacting intermolecular forces. This involves focusing long-wave laser beams (800 to about 1000 nm) through microscope objectives with a high numeric aperture. The photons of these highly concentrated laser beams transfer impulses on objects that refract the light more than their surrounding aqueous medium, such as microscopic beads (diameter between 0.1 and 1 μm, consisting of glass or polystyrene), organelles, or (small) cells. This produces a parabolic potential well in the focus of the objective, creating a force pointing towards the center of the laser spot. This optical force can be used to set the position of the trapped objects or for indirect force measurements on **motor proteins**. Figure 20.3 shows the effect of optical tweezers or trap on an object. A polystyrene bead, 6 μm in diameter (marked with an arrow), is attracted towards the center of the trapping laser. Note that none of the other objects move! The images of the image sequence shown here are captured at a rate of 25 Hz.

20.4.3
Laser Microdissection

Laser beams with high energy density, sharply focused through a microscope, can perforate cell membranes and be used generally for the microdissection of cells or tissue. Although almost all lasers used in microscopy have the ability to

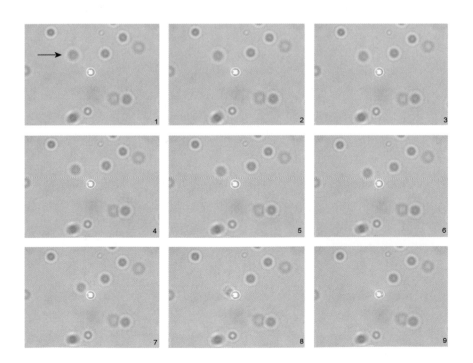

Fig. 20.3 Effect of optical tweezers or trap on an object.

damage cells at high energy, short-wave lasers such as pulsed **nitrogen lasers** (332 nm) are mostly used for precision work (e.g., the excision of single cells from tissue material). Their wavelength corresponds approximately to the actual section width. This method of cell preparation without touching the cells or their components can be combined very effectively with molecular biological or cell physiological methods.

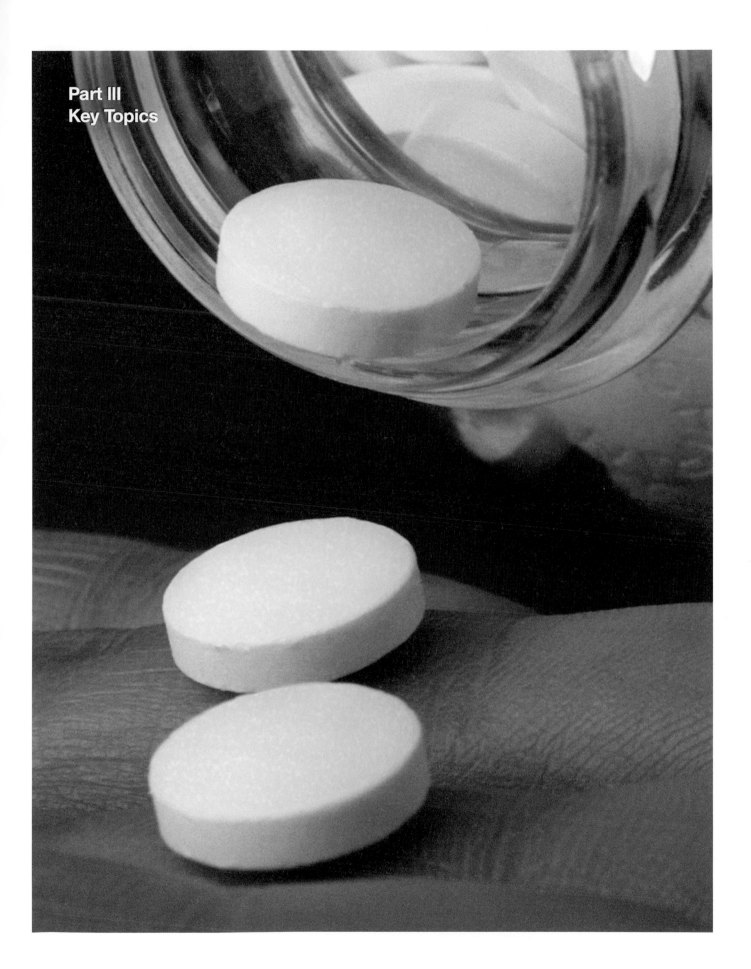

Part III
Key Topics

21
Genomics and Functional Genomics

Learning Objectives

This chapter introduces the foundations of genomics. It describes the methods of DNA and genome sequencing, and the sorting methods associated with them. Additional methods, from cDNA libraries to expressed sequence tag (EST) libraries and full-length clone libraries, are also described in detail. The development and state of the Human Genome Project will be addressed specifically. Then, the next part focuses on functional genomics. This part will begin by discussing methods by which individual genes can be identified or the function of individual genes analyzed. This leads on to methods used to investigate either a large set or all the regulated genes in a system with respect to their transcriptional activity. We then introduce cell-based assays, which takes us into a field that borders on cell biology and proteomics. Finally, we look at functional genomics from the point of view of genome-wide screens, taking as examples directed screening in yeast and undirected mutagenesis screening in the mouse.

21.1
Introduction

Today we live in a world of science dominated by DNA. This is already reflected in the generic term **genomics**, but it is also found in the term **functional genomics**. Genomics is derived from the word genome, which in the form of DNA is the carrier of the genetic information of most organisms. Our repertoire of genetic information is encoded in chromosomes and in genes, but mRNAs (referred to as the **transcriptome** in analogy to the genome) and proteins (the **proteome**) are the executing material of this information. The chromosomal DNA is virtually identical in all somatic cells. The differences in the approximately 200 different types of cells are based on processes mainly controlled by mRNAs and proteins. Furthermore, differences between individual genomes are sometimes minor. Therefore, only the study of mRNAs and proteins will allow an analysis of what makes humans and other organisms, and to what the differences can be attributed to. The once postulated **central dogma of molecular genetics** that "DNA makes RNA makes protein" has more recently been supplemented and softened by the discovery of the enzyme **reverse transcriptase** (Fig. 21.1). Other hypotheses, such as the **one geneone enzyme hypothesis**, have proven to have been formulated too tightly, which can be demonstrated impressively by the term **alternative splicing** (see Chapter 4). Thus, it really does make sense to rethink the frequently accepted postulate of the dominance of DNA.

Although RNA and proteins are functionally very important, DNA-based approaches are of high relevance. For example, without knowledge of the genetic information, it is not possible to identify and characterize an organism's genes. With the genome as a detour, **transcriptomes** and **proteomes** are easier to process and analyze. The **chromosomal localization** of a gene can often give information about possible relevance with regard to diseases, if a disease shows asso-

An Introduction to Molecular Biotechnology, 2nd Edition.
Edited by Michael Wink
Copyright © 2011 WILEY-VCH Verlag GmbH & Co. KGaA, Weinheim
ISBN: 978-3-527-32637-2

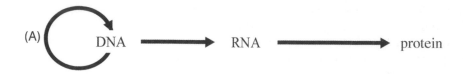

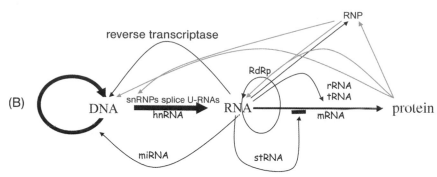

Fig. 21.1 Information flow in the cell.
(A) Originally developed picture of the information pathways (central dogma of molecular genetics). (B) An incomplete description of the connections between DNA, RNA, and protein. The arrow drawn between DNA and RNA is thicker than the one between RNA and protein, since only a small amount of RNA is actually translated to protein. The gray arrows describe the role of proteins and RNA in cellular replication processes, transcription, and translation. hnRNA, heterogenic nuclear RNA (transcription product, still includes introns); snRNP, small nuclear ribonuclear protein (in splicosomes U-RNAs are small RNAs, which are involved in splicing); rRNA, ribosomal RNA; tRNA, transfer RNA; mRNA, messenger RNA (fully processed messenger between gene and protein); miRNA and stRNA, micro RNA and small temporal RNA (small, about 21-nucleotide long functional RNAs, which play a part in cell cycle regulation, translation activity of mRNAs, and the degradation of RNAs); RdRp, RNA-dependent RNA polymerase; RNP, ribonucleoprotein (complex consisting of protein(s) and RNA(s), e.g., RNase, telomerase, spliceosome).

ciation (see below) to this region. Differences (**mutations, polymorphisms**) in the genomes of individuals and/or species not only allow the analysis of **evolutionary processes**, but also the verification and diagnosis of genetic diseases. The research on so-called monogenic illnesses (i.e., illnesses which appear due to one defective gene) was often successful without knowledge of the full genome sequence. The triggers, such as the genes responsible for the processes in redgreen color blindness or hemophilia A (blood clotting Factor VIII), have been known for a long time. **Complex diseases** such as schizophrenia, autism, or Down's syndrome, for whose appearance a variety of genes are probably responsible, can be approached much more effectively with knowledge of the genome sequence and of all genes. With the growing amount of completely sequenced genomes, including the **human genome**, the importance of approaches, which can be summarized in the term **functional genomics**, is increasing. These approaches allow the continuation of high-throughput strategies – which go beyond sequencing and towards a complete characterization of genes, the encoded RNAs, and proteins – as well as the interactions between DNA, RNA, and proteins in the context of cells and organisms. The majority of information gained with high-throughput methods, such as **genome sequencing**, makes the parallel development of **bioinformatic software** and other tools necessary in order to filter the relevant data out of the large amount present and put it into a larger context. The development of fast computers and cheap memory space were a necessary requirement for the development of high-throughput strategies. Later in this chapter this will be emphasized by the example of the **whole-genome shotgun approach**, but it concerns all areas of modern biology. A **main aim of genomics** and **functional genomics** was, and still is, to know and to understand the function of all human genes and all genes of other model organisms. Although the human genome has already been sequenced, this target has not yet been achieved. As the exact number of human genes is still not certain, we do not even know how far away from the aim we are.

In the first following paragraphs genomic approaches will be described, which create a basis for functional genomic approaches. These include technological developments in the fields of DNA and RNA analysis, a review of the human genome sequencing project, which can already be regarded as a historic process, and a forecast of the future goals of genomics with respect to diversity and disease.

21.2
Technological Developments in DNA Sequencing

About 35 years after the discovery of the structure of DNA, two technologies were created in parallel in 1977, both of which led to improvements in the sequencing technology of the time. However, the two technologies, created respectively by Maxam and Gilbert and by Sanger, which are still used today, clearly differ in their approach (see Chapter 14). Due to its automation, the quality of the sequencing and the longer read lengths, the **Sanger sequencing method** has prevailed. Further development of the **Sanger chain terminators** has taken place in all areas of biochemistry, equipment development, and analysis, which has led to a rise in the performance from a few bases to many megabases (millions of bases) that can be decoded per day by an individual. In recent years new **high-throughput sequencing technologies** (such as 454 sequencers (www.454.com) or Illumina sequencers (http://www.illumina.com/technology/sequencing_technology.ilmn)) have been developed with different methodologies (see Chapters 14 and 27). Using these new sequencers, complete genomes and transcriptomes can be analyzed in a comparably short time.

Developments in three areas were necessary to produce, administer, and utilize the large amounts of data and make them available in databases. This job is seen as a branch of bioinformatics (see Chapter 24). One must also not forget the **development of databases** (e.g., EMBL, GenBank, DDBJ, NCBI) and search systems (e.g., BLAST), which are essential for the users of the available sequence information.

All of the described developments, together with the developments in mapping discussed below, were necessary to make the sequencing of larger genomes, such as that of the human genome, possible. For example, it was only in 1997 that the sequencing of baker's yeast *Saccharomyces cerevisiae* was finalized. This genome consists of about 14 million base pairs. Many hundred scientists in many institutions were involved in the project, which took about 5 years. A genome of similar size can today (2010) be decoded in a suitably equipped sequencing laboratory within in a week. DNA sequencing has moved from a science to a routine, which is performed on a large scale by specialized institutions and companies.

The future of sequencing lies in a further increase in performance combined with lower cost. The aim is to decode complete genomes within a few hours (e.g., to make genome sequencing a part of routine diagnostics in human medicine). In the area of tumor research the plan is to sequence the genome and transcriptome of every tumor that has been surgically removed. The hope is to gain a better understanding of tumor development and to advise on more specific therapies. The ethical and legal aspects connected to this progress are not yet clear.

21.3
Genome Sequencing

21.3.1
Mapping

Depending on the strategy of a genome project (see below) it is necessary to perform a more or less extensive **genome mapping**. In closer context the term mapping is used for the definition or exact description of a place in the genome (i.e., a gene is "mapped"). In genomics, the term mapping describes the creation of maps with variable depths of information, in order to analyze larger areas of a genome, or even an entire genome, more exactly. A larger area can be an entire chromosome, a chromosomal band, or even an individual gene. Mapping makes orientation amidst the flood of information of a large DNA section easier and plays a part in performing sequencing economically (i.e., without too high redundancy). Furthermore, important **structural properties** (e.g., the position of centromeres, telomeres, deletions, amplifications, etc.) can be uncovered through mapping.

To explain the necessity of creating **maps in genomics**, we shall consider historic seafaring. Assuming an explorer of earlier centuries found a new island, he would first outline it on his map. Further expeditions then roughly investigated the area, and the original sketched map was enhanced with geographical features such as rivers and mountains, which functioned as landmarks for orientation and estimating distances. More detailed maps of interesting areas were then created. Later mapping may concentrate on vegetation and treasures of the soil, and further information was gained through, for example, questioning natives. In a similar way, genome maps first concentrate on rough maps, and keep going into more and more detail until, finally, sequence information is acquired. This procedure is defined as a top-down strategy (Fig. 21.2).

Whole-genome shotgun sequencing can, however, be seen as a jigsaw puzzle. The correct pieces are put together in such a way that a picture appears. In this

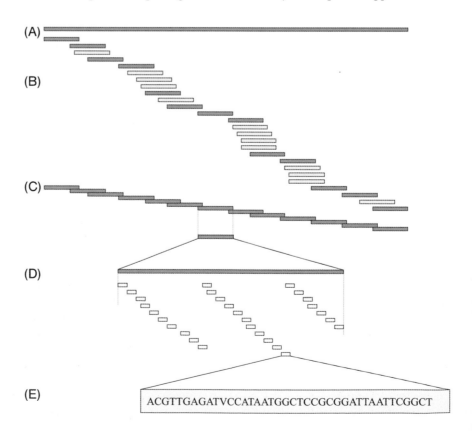

Fig. 21.2 Principles of genome mapping.
(A) Genome (e.g., a chromosome) and its representation in a clone library (fragments) (B). (C) The dark gray fragments represent the minimal tiling path (e.g., the genome can be completely covered by this minimal amount of clones).
(D) One of these clones is subcloned to plasmids.
(E) By analyzing the order of overlapping sequences the entire sequence can be reconstructed

ACGTTGAGATVCCATAATGGCTCCGCGGATTAATTCGGCT

process there are pieces that fit together but do not lead to a sensible picture at the end and others that contain only very little or redundant (repetitive DNA) information (i.e., the blue sky in a landscape puzzle). With as much advanced information as possible, mistakes can be avoided during assembly. For example, in a puzzle you can sort the pieces in advance according to color, use a template, and better define difficult areas (blue sky) by starting with structured areas. In genome sequencing one also tries to acquire as much additional information as possible. For this reason maps, as a frame for sequence information, are very helpful – if not even essential – for many whole-genome shotgun projects. From the sequence information to the final aggregation of a genome one talks of a **bottom-up strategy**.

As described above, for the following sequencing it is convenient to have larger parts of a genome present as smaller manageable pieces cloned in bacteria (cosmids, bacterial artificial chromosomes (BACs), etc.) (see Chapter 15). One or more such clone libraries are used to create a **physical map**. This map shows the order of the fragments on the next larger DNA piece (e.g., a chromosome). The aim can be to define clones, which cover the entire genome with a **minimal tiling path**. Only these are used in further cloning steps, in order to be finally sequenced. In this way a higher level of effectiveness can be achieved. Physical maps represent real distances, measurable in base pairs, which is why they are more suitable as starting points for sequencing projects than **genetic maps**, which show genetic distances, such as the frequency of recombination events.

A common problem in mapping is the **loss of position information** (e.g., during cloning). As well as the mixing of fragments during the biochemical procedures, which at first does not enable the allocation of a particular DNA area to a bacterial clone, other additional factors also play a part. The DNA is cut in many different places, whereas (according to the method applied) neither the exact location of the interface nor the length of the fragment is determined. In this process overlapping fragments are created, whereby the redundancy of the individual DNA pieces in the DNA library is not known. The **aim of mapping** can be to identify neighboring fragments based on mutual characteristics and to finally cover the entire mapped DNA (genome, chromosome, etc.) with clones from the library. Depending on the size of the DNA concerned, advanced information available, and the scientist's preferences, different systems, which are partly combined, are available for this process.

The procedures can sometimes be differentiated into **top-down** and **bottom-up methods**. Top-down describes an approach that uses a DNA unit larger than the one from which the positional information is to be gained (e.g., the use of larger fragments for the grouping of plasmid clones in a whole-genome library). If one sorts these plasmids according to analysis of overlapping areas within a **plasmid library** (i.e., through overlapping sequences), in order to later define a longer section, it would involve a bottom-up method. This works from the bottom upwards – from the small fragments to larger, ordered areas. In practice, both strategies are often applied either in parallel or sequentially.

Techniques working with **fingerprints** and/or **landmarks** play an important role among the methods with which a library of DNA fragments is sorted. In fingerprint methods, typical characteristics of the fragments are identified and compared with one another. If a clone library has been built in such a way that individual clones can overlap, their fingerprints will display mutual characteristics and the fragments are probably neighbors or may be identical. A landmark such as a molecular marker can be used to determine the position of a certain fragment at a particular place (e.g., that of a gene on a chromosome). However, a statement about the degree of overlap of two fragments cannot be made at this point in time – several landmarks are necessary for this. Many landmarks on a DNA fragment can therefore also be used to identify a fingerprint.

Apart from this rather theoretical division, the methods also differ in the practical procedures concerned.

The oldest physical mapping method is **restriction mapping**, during which a map is created using partially **overlapping restriction fragments**. Traditionally gel electrophoresis and Southern blotting are used. New approaches successfully use microscopic techniques to make restriction fragments visible through **optical mapping**.

Methods based on hybridization, such as the binding of complementary DNA (see Chapter 11), describe a different approach. The library that is to be ordered can, for example, be attached to a membrane in an array. The probes that are hybridized can be short, labeled oligonucleotides, with which characteristic fingerprinting can be performed. **Genomic markers** (DNA sequences) or clones from other clone libraries can also be used. The **reverse procedure** – the use of arrays with oligonucleotides, **polymerase chain reaction (PCR)** products, or other markers – on which labeled clones of the library to be sorted, are hybridized to create a characteristic pattern, is also used.

Instead of hybridization of markers, a PCR with specific primers can also verify whether or not a marker is present in the DNA sequence to be mapped. A unique sequence addressed in such a way is a **sequence tagged site (STS)**.

A further possibility lies in the complete sequencing of only a few clones at first (such as a BAC or cosmid library) and, at the same time, sequencing on as many as possible clones from their ends. In this case, the clone map is created in parallel to the sequence map and is used to select the nearest, most favorably placed clone for further sequencing.

Radiation hybrid mapping and **HAPPY mapping**, however, deliver very large fragments that are not based on clone libraries.

Often a combination of several approaches is necessary in order to completely map a genome or chromosome. Occasionally optical methods such as **fluorescent *in situ* hybridization (FISH)** (see Chapter 11) can support a mapping project.

21.3.1.1 Restriction Mapping and Restriction Fingerprinting

Restriction enzymes can be applied in many different ways during mapping. In classical molecular biology the simplest method lies in creating a **restriction map** of a DNA section (i.e., of a clone) with the help of several restriction enzymes (see Chapter 12). Through various combinations, different sized fragments are generated, which are separated through gel electrophoresis and made visible. The combinational analysis then models the restriction sites onto the target DNA (Fig. 21.3 A–C).

During restriction fingerprinting a library's clones are digested by one or more restriction enzymes. Each clone has an individual pattern in the gel electrophoresis, which matches in the overlapping areas with other clones. In this way **contigs** (from "contiguous") can be defined – these consist of a DNA strand of continuous sequence.

Restriction fingerprinting was first put to use in the 1980s based on cosmid libraries to map the model genomes of the nematode *Caenorhabditis elegans* and the yeast *S. cerevisiae*. Here, however, consecutive digestions with various enzymes and radioactive labeling of the ends were still used. Major improvements and combined with this the parallel and reproducible applicability, even for larger genomes, were the introduction of much larger BACs as starting material, fluorescent labeling, and improved software for the identification of similar fingerprints (Fig. 21.3 D and E).

21.3.1.2 BAC End Sequencing

This method is an example of how sequencing and mapping can interlock, and a **physical map** is developed in parallel to the sequence. The starting material for this method, suggested by **J. Craig Venter** in 1996, is a **BAC library**, in which

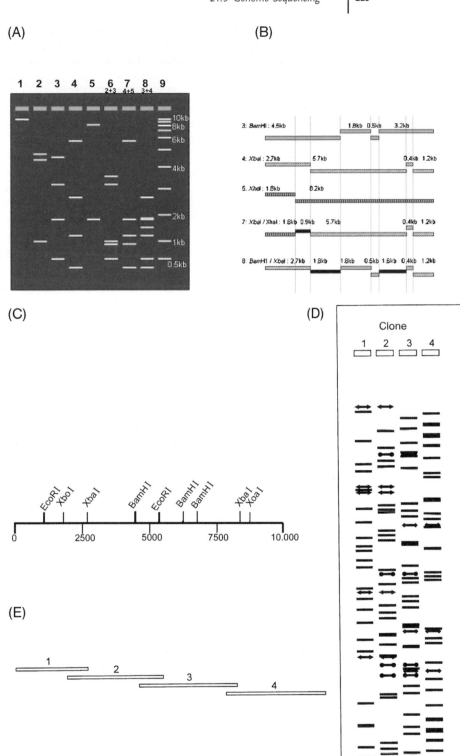

Fig. 21.3 Overview of restriction mapping (AC) and restriction fingerprinting (D and A–C). (A) Gel electrophoresis of 10-kb long DNA cut with various restriction enzymes (Track 1, uncut DNA; 2, EcoRI; 3, BamHI; 4, XbaI; 5, XhoI; 68, double digestions; 9, distance marker). (B) Through skillful combination a map can be created. Fragments that were only cut by one enzyme are at first placed randomly. In double digestion 7 one can already see that the length of the original 8.2-kb fragment (from 5) is exactly the sum of the fragments 0.9+5.7+0.4+1.2 kb.

New fragments are seen in black. For each fragment it can be determined by which restriction interfaces it is defined and a restriction map (C) can be made. (D) Restriction fragments (schematic) of four larger inserts (e.g., from BACs with approximately 150 kb) after digestion with a six-cutter enzyme (recognizes six base pairs) and gel electrophoresis. Restriction fragments that are the same in two clones are indicated (e.g., with arrows) and through analysis of overlapping areas a clone contig (i.e., a continuous section) is created.

Fig. 21.4 Creation of a BAC library.
(A and B) A BAC library is characterized by fingerprints and clone end sequencing (black/dark gray). (C) A starter clone is chosen from which a plasmid library is created (D), which is sequenced and assembled until the full sequence of the clone is gained (E). (F) Clones are chosen by their end sequences with the help of fingerprints showing common sequences with the starter clone (black ends). (G) Two suitably placed BACs are chosen for the next sequencing process. Once again plasmid libraries are set up, sequenced, and assembled (H and I). (J) The sequences of the BACs are put together and new overlap areas are defined at the ends.

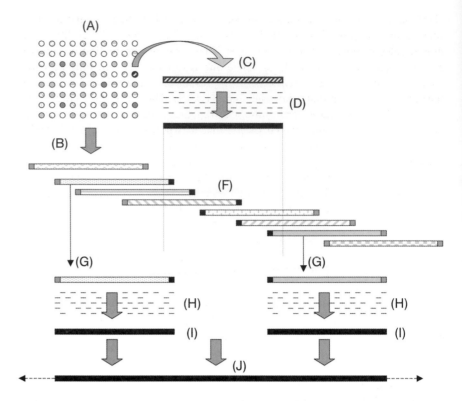

the clones are added to the sequence from both ends and are characterized by a fingerprint (Fig. 21.4). A clone is now chosen as a starting point and completely sequenced. As already described, a **plasmid sublibrary** is created from the clone's DNA, from which a representative amount of clones are sequenced and assembled. With the help of the known end sequences of the BACs, those that overlap with the starting clone can be chosen. This can be verified with their fingerprints and two BAC clones, which only have a minimal overlap, can be chosen at each end for the next sequencing round. Once these have been sequenced, best possible overlapping BACs are again searched for in the next round. The aim is to make the sequencing process as **economic** as possible and to **minimize redundancy**. The procedure is also known as **BAC walking**, since one "walks" from a starting point along the genome.

21.3.1.3 Genetic Mapping

Unlike a physical map, a genetic map is not based on distances, but on the **probability of recombination events** during meiosis. Two characteristics, which are normally inherited together, are separated from one another with a certain probability. The positions of the characteristics in the genome are referred to as **loci** and their **genetic distance** is given in centimorgan (cM), where 1 cM represents a recombination frequency of 1% (for small distances). Simplified, this means that two loci that are further apart are more likely to be separated by a recombination event. There are, however, limitations to this, given by frequently appearing, irregularly distributed **recombination events**. This is why the genetic and physical distances are not the same.

In spite of this, **genetic maps** are important tools, especially in order to identify **disease genes**. As crossing experiments with humans are not possible, in order to create pedigrees as done with fruit flies, family trees can be utilized. Often maps are created with the help of genetic markers. These are characteristics that follow a hereditary path according to Mendel and usually appear heterozygously. Classical examples are the **blood groups** A, B, AB, and zero. However, for a genetic map of sufficiently high resolution these are nowhere near sufficient, so other marker systems have been established. Initially, **restriction fragment length polymorphisms (RFLPs)** were used; these are individual differences due to polymorphisms of restriction sites. Their level of information is, however, not sufficient for all our analyses, since one restriction site can only have two alleles (i.e., two possible conditions): it is present or not. **Mini-** and **microsatellites** play a more important part (see Section 4.1.1). These are short **tandem-like repetitive DNA sequences**. The amount of repetitions differs among individuals depending on the satellite, which means that it is polymorphic, which is highly advantageous with reference to the amount of information available. Their quantity even allows mapping (at least of the human genome) with an acceptable resolution. Together with the RFLPs, the **satellite markers** have the additional advantage that they can be easily physically mapped as DNA characteristics are addressed directly. These are easy to analyze with PCR (unlike the blood groups, for example).

The aim of mapping is to cover a chromosome or the entire genome with markers in order to determine the **length and position of disease genes**, for example, more easily. For this, linkage analyses have to be performed (i.e., an analysis is made of how often the observed markers are inherited together in a certain family). From here, maps can be created, with the help of statistical methods, which show the genetic succession of markers. If necessary, further data can be integrated at this point, such as the length of genes.

21.3.1.4 Radiation Hybrid Mapping

Radiation hybrid mapping is a method (Fig. 21.5) that includes components of genetic as well as physical mapping. The aim is to create very large fragments, which can, for example, be used as a backbone for the ordered layout of the BACs, which are still very big with a size of 100–200 kb. Hybrid cell lines are the starting material. These are rodent cells (usually hamster), which include a single human chromosome. These cells are then subjected to a lethal dose of X-rays, during which strand breakage appear in the human and recipient DNA. The radiation dose determines the average length of the fragments. This way a dose of 3000 rad, for example, creates an average fragment length of 0.25 Mb,

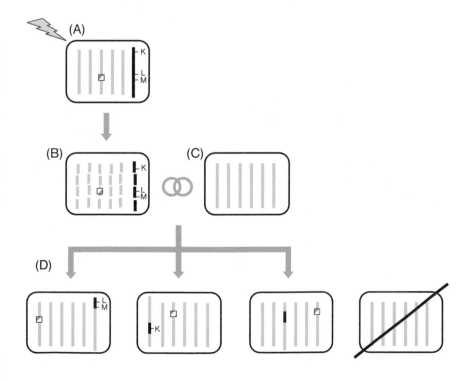

Fig. 21.5 Schematic graph of radiation hybrid mapping. (A) A hybrid cell with a human chromosome (black) is irradiated with X-rays. The cell also contains a selectable marker (box). Three loci K, L, and M are marked on the human chromosome. (B) The DNA is fragmented by the radiation. (C) A new hybrid cell is created by fusion with an untreated cell. (D) Only hybrid cells survive cultivation in a selective medium. These contain fragments of the human chromosome. A hybridization test performed with the three loci K, L, and M shows that L and M frequently appear together, and therefore lie near to one another, while K lies further away.

while a dose of 50 000 rad creates fragments with an average size of 4 kb. The lethally irradiated donor cells are themselves fused with intact cells and regenerate. A selection marker from the donor cell only lets (new) hybrids survive. These integrate parts of the fragmented DNA, and therefore also the human DNA, in the genome permanently.

An entire chromosome can be covered by a panel of 100–200 of such hybrid cells. With their DNA, the linkage of loci and, therefore, their distances can be determined by PCR, for example. Compared to genetic mapping, this method has the advantage that the fragments induced by radiation appear coincidentally (unlike recombination events) and that an approximately 10 times higher resolution can be achieved. Furthermore, polymorphic markers are not necessary. Short, known sequence pieces are sufficient for mapping (so-called STS).

21.3.1.5 HAPPY Mapping

HAPPY mapping is, in principle, related to **radiation hybrid mapping**, but can operate completely without cell systems or vectors. For this reason this technology is especially suited for genomes, which are troublesome due to a **high AT content** such as the slime mold (*Dictyostelium discoideum*) or the malaria vector *Plasmodium falciparum*. In this technology, the genome (or a chromosome) of the organism which is to be analyzed is broken up by radiation or mechanical shearing. A dilution step follows and many aliquots (preferably in microtiter plates) are prepared, each containing a haploid genome equivalent. In a test for cosegregation (i.e., mutual appearance) of two markers with PCR, these will be statistically amplified together more frequently if they are found close to one another. The use of a *hap*loid genome, as well as the *p*olymerase chain reaction, gave this method its name: HAPPY (Fig. 21.6).

21.3.1.6 Mapping Through Hybridization

In **mapping through hybridization** it is necessary to work with DNA libraries, which are bound to a **solid carrier**. This is usually a membrane (also called a "filter") made of nylon or nitrocellulose, or a modified glass surface in the form of a microscopic slide. As the DNA is applied in an ordered pattern, one speaks of an **array**, depending on the size, referred to as a **macro- or a microarray**. Microarrays are often referred to as **DNA chips**, although the term can be confusing due to its use in electronics.

During hybridization a labeled DNA in solution is bound to its complementary counterpart on the carrier. The labeling then shows at which points binding has taken place. Starting material is often a **library of clones with large inserts (i.e., cosmids or BACs)**, which are arranged in such a way that overlapping areas (contigs) appear. Different principles can be applied in this process (Fig. 21.7). It can be determined which clones show similar fingerprints, in a similar way to restriction fingerprinting, by hybridizing these clones separately on the array. The **target DNA** may consist of relatively short oligonucleotides, whose sequences were chosen so that they are more or less universally applicable. Each clone will create an individual pattern on the **oligonucleotide array**. It is then possible to determine overlapping clones through the oligonucleotides which hybridize with two clones.

It is economic if the library that is to be sorted is itself bound to the array and then individual clones from this library are hybridized. Each clone creates a signal where there is an overlap with a clone on the array (Fig. 21.7). After each hybridization, a record is generated for the hybridized sample that displays with which other clones this sample clone generated a signal. From here, a map can be established (Figs 21.8 and 21.9). In principle, other DNAs such as restriction fragments, PCR products, or clones from another library can be used as labeled samples in hybridization. The result shows which clones on the array have the

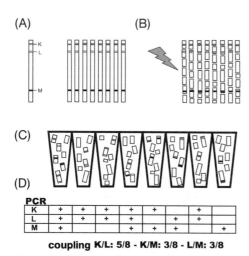

coupling K/L: 5/8 - K/M: 3/8 - L/M: 3/8

Fig. 21.6 Schematic illustration of HAPPY mapping. (A) Scheme of the genome that is to be analyzed with three markers. (B) Genomic DNA is fragmented by radiation. (C) The DNA is diluted to one genome equivalent per reaction. (D) The following PCR, with primers for the individual markers, shows that K and L amplify together more frequently, so they locate closer to one another than K and M or L and M.

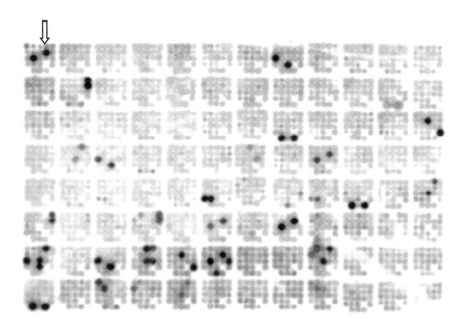

Fig. 21.7 Hybridization with a labeled cosmid (arrow) on a cosmid array. The DNA of all clones in a library was applied to a nylon filter in duplicates (for better evaluation). The cosmid of a randomly chosen clone was labeled radioactively and applied to the array. The sample hybridizes with complementary sequences from other (and also the same) cosmid. The signals are assigned to the clones and a hybridization map is calculated from them, which shows which samples hit the same clones.

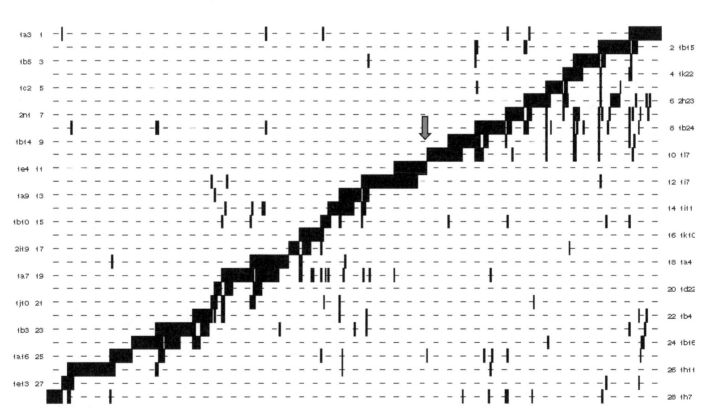

Fig. 21.8 Cosmid hybridization map of a chromosome (680 kb of *Trypanosoma cruzi*). Twenty-eight overlapping clone samples, which cover the entire chromosome, apart from the gap between samples 10 and 11 (arrow), are displayed on the ordinate. The clone names on the abscissa are not displayed. Positive hybridization events are marked with a black bar between sample and clone position at the overlapping area.

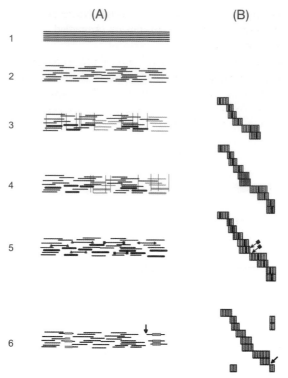

(A)   (B)

1

2

3

4

5

6

Fig. 21.9 Procedures during cosmid hybridization mapping: (A) DNA fragments and (B) the resulting maps. In the maps, the samples used are pictured horizontally (e.g., six samples in map B) and the clones hit are pictured vertically (e.g., the top sample in map B3 hits seven clones, the second sample hits four, etc.). A1: uncut genomic DNA in many identical copies (e.g., a chromosome). A2: unsorted genomic cosmid library – the relative position of the cosmids is unknown before mapping, but pictured here for emphasis. A3: six clones (fat) were chosen as samples and hybridization took place with the entire population. The thin vertical lines are to illustrate overlapping areas between the cosmids. Clones that were hit by none of the six samples during hybridization are pictured in gray. B3: first hybridization map. The six samples each recognize a different number of clones. The sample clone is always marked in bold for emphasis. Several overlapping areas can be defined. In the left part of the map only the first and second sample overlap with one clone. A4: From the entity of clones that have not yet been hit (gray) four clones are (randomly) chosen as hybridization samples (bold, gray). B4: second hybridization map. The four new hybridization clones open up new areas of the chromosome (shaded clones). All clones present in the library have now been hit at least once by a sample. A5 and B5: the map still displays one gap. The selection of two samples at the end of the respective contigs (arrows in B5) could close the gap. In A5 these two clones are marked with a box. The map displayed in B5 is sufficient to define a minimal set of clones, which can cover the entire genome. The samples used up until now are, however, not sufficient and more clones have to be chosen, as the sample clones in B5 do not actually overlap. The illustration only shows that two samples hit the same clone (shaded). The respective clones to be chosen in A5 are marked with a triangle at the end of the line. A6 and B6: the example has now been constructed in such a way that a real gap is present, which cannot be closed by clones (arrow). Apart from that, there is a repetitive area (gray boxes in A6), which appears twice in the genome. This enables cross-hybridization (i.e., a cross structure appears in the map).

same sequences, as they deliver a signal with the same sample. However, areas that generate so-called **cross-hybridization** cause problems. This means that there are various sections in the genome that hybridize with the sample.

21.3.1.7 Sequence Tagged Sites, Expressed Sequence Tags, Single Nucleotide Polymorphisms, and Sequence Length Polymorphisms (Amplified Fragment Length Polymorphisms)

The basis of many mapping strategies is the use of known sequence pieces, which one tries to place at the correct position in the genome or on the chromosome. From such markers clone maps can then be created, which give a base for complete sequencing. The check as to whether a clone contains a certain sequence is performed through PCR or hybridization. A requirement is that the sequence is unique in the genome, as otherwise cross-amplification or hybridization appears and the clone may be assembled incorrectly. Short fragments with a few hundred base pairs are called **STSs** and defined by an **individual primer pair**. This is also the major advantage of an STS – the primer sequence can be handed on in an electronic format and every interested scientist can quickly get hold of the DNA. The costs of primer synthesis and the many PCR reactions necessary are, however, not insignificant during large projects. It is an advantage if the region that is to be analyzed is uniformly covered by STSs. STSs are well suited for mapping **deletions** (e.g., in tumor cell lines). STSs that are in the deletion area within a sample will not be amplified during PCR.

While STSs stem from genomic DNA and therefore usually include noncoding sequences, **expressed sequence tags (ESTs)** are derived from transcribed sequences, often from cDNA libraries (see Section 21.4.3). This can be an advantage or a disadvantage, depending on the approach. On the one hand, you are focusing directly on the genes; on the other hand, you are also limited to them and therefore have problems in areas of low gene density. Furthermore, it must be considered that not all sections of a gene are convenient, but only those that show a unique sequence, which is often not the case in translated regions.

Where sequences show **polymorphisms** (i.e., individual differences amongst different people) they have particular significance, especially if they are located at the gene level. Such areas are especially suited for a combination of genetic and physical mapping, and furthermore, individual genetic differences, such as susceptibility for diseases (see Chapter 27), are of increasing importance in functional genomics.

Individual differences in a base pair, which should, however, appear within a population to a certain degree (i.e., the least common allele with a probability of 1%), are known as **single nucleotide polymorphisms (SNPs)** (see Chapters 4 and 27). Length polymorphisms of fragments are created if there is an area within the individual primers that genetically characterizes an individual by di- or trinucleotide repetitions of variable length (**microsatellite DNA**).

Such areas are not only of interest in human genetics or functional genomics, but also in plant research. Here, for example, a characterization of types with certain traits is desired with the help of relatively little sequence information (in the sense of variety, subspecies). **Polymorphic areas** are mainly identified with the help of **amplified fragment length polymorphism (AFLP)**, and **inter simple sequence repeats (ISSR)** for further characterization (Fig. 21.10).

In the case of AFLP adaptors with primer binding areas are ligated to the cut DNA after digestion with various restriction enzymes. Through the choice of selected primers for the following PCR, individual genomic fingerprints are created. AFLP fragments that prove to be unique for a certain type are then analyzed in more detail.

ISSR analysis is easier to handle than AFLP. It delivers the same or more information. For ISSR one uses one single primer whose sequence conforms to a microsatellite domain, e.g., $(CA)_{10}$. If a neighboring microsatellite is present in reverse orientation within a distance of up to 2000 bases, a PCR product can arise. Due to the variability in length among the microsatellite sections, one often receives a large number of polymorphic DNA markers with ISSR. Another advantage of ISSR is that the primers can be used universally, as microsatellites are built in the same or in a similar way in many organisms.

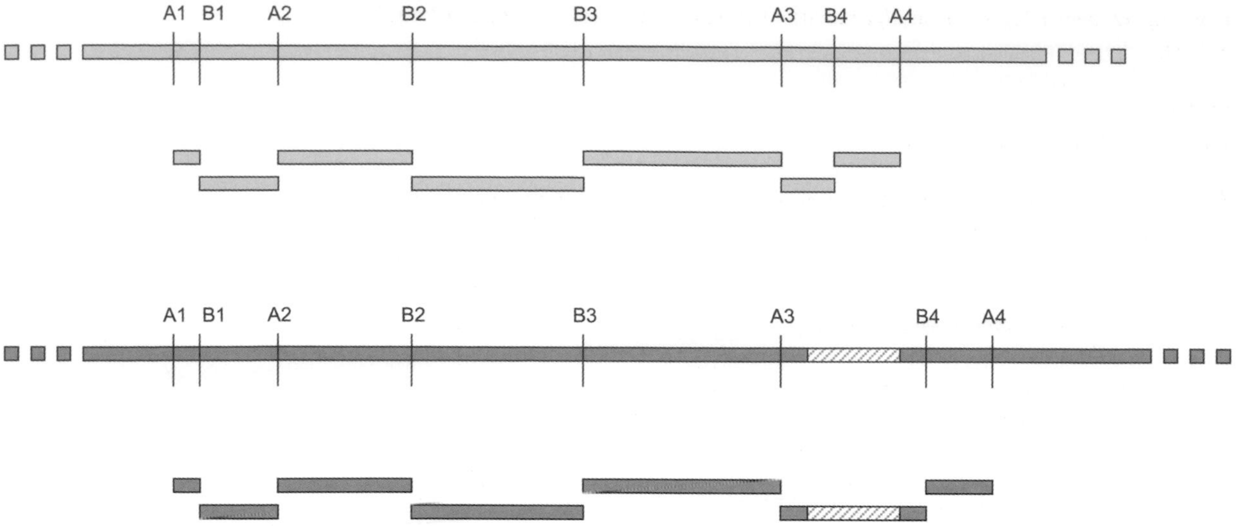

Fig. 21.10 General scheme of RFLP and AFLP analysis. The two genomes that are to be compared (light and dark gray) are fragmented (e.g., with the restriction enzymes A and B). The polymorphism lies between A3 and B4. After PCR amplification with respective primers, this is emphasized with the help of fingerprinting after gel electrophoresis.

21.3.1.8 Fluorescence *In Situ* Hybridization, Fiber Fish, Optical Mapping, and Comparative Genome Hybridization

In **FISH**, DNA with a fluorescent dye is hybridized with metaphase chromosomes. The position at which the sample binds can now be seen under a fluorescence microscope. In this way, the position of genes can be determined with an accuracy of up to 1 Mb. An improved resolution can be gained through mechanical stretching of the condensed chromosomes. In the mid-1990s, the technology was improved in a way that one could also work with the chromatin in interphase cell nuclei (**fiber fish**), which improved the resolution by a factor of up to 1000. This method is very suitable for the localization of individual samples, although not for a high throughput, as it is technically demanding and consuming.

A promising connection to mapping is provided by **optical mapping**. This method proved to be especially successful when other methods were not, due to high AT levels or large amounts of repetitive elements within the DNA to be mapped. DNA labeled with a fluorescent dye is stretched on a surface of molten agarose and then treated with a restriction enzyme. Under a microscope it can be seen how restriction fragments appear. As the fragments remain more or less stretched at their original position, their length (with the help of an added standard) and the positions of the restriction sites can be determined.

Finally, a method especially suitable for identifying amplified or deleted areas of, for example, tumors in genomic DNA, will be briefly discussed. **Comparative genome hybridization (CGH)** always compares two DNA populations (e.g., the genomic DNA from a tumor and that of a healthy tissue). The DNA is amplified and labeled with different fluorescent dyes. Then both reactions are hybridized with metaphase chromosomes and analyzed under the fluorescence microscope (see Chapters 19 and 20). DNA from an amplified area binds stronger to the complementary chromosomal DNA. For this reason this sample's dye will appear more intensely in the comparative picture analysis of the position of interest. With a deletion it is the other way around. In this way, changes to the number of copies of genomic areas on a chromosome can be mapped in an approximate area of 2–10 Mb. A further development of this method – **matrix CGH** – uses DNA chips (microarrays) with large DNA fragments (e.g., BACs or cosmids) instead of metaphase chromosomes. The resolution then depends on the covering of the area to be analyzed, and the clone size lies between 0.1 and 1.0 Mb.

21.3.2
Timeline of Genome Sequencing

Advances in genome sequencing have always been linked to technological developments and have depended on them. The first publications in which sequences were described contained information on only a few base pairs. A small genome such as the 53–86 bp of the bacteriophage *ΦX174* could already be decoded in 1978 with the comparatively simple technical solutions available at that time. The first fully sequenced **human gene** was published in 1990. The 57 kb were decoded exclusively using automatic DNA sequencing. The number of entirely decoded genomes has literally exploded, particularly in the last 5 years (Table 21.1). By 2010, more than 1000 complete genomes were sequenced (among them more than 100 from eukaryotes). This has been due to the developments in DNA sequencing described above and to the advances in cloning technology and sequencing strategies. In the "Genome 10k" project it is planned to sequence 10 000 genomes of vertebrates in order to obtain a fine resolution of the tree of life of vertebrates (www.genome10k.org) The aspects of cloning and sequencing strategies will be discussed in this section.

Table 21.1 List of eukaryotic large-scale genome projects (in addition, over 1000 bacterial and over 1000 viral genomes have been completely sequenced; http://www.ebi.ac.uk/genomes/eukaryota.html).

Unicellular life forms, algae, fungi
Aspergillus niger, A. fumigatus, A. terreus (filamentous fungi)
Candida albicans, C. glabrata (pathogenic yeasts)
Cryptococcus neoformans (pathogenic yeast)
Cryptosporidium parvum (enteroparasite)
Dictyostelium discoideum (slime mold; model organism)
Eimeria tenella (sporozoon, poultry parasite)
Encephalitozoon cuniculi (intracellular parasite)
Entamoeba histolytica (causes amebiasis)
Fusarium oxysprum (phytopathogenic mold)
Giardia lamblia (cause of giardiasis)
Gibberella zeae PH-1, G. moniliformes (filamentous fungi)
Guillardia theta (flagellate, nucleomorph organelle)
Leishmania major, L. braziliensis, L. infantum (cause leishmaniasis)
Neurospora crassa (filamentous fungus)
Ostreococcus lucimarinus (green algae)
Paramecium tetraurelia (ciliate)
Phanerochaete chrysosporium (filamentous fungus)
Phaeodactylum tricornutum (diatoms)
Pichia stipitis (yeast)
Plasmodium falciparum, P. yoelii, P. knowlesii, P. vivax (cause malaria)
Pneumocystis carinii (pathogenic fungus)
Saccharomyces cerevisiae (baker's yeast)
Schizosaccharomyces pombe (fission yeast)
Thalassosira pseudonana (diatoms)
Theileria annulata, T. parva (flagellates, bovine parasites)
Toxoplasma gondii (pathogenic sporozoan)
Trypanosoma brucei, T. cruzi (causes sleeping sickness and Chagas disease, respectively)
Yarrowia lipolytica (yeast)

Invertebrates
Anopheles gambiae (mosquito)
Brugia malayi (pathogenic nematode)
Caenorhabditis elegans, C. briggsae (nematodes)
Drosophila melanogaster, D. jacuba, D. simulans (fruit fly)
Hydra magnicapillata (coelenterate)
Schistosoma mansoni causes bilharziosis (trematode)

Vertebrates
Bos taurus (cattle)
Callithrix jacchus (primate)
Canis lupus (dog)
Ciona intestinalis (sea squirt)
Danio rerio (zebra fish)
Equus caballus (horse)
Fugu rubripes (fugu fish)
Gallus gallus (chicken)
Homo sapiens (human)
Macaca mulatta (primate)
Mus musculus (mouse)
Pan troglodytes (chimpanzee)
Pongo abelii (orangutan)
Rattus norvegicus (rat)
Ornithorhynchus anatinus (platypus)
Oryctolagus cuniculus (rabbit)
Oryzias latipes (rice fish)
Ovies aries (sheep)
Sus scrofa (pig)
Taeniopygia guttata (zebra finch)

Higher plants and algae
Arabidopsis thaliana (model plant)
Brachypodium distachyon (a grass)
Glycine max (soy bean)
Medicago truncatula (barrel medic, model plant)
Oryza sativa ssp. indica, O. sativa ssp. japonica (rice)
Porphyra yezoensis (edible red algae)
Solanum lycopersicum (tomato)
Sorghum bicolor (a grass)
Zea mays (corn)

Several steps take place upstream or downstream of the decoding of a sequence that have to be automated in order to keep up with the increasing sequencing capacity. Examples are the production of appropriate templates to be sequenced and the bioinformatic processing of large amounts of sequence information. The latter problem is solely of a bioinformatic nature (see Chapter 24), the former, however, required developments in technology, equipment, and processing technology.

For the sequencing of the **human genome**, compared to earlier projects, a much higher throughput in sequencing technology was necessary, which led to developments in the accompanying technologies. These sequence-accompanying technologies can be emphasized with the sequence strategies that were applied and developed in order to decode the approximately 3 billion bases (Gb) of the human genome. After the clone-based strategy had been practiced for many years (conventional approach) – with an intermediate step of genomic subclones – a new strategy has been developed in the last few years (whole-genome shotgun). This triggered a revolution and great controversy in this sector. This is partly due to Craig Venter, who developed the whole-genome shotgun method and applied it to human genome sequencing in a company (Celera Genomics). In order to understand the differences between the strategies, the core aspects of the two approaches will be illustrated.

21.3.3
Genome Sequencing Strategies

21.3.3.1 **Conventional Approach: Random Shotgun Strategy**

The aim of genome sequencing is to determine all the building blocks of a genome without any gaps. When a genome has more building blocks than can be detected in a sequence reaction, the problem of having to assemble the entire genome from individual sequences appears. If the genome can be fully cloned to provide a vector that is also suitable for sequencing, the genome can also be analyzed by directed sequencing (e.g., primer walking). Most genomes are of a size that far surpasses the upper limit of what can be cloned into an available vector at the moment.

The bioinformatic analysis, necessary to assemble individual sequences to contigs, also proved to be a problem at the beginning of the **Human Genome Project**. With a linear increase in the number of sequences, the calculation effort rose exponentially in the comparing of "all to all". Therefore, it was not possible for a long time to process projects with several thousand sequences with the computers and programs available.

One solution was to split the sequenced genome into **sections**, which were then processed with the technical solutions available. This was achieved in two consecutive steps. First, the genome was split into small fragments and cloned into a **genomic library**. Next, clones from this library were chosen and further fragmented (see Section 21.3.1). These fragments were cloned into sequence vectors and, by doing so, **shotgun sublibraries** were produced.

The size of genomic DNA fragments that could be cloned into genomic libraries was, for a long time (until about 1995), about 40 kb. Fragments of this size could be cloned in vectors, which were derived from the **phage λ system (cosmids)** (see Chapter 15). Cosmid vectors created the basis for sequencing of the yeast genome. For the 3.3 billion bases of the human genome these vectors were, however, not practical.

As a further development, **yeast artificial chromosomes (YACs)** were developed, which could be propagated in yeasts relatively stably and with one copy per cell. They could incorporate over 1 million base pairs. YACs clearly showed that organisms do not like foreign DNA and the YACs were therefore often subject to mutations. It was deletions, in particular, that made the YACs relatively unsuitable for genome sequencing. In mapping, however (see below), YACs were still very successfully employed. **BACs** served as a basis for sequencing the human genome, in the sequencing project run by the International Consortium and finally brought a major improvement.

Due to sequencing technologies, there were limitations in regard to the choice of sequencing strategy. Until the mid-1990s, a fluorescent marker could only be applied to the sequence primers in use (**dye primer** sequencing). Labeling primers was expensive and therefore the same primer was used for all sequencing reactions. This made the use of **vector-based primers**, which would fit any clone, necessary. Thus, it appeared sensible to further fragment the 40-kb sections of a genome, which were cloned in cosmids (or 150–500 kb sections in BACs), and clone them into sequence vectors as 1- to 2-kb fragments. If the fragmentation was achieved by shear force, the double-strand breaks appeared randomly spread across the genomic section, due to which overlapping fragments were to be cloned and sequenced. The available sequencing capacities were sufficient for sequencing a large amount of sequence clones (approximately 200 per cosmid, approximately 800 per BAC) in order to statistically achieve a complete coverage of the original cosmid, if their amount was high enough. This sequencing strategy is known as **random shotgun sequencing**, as the clones to be sequenced are randomly selected from the shotgun sublibrary.

A complete coverage of the DNA with individual sequences was, however, still not possible with random shotgun sequencing. Gaps without a sequence always remained as well as areas with sequences of low quality and therefore resulting in inaccuracy. For this reason, a directed sequencing process (finishing) was always added, in order to finally receive the complete sequence of the DNA section cloned in the genomic clone (cosmid or BAC) with high quality.

The above describes how in genomic sequencing, first, a genomic library was produced in cosmid or BAC vectors and afterwards individual clones were used to generate shotgun libraries for sequencing. The selection of appropriate clones from the genomic library was vital for finishing genome sequencing. In order to have acquired all of the genome's bases in the genomic library with high statistical probability, the library had to display a certain redundancy. One also speaks of *x*-fold coverage, in which each base is statistically captured. A good genomic library had a 10-fold coverage. This meant that each base in the genome should theoretically be included in 10 independent genomic clones. However, it also meant that a genomic library for the human genome would have to consist of about 150 000 BACs (with an average fragment length of about 200 kb per BAC and a genome size of three billion bases; 3×10^9 bp (genome size) $\times 10$ (fold covering)/200 000 bp (length of clone) = 150 000 BACs). It was, however, impossible to sequence this amount of BACs. Thus, it was necessary to sort the individual clones in the library and finally select only a minimum number of clones for the production of shotgun libraries and for sequencing (Section 23.3.1).

21.3.3.2 Whole-Genome Shotgun Strategy

In 1996, **The Institute for Genome Research (TIGR)** finalized the 1.8 million bp long sequence of the bacterium *Haemophilus influenzae*. In this case, a method in which the genome was no longer subcloned in a cosmid or BAC-based library was used for the first time. Instead *H. influenzae*'s genome was split into very small sections and random shotgun libraries were produced directly. Through the sequencing of an adequately high number of shotgun clones, one hoped to receive the entire sequence of the original genome, preferably gapless and overlapping. The step of mapping genomic subclones, which was a requirement up until then, became unnecessary.

The sequencing took place in six steps:

i. *Production of random shotgun libraries.* For the assembly of the single sequences to contigs, it not only seemed necessary to create very small clones with a fragment size of 2 kb, but also a library with an average fragment length of 15–20 kb. The first library was supposed to serve for the mass production of sequences. The sequences in the second library were to help with the assembly of the individual sequences. For both libraries it was important that the length of the cloned DNA fragments only marginally varied. Only then was it possible to make use of length information as well as the sequence itself during assembly. Furthermore, the long clones made templates available that could be used for closing many sequence gaps (see below).

ii. *Selection of the clones to be sequenced.* It was necessary to ensure that, during creation of the libraries, no sections of the genome were preferably cloned, but that it really was a random library. Furthermore, the length distribution of the fragments in the library had to be checked.

iii. *High-throughput sequencing.* The redundancy in sequencing was to be so high that statistically each base of the genome was to be sequenced 6-fold. Therefore, with an average length of 460 bases for each sequence and a genome size of 1.83 million base pairs, almost 24 000 individual sequences were theoretically necessary. In fact, 24 300 sequences were actually produced.

iv. *Assembly of the individual sequences* (Fig. 21.11). Based on overlapping areas in sequences (i.e., areas with identical areas in two separate sequences) and distance information from the 15–20 kb clones, an attempt was made to assemble the genome from the individual sequences. Up to this step almost complete automation of the processes was possible.

v. *Closing the clone gaps and sequence gaps.* A 6-fold coverage of the starting sequence is, for statistical reasons, not sufficient to reconstruct a genuinely continuous sequence. The assembly of the 24 300 sequences resulted in 140 contigs, so 140 sections of the genome, each including a large or small number of individual sequences that did not overlap. Two kinds of gaps especially needed closing. First, there were **clone gaps**. In these sequence sections there were no clones whose sequencing could close the gaps in the library. Suitable alternative strategies (e.g., PCR) had to be employed to close these gaps. Furthermore **sequence gaps** existed, where one or more available clones from the library simply had to be sequenced further to close the gap. In both cases, however, human interactions were necessary (i.e., automation of these processes was not possible).

vi. *Editing the contigs.* By closing the final gaps, the sequence of the genome had been made contiguous. The quality of the sequence was, however, not sufficient at every position. There were areas in which after analysis of several sequences, different bases were found to be present at the same posi-

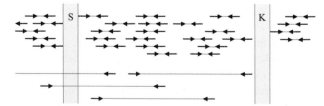

Fig. 21.11 State of the project at the end of the random shotgun phase (step (iv) in the whole-genome shotgun project). A large number of clones from the shotgun library with short (2 kb) fragments (above) and a small number of clones from the shotgun library with long (15–20 kb) fragments (below) have been sequenced. The individual sequences (arrows) were assembled and the distance information (lines between the arrows) was incorporated in the assembly. In this way, two of the three contigs could be related to one another. The sequencing gap (S) is covered by two independent long clones (below) and can be closed by sequencing these clones. On the right a clone gap (K) is illustrated, which cannot be closed by any available clone. No relation can be made between the two contigs on the left and the contig to the right of the gap. This problem can only be solved by the following steps.

tion ("ambiguities"). As a result of the assembly, it could happen that individual sections had been put together incorrectly and that the results of the distance analysis could be contradictory. These problem positions were to be analyzed. Only experienced staff were capable of doing this.

With this relief, however, an increased amount of sequencing was achieved, which was statistically necessary in order to leave no gaps in the sequence. To sequence the 1.8 million bases of *Haemophilus* conventionally, 15 BACs (each with a fragment length of 200 kb and 800 sequence reactions per BAC) would have been necessary. This would have meant about 12 000 sequencing reactions, compared to the 24 000 reactions in the whole-genome shotgun approach. The reason for the success of the whole-genome shotgun approach was that sufficient sequencing capacity was available. The developments in **automation** of sample preparation and in sequencing technology were fast enough to keep up with the rising demand for sequence information. Furthermore, TIGR had a bioinformatic department, which in the TIGR assembler developed an algorithm capable of processing the large number of individual reactions taking place. Its success reinforced TIGR. The whole-genome shotgun approach was superior to conventional approaches – in regard to the speed at which a genome could be decoded, at least for small genomes. For small genomes (up to 30 Mb) this method has prevailed. It was, however, questionable whether the whole-genome shotgun strategy could also be used for larger genomes such as that of humans.

21.3.3.3 Sequencing of the Human Genome

Compared to *H. influenzae,* the human genome is a factor of about 2000 larger. In 1998, the company **Celera** was founded with the aim of using the whole-genome shotgun strategy for sequencing the human genome and finishing this within 3 years. That of the **fruit fly *Drosophila melanogaster*** was supposed to serve as a test genome for Celera. The genomic sequence of the fruit fly was actually published in 2000. Competition with a project within the international genome project, which had been making progress for several years, had been established. In 1993, this project had already stated it wanted to finish the first 80 million base pairs by 1998 and decode the entire human genome by 2005. The constantly rising sequencing capacities were beneficial for sequencing the **nematode *Caenorhabditis elegans***, whose genome, which was completed in 1998 (see Table 21.1), was used to test strategies and technologies.

However, even after the human genome had been published by the two groups, the arguments about the respective share of the work done went on and are still unfinished. The usefulness of both projects is not in doubt, perhaps due to the competitive situation, meaning that finishing the project was quicker than expected or that the quality of the genome sequence is higher than could have been achieved by only one project.

The applicability of the whole-genome shotgun method was later accepted by the International Consortium and, eventually, the **mouse genome** was decoded using this strategy. A vital comment is appropriate in this context. Further on in the text, three very different meanings of the phrase "full-length DNA" will be explained. For the genomic sequence, the term "finished" is also associated with different meanings.

i. If DNA has a length *n*, "finished" means that the sequence of this DNA is in correct order and clearly ascertained at every position from base 1 to base *n*. The sequencing centers have set a target of only tolerating sequences with less than one mistake per 10 000 base pairs.

ii. Many DNAs consist of euchromatic and heterochromatic areas. The **euchromatin** contains the large amount of genes; the **heterochromatin**, which is very repetitive and mainly situated in the areas of the centromeres and telomeres on the chromosomes, only contains a few known genes. Heterochromatin is usually not sequenced. As long as the euchromatin is completely sequenced and the sequence is of high quality (less than one mistake in 10 kb) the genome can still be regarded as "finished".

iii. In euchromatin there are also areas with a high degree of repetitive sequences, which often cannot, or only with great difficulty, be cloned. As long as the existence and approximate length of these sections is known, a sequence of the surrounding DNA can be regarded as "finished", as long as it only displays few sequencing errors.

The genome sequences published in 2001 did not even comply with the third standard. There were a variety of clone and sequence gaps, and the assembly of the genome had many errors. In the meantime, however, the quality has been much improved, so that the human genome should at least meet the criteria for the third definition of quality. Also, in 1998, when the genome of *C. elegans* was published, it was in a state which conformed to the third standard. Particularly due to the involvement of John Sulston, most gaps have been closed in the past few years. It will still take some time until the human genome that meets the first quality standard is available.

21.3.4
Outlook for Genome Sequencing

Depending on the infrastructure available, genome projects used to be performed in individual laboratories, for whom sequencing was a necessary evil, in order to work later on with the product. This strategy meant that the sequencing of *Escherichia coli* took many years. In the yeast genome project (*S. cerevisiae*) too, a major part of the sequencing was performed by nonspecialist laboratories. In the mean time, centers which specialize in DNA sequencing and achieve high performance have appeared. Recently, high-throughput sequencers (454 or Illumina sequencers) have been replacing the former capillary sequencers. One advantage is that, due to specialization, much effort has been invested into optimizing processes. In this way costs have been lowered significantly, while the product (i.e., the sequence) is of consistently high quality. The list of genomes sequenced is extended almost daily (http://www.ncbi.nlm.nih.gov/Genomes/index.html). A variety of genomes (e.g., of pathogenic bacteria) have been decoded and the sequence information is used for the development of methods to combat these bacteria. The sequences of *Plasmodium* and the vector *Anopheles* could show points of attack when combating malaria. Comparison genome analysis (e.g., between humans and chimpanzees) for evolutionary studies will be possible for the first time. Similarities and differences between species will become apparent and interpretable.

Genome projects are now often performed by companies. The whole-genome shotgun procedure is usually the chosen approach. The time between the start and end of a project has been dramatically shortened in the past few years. Experimental work in automated labs and improved computer capacities have had a great impact. The next stated target is the sequencing of an entire genome in a day. In this way, the human genome could be put to use in diagnostics.

21.4
cDNA Projects

21.4.1
cDNA Libraries Represent the Cell's mRNA

During the expression of a gene, the chromosomal DNA in the nucleus is first transcribed to **heterogeneous nuclear RNA (hnRNA)**. This hnRNA is a full copy of the transcribed section of the gene concerned. Control elements (**promoters**) are not transcribed and are therefore not included in these RNAs (see Chapter

4). The hnRNA, however, includes the gene's introns as well as its exons. Still in the nucleus this hnRNA is processed to mRNA by **riboprotein complexes (splicosomes)**. This processing (also known as maturation) includes several steps that are partially performed in parallel. The **5'-end of the RNA** is modified, by receiving a so-called cap. At the **3'-end of the RNA**, a nuclease specifically splits off a part of the RNA and a second enzyme adds (without a template) a row of A nucleotides to the RNA. This is known as the **poly(A) tail**. In a third process, in which again several proteins and RNAs are involved, the introns are cut out (spliced) and the exons are connected to one another. Only then does the continuous sequence of the **open reading frame (ORF)**, which later in the cytoplasm is translated into protein, appear in protein-encoding RNAs. Therefore, an mRNA no longer directly represents the initial gene sequence. Nonprotein encoding RNAs such as ribosomal RNAs and transfer RNAs are also processed.

The analysis of mRNA has advantages as well as disadvantages compared to the analysis of genomic sequences.

Advantages:
- As the exons can be found in uninterrupted order in mRNA, the sequence of an encoded protein can be directly derived from the mRNA sequence.
- If the mRNA is written back to DNA, this DNA can be used to produce an encoded protein *in vitro* as well as *in vivo*. Due to the exonintron structure of higher organisms' genes, this is not possible with genomic DNA.
- With the mechanisms of **alternative splicing** (see Chapter 4), different mRNAs can often be produced by the same hnRNA. Alternative splicing is not predictable from the genomic sequence at the moment, so that the analysis of cDNA, produced from mRNA, is indispensable for finding such varieties.

Disadvantages:
- In comparison to DNA, RNA is a very labile molecule. The hydroxy group on the C2 atom of the ribose makes the RNA unstable in an alkaline milieu. At pH values above 8, RNA quickly degrades, while DNA, whose C2 atom possesses two hydrogen atoms, is still stable under these relatively mild conditions. Furthermore, all organisms possess very effective enzymes that degrade RNA. RNases also offer a protection mechanism against RNA viruses and are present in large quantities on the skin,for example.
- The analysis of DNA is simple compared to that of RNA. Double-stranded DNA can be manipulated with conventional molecular biological methods. Restriction enzymes allow DNA to be cut specifically according to its sequence, ligases connect ends, and vectors have been developed to recombinantly multiply DNAs. The already explained task of DNA sequencing has extensively been developed. Due to the characteristics of RNA, this molecule is not suitable for these established methods.
- The fraction of a specific mRNA in the entire population of mRNAs is extremely variable. Great fluctuations in the amount of mRNA produced are possible in gene expression. It depends on the gene concerned, the amount of RNA, or the amount of the encoded protein, which is required in each individual case by the tissue in which the gene is expressed, or not, and on the state of differentiation in which the organism finds itself. The following serve as examples. The pepsin gene is related to a specific organ – the stomach – and can therefore only be isolated from stomach tissue in form of mRNA. At the same time, pepsin-producing cells produce a large amount of this enzyme and therefore also its mRNA. Thus, pepsin mRNA is present in these cells in large quantities. However, other proteins, such as many transcriptional factors and surface receptors, are only required in small quantities, so the associated mRNAs are only present in small quantities within these cells. Many genes specific to development are only expressed at a certain time during embryonic development and are **silenced** at later stages. In order to study the mRNA of

these genes, embryonic tissue must be prepared, which, for some species such as humans, is not easy for ethical reasons. In a genome, however, the genes are normally present in an equal number of copies. In humans there are two copies of every chromosome; each gene therefore has two alleles. For this reason, if the complete genome of a species has been sequenced, all genes have been captured. Therefore, in the analysis of genomic sequences, only the problem of actually recognizing these genes remains. This is where mRNA and cDNA play a key role.

As mRNA is unstable compared to DNA and cannot be cloned or directly analyzed with regard to the sequence, the mRNA is rewritten into DNA – so-called **cDNA or complementary DNA** – with the help of the enzyme **reverse transcriptase** (see Chapters 4, 12, and 15). The cDNA is made double-stranded by a DNA polymerase and is then cloneable. If the entire mRNA of a cell or a tissue is isolated and then rewritten to cDNA, packed in vectors, and cloned in bacteria, a so-called **cDNA library** is finally obtained. Such a cDNA library in the best case contains cDNAs that represent all mRNAs present in the cell, or the tissue, or the organism concerned at the time of the mRNA isolation. Therefore, in the ideal case only the production of one single cDNA library should be necessary to gather all genes of an organism. Depending on the abundance of mRNAs of individual genes in the entire population of mRNAs of the tissues, the problem of actually gathering all genes expressed arises.

It was previously described that theoretically a single cDNA library should be sufficient to gather all genes of an organism and to isolate the respective cDNA clones for each gene. The problem of varying gene expression in the individual tissues has already been discussed. A second problem results from the technology of cDNA synthesis and cloning. Therefore, in the following, the process of cDNA generation will be illustrated, from which many difficulties in cDNA-based analysis become clear.

21.4.2
Production of cDNA Libraries

The basis for the production of cDNA is mRNA, which serves as a template. It has already been described above that mRNA is an unstable molecule, which quickly degrades. First, the scientist is responsible for this degradation, as he/she has **RNases** on his hands and also perhaps on the equipment. This problem can be minimized by wearing gloves and carefully using plastic containers. A second source of RNases are the tissues from which the RNA is to be isolated. Each cell has RNases at its disposal in order to keep up the unstable balance of gene expression. To keep the time available for cellular RNases to degrade the short RNA, the acquired material is either prepared straight away or shock-frozen for storage. The ultimate aim is to isolate high-quality RNA from the tissue.

The mRNA amounts to a maximum of 5% of the entire RNA within a cell. Therefore, an enrichment of this fraction is usually performed to keep the amount of contaminated ribosomal or other RNAs in the cDNA library low. Two characteristics differentiate mRNA from other RNA species. These are the cap structure at the 5'-end of the mRNA and the poly(A) tail. Only mRNAs, that have a very short lifespan (e.g., **histone RNAs**), have no poly(A) tail and are not present in conventional cDNA libraries. The cap structure is only small and does not play a part in the enrichment of mRNA, although it is essential for the selection of full-length cDNAs. However, the **poly(A) tail**, which is often over 100 nucleotides long, is suited for the selection of mRNAs with this structure. Oligo(dT) poly nucleotides (known as **oligo(dT) primers**), which are coupled to a carrier matrix (e.g., agarose), are incubated with the complete RNA (see Chapter 15). mRNAs, which have a poly(A) tail, are taken up by the oligo(dT) primers and can be isolated (eluted) from the matrix after nonbound RNAs have been washed away.

Finally, the mRNA, in fact the poly(A)-containing RNA (or poly(A)+ RNA), is reverse transcribed to DNA as the first step in creating a cDNA library. Like all currently known DNA polymerases, the **reverse transcriptase** also needs a short **DNA starter molecule (or primer)**, on whose OH group of the 3'-C atom of the ribose, the polymerization can start. For cDNA production two different primer species are usually used.

- These are especially **oligo(dT) primers**, which consist of a row of desoxythimi-dine bases (usual lengths are between 16 and 25 nucleotides) and primarily bind to the poly(A) tails of mRNAs. As these poly(A) tails are, respectively, on the 3'-ends of the mRNAs, the reverse transcriptase can synthesize the entire complementary strand to the mRNA (Fig. 21.12).
- Instead of oligo(dT) primers, **random hexamer primers** are also used in cDNA synthesis. These hexamers consist of a succession of six nucleotides, which are produced in random order by a synthesis system so that they contain many various base combinations. Since a mixture of different hexamers is always used, these associate to various unpredictable positions on the RNA and then serve as starter molecules for the reverse transcriptase. Reverse transcriptase is **single-strand specific**. This means that it uses single-stranded RNA as a template, but not a double strand consisting of RNA and DNA. So if, for example, two hexamers are present on a RNA, two independent cDNA products are produced, which do not overlap (Fig. 21.12). The synthesis of a cDNA that displays a complete copy of the mRNA template is therefore only possible if oligo (dT) primers are used.

An additional problem in using hexamers is that these primers have no selectivity for mRNA as a template. If an RNA isolation is contaminated with other RNAs, this RNA is also rewritten into cDNA. With the resulting cDNAs, finding the desired cDNA becomes complicated.

The cDNA produced by the reverse transcriptase cannot, however, be cloned in this state, as it is, like the mRNA, single-stranded (Fig. 21.13). Therefore, it must first be changed into a double-stranded form. This takes place with the help of a **DNA polymerase**. At this point there is a dilemma, which is intrinsic to cDNA synthesis. For the synthesis of the first cDNA strand an oligo(dT) primer could be used, which is able to associate to the mRNA independently

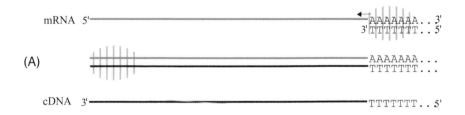

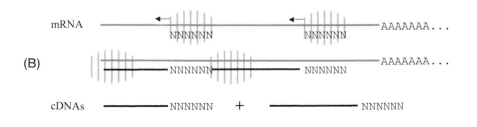

Fig. 21.12 Production of cDNA with oligo (dT) primers (A) or random hexamer primers (B). The primers associate to complementary sequences within the mRNA. The enzyme reverse transcriptase (gray oval) recognizes the OH group on the 3'-C atom of the primer's ribose and starts adding sNTP blocks and synthesizing a cDNA, which is complementary to the template RNA. Reverse transcriptase cannot displace cDNA from an existing duplex. Therefore the synthesis is terminated when the enzyme meets a synthesized strand (in B). The result is two partial cDNAs, which both represent only a part of the initial mRNA.

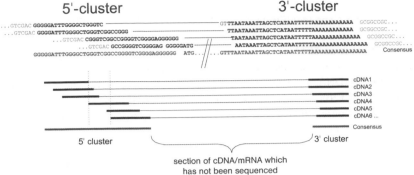

Fig. 21.15 Cluster creation of EST sequences. Individual EST sequences are grouped together if their sequences are identical (i.e., the cDNA comes from the same genes). From the sum of the sequences, the consensus sequence, which may be longer than each of the individual sequences, can be calculated. The 3′-ESTs should all begin at the poly(A) tail of the mRNA and therefore allow reliable clustering. The position of the 5′-ESTs relative to the initial mRNA depends on how much of the mRNA sequence was covered by the cDNA. In the top area of the picture the individual ESTs are shown as sequences. To illustrate the term clustering the usually 400–600-base long ESTs have been shown in a form shortened to a few bases. The cloning sites (CTCGAC for the 5′-EST and GCGGCCGC for the 3′-EST) are colored gray, as these restriction sites do not stem from the mRNA, but are in every cDNA. Areas of the cDNA that are not covered by ESTs are marked with a line drawn through. A consensus sequence is calculated from the overlapping areas. At the bottom of the illustration the ESTs are symbolized with bold lines. The consensus is derived from the overlapping sequences and is longer than the individual sequences. While the 5′-ends of the cDNAs 16 do not overlap (indicated by the vertical dotted lines), the connection is made by the other ESTs. Thin lines between the 5′- and 3′-ESTs mark the section of cDNA that is not covered by EST sequences. After the 5′ and 3′ clusters have been formed and the two consensus sequences have been calculated, a section of the cDNA remains that has not been sequenced and about whose sequence therefore no statement can be made.

large quantities and identical sequences were to be summarized (clusters) (Fig. 21.15). In this way redundancy, which was inevitable with an amount of ESTs larger than the number of genes, was supposed to be reduced. Finally, a resource was to be produced, which was to cover all genes. In the meantime, this approach has been more or less realized and this sequence resource has been made publicly available as the **UniGene collection**.

In contrast to genomic DNA sequencing, in **cDNA sequencing** the point in time at which all mRNAs have been gathered is difficult to estimate. The genome has a finite number of building blocks and once these have been decrypted the sequencing of the genome is complete. In April 2003 this state was announced for the human genome. With cDNAs, the number of possible sequences may also be finite, but because of different concentrations of the individual mRNAs within the entire population of mRNAs in the wide variety of cell types, it is not possible to actually gather all RNAs (Fig. 21.16). The same effect can be seen in the analysis of full-length cDNAs.

The possibility of moving the point at which saturation is reached, and thus identifying more genes with the unchanging sequencing effort, is given by normalizing. cDNAs that were produced by low expressing mRNAs are enriched in the library. However, artifacts are also, in comparison, rare events and the cDNAs resulting from them accordingly rare, and are also enriched during normalization. The new genes projected on the *y*-axis therefore, depending on the quality of the initial cDNA library and the amount of artificial cDNA present, contain a smaller or larger fraction of artifacts.

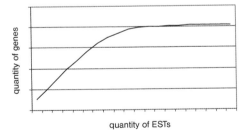

Fig. 21.16 During cDNA analysis, a point of saturation is reached where the identification of new genes is concerned. With a rising number of ESTs, the number of newly identified genes initially increases in an almost linear mode. Due to the redundancy of mRNA in the RNA population and therefore also of cDNA in cDNA libraries, this saturation appears. The sequencing effort would have to be increased greatly in order to find just a small number of new genes.

21.4.4
Full-length Projects for the Production of Resources for Functional Genomics

The production of cDNA libraries was discussed above. It was explained that when using oligo(dT) primers each time an entire copy of the template mRNA is received. Unfortunately, this does not reflect reality. The processivity (i.e., the adding rate and the residence time of the reverse transcriptase on the template) is not high enough so that the polymerase can synthesize the cDNA of long templates sufficiently to their 5'-end. The fraction of partial cDNA for each mRNA therefore depends on the length of the mRNA and furthermore also on its sequence. The reverse transcriptase cannot resolve double-stranded areas (lacking strand displacement) and terminates polymerization when such an area is reached. This effect was discussed in the production of randomly primed cDNAs. Many mRNAs (also) have sequences at the 5'-ends that can participate in base pairing. Furthermore, at their 5'-ends mRNAs often display large quantities of G and C bases, through which double-stranded areas are, in comparison to sections with large quantities of A and T bases, thermally stabilized. *In vivo* double-stranded areas are especially areas to which *in vivo* regulatory proteins can bind. A smaller or larger fraction of the cDNAs in a library is for this reason only partial (i.e., an often considerable part of the mRNA sequence information is not present in the cDNA). For cloning and sequencing full-length **cDNA** there are, however, further difficulties. On the way from gene to mRNA there are intermediate stages (**hnRNA**) that do not represent a homogeneous population but, for example, consist of several interstage products in the splicing process. Introns are not at the same time cut out of the hnRNA. In addition, the cell seems to make mistakes in this splicing process, which lead to falsely processed RNA. Normally, falsely spliced mRNAs should be degraded by a nonsense-mediated decay mechanism. The large number of artificial cDNAs in the databases, however, makes it probable that at least some of these are stable in the cell after a certain time.

Still, in complex organisms with distinctive exonintron structures, there is currently no alternative to the analysis of cDNAs in order to get complete copies of the mRNAs or even to be able to identify splicing variations. Full-length cDNAs are also needed for the production (expression) of proteins, as described later.

Worldwide, **four large projects** have so far been performed with the aim of gathering as large as possible fraction of human RNA in the form of full-length cDNA. First, however, the term "full-length" must be defined, as there are several interpretations of it. A mRNA has defined 3'- and 5'-ends. Two mRNAs that are transcribed by the same gene can, however, differ at their 5'- as well as at their 3'-ends. This has several reasons. At the 5'-end in the promoter, different nucleotides can be used as the transcription start or the transcription can start at different promoters. At the 3'-end, different **polyadenylation signals** can be put to use, which means that different **3'-untranslated regions (3'-UTRs)** appear. Furthermore, the length of the poly(A) tail varies depending on the mRNA species and the age of the mRNA. The length of the poly(A) tail is related to the mRNA's half-life. Alternative splicing has no significance here, as internal exons are not affected in this procedure.

The following definitions for a full-length cDNA are in use:
* *1:1 copy of the mRNA*. This means that the entire sequence information from the 5'-cap structure to the poly(A) tail is represented in the cDNA. This is the ideal state. The encoded protein can be expressed from the cDNA. Furthermore, the sequence of the 5'- and the 3'-UTRs allow the analysis of sections in these areas, which, for example, regulate translation, localization, and stability of the mRNA.
* The *protein encoding area is completely included in the cDNA*; sequence information may, however, be missing at 5'- and 3'-position (i.e., called on by the

cDNA production process). This cDNA can therefore also be used for applications in functional genomics.

- A *fully sequenced cDNA*. This connection to the term "full-length" cDNA should be avoided, since a fully sequenced artificial cDNA remains an artifact.

Depending on the method of cDNA production, **cDNAs of various qualities** can be found in a library. Through the use of suitable technology and good RNA preparation, the fraction of high-quality cDNAs can be increased; for each individual cDNA and cDNA sequence, however, the status is only accessible after in-depth bioinformatic analysis.

The first project performed in sequencing long cDNAs was started in Japan at the Kazusa DNA Research Institute in 1994. In 1997, a project was initiated in Germany (German cDNA Consortium (GCC)). In 1999, two projects were initiated: one was the Mammalian Gene Collection (MGC) in the USA and the other was the NEDO project in Japan.

None of these projects is sufficient to gather all human genes. Together, however, a large number of human genes are covered by cDNAs, although it will be difficult to decide when all genes have actually been gathered. Of particular value are cDNAs that, new genes apart, represent splicing variations of possibly already studied genes and therefore present the opportunity of being analyzed themselves. The interest in splicing alternatives has increased greatly since the quantity of human genes has been estimated to be little higher than that of low organisms (such as the nematode *C. elegans* or the mouse ear cress *Arabidopsis thaliana*). Alternative splicing offers a possible solution to the problem, which arises from the difference in complexity between humans and other model organisms (animals or plants) with a similar number of genes.

Apart from that of humans, in an equivalent way of genomic sequencing, the cDNA of other model organisms is also analyzed systematically. In this context the mouse must be especially mentioned, which was extensively processed in a Japanese project performed at the RIKEN Institute. The focus of this project is on complete cDNAs (including 5′-ends) and should gather as many genes as possible. Many more than 20 000 cDNAs have been sequenced up until now. There are further projects for the rat, the pipid frog (*Xenopus*), the zebra fish, but also for several plants (i.e., *Arabidopsis*, rice, wheat). The **National Center for Biotechnology Information (NCBI)** at the National Institutes of Health in the USA runs a website that details the most important cDNA projects (http://www.ncbi.nlm.nih.gov/genome/flcdna/).

21.5
Functional Genomics

The concept of functional genomics describes the **functional analysis** of the genome and the genes encoded in it (including regulatory elements) and their gene products (functional RNAs, proteins). The functional units we shall focus on are almost entirely the genes and gene products. We do not propose to discuss here structural functional units such as the telomeres, the centromere, or elements required for the higher organization of the chromosomes in the nucleus. "Function" is usually understood to mean the way that gene products operate in the cellular context. Gene products comprise proteins and functional RNAs. Increased importance has been ascribed to the latter, especially in recent times.

The term **proteomics** is now frequently used for the concept that covers proteins in their functional entirety – the proteome comprises all the proteins in a system. Functional genomics and proteomics comprises complementary, comprehensive attempts to characterize completely the functional units that carry out the processes of life. Functional genomics in the narrower sense is concerned with the question of which genes are regulated, how and under what conditions, and what function the gene products have in the cell. This regulation cannot be seen in isolation from its function. Digestive enzymes (e.g., trypsin) fulfill specific functions and their expression has to be strictly regulated in order to avoid undesired effects, as might occur, for example, if trypsin were to be expressed in tissues other than the pancreas.

Thus, functional genomics investigates the nucleic acids DNA and RNA, but also proteins in their **functional aspect**. This section introduces the fundamental technologies and developments that are applied. The distinction between this and proteomics is that we shall not look at any techniques that work with or aim to identify isolated proteins (see Chapters 7 and 8).

The first general question as to whether a gene is switched on or off can be decided by looking to see whether the corresponding mRNA is present. The information that is transmitted from the genome (DNA) via the RNA and, in particular, the mRNA is also called the **transcriptome**. RNA is a carrier of information – a unified class of molecule that in many ways resembles DNA. The physical and chemical characteristics of the various **RNA molecules** are very similar. They primarily consist of four bases that make up the code; there are certain regulatory regions and the information carried by the RNA can be transcribed with the aid of enzymes either in the direction of the protein or in the reverse direction, back to the DNA (cDNA) (see Chapters 4, 12, and 13). It is therefore comparatively easy to manipulate and this can be done almost irrespective of the source (i.e., the organism or the gene). **Proteins**, on the other hand, are very diverse in their characteristics; they have 20 or more coding information units (amino acids) (see Chapter 2). There is no flow of information copied from protein to protein nor is there a reverse flow of information that could be used to recreate the RNA from the protein. Finally, in transcriptional analyses it should not be ignored that the presence of a particular mRNA cannot be taken to mean that the relevant protein is also present nor does it follow that a large quantity of protein necessarily requires a large quantity of mRNA.

Functional genomics can also be seen as a **logical extension of genomics**. The first step was to sequence genomes; now the task is to understand the information we have in our hands (or rather, in our computers). One approach to acquiring this understanding is via **gene expression** – in the first instance via the transcriptome through to the proteome. Another approach is to examine changes at the genome level from the point of view of their functional effects.

The classic procedure in molecular biology is to identify the gene, sequence it, and investigate its function. This is usually done with a particular end in view (e.g., to find a gene for a particular disease). In genomics the first stage is often to collect, integrate, and describe large amounts of information to obtain an overall picture (e.g., about the **regulation of metabolic processes**).

With genomics, and functional genome analysis in particular, a number of important changes have taken place in the biological sciences. On the one hand, the need for **automation** and **high-throughput procedures** has given rise to an industry that has quickly developed the relevant procedures and made them available. On the other, we have also seen the emergence of a scientific industry that has produced large quantities of genomic data out of commercial interest. This has led to a division of labor ("specialization"). Whereas previously the scientists applied a great variety of methods when addressing a problem in order to generate their own data, now increasingly we find scientific analysts using resources or data that were produced elsewhere. This arrangement lends special weight to the producers on account of the particular equipment and funds at their disposal, their access to special resources, an advantage of time, and possibilities of distribution and direction. There are also political and financial facets. The size and complexity of projects has contributed considerably to an increased forging of links across the scientific world, and so greatly accelerated the acquisition of scientific knowledge.

The initial reaction was for genomics to tend away from the focus on matters of detail to a very broad but less profound examination (this might also be described as a **holistic approach**). If we look at this development from the point of view of a theory of knowledge, we might even say that the dogma of a **hypothesis-oriented scientific endeavor** has been abandoned and to some extent replaced by the **pure collection of data**. Whereas previously there would have been a particular hypothesis prior to experiments being done and the sum of the results leading to a theory, genomics can often initially manage without a hypothesis and proceed purely descriptively.

21.6
Identification and Analysis of Individual Genes

Methods of identifying and analyzing individual genes can be classified as part of molecular biology (in the classical sense), genomics, or functional genomics, according to the particular approach. The targeted cloning of a gene can certainly be seen as marking the transition from genomics to functional genome analysis, as the hunt for a particular gene is usually linked to a particular question about its function.

In this section we shall address techniques that help us understand the function of genes we already have (i.e., cloned or sequenced genes). Function here does not only mean the immediate effect of the expression or nonexpression of a gene, but also the **patterns of expression** in terms of time and location that place a gene in an important functional context. One of the most interesting techniques in this respect is **RNA interference** (see Chapters 2, 21, and 31), which in the long term will replace knockout studies (see Chapter 28).

21.6.1
Positional Cloning

Before the high-throughput sequencing of the human genome (see Section 21.3) and that of various model organisms was undertaken, **positional cloning** was the most promising method of isolating genes in a targeted way on the basis of their activity or disease association. The positional cloning process falls into two parts – coupling analysis and candidate gene analysis.

The first step is to link a disease to a chromosome or subchromosomal region by **coupling analysis**, which involves comparing the DNA of patients with that of healthy individuals. Coupling is the inheritance of two loci together (these may be two markers or one marker and one disease locus) on the same chromosome. A marker is an element of known sequence that can clearly be assigned to a particular region of the chromosome (e.g., a STS or a polymorphic microsatellite).

During meiosis, **homologous recombination** occurs between sister chromatids of the maternal and paternal chromosomes (see Chapter 4). The further apart two loci are located on a chromosome, the greater the likelihood that recombination events will take place in between these loci (Fig. 21.17). With increasing distance it becomes less and less likely that these two loci will be inherited together. If, however, two loci are close neighbors, there is a great probability that no recombination of the chromosomes will occur in between these particular loci. As a result these two loci will be inherited together. Coupling analyses therefore require families in which the diseases being studied occur frequently and are inherited from one generation to the next (predisposition and/or actual disease).

Coupling analysis is used to find the frequency with which two loci are inherited together (Fig. 21.17). Therefore, the closeness of two loci is defined in terms of the recombination frequency h. Noncoupled gene locations on different chro-

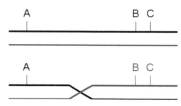

Fig. 21.17 Coupling analysis. In meiosis, homologous recombination occurs between the chromosomes. The greater the distance between two loci on the chromosomes, the greater the likelihood that a recombination event will occur between them (loci A and B). If, on the other hand, the loci are very close together (B and C) the frequency of recombination is very low and there is a corresponding likelihood that the two loci will be inherited together. Coupling analysis uses known markers (e.g., A and C) to look for unknown disease loci (e.g., B) that are inherited together with a marker. It begins by trying to define the position of the disease gene more closely and then to identify it. As a general rule, the more densely positioned the known markers are, the more closely the region can be defined.

mosomes have a value for h of 0.5. Coupled loci show a lower recombination frequency, which gives them a higher value for h. A **maximum likelihood analysis** is carried out. A **logarithm of odds (lod)** then provides information on the coupling of the two loci. In genome-wide coupling analyses, positive lod values above $Z=3.6$ (with an acceptable statistical error rate of 5%) are considered significant.

Such coupling studies, of course, are only possible in the case of **genetically determined diseases**. Coupling analyses offer a promising approach mainly for monogenic diseases, which can be traced to a defect in just one gene. The study of polygenic diseases, which are due to a malfunction of more than one gene, is much more complex, depending on the number of genes affected. Examples of monogenic diseases that have been identified by positional cloning are redgreen color blindness (the gene is located on Xq28) and fragile X syndrome (also Xq28). **Polygenic diseases** for which the mechanism is not yet fully understood because there is insufficient knowledge available about the genes involved include Down's syndrome (trisomy 21) and various psychological disorders, such as schizophrenia and autism. Although attempts are also being made to define the chromosomal regions affected in these diseases by means of coupling analyses, these regions are at present still very large (greater than 20–100 Mb with more than 100 genes on different chromosomes). In the case of Down's syndrome it is now assumed that not only genes on chromosome 21, but also genes on other chromosomes are involved in causing the disease and determining its severity.

Two basic conditions have to be fulfilled to enable coupling analyses to be done:
- There must be a **sufficient number of markers** that can be located with certainty to specific regions of the chromosomes, and
- The **number of families and family members** recruited must be sufficient to make it possible (i) to perform a statistical analysis of the coupling analyses and (ii) to define the chromosome region bearing the disease gene or genes to as small an area as possible.

When the human and murine genome sequences were still unknown, efforts were made to produce maps of markers that were as detailed as possible. In the 1990s these markers were mainly sections of DNA whose sequences had been found and whose position in the genome had been discovered as precisely as possible. A distinction is made between **STS markers** and **polymorphic microsatellites**.

An STS is generated, for example, by sequencing a genomic clone (cosmid or BAC) from the ends, so that a piece of the sequence is known. The position of the genomic fragment in this clone is then localized in the genome by **hybridization methods** (e.g., *in situ* hybridization, FISH, or radiation hybrid mapping). **Polymorphic microsatellites** are short repeats, usually of dinucleotides (CA repeat, e.g., $NNNNCA(CA)_nCANNNNN$). The polymorphism at these positions (see Chapter 4) is due to errors in DNA replication that occur when DNA polymerase "stutters" on encountering repeats and synthesizes more or fewer copies. The greater the number of repeats in a section, the higher the degree of polymorphism and the more powerful that section will be as an indicator in the coupling analysis.

For **coupling analyses** the corresponding region of a polymorphous microsatellite is obtained from patients and healthy individuals, and **amplified by PCR**; the number of repeat units in the microsatellite can be determined. If a particular number of the repeat units is inherited, coupled with a disease, the marker is associated with the disease and the disease gene concerned is localized on the chromosome near this marker. Near is a relative term in this case, however, since **recombination frequency** is not equally distributed over the chromosomes (there are hotspots and deserts). Also, the quality of the coupling analysis depends considerably on the density of the markers being investigated. In general, the more markers per chromosomal region, the better the coupling that can be achieved and the smaller the chromosomal region to which the presumed disease gene can be defined.

Since the number of microsatellites in the genome is not unlimited, although it is large, alternative markers were sought and eventually found with the discovery of **SNPs** (see Section 21.3.1.7, and Chapters 4 and 27). Almost every 100–1500 bases there will be a difference at one base position between the genomes of two human individuals. This fact can be exploited to identify these differences and any coupling with diseases, using high-throughput methods such as **mass spectrometry**, and thus find disease genes.

The second step of positional cloning is to investigate genes in the candidate region for any mutations that might be present in patients. The candidate genes are first examined as thoroughly as possible for any available information. The expression patterns of these genes in the relevant tissues make it possible to further limit the range of candidates, so that fewer genes need to be examined for mutations, for example, in the experimental analysis. There are in principle three types of pathogenic change:

- **Mutations.** Point mutations in the genomes of patients, or short insertions/ deletions that affect the protein sequence (see Chapter 4), are usually an initial indication that the gene concerned does have an effect on the disease under investigation. The usual method is to amplify the exons of the gene to be studied from the genomic DNA of patients and to analyze it for mutations by comparison with a healthy control. Techniques such as sequencing, denaturing high-performance liquid chromatography, or mass spectrometry are normally used for this. One problem is that only about half the mutations leading to a pathogenic change lie in the protein-coding region. The other half is distributed over introns and regions responsible for the regulation of gene expression. Mutations in introns can produce changes in splicing patterns, for example, perhaps by creating new splice sites or causing the disappearance of sites that would normally be used. In this case the protein sequences are directly affected. Changes in regulatory regions alter the pattern of gene expression. The gene is over or underexpressed or expressed in the wrong organs or at the wrong stages of development, and this may also lead to the emergence of a phenotype.

- **Deletions or amplifications** of genomic regions are a second type of change that can be a possible disease cause. Most of the diseases associated with this type of change are cancerous. **Amplification of an oncogene** causes increased expression of this gene (dose response effect) and leads to the increased proliferation of the affected cells. The **deletion** of a region in which a tumor suppressor gene is located also leads to increased proliferation, because insufficient protein is present to block this effect.

- In addition, the **translocation of chromosome regions** to other chromosomes can be responsible for certain effects, in particular for some cancers and also for mental retardation. For these diseases too a number of affected genes were cloned by positional cloning. These had either been destroyed (loss of function) or acquired a new function by the fusion of sections with another gene (gain of function) (e.g., *bcr-abl* fusion gene and the encoded protein in the case of the Philadelphia chromosome).

Techniques such as **FISH**, **CGH**, and **matrix CGH** (CGH using DNA chips) are especially useful (see Section 21.3.1.8) for the analysis of deletions, amplifications and translocations.

21.6.2
Gene Trap

The gene trap is a system that makes it possible to identify genes that give rise to a phenotypic effect when they are switched off and also to analyze the effects of these genes in cells or complete organisms. The first step in the gene trap is to integrate an artificial DNA construct into the genome of a cell line, usually an embryonic stem cell line. For the success of this step, it is important for the gene to be integrated as randomly as possible rather than at hotspots (locations with an increased frequency of integration). Success of the gene trap requires the location where the construct is integrated to be in the intron of a gene and, if the intention is to switch off the functionality of the gene product, the integration position should ideally be in one of the upstream introns. The integration can then have the effect of preventing the exons of this gene that are located downstream from being translated into protein, so that important sections of the protein are missing. Instead, a fusion protein results, made up of the natural N-terminus of the gene concerned and a **reporter protein**. This reporter protein forms a considerable part of the DNA construct that has been integrated into the genome. It is used to demonstrate that gene expression has occurred (e.g., blue coloration by the *lac*Z gene product). A signal (i.e., the time and place of expression of the reporter gene) will also be specific for this gene, since the expression is under the control of the regulatory region (promoter/enhancer/silencer) of the gene concerned. The second component needed for this construct is a **splice acceptor sequence**. This sequence enables the integrated, cotranscribed construct to be recognized as an exon during processing of the hnRNA to mRNA, so that it will be contained in the mRNA. A third vital component of the construct is a **polyadenylation signal**. In the cell, this causes the mRNA to break off at this point during processing, instead of extending into the following exons of the gene concerned. This prevents degradation of the mutated mRNA, such as through nonsense mediated decay (i.e., stop codon not in the terminal exon). The artificial mRNA is then stable, and the fusion protein can be translated and its presence demonstrated.

If the integration of the construct is linked to a **change in phenotype**, the next step is to identify the gene concerned. Since the sequence of the reporter gene is known, suitable methods (**rapid amplification of cDNA ends (RACE)**) can be employed to find out the sequence of the gene (see Chapter 13). The entire gene sequence can then be reconstructed by comparing sequences, at least for those species whose genome sequence is known. The gene trap method has particularly been used to identify **disease genes**, because it is possible to identify the integration location of the construct directly and associate it with a phenotype. However, the advent of systematic cell-based assays, in which the effects of protein overexpression or **small interfering RNA (siRNA)-mediated underexpression** can be investigated, means that the importance of the gene trap method will decline in the comparatively near future.

21.6.3
DNA/RNA *In Situ* Hybridization

It was the **Southern blotting method**, developed by Ed Southern, that first made it possible to investigate the presence of specific sequences in a complex mixture of DNA. This method uses DNA probes that are hybridized on membranes on which genomic DNA has been immobilized (see Chapter 11). Later, the **Northern blotting method** was also developed, in which the object of the analysis is not DNA, but RNA, especially mRNA. The aim was to identify expressed genes by using specific probes to produce hybridization between the probe used and the immobilized RNA on membranes. The biological material (DNA or RNA) needed for hybridization by Southern and Northern blotting had to be extracted from the cells or tissue, and further processed by gel electrophoresis and transfer onto a membrane.

In situ **hybridization** made it possible to carry out the analysis of the DNA or RNA in the cell nuclei, cells, or tissues, as the term *in situ* implies. There are several variants of *in situ* hybridization, each developed to address a specific type of question.

FISH analysis was developed to localize genomic regions (i.e., sections of DNA) in the chromosomes. Cells are mounted and fixed on slides. The probes used are genomic clones (YACs, BACs, or cosmids) or, with a lower success rate, cDNAs. In FISH there is a lesser chance of success when using comparatively short cDNAs (3–5 kb) (which are also represented on various exons in the genome) than using genomic clones with a length of greater than 40 kb. The DNA is labeled using fluorescence dyes and hybridized. **Condensed chromosomes** (metaphase) or interphase nuclei are then examined for signals. The number of signals provides information about possible amplifications or deletions of genomic regions, breaks in chromosomes, translocations, and so on. These abnormalities are often associated with tumors and diagnostics is one area where investigations are performed to look for such abnormalities.

RNA-ISH (*in situ* hybridization) enables the extent of mRNA expression of particular study genes to be analyzed in an association of cells. An **antisense probe** specific for a particular gene, which is complementary to an mRNA species, is radioactively or fluorescently labeled and hybridized on a tissue section, for example. The probe makes visible those locations where the corresponding mRNA is present. A concomitant histological examination of the section enables the gene expression to be assigned to individual cell types, so that a very specific gene expression pattern is obtained. **The specificity of RNA-ISH** is far superior to Northern blotting or even quantitative PCR (TaqMan®), as the locational resolution of this method extends to the individual cell, even to subcellular structures. However, RNA-ISH is considerably more demanding to perform than Northern blotting or quantitative PCR, although many of the steps involved have now been automated.

21.6.4
Tissue Arrays

Tissue arrays are a very promising modern method that has become available recently to investigate the expression of individual genes. With these, unlike conventional *in situ* hybridization or immunohistochemistry with a tissue sample, many tissues can be studied in parallel. To do this, many small pieces of tissue (stamped out and cut with a microtome) are fixed on a glass slide in the form of an array (compare array-based techniques for the investigation of transcriptional activity (Section 21.7.4)). These tissue patches have a minimum diameter of 0.6 mm and more than 3000 can be arranged on an array. A probe (i.e., a labeled oligonucleotide representing a gene) or an antibody to the protein, is then used to demonstrate expression at transcriptional or protein level. This has a wide spectrum of application in cancer research, as tissue arrays offer the possibility of investigating many individual tumors in a single experiment. This can be an advantage in characterizing the individual molecular features of different tumor entities that cannot be distinguished histologically or when testing new biomarkers against a broad panel of tumors. They can also be used to investigate the expression of a gene in quite different tumor types and the corresponding normal tissues (Fig. 21.18).

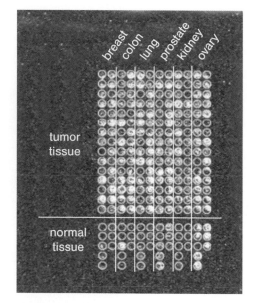

Fig. 21.18 Hybridization with an oligonucleotide probe on a tissue array. The aim is to investigate the expression of a candidate gene (at the transcriptional level) in different tumors and in corresponding healthy tissue. A labeled oligonucleotide probe represents a candidate gene, and is hybridized with 180 tumor sections and 48 sections of normal tissue on the array. The evaluation shows that the gene is clearly being expressed in most of the tumors, but hardly at all in the corresponding healthy tissues (except the ovarian tissue).

21.7
Investigation of Transcriptional Activity

In this section we shall introduce developments that look less at individual genes and more at the **transcriptome** in its entirety; in other words, they examine all the genes expressed as mRNA – or a large subset of these. A subset is arrived at either by prior selection, where it is limited to genes of a particular group, or by looking only at genes that show a change in their regulation in the context of a particular question that is being investigated.

The basic distinction to be made here is between two methodological approaches. On the one hand are methods that investigate the **change in gene expression** directly in a molecular biological or biochemical way, producing a result that consists just of the regulated genes. On the other is the approach that uses **global analyses** to generate what are sometimes enormous quantities of data. Filter mechanisms are then used to extract the relevant data, such as the change in expression.

The first group includes **subtractive methods** such as the generation of subtractive cDNA libraries, representational difference analysis (RDA), suppression subtractive hybridization (SSH), and differential display. These techniques investigate changes at the transcriptome level and reduce the quantity of data at the test tube stage, so that extensive bioinformatic analyses are usually unnecessary. However, the quantification of changes is only very limited. The great advantage of these methods is that they require no *a priori* information, so they are especially suitable for systems that have not yet been much investigated or where there is a wish to find new genes.

The second group includes **array-based techniques** that measure gene expression by hybridizing labeled cDNA on a surface with DNA as the analyte (DNA chip or array). This approach normally requires a high level of *a priori* information. This means that usually the genes used on the array are known and that the investigation is limited to these. Therefore, in contrast to the methods mentioned above, no new genes can be found. Techniques that operate on a sequence basis, such as **serial analysis of gene expression (SAGE)**, begin by generating as large a quantity of sequence data as possible. They then select to arrive at a result stating the extent to which transcription in the system under investigation is taking place or is changed.

Another way of describing the various systems is with terms like **open and closed architecture**. An open architecture allows a process without *a priori* information (e.g., in the form of a sequenced genome), and the discovery of new genes, splicing variants, SNPs, and so on. A closed architecture is based on the information already available (e.g., on the sequenced genome or the precise knowledge of a gene family). Systems with an open architecture include the subtractive methods RDA and SSH, differential display, and the like. Those with a closed architecture include all array techniques and **reverse transcription (RT)-PCR**-based techniques with specific primers, since the analysis here is limited to the sequences available. **SAGE** and **massive parallel signature sequencing (MPSS)** occupy a special place in that the method enables them to generate new sequence information, but the reasoned interpretation of this requires the use of sequence information that is already available (MPSS will not be further elaborated upon here).

21.7.1
Serial Analysis of Gene Expression (SAGE)

In addition to the methods of quantifying mRNAs that can only investigate known genes, because they rely on specific probes, there are three widespread methods that can also be used to capture hitherto unknown mRNAs. One of these is the **high-throughput EST sequencing** of cDNAs, in which large numbers of cDNAs belonging to a cDNA library are systematically sequenced from the ends. This technique was used especially in the 1990s with the aim of gene discovery (identifying if possible all the genes of humans and those of other organisms). In these libraries, rarely expressed mRNAs and the cDNAs derived from them were often enriched (normalization) in order to increase the number of the found genes. Another consequence of this, however, is that no conclusions can be drawn from the number of cDNAs sequenced for a gene about the degree of expression of the underlying mRNA.

SAGE is a method invented by Victor Velculescu in 1995. The number of mRNAs that are analyzed and identified for each sequence reaction is greater by this method, compared with EST sequencing. It also depends on sequencing. The aim was to reduce the number of sequence reactions and gels needed. In SAGE, only a short section (typically 10 bases long) of each mRNA, or the cDNA derived from it, is sequenced. Assuming that the genome is sequenced, the underlying gene can be clearly identified by sequence comparison.

First, a biotinylated oligo(dT) primer is used to produce cDNA from the mRNA. This is then cut with a **restriction enzyme** that has a four-base recognition sequence (NlaIII recognition sequence: CATG) and so cleaves frequently. Two different linkers are then ligated to the fragments in separate batches. The fragments, which contain the poly(A) tail of the original transcript, are purified via the biotin labeling. A second restriction digestion follows, using a type III restriction enzyme (BsmFI). Type III restriction enzymes, unlike type II, do not cut in the recognition sequence, but at a certain distance – one that is specific for each enzyme. The fragments obtained therefore contain 10 nucleotides that are specific for the original mRNA, as well as the linker sequence and the four bases of the *Nla*III sequence. The fragments with the different adaptors are mixed and ligated to produce ditags, in which two fragments are randomly linked together to form a longer product. These products are amplified, cut again with the NlaIII enzyme, and finally ligated into long chains and cloned into suitable vectors. Sequencing reveals the sequence of bases of the individual fragments. There are fragment pairs consisting of two independent mRNAs directly linked to each other and separated from the other fragments by NlaIII restriction sites. The entire sequence can be broken down into the individual fragment sequences using suitable software, and these can then be clearly assigned to the genes by sequence comparisons.

A variant of the SAGE technology, called **CAGE** (C stands for 5′-cap structure), has recently been developed. Whereas SAGE clones and analyzes fragments covering the 3′-end of cDNAs or mRNAs, CAGE seeks to characterize the start of transcription of mRNAs. Thus, CAGE uses an oligonucleotide attached to the 5′-end of the mRNA, instead of the oligo(dT) primer, to immobilize the fragments. Both methods have in common the complexity of the genes expressed in any cell. Two parameters are considered here: (i) at least 10 000 different genes are expressed in each cell and (ii) these genes are present in very different numbers of mRNA copies; the dynamic range of expression extends from about one mRNA per cell to 100 000 mRNAs of a gene per cell. A very large number of analyzed SAGE or CAGE fragments (10 000 genes×100 000 copies=10^9 fragments) would be needed to account completely for this complexity. In most SAGE projects however, only 100 000 to 1 million fragments are analyzed. Statistically useful results can therefore only be achieved for strongly expressed genes.

21.7.2
Subtractive Hybridization

The precursor of the subtractive techniques was the construction of normalized or amplified **cDNA libraries** (Fig. 21.19). The starting point for this technically demanding method is the fact that very great differences exist as regards the proportion of the various mRNA species in a cell. **Three classes of mRNA** can be distinguished in an average somatic cell: (i) the highly abundant (superpreva-

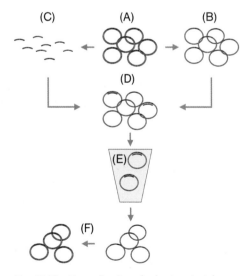

Fig. 21.19 Normalization: the basic principle (simplified). A cDNA library (double-stranded plasmids) is used to produce single-stranded plasmids (B) and the single-stranded DNA of the erts (C). These are hybridized with each other (D) and passed through a hydroxylapatite column. Double-stranded DNA (which mainly comes from the most abundant inserts) is bound (E). The single-stranded plasmids are remade into double-stranded form (F) to be transformed in cells. The effect of normalization is to reduce the proportion of abundant sequences in the starting library.

lent) class comprises just 10–15 different species, which make up 10–20% of the mass, however; (ii) the **moderately abundant** (intermediate) group comprises 1000–2000 species, which together add up to 4045% of the mass of mRNA; and (iii) the **rare (complex)** ones comprise 10000–20000 species, also with 40–45% of the mass. That means that there will be considerable redundancy in a representative cDNA library (among 10000 analyzed clones there will be some that appear more than 100 times, while others will not be represented). Normalization attempts to reduce these effects. The starting point is a **representative cDNA library** from which single-stranded plasmids and single-stranded DNA or RNA are obtained from the total number of the cloned genes. If these two groups are allowed to hybridize, the most frequently occurring sequences will be the first to form double strands. **Double-stranded DNA** binds easily to hydroxylapatite columns and this provides a means of separating it from the **single-stranded plasmids**. These are then made up into double strands and used to transform bacteria once more to produce a library that is depleted in the frequently occurring fragments. Subtractive libraries are a development of this approach.

Functional genomics uses the principle of **molecular subtraction** to subtract DNA populations from one another, ideally leaving just the differences as the remainder (Fig. 21.20). If the experiment has been drawn up appropriately these are precisely the genes that are of interest, because these are differentially regulated in one set of investigation conditions (as opposed to another). DNA is comparatively well suited for molecular subtraction on account of its propensity to form double strands. For this process the two DNA populations to be compared are first usually made into single-stranded form (denatured) by heat, mixed, and then hybridized – making them double-stranded again. The experiment can be set up in such a way that one population is present in excess (this is usually called the driver or competitor). Its purpose is to mop up (compete out) the other (the tester). The effect is to subtract the driver from the tester and the remaining DNA will be parts of the DNA competed against (i.e., sequences that are present in the tester but not in the driver). These then have to be isolated by appropriate methods. **RDA** and **SSH** operate on this basic principle. Both methods take as their starting point cDNA made from complex mRNA; in other words, the transcriptome of the system to be investigated (e.g., a cell culture, a tumor, or an organ).

A system is always looked at in comparison to another state, so that the question being asked is always: which genes are switched on (or upregulated) in system 1 as opposed to system 2? Or the reverse (reciprocal) question might be asked: which genes are switched on in system 2 as opposed to system 1? The second question would then be equivalent to asking which genes in system 1 are switched off (downregulated) as compared to system 2.

RDA was originally developed for genomic DNA to identify chromosomal rearrangements. However, this method acquired greater importance in the identification of differentially regulated genes. Since a transcriptome (still more a genome) is very complex with regard to the quantity of DNA sequences present, work is done in the first instance on just a representative part of these – the representation. These are fragments of the transcribed genes that are much less complex than the complete mRNA (or cDNA) sequences. This is done by cutting the cDNA with a restriction enzyme and ligation of primer adaptors followed by PCR. Only fragments of suitable length are amplified and contribute to the representation. The two representations to be compared (e.g., from a tumor and equivalent healthy tissue) are then, after further modification, mixed and denatured (i.e., made single-stranded) (Fig. 21.21). If they are allowed to hybridize once again, strands from both DNA populations reanneal into double strands. Some are unique to the tester population, as they come from genes that are not present in the other cDNA (the driver), not having been transcribed. These strands (only) bind exclusively to complementary strands from the same population. If the tester DNA has been modified in such a way that only the tes-

10 − 7 = 3
ABCD − BCD = A

Fig. 21.20 Subtraction and tester/driver explained. Just as we say subtract 7 from 10, leaving a remainder of 3, so in this experiment we subtract BCD from ABCD, leaving A. ABCD is the tester and BCD the driver or competitor. For the purposes of the experiment the driver is added in excess.

Fig. 21.21 Schematic diagram to illustrate RDA of cDNA (simplified). Two complex cDNA populations are cleaved with a restriction enzyme, primer adaptors added, and they are amplified (representation). Only one of these populations (the tester) has primer adaptors. The driver has no primer and so cannot be amplified. The two sets of fragments are mixed together, denatured by heat (making them single-stranded), and allowed to reanneal. The driver, which is present in excess, competes with the tester and tester-specific fragments are then preferentially exponentially amplified by PCR. Several rounds of this enrichment procedure can be performed. Finally, a difference product can be analyzed.

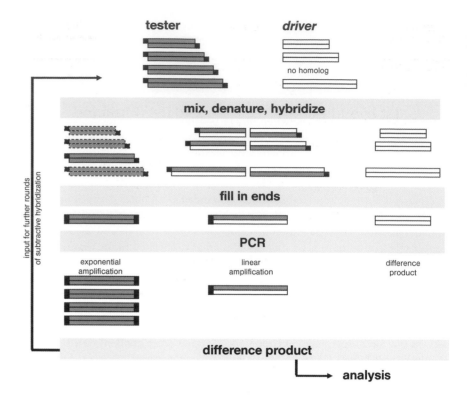

ter DNA and not the driver DNA can be amplified in the ensuing PCR amplification, enrichment for these unique tester sequences will occur. At the end of the procedure these can be cloned or otherwise analyzed, providing an indication of which genes were switched on. The disadvantage here is that only fragments are involved and the search for the entire gene can be demanding. Also, it is basically not possible to carry out a differentiated quantitative study of the extent of overexpression of a gene, since in most cases the situation is not as simple as a gene simply being switched on or off. Rather, there is often a certain basic level of transcription, which is then increased by some factor. The advantage of this method is its robustness and the possibility of discovering new genes without a priori information. Normally results of methods such as cDNA-RDA need to be validated, such that a gene found by this type of method has to be checked using another method, such as **Northern blotting**, **RT-PCR** or in situ **hybridization.**

The strategy followed by **SSH** is similar. However, there is an additional element to this method: the tester is normalized in the course of the procedure (Fig. 21.22). This means that differences in the abundance of differentially expressed genes are evened out. This avoids producing an unhelpfully large redundancy, since there is usually the same process of cloning and analysis of the clones at the end of the procedure. The name of this procedure derives from a special feature of the primer adaptors. In SSH the tester population is divided and given two different primer adaptors. These compete with the driver in two separate reactions and the two groups are then mixed. Fragments of both testers that have not yet hybridized can now form double strands. These have two different adaptors at each end and are then preferentially amplified by PCR. Tester fragments that annealed quickly to form double strands in the first hybridization reaction because they were abundant have the same primer adaptors at both ends. These form internal stemloop structures in the single-stranded DNA molecule, and the effect of this is to suppress (hence the name SSH) amplification, so helping to achieve normalization.

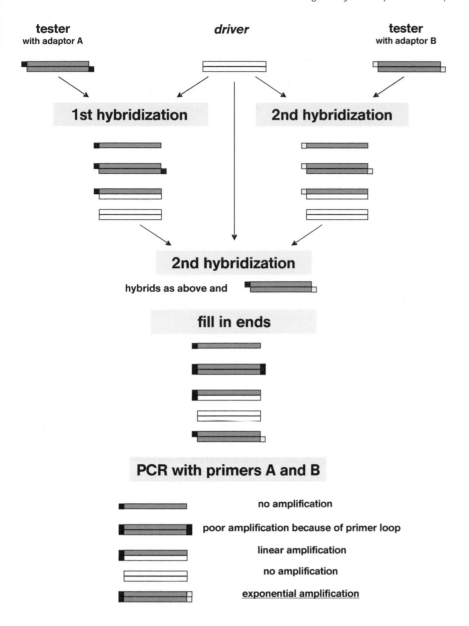

tester
with adaptor A

driver

tester
with adaptor B

1st hybridization

2nd hybridization

2nd hybridization

hybrids as above and

fill in ends

PCR with primers A and B

no amplification

poor amplification because of primer loop

linear amplification

no amplification

exponential amplification

Fig. 21.22 SSH. The principle of SSH is similar to RDA, but the primer adaptors are constructed in such a way that hybrids of fragments that have the same adaptor form an intramolecular stem-loop structure. As a result, they are poorly amplified by PCR. This provides a means of reducing the abundance of frequently occurring fragments (normalization).

21.7.3
RNA Fingerprinting

RNA fingerprinting techniques follow a different strategy. These were developed from 1992 onwards, shortly before the methods described above. They have become even more important than the subtractive techniques, although they are no easier to establish, and there has been much controversial discussion about their specificity and sensitivity (further details below). A great advantage of RNA fingerprinting and the procedures deriving from it is that several samples can be examined in parallel. The subtractive techniques only allow comparison of one pair at a time. The two basic techniques of RNA fingerprinting are often called **differential display (DD)** and **RNA arbitrary primed (RAP)-PCR**, with DD sometimes also being used as an umbrella term to include RAP-PCR.

The basic scheme is first to reduce the complexity of the RNA samples being compared during the cDNA synthesis or subsequent PCR. The choice of primers creates various subsets. The samples within each subset that are to be compared are separated side by side on a polyacrylamide gel. The reduced complexity makes it possible to see bands representing cDNA fragments. There may be a few dozen of these, depending on the choice of primer. When two samples are directly compared, individual bands of different strength may be found or a band may appear in only one of the two samples. The bands can be cut out and amplified by PCR for further analysis to identify differentially expressed genes.

In DD, the cDNA is produced using oligo(dT) primer that binds to the 3'-poly(A) tail of the mRNA and enables reverse transcription starting at that point. If this primer has one or two additional bases at its 3'-end, it is called an anchor primer, and binds preferentially to the transition zone between the transcribed sequence and poly(A) tail. Various anchor primers can be designed for PCR (3'-GGTTT..., 3'-GCTTT..., 3'-GATTT, 3'-CGTTT..., etc.), which produce different subsets on amplification, because they can each reversely transcribe or amplify only one particular subgroup of mRNAs. The subsequent PCR requires another, second primer that will bind upstream on the transcript. However, since there is a mixture of several transcripts, the primer sequence has to have a composition that allows the primer to bind specifically, but to many cDNAs, to form amplifiable fragments. This type of primer is termed **arbitrary**. Its arbitrary character is achieved either by the specific sequence being short (so that it occurs frequently) or by using it at low temperatures (so that it binds less specifically). A **random primer**, on the other hand, has a random, mixed sequence (e.g., a random 6mer has any one of four possible bases at six base positions, leading to a possible $4^6 = 4096$ different molecules in the mixture). For DD, arbitrary primers with a length of 10–12 bases are often used. These can bind imprecisely at slightly lower temperatures. As long as binding at the 3'-end, where the polymerase will attach the first nucleotides, is specific, a binding mismatch has no further effect. The 5'-end does not have to fit precisely. All the newly synthesized DNA fragments in a PCR of this type have the new primer sequence and so subsequent PCR cycles can be carried out with greater specificity at higher temperature.

The only essential difference between **RAP-PCR** and the method described is that it does not use an oligo(dT) primer, but an arbitrary primer, even at the reverse transcription stage. As a result, priming takes place anywhere, instead of the 3'-ends of the genes being amplified as in DD. One advantage of this is that in this way transcripts without a poly(A) tail can also be reversely transcribed (e.g., as in bacterial systems).

There are several developments of this methodology that can be grouped together as **systematic DD**. Some of these try to avoid the problem that a gene can be represented by several amplicons, which leads to a certain redundancy in what is already an extensive analysis. The individual methods cannot be presented in detail here nor have they (yet) achieved the same degree of importance as the original method. The following list may inspire the interested reader to further individual research:

- Gene expression fingerprinting
- Ordered DD
- RNA fingerprinting by molecular indexing
- Restriction landmark cDNA scanning
- AFLP-based mRNA fingerprinting
- Targeted RNA fingerprinting
- Total gene expression analysis (TOGA).

21.7.4
Array-based Techniques

A **DNA array** (also called a DNA chip) is an arrangement of various DNAs, usually of known sequence, in an ordered grid on a solid surface (Fig. 21.23). The individual positions that can be addressed are called **probes**, **spots**, or **features**. They act as analytes; that is, other DNA molecules bind to them by **Watson-Crick hybridization**, depending on the particular sequence. These molecules usually come from a solution over the spots (this solution is usually called the target,

(A)

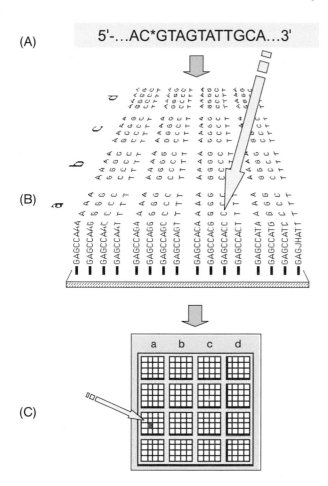

(B)

(C)

Fig. 21.23 Basic principle of array hybridization. (A) A labeled nucleic acid (either radioactively or by fluorescence labeling) is applied to an array (B) consisting of ordered DNA sequences, which serve as analytes (probes). The labeled nucleic acid binds specifically (hybridizes) to the probes (arrow, C). A signal can be detected at the position where this binding has occurred. It can then be attributed to a particular sequence.

although it is also called the sample and the term probe has been applied to this too). Here, we will not discuss protein arrays or DNA chips that bind proteins or other molecules.

Arrays have received a great deal of attention in the past because they fit so well into the overall concept of genomics and have a whole range of advantages. Arrays were sometimes used in structural genomics (see Section 21.3.1.6) to create genome maps, so their use in functional genomics corresponds to the onward developmental history of the technology. The sequencing projects also created the resources for DNA chip technology, in that it was possible to refer directly to available clone libraries or sequence databases when producing arrays.

The terms **DNA chip and array** are often used synonymously. A brief explanation is appropriate here. Unfortunately, the term "chip" is already used in electronics for a semiconductor component. This fact often leads those unfamiliar with what is meant to make false associations at first when DNA chips are discussed. On the other hand, the widespread awareness that (semiconductor) chips are used in computers has probably played a part in creating greater public interest in the concept of DNA chips. Moves are indeed already being made to integrate DNA into electronic components. At present this technology is still very much in the development stage and will not be presented here.

A **DNA chip** is usually understood to be DNA mounted on a flat, nondeformable surface. There is a reason for this distinction: the first DNA chips were produced on nylon or similar flexible materials. There is the further fact that DNA can also be bound to beads in suspension. Like an electronic component (a chip), a DNA chip is therefore something that is quite small and has a solid structure. "Array" in this context simply describes the fact that biomolecules are arranged in an ordered form on a surface. A distinction is sometimes made between **macroarrays** and **microarrays**; DNA chips in the narrower sense are microchips. Macroarrays can be defined for our purposes as not necessarily requiring any means of image enlargement in order to interpret them, whereas microarrays usually require enlargement of the image when they are being evaluated. Macroarrays are often produced on membrane supports such as nylon or nitrocellulose. The 3D structure of these surfaces means that the DNA is able to penetrate to a limited extent when it is applied to them (called spotting). In size, a macroarray has a side length of approximately a few centimeters (e.g., in microtiter plate format). The surface used for microarrays is often chemically modified glass and their dimensions are in the centimeter range. The density of the spots is up to about $100\ cm^{-2}$ in the case of macroarrays. Special microarrays, for which the DNA is synthesized on the surface, may achieve a density of greater than $100\,000\ cm^{-2}$.

Another basic distinction is between a procedure in which the DNA is chemically synthesized directly onto the surface, and one in which it is first manufactured and then transferred onto it. In the case of the latter the density that can be achieved is limited by the technical apparatus used to transfer the DNA. However, this approach enables longer DNA fragments to be used, whereas synthesis on the chip successively builds up a relatively short oligonucleotide.

The basic principle of **transcription analysis** using arrays is that complex mRNA is transcribed into cDNA, labeled either radioactively or with fluorescence dyes, and applied to arrays for comparison, so that the labeled (target) cDNA can hybridize with complementary (probe) DNA on the array (Fig. 21.24). The signal intensities from the labeling at the various spots are measured. Usually the signals from two complex cDNAs are compared with each other and the ratio between them found. The aim is to state the ratio by which a given gene (corresponding to one or more spots (features) on the chip) is upregulated

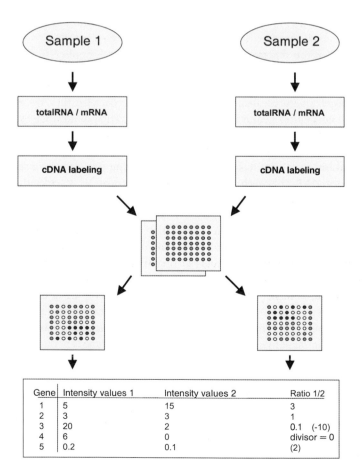

Fig. 21.24 From array hybridization to transcription analysis (simplified). RNA is isolated from two samples that are to be compared (tissue, cells, etc.). cDNA is produced by reverse transcription and labeled, either radioactively or by fluorescence labeling. Hybridization can be carried out on one or two arrays, depending on the labeling. Two images are obtained when the image has been recorded. Special software is used for the recognition of the individual spots and intensity values assigned to the positions. These are used to arrive at ratios, the aim of which is to reflect changes in gene expression. The table shows some examples and problems. Gene 3, for example, is downregulated by a factor of 10, gene 4 upregulated by a factor of infinity, and gene 5 is hardly expressed (it lies below an arbitrary threshold value), so that the stated ratio would need to be taken as imprecise.

Gene	Intensity values 1	Intensity values 2	Ratio 1/2
1	5	15	3
2	3	3	1
3	20	2	0.1 (-10)
4	6	0	divisor = 0
5	0.2	0.1	(2)

or downregulated when two states are compared (corresponding to two cDNAs). The more genes there are on an array, the more the isolated consideration of a single gene slips into the background, to be replaced by the consideration of whole groups of genes. This applies all the more if several sets of experimental conditions are being studied, such as a development over time, when genes with a similar transcription profile are grouped together. The meaningful interpretation of such data or sets of data and ultimately the integration of already known gene functions into a meaningful whole is one of the main challenges of **bioinformatics**.

Below is a brief introduction to the most important developments in the field of array technology.

21.7.4.1 Macroarrays

Macroarrays were first produced on materials usually used for the various blotting techniques (**Southern**, **Northern**, etc.). They probably derive from **colony lifts**, in which a replica of an unordered clone library was made by transferring an impression from the agar plate to a membrane (also called a filter). The attached DNA was used to screen for a particular sequence using a radioactive probe and finally identify the relevant clone. **Dot blots** can be seen as the first simple arrays, in which DNA was applied to a surface in an ordered manner. Here, DNA is pipetted onto a membrane – a rather imprecise method as regards the dimensions of the individual spots. If a certain number of DNA samples are applied in an ordered fashion, this produces a simple array. **Slot blots** were a further development. Here, apparatus was used to apply the DNA to the membrane through slots, usually by aspiration. **Colony filters** exploit the fact that nutrients can also diffuse through the membranous surface of the support material, so that, if a membrane is placed on the surface of an agar plate, bacteria or yeasts can be cultured on it. The bacteria can be transferred from microtiter plates by means of a **replicator**. This is a tool that works rather like a stamp, but using pins that each transfer small quantities of bacteria onto the membrane in an ordered grid on the membrane. The bacterial colonies are grown on and the filter is then processed. This involves opening up the cells so that the DNA can bind to the filter. These colony arrays are used in screening for individual genes or in genome mapping by hybridization. They are less suited for transcriptional studies, as they contain the entire *E. coli* DNA in addition to the plasmids with the DNA fragments of interest and this gives rise to a fairly high number of background signals. PCR products are better for transcriptional studies. These can be transferred onto the membrane with the aid of machines, robots, or similar, designed for spotting DNA from microtiter plates onto the surface of the array-to-be using a pin tool (like the replicator mentioned above) (Fig. 21.25).

Radioactively labeled cDNA is often the target used on membrane surfaces such as nylon or nitrocellulose, since there are usually disadvantages to the use of dyes for labeling on these surfaces as regards sensitivity, the background signal ratio, and ease of manipulation. This labeling is done with nucleotides containing **radioactive isotopes** (e.g., ^{33}P) that have been incorporated during the reverse transcription of the mRNA to cDNA. Two membranes are hybridized with two cDNA targets, separately from each other and under stringently identical conditions, meaning that as far as possible only specific or homologous binding will be stable (achieved for example by using a sufficiently high temperature). Signal detection used to be done using **X-ray films** (autoradiography) that reacted on exposure to the radioactivity. Now this is done using **phosphoscreens**. These work by using an effect in which the radioactivity excites molecules in the surface of the screens ("phosphorescence"). These then give out radiation as they fall back to their ground energy state. Emission excitation in a special scanner finally produces an image very much like an exposed X-ray film, but span-

Fig. 21.25 Macroarray. Section of a macroarray with 240×240=57600 positions over an area of 22×22 cm (119 cm^{-2}). Here, hybridization was carried out using a complex cDNA radioactively labeled with ^{33}P. PCR products represent the genes and have been applied as double probes. The circular marking of the individual spots indicates that spot recognition software has been used following image recognition. A difference in signal intensity indicates a difference in the abundance of the corresponding cDNA in the complex target. Comparison with another array allows conclusions to be drawn about differences in gene expression in the two targets.

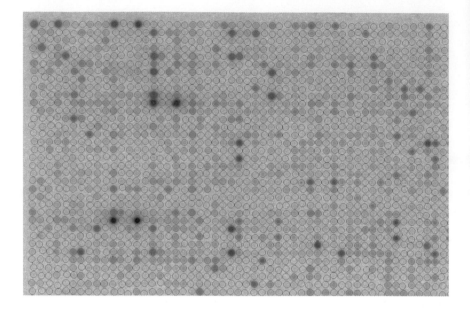

ning a much greater dynamic range. This means that weak signals can still be detected, and strong ones do not exceed saturation point. Appropriate software recognizes the individual spots, assigns intensity values to them, and also offsets any background signals.

The **signal strength** at each individual spot position is dependent on the quantity of DNA on the membrane and the concentration of the cDNA that binds to it in the complex target. The first variable is largely removed by the binding of a large quantity of DNA to the membrane. The exact amount is not allowed to differ between two filters that are being compared. This ensures that the signals measured at the spots will be proportional to the concentrations of individual cDNAs in the targets.

A disadvantage of these arrays is that they require comparatively large amounts of PCR products, as the 3D structure of the membranes adsorbs a lot of fluid. Also, the size of the membranes requires a relatively large hybridization volume, so that a comparatively large amount of target needs to be used. Work with radioactive isotopes also requires more elaborate protection measures and some people find this unpleasant.

21.7.4.2 Microarrays

Microarrays can be seen as a further development of macroarrays. Not only do these take account of a trend toward **miniaturization**, their smaller size offers advantages, including the fact that smaller quantities of DNA need to be used. This comes at the price of making the production and application more demanding, so that special laboratory equipment is usually needed. Slides are used; these are glass slides similar to those used in microscopy. In fact, microscope slides were (and still are) used, as they were easily available. Their size was not ideal, but the apparatus has developed on the basis of these glass slides. When used as a surface, glass has the advantage of being translucent, but it also requires chemical activation of the surface. This involves the introduction of functional groups that enable DNA binding on the surface and spacer molecules are often also inserted.

This methodology faces quite high intrinsic demands as to sensitivity. Macroarrays have the advantage of signal accumulation, on account of the 3D surface and their use of radioactivity. The structure of the surface enables a large amount of radioactive cDNA to be bound in one place and the signal can be recorded cumulatively. Longer exposure therefore means a stronger signal. These advantages are lost in microarrays, because they use a completely flat surface and employ fluorescence dyes to provide the signal.

Glass microarrays offer the advantage that fluorescence dyes can be used. These absorb light of one particular wavelength and emit another. The use of two or more dyes in combination enables two or more experiments to be carried out in parallel on the same chip. This in turn avoids experimental errors such as might arise from individual differences between two arrays when using macroarrays. In practice this means that two targets that are to be compared are labeled with two different dyes during or following reverse transcription of mRNA to cDNA. They are hybridized on the chip together and the labeled individual cDNA molecules compete with each other for binding sites on the chip. An equilibrium is eventually established in which the ratio of the DNA molecules labeled with the two dyes on the chip corresponds to that in the solution. The **cyanin dyes** Cy-3 and Cy-5, **Alexa Fluor dyes**, and **phycoerythrin** are widely used. Almost all dyes consist of relatively large, complex molecules, because they contain conjugated ring systems. Thus, fluorescence labeling is a considerable chemical modification of a nucleic acid, in contrast to radioactively labeled DNA. Polymerases may operate with varying efficiency when dyes are involved, depending on the type and quantity, so that it can be helpful to couple the dye to the cDNA afterwards, instead of incorporating it directly during reverse transcription. This can be done either directly or via reactive groups introduced into the DNA. In practice the two labeled cDNAs that are to be compared are then purified, the concentration and fluorescence (and thus the rate of incorporation) measured, and equivalent quantities of cDNA applied together to a chip. From the technical point of view this is sometimes done using quite simple constructions: a glass coverslip like the ones used in microscopy, to ensure the distribution of the target, a chamber in which the chip is fixed, to ensure a constant humidity of the atmosphere and prevent drying out, and an oven or waterbath to maintain the required temperature. The disadvantage of this experimental arrangement is that only slight mixing of the solution occurs under the coverslip. Technical remedies on offer use sound waves or pump mechanisms to achieve homogeneous mixing of the target. Without this mixing, only a portion of the specific cDNA in the solution would come in contact with its complementary probe within an acceptable length of experimental time; only then is it able to bind to it and produce a signal. After hybridization the chip is washed, dried, and optically read. Larger throughput amounts can be processed more quickly if automated stations are used instead of individual hybridization chambers. These carry out the washing process, for example. There are camera systems to record the signals, although **laser scanners** are usually used. If two dyes have been used for labeling, they will have two different excitation and emission wavelengths, and two images of the same chip can be made. Each represents a particular hybridization with a complex cDNA. The parameters of the recording system with regard to signal generation (e.g., intensity of excitation or photomultiplication) can be changed and these can be aligned with each other as required. Redgreen representation is often used for quick examination of the image; for this, one image is generated in red and one in green, and an image-processing program used to produce a superimposed image. Signals from spots that appear equally strong in both images are then represented in yellow.

Spot recognition software creates a model grid of spots corresponding to the array over the signals (spots) of the recorded image. The intensity values of all the spots are obtained. A number of parameters and filters can be applied at this stage (e.g., to mark spots of inadequate quality, to measure the local or global background, to integrate the signal intensity in various ways, etc.). The end result is a table assigning a position and so a name to each signal (e.g., the name of the clone that was used for PCR) and assigns a measured intensity to the dyes used (and so the cDNA targets used). Background values and/or other statistical values may also be given as appropriate. See Chapter 24 for information on the further analysis of array data. In essence, two types of microarrays are used:

- **Print arrays**, in which conventionally produced DNA is applied to the chip using a printing process (similar to macroarrays).
- *In situ* **synthesized arrays**, in which the DNA is chemically synthesized on the chip.

Any type of DNA can be used in print arrays, although these are usually **PCR products** representing the genes of interest. Amplification of these genes is often done by high-throughput processes and the template is provided either by genomic DNA, amplified using gene-specific primers, or by using clone libraries (gene collections). These last enable the use of universal primers that prime on the plasmid vector in question. This brings cost advantages, but a small piece of the vector is always amplified and so finds its way onto the chip. The transfer from the mictrotiter plate onto the chip is done using a precise spotting robot. The most widely used procedure at present uses **split pins** that take up a small quantity of DNA solution and can then release it in spots rather like a fountain pen. The quantities spotted onto the chip by this method are in the nanoliter range. Only a few scientific laboratories have the equipment to produce such chips themselves, as it is very expensive to acquire. **Print arrays** for a whole range of applications are therefore available to the end-user from various commercial manufacturers.

For arrays that are synthesized *in situ*, oligonucleotides are chemically synthesized on the chip. For these, only considerably shorter sequences can be produced. The length is determined by the quality of the chemical synthesis (with a synthesis accuracy of 99%, as many as more than 22% of all molecules contain errors after 25 synthetic steps, corresponding to 25 bases). Chips of this type are mainly available ready configured from the company Affymetrix (see also Chapter 24).

21.7.4.3 Global and Specific Arrays

Arrays can vary enormously in their dimensions with regard to the number of features (genes) on the chip. A number of considerations govern these dimensions: experimental purpose, available apparatus, or financial resources, as well as data evaluation and processing capacities. For the purposes of explanation, we shall distinguish here between arrays for the global analysis of an entire transcriptome (**transcriptome arrays**) and arrays for the focused examination of a particular question (specific arrays). In simple terms, we are making a distinction between arrays for the analysis of thousands of genes and arrays that cover dozens or hundreds.

The starting point for a specific array is a question that requires a particular gene set such as a collection of available **cDNA clones**, all the genes associated with particular inflammatory processes, those expressed in a particular tissue, all the members of a particular gene family, or all those involved in a particular pathway. This means that the experimenter must work out precisely in advance what should be put on the chip as the probe and this set may have to be kept fairly small for economic reasons. A very high standard of quality will need to be set for the production of the single defined probe, as it may be the intention to carry out the evaluation in a probe-oriented way. In other words, the question is being posed of a particular probe on the chip, to find out if it can yield information relating to the experiment. A global analysis, in contrast, will tend to ask which probes offer information relating to particular experimental conditions. These two questions may seem at first sight to be very similar, but a global transcriptome chip will have many sequences on it whose precise function is not yet known. Depending on the resources, a lower level of quality may be demanded of any given probe on a global array in respect of the sequence selection, and its

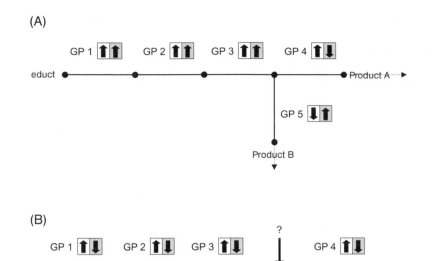

Fig. 21.26 Study of a known metabolic chain (A) and another that is not yet fully elucidated (B). In (A), an educt passes through four steps involving four gene products (GP 1–4) that modify the educt to create product A. Under the conditions shown as white, all the gene products are present (arrow pointing upward). Under the conditions shown as gray, gene product 4 is not formed. Instead, gene product 5 is made. The metabolic pathway branches off and leads to product B. In (B), the metabolic chain that leads to a particular product is not yet fully known. The regulation of gene products X and Y is similar under conditions white and gray, so that both are potential candidates that might be involved in this metabolic pathway. Z, however, is probably not suitable.

validation and production. Some compensation for this may be achieved by redundancy and filtering of the data. **Redundancy** arises from the multiple representation of a gene on the array (e.g., by several representative oligonucleotides or different original clones). Also, several redundant data records will be produced for an experiment, to ensure the statistical validity of the results. **Filtering** of the data can help to reduce experimental inaccuracies that can arise in a very large array (e.g., weak signals resulting from characteristics of the arrays can be rejected for further analysis). A global transcriptome array therefore offers the possibility of placing the expression of genes that are on the array and have so far been little annotated, if at all, in a functional relationship. For example, it might be possible to postulate that a gene that has so far not been described, but is always regulated together with genes of a particular metabolic pathway, is also associated with this pathway (Fig. 21.26).

21.7.5
Specificity and Sensitivity

The **specificity and sensitivity** of the various methods, in other words the **proportion of false-positives and false-negatives**, poses an interesting general question. False-positives are genes that have been identified in the experiment, but are unchanged in their expression. False-negatives are those that wrongly remain unidentified. A maximum of specificity and sensitivity is of course desirable. In experimental practice, however, these demands are sometimes mutually contradictory. If the specificity of an experiment is increased in order to reduce the number of false-positives, this often happens at the expense of sensitivity, so that a number of genuine positives are not found. A large part of practical methodological development is therefore concerned with these issues. A simple example will illustrate this. A blot or array is hybridized with a labeled DNA, and produces both the expected signals and background signals. If the conditions are altered to reduce the background signals, such as by introducing more **stringent hybridization conditions** or additional washing, the art is then to achieve this as far as possible without reducing the specific signals.

Several comparisons of the various methods presented have been published, but there is an ever-increasing number of publications with improvements and further developments. No current recommendation will therefore be made here.

From a critical look at the literature, one particular fact can be noted: the various authors are usually at home in one particular method and this method then duly emerges favorably from the comparison. A decisive factor in the choice of the right method for the investigation of transcriptional activity, in addition to sensitivity and specificity, is the availability of instrumentation and staffing resources. All methods are comparatively demanding to perform, so that the ability to call upon the relevant experience can be an important criterion for the choice of a particular method.

21.8
Cell-based Methods

21.8.1
Green Fluorescence Protein Techniques

A central problem when analyzing the activity of proteins in transiently transfected mammalian cells is that the transfected cells have to be distinguished from nontransfected cells so that observed effects can clearly be ascribed to the relevant proteins. This distinction can only be made indirectly, as most proteins cannot be demonstrated directly. For fixed cells it is possible to use antibodies against the protein being investigated. For observations *in vivo* and for proteins for which there are no specific antibodies available, alternative methods need to be selected. The usual choice for these in the first instance was to use short peptide sequences that were expressed together with the protein (**fusion proteins**). Antibodies to these peptide sequences are available, making it possible to specifically detect them. Common peptide sequences are derived from *myc* **protein** or are artificial (**FLAG tag**). However, with these too, investigations in living cells (lifetime imaging) are not possible. An alternative first emerged with the arrival of **Green Fluorescence Protein (GFP)**. This is a protein from coelenterates, Pacific jellyfish such as *Aequorea victoria*. The protein spontaneously (ATP-dependently) emits fluorescence in the green part of the spectrum. The gene was cloned and can be expressed recombinantly. Since all cells contain ATP, the protein produces a green coloration in the cells labeled with it, even in entire organisms. As a result there are green zebra fish and green mice.

If the ORF of GFP is fused with that of a protein under investigation, a fusion protein will eventually be translated. Ideally this will carry the characteristics of both proteins, independently of each other: (i) the fusion protein should fluoresce green and (ii) it should perform the original function of the other protein. The simplicity of the GFP system has ensured that this protein is used in many projects, often extensively.

However, all tags produced by fusion in this way involve a problem that needs to be mentioned. Proteins usually consist of several domains, which determine both the activity/function and the localization of the protein. Proteins that are secreted by the endoplasmic reticulum (ER)/Golgi apparatus route normally have an **N-terminal signal sequence** that directs the mRNAribosome complex to the rough ER to be translated so that the protein passes into the ER lumen (see Chapter 5). This sequence is made up of hydrophobic amino acids and the position of the sequence in the protein is particularly important for its recognition. If a tag is attached N-terminally to the protein sequence (i.e., C-terminal fusion), then the signal sequence disappears into the total sequence and is masked by it. As a result the fusion protein is not translated in the rough ER, does not pass into the ER/Golgi apparatus, and is not secreted. The protein will be wrongly localized. The same applies to many proteins that are nuclear coded, but carry out their function in the mitochondria, such as the mitochondrial ribosomal protein L18 (Fig. 21.27, top). Other proteins, such as **Rab proteins**, have their localization signal at the C-terminal end, and this could be masked by N-terminal fusion with GFP or another peptide. They too are wrongly localized (Fig. 21.27, bottom). In the middle example shown in Fig. 21.27, the orientation of the tag has no effect on the localization of the fusion protein. The GFP, if expressed alone, can localize either in the cytoplasm or in the nucleus. This can be seen in the bottom example on the left in N-terminal fusion column.

Several conclusions can be drawn from these observations. (i) All results achieved using fusion proteins should be treated with caution and independently verified, ideally using specific antibodies. (ii) It highlights the importance of alternative investigation methods such as bioinformatics. It was the homology of the mitochondrial **L18 protein** with the cytoplasmic ortholog that made it possible to conclude that the localization of the N-terminal fusion was correct. The **Rab protein** was unambiguously identified on the grounds of its homology with known members of this protein family and this made it possible to recognize the Golgi localization of the C-terminal fusion as correct.

Despite these problems, GFP-based technologies have become established in cell biology (see Chapter 19). Further methods have been developed in addition to protein localization, which also require a fluorescence tag in proteins.

21.8.2
Alternatives to Green Fluorescence Protein

For fluorescence dyes to be applicable in practice, they must fulfill a number of prerequisites:
- The dyes must not be toxic or interfere with biological function. As an example, GFP has a slightly inhibiting effect on cell division and this must be taken into account in the analysis of results.
- The dyes must be stable and must not lose color too quickly in daylight or illumination by laser (**photo bleaching**). It would otherwise be difficult to achieve reproducible results.
- The dyes should have an **excitation** and **emission spectrum** that is as sharply defined as possible, so as to avoid spectra overlapping with those of other dyes, unless this is desired (e.g., for **fluorescence resonance energy transfer (FRET)**, see below). Suitable excitation (laser) and filter methods (long pass or band pass) should be available to measure the emitted radiation.

Following the discovery and establishment of the naturally occurring GFP, alternatives were developed. The first of these were produced by means of mutations in the GFP sequence. The spectral characteristics are altered if the amino acids were exchanged at a few positions in the protein sequence. In this way a **Yellow Fluorescent Protein (YFP)** and a **Cyan Fluorescent Protein (CFP)** emerged. Other variants such as Blue and Red Fluorescence Proteins, and a photoactivated form of GFP were developed, and these are used in cell biology. Common to all these protein-based fluorescence dyes is the fact that, as proteins, they introduce a considerable molecular weight and a particular structure into the fusion protein concerned. Myc and FLAG tags had the great advantage of simply consisting of short peptides made up of only a few amino acids.

A recent development has been that of **FLASH markers**. These consist of two components. The first of these is a short peptide with the sequence CCXCC (where X can be any amino acid except cysteine), which is expressed as a fusion tag along with a protein that is to be investigated. The second component is a small, membrane permeable molecule, produced on the basis of fluorescein with arsenic substituents, which can be complexed by the cysteines of the fusion protein. This reaction is described as being specific and affine, so this system offers an alternative to the GFP, bearing in mind that the GFP protein has a size of about 35 kDa.

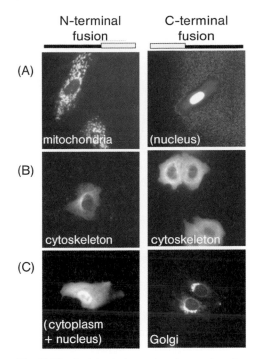

N-terminal fusion	C-terminal fusion

(A) mitochondria | (nucleus)

(B) cytoskeleton | cytoskeleton

(C) (cytoplasm + nucleus) | Golgi

Fig. 21.27 Proteins to be analyzed were cloned N-terminally to the ORF for GFP (N-terminal fusion) or C-terminally to GFP (C-terminal fusion). In order to make clear the orientation of the two protein components relative to each other, the ORF of the protein to be analyzed has been shown as a black bar and GFP as a gray bar. Three different proteins were investigated to find out the effect of the GFP component on the localization of the fusion protein (L18). The top photographs show a mitochondrial protein. This localizes correctly when the protein is oriented ORFGFP, but incorrectly when oriented GFPORF. The middle photographs show a protein that localizes correctly in both orientations, and the bottom example (Rab) only localizes correctly in the Golgi apparatus if it is oriented GFPORF.

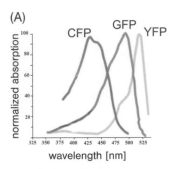

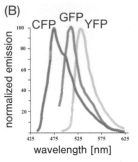

Fig. 21.28 Absorption (left) and emission (right) spectra of GFP and two of its derivatives: CFP and YFP.

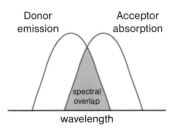

Fig. 21.29 If the emission spectrum of the donor dye overlaps with the absorption spectrum of the acceptor dye (spectral overlap), the energy will pass directly from the donor to the acceptor without any emission of energy from the donor dye in the form of light.

21.8.3
Fluorescence Resonance Energy Transfer

The acronym **FRET** stands for the physical phenomenon of fluorescence resonance energy transfer. This phenomenon is based on a property common to all fluorescence dyes – that of absorbing light (energy) of one wavelength and then emitting energy again as light of a longer wavelength. In GFP that has not been externally excited, the excitation of the fluorescence dye is brought about by intracellular ATP. Alternatively, the excitation can be done with laser light. The advantage of using a laser is that it emits monochromatic light in laser lines, which each have a discrete wavelength. An **argon-ion laser** emits light at 488 nm (main line) and an additional line at 524 nm. The absorption spectrum of GFP reaches its maximum at just about 500 nm, so an argon laser will bring about optimum excitation of this molecule to fluoresce. Derivates of GFP produced by amino acid exchange have modified spectral properties. CFP emits in the blue part of the spectrum; YFP in the longer wavelength yellow part of the spectrum (Fig. 21.28).

If the absorption spectrum of a dye overlaps with the excitation spectrum of another dye (spectrum overlap), the energy emitted by the excited dye can be directly transferred to the second without any emission of light by the first, in a dark reaction (Fig. 21.29).

One essential requirement for this process is **compatibility of the two dyes** (overlapping absorption and excitation spectra); the other is their **proximity in space**. The direct transfer of energy from one dye to the other only operates if they are no further than about 50 Å apart. This distance depends on the particular molecules being used and can be calculated accordingly. The Förster radius (R_0) gives the separation distance between the two molecules at which FRET operates at 50% efficiency.

The necessity for the two dyes to be immediately adjacent brings with it a huge opportunity. If, for example, the aim is to confirm **proteinprotein interactions** in the living cell, FRET can be used (Fig. 21.30). Even if both dyes are present in the same solution (they may, for example, be linked to protein A and protein B), FRET only occurs if these two proteins have an affinity to each other and bring the attached dyes into immediate proximity as well. If the proteins remain more than about 50–100 Å apart, no FRET signal will be measured. This provides a means of measuring proteinprotein interactions directly in living cells.

Measurement is done in two ways: (i) the emission of the acceptor dye is measured after excitation of the donor dye at a wavelength specific to it and (ii) the

Fig. 21.30 Basic principle of FRET. Assuming a compatible pair of dyes, the energy emitted by the donor can be transferred directly to the acceptor in a dark reaction. The acceptor then emits light of a wavelength specific to itself. This FRET energy transfer requires the distance between donor and acceptor to be very small. The Förster radius (R_0) gives the distance between two dyes at which FRET takes place at an efficiency of 50%. The value of this radius is different for the various possible dye pairs, and lies between about 40 and 80 Å. If the distance between them is greater, FRET does not take place.

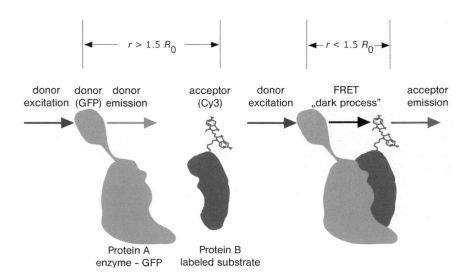

half-life for the continuation of the donor dye at the higher energy level is measured. During normal fluorescence the dye has a characteristic half-life, at which 50% of its electrons have returned to their lower energy level, emitting the corresponding fluorescence radiation. However, if there is a FRET partner, the donor dye can emit its energy more quickly and this is expressed in a reduced half-life. For GFP, the half-life without an acceptor is about 3 ns, but with an acceptor, FRET can reduce this time to about 2.4 ns. The choice of acceptor dye determines the difference in half-life.

21.8.4
Fluorescence Recovery After Photobleaching

Fluorescence recovery after photobleaching (FRAP) makes it possible to follow the movement of molecules in living cells. A focused, strong pulse of laser light in the excitation wavelength of the fluorescence dye is used to destroy the dye molecules at one position in the cell. A record is made of whether fluorescent molecules migrate into the photobleached area and the time taken for this to happen (Fig. 21.31). This process can be followed using an **epifluorescence microscope.** If active transport of the molecules is occurring, this process will take place quickly and in a directed way. If, however, movement is only happening by diffusion, the process will take longer. As well as depending on the type of transport, FRAP is dependent on the size of the molecules under investigation (e.g., monomers or multiprotein complexes) and on any possible anchoring to cellular structures (e.g., membranes). The time needed before saturation of the signal occurs is determined and also the degree of saturation relative to the signal that was measured before the pulse.

Figure 21.32 shows an example of a FRAP experiment. A protein was labeled with GFP and expressed in cells. The fusion protein localizes in vesicular structures of the Golgi apparatus. A strong pulse from an argon laser has caused local destruction of the fluorescence and the area is then observed to see how the signal is re-established from neighboring regions.

21.8.5
Cell-based Assays

The advent of high-throughput projects for the identification of genes and the preparation of cDNA clones made it possible to establish and carry out cell-based assays at higher throughput. In this type of assay a fairly large number of proteins is investigated for particular functions in the cell by targeted destruction of the cellular balance. A high degree of automation is needed to achieve the high throughput. This in turn requires the assays to be simple in structure, meaning that they can be carried out using automatic pipettors and the data captured by automatic systems. Finally, they then have to be evaluated by automatic means.

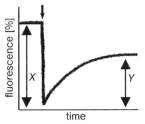

Fig. 21.31 FRAP. The fluorescence signal is first measured in its original state (*X*). Then the signal is destroyed with a strong pulse of laser light (black arrow). Then the time taken for the signal to reach saturation (*Y*) is measured. The ratio of *X* to *Y* gives the degree of saturation. The greater the freedom to move that the molecules have in the medium, the more quickly saturation is achieved.

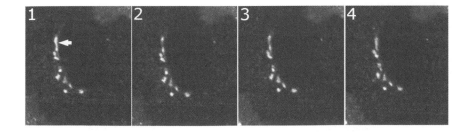

Fig. 21.32 Example of FRAP. The fluorescence molecules are destroyed by a laser pulse at the position marked by the arrow (1), causing the signal to disappear accordingly (2). With time, fluorescence molecules are transported back into that position (3+4). The areas that are fluorescently marked in this figure are vesicular regions of the Golgi apparatus.

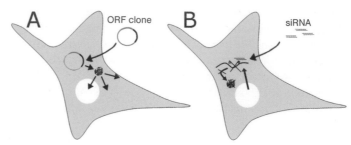

Fig. 21.33 The disruption of the cellular system is carried out either through the over-expression of a recombinant protein introduced into the cell (A) or by switching off the expression of a protein that is otherwise endogenously expressed, by means of RNA interference (RNAi) (B). Overexpression is achieved using expression constructs that usually carry a marker (e.g., GFP green) as well as a suitable promoter and the ORF to be analyzed (black). The marker enables the progress of the protein in the cell to be followed. The arrows indicate that, in the ideal case, the protein is transported intracellularly to the locations where it would carry out its normal function. RNAi operates by importing small, double-stranded siRNAs into the cell. These work together with a protein complex found in every eukaryotic cell so that one strand of the siRNAs attaches to complementary sites on an mRNA and initiates the degradation of this mRNA (symbolized by scissors). This suppresses translation of the mRNA species and the concentration of the protein that it encodes will be reduced. The speed with which this reduction occurs depends on the half-life of the protein and the efficiency of the siRNAs used. In instance A, more than the normal quantity of the protein under investigation is present in the cell; in instance B there is less.

21.8.5.1 Assay Design

Two types of molecules can be considered for targeted investigations of protein function. These are **cDNA clones**, which contain the regions coding for the proteins in an expressible form (**expression vectors**), and **siRNAs** (see Chapters 2, 21, and 31), which can be used to switch off the endogenous expression of the protein under investigation and thus also switch off its function (Fig. 21.33).

A number of crucial factors have to be considered in the planning and execution of **high-throughput assays**. A suitable cell system has to be selected, in line with the biological question at issue. This may seem a trivial matter at first sight, but, to give an example, **HeLa cells** are not suitable for an **apoptosis assay**, because they do not have p53 protein and so an essential branch of the apoptosis pathway in these cells is inoperative. If HeLa cells were used, it would not be possible to record any information about proteins that carry out their function using this pathway.

21.8.5.2 Pipetting Systems

The reproducibility of results is increased if the assays are set up using **pipetting** robots instead of by hand. There are suitable systems available that operate using either cannulae or plastic tips. For cellular assays these systems should be enclosed in a housing to create at least a semisterile working environment. It is important for the cells to be handled gently, to avoid problems such as loss of cells during the various steps of the experiment. The performance of the experiment must also be adapted to the capabilities of a pipetting robot. If reproducibility is high, the human hand is usually quicker and a human is flexible enough to react to changes (e.g., flexible adaptation of reaction times). A robot can do none of this. Assays are therefore usually developed for manual operation in the first place and then adapted for the robot.

21.8.5.3 Reading and Recording of Data

A variety of systems exists to choose from to capture the assay data produced, with three types in particular being used. These are microtiter plate readers, automatic microscopes, and cell sorters for **fluorescence activated cell sorting (FACS)** analysis. These all have in common the use of lasers for excitation of fluorescence, and a detector to receive the emitted fluorescence and direct it to the attached computer. These systems differ considerably both in format (individual analysis or high throughput) and in their different specifications.

- *Plate readers.* Various companies offer devices to read **microtiter plates** of 96-, 384-, and/or 1536-well size. These readers are suited to high and ultrahigh throughput, as the time taken in each case to record the data is short (from seconds to minutes for all the wells of any one plate). Excitation is usually done using a lamp with monochromator or using a laser. Photodiodes or **charge coupled device (CCD)** chips are used for signal detection. The advantage of speed is balanced by the disadvantage of the lack of spatial resolution, which is necessary for various applications. However, if the aim is to measure the influence of a number of chemical compounds on a cell system, for example, and if the device being used is capable of measuring a particular reaction, then plate readers are the method of choice. It is a prerequisite that the number of cells in which a signal can be measured should be as close as possible to 100% of all the cells in the well. If they only number 70%, the proportion of uninvolved cells (in effect control cells) creates an increased background signal (noise) which can swamp weak but genuine signals.

- *FACS* (see Chapter 18). In the FACS cell sorter, parameters are set for individual cells; these parameters are usually fluorescence intensities, which can be measured in various wavelength regions. FACS instruments also have two or more lasers and filters that cover regions of the spectrum that correspond to fluorescence dyes. This enables different parameters to be measured in parallel in each cell. Throughput is considerably lower than with a plate reader; between 30 min and 2 h are needed to complete the measurement of a 96-well microtiter plate (depending on the manufacturer and model). As against this, the parameters measured are recorded for each cell individually.

- *Microscopes (high content screening microscope)* (see Chapter 19). The highest resolution is achieved by fluorescence microscopes, which can also analyze subcellular structures (see **FRAP**). Automation of microscopes requires automatic *XYZ* object tables, comprising an autofocus and means of cell recognition. The photographs also have to be taken automatically and the fluorescence intensities for the automatically identified structures have to be produced in tabulated form. Image recording needs to have functioning image analysis attached to it. Various manufacturers (e.g., Zeiss and Olympus) now offer systems suitable for high throughput. The problems associated with these systems are the time needed to record the data from a 96-well microtiter plate, for example, and also the number of cells analyzed. For many applications, cell numbers of 500 are too few to enable any significant results to be stated.

Overall, there is no ideal system for registering the data from **cell-based assays**. Instead, the best system should be selected for each particular biological and biomedical question.

21.8.5.4 Data Analysis

Data analysis is crucial for every experiment. This is particularly true of cellular assays being performed by a high-throughput method. The rate of false-positive and false-negative results needs to be a low as possible, which often calls for statistical analysis. Depending on the assay, the results to be expected (and observed) may be either of two extremes, as well as anywhere in between. One extreme is the situation where a yes/no decision can be deduced from an assay. This may mean, for example, that interaction is or is not occurring. Large numbers of cells or sophisticated bioinformatics and statistics are unnecessary to evaluate an assay that produces digital information of this sort. The other extreme is an assay or biological question that aims to produce a gradation of differences, but no clear, unambiguous results. The example of a proliferation assay, discussed in Section 21.8.5.1, comes close to this second possibility. Not all nontransfected cells have passed through the entire cell cycle and, in addition, activation by the protein under investigation is not so strong that all the transfected cells have gone through an S phase. Further, the measured effect is dependent on concentration, so that a decision has to be made as to which area is capable of having most confidence placed in it. For this type of assay, it is necessary to increase the number of cells measured and the number of repeats in order to arrive at significant conclusions. Both specificity and sensitivity are increased, so that a greater number of candidate proteins can be filtered out of the general noise (background signal), and the genuine signal brought closer and closer to the noise threshold. The same problem also exists in the case of DNA arrays, for example, in expression pattern analysis.

21.9
Functional Analysis of Entire Genomes

To conclude this chapter we present the example of two methods that each look systematically at the entire genome of an organism.

21.9.1
Genotypic Screening in Yeast

Baker's yeast has been used as a model organism for many years, because it is as easy to culture as bacteria, yet is a eukaryote. It is of economic importance in bread baking and beer production, and its small genome (12 Mb) makes it an attractive choice. This is also the reason that *S. cerevisiae* was the first eukaryote to be sequenced. It has only slightly more than 6000 genes, and possesses other properties such as its unicellular structure, ease of culture in liquid and on agar, ease of manipulation from the molecular genetic and biochemical point of view, and especially the ability to work in both the haploid and diploid state. All these meant that yeast emerged early on as a subject for studies whose focus was on all genes. The screens take the **genotype** as their starting point (in contrast to the mouse, in the example below, where the focus is on the **phenotype**). This means that the experiment tries to switch off or label the genes. It would be better to speak of ORFs rather than genes here, because a third of the more than 6000 potential genes have not yet been precisely characterized. They have simply been determined via an ORF on the genomic DNA; in other words they have had their start and end points defined. First, suitable DNA fragments are produced by molecular genetic means outside the yeast, using either cloning techniques in *E. coli* or PCR. Then these can fairly easily be inserted into the yeast genome by homologous recombination. An older strategy proceeds in an undirected way and targets only transcribed sequences by means of **a minitransposon**, in a quite general manner. Newer methods delete or label the annotated ORFs in a targeted way (Fig. 21.34). Markers are often introduced, which allow a color reaction or selection advantage when grown on special media and so enable the identification of the desired cells. Automated high-throughput techniques are used upwards of 6000 ORFs, since a multiple of more than 6000 clones will need to be examined. Deletion mutants are used to investigate the loss of function of a gene. Here it is very helpful to have the yeast available in both the haploid and the diploid state, as some mutants are homozygous, so a deletion in the haploid genome will be lethal in that case, but compatible with life if heterozygous. Screening can then be carried out according to morphological and metabolic traits such as cell form or growth on particular media. Labeling, positioned at the end of an ORF, enables the protein to be purified or demonstrated in the

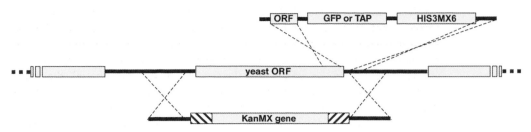

Fig. 21.34 Targeted labeling or deletion of a yeast gene. In the middle is a section of the yeast genome with three genes (or ORFs). The central ORF is the one under investigation and is labeled or deleted by homologous recombination (dotted lines). The top example shows labeling by means of a cassette at the end of the ORF. This contains a GFP or tandem affinity purification (TAP) tag and also a selection marker (HIS3MX6). GFP makes it possible to follow the expression of the protein in the cell and TAP enables purification. In the bottom example the whole gene is replaced, creating a loss-of-function deletion mutant. The KanMX gene enables the recombinants to be selected and the two hatched cassettes represent a molecular bar code that individually identifies the deleted gene.

living cell. Demonstration of its presence in the cell enables subcellular ascription to various compartments or the demonstration of expression under certain conditions (e.g., during the formation of daughter cells (budding) or under the influence of certain inducers, inhibitors, minimal media, etc.).

21.9.2
Phenotypic Screening in the Mouse

Among the higher eukaryotes, the **mouse** is, along with human subjects, without doubt the model organism with the most significance since it is a mammal. Given its significance, the mouse is the example given here. Although other models such as the **fruit fly** (*D. melanogaster*), **zebra fish** (*Danio rerio*), **nematode** (*C. elegans*), and others have also markedly extended our horizons in life sciences, the relative similarity of the mouse with humans and its importance in medical research make it a particularly interesting model from the point of view of functional genomics.

Experimental approaches like those above, which take the genotype as their starting point, are clearly difficult to perform when dealing with around 30000 mammalian genes; however, several thousand gene sequences were isolated by **gene trapping** in embryonic stem cells in particular.

Approaches that investigate gene function by taking the phenotype as their starting point are promising when dealing with complex organisms. In the mouse system intensive work is being done using this route, by a method called **N-ethyl-N-nitrosourea (ENU) mutagenesis** (Fig. 21.35). ENU is a very potent mutagen that produces point mutations as a result of **alkylation** and consequent mispairing or substitution. Administered by injection in the right concentration, it is not too toxic or carcinogenic and has the advantage of affecting mainly spermatogonia, so that the offspring of a treated male will exhibit mutant phenotypes. Pure lines are bred from these by performing the appropriate crosses. Statistically, about 2540 different genes are affected, but most mutations show no effect or become lost. The probability that any eventual phenotype that is observed is the consequence of a single gene is therefore sufficiently high.

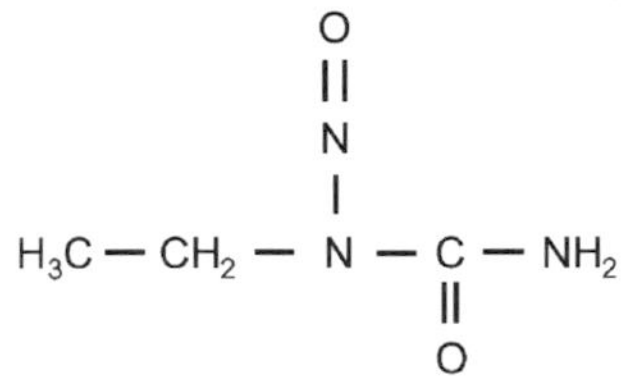

Most ENU mutagenesis screens are designed to investigate dominant inherited traits, so the direct progeny of the treated male exhibit a heterozygotic phenotype. This strategy is the simplest, but is not capable of recognizing the many traits that are only apparent in homozygotes, which are, in other words, recessive. A recessive screen calls either for a high degree of backcrossing to arrive at homozygotic traits or for the use of females with a deletion of a chromosomal region when the cross with the treated male is performed. This then means that

$$H_3C - CH_2 - \overset{\displaystyle \overset{O}{\underset{\displaystyle}{\|}}{\underset{\displaystyle N}{\mid}}}{N} - \overset{\displaystyle}{\underset{\displaystyle \underset{O}{\|}}{C}} - NH_2$$

Fig. 21.35 Structural formula of ENU.

a mutated allele from the father in this part of the chromosome will show an effect in the first generation. Various other strategies are also used, such as chromosomal markers and balancers or pairing of the treated males with females that already exhibit a certain phenotype.

The **screening of the offspring looks** for particular traits and often aims to find a model for a particular human disease. Thus, various medical parameters will be studied to find any deviation from the norm, depending on the screen. The most obvious parameter for such an investigation is the individual's appearance. A variety of study traits will be looked for (e.g., color of fur, number of digits on the feet, ear shape, skeletal deformation, length of teeth, weight, etc.). Mice with altered behavior are tested and selected out in tests for sensitivity to pain, preferred periods of activity, sense of hearing, learning capacity, aggressiveness, and so on. Laboratory clinical parameters such as blood lipids, enzymes, salts, and so on, are measured to find metabolic irregularities. Levels of immunoglobulins and leukocytes can serve to identify mutants with a relevance from the immunological point of view.

Finally, the next step is to map or identify the gene responsible for the phenotype that was found. The preferred method of mapping is via **linkage analysis** after backcrossing of the mutant with a genetically different strain, using **polymorphic genetic markers** (see Section 21.3.1). When a chromosomal region is found that seems a likely candidate for the site of the mutation, possible candidate genes can be found fairly quickly using the sequence of the mouse genome, which has now been established.

There is another aspect of ENU mutagenesis that is of particular interest for the understanding of gene functions. This is the possibility of observing very similar phenotypes that can be traced back to different genes and also that of observing mutations in different locations on a gene that do not necessarily produce the same phenotype. However, one cannot expect to find a phenotype for every mutated gene, since many gene functions can be compensated for in the body when gene defects are present. It must also be borne in mind that the observed effects represent the outcome of very complex metabolic processes involving a multiplicity of gene products.

Finally, the point should be made that **large-scale mutagenesis** screens such as those carried out in the GSF (National Research Center for Environment and Health) in Neuherberg, Germany, involve enormous logistical demands related to the keeping of the mice. For this reason, mouse lines are sometimes simply archived in the form of deep-frozen sperm and revived only when required, by *in vitro* fertilization.

One promising perspective is the identification of mutant individuals that react differently to pharmaceuticals. These may be able to serve as models for **pharmacogenomics** or **toxicogenomics**. The basis in this regard is (human) genetic individuality and the individualized treatment for disease that may arise from it. It demands a knowledge of individual genetic features that is as comprehensive as possible. Aspects of the **genotyping** and **SNP analysis** that this demands are addressed in other chapters.

22
Bioinformatics

Learning Objectives

Bioinformatic methods hold the key to the analysis and understanding of large amounts of data gathered from genomics, functional genomics, proteomics, and molecular diagnostics, to name but a few. This chapter introduces the different methods and problems of bioinformatics.

22.1
Introduction

Bioinformatics as a discipline arose from the necessity to process and analyze sequencing data. The availability of large amounts of data provided by molecular biological techniques consequently led to the development of computer programs to store and compare this data. The process repeated itself about two decades later with the development of DNA chips, which promised insights into the **transcriptome** (after the genome had already been investigated). Ironically, bioinformatics was not meant to be more than an auxiliary discipline at first. However, it quickly rose to a **full discipline** in its own right (especially in the field of sequence analysis) and has significantly contributed to biological knowledge ever since. For instance, the investigation of evolutionary processes, which are not accessible to conventional experiments, has only become possible through the help of mathematical methods and statistical analysis of sequence data. In fact, today's arrangement of the tree of life is based on molecular similarity instead of morphological criteria (Fig. 6.1).

Bioinformatics can be subdivided based on fields of application or methods employed. Selected applications of **bioinformatic methods** (in temporal order) would be sequence alignment, database search, motif recognition, phylogenetic analysis, structure prediction of RNA and proteins, gene prediction, promoter analysis, transcriptome analysis, proteome analysis, and modeling of complex biological systems. These methods contain algorithms to determine similarities in series of characters (including addition, deletion, and alteration of letters), methods from graph theory, statistical procedures (e.g., maximum likelihood estimation), methods of machine learning (e.g., artificial neural networks (ANNs) and hidden Markov models (HMMs)),and methods of system theory (e.g., Boolean or stochastic networks).

The scope of this textbook allows only for a conceptual introduction to these procedures; please refer to Further Reading for a more thorough study of bioinformatics. Prediction and analysis of protein structures as well as analysis of biological networks is treated in Chapters 23 and 24.

An Introduction to Molecular Biotechnology, 2nd Edition.
Edited by Michael Wink
Copyright © 2011 WILEY-VCH Verlag GmbH & Co. KGaA, Weinheim
ISBN: 978-3-527-32637-2

22.2
Data Sources

Bioinformatics would not be possible without readily available molecular data. The first sequencing projects provided all determined sequence data to the international public; the same open conduct is striven for regarding microarray data. Development of the internet as a medium of electronic communication has also contributed to the further enhancement of bioinformatics. Available online databases store primary results as well as derived data. We now introduce several important databases.

22.2.1
Primary Databases: EMBL/GenBank/DDBJ, PIR, Swiss-Prot

Primary databases store DNA, RNA, and protein sequences. Two concepts exist for this type of database. The first concept allows every scientist to file sequence data in the database after only minor plausibility checks. **GenBank** is a prominent example of this approach; today, international data comparison makes its entries identical to those found in **EMBL (European Molecular Biology Laboratory)** and **DDBJ (DNA Database of Japan)**. The strength of these databases is the fast public accessibility of sequences, which are often available only hours after the actual sequencing. The lack of quality control is a disadvantage, however, as the database sometimes contains hundreds of redundant sequences for the same gene that may carry totally different names. Furthermore, the database operators do not correct flawed sequences once they have been saved, so incorrect sequences remain in the system permanently. There are efforts to establish databases of nonredundant sequences from information in primary databases (e.g., the **RefSeq** database).

The diversity of noncoding transcribed sequences (**noncoding RNA (ncRNA)**) has been only discovered recently – they are believed to have a central role in gene regulation. **RNAfam** (Sanger Institute) is probably the best known of the databases storing ncRNA information; in addition there are databases like **ncRNA-DB**, **RNAdb**, and several specialized databases (e.g., for **small nucleolar RNAs** or **microRNAs**).

The second concept is a **curated database**; entries into this kind of database are closely supervised and carefully checked for consistency with existing data and compliance with quality standards. Bairoch from **Swiss-Prot** and the **Protein Information Resource (PIR)** follow this approach. The obvious advantages are juxtaposed to lacking up-to-date information, which has become an increasing deficit in the days of high-throughput sequencing. In consequence, **Swiss-Prot** and the translated coding sequences from **EMBL (TrEMBL database)** are often summarized in a meta database.

Primary databases often contain additional information beyond name, a short description, and the actual sequence, such as information on the author, literature citations, sequence properties (e.g., exons/introns for genomic sequences), or – in the case of curated databases – function, cellular localization, and more. Cross-references to other databases are also important for practical work; in some cases it is still not a trivial effort to find a protein sequence based on the access number of a nucleic acid database.

22.2.2
Genome Databases: Ensembl, GoldenPath

Information from completely sequenced organisms needs to be presented in an integrative fashion. This not only includes the genome sequence itself, but also information on mRNAs, tRNAs, rRNAs, microRNAs, protein, and sequence polymorphisms. Genome databases allow a consistent view on all of this data in a single browser. For eukaryotic genomes, **Ensembl** or the **GoldenPath Browser** (UC Santa Cruz) are most frequently used. They include not only information on genome-related information and homology between organisms, but also allow researchers to include their own data and show them alongside the genome (e.g., on the location of transcription factor binding sites).

22.2.3
Motif Databases: BLOCKS, Prosite, Pfam, ProDom, SMART

Proteins – the most important and versatile building blocks of the cell – have a modular nature. Some estimate that the number of possible proteins in higher organisms (like humans) exceed the number of available protein domains by one or two orders of magnitude. At the same time, the term **protein domain** is not defined precisely – protein domains are thought of as structural or functional modules that present practice usually defines as blocks of conserved amino acids in multiple sequence comparisons. Such precalculated blocks are archived in **motif databases**, which can often serve to understand an unknown protein's function when there are no sequence homologies to proteins whose functions are understood. Furthermore, such databases are indispensable for the understanding of protein evolution through new combination of domains. The databases differ in the mode of calculation as well as in the representation of domains. For instance, **Prosite** (the earliest motif database) uses regular expressions, **ProDom** uses position-specific score matrices, and **Pfam** (protein families database of alignments and HMMs) uses HMMs. Automatically generated databases like ProDom have a considerably larger number of entries (about 1 700 000 at present) than curated databases like Pfam (about 10 000).

22.2.4
Molecular Structure Databases: PDB, SCOP

The main database for **protein structures** is **PDB (protein database)**. Molecular structures of proteins and nucleic acids are determined by x-ray diffraction of single crystals or nuclear magnetic resonance (NMR) methods. Atomic coordinates, crystallographic parameters, and quality factors are all saved here. Keep in mind that structures derived from x-ray diffraction patterns do not contain hydrogen atoms, which show practically no interpretable diffraction due to their low atomic mass. This can be relevant for the detection of hydrogen bonds.

Derived structure databases classify structures based on characteristic features. **SCOP (structural classification of proteins)**, for instance, subdivides structures into all-α, all-β, α/β (antiparallel sheets), α/β (parallel sheets), complex structures, and small structures. The number of saved protein structures (about 60 000 at present) belies the fact that many of these structures display identical folding patterns. Recombinant forms of the same protein are frequently saved as different entries. At present, the number of truly different folding patterns is probably no greater than a few thousand.

22.2.5
Transcriptome Databases: SAGE, ArrayExpress, GEO

DNA microarray technology and **serial analysis of gene expression (SAGE)** allow us to examine the expression levels of thousands of genes at the same time. After a rapid development of technology, the desire has increased to publish the gathered data in a uniform manner, as has previously been done with sequence data. However, the understanding and interpretation of **microarray data** requires information regarding a number of technical parameters (e.g., information on the system used, labeling schemes of the nucleic acids, hybridization conditions, etc.). A correct description of the samples, whose transcriptome has been examined, requires systems of description that do not exist yet. Without such systems, problems arise through synonyms (i.e., different terms describing the same thing), ambiguous expressions, and misspelled words. Current approaches to avoid these problems are the use of controlled vocabularies or so-called ontologies. Ontologies are hierarchical systems of nomenclature for the consistent description of biological entities. Unfortunately, many fields (e.g., the histologically correct description of cell type) still lack universally acknowledged ontologies.

For a meaningful description of microarray experiments, a minimal set of information has been agreed upon, which needs to be provided along with the experimental data to ensure a correct interpretation. This standard has since been called **MIAME** (minimal information about a microarray experiment). An XML-based data format, **MAGE-ML** (microarray gene expression markup language), was developed to ensure a uniform data exchange between different databases. Important public-access databases for gene expression data are **GEO (Gene Expression Omnibus**, National Center for Biotechnology Information, USA) and **ArrayExpress** (European Bioinformatics Institute, UK). SAGE data is saved in an independent project. So far, no method exists to convert expression data gathered from SAGE, cDNA microarray, or oligonucleotide microarray into a comparable format. Only SAGE data can be translated into an absolute value (number of mRNA copies per cell or by 100 000 transcripts).

22.2.6
Reference Databases: PubMed, OMIM, GeneCards

Reference databases establish a relationship between a sequence database entry, the original scientific literature, and the respective gene or protein. Certainly, the most important database is **PubMed**, which contains the **MEDLINE** abstract information of about 4500 bioscientific and medical journals. It also contains features that connect to sequence databases. Furthermore, there is **OMIM (Online Mendelian Inheritance in Man)** that originally listed genes associated with inheritable diseases but also contains other disease relevant genes today. Every gene is listed with literature information on the respective diseases. GeneCards is a meta database that concisely compiles the most important information on human genes from a number of other databases (e.g., GenBank, Locus Link, OMIM, Swiss-Prot, etc.).

Open-access publications (i.e., electronic journals that provide free full-text access to their articles and instead charge authors for publishing) allow for novel bioinformatics methods to fully search the complete text of articles by computerized methods. The best known sources of full-text research articles are **BioMed Central (BMC)** and the **Public Library of Science (PLoS)**.

22.2.7
Pathway Databases and Gene Ontology

"Pathway" in this context means a functional biological module like a metabolic pathway, a signal transduction pathway, or a gene regulatory network. Pathway databases try to collect and structure the information on such modules. Pathways are often represented as mathematical graphs, where nodes are the components (genes, proteins, reactants) and edges mean functional interactions; sometimes edges may have a different meaning depending on context (activation or inhibition) (see Chapter 21). The **Kyoto Encyclopedia of Genes and Genomes (KEGG)** represents such a database with special emphasis on metabolic processes. EcoCyc and related projects provide organism-specific information. Other pathway databases include **Reactome**, **Transpath**, or **Biocarta**. In all cases, the division into single modules ("pathways") is done manually based on prior biological knowledge. Automatic methods to modularize proteinprotein interaction networks are at an experimental stage and not generally accepted.

In contrast to pathway databases, the **Gene Ontology** project provides a terminology to consistently describe the function of gene products. The descriptive terms are taken from an ontology (i.e., a hierarchy of linked terms). The database with gene-to-GO or protein-to-GO annotations can be used to identify all genes or proteins that are associated with a special term from this hierarchy. This is formally equivalent to the list of components of a pathway from one of the pathway databases listed above and is frequently done for over-representation analysis of functional modules (see Section 22.6.6).

22.3
Sequence Analysis

This section summarizes all of the methods used to analyze or compare the **sequences of building blocks** in nucleic acids or proteins. The first methods introduced are based solely on the structure and composition of a peptide chain. Most (and the most relevant) methods also examine similarities to other sequences. Determining these similarities proves quite challenging for the informatic algorithms employed a challenge that is only recently being met by bioinformatics. Sequences are treated as a series of letters from an alphabet; the alphabet for nucleic acids, for instance, would be $\aleph = \{A, C, G, T\}$. Finally, statistical methods of inference can also be used to examine hypotheses on the degree of relatedness; this has provided significant information toward the understanding of evolution on a molecular level.

22.3.1
Kyte-Doolittle Plot, Helical Wheel Analysis, Signal Sequence Analysis

This section covers three methods that are based solely on the amino acid sequence of a polypeptide chain. The **Kyte-Doolittle plot** involves determining a peptide's hydrophobicity in a sliding window of usually five to seven amino acids in range (Fig. 22.1). The **hydrophobicity** is calculated from increments for the individual amino acids and is linked to the amino acids' solvation enthalpy. Hydrophilic amino acids (e.g., serine, threonine, aspartic acid, lysine) receive negative values. The calculated score for a peptide of the given window length is plotted against the window's position, which results in a hydrophobicity profile of the protein. Choosing a suitable window size can smooth the profile, although care should be taken not to **smooth out** the observable effects. This method is most frequently used in the search for protein transmembrane domains. These α-helical structures have a length of 17–21 amino acids, which corresponds to the 3 nm thickness of a lipid bilayer's lipophilic part. A **hydrophobicity diagram** displaying peaks with this length of amino acids is a strong indicator for a transmembrane region.

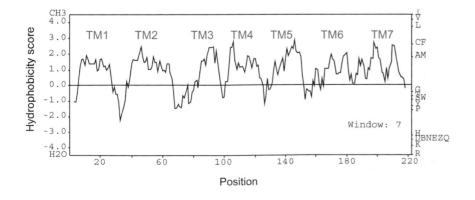

Fig. 22.1 Kyte-Doolittle plot of bacteriorhodopsin from *Halobacterium* spp. The hydropathy index is plotted against the position of the amino acid window (length: seven amino acids). The second ordinate shows the individual amino acid positions. The location of the seven transmembrane helices, as derived from the protein's crystal structure, are indicated (TM1–TM7).

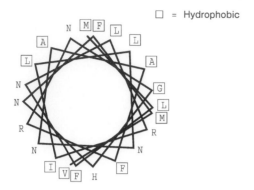

□ = Hydrophobic

Fig. 22.2 Helical wheel plot of the first 22 amino acids of ornithine transcarbamylase (a mitochondrial enzyme) at a rotation angle of 95°. Hydrophobic side chains are boxed; hydrophilic side chains, especially the two positively charged arginine residues, are located on one side of the helix; the structure forms an amphiphatic helix.

Helical wheel analysis focuses on periodic structures occurring in certain protein areas. They are found in **transmembrane areas**, but also in signal peptides' so-called **amphipathic helices**. Amphipathic helices are polar on one side and hydrophobic on the other. Plotting an amino acid sequence in a staggered fashion and with a twist of 100° creates a structure that is quite similar to looking upon a helix along its longitudinal axis (Fig. 22.2). Amphipathic helices can easily be recognized by different coloring of the polar and hydrophobic amino acids. Note, however, that the rotation angle between amino acid side chains is not always exactly 100°. A deeper type of analysis determines the amphipathic moment (i.e., the permanent dipole moment perpendicular to the helix axis) against the rotation angle; a maximum in the range of 85–115° along with sufficient moment size strongly suggests an amphipathic helix.

Signal sequence analysis is based on the presence of certain patterns and deviations from the average amino acid composition in signal sequences that direct protein localization within a cell. The signal sequences for, say, mitochondrial matrix proteins are located at the protein's N-terminal end and have a length of about 25–75 amino acids. Positively charged amino acids are more frequently found in these sequences than on the average; the cleavage site, where the signal sequence is cleft off after import, often features an arginine in position −2 or −10. The signal peptide often forms an amphipathic helix. A number of such observations have been documented and can now be useful in the prediction of protein localization. The program **PSORT**, for instance, performs about 20 single analyses, calculating a score for each of them. Predictions are made using the k-nearest-neighbor method and a training data set of 1500 proteins of known localization. This involves comparing the calculated scores with those of the proteins in the training data set in order to find the k (the standard is $k=9$) proteins with the best matching scores. If a significant portion of these are located in a single target compartment, this compartment is considered a valid prediction regarding the protein's localization.

22.3.2
Pairwise Alignment

Pairwise alignment of two sequences means aligning the sequences to each other in such a manner that a previously defined score is maximized or minimized. The most commonly used target parameters measure the distance of two objects. The Hamming distance measures the number of exchanges that need to be made to Sequence 1 in order to receive Sequence 2. The Edit distance also knows the operations **insert** and **delete** in addition to **exchange**, and is thus better suited for biological sequences, which generally tend not to be of equal length. Since the order of comparison is arbitrary in sequence alignments, delete and insert must be treated equally — after all, deleting a character in Sequence 1 is the same as inserting it into Sequence 2. **Score matrices** define the increments that need to be added to the score for adding or retaining a character in a sequence. Simple scoring schematics are often used for nucleic acids (e.g., retain: +4; exchange: 0), whereas the protein examination takes the different evolutionary pressures on amino acid exchanges into account. The exchange of, for example, leucine for alanine is much more common than, say, arginine for tryptophan. Common score matrices (e.g., the **PAM**, **BLOSUM**, and **Gonnet** series) calculate score increments from observed frequencies of amino acid exchanges in multiple alignments of protein families. The PAM series (see Section 22.4.2 can furthermore be extrapolated upon evolutionary distances that would make multiple alignments impossible due to lacking sequence similarity, simply by multiplying the transition probability matrix, upon which the score matrix is based, with itself.

The score usually serves as a **measure of similarity** rather than distance. Increments are determined as entries in the score matrix for retaining or exchanging a character or simply by predefined values for insertion or deletion. As an enhancement of the Edit distance mentioned above, different scores can be assigned to the initial occurrence of an insertion or deletion and the extension of such an event (**affine gap cost model**). The biological background for this is that the length of an inserted sequence is not subject to conservation; it is thus easier to elongate a sequence than to insert it in the first place. Two increments exist in this kind of model: one for opening an insert or gap (gap opening cost), the other for extending it (gap extension cost). Usually, the former score is rated much higher (up to 10-fold) than the latter. These values need to be adapted when choosing a different score matrix, as they are not independent from the matrix used. The choice of score matrix and gap cost model crucially influences the alignment and should be chosen based on all biological expertise.

Novel methods of **ultraparallel sequencing (next-generation sequencing)** are now producing large volumes of data that need to be mapped by sequence alignment (see Chapter 14). A single run on a new sequence can easily produce more than 30 million sequence reads of 30–150 nucleotides long. These sequences need to be assembled to complete genomes (for *de novo* sequencing or in metagenomics) or be mapped to a unique position in a reference genome (for resequencing studies). The high number of sequences requires special algorithms and hardware to be able to deal with the massive amount of data. In fact, sequencing technology today seems to be more limited by Moore's law, which describes the growth of computer capacity (i.e., it doubles every 8 months) than by the sequencing technology itself. Further bioinformatic developments in this field include algorithms for detection of **single nucleotide polymorphisms (SNPs)** or somatic mutations, but these are still in an early stage.

22.3.2.1 Local/Global

Sequences can be aligned over their entire length or only over those segments with the highest sequence similarities. These **high-scoring segment pairs (HSPs)** are defined as alignments that cannot be expanded in either direction without lowering the achieved score. A certain minimum length has to be exceeded, however. Alignments of the entire lengths of sequences are referred to as **global alignments**, those only over segments with high similarities as **local alignments**.

22.3.2.2 Optimal/Heuristic

At first glance, the number of possible alignments that need to be evaluated in order to find optimal alignments seems to grow exponentially with the number of characters within the aligned sequences. Algorithms whose running time or storage space increases exponentially with input size are generally considered impractical. However, at closer look, a large number of alignments do not need to be re-evaluated because they share many subalignments that only need to be calculated once. The full algorithmic solution to this problem is called **dynamic programming**; it can be applied to all problems that involve optimizing an additive score as a function of the alignment of two sequences. The general solution increases with the length of the sequences to be aligned to the third power; appropriate score schematics can lower this increase to second power complexity. Such algorithms solve the alignment problem optimally in all cases, since they consider all possible arrangements of the sequences. Global alignments are calculated with the **Needleman-Wunsch algorithm**, local alignments with the **Smith-Waterman algorithm**.

When aligning only two sequences, the running time of these **optimal algorithms** is absolutely sufficient. They are impractical, however, for database searches for homologous sequences (with high sequence identity) from hundreds or thousands of entries, or for one-on-one comparisons of a great number of sequences, which are common for sequence clustering. These applications require so-called **heuristics**, which solve a problem well in the majority of cases (i.e., the identified alignments have scores close to the optimal value). Note, however, that heuristic algorithms can fail in some cases.

The most important heuristic algorithms are **FASTA** and **BLAST (Basic Local Alignment Search Tool)**. They are used almost exclusively in database searches and there are several modifications that increase the sensitivity when tracking remotely related sequences (PSI-BLAST, PHI-BLAST).

22.3.3
Alignment Statistics

Especially with database searches, a score is often not enough to evaluate a received alignment's significance. It has been shown theoretically that the scores of databases containing randomly generated sequences of equal length follow an extreme value distribution. The parameters of this extreme value distribution can be determined via a simulation with a small number of random sequences (1000–5000). Two values can now be assigned to each score:

1. The probability of finding an equal or greater score in a database of random sequences (*P*-value).
2. The expected value for the number of alignments with a database of random sequences that have an equal of greater score (*E*-value).

Both values are linked by the following equation:

$$P = 1 - \exp(-E). \tag{22.1}$$

At very small values, *E* and *P* become equal to each other:

$$P \approx E \text{ for } E \ll 1. \tag{22.2}$$

In practice, sequences are considered identical if they have *E*- or *P*-values below 10^{-30}; scores below 10^{-8} indicate related sequences. *E*-values of 0.5 or greater do not indicate any relationship between the examined sequences. Evaluation of sequence alignments is a complex thing, however, and should always also draw upon biological expertise (e.g., on conservation of structural elements).

22.3.4
Multiple Alignment

If more than two sequences are aligned with each other in a manner that optimizes a score, the procedure is referred to as **multiple alignment** (Fig. 22.3). A common-use score function for these cases is the **sum of pairs** function:

$$S(m_i) = \sum_{k<l} s(a_{k,i}, b_{l,i}) \tag{22.3}$$

with $S(m_i)$ being the score for a column *i* in an alignment *m*, and $s(a,b)$ being the increment value for the pair *a,b* in sequences *k* and *l* at position *i* as defined by the score matrix. Summation is performed over all sequence combinations of *k* and *l*.

Multiple alignments can be written as multidimensional dynamic programs. Again, the algorithm's running time increases exponentially with the number of sequences aligned, so they tend to be too slow for most practical applications. A number of heuristic algorithms are used instead. One algorithm that is implemented in the program **ClustalW**, for instance, initially calculates all pairwise alignments and from these draws a tree that reflects the respective sequence similarities. The tree structure is then followed outside-in by aligning sequences to each other, summarizing them as profiles (with ambiguous positions), aligning sequences to profiles, and finally aligning the profiles themselves (progressive alignment).

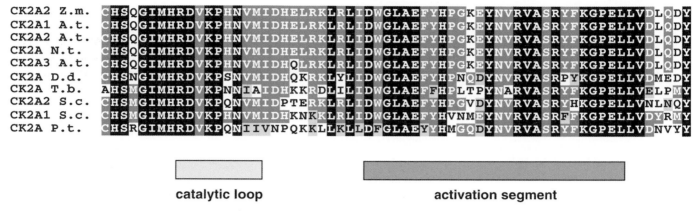

Fig. 22.3 Part of a multiple alignment of sequences of the a subunit of casein kinase II. The abbreviations denote the species: Z.m., *Zea mays*; A.t., *Arabidopsis thaliana*; N.t., *Nicotiana tabaccum*; D.d., *Dictyostelium discoideum*; T.b., *Trypanosoma brucei*; S.c., *Saccharomyces cerevisiae*; P.t., *Paramecium tetraurelia*.

DCA is an algorithm that follows the **divide and conquer** strategy. It involves breaking sequences down into shorter fragments that can be aligned much more quickly. The final alignment is composed from the single solutions. Of course, finding the optimal break points is as complex a problem as the original multiple alignment, so a **heuristic algorithm** is introduced at this stage. These algorithms prove advantageous if the multiple alignment serves as the basis for calculating phylogenetic trees (see Section 22.4.3); performing such calculations with an algorithm that is itself based on a tree (such as ClustalW) would be a circular argument. The resulting tree would probably strongly resemble the tree on which the algorithm is based initially.

Choosing the right score matrix and appropriate gap costs is crucial for practical results. When evaluating multiple alignments, biological knowledge on the aligned sequences should be consulted whenever possible, such as information on the position of an enzyme's active center, single amino acids important for activity or structure (point mutation experiments), or the position of other functional domains (Fig. 22.3).

22.4
Evolutionary Bioinformatics

Bioinformatics has crucially contributed to the study of evolution and evolutionary processes in biology. Since the periods for evolutionary changes, at least in speciation, are much too long to be studied experimentally, and since molecular or even fossil information on common ancestors is most often not available, the only way of getting information on molecular composition of these ancestors is by statistical inference. The underlying theory of evolution is the so-called **Neo-Darwinian Synthesis** – a combination of Darwin's theory of evolution with Mendelian genetics and theories from modern molecular biology. To put it in a nutshell, this theory predicts that evolution is based on random mutagenesis of genetic material and the selection of individuals that are best adapted to their environments. The generation of variation happens on the level of nucleic acids, which are changed by point mutations, insertions, deletions, and genetic rearrangements.

More recent insight into epigenetic inheritance (e.g., by DNA methylation) have not yet been incorporated into a systematic theory of evolution since they are only incompletely understood. It should be mentioned that recent criticisms of Darwinian theory of evolution, in particular the **"intelligent design"** hypothesis encountered in the United States, do not comply with scientific standards;

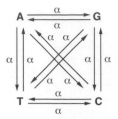

Fig. 22.4 Jukes-Cantor model. Each single nucleotide changes to any other nucleotide with rate α; this is the sole parameter of the model.

for example, by definition, they can neither be falsified nor verified and should be regarded as outside of any natural science.

22.4.1
Statistical Models of Evolution

The most simple statistical model of evolution assumes one common substitution rate α. It is called the **Jukes-Cantor model** (Fig. 22.4). The probability of a single base (e.g., adenine) mutating in unit evolutionary time to another base is 3α; the probability of not mutating is $(1-3\alpha)$. The probability of mutation or retention of a base after a time t is obtained by integration of the associated differential equations. This model allows us to relate the estimated number of mutations to the number of mutations observed in reality, even if the evolutionary time is so long that the same base has been mutated repeatedly.

Solutions of differential equations under the JukesCantor model:

$$p_{ii}(t) = \frac{1}{4} + \frac{3}{4}e^{-4at} \tag{22.4}$$

and

$$p_{ij}(t) = \frac{1}{4} - \frac{1}{4}e^{-4at} \tag{22.5}$$

give the probability for retention (p_{ii}) or mutation (p_{ij}) of a base within a nucleic acid or of an amino acid within a protein in evolutionary time t.

The Jukes-Cantor model makes important assumptions on evolution that also underlie other models:

1. Evolution at one position of a nucleic acid sequence is independent of events at other positions.
2. Evolution is **time-reversible** (i.e., the mutation rate A→C is the same as that for C→A). This allows us to compare two actual sequences that have evolved from a common ancestor in evolutionary time t_1. Based on time reversability, this can be seen as an evolutionary process from Sequence 1 to Sequence 2 over the time of $2t_1$ (Fig. 22.5).
3. Mutation rates are regarded as constant. As a consequence, evolution is analyzed on time scales where genetic changes occur at constant rates (**molecular clock**). This scale is not related to real time. In contrast, theoretical studies as well as experiments on micro-evolution have shown that increase of fitness or improvements in adaption to new environmental conditions occur in a jump-like fashion, not continuously.
4. The majority of changes in protein sequences is neutral (i.e., not under selective pressure) (**Kimura hypothesis**). This means in turn that an observed conservation of an amino acid in long evolutionary time is likely to be under selective pressure (i.e., this amino acid has an important functional or structural role). The neutral distance (i.e., the time in which all amino acids of a protein are expected to have mutated at least once) is estimated to be 320 million years. The most distantly related species compared to humans, whose com-

Fig. 22.5 Consequences of time reversability. Two actual sequences, 1 and 2, have evolved from a common ancestor in evolutionary time t_1. Time reversability allows us to change one arrow and look at this as an evolutionary process from Sequence 1 to Sequence 2 in evolutionary time $2t_1$, passing the (unknown) ancestral sequence as an intermediate step. Thus, a distance between any pair of actual sequences can be calculated.

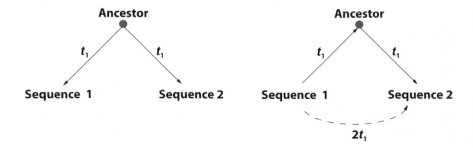

mon ancestor is dated back to this time, are the marsupials. More closely related species (e.g., rodents, which include some prominent model organisms) might show conservation of amino acids just by chance; the time for changing these amino acids by evolution from the least common ancestor has simply not been long enough to change them by neutral evolution. Thus, comparing the genomes of mice and men, one cannot simply conclude on the functional importance of a residue when this residue is conserved.

Extended models of DNA evolution exist. One example would be the **Kimura model**, which has different mutation rates for transitions, a, and for transversions, $\beta\,(a>\beta)$. Models that include more than six different mutation rates are not time-reversible. Those models are rarely found in applications, which is due to the practical difficulties in reliably estimating multiple mutation rates.

22.4.2
Relation to Score Matrices

The matrix of mutation rates, $\mathbf{Q}$, can be transferred into a matrix of **transition probabilities**, $\mathbf{P}$, by either matrix operations or solving the associated differential equations. The logarithm of the calibrated transition probabilities (log odds score) is the score in a score matrix that is used to align protein sequences. This score matrix is calibrated to certain evolutionary distances, which are usually given in **percent accepted mutation (PAM)** units (i.e., the percentage of observed mutations). The real number of mutations, which includes unobserved multiple changes, is given as **percent expected mutation (PEM)** units. Due to sequential mutations at the same position (Fig. 22.6), PEM is larger than PAM, but can only be estimated since these sequential mutations have never been observed.

The matrix $\mathbf{P}$ of transition probabilities can be multiplied by itself to extrapolate to longer evolutionary distances. For example, $\mathbf{P}(2)$ that is the basis for PAM2 can be obtained from $\mathbf{P}(1)$ from which PAM1 is calculated: $\mathbf{P}(2)=\mathbf{P}(1)\times\mathbf{P}(1)$, or more generally: $\mathbf{P}(s)\times\mathbf{P}(t)=\mathbf{P}(s+t)$ (**Chapman-Kolmogorov equation**).

22.4.3
Phylogenetic Analysis

Phylogeny means the evolutionary development and history of a species as opposed to ontogeny, the development of an individual. This section introduces methods to examine hypotheses on phylogenesis with methods of molecular biology and statistical inference.

When looking at a **phylogenetic tree** that describes the kinship between species (Fig. 22.7), one should always keep in mind that the sequences of the last **common ancestors** are unavailable. Barring very few exceptions, only sequences from living organisms can be drawn upon for the construction of such trees. Sequences corresponding to the inner nodes of the tree can thus only be concluded. Different assumptions on the course of evolution are formulated as a model upon which the most plausible progenitor sequence can be calculated. An example of a common principle is **maximum parsimony (MP)**, which states that those two sequences that require the least changes to transfer one into the other are the most closely related. A more complex calculation formulates and optimizes a **likelihood function** in order to find an optimal solution (**maximum likelihood principle (ML)**). ML methods include statistical models of evolution (e.g., the Jukes-Cantor model in Section 22.4.1) to estimate branch lengths in phylogenetic trees.

Calculating phylogenetic trees does not only consider sequence similarity, but also that an observed similarity might be due to random effects. It thus calculates the ratio of probabilities under the assumption of kinship and totally random changes, referred to as the **odds**. For technical reasons, calculations are per-

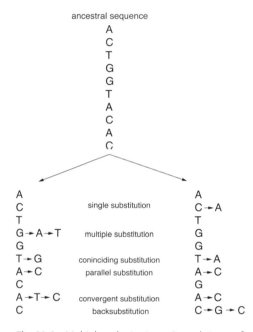

Fig. 22.6 Multiple substitutions. Several times of multiple substitutions in comparisons of two actual sequences are shown. These can only be concluded by statistical methods since only simple differences between the two sequences can be observed. Note that even an apparent observation of a base (or of an amino acid in a protein sequence) can result from multiple substitutions (i.e., backsubstitution).

Fig. 22.7 Phylogenetic tree for sequences of the α subunit of casein kinase II. It can be observed that a gene duplication must have occurred very early in vertebrate development. Independent duplications have also taken place in fungi, while plants have multiple isoenzymes. Abbreviations of species (also see Fig. 22.3): T.p., *Theileria parva*; T.a., *Triticum aestivum*; O.s., *Oryza sativa*; G.g., *Gallus gallus*; M.m., *Mus musculus*; B.t., *Bos taurus*; H.s., *Homo sapiens*; D.r., *Danio rerio*; D.m., *Drosophila melanogaster*; C.e., *Caenorhabditis elegans*; S.f., *Spodoptera frugiperda*; H.p., *Hemicentrotus pulcherrimus*; X.l., *Xenopus laevis*; O.c., *Oryctolagus cuniculus*; R.n., *Rattus norvegicus*; Y.l., *Yarrowia lipolytica*; S.p., *Schizosaccharomyces pombe*.

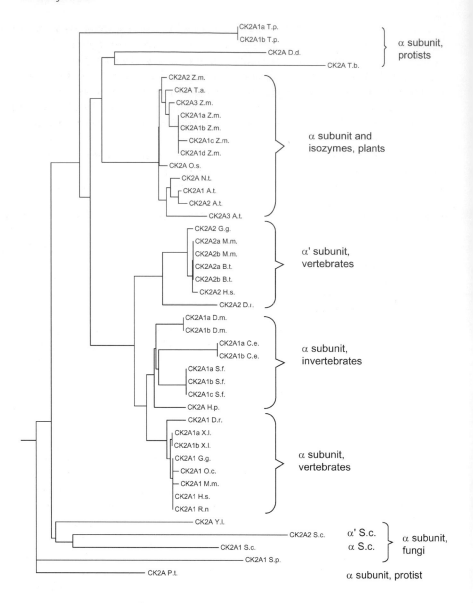

formed with the logarithm of these odds, so these score functions are often known as **log odds scores**. In order to select the most probable tree for a given evolutionary model, all possible trees need to be evaluated. Unfortunately, the number of possible trees grows exponentially with the number of sequences whose relationship the tree is to describe. Again, **heuristic algorithms** are used instead (e.g., **neighbor joining (NJ)** or the **MP model** (see above)). Another approach analyzes all possible four-branched trees (which can still be calculated in acceptable time) and then composes a full tree from these (**Quartet-PUZZLE**). Many of these heuristic methods assume that a tree is globally optimal (with a maximum score) if it is locally optimal at all places (i.e., the subtrees all have maximal scores); the best tree can then be composed from the best subtrees. This assumption is not always true, however; there is still great need for improved algorithms that will help derive phylogenetic trees. Important phylogeny programs are **PAUP**, **MEGA**, and **PHYLIP**.

22.5
Gene Prediction

The problem of gene prediction is differentiating coding regions of the genomic sequence from noncoding ones. This difficulty grows with increasing genome complexity. While it is generally possible to differentiate coding regions from intergenic regions in bacteria, the problem is only unsatisfactorily solved in mammals with about only 1% of coding DNA. If an mRNA sequence is known, the gene structure (transcription start site 5'c-untranslated region (exon 1) – start of translation (ATG) – intron 1 – (…) – exon n – stop codon – 3'-untranslated region – stop of transcription) can be derived from comparisons with the genomic sequence. In the human genome, this is possible for only part of the estimated 25 000 genes. Automatic methods of gene prediction are thus necessary. All possible statistical sequence features can be used for this, from the composition of di-, tri-, and hexanucleotides, the presence of characteristic patterns (**splice sites**), the distribution of **stop codons** in different reading frames, and more. In order to train statistical learning procedures, these are used on genes of known structure in order to predict genes in other genomic areas. Presently, prediction of the **untranslated regions**, and thus the first and last exon, is not possible with satisfactory accuracy; this is increasingly true for the regulatory regions upstream from the origin of transcription. There are also difficulties with protein coding regions; estimates say, for example, that half of the 19 000 genes from *Caenorhabditis elegans* proved faulty in the genome project.

22.5.1
Neural Networks or HMMs Based on Hexanucleotide Composition

The statistical learning procedures used for gene prediction are **ANNs** or **HMMs**.

ANNs consist of a layer of input nodes, a so-called hidden layer, and an output node. The input nodes transfer inputs (e.g., relative frequencies of the possible hexanucleotides within a given frame) to the hidden layer as numerical values. A transfer function defines the connection between a node's input and output value; usually these are sigmoidal functions (e.g., tanh). Every internodal link within the network is assigned a weight factor. During the training process, these are iteratively adjusted in such a manner that the difference between the output node's prediction (here the probability that the region in question is a coding region) and the actual known fact about the coding sequence (coding or noncoding) is minimized.

HMMs are based on Markov models in which a random event only depends on the preceding event. A Markov model with probabilities for the succession of two letters in a chain of characters could be understood as a generating model for this chain of characters. HMMs furthermore introduce additional (hidden) states from which the next characters can be generated. This could be, for example, the states **coding** and **noncoding** ("c" and "nc", respectively). In this respect, hidden means that the generated sequence does not give away from what states the letters were generated. There are different transition probabilities for the same letter in different states. For nucleic acids, for instance, a simple Markov model with four letters would have $4 \times 4 = 16$ transition probabilities (e.g., $P(A|A) = 0.015$); this would be read as the probability that an A follows an A is 1.5%. A HMM with four letters and two states c and nc would have $8 \times 8 = 64$ transition probabilities (e.g., $P(A_c|A_c)$) for the probability that an A in the "c" state would follow an A in the same state, while the probability $P(A_{nc}|A_c)$ states how often an A in the "nc" state will follow an A in the "c" state and would thus change its state. Additional pseudocharacters are also introduced, which model the beginning and end of the sequence (Fig. 22.8).

The complete set of probabilities is determined by training with sequences for which coding and noncoding regions are known (**Baum-Welch algorithm**). In-

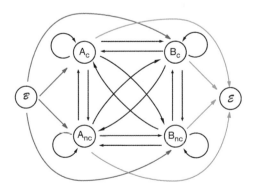

Fig. 22.8 Operating principle of a simple HMM. In this case, there are only two characters, A and B, as well as two hidden states, "c" (coding) and "nc" (noncoding). Additionally, pseudocharacters B and E for beginning and end have been added to the sequence. Each arrow in the schematic is assigned with a transition probability, which is calculated through training with sequences for which the coding and noncoding regions are known.

stead of using the HMM as a generating model for the sequences to be analyzed, the inverted problem is solved – we calculate the probability that the analyzed sequence was generated by the Markov model. This allows for a prediction of coding and noncoding regions within the sequence. The decoding of hidden states that shows, for example, which regions are coding or noncoding, is done by the **Viterbi algorithm**. HMMs are also used in other areas of bioinformatics (e.g., predicting secondary structure elements from the sequence).

22.5.2
Comparison with Expressed Sequence Tags or other Genomes (*Fugu*, Mouse)

Since the prediction accuracy of the statistical learning procedures mentioned above is unsatisfactory in many cases, they are often combined with information on the degree of sequence similarity between closely related organisms. This method is limited by the availability of fully sequenced genomes; **model organisms** for human sequences include the mouse, domestic or farm animals (dog, chicken, pig, or cattle), or the pufferfish *Fugu*. The idea behind this approach is that coding sequences are more strongly conserved than introns, promoter regions, or intergenic regions. Insight can also be gained from **mRNA fragments** – so-called **expressed sequence tags (ESTs)** – whose sequences are gained from high-throughput procedures, but tend to be of low quality. This again poses new challenges for alignment algorithms. On the one hand, they need to be able to perform calculations with very long genomic fragments (several hundreds thousand base pairs); on the other hand, they need to handle the frequent sequencing errors and frame shifts in the ESTs.

Comparative genomics, which compares entire genomes of related organisms, has also been used, in combination with other methods, for the prediction of regulatory sequences (**Comparative Regulatory Genomics (CORG) database**).

22.6
Bioinformatics in Transcriptome and Proteome Analysis

Since 1995, several procedures have been established to examine the identity and frequency of transcripts and proteins on a whole-genome scale. The most important ones are **DNA microarray technology**, **SAGE** and **mass spectrometry** methods in conjunction with 2D gel electrophoresis or column chromatographic procedures (see Chapters 7 and 8). The mere volume of generated data makes bioinformatic storing and processing unavoidable, but it also poses a new challenge for the statistical algorithms used for analysis. This challenge can be summed up in the curse of dimensionality and results from the high number of variables determined at the same time in comparison to the number of examined samples. There are always some gene-specific fragments among the 10 000 on a DNA chip whose signal values correlate with the behavior expected over 40 samples merely for stochastic reasons. Procedures of feature selection are often employed to avoid this dilemma and to make an informative selection from the initially high number of variables.

22.6.1
Preprocessing, Normalization

Data preprocessing depends on the method used for transcriptome or proteome analysis. With **SAGE**, the received tag – a short piece of sequence characteristic for one mRNA – needs to be assigned to the corresponding gene. The number of tags of one type found within a single analysis is then scaled to 100 000 transcripts. The resulting values are easily comparable, even though the real frequency of rare tags can only be estimated with large errors (the error is proportional to $n^{-1/2}$, with n being the number of found tags).

When performing proteome analyses, quantification is essential for the type of preprocessing. Staining proteins in 2D gels with subsequent image analysis uses similar procedures as DNA microarrays (see below). The proteins are identified with **matrix-assisted laser desorption/ionization (MALDI) mass spectrometry**, which compares the received fragment sizes with a database. In combination with a second mass spectrometer (**MS/MS**), a short piece of sequence can also be generated, which facilitates identification in combination with the fragment mass (see Chapter 8). The bioinformatic methods involved in this procedure cannot be covered in the framework of this book, however. If proteins are quantified chromatographically, the chromatograms need to be evaluated by identifying peaks and integrating the areas beneath them; the proteins are then identified with the mass spectrometric methods mentioned above.

DNA microarrays can be used in two modes of analysis. The first mode involves competitive hybridization of two cDNA representations (each of which is marked with a different fluorochrome) to the DNA fragments robotically spotted onto the chip (**spotted** or **printed** chip). After hybridization and washing, a color mixture can be read out from every spot, which is determined by the different fractions of complementary mRNA in the two cDNA representations. One of these two preparations stems from control conditions and is used in all hybridizations of a study. The other preparation stems from the condition to be examined. The control **cDNA preparation** serves as an internal standard. Preprocessing is image analysis with the steps segmentation (localizing spots), addressing (assigning positions to known fragment arrangements), and quantification (discriminating foreground and background, pixel readout). Since this type of microarray works with an internal standard, results are recorded in reference to this standard as the ratio of the signal strengths of sample and control or its logarithm (**log ratio**).

The second mode works with *in situ* **synthesized oligonucleotides** as fragments on the chip. Since the resulting hybridization is not as specific as with the longer DNA fragments used in the other method, 10–20 oligonucleotides are used for each gene and each oligonucleotide is neighbored by another oligo with a single base exchange in the center, functioning as control. Hybridization is performed with a single cDNA or cRNA representation that is coupled with **biotin** and can be recognized by a **streptavidin fluorescent dye reagent**. Calculating the individual values into a single value per gene, whose expression strength was to be measured, is quite complicated and presently subject to controversial debate.

The received signal values now need to be normalized in order to eliminate technical influences that are hard to control and lead to systematic errors. When using chips that present a genome-wide selection of genes, it is often assumed that most of the presented genes will not change their expression based on minor condition alterations; if this assumption is no longer valid, heterologous control fragments on the chip have to be used in order to determine correction factors.

Since the largest number of studies have been performed with DNA microarray technology, further data evaluation will be covered from this angle. Results of a microarray study are presented in a table with rows displaying the signal values of the examined genes and columns showing the values of an examined sample. Since signal values are assumed to be proportional to a gene's expression strength (the number of copies per cell), they are often referred to as gene expression values. One should keep in mind that microarray results depend on the special procedure and cannot simply be compared with results gained from different procedures.

22.6.2
Feature Selection

The intrinsically high number of codetermined variables can be decreased with procedures that directly select informative variables. These procedures correspond to the biological insight that any specific cell type expresses only a fraction of the available genes and that the expression strength of many genes does not change with the observed conditions (e.g., because they are constitutively expressed, unregulated genes). Feature selection includes, for instance, excluding all genes whose microarray signal levels are so low that they can be considered unexpressed. The respective threshold can only be estimated, however, since the signal strengths of weakly expressed genes and unexpressed genes can be of the same magnitude. Similarly, a minimum change during a series of microarray experiments can be required in order to exclude genes whose expression does not change. This method presents some problems, however, as a moderate but highly reproducible change can be more interesting than a strong but variable change; the explained method would only accept the latter gene expression values for further analysis.

Another approach of feature selection uses information about the **biological differences** between the examined samples. If they are, for example, samples from defined classes (like bone marrow samples from two forms of leukemia), genes that are expressed differently by these forms can be looked for specifically. This often involves calculating a statistical value that takes the differences between the classes into account as well as variability within the classes. Some measurement distributions can be predicted theoretically (with certain assumptions, such as normal distribution of the measurements); a significance threshold is then derived from these predictions. This is true, for instance, for t-statistics, which are used for t-tests. Note that one t-test is performed per examined gene, which means 10 000 t-tests for 10 000 genes. The calculated significance thresholds must be adjusted for multiple tests, which can be achieved through several methods. These problems can be avoided by simulating the statistical distribution. This simulation is performed by random switching of the samples' class notations and calculating the statistics from a great number of these switches (usually several thousands). A significance threshold can be set by comparing observed statistics with those gained through permutation.

If the examined parameters cannot be divided into classes, but change continuously instead (concentration or time series, cell cycle), a statistical model can be established to extract informative genes. When dealing with the cell cycle, for instance, **Fourier transformations** can help select genes with cyclic changes of expression and with a period that roughly equals one passage of the cell cycle.

22.6.3
Similarity Measures: Euclidean Distance, Correlation, Manhattan Distance, Mahalanobis Distance, Entropy Measures

Some of the further methods of analysis require a quantification of the similarity between two gene expression profiles. Of course, this requires a means to measure similarity or lack thereof, which is mathematically realized as a measure of distance. A gene expression profile can be understood mathematically as vector – an ordered list of numeric values. Frequently used distance metrics include the **Euclidean distance** (i.e. the geometric distance between points defined by gene expression vectors), the **Manhattan distance**, which is more robust to outliers than the Euclidean distance, the **Mahalanobis distance**, which accounts for the covariance between gene expression profiles, the **correlation distance**, which is scale invariant and is given as 1 minus the correlation coefficient, and the **mutual information** measure from information theory, which is calculated from relative entropies.

22.6.4
Unsupervised Learning Procedures: Clustering, Principal Component Analysis, Multidimensional Scaling, Correspondence Analysis

Unsupervised procedures are used to recognize **patterns** independent from other information pertaining the data. They recognize every kind of pattern, including those that are caused by technical influences (such as processing samples in different laboratories). Unsupervised procedures help identify subgroups in a data set. Applied to genes, clusters with similar expression can be identified. Unsupervised learning procedures are less suited for the analysis of known class divisions as they do not use this special information.

Clustering procedures create a series of groups based on a similarity or distance matrix and thus cluster similar objects. In transcriptome analyses, these objects could be the examined samples that are grouped based on their expression profiles as well as the genes that are grouped based on their expression patterns. The procedures in use either operate agglomeratively (i.e., they gradually cluster objects to higher and higher groups), or in a partitioning manner (i.e., dividing into a number of clusters that is often preset). Methods of hierarchical clustering belong to the former type; *k*-**means** and clustering with **Kohonen networks** (self-organizing maps) belong to the latter.

In contrast to clustering methods, **principal component analysis (PCA)**, **multidimensional scaling (MDS)** and **correspondence analysis (CA)** attempt to project the data into a low-dimensional space (plane, 3D space). A visual analysis of the data structure is thus made possible. PCA attempts to maximize the data dispersion along the axes, while MDS tries to conserve the distance of the individual data pieces as much as possible. CA is special in the way that it tries to display the objects from columns and rows of a data matrix together in a low-dimensional space. The distances between the individual objects (e.g., points representing genes or the samples of a transcriptome analysis) can then be interpreted as a measure of correspondence.

All the unsupervised procedures described are explorative in nature and do not provide definitive, statistically sound results. Their validation is difficult and can at best provide a measure of an observed cluster's robustness. Different procedures will generally provide results that diverge in the details, and no way exists to tell which result is closest to reality. Still, they represent an important source of biological understanding, as has been shown by a number of studies.

22.6.5
Supervised Learning Procedures: Linear Discriminant Analysis, Decision Trees, Support Vector Machines, ANNs

In contrast to their unsupervised counterparts, **supervised procedures** do utilize additional information about the examined samples. If these can be divided into defined groups (classes), methods of classification can be employed; if the subdivision is based on a continuous parameter like concentration or time, methods of multivariate regression are used. Periodic processes can often be accessed through Fourier or wavelet transformations. We now introduce a few classification procedures.

The **classification process** can be divided into the following steps. First, a suitable learning procedure, the classifier, is selected. The data set is then divided into three parts: one part for training the classifier, one to optimally adapt the algorithm parameters, one part for validation. Ideally, about half of the data set should be used for training, the rest for tuning and validation. Unfortunately, microarray studies are too small, at least so far, to discard half of the data for training. Instead, studies often resort to cross-validation, which involves dividing the data into *n* subsets of roughly equal size. The *n* cycles each omit one (changing) subset, train the classifier with the remaining data, and test its predictions on the omitted subset. The measured error is then averaged over the *n* cycles. Once the classifier procedure is optimally adjusted, it is trained with all the data. The algorithm can then be used on new samples for prediction in the same class. Note that when estimating the error of cross-validation, feature selection (if used) needs to be reperformed for each cycle of the cross-validation and only under consideration of the training data. Otherwise, the error is estimated too optimistically (selection bias).

Classification procedures stem from methods from statistical learning theories or from procedures of machine learning. A simple method from statistics is, for example, **linear discriminant analysis (LDA)**. Like PCA, this method is about finding a linear coordinate transformation that maximizes the division into (two) classes along the axes. The single variables can then be ordered by their influence on the first discriminants. The procedure is only suited for the analysis of gene expression when a high number of genes clearly supports the class division, otherwise it can only be used after drastic feature selection. **Decision trees** are another method of classification. They have the advantage of implicitly selecting features and classifying along easily formulizable rules that can be interpreted biologically. Unfortunately, decision trees are not robust and can easily produce very different trees upon minor changes in the input data. This can be overcome by using several trees at the same time; such classifier ensembles can also be constructed with other algorithms. Classes can also be defined by ANNs; their working principles have already been outlined (see Section 22.5.1). Before applying **neural networks** to gene expression data, note that they can easily be overtrained; they often demand a significant reduction in the number of entry parameters as well as special procedures to prevent overtraining. The last method to be mentioned is based on **support vector machines (SVMs)**; please see "Further Reading" for more details. SVMs have proven to be robust classifiers that also get along well with higher numbers of features used for classification. Unfortunately, said classification is not always easily understood; only with difficulty is it possible to name the features most important for classification.

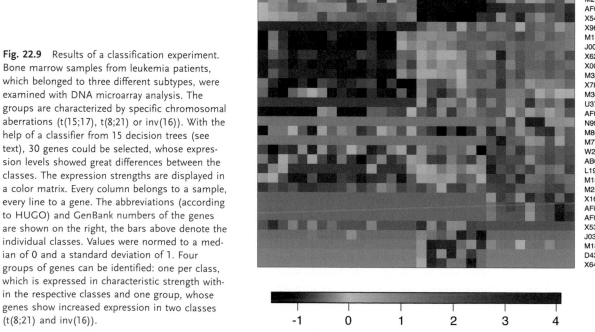

Fig. 22.9 Results of a classification experiment. Bone marrow samples from leukemia patients, which belonged to three different subtypes, were examined with DNA microarray analysis. The groups are characterized by specific chromosomal aberrations (t(15;17), t(8;21) or inv(16)). With the help of a classifier from 15 decision trees (see text), 30 genes could be selected, whose expression levels showed great differences between the classes. The expression strengths are displayed in a color matrix. Every column belongs to a sample, every line to a gene. The abbreviations (according to HUGO) and GenBank numbers of the genes are shown on the right, the bars above denote the individual classes. Values were normed to a median of 0 and a standard deviation of 1. Four groups of genes can be identified: one per class, which is expressed in characteristic strength within the respective classes and one group, whose genes show increased expression in two classes (t(8;21) and inv(16)).

A number of further classification procedures are currently in use. These procedures are for used to create a means of diagnostic prediction based on gene expression measurements in problematic cases. Another aim is to select the most important genes whose expression changes between defined groups (Fig. 22.9). Furthermore, these methods can be used to identify samples that do not fit into a defined class and whose class can only badly be predicted. Such samples can be hints for the presence of new disease subgroups.

22.6.6
Analysis of Over-Representation of Functional Categories

A prime goal of analysis of proteomic or transcriptomic data is the identification of differentially expressed proteins (or their transcripts). The result consists of a list of such proteins or mRNAs and of a measure of the significance attached to them. Sometimes these lists tend to be very long, so that researchers sought to summarize these lists with respect to the biological functions covered by the proteins (or mRNAs) in the list. They were implicitly hoping that the list of over-represented cellular processes was much shorter than the list of proteins itself. As a source of information on biological processes, pathway databases (see Section 22.2.7) are used that list metabolic or signal transduction pathways, or gene regulatory networks. The interaction information in pathway databases is, however, never used in over-representation analyses. The individual modules are taken as a set of entities to be tested against the list of differential proteins or mRNAs from an experiment. As an alternative, Gene Ontology (see Section 22.2.7) can be used. Here, functional categories are embedded in a hierarchical system of descriptors such that for any particular term there is one or several more general terms attached to it. For the purpose of over-representation analysis, the entities associated with a special term are supplemented by those attached to any category that is more downstream of the category under study. In this way, hierarchically nested lists of mRNAs or proteins are created, which are annotated with terms from the Gene Ontology hierarchy, and each of which is tested whether it is over-represented in the list of differentially expressed entities in an experiment.

The components from pathway databases, or Gene Ontology associations, have to be mapped to the list of studied entities, where it is common that a large portion of the latter has no counterpart in pathway databases. To study whether any of the categories is over-represented, statistical testing is used, most often Fisher's exact test of a two-way contingency table. The results need to be corrected for multiple testing, since one test per functional category is performed and individual p-values do not account for this. Alternative methods are in use, two examples being the **globaltest** and **gene set enrichment analysis**.

The underlying hypothesis of over-representation testing is that a biological process or a functional category might be relevant if several of its components (i.e., more than expected by chance) are subject to differential expression. This hypothesis may hold true, but not necessarily. It is, for instance, possible that by changing the expression of a transcription factor even those target genes are changed, too, that are not at all related to the studied experimental conditions (**bystander or passenger effect**). Thus, a functional category found by over-representation analysis needs to be validated by independent cell biological experiments.

22.7
Bioinformatic Software

Bioinformatic analyses have first been carried out on mainframe computer systems. Thus, software has been primarily developed for Unix-based operating systems. One of the packages still in use is **GCG (Genetics Computer Group)**, which comprises some 300 programs for sequence analysis of proteins and nucleic acids as well as for database search. It needs to be licensed and is nowadays most often maintained by core computing facilities. In contrast, the **EMBOSS** project wants to provide similar programs for private users under an open-source license (i.e., royalty free); it is available for Linux-based computer systems. Database searches by **BLAST** or profile searches by means of HMMs models (see Section 22.5.1) can be triggered from web-based forms on the internet.

Statistical analysis of microarray data can be performed using the programming environment **R**, which has evolved from the statistical programming language S. The **Bioconductor** project provides some 280 extension libraries for different data analysis tasks in functional genomics. In addition to this, more specialized software exists for some types of microarray data (e.g., **MTEV** or **dCHIP**).

23
Cellular Systems Biology

Learning Objectives

This chapter introduces exemplarily basic concepts and methods to gain an understanding of cellular processes on a systems view. Two approaches will be explained: Top-down approaches examine the cell first on a global level using high-throughput data of a larger portion of all cellular genes, proteins and their interactions to discover the relevant pathways of the disease or treatment under examination. In turn, bottom-up approaches use much more detailed mechanistic models for a well defined pathway to identify physiologically relevant cell properties emerging from the complex interplay of the molecules of this pathway.

23.1
Introduction

The regulation of the cell is a highly complex process. It comprises many different molecular species (e.g., genes, nucleotides, proteins, and metabolites), multiple layers of regulation (e.g., transcriptional, posttranscriptional, and posttranslational), and ample feedback regulation. Therefore, systems approaches at the interface of biology, informatics, and mathematics are needed to integrate the large body of biological knowledge and to understand its functioning. Analyzing the cell on a systems view can be done by **top-down and bottom-up approaches**. Top-down approaches examine the cell first on a global level at which each signaling event or metabolic flux is regarded at the same level of complexity. Typically, the cells are experimentally screened with **high-throughput methods (gene expression profiling, proteomics, knock-out and knock-down, sequencing, affinity assays)** with respect to the treatment or disease under study. The experimental data then needs to be compiled into a model that explains the orchestrated behavior of the cellular components. Therefore, networks are set up connecting proteins, complexes, genes, and metabolic compounds. Conceptually, the cellular network can be divided into three parts: the **metabolic network**, the **signaling network**, and the **transcriptional regulatory network** (Fig. 23.1).

In a typical scenario of a human cell in a tissue, an extracellular signal for growth, apoptosis, or food uptake is passed through membrane receptors. It is processed in signaling cascades down to the transcription factors that are then activated or deactivated. This changes the transcriptional program within the regulatory network. A new composition of proteins is built, which changes the metabolism. Additionally, direct signals may be passed to the metabolism, multiple feedback loops are possible between the networks, and signals may also be processed and passed to the neighboring cells or into the blood system. The knowledge and temporal changeability of the three networks is very different. This has led to different modeling approaches for these networks. Metabolism is the best observed and described part of the network. This is due to the fact that metabolic reactions typically involve enzymatic conversion and mass flow of small molecules (e.g., sugars) the have been studied for several decades using

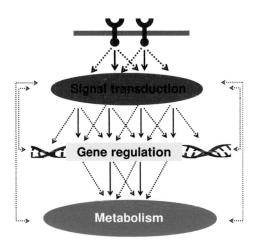

Fig. 23.1 The three networks of a cell.

An Introduction to Molecular Biotechnology, 2nd Edition.
Edited by Michael Wink
Copyright © 2011 WILEY-VCH Verlag GmbH & Co. KGaA, Weinheim
ISBN: 978-3-527-32637-2

enzyme kinetics and tracer experiments. Hence, elaborate qualitative and quantitative flux models have been developed on a hard-wired well-defined network structure. In contrast, knowledge about signaling interactions is much less established on a general level, and models often obtain functional context from potential wiring and rewiring aspects. Finally, the regulatory network contains the less-conserved topology. It adapts broadly and often very dynamically to the physiological situation. It operates on a much slower timescale than metabolism and signaling, and can best be characterized by an integrated approach including all three (sub-) networks using statistical models. Bottom-up approaches focus on well-characterized parts of the network, and are typically based on the assumption that the properties of these subnetworks (**modules**) can be studied in isolation. Detailed mechanistic mathematical models constructed from the molecular characteristics of individual proteins (**bottom-up models**) have so far only been developed for metabolism and, in part, for the signaling network. Based on prior knowledge and on time-resolved experimental data, mechanistic mathematical models are constructed that describe the interactions of individual proteins in the module (e.g., by using sets of coupled differential equations). The goal of bottom-up modeling is to identify physiologically relevant systems-level properties emerging from complex interactions within the network and to understand the underlying molecular mechanisms. Section 23.2 describes the basic ideas and principles for network analyses in top-down approaches and Section 23.3 explains the bottom-up approaches.

23.2
Analysis of Cellular Networks by Top-Down Approaches

23.2.1
Motivation

High-throughput methods such as gene expression profiling by microarrays or deep-level sequencing typically come along with larger sets of genes being upregulated, mutated, or special in some way to some treatment or disease. We need to get an understanding of how these genes act as a whole. This can be approached by gene set enrichment tests (Section 23.2.3). A further challenging goal in the analysis of cellular networks is to define drug targets. For this, typically, a node in the network model is discarded, mimicking specific drug treatment that inhibits the corresponding protein. Such simulations have been successfully applied to bacteria or, in general, single-cell pathogenic microorganisms to define drug targets for antibiotics. We describe two methods for this – one that uses network topology features (Section 23.2.6) and one based on qualitative decompositions of the stoichiometry of metabolism (Section 23.2.7). It should be noted that a specific treatment of cancer cell networks is more difficult as cancer cells are in principle very similar to cells of their host tissue. Modeling approaches for specific cancer treatment may need to focus more on signaling networks as their topology is often more specific. Apart from this, network analysis is successfully applied to optimize bacterial strains for the production of vitamins, amino acids, and other nutrient additives. Finally, compiling the functionality of sets of genes and proteins by assembling them into consistent global network models may produce new concepts to describe complex principles of nature in a systematic way.

23.2.2
Definitions and Reconstruction of the Networks

The terms "network" and "graph" will be used synonymously in the following. A graph $G = (V, E)$ consists of vertices $u, v \in V$ and edges $(u, v) \in E$ connecting these vertices. Edges (u, v) can be undirected or directed. Directed edges are represented by ordered pairs of nodes (u, v) and lead from source u to sink v. They are graphically depicted by arrows. Undirected edges are represented by unordered pairs of nodes (u, v), and are depicted by a line between vertices u and v. They are used if information about the direction is lacking or not needed. Bidirectionality between vertices u and v is represented by two edges – one leading from u to v and the other in the opposite direction. Metabolic networks are represented as bipartite graphs consisting of two disjoint sets of vertices $m \in M$ and $r \in R$ representing metabolites and reactions. Directed edges lead from the substrates of a reaction to the reaction and from the reaction to its products. Doing this for every reaction yields a network that consists of alternating nodes of metabolites and reactions. For some applications a reaction-based representation is needed in which the vertices of the network are the reactions and edges are set if a product of one reaction is the substrate of the other. Similarly, in a metabolite-based representation, the vertices are the metabolites that are connected by reactions (e.g., see Fig. 23.4). Commonly, ubiquitous metabolites like water, oxygen, ATP, and cofactors are discarded to model only the most relevant metabolic fluxes. Reconstructing signaling networks is much more demanding and several different approaches have been reported. In the simplest and most commonly used case, data of known protein-protein interactions is used as edges forming an undirected graph. Many protein complexes are not known or are described differently in different databases. Therefore, protein-protein interactions are often described just by their coding genes (e.g., the Human Protein Reference Database). This is the most simplified description of protein-protein interactions and has the advantage that interaction information can easily be integrated from several databases. One of the most elaborated approaches was suggested by Kohn in 1999, in which a detailed signaling flow was reconstructed similar to maps for electronic circuits. Regulatory networks can be reconstructed by linking transcription factors and their regulating genes. The interaction information for this can be inferred experimentally from **chromatin immunoprecipitation (ChIP)** and **ChIP on microarrays (ChIP-chip)**.

23.2.3
Gene Set Enrichment Tests

High-throughput methods such as gene expression from microarrays yield quantitative data for a major portion or all genes of a cell and therefore enable the discovery of parts of the whole network (**pathways**) relevant for a certain disease or treatment. For this, **gene set enrichment tests** have been developed. The method will exemplarily be explained for studies with gene expression microarray profiles of two different sample entities (e.g., normal and tumor samples). The basic idea behind these enrichment tests is to screen groups of genes with common functionality and select groups whose genes show significantly more differential expression than randomly selected genes. A group of differentially expressed genes is defined by a significance test (e.g., Student's t-test) of each gene to be differentially expressed in one class of the samples (e.g., the tumors). A reasonable significance threshold needs to be defined yielding two groups: genes with differential expression and genes without differential expression. We now want to detect common functions for the group of differentially expressed genes. For this, groups of genes are defined with common function, such as genes with common **Gene Ontology** terms.

The Gene Ontology project (www.geneontology.org) is a collaborative effort that addresses the need for consistent descriptions of gene products in different databases. The Gene Ontology Consortium includes many databases, including several of the world's major repositories for plant, animal, and microbial genomes. The Gene Ontology project has developed three structured controlled vocabularies (**ontologies**) that describe gene products in terms of their associated biological processes, cellular components, and molecular functions in a species-independent manner.

Gene Ontology terms are organized in a directed acyclic graph starting from very general terms like "metabolic process" and leading to very specific terms like "sphingosine biosynthetic process". Once the gene groups are defined, gene set enrichment tests can be applied, which in the simplest case are χ^2 or Fisher's exact tests. The differentially expressed genes are now mapped onto one of the defined gene groups from Gene Ontology, something like "sphingosine biosynthetic process". Altogether, this leads to four sets of genes that need to be counted: differentially expressed genes in the group, all genes (with any expression value) in the group, all differentially expressed genes, and all measured genes that can be mapped to any Gene Ontology term. These four values are taken for the statistical test (χ^2 or Fisher's exact test) and tested if in the group under consideration there are significantly more or significantly less genes than expected by a random selection. The test yields a significance value (p-value) for enrichment of differentially expressed genes in this group. The same is done for all groups and significantly enriched groups are given out. As these tests are done for all gene ontology terms, the probability that a p-value becomes low for one of these gene ontology terms just by chance is increased. Therefore, a correction needs to be applied for all p-values, known as the **multiple testing correction**. The simplest but most conservative correction is to multiply all p-values by the number of tests that were performed (Bonferroni correction). There exist several web-based servers for gene set enrichment tests with Gene Ontology terms. For example, **GOstat** (http://gostat.wehi.edu.au) is applicable in a straightforward way by copying and pasting Gene Ontology terms of gene lists into a graphical user interface of a browser. However, the tree-like (**directed acyclic graph (DAG)**) architecture of Gene Ontology causes some problems when using these tests. Specific terms are subsets or partial subsets of their less specific parents. Therefore, the parents are comprised of more genes and often yield higher significance of enrichment. This can lead to rather unspecific terms like "protein binding" or "metabolic process", which is insufficient information for the investigator. **TopGO** solves this by discarding genes in parental terms if the corresponding children already showed significant enrichment. In practice, this may solve the problem. However, sometimes still only unspecific terms show up. Therefore, cellular networks can also be decomposed into a rather flat hierarchy of gene groups with low overlap using well-defined subnetworks of the whole network, which we also call "pathways" in the following. The **Kyoto Encyclopedia of Genes and Genomes (KEGG)** database provides approximately 200 pathways ("KEGG maps"), half of which describe the metabolism and the other half signal transduction. The above-described gene set enrichment method can be applied for each of these pathways like as for Gene Ontology terms. As well as χ^2 or Fisher's exact tests, more sensitive methods can be used, such as Significance Analysis of Microarrays for Gene Sets and the Global-Test. Elaborated methods have been developed like PathWave that take network topology into account for finding significant patterns of differentially expressed genes in the networks.

23.2.4
Network Descriptors

In the following, the most relevant network descriptors are explained mostly for defining the substantiality (essentiality) of nodes in a network. In Section 23.2.5 we describe a machine learning concept that integrates these features for an *in silico* prediction of essential proteins in metabolism.

23.2.4.1 Scale-Free Networks

Barabasi and Albert (1999) investigated a wide variety of different networks, such as power grids for supplying electricity to a state, the world-wide web, a network of mating students in a college, and metabolic networks of several organisms. In contrast to regular lattice grids and more randomly organized glass structures of condensed matter, they observed a so-called **power law distribution** for these networks. Qualitatively speaking, these networks are comprised of many nodes with only a few neighbors (**orphans**) and a few (central) nodes with many neighbors (**hubs**). In a metabolic network, hubs are metabolites like water, oxygen, and ATP; orphan nodes are specific metabolites like δ-8,14-sterol. These observations can be formulated by a probability distribution:

$$P(k) \sim k^{-\gamma} \tag{23.1}$$

where $P(k)$ is the probability of obtaining a node in the network with connectivity k. The connectivity is the number of neighbors of the node. γ is a decay constant and was observed to be 2.2–2.4 for metabolic networks.

23.2.4.2 Triangle Motifs in Networks

Local network topology can also be described by simple triangle motifs. Thirteen motifs are possible (Fig. 23.2) for a directed graph. These motifs were counted for a large variety of different networks, such as the signal transduction networks in *Drosophila* and sea urchin, regulatory networks in *Escherichia coli*, yeast and *Bacillus subtilis*, human signal transduction, parts of the world-wide web, social networks (consisting of people as nodes and the relation "having a positive opinion" as edges), and ordered word-pairs, one word of which followed by the other in English, Spanish, French, and Japanese literature. Motif abundances were compared to networks with random links but remaining connectivity distribution of the nodes. A Z-score was defined for each motif and network by:

$$Z_i = \frac{N_{\text{real},i} - \overline{N_{\text{rand},i}}}{\sigma_{N_{\text{rand},i}}} \tag{23.2}$$

where $N_{\text{real},i}$ and $N_{\text{rand},i}$ are the number of motifs in the given (real) network of type i and an average of randomly reconnected networks of type i, respectively. The relative significance for each profile was observed by normalizing the profiles to a length of one by $SP_i = Z_i / \sqrt{\Sigma Z_i^2}$. The regulatory networks of the investigated microorganisms showed high abundance of motif 7, which is a typical feed-forward motif making the signal more robust against noise and misleading short sequences of wrong signals. In motif 7, the signal is passed directly from above to right and via a second node confirming the signal from top to left to right. This tendency was also observed in signal transduction networks. These signal transduction networks also showed a significantly high score of motifs 9 and 10 representing the same family of motifs called two-node feedbacks that regulate or are regulated by a third node. Common to all these signal flows is that they do not need short response times, but rather slower and more robust signals. Word-pair networks showed high abundances of motifs 1–3 representing sequential cascades of alternating word entities (preposition and noun, article and noun, etc.). Social networks and the world-wide web showed high abundances of mutual relationships (motif 13), such as if A has a positive opinion of

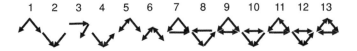

Fig. 23.2 Triangle motifs in networks.

B, there is a probability higher than in a random selection that B has a positive opinion of A.

23.2.4.3 Centrality and Further Topology Features

Features describing the location and the local vicinity of a node in the network are well suited for finding out how necessary a node in a cellular network is. We start with the centrality features. They describe how central a node is placed in the network. Let $G(V, E)$ be an undirected graph with n vertices. **Betweenness centrality** measures how often a node is part of the shortest paths between all other nodes. The betweenness centrality $C_b(v)$ for a vertex v is given by:

$$C_b(v) = \sum_{i \neq j \neq v \in V} \frac{d_{ij}(v)}{d_{ij}} , \tag{23.3}$$

where d_{ij} is the number of shortest paths from i to j and $d_{ij}(v)$ is the number of shortest paths from i to j that pass through a vertex v. **Closeness centrality** $C_c(v)$ defines the inverse of the average length of the shortest paths to all the other nodes:

$$C_c(v) = \frac{n-1}{\sum_{i \neq v, i \in V} d_{vi}} . \tag{23.4}$$

Eccentricity $C_e(v)$ is the longest distance from the given node to any other nodes:

$$C_e(v) = \frac{n-1}{\max_{i \neq v, i \in V} (d_{vi})} . \tag{23.5}$$

Eigenvector centrality is based on the assumption that the utility of a node is determined by the utility of the neighboring nodes. It scores a node higher if it is connected to high-scoring nodes. It is defined as the first eigenvector of the adjacency matrix of the network. Let x_i denote the score of the ith vertex. Let A_{ij} be the adjacency matrix of the network. Hence, $A_{ij}=1$ if there is an edge between the ith vertex and the jth vertex, and $A_{ij}=0$ otherwise. For the ith vertex, the centrality score is proportional to the sum of the scores of all vertices which are connected to it. Hence:

$$x_i = \frac{1}{\lambda} \sum_{j \in M(i)} x_j = \frac{1}{\lambda} \sum_{j=1}^{n} A_{ij} x_j , \tag{23.6}$$

where $M(i)$ is the set of vertices that are connected to the ith vertex, n is the total number of vertices, and λ is a constant. This leads directly to the well-known eigenvector equation, $Ax = \lambda x$. In general, there are different eigenvalues λ for which an eigenvector solution exists. However, by the Perron Frobenius theorem only the eigenvector of the largest eigenvalue is the eigenvector centrality.

A reaction that uniquely consumes or produces a certain metabolite in the metabolic network is considered as a **choke-point**. Finally, the **clustering coefficient** is given by:

$$C_i = \frac{n_{edges}}{k \cdot (k-1)} , \tag{23.7}$$

where n_{edges} is the number of edges connecting the neighbors of node i. k is the number of neighbors. This feature describes how well the neighbors are connected within each other. If they are fully connected, the clustering coefficient is 1; if they are not connected at all, the clustering coefficient is 0. This attribute may hint at alternative pathways or deviations, but can also be used to detect clusters in networks.

23.2.5
Detecting Essential Enzymes with a Machine Learning Approach

Screening pathogenic microorganisms for drug targets usually starts with a genome-wide knock-out screen of all open reading frames. Positive hits are knock-out mutants with considerably reduced viability and proliferation. The genes and the corresponding enzymes can then be considered as candidates for further, more detailed investigations. We will describe how to use and refine this kind of screening data with a systematic machine learning approach for defining essential reactions or enzymes in a metabolic network. The method and algorithms for supervised machine learning have been introduced in Section 24.5.5 in which they were applied to design a diagnosis method with gene expression data. The basic idea is to perform an experimental knock-out screen, which is then computationally validated with a systematic network analysis. The workflow is depicted in Fig. 23.3.

An experimental knock-out screen is performed for the strain under consideration and the viability of every knock-out mutant determined. This defines a class label for each gene to be essential or nonessential. The class labels of the gene knock-outs are transferred to the corresponding enzymes and reactions. Enzymes that consist of several peptides (complexes) are used if all coding genes show the same class label, otherwise they are not considered for classification. This data is taken for training and validation. The aim is now to predict these classes with a classifier that uses the above-described network topology features (Section 23.2.4) for the reactions. Additionally, features describing the likelihood of homologous (BLAST hits from a genome-wide alignment screen) and analogous (coregulated) genes can be added for improving the prediction results as these genes may take over the function of the knocked-out gene. The machine is then trained and validated with a cross-validation. The final prediction results are compared to the original experimental data of the knock-out screen. Predictions that are inconsistent with the screening data are candidates for a selection of genes that need further refinement in the lab by, for example, a second, more elaborate smaller screen. For many pathogenic organisms constructing knock-out strains is too demanding or hazardous. For such organisms, to some extent, the machine learning technique can be applied by inferring essential enzymes with a machine that has been trained with an organism for which knock-out screening data is available. These studies have also been performed with proteinprotein interaction networks.

23.2.6
Elementary Flux Modes

Elementary flux modes are based solely on the stoichiometry of metabolism and do not need any experimental data, such as turnover rates, binding constants, or gene expression. They have been astonishingly successful in predicting bacterial knock-out strains or carbon sources for optimizing the production of a metabolite of interest such as vitamins and amino acids. In Fig. 23.4, a simplified scenario is sketched in which substrate A is processed into two metabolites C and Vitamin B. When only considering the stoichiometry, knocking out reaction R_3 theoretically doubles the yield of Vitamin B.

However, cellular metabolism is more complex. In Fig. 23.5, a larger, more complex section of metabolism is depicted. Phosphoglycerate (PG) from glycolysis is processed via the tricarboxylic acid (TCA) cycle into several amino acids

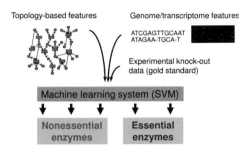

Fig. 23.3 The machine learning system needs features of the network, genome, and transcriptome to predict essential and nonessential reactions. It is trained and validated with experimental data from a genome-wide knock-out screen. Finally, for every enzyme class labels are assigned for being predicted as essential or nonessential.

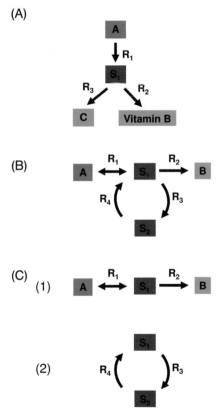

Fig. 23.4 (A) Example of a simple network. Knocking out reaction 3 (R_3) enhances vitamin B production, A is an external source, C and Vitamin B are external sinks; R_1, R_2, and R_3 are irreversible, S_1 is an inner metabolite. (B) The network is decomposed into its elementary flux modes (C(1) and C(2)).

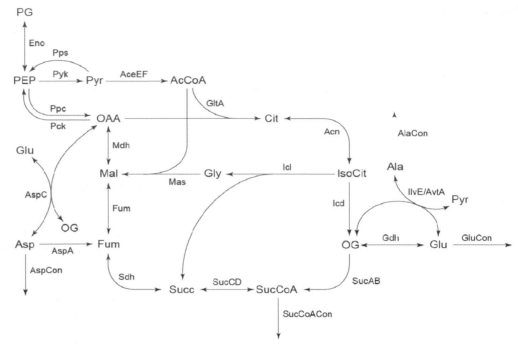

Fig. 23.5 TCA cycle and glyoxylate shunt of *E. coli* for the example in the text. Abbreviations: AcCoA, acetyl-CoA; Ala, alanine; Asp, aspartate; Cit, citrate; Fum, fumarate; Glu, glutamate; Gly, glyoxylate; IsoCit, isocitrate; Mal, malate; OAA, oxaloacetate; OG, 2-oxoglutarate; PEP, phosphoenolpyruvate; PG, 2-phosphoglycerate; Pyr, pyruvate; Succ, succinate; SucCoA, succinyl-CoA. Abbreviations of enzymes: AceEF, pyruvate dehydrogenase; Acn, aconitase; AspA, aspartase; AspC, aspartate aminotransferase; Eno, enolase; Fum, fumarase; Gdh, glutamate dehydrogenase; GltA, citrate synthase; Icd, isocitrate dehy- drogenase; Icl, isocitrate lyase; Mas, malate synthase; IlvE/AvtA, branched-chain amino acid aminotransferase/valine-pyruvate amino- transferase; Mdh, malate dehydrogenase; Pck, PEP carboxykinase; Ppc, PEP carboxy- lase; Pps, PEP synthetase; Pyk, pyruvate kin- ase; Sdh, succinate dehydrogenase; SucAB, 2-oxoglutarate dehydrogenase; SucCD, succi- nyl-CoA synthetase; AlaCon, AspCon, Glu- Con, and SucCoACon, consumption of ala- nine, aspartate, glutamate, and succinyl-CoA, respectively (with permission from Schuster *et al.*, 1999).

and succinyl-CoA. Let us try to optimize the production of glutamate by predict- ing a suitable knock-out strain. We will use elementary flux modes for this. The network needs to be decomposed into subnetworks that (i) consist of a minimal set of reactions and (ii) can "exist" on their own.

A decomposition of the network of Fig. 23.4 B is depicted in Fig. 23.4 C. In Fig. 23.4 B, the network comprises of an external source A and an external sink B. Decomposing the network into subnetworks yields two elementary flux modes (Fig. 23.4 C(1) and C(2)), which do not need further sources for sub- strates and sinks for products, and cannot be further decomposed.

Applying this method to the network of Fig. 23.5 leads to 16 elementary flux modes (taken from (Schuster *et al.*, 1999)):

Mode 1: $ATP \rightarrow ADP$
(Pck Ppc)

Mode 2: $ADP \rightarrow AMP$
(Pyk Pps)

Mode 3: $NH_3 + NADPH + CO_2 + PG \rightarrow Asp_{ex} + NADP$
(Eno AspC AspCon Gdh Ppc)

Mode 4: $ADP + NH_3 + NADPH + PG \rightarrow Ala_{ex} + ATP + NADP$
(Eno Pyk Gdh IlvE/AvtA AlaCon)

Mode 5: $NADPH + NAD \rightarrow NADP + NADH$
(Fum Mdh AspC AspA Gdh)

Mode 6: $ADP + FAD + 4 NAD + PG \rightarrow ATP + FADH_2 + 4 NADH + 3CO_2$
          (Eno 2Pyk 2AceEF GltA Acn Sdh Fum 2Mdh Icl Mas Pck)

Mode 7: $2 ADP + NH_3 + FAD + NADPH + 4 NAD + 2 PG \rightarrow 2 ATP + Asp_{ex} + FADH_2 + NADP + 4 NADH + 2 CO_2$
          (2Eno 2Pyk 2AceEF GltA Acn Sdh Fum 2Mdh Icl Mas AspC AspCon Gdh)

Mode 8: $ATP + FADH_2 + NADPH + CO_2 + PG \rightarrow Suc_{ex} + ADP + FAD + NADP$
          (Eno – SucCD – Sdh AspC AspA Gdh Ppc SucCoACon)

Mode 9: $ADP + 3 NAD + 2 PG \rightarrow Suc_{ex} + ATP + 3 NADH + 2 CO_2$
          (2Eno 2Pyk 2AceEF GltA Acn – SucCD Mdh Icl Mas SucCoACon)

Mode 10: $ATP + FADH_2 + NADH + CO_2 + PG \rightarrow Suc_{ex} + ADP + FAD + NAD$
          (Eno – SucCD – Sdh – Fum – Mdh Ppc SucCoACon)

Mode 11: $FADH_2 + 2 NAD + 3 PG \rightarrow 2 Suc_{ex} + FAD + 2 NADH + CO_2$
          (3Eno 2Pyk 2AceEF GltA Acn – 2SucCD – Sdh – Fum Icl Mas Ppc 2SucCoACon)

Mode 12: $ADP + NH_3 + NAD + 2PG \rightarrow Glu_{ex} + ATP + NADH + CO_2$
          (2Eno Pyk AceEF GltA Acn Icd Gdh Ppc GluCon)

Mode 13: $ADP + NADP + 2 NAD + 2 PG \rightarrow Suc_{ex} + ATP + NADPH + 2 NADH + 2 CO_2$
          (2Eno Pyk AceEF GltA Acn Icd SucAB Ppc SucCoACon)

Mode 14: $3 ADP + NH_3 + FAD + 5 NAD + 3 PG \rightarrow Glu_{ex} + 3 ATP + FADH_2 + 5 NADH + 4 CO_2$
          (3Eno 3Pyk 3AceEF 2GltA 2Acn Icd Sdh Fum 2Mdh Icl Mas Gdh GluCon)

Mode 15: $2 ADP + FAD + NADP + 3 NAD + PG \rightarrow 2 ATP + FADH_2 + NADPH + 3 NADH + 3 CO_2$
          (Eno Pyk AceEF GltA Acn Icd SucAB SucCD Sdh Fum Mdh)

Mode 16: $3 ADP + FAD + NADP + 6 NAD + 3 PG \rightarrow Suc_{ex} + 3 ATP + FADH_2 + NADPH + 6 NADH + 5 CO_2$
          (3Eno 3Pyk 3AceEF 2GltA 2Acn Icd SucAB Sdh Fum 2Mdh Icl Mas SucCoACon)

The enzyme names given in brackets indicate the enzymes used in the respective mode weighted with their fractional flux (unity if no number is given). Negative values indicate that the reaction is used in the reverse sense. Abbreviations are as in Fig. 23.5; consumption of alanine, aspartate, glutamate, and succinyl-CoA are represented by AlaCon, AspCon, GluCon, and SucCoACon, respectively. Modes 12 and 14 both lead to the production of glutamate. Mode 14 needs three PG to produce one glutamate. In comparison, mode 12 needs only two PG and is therefore advantageous. To improve glutamate production, mode 14 may be discarded by knocking out genes for reactions that are specifically needed for this mode, such as malate synthase (Mas). Elementary flux modes have also been successfully applied to optimize the choice of nutrients for the production of specific metabolites. As an alternative to the calculation of elementary flux modes, the method of **flux balance analysis (FBA)** can be used to optimize the production of biomass and compounds of interest. It also considers stoichiometric constraints and also takes nutritional availability into account. FBA is based on linear equations of in-coming and out-going fluxes for each inner metabolite considering steady-state conditions similar to the Kirchhof law for electric circuits. This system of linear equations is solved by a linear integer algorithm. However, we will not explain the flux balance method in more detail as numerous references exist for this.

23.2.7
Inference of Regulatory Networks: Boolean and Bayesian Networks

Boolean networks were used for the description of gene regulation since as early as 1969. Their attractiveness lies in their simplicity, as they know only two states for each gene ("on" and "off"), and reduce interactions between genes or genes and gene products to logical rules in Boolean algebra. Reaction kinetic parameters, for example, are not needed for the construction of such networks. Boolean networks have no stochastic component; a given state of the network will always pass into the same second state. Since stochastic fluctuations do occur in biological systems and can also play a substantial role in the system's transition from one state into another, enhancements of Boolean networks introduced probabilities for every effect in the network. **Bayesian networks** are graphs that plot the conditional probabilities between events. They are used for the modeling of networks from experimental data and can be used, for instance, on gene expression data. The nodes of the Bayesian network then correspond to genes whose expression is measured and the connections between the nodes state whether an event (e.g., gene 1 expressed) often or always coincides with another event (gene 2 expressed). The edges are directed and weighted with the value of the conditional probability. To reduce complexity, each event in the model can only depend on a limited number of other events and needs to be independent of all the others. Efficient algorithms exist for the construction of Bayesian networks, which find the optimal solution when provided with a score function.

23.3
Overview of Bottom-Up Modeling of Biochemical Networks

The top-down modeling approaches discussed above allow us to integrate and interpret large experimental data sets even without much prior knowledge; however, in most cases top-down analysis does not provide mechanistic insights into the system dynamics. This is in contrast to bottom-up approaches, where detailed knowledge about isolated biological subsystems is used to implement mechanistic mathematical models of molecular mechanisms underlying complex cellular behavior. The aims and requirements of bottom-up modeling approaches based on **ordinary differential equations (ODEs)** will be described in the following.

23.3.1
Motivation

Bottom-up modeling is based on the premise of modularization of biological function. Modules are small subsystems whose dynamics are assumed to mainly depend on reactions within the module and thus can be studied in isolation. Mathematical modeling (i.e., the translation of these interactions into sets of equations) is then used to understand the network behavior arising within these modules of interacting cell components, when mere intuition is no longer sufficient.

Functional properties arising from complex interactions are, for example, oscillations of protein levels as they occur in the cellular module that constitutes the circadian clock. Here, the oscillating abundance levels of certain proteins determine an organism's daynight rhythm. Another example is the transition of cells between mutually exclusive behavioral states including proliferation, differentiation (the transition to a different cell identity, such as during development), and apoptosis (a suicide program that can be invoked in cells). Here, modeling can help to understand how such cell fate decisions can be realized in an all-or-none manner, so that a program is either fully initiated in an irreversible manner or completely suppressed. Modules that show such discrete activity states are termed network switches. Biological systems are subject to fluctuations, such as in the concentrations of their constituents. Such biological noise and its propagation in biological networks can be studied using stochastic modeling approaches (Raj and van Oudenaarden, 2008). In particular, modeling often provides insights into how biological networks maintain reliable decision making even in the presence of uncertainty. However, such biological reliability, often also referred to as **robustness**, can never be simultaneously established for all possible network perturbations. Thus, each biochemical regulatory system exhibits certain weak nodes (e.g., proteins) that are sensitive to perturbations (e.g., pharmacological intervention or knock-down). Modeling approaches such as sensitivity analysis allow us to systematically identify such weak points, and might therefore unravel new disease mechanisms and sites of effective pharmacological intervention.

23.3.2
Choosing Model Complexity

Typically, the first step in modeling is to identify how experimentally measurable inputs, perturbations, and readouts are best represented by a model. In general, a trade-off exists, as simpler models are easier to tackle mathematically, while more complex models are often required to adequately represent biological phenomena.

Whether certain biochemical processes are explicitly considered in a model typically depends on their time scale and on the stoichiometry of molecules involved. Consider the case of a signal transduction pathway where ligand binding to a receptor induces phosphorylation of an intracellular kinase, which in turn triggers gene expression in the nucleus. In many cases, it is reasonable to assume that the ligand in the extracellular medium is in stoichiometric excess over its cognate receptor, implying that we can neglect ligand depletion in the model and assume the ligand concentration to be constant. Similar arguments also hold for ATP, which is an important component in kinase-mediated phosphorylation, but is usually not limiting within the cell. Gene expression operates on a time scale of hours and is thus a slow process relative to upstream signaling, which typically occurs within minutes after stimulation. If we seek to describe slow gene expression dynamics with our model, we might therefore assume that receptor-ligand binding and receptor-mediated binding reach a steady state very rapidly (i.e., that they are not rate limiting), and do not need to explicitly govern the dynamics of these fast processes.

Another critical decision in model implementation is the level of abstraction. Bottom-up modeling of biological systems has been applied with various levels of abstraction, ranging from intra-molecule-scale modeling of single enzymes and their allosteric states to a mesoscale description of complex intercellular communication at the organ level. In general, detailed mechanistic models quantitatively describing events at the level of protein-protein interactions are most reliable in predicting new experiments. However, such detailed models typically require a lot of knowledge about intracellular protein concentrations and kinetic parameters, either from the literature or from one's own quantitative measurements, so that they have only been applied to small biological subsystems so far. Mathematical models of larger biological systems therefore typically include net reactions lumping many molecular steps into a single rate.

23.3.3
Model Construction

Once a level of abstraction has been chosen, the biological system is formulated as a system of ODEs. This is often done using the formalism of mass-action kinetics. Consider the example in Fig. 23.6 A, where a protein exists in an unphosphorylated form (X) and in a phosphorylated form (X^*). The protein is subject to interconversion between both forms due to kinase-catalyzed phosphorylation (v_k) and phosphatase-catalyzed dephosphorylation (v_p). For simplicity, we shall assume that protein X is present in vast excess over the catalyzing enzymes. According to biophysical textbooks, the rates of phosphorylation and dephosphorylation are then given by the MichaelisMenten equations:

$$v_k = \frac{V_{\max,k} \cdot X}{K_{M,k} + X} = \frac{k_{cat,k} \cdot K \cdot X}{K_{M,k} + X} \quad \text{and} \quad v_p = \frac{V_{\max,p} \cdot X^*}{K_{M,p} + X^*} = \frac{k_{cat,p} \cdot P \cdot X^*}{K_{M,p} + X^*},$$

where $V_{\max}$ is the maximal velocity of the enzyme (given by the catalytic rate constant times the enzyme concentration) and K_M is the Michaelis-Menten constant. In other words, we have lumped the molecular mechanism "enzyme + substrate ⇔ enzyme substrate complex ⇔ enzyme + product" into a single reaction rate. In order to describe the dynamic behavior of the full phosphorylation-dephosphorylation cycle, we write down the following system of differential equations:

$$\frac{dX}{dt} = -v_k + v_p$$

$$\frac{dX^*}{dt} = v_k - v_p$$

(23.8)

Here, the differential equations describe the change of the concentrations of the dynamic variables, X and X^*, with time. The differential equation system (23.8) can be extended to describe the kinase cascade depicted in Fig. 23.6 B:

$$\frac{dX_1}{dt} = -v_{k,1} + v_{p,1}$$

$$\frac{dX_1^*}{dt} = v_{k,1} - v_{p,1}$$

(23.9)

Here, v_k and v_p are still the same as above, while:

$$v_{k,1} = \frac{V_{max,k1} \cdot X_1}{K_{M,k1} + X_1} = \frac{k_{cat,k1} \cdot X^* \cdot X_1}{K_{M,k1} + X_1} \quad \text{and} \quad v_{p,1} = \frac{V_{max,p1} \cdot X_1^*}{K_{M,p1} + X_1^*} = \frac{k_{cat,p1} \cdot P_1 \cdot X_1^*}{K_{M,p1} + X_1^*}.$$

23.3.4
Model Simulation

The differential equation system of the kinase cascade (equations 23.8 and 23.9) can be considered as a simple model for the three-tiered mitogen-activated protein kinase (MAPK) cascade. MAPK signaling is stimulated by extracellular growth factors and the cascade output kinase, extracellular regulated kinase (Erk), is involved in cell fate decisions (e.g., between cell proliferation and differentiation). As indicated in Fig. 23.6 B, the upstream kinase K in our model might represent the growth-factor-regulated kinase Raf, which mediates phosphorylation of MAPK kinase (Mek) ($X \Rightarrow X^*$) and finally the phosphorylation of Erk ($X_1 \Rightarrow X_1^*$). Using the model, we can now simulate how the MAPK cascade responds to extracellular stimulation by growth factors. The model consists of four dynamical variables (X, X^*, X_1, and X_1^*) and eight kinetic parameters ($V_{max,k}$, $K_{M,k}$, $V_{max,p}$, $K_{M,p}$, $V_{max,k1}$, $K_{M,k1}$, $V_{max,p1}$, and $K_{M,p1}$). In order to simulate serum starvation (before stimulation) we assume all proteins to be unphosphorylated at $t=0$ and set the initial concentrations of the dynamical variables to $X(0)=X_{tot}$, $X^*(0)=0$, $X_1(0)=X_{1,tot}$, and $X_1^*(0)=0$ (where X_{tot} and $X_{1,tot}$ are the total concentrations of X and $X1$, respectively). As depicted in Fig. 23.6 C, we can now simulate the temporal evolution of the dynamical variables of interest (X^*, X_1^*) for a given set of kinetic parameters using numerical integration techniques. How does this work? The simplest numerical integration method, known as the Euler method, approximates the solution of the differential equation $dx/dt=f(x)$ by the following relationship:

$$x(t_{i+1}) = x(t_i) + f(x(t_i)) \cdot \Delta t,$$

(23.10)

where $x(t_i)$ is the solution at time point t_i and $\Delta t = t_{i+1} - t_i$. In practice, the solution of the differential equation at any time point is obtained by iteratively applying equation (23.10) starting from the initial conditions at $t=0$. The smaller the time increment δt is chosen, the more accurate the solution. However, in general, the numerical error of the Euler method increases with increasing number of numerical integration steps. Thus, more accurate algorithms are usually ap-

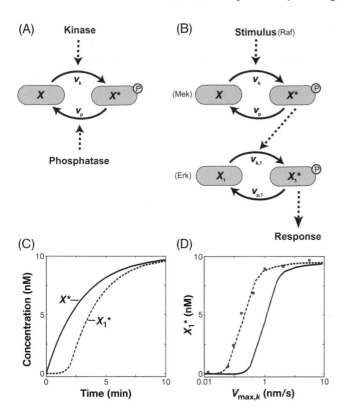

Fig. 23.6 Bottom-up modeling of signal transduction pathways. (A) Schematic representation of a phosphorylation/dephosphorylation cycle with interconversion between phosphorylated and unphosphorylated forms X and $X*$, respectively. (B) Schematic representation of a kinase cascade comprising modification of two consecutive kinases, X and X_1. (C) Numerical simulation of the temporal behavior of phosphorylated species in the kinase cascade. (D) Simulated doseresponse curves (lines) and comparison with hypothetical experimental data (circles). The dashed line represents a quantitative match between model and experiment, while the solid line corresponds to a qualitative model (see text for details).

plied, such as the RungeKutta method that uses a weighted average of slopes instead of a single slope $f(x(t_i))$ in equation (23.10). Corresponding numerical integration algorithms are integrated in standard mathematical computing software packages (e.g., MATLAB, Maple, Mathematica). Based on the time-course calculations depicted in Fig. 23.6 C, one can now simulate all kinds of experimental conditions *in silico*. For the MAPK cascade one might, for example, analyze the dose-response behavior towards growth factor stimulation at a particular time point. In order to do so we assume that growth factor stimulation alters the catalytic activity of Raf kinase ($V_{max,k}$) and calculate time courses for different values of $V_{max.k}$. The numerical results obtained for the desired time point are then used to construct an *in silico* doseresponse curve (Fig. 23.6 D, solid line) that depicts the cascade response (X_1^*) as a function of the stimulus parameter ($V_{max.k}$).

23.3.5
Model Calibration

Two fundamentally different approaches exist for comparing bottom-up models with experimental data: (i) qualitative modeling where the model represents the essential features of the data, but does not aim for a quantitative description, and (ii) quantitative modeling, which aims for an optimal match between quantitative experimental measurements and the model. Consider again the dose-response simulations depicted in Fig. 23.6 D: hypothetical experimental data points (circles) and two simulated dose-response curves (solid and dashed lines) are shown, both of which were obtained for the cascade model, but with different kinetic parameters. We see that the dashed curve overlaps well with the data and thus represents a quantitative description of the biological system. In contrast, the solid curve deviates strongly from the measurements. Still, it reflects one essential feature of the data – the highly nonlinear switching between on and off states at a certain threshold stimulus – and thus represents a qualitative description of the system. Obviously, qualitative modeling is applied if only qualitative and noisy experimental data is available or if the data is collected from heterogeneous sources (e.g., from different laboratories). Qualitative modeling can also be helpful to systematically analyze what kind of qualitative behavior (e.g., oscillations) a system can, in principle, show. A wide variety of methods have been developed to analyze the qualitative features of biochemical network models (e.g., bifurcation theory, phase-plane analysis). Finally, qualitative modeling is helpful to analyze simple toy models that are used to understand the minimal requirements for a system property to be observed. In many cases, the differential equations of these toy models can be solved analytically, and the analytical solutions provide deep insights into the functioning and parameter dependency of biochemical systems. For qualitative modeling, the values of kinetic parameters are chosen arbitrarily, but it is important to ensure that they lie within the physiological range. In contrast, for quantitative models the parameters are optimized using computer algorithms in order to minimize the deviation of the model from the data (i.e., to fit the model to the data). A popular minimization criterion is the χ^2-value:

$$\chi^2 = \sum_{i=1}^{n} \frac{(M_i - S_i)^2}{E_i^2} , \qquad (23.11)$$

where M_i is the ith measurement value, and S_i and E_i are the corresponding simulated value and the measurement error, respectively. Thus, the χ^2-value sums up the deviations between model and experiments over all n measurements, with each deviation being weighted by the corresponding experimental error. In other words, the larger the experimental error of a measurement i, the less its weight. A good model fit to experimental data is characterized by $\chi^2 < n$ (i.e., on average all simulated values lie within experimental error). It should be noted that experimental measurements (e.g., Western blots) can often only be obtained in arbitrary units, while the model is typically formulated in absolute concentration units. Therefore, scaling factors have to be introduced in order to allow for a comparison between experiment and model (i.e., to calculate the χ^2-value). More specifically, the simulated values S_i in equation (23.11) need to be modified to $S_i = a \cdot S_{i,real}$ where a is the scaling factor and $S_{i,real}$ is the model simulation in absolute concentration units. Thus, in addition to the kinetic parameters of the model the scaling factor has to be optimized in order to fit the model to the data. Optimization algorithms (e.g., LevenbergMarquardt or Simulated Annealing) are implemented in commercially available software packages such as MATLAB or can be downloaded as freely available toolboxes (e.g., PottersWheel, COPASI, AsaMin). However, when using these packages one has to keep in mind that model fitting can yield misleading results, especially if the model is comprised of many kinetic parameters. For example, the optimization algorithm might get stuck in local minima and thus be unable to find the globally optimal solution. Moreover, even if a global optimum can be found it might not be unique in the sense that many different parameter combinations yield a similar match between model and experiment. In other words, the parameters cannot be unambiguously determined from the data (i.e., the parameters are not **identifiable**). Finally, the fitting results may be strongly dependent on the model topology (i.e., on the biochemical mechanisms assumed during model construction). In order to investigate such topology dependencies and to discriminate biochemically feasible model variants, some research groups have applied a so-called ensemble modeling approach, where different model topologies are systematically compared with respect to their ability to fit experimental data.

23.3.6
Model Verification and Analysis

Any mathematical model of biological systems must be able to generate experimentally testable predictions. Biological modeling is therefore often described to involve an iterative cycle between experiment and theory: an initial experimental data set is used to construct the model and/or to calibrate its kinetic parameters. Once a model successfully describes the data, its predictions need to be verified by independent experiments which were not part of the construction/calibration data set. Eventually, the verification data set is then used to further refine the model (and so on). The design of model verification experiments is not trivial. Therefore, optimal experimental design strategies were proposed for planning verification experiments such that they optimize the discrimination of model variants or the accuracy in parameter estimates. However, these algorithms are not fully developed, so that, in most cases, the design of verification experiments is led by the experience of modelers and experimentalists.

23.4
Biological Examples

Having introduced bottom-up modeling of biochemical regulatory networks, we will now discuss two specific biological systems where qualitative mechanistic modeling provides valuable insights into the system's underlying function principles. First, we will analyze a simplified model describing the initiation of programmed cell death by proteolytic enzymes called caspases. In particular, we will focus on a positive feedback loop involved in caspase activation, which ensures that the system exhibits two discrete activation states – one with low caspase activity (life state) and one with high caspase activity (death state). This minimal model of a network switch can be understood qualitatively without explicitly solving the differential equations. In the second example, we analyze a genetic toggle switch of mutual inhibition, where gene X inhibits the expression of gene Y and *vice versa*. Numerical analysis of the toggle switch shows that, depending on the parameter values, this system again shows two different activity states – one characterized by exclusive expression of gene X, the other by exclusive expression of gene Y.

Generally, bistable systems (i.e., systems with two distinct stable steady states) possess two main properties that make them a qualitative feature of many signaling modules involved in cell-fate decisions. First, they show switching behavior, whereby a graded signal is translated into a sharp, all-or-nothing response. The second property is irreversibility, meaning that after a system has switched to a different steady state by application of a stimulus, the stimulus does not need to be continuously present to sustain the new state.

Bistability relies on a positive feedback loop in which a component enhances its own production. This positive feedback can be either in the form of direct enhancement (as will be discussed in the example of self-enhanced caspase-3 production) or by repressing a factor that acts negatively on the component ("repress the repressor"; as discussed in the toggle switch example of mutual inhibition).

Figure 23.7 depicts a simplified illustration of core events leading to activation of caspase-3, one of the main executioner proteins of the cell suicide program apoptosis. Caspase-3 exists as an inactive proenzyme and undergoes proteolytic processing upon activation. Activation is stimulated through two different sources (Fig. 23.7): (i) through signals emanating from death receptors that reside in the cell plasma membrane and are activated by different ligands, and (ii) through signals from permeabilized mitochondria. Active caspase-3 in turn induces permeabilization of mitochondria, thus giving rise to a positive feedback

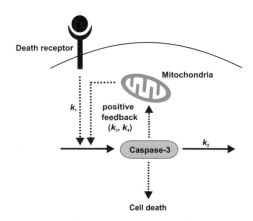

Fig. 23.7 Schematic representation of signaling pathways leading to caspase-3-mediated programmed cell death. Death receptors activated by extracellular ligand induce formation of caspase-3, which in turn amplifies its own production via a mitochondrial positive feedback loop.

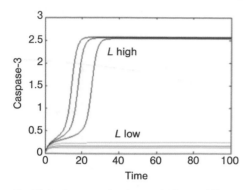

Fig. 23.8 Caspase-3 levels can reach two different steady states, depending on stimulus *L*. Plotted are the trajectories for caspase-3 of differential equation (23.12) for three different values of high (blue) and low (green) stimulus *L*.

loop, where caspase-3 amplifies its own activation. Activation of caspase-3 could thus be described by the equation:

$$\frac{\mathrm{d}[C3]}{\mathrm{d}t} = f(C3) = k_1 \cdot [L] - k_2 \cdot [C3] + k_3 \cdot \frac{[C3]^2}{k_4^2 + [C3]^2} \,. \tag{23.12}$$

The first term denotes activation of caspase-3 (*C3*) by binding of ligand *L* to death receptor. The second term denotes caspase-3 degradation, while the last term models the positive feedback of caspase-3 on its own production via mitochondria permeabilization.

Figure 23.8 shows the temporal evolution of caspase-3 according to equation (23.12) for different input levels of stimulus *L*. We see that for high stimulus levels (blue curves), caspase-3 reaches a high steady-state level, whereas for low stimulus levels (green curves), caspase-3 remains close to zero. Thus, the steady-state value of caspase-3, whether high or low, depends on the stimulus *L* and switches to the high steady state (death state) if *L* increases above a certain threshold. To understand how this is explained by equation (23.12) we study the graph $f(C3)$ (i.e., the rate of change of caspase-3 as a function of its own abundance level).

Figure 23.9 shows the graph $f(C3)$ for three different levels of stimulus *L*, with *L* increasing to the right. First note that whenever the graph intersects the C3 axis ($f(C3)=0$) the rate of change is zero (i.e., the system has reached a steady state). The flow of *C3* is indicated by black arrows on the C3 axis in Fig. 23.9: for $f(C3)>0$, the caspase-3 level increases, while for $f(C3)<0$ the caspase-3 level decreases until a steady state is reached.

Two different types of steady states are indicated in Fig. 23.9. A stable steady state is characterized by the fact that if small perturbations of the level of caspase-3 occur, the system is forced back into the steady state. This is the case if the flow of *C3* is increasing for values smaller than the steady state, while the flow of *C3* is decreasing for values higher than the steady state. Consequently, a steady state is stable if the graph $f(C3)$ intersects the C3 axis while it is decreasing. If $f(C3)$ is increasing at the intersection of the C3 axis the steady state is unstable. In Fig. 23.9, stable steady states are marked by filled circles, unstable steady states are marked by open circles.

How does an increase of stimulus *L* switch caspase-3 levels to a high steady state? To see this, note that the graph of $f(C3)$ always exhibits the S-shaped form shown in the three plots in Fig. 23.9 and that at C3=0, $f(C3) = k_1 \cdot L$, independently of the values for the parameters k_1 to k_4. If the value of *L* is very small, that is, if the death receptor is activated by a small number of ligands, $f(C3)$ will intersect the C3 axis 2 times, where the first and third steady states are stable (Fig. 23.9, left panel).

Accordingly, if we consider a cellular situation in which, prior to encounter with the stimulus *L*, cells possess no or very little active caspase-3, as production of caspase-3 starts and increases, it will hit the first stable steady state. This means that for low stimuli *L*, the production of caspase-3 will reach a steady state precisely at that low value of C3 corresponding to the first intersection of

Fig. 23.9 Equation (23.12) plotted as a function of C3 for three different levels of ligand *L*. Stable steady states are marked by filled red circles. As *L* increases the graph is shifted upwards so that in the right panel only one stable steady state remains.

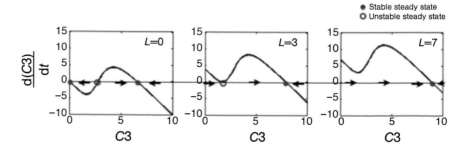

the $C3$ axis. As stimulus L increases, however, the graph $f(C3)$ is shifted upwards (Fig. 23.9, middle and right panel), it will intersect the $C3$ axis only once and the system exhibits only one steady state, corresponding to high caspase-3 levels. Thus, for high stimulus levels L, caspase-3 production will come to a rest in the high state.

In Fig. 23.10, the steady-state value of caspase-3 is plotted against increasing stimulus L. The two stable steady states are plotted in solid, the unstable state in a dotted line. We see that for low values of L, two stable and one unstable steady state exist. When L passes a threshold, the steady-state level of caspase-3 "switches" from the low to the high steady state ("going up"). In contrast, if the system starts in the high steady state, it will stay in this state ("coming down"). In fact, the mechanism described by equation (23.12) holds the explanation for yet a different way to achieve a switching mechanism between distinct steady states. So far, we assumed that cells possess no caspase-3 prior to receptor activation. This assumption is correct in our example, since in the absence of apoptosis stimuli, cells do not possess active caspase-3. However, equation (23.12) could be applied in other contexts, to describe the dynamics of different molecules in place of caspase-3 where it might be reasonable to consider cells with different (i.e., nonzero) initial values of the molecule of interest.

As mentioned earlier, each stable steady state has a certain regime associated with it, from which values that fall within these regime will tend towards the particular steady state (indicated by black arrows in Fig. 23.9). If we now consider a case of parameters, where two stable steady states exist (such as depicted in Fig. 23.9, left panel) we can see that **different initial conditions will be sorted into distinct steady states**, provided they fall in the regimes of the respective steady states. That is, if the parameters are such that two stable steady states exist, it will depend on the initial values, which of these steady states is reached. We will encounter this situation again in the example of mutual inhibition discussed below.

Becskei *et al.* (2001) implemented a transcriptional positive feedback mechanism in *Saccharomyces cerevisiae* using a plasmid-encoded tetracycline-responsive transactivator (rtTA, Fig. 23.11). In the presence of inducer doxycycline rtTA is activated and binds DNA containing appropriate binding sites. Figure 23.11 illustrates the architecture of an autocatalytic positive feedback of the rtTA system. **Green Fluorescent Protein (GFP)** expression was used to assess activity. Upon induction, the population of cells split into two distinct subpopulations with cells either fully expressing GFP or not at all indicating the existence of two distinct steady states as predicted by equation (23.12).

In this section we derive how negative interaction of two molecules can give rise to a bistable system. This motif of mutual inhibition is a positive feedback circuit too and can be viewed as a more detailed description of the processes giving rise to the dynamics of the earlier example.

Consider two molecules X and Y, which directly or indirectly repress each other. For example, X and Y could be transcription factors, which bind and block each other's promoters, thus preventing transcription (Fig. 23.12 A). This negative effect does not need to be direct: protein X, for example, could be involved in the activation of a repressor Z of Y's promoter and *vice versa*. On the protein level, the inhibition could be envisioned as, for example, part of a signaling cas-

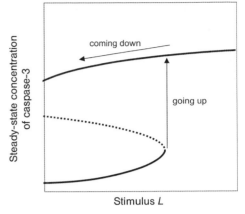

Fig. 23.10 Steady states (stable, solid line; unstable, dotted line) of caspase-3 as a function of input stimulus L (equation 23.12) (see text for details.)

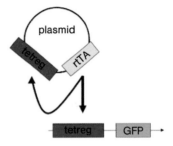

Fig. 23.11 Architecture of an autocatalytic positive feedback of the rtTA system. rtTA binds and activates its own promoter as well as that of a reporter GFP gene (tetreg).

(A)   (B)

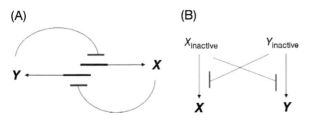

Fig. 23.12 Mutual inhibition of two molecules on transcription (A) and protein level (B).

cade where X and Y stand for the activated form of two proteins, and where the repression takes place at the level of the conversion from precursor to active state (Fig. 23.12 B). Generally, the negative interaction between X and Y is described by the system of equations:

$$\frac{dX}{dt} = f(Y) - k_1 \cdot X$$

$$\frac{dY}{dt} = g(X) - k_2 \cdot Y,$$

(23.13)

where $f(Y)$ and $g(X)$ are decreasing functions, and the second terms represent degradation, respectively. We will discuss one specific form of these equations, which can arise in cooperative binding events:

$$\frac{dX}{dt} = f(Y) - k_1 \cdot X$$

$$\frac{dY}{dt} = g(X) - k_2 \cdot Y,$$

(23.14)

The qualitative results, however, can be achieved for different implementations of mutual inhibition, such as, for example, in describing competing species. To understand the behavior of the dynamic system of two components repressing each other, we can apply knowledge from the previous example. As in the 1D case of caspase-3 activation, we will see that it depends on the relative strengths of interaction parameters (mutual repression: k_3 and k_4; degradation: k_1 and k_2) whether the system exhibits one or two stable steady states, and what the nature of these steady states is.

Investigating equation (23.12), we had seen that it can either possess one or two stable steady states. In the latter case which of these two steady states would be reached depended on the initial values. We find the same aspects in the analysis of equations (23.14), only with slightly more possibilities. We refer the reader to Edelstein-Keshet (1988) for a detailed mathematical analysis and here will only outline the different results that can be achieved.

Four different scenarios can occur:
1. The system exhibits exactly one steady state, with X always dominating and Y fully repressed, if repression of X on Y is much higher than Y on X.
2. The system exhibits exactly one steady state, with Y always dominating and X fully repressed, if repression of Y on X is much higher than X on Y.

In cases (1) and (2) one stable steady state will be reached irrespective of the initial conditions. These scenarios correspond to the case of high ligand stimulation in the caspase-3 example (Fig. 23.9, right panel).

3. The system exhibits two stable steady states, where either X or Y is fully repressed and the other molecule dominates, if the mutual repression is both strong and similar for both molecules. Which of the two molecules will dominate depends on the initial values: if initially more X than Y is available, then X will fully repress Y and *vice versa* (Fig. 23.13). This scenario corresponds to the case of low ligand stimulation in caspase-3 activation, where, as discussed, in the presence of two stable steady states the outcome depends on the regime the initial value lies (Fig. 23.9, right panel).
4. Finally, if the mutual repression strengths are similar and weak, stable coexistence of the two molecules may occur. In this case one stable steady state exists, which is reached irrespective of the initial values.

Gardner *et al.* (2000) constructed a switching mechanism of two genes in *E. coli* that implements the architecture shown in Fig. 23.12 A. The basic form of mu-

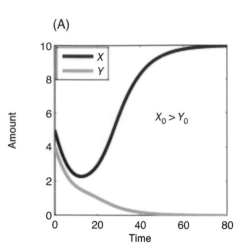

(A)

$X_0 > Y_0$

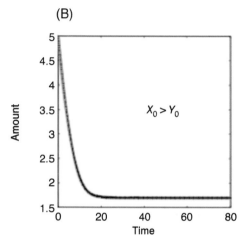

(B)

$X_0 > Y_0$

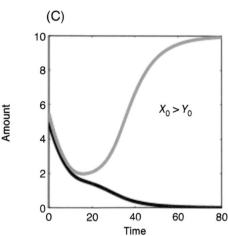

(C)

$X_0 > Y_0$

Fig. 23.13 Simulation of the mutual inhibition mechanism (equations 23.14). Parameters are chosen such that the system has two stable steady states in which small differences in the initial values (X_0, Y_0) are sorted by the system into two distinct steady states (left and right).

tual inhibition on the transcription level is two promoters, each controlling the expression of a repressor of the opposing promoter. In addition, two inducers are needed that specifically block the interaction of one repressor-promoter pair – inducers can then be added to the reaction to shift the system between the two steady states.

24

Protein–Protein and Protein–DNA Interaction

Learning Objectives

Virtually all biological processes rely on protein–protein interactions. Most of these interactions are mediated by protein domains, of which the human genome alone codes for at least 1000 different kinds. While all domains are assumed to have defined interaction partners, these are known for only a fraction of all domains. Protein interactions are often organized in stable complexes. On average, a single interaction involves 20 amino acids; this corresponds to a surface area of about 800 $Å^2$. The most important forces involved are hydrophobic interactions, hydrogen bonds, and ionic bonds. The mass action law can describe protein interactions as bimolecular reactions; an average energy of around 10 kcal is needed to break 1 mol of dimers. Essential techniques for the examination of interactions are protein purification and characterization of complexes with mass spectrometry, but the two-hybrid method, fluorescence resonance energy transfer, and in vitro binding assays are also important tools. Protein interactions are regulated by the expression of proteins, but also their localization, stability, and covalent modifications (and noncovalent modifications, such as bound ligands). Protein interactions can sometimes be predicted theoretically by molecular docking or homology-based methods; however, experimental validation remains essential. Protein–DNA interactions play essential roles in all aspects from gene regulation to inheritance. Enzymes such as DNA polymerase are responsible for the duplication of the genetic material. Other DNA-binding enzymes recognize and repair DNA damage. Transcription factors regulate gene expression and thus ensure that depending on the current cellular environment, the appropriate proteins are produced at the optimal level. Transcription factors need to specifically recognize and bind gene sequences located in the promoter region of the regulated genes. In spite of recent advances in computational biology, predicting the sequence specificity of a given DNA-binding protein remains a formidable challenge. Protein–DNA interactions can be examined with a number of methods stemming from biophysics and molecular biology. X-ray structure analysis is the only method capable of exploring interactions in atomic detail. Protein interactions and protein–DNA interactions are at the center of much research in medicine and biotechnology, such as when developing cancer therapeutics that block protein–protein interactions or ligands that activate or inhibit certain protein–DNA interactions.

24.1

Protein–Protein Interactions

Almost all cellular processes feature protein–protein interactions in prominent roles. For instance, all of the structural elements such as actin filaments or microtubules consist of protein complexes held together by protein interactions. Furthermore, a very large number of enzymes are composed of subunits that develop their full activity only in concert. RNA polymerases are an arbitrary exam-

An Introduction to Molecular Biotechnology, 2nd Edition.
Edited by Michael Wink
Copyright © 2011 WILEY-VCH Verlag GmbH & Co. KGaA, Weinheim
ISBN: 978-3-527-32637-2

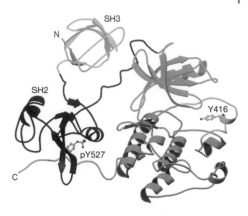

Fig. 24.1 Protein domains of the Src oncoprotein. The Src protein has three major domains: SH3, SH2, and SH1, the latter of which is the kinase domain. All three enter several well-defined interactions. The smaller domains do not only interact with other proteins, but also with sequences within Src: SH3 binds a proline-rich sequence between SH2 and the kinase domain; SH2 binds to a phosphorylated tyrosine at position 527, close to the C-terminus (pY527).

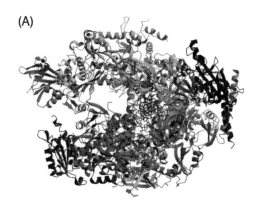

(A)

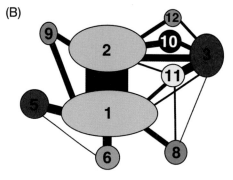

(B)

Fig. 24.2 RNA polymerase II: a multimeric protein complex. (A) Structure of yeast RNA polymerase II. (B) Schematic diagram showing the interactions between the 10 subunits. The thickness of the connecting lines corresponds to the size of the contact area between the individual subunits. The colors correspond to those in (A). (From Cramer *et al.*, 2001.)

ple for this principle, taken from hundreds of protein complexes within a cell. Their subunits need to enter numerous protein interactions among themselves, but also with nucleotides, DNA, and RNA (i.e., their enzymatic substrates and products). Proteins further interact with practically all other low-molecular-weight substances like sugars, fats, and salts. Due to limited space these aspects will not be covered in this chapter; however, they should still be kept in mind since they are of considerable importance to cell metabolism.

24.1.1
Classification and Specificity: Protein Domains

Owing to their great variety, it is almost impossible to classify protein interactions in a straightforward way. Arbitrarily, interactions can be divided into **strong (stable)** and **weak (transient) interactions**, but there is no clear-cut border between the two types. Many protein complexes are assembled quite stably, as their integrity is essential to their functions; ribosomes, for example, are largely stable as protein–RNA complexes. On the other hand, even form-giving structures like actin filaments are constantly being assembled and disassembled.

Although protein interactions need to be extremely specific (e.g., the binding of a peptide hormone like insulin to its receptor), many weak interactions appear to be relatively unspecific and thus without obvious significance. They are tolerated without consequences as long as they do not set the organism at a disadvantage. Unspecific interactions should not be confused with chance collisions caused by **Brownian motion**, as the latter does not create cohesion. Weak interactions might have played an important part in evolution, as they can be enhanced through mutation and selection, and thus be made useful.

From the biologists' point of view it may be more meaningful to classify interactions by protein domains involved. **Domains are the structural and functional units of protein interaction.** They fold independently of other protein areas and tend to be globular with a length of 40–150 amino acids (Fig. 24.1). Many domains have defined interaction properties (e.g., **SH3 domains generally** bind proline-rich sequences, **SH2 domains** bind peptide sequences that contain phosphotyrosine, etc.). These two domains have been named after their homology to the **oncoprotein Src**, which may cause sarcomas when its inhibitory tyrosine 527 residue is mutated. All Src-related proteins have such **Src homology (SH)** domains; the **SH1 domain** represents a **kinase domain**. However, there are still numerous examples among the more than 1000 protein domains in the human proteome whose binding qualities are barely, if at all, known. Even when a domain's principle qualities have been determined (e.g., binding proline-rich sequences), it is still impossible to exactly predict its interaction partners as numerous proline-rich proteins are encoded by most genomes. Predicting such interactions remains an important challenge to structural biologists and bioinformaticists.

24.1.2
Protein Networks and Complexes

Eukaryotic cells are known to contain hundreds of different discrete protein complexes. Many of these complexes contain dozens if not hundreds of proteins (ribosomes, spliceosomes, sarcomere elements in muscles, RNA polymerases; Fig. 24.2). However, even well-defined complexes interact with other, transiently associated proteins (e.g., translation factors with ribosomes). The proteins within a cell can thus be thought of as nodes within a giant protein network that links most of a cell's proteins (Fig. 24.3). Although systematic protein interaction analyses have been performed for only a few organisms (like several viruses, some bacteria, yeast, and a few other model organisms), **proteins have been found to interact with at least three other proteins on average**. It is estimated that the

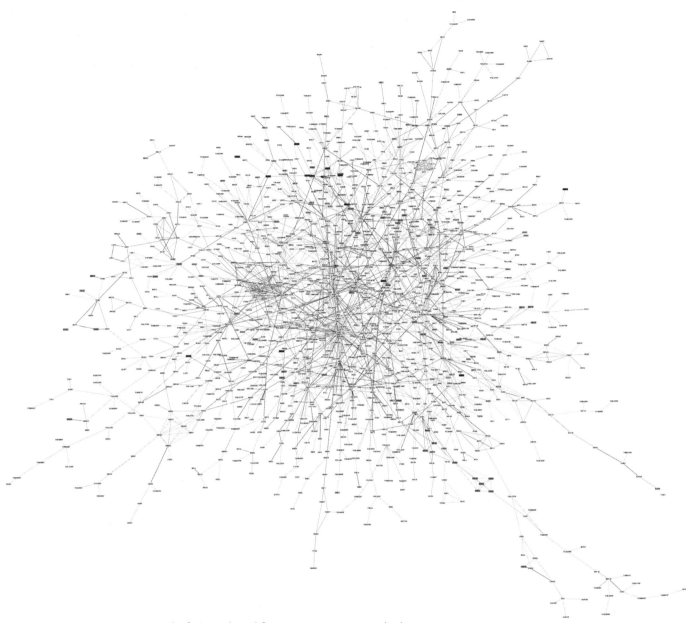

Fig. 24.3 Protein interaction network of a yeast cell. This map was reconstructed from published interaction data and contains 1548 proteins linked by 2458 interactions. The proteins are colored according to their bio- logical function (e.g., proteins involved in membrane fusion are blue, chromatin pro- teins are grey, etc.). (After Schwikowski *et al.*, 2000.)

25 000 or so proteins of the human body interact with each other in more than 100 000 ways. So far, only several ten thousand of these interactions have been experimentally proven and catalogued in databases (Table 24.1). However, mod- ern high-throughput methods will quickly increase our knowledge of such inter- actions. Three important problems remain to be solved. First, the quantitative measurement of interactions (i.e., finding out which ones are weak or strong en- ough to hold together a stable complex). Many proteins in complexes are also part of a less-well defined network of interactions (Fig. 24.3) so that the bound- ary between stable complexes and more transient associations is blurring. Sec- ond, and even more important: what are the biological functions of all these in- teractions? Third, what is the atomic structure of these complexes?

Table 24.1 Selected databases and internet resources.

Database	URL
Database of Interacting Proteins	http://www.ebi.ac.uk/intact
IMEx consortium	http://imex.sourceforge.net
MPIDB	http://www.jcvi.org/mpidb
PDB (3D structures)	http://www.rcsb.org
NDB (nucleic acids and proteins)	http://ndbserver.rutgers.edu
Protein domains	http://smart.embl-heidelberg.de

24.1.3
Structural Properties of Interacting Proteins

Several hundred protein complexes have already been examined by x-ray crystal structure analysis and other methods; their structural data is available from the Protein Data Bank (PDB, Table 24.1). The following statements regarding the geometry and energetics of protein interaction have been derived from the analysis of several dozen to about 100 crystallized protein pairs.

In stable complexes the **contact area** between two proteins is almost always greater than 1100 $Å^2$, with each interaction partner contributing at least 550 $Å^2$ to the entire surface. On average, every partner loses about 800 $Å^2$ of **solvent contact surface** area per interaction, which corresponds to about 20 amino acids per partner. In other words, every amino acid residue involved in an interaction covers about 40 $Å^2$.

On average, dimers contribute 12% of their surface to an interaction, trimers contribute 17.4%, and tetramers contribute 21%. There are considerable differences between individual complexes, however; the entire contact surface ranges from 6% for dimers of inorganic **pyrophosphatase** up to 29% for **Trp repressor** homodimers. This also means that protein surfaces almost always allow for interactions with several proteins at a time.

About 80% of contact surfaces are more or less flat. With a few exceptions, contact surfaces are almost round areas on the surface of stable or transient complexes. Contact surfaces in stable interactions tend to be larger, less plane, more strongly segmented (on the sequence level), and more densely packed than contact surfaces of unstable interactions.

Concerning **secondary structure**, one investigation showed that loop interactions on average constitute about 40% of the contact area. Another study of 28 homodimers showed 53% of contact surfaces to be *α*-helical, 22% to be *β*-sheets, 12% to be *αβ*, and the remaining 11% to be coils.

Complementarity can be defined as fitting surface shape. Contact areas in homodimers, enzyme-inhibitor complexes, and stable heterodimers tend to be the most complementary. Antigen-antibody complexes and unstable heterodimers appear to possess the weakest complementarity.

Concerning **amino acid composition**, contact surfaces between proteins tend to be more hydrophobic than their outsides, but less hydrophobic than the protein interior. One study showed 47% of interacting amino acid residues to be hydrophobic, 31% to be polar, and 22% to be charged. Stable complexes have contact surfaces with hydrophobic residues while unstable complexes tend to prefer polar residues. Mutagenesis experiments have shown that often more than half of a contact surface's amino acids can be changed to alanine without significantly altering the affinity constant (K_d). This means that the *functional epitope* is only a fraction of the structural epitope.

24.1.4
Which Forces Mediate Protein–Protein Interactions?

Remarkably, the average contact area of two interacting proteins is hardly any more polar or hydrophobic than the rest of the protein that is in contact with the solvent. **Transient complexes**, however, tend to have more hydrophilic contact areas, which makes sense, as both components need to exist separately in the cell's aqueous environment. Water is usually excluded from the contact site.

Some authors have proposed that **hydrophobic interactions** provide the energetic basis for the interaction while **hydrogen and salt bridges** ensure specificity.

Although **van der Waals forces** affect all neighboring atoms, these interactions are by no means stronger between two proteins than between a protein and the solvent. Still, they contribute to protein interaction energetically, because they are more frequent on the densely packed contact sites than on the solvent interface.

Hydrogen bonds between proteins are often energetically favored over those with water. The number of hydrogen bonds roughly equals 1 per 170 Å^2 of surface area. The average interaction area (around 1600 Å^2) thus contains about 900 Å^2 of unpolar surface, 700 Å^2 of polar surface, and about 10 (±5) hydrogen bonds. A random sample of relatively stable dimers featured 0.9–1.4 hydrogen bonds per 100 Å^2 on average (with entire contact surfaces usually above 1000 Å^2). However, the range from 0 (e.g., uteroglobin) to up to 46 (variant surface glycoprotein) was considerable. The amino acid side chains are involved in about 77% of hydrogen bonds. Only 56% of homodimers even possess salt bridges; those that do can have up to five.

24.1.4.1 Thermodynamics
Protein interactions can be described as simple chemical reactions of the following form:

$$A + B \underset{k_d}{\overset{k_a}{\rightleftarrows}} AB \tag{24.1}$$

A and B represent two proteins that form the complex AB. Multiprotein complexes are assumed to form through successive binding of subunits.

Protein–protein interactions can be very weak and short-lived as well as strong and permanent. The former are referred to as **transient**, the latter as **stable**, although all numbers of intermediate grades exist. For example, an enzyme can bind its substrate, phosphorylate it, and dissociate afterwards in less than a microsecond. On the other end of the scale, some protein complexes like the collagen triple helix stably persist in bones or other tissues for weeks or even years without dissociating.

The interaction between two proteins can be described quantitatively with the **mass action law**:

$$\frac{[A][B]}{[AB]} = \frac{1}{K_a} = K_d = \frac{k_d}{k_a} \tag{24.2}$$

with k_a being the reaction constant of second degree for the biomolecular association, k_d equaling the reaction constant of first degree for the unimolecular dissociation, and $K_d = k_d/k_a$ being the reaction constant of dissociation (K_a for association).

K_d depends on the concentrations of A, B, and AB at thermodynamic equilibrium; K_d has the dimension of concentration (mol L^{-1} or M). K_a and K_d values for protein–protein interactions vary widely, and range over 12 orders of magnitudes from 10^{-4} to 10^{-16} M.

Interactions with K_d values in the millimolar range are considered weak; values in the nanomolar range and below are strong. The interaction between **trypsin** and **pancreas trypsin inhibitor**, for example, has a dissociation constant in the range of 10^{-14} M, and the binding is thus very strong and stable. Biological interaction strength can also depend on other factors, such as **cooperativity**. Several weak interactions between the subunits of a complex can form a very stable complex.

24.1.4.2 Energetics

K_d values between 10^{-4} and 10^{-14} M correspond to free enthalpies ΔG_d of 6–19 kcal mol^{-1} (i.e., 19 kcal are required to dissociate 1 mol of the complex). Dehydration of the nonpolar groups on the contact surface is definitely decisive for stable association. K_d values for protein–protein interaction can be looked up in special databases (Table 24.1).

Interactions between single amino acids can contribute up to 6 kcal mol^{-1} to a single protein–protein interaction. The greatest energy gain, however, is provided by salt bridges and hydrogen bonds between charged amino acids. The strength of neutral hydrogen bonds lies in the range of 0–3 kcal mol^{-1}. This amount is significantly below a normal hydrogen bond's energy and means that the interaction between two amino acid residues within a complex is hardly stronger than the interaction of a soluble protein with the surrounding water molecules. In complexes of known 3D structure, the peptide bonds form at least half of the hydrogen bonds between interacting proteins. Bonds between side chains and primary chains are especially common, although bonds between both primary chains are also observed at times.

It is estimated that the nonpolar contact areas of hydrophobic interactions provide an energy gain of about 25–70 cal Å^{-2}. Sometimes protein–protein interactions can be so strong (i.e., with a K_d value lower than 10^{-16} M^{-1}) that the components can only be separated by denaturing them.

24.1.5
Methods to Examine Protein–Protein Interactions

Methods for the analysis of protein–protein interactions are based on only a few fundamental principles.

One of the dominant methods is the purification of proteins that have been fused to a foreign protein such as **glutathione S-transferase (GST)**. The fusion proteins and associated proteins can then be isolated on a glutathione-linked matrix and identified with mass spectrometric methods (see Chapter 8) or Western blotting (if antibodies are available).

Several methods allow the analysis of proteins without their purification. Most of these methods use two fusion proteins (Fig. 24.4). *In vivo* methods involve expressing genes in such a manner that their interaction activates a so-called **reporter gene** (e.g., the **two-hybrid system**; Fig. 24.4). Today, interaction screens are routinely performed on a genome-wide scale using robotics to test all possible protein pairs of a proteome. Some reporter systems use light as read-out (e.g., **fluorescence resonance energy transfer**). This method uses two fluorescent proteins to detect their spatial proximity. One of the proteins to be examined is fused with the **Cyan Fluorescent Protein (CFP)** – a candidate interaction partner protein with **Yellow Fluorescent Protein (YFP)**. When these proteins interact or otherwise come into close proximity with each other (at least 100 Å; 30 Å at best), the colocalization can be detected by irradiation with blue light with a wavelength of 434 nm. This wavelength is absorbed by CFP, which immediately transfers the absorbed energy to YFP; YFP then emits its characteristic yellow light with a wavelength of 527 nm. A yellow signal in the fluorescence microscope thus indicates protein interaction (or close proximity).

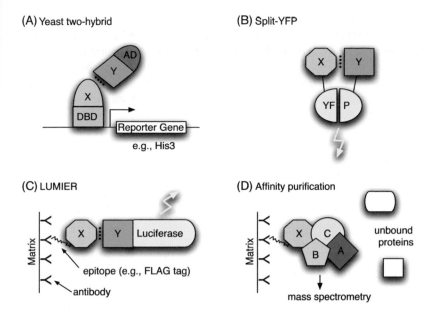

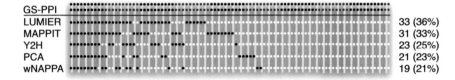

Fig. 24.4 Selected methods for the study of protein–protein interactions. (A) The yeast two-hybrid (Y2H) system is based on the expression of two fusion proteins within a cell. One of the proteins contains a DNA-binding domain (DBD), which can bind to the promoter of a reporter gene (here, His3), and a second protein X, the bait. The second fusion protein consists of a transcription activation domain (AD) and a second protein, Y. If proteins X and Y interact, a transcription factor is formed and the reporter gene is activated. In this case, this means that the cell can grow on histidine-free medium. A yeast colony growing on such medium thus indicates an interaction of the two inserted proteins. (B) Protein complementation assay, e.g. split-YFP. As in the Y2H assay two interacting proteins bring together two protein fragments that are inactive when separate but active when in close proximity. Here, fragments of YFP reassociate and fluoresce when reassembled. Other fluorescent proteins such as GFP have been used in a similar way. (C) LUMIER (LUMInescence-based mammalian intERactome). Two fusion proteins are purified by means of an epitope tag

(here, FLAG tag), usually on an antibody-coated matrix. The interactions between X and Y can be detected using luciferase that is fused to Y and that emits light when luciferin is added. (D) Affinity purification. Protein complexes can be purified from cellular lysates using an affinity epitope, as in (C) with a FLAG tag. The components of the complex can then be identified using mass spectrometry, using the unique mass of peptides when the protein is digested by trypsin. (E) Comparison of various protein interaction methods. MAPPIT and wNAPPA are not described here in any further detail due to space limitations. GS-PPI are "gold standard interactions" (i.e., protein pairs that are known to interact). Red and black dots indicate groups of 10 without further meaning. When these protein pairs are tested, each of the methods shown detects only a small subset (black dots). White dots indicate negatives tests. The success rate of each method is shown on the right (i.e., the number of interactions (out of 92 total) detected with each method). Only eight interactions were found with all methods, but 37 with none of them. ((E) Modified after Braun *et al.*, 2009.)

Comparisons of various methods have shown that no single method is superior to all of the others – all of them are able to detect only a subset of all interactions and no method is able to detect more than about a third of known interactions (Fig. 24.4 E). The complete analysis of all of the interactions of an organism therefore requires a combination of a broad spectrum of methods.

The atomic structure is required for a detailed functional understanding of a complex. Ideally, the structures of the interacting proteins are determined both individually and in the complex. The methods used are **nuclear magnetic resonance (NMR) spectroscopy** (for smaller proteins) or, especially for larger complexes, **x-ray crystallography.**

Protein interactions can also be measured quantitatively. **Dissociation constants** are determined on a micromolar scale through equilibrium centrifugation or **microcalorimetry.** More accurate measurement on a nanomolar scale requires radioactive markers or antibody reactions. These methods are not frequently employed, however, and will therefore not be covered in this book. Several other methods of examining protein–protein interactions are described in Chapters 8 and 27 (**recombinant antibodies** and **phage display**).

24.1.6
Regulation of Protein–Protein Interactions

Biologically relevant protein interactions are subject to tight regulation; if this regulation is disturbed, important diseases such as cancer may be the result.

The most important regulator of protein–protein interactions is **expression control**, because, naturally, proteins can only interact if they are expressed in the same place at the same time. The central control mechanisms are those for transcription and translation (see Chapter 4). For example, most **growth factors**, like some fibroblast growth factors (FGFs), are expressed only in certain tissues (e.g., limbs, brain, or kidneys). Some FGFs are given off into the bloodstream, from where they can reach and bind to receptors that are expressed only in certain tissues. FGFs are also strongly regulated in a temporal dimension (e.g., FGF4 and FGF8 are only expressed in the embryo, while the other FGFs are primarily found in adult animals). The same is true for the respective receptors. **Protein localization** within a cell is also of great importance. Some transcription factors like **NF-κB** (composed of two subunits: RelA and p50) are normally found as inactive protein complexes in the cytoplasm (Fig. 24.5). NF-κB is bound to its inhibitor IκB that dissociates after phosphorylation and is then degraded. The liberated NF-κB protein then enters the nucleus where it regulates the activity of target proteins. **Protein stability** is similarly important as expression, as the final concentration is determined by the equilibrium between synthesis and degradation. Numerous proteins are regulated at this aspect. **Cyclins**, for example, are specifically degraded during certain phases of the cell cycle and thus can no longer interact with their partners – the **cyclin-dependent kinases (CDKs)**.

Covalent modifications are essential regulators for many protein–protein interactions. An important example in addition to phosphorylation is acetylation of histone proteins, which allows for the association of so-called **bromo domains** (see Chapter 4). These protein domains can bind only to acetylated histones. It has been estimated that there are hundreds of different chemical modifications of proteins. About a third of all human proteins are phosphorylated, even if not all of these modifications have a biological function. However, the myriad of **posttranslational modifications (PTMs)** provides a glimpse of the complexity of regulatory interactions that are required to coordinate thousands of enzymes and their substrates of the human proteome!

Ligands are another important form of regulation. GTP, as a prominent example, binds to the a-subunit of **trimeric G-proteins** and causes the dissociation of the $\beta\gamma$-subunit (see Chapter 32). The unbound subunits bind other proteins in

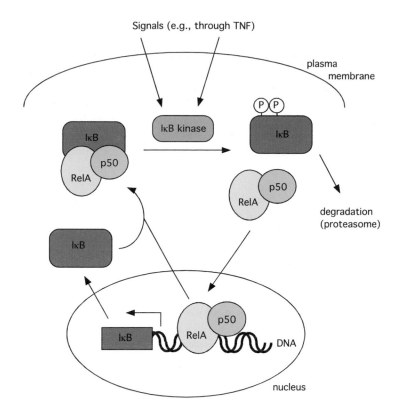

Signals (e.g., through TNF)

plasma membrane

IκB kinase

degradation (proteasome)

DNA

nucleus

Fig. 24.5 NF-κB signaling pathway as an example of protein–protein and protein–DNA interactions. Several signals influence the activity of the IκB kinase complex (IKK), such as those coming from the tumor necrosis factor (TNF) receptor. When induced, IKK phosphorylates IκB. This phosphorylation triggers the recognition of IκB by ubiquitination enzymes, which modify it so that it is recognized and degraded by the proteasome. The removal of IκB exposes a previously covered nuclear localization signal on the NF-κB complex, which can now migrate into the nucleus and bind to specific DNA sequences there. It functions as a transcription factor that activates several target genes, among them the gene for IκB. The now newly expressed IκB in turn binds the NF-κB complex at the promoter and deactivates the gene once more. Many other target genes and interaction partners of NF-κB exist beyond IκB. This example illustrates how complex regulatory networks can be and how many layers of regulation they are composed of (in this case, transcription, localization, modification by phosphorylation, and ubiquitination).

(A)

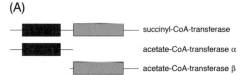

succinyl-CoA-transferase

acetate-CoA-transferase α

acetate-CoA-transferase β

(B)

Fig. 24.6 The Rosetta Stone method. (A) Some proteins, like human succinyl-CoA transferase, are distributed over two proteins in other organisms. In *E. coli*, the protein's function is fulfilled by the α- and β-subunits of the enzyme acetate-CoA transferase. (B) The historical Rosetta Stone from Egypt. Further details in the text.

turn and regulate their activity. Exchanging GTP for GDP triggers the reassociation of the subunits.

24.1.7
Theoretical Prediction of Protein–Protein Interactions

The experimental analysis of protein–protein interactions is quite costly, especially if the structural details need to be investigated. Luckily, many interactions can be predicted even though such predictions are usually based on experimental data, namely known structures or sequence similarities between proteins. So-called "docking" methods can scan 3D structures for compatible potential binding sites. This computational method works reasonably well for rigid proteins with relatively large and hydrophobic interaction surfaces. However, many proteins change their conformation upon interacting and hence *in silico* docking will not work well in such cases. For that reason enzyme–inhibitor complexes can be docked more easily than interactions among signaling proteins, which often change their conformation due to posttranslational modifications.

24.1.7.1 Predicting Interacting Proteins by their Genome Sequence
Several attempts have been made to predict protein–protein interactions *de novo*. One of these approaches is the **Rosetta Stone method** (Fig. 24.6) that uses the observation that some proteins are split up in one organism but not in another. From this, it was concluded that the respective protein halves must interact, since they interact similarly in the fusion protein. These fusion proteins are hence called Rosetta Stone proteins after the famous Rosetta Stone found in Egypt in 1799, engraved with the same text in Greek, Demotic (an early Egyptian common writing), and Hieroglyphs. The stone marked a breakthrough in the translation of hieroglyphs. One example of a Rosetta Stone protein is **human succinyl-CoA transferase** that is found split in two halves in *Escherichia coli*, namely the α- and β-subunits of acetate-CoA transferase (Fig. 24.6).

24.1.7.2 Phylogenetic Profiles

Some gene combinations are retained stably over the course of evolution, which means that these genes apparently cannot exist in isolation. It is concluded that these genes code for components of metabolic pathways or protein complexes that require all respective subunits or otherwise lose their function. Although such phylogenetic profiles do not necessarily allow for conclusions regarding physical interactions, a functional cohesion of the respective gene groups has been found in many cases. Good examples are **ribosomal proteins** or the **yeast proteins Hog1 and Fus3** – two kinases in the yeast mitogen-activated protein kinase (MAPK) signaling pathway.

24.1.8
Biotechnological and Medical Applications of Protein–Protein Interactions

Substantial areas of biotechnological research are concerned with the production of **interacting proteins** such as antibodies or peptide hormones. For example, one therapeutically used antibody is **Herceptin®** (trastuzumab), which binds to the cancer protein HER2 that is overexpressed in 25–30% of all cases of breast cancer. **Erythropoietin** is an example of a peptide hormone that stimulates the formation and maturation of red blood cells (erythrocytes) in the bone marrow. It has been produced biotechnologically for several years now and has become infamous for its involvement in doping cases in professional sports. **With detailed knowledge of protein–protein interactions, substances that specifically block those interactions can be identified.** For example, it is desirable to block the binding of **HIV** to its **target receptors** CD4, CCR5, and CXCR4. Meanwhile, there are also a number of substances available that develop their effect *within* cells by blocking specific protein interactions. The **immunosuppressant FK506**, for instance, binds the **FK506-binding protein (FKBP)**. The resulting complex in turn blocks the activity of the phosphatase **calcineurin** through direct interaction, upon which calcineurin activates T cells of the immune system. Blocking calcineurin thus triggers the actual immunosuppressant effect of FK506.

Occasionally, protein–protein interactions prove disadvantageous, like with insulin, which tends to form dimers or hexamers that are less active than monomers. This tendency toward oligomerization can be suppressed via genetic modification and thus insulin of higher activity can be produced.

24.2
Protein–DNA Interactions

As outlined above, protein–DNA interactions play a major role in all fields of genetics from regulation and transcription of individual genes to repair of damaged sequences, even to the stabilization of DNA in chromatin and the replication of entire genomes. It is estimated that 2–3% of prokaryotic and 6–7% of eukaryotic genes code for **DNA-binding proteins**. Additionally, many of these proteins do not merely bind DNA, but also interact with other proteins and sometimes, as is shown in the example of RNA polymerase, only display their full activity when organized in multimeric complexes.

24.2.1
Sequence-Specific DNA Binding

Protein recognition of specific sequences on the DNA double helix is of critical importance to fundamental genetic processes such as gene regulation and transcription (see Chapter 4). The specific binding occurs on the atomic level via interactions of certain side chains of the protein with nucleotides of the DNA. In spite of numerous attempts to formulate them, no simple or general rules to explain or predict sequence specificity exist to date. Instead, the same basic interactions known from **protein–protein** or **protein–ligand** interactions are also found in **protein–DNA** complexes. The sequence-specific binding results from an individual combination of the different possible interactions. The first aspect of this specificity arises from the different hydrogen bond patterns of the base pairs **adenine–thymine** (A–T) and **guanosine–cytosine** (G–C), that are presented in the minor or major groove of the DNA double helix (Fig. 24.7).

Statistical analyses of known protein–DNA complex x-ray structures showed that the positively charged side chains of **arginine and lysine** preferably form hydrogen bonds with guanine; **asparagine and glutamine** mainly form hydrogen bonds with adenine. Not quite as specific, the shorter side chains of **serine** and **threonine** have been shown to bind to the sugar-phosphate chain and are probably concerned with overall stability rather than specificity. In addition to hydrogen bonds, hydrophobic interactions are also important: although van der Waals contacts tend to be less specific than hydrogen bonds, certain preferences could still be observed. **Arginine** shows a preference for guanine while threonine prefers methyl–methyl van der Waals interaction with thymine. **Phenylalanine, proline**, and **histidine** form hydrophobic stacks with the planar bases, especially in those structures in which the DNA is sufficiently deformed (e.g., in **TATA box-binding proteins**). A third possibility for interaction is indirect contact mediated by water molecules. The importance of these interactions has become apparent in recent years with the growing number of high-resolution x-ray structures (with resolutions better than 2 Å). Unfortunately, the resulting binding patterns are still too complex to formulate general rules.

24.2.2
Thermodynamic Considerations Regarding Protein–DNA Complexes

Although only a small number of protein–DNA complexes have been characterized thermodynamically (much fewer than are structurally known), some important and basic considerations can be made and a few conclusions be drawn. **Sequence-specific proteins** generally have a high DNA affinity with association constants of $K_a > 10^7$ M. In contrast to this, the respective values for **unspecific binding** are lower by up to three orders of magnitude, to ensure sufficient discrimination. The upper limit for K_a is estimated to be in the range of 10^{12} M, because at higher values complex formation would not be reversible under physiological conditions and would react too sensitively to minor concentration fluctuations within the cell. Furthermore, it could be shown that the association constants (and thus the free binding enthalpies ΔG^0) of several protein–DNA complexes tended to be quite similar, while the individual contributions of ΔH^0 and $T\Delta S^0$ varied greatly. Stabilizing enthalpy contributions result from the formation of hydrophilic and hydrophobic interactions, whereas the loss of hydrogen bonds to solvent molecules has a destabilizing effect. Still, the liberation of water molecules constitutes the most important contribution to $T\Delta S^0$; the actual complex formation lowers entropy. Further destabilizing enthalpy contributions result when complex formation forces one of the partners into a disadvantageous conformation. In many protein–DNA complexes the DNA double helix is bent from its canonical **B-conformation** in this manner.

24.2.3
Methods to Study Protein–DNA Interactions

The methods used to examine protein–DNA interactions are similar to those that have already been described in the first part of this chapter (Fig. 24.4). The special methods used in addition to these are summarized in Fig. 24.10 and Table 24.2.

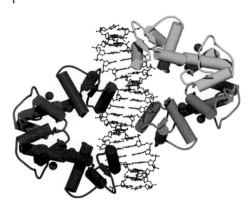

Fig. 24.7 Crystal structure of the activated diphtheria toxin repressor (DtxR) from *Corynebacterium diphtheria* in complex with DNA. DtxR is a global repressor controlling a wide variety of genes including the *tox* gene encoding diphtheria toxin. *In vivo* DtxR is activated by Fe²⁺ (red balls) and once activated binds its DNA target sequence as double dimer. The DNA recognition helix located in the major groove of DNA is part of an N-terminal helix-turn-helix motif. (From Pohl *et al.*, 1999.)

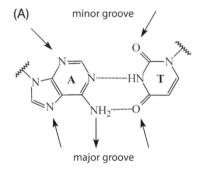

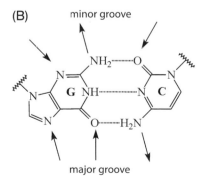

Fig. 24.8 Watson–Crick hydrogen bonds of the base pairs A–T (A) and G–C (B). The arrows indicate potential hydrogen donors or acceptors.

Table 24.2 Important methods to examine protein–DNA interactions. (See also Luscombe *et al.*, 2001; Mandell und Kortemme 2009; Moss, 2001.)

Method	Experiment	Results
DNase footprinting	Binding to a protein protects DNA. The DNA is first either radiolabeled or chemically marked and then exposed to DNase (or *in situ* generated OH radicals) and then analyzed by gel electrophoresis.	The DNA target sequence can be determined. Varying the concentration leads to binding constants. In addition, the influence of activating or inhibiting chemical can be assessed.
Electrophoretic mobility shift assays (EMSA or band shift)	Gel electrophoresis under native, nondenaturing conditions of the protein–DNA complex. Pure DNA serves as the standard. If the protein is bound the complex will travel at lower velocity and hence the band will appear *shifted* compared to the DNA alone.	Using different DNA oligonucleotides enables the determination of target sequences. Varying the concentration leads to binding constants. In addition, the influence of activating or inhibiting chemical can be assessed.
Capillary electrophoresis	Here, the gel is substituted by a capillary of 50–250 μm diameter and 5–20 cm length. Protein–DNA complexes and DNA travel at different velocities through the capillary. At the end they are detected by an UV monitor.	This methods is similar to EMSA with the advantage that much less material is needed and the results tend to be more reproducible. However, capillary electrophoresis equipment is expensive.
Fluorescence resonance energy transfer (FRET)	This method is based on the energy transfer observed from one fluorescently labeled biomolecule to another that is labeled with a different fluorescent marker. The energy transfer is a function of direct distance.	By labeling different parts of the DNA and the protein in separate experiments, structural information such as approximate distances and possible conformational changes upon binding can be obtained.
Isothermal titration calorimetry (ITC)	ITC measures the heat generation upon binding. Here one component is carefully titrated in small volumes to the second component.	ITC allows the determination of all thermodynamic parameters such as ΔH, $T\Delta S$, and ΔG as well as binding constants. The method requires relatively large amounts of sample and the ITC equipment.
Surface plasmon resonance (SPR)	SPR is an optical method based on the refractive index of an aqueous solution floating over a mono-molecular layer immobilized on a glass plate. One component is immobilized (i.e., the DNA labeled with biotin and bound to a streptavidin-coated surface). Binding of the protein causes a change of refractive index that is detected.	This method also allows the determination of thermodynamic parameters and only requires small sample amounts. However, the equipment is expensive and the analysis sometimes less than trivial.
SELEX (Systematic evolution of ligand by exponential enrichment)	SELEX is an experimental in vitro procedure to identify and optimize DNA or RNA molecules that bind a target protein.	This method is used to find new DNA sequences that bind a target protein. These sequences can be used to find the genes regulated by this protein.
ChIP-seq (Chromatin immuno-precipitation plus sequencing)	ChIP is a method to study protein-DNA interactions in vivo. Antibodies are used to isolate proteins that are still bound to genomic DNA. The bound DNA can be partially digested to reduce fragment size, separated and sequenced.	The main advantage of this method is that protein-DNA interactions can be studied in living cells. In addition, the whole genome and its regulatory networks can be analyzed.

The most powerful method capable of unraveling protein–protein as well as protein–DNA complexes in atomic detail is **x-ray crystallography**. The importance of this method is exemplified by recent Nobel Prizes in Chemistry awarded for the structure determination of the ribosome (2009), RNA polymerase (2006), trans-membrane "in" channel (2003), and ATPase (1997).

24.2.3.1 Structural Classification of Protein–DNA Complexes

DNA-binding proteins can be divided into eight groups based on their structure and function; each of these groups uses similar motifs to recognize and bind DNA (one example is shown in Fig. 24.7). Note that this classification is based on the several hundred complex crystal structures solved so far. Table 24.3 gives an overview respective frequency of the classes in the human genome.

Domain	Human	Fly	Worm	Yeast	Plant
Histone core domain	75 (81)	5	71 (73)	8	48
Helix-loop-helix DBD	60 (61)	44	24	4	39
Homeobox domain	160 (178)	100 (103)	2 (84)	6	66
Myb-like DBD	32 (43)	18 (24)	17 (24)	15 (20)	243 (401)
Leucine zipper domain	114	55	36	16	134
RFX DBD	7	2	1	1	0
TATA-binding protein (TBP)	2 (4)	4 (8)	2 (4)	1 (2)	2 (4)
Other zinc finger domains	77 (100)	34 (37)	50 (72)	19 (21)	87 (102)
Zinc finger, C2H2 type	564 (4500)	234 (771)	68 (155)	34 (56)	21 (24)
Zinc finger, C3HC4 type (RING finger)	135 (137)	57	88 (89)	18	298 (304)
Other DBDs	46 (47)	26 (27)	19	6	7
DEAH box helicase (RBD)	63 (66)	48 (50)	55 (57)	50 (52)	84 (87)
KH domain (RBD)	28 (67)	14 (32)	17 (46)	4 (14)	27 (61)
RNA recognition motif (RRM zinc finger) (RBD)	224 (324)	127 (199)	94 (145)	43 (73)	232 (369)

Numbers in parentheses indicate the number of domains. For example, 28 (67) in the KH domain line means that the human genome contains 28 proteins with a total of 67 domains. Several proteins thus contain more than one KH domain. Other DNA-binding proteins contain the so-called ARID and *forkhead* domain. The RFX domain is an uncommon helix-turn-helix motif and thus related to the homeobox proteins. The helix-turn-helix DNA-binding domain resembles the leucine zipper group as well as the Myb domain. (Modified from Venter *et al.*, 2001.)

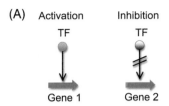

Table 24.3 DNA- and RNA-binding domains in the human genome as well as in the genomes of fully sequenced model organisms (the classification is based on sequence; RNA indicates RNA-binding domains, all other domains are DNA-binding domains).

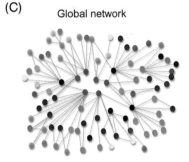

Fig. 24.9 Schematic representation of regulatory networks. (A) Transcription factors (TF; depicted as circles) can act as gene activators (left) or repressors (right). Interaction with the promoter is represented as arrow. (B) Examples of simple network motifs: single-input module, feed-forward loop, and dense overlapping regulon. Transcription factors are depicted as circles, genes as boxes, and interactions (activating or inhibiting) as arrows. (C) Simplified representation of a theoretical network. Transcription factors are shown as circles with interactions to genes of other transcription factors as arrows. The red circles in the center represent global regulators.

24.2.4
Regulatory Networks and Systems Biology

Over the last decade a large number of transcriptions factors have been identified, their activation mechanism by small molecules has been studied, and their DNA target sequences determined. One of the great challenges of **systems biology** is to translate and summarize this vast amount of information in so-called **regulatory networks** that can be simulated in computational models (s. Chapter 23). The long-term goal is to be able to simulate and ultimately predict the responses of cells and organism to a changing environment.

In order to construct regulatory networks, the system can be divided into three organizational levels. The lowest level is represented by the interactions of one single element (transcription factor) with its target DNA promoter (Fig. 24.9 A). This interaction can either activate or repress the expression of the downstream gene(s). In many, but not all, cases is the DNA-binding activity of the transcription factor itself that is regulated, for instance by binding of a small molecule or even a single atom such as iron (Fig. 24.7). The next level is composed of certain network motifs where the single transcriptions factor is part of a regulatory module. Three examples of such motifs are depicted in Fig. 24.9 B. The **single-input module (SIM)** consists of one transcription factor that controls a number of genes. The **feed-forward loop (FFL)** is represented by one transcription factor that activates the expression of another transcription factor. **Regulons** are characterized by several transcription factors that regulate a number of genes in concerted action. These and a limited number of other motifs can be used to construct genome-wide regulatory networks. It should be noted that in spite of all recent progress current networks are still severely limited by the lack of complete knowledge of possible interactions. Nevertheless, network analysis, particularly of prokaryotic organisms, has led to important conclusions. It was

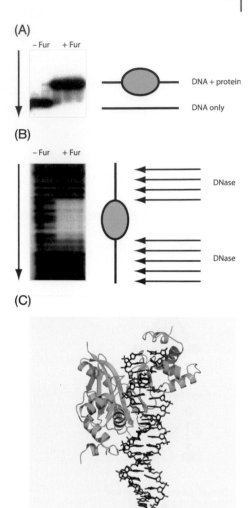

Fig. 24.10 DNA band shift assay and DNase I protection assay of the ferric uptake regulator (Fur). Fur is an alternative Fe^{2+}-dependent regulator not related to DtxR. (A) DNA band shifts are based on the fact that free DNA and free proteins migrate through a gel faster than a DNA–protein complex. (B) DNase I footprinting involves digesting DNA incompletely (with or without bound protein), resulting in fragments of different length. DNA-binding proteins protect DNA from digestion. If the (radioactively marked) fragments are loaded onto a gel, a gap between the bands corresponding to the binding site can be seen. (With kind permission from M.L. Vasil). (C) Computer model of Fur bound to DNA. Whereas the crystal structure of Fur has been determined by x-ray crystallography, the complex with DNA has been elusive. This model, however, shows how the putative DNA recognition helix, which is also part of a N-terminal helix-turn-helix motif, could bind into the major groove of DNA. It is remarkable that DtxR (Fig. 24.7) and Fur employ similar DNA-binding motifs but completely different dimerization domains. Note that the DNase I protection assay shows a much longer region to be protected than indicated by the computer model.

shown, for example, that a small number of transcription factors play a global role controlling a large number of genes, including many other local transcription factors. These in turn control only a small number of genes. One example of such a global regulator is the **ferric uptake regulator Fur** (band shift depicted in Fig. 24.10). In *Pseudomonas aeruginosa*, Fur is an Fe^{2+}-dependent regulator responsible for activating or inhibiting the expression of hundreds of genes responsible for iron uptake and storage, and for the oxidative stress response.

24.2.4.1 Medical Relevance of Protein–DNA Interactions

Numerous **diseases** are caused by **incorrect protein–DNA interactions**. The great importance of these interactions stems from the fact that DNA-binding transcription factors are central switches of a cell's regulatory network. The **transcription factor SRY**, for instance, is sufficient to trigger the development of the male sex. Mutations in the protein can lead to a sex change of affected embryos. A number of **hormone receptors** like the glucocorticoid or estrogen receptor are zinc finger proteins, which play an important part in hormone-controlled metabolism. Finally, cancer can also be caused by mutated transcription factors. The proteins Jun and Fos are well-studied **leucine zipper proteins** that do not only have to bind the correct promoter sequences, but can only fulfill this physiological function as a Jun/Fos protein complex. These proteins are of great medical importance since mutations in the underlying genes have been linked to the development of certain cancers.

24.2.5
Biotechnological Applications of Protein–DNA Interactions

The detailed knowledge of **protein–protein** and **protein–DNA** interactions allows us to specifically manipulate them for different purposes. In some cases, DNA-binding proteins can be manipulated in such a manner that they recognize specific DNA sequences. One aim is, for example, the creation of DNA-binding proteins that specifically recognize and bind to defined target genes to activate or deactivate them. If specific promoters were to be placed before such genes, they could be manipulated in a tissue-specific manner. Another aim is the specific activation of DNA binding through added substances. One example for the latter is the **Tet repressor** that binds to the *tet* operator DNA sequence as well as to the antibiotic **tetracycline**. This system can be modified in such a way that the addition of tetracycline or related substances can induce or inhibit the DNA binding. It is possible to insert the *tet* repressor gene and a target gene under control of the *tet* operator into mammalian cells and then switch the target gene on or off simply via the addition of tetracycline (Tet system). Many laboratories work on the manipulation of DNA-binding domains that recognize specific sequences (even unnatural sequences when required). In the case of restriction enzymes, proteins could be created that cleave DNA on predefined sites. With an increase of detailed knowledge, numerous applications are imaginable that not only allow for the manipulation of bacteria, animals, and plants, but also make alterations of the human genome possible. While therapeutic applications are desired, it will be a great challenge to avoid unintentional side effects and misuse.

24.2.5.1 Synthetic Biology

The technology described in this chapter aims at the manipulation of individual genes or proteins. However, within the past few years it has become possible to chemically synthesize complete viral or bacterial genomes. Even if we do not know the exact function of all proteins or their interactions, we can synthesize organisms with completely new properties by adding, removing, or replacing genes at will. Synthetic organisms can only work with functional protein and gene interaction networks. However, these networks can be tailored to specific properties or needs. We expect many new biotechnological or medical applications to develop in this area in the near future.

25
Drug Research

Learning Objectives

Most drug targets are proteins and should feature two properties: they should be responsive to low-molecular-weight substances and be associated with a medical condition. In the target validation process, the modulation of the target is simulated by modifying their activity by genetic and molecular biological means. Biologicals are a special kind of therapeutically usable proteins. Here, the protein itself is the active substance. The use of a gene or its product as a target or the active substance can be patented. The search for active substances relies on large compound libraries. Assays with a readout on target activity must be developed for each target. They must have good statistical parameters, and be able to be miniaturized and automated. Substances found in assays are characterized and optimized regarding their dose–response relationship and their pharmacological as well as toxicological properties. Animal trials are carried out to demonstrate the efficacy and innocuousness of the substance in question in a living organism before clinical trials can begin. Clinical trials of compounds in humans are time-consuming, labor-intensive, and costly. Even after the market admission of a drug, its medical effectiveness remains under observation.

25.1
Introduction

Although enormous progress has been made in **drug research** over the last century, and life expectancy has gone up considerably, many medical conditions are still not treatable or there is scope for improvement in existing therapies. Furthermore, **new conditions** are being discovered or well-known diseases such as Alzheimer's are reaching new proportions due to changes in living conditions. The way in which drug research is carried out nowadays has been revolutionized by the further development of molecular biotechnology, chemistry, chemical analysis, and information technology. This chapter gives an overview of the processes involved in the development of a new drug.

25.2
Active Compounds and their Targets

How do therapeutic agents act on the human body? Healing substances have been known for thousands of years, but nobody knew what the underlying principles were that made them effective. In the nineteenth century, Ehrlich postulated the existence of what he called chemoreceptors on parasitic microorganisms, to which compounds with anti-infective properties would bind. Similarly, in the first descriptions of hormones in the early twentieth century, the existence of such receptors in the human body was also postulated. When biochemical and molecular biological methods became more refined during the 1970s, it

Fig. 25.1 Distribution of targets of known therapeutic agents over protein categories.

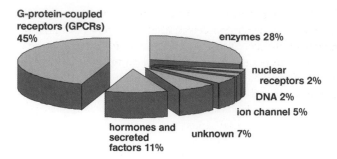

was possible to isolate large numbers of these receptors and describe them. It turned out that among the four predominant compound classes, proteins were particularly suited as **targets** for therapeutic agents. Lipids and sugars, by contrast, hardly ever act as receptors, while nucleic acids take on this function in some exceptional cases. Since the human genome has been decoded and 22 000 coding genes have been described, all human proteins should soon be known and could be assessed for their suitability **as targets**.

Not all proteins make suitable targets. When the list of 500 known targets of today's drugs is grouped in terms of protein categories, it turns out that the vast majority of targets for low-molecular-weight compounds belong to just a few categories (Fig. 25.1).

These predominant categories include various **enzymes, membrane receptors, ion channels**, and **nuclear receptors**. There are many protein categories that have never been described as targets for a therapeutic agent. For example, there is no active agent acting directly on **transcription factors** that are not members of the nuclear receptor family, although such proteins occur in large numbers. Why this is can be explained in fairly simple terms – active agents that bind specifically and with sufficient affinity to a certain protein attach to pockets on the surface of the protein. The hydrophobicity and charge distribution of the binding pocket enable the compound to dock into the pocket in an energy-efficient way. Proteins that do not carry a binding pocket on their surface are usually not suitable as targets for low-molecular-weight chemical compounds.

This also explains why a fairly high number of targets are enzymes. They naturally carry binding pockets in order to bind their natural substrates. Therapeutic agents also bind to these pockets, which, among other things, enables them to inhibit the activity of the enzyme by preventing the natural substrates from binding. Nuclear receptors and membrane receptors also carry binding pockets that recognize the organism's own messenger substances or metabolites and initiate changes in receptor activity. The ability of a protein to bind to low-molecular-weight compounds has been coined "**druggability**". The choice of proteins suitable as targets is further limited by the necessity of developing a **robust and cost-effective testing system**, which allows the assessment of the interactions of the protein with low-molecular-weight compounds. Such a testing system is called an **assay** and the search for active compounds is known as **screening**.

An exception to the rule that active agents bind to proteins are those therapeutic agents that exhibit a DNA-modifying activity. These include chemotherapeutic agents used in cancer therapy (e.g., cisplatin). Further exceptions are some compounds that attack biomembranes or act as anti-infectives, killing microbial infective agents. Some antibiotics, for example, bind to the RNA backbone of ribonucleoproteins.

25.2.1
Identification of Potential Targets in the Human Genome

Molecular biological research on an industrial scale, as it evolved in the process of the Human Genome Project, has also resulted in profound changes in the way preclinical research is carried out. It used to be the case that active compounds were identified by testing tissue or animal models. Research was focused on the chemical questions, while the detection of the actual site of action was more or less a matter of chance. Often, the molecular site and the mechanism behind the action of a compound was only discovered many years or even decades after the approval of a medication. Today a large number of proteins are known to be the targets of approved drugs. If proteins are druggable, there is a likelihood that this applies to the other members of the protein family as well and these proteins are defined as potential targets.

Unknown sequences can be assigned to functional categories on the basis of bioinformatically established similarities (Box 25.1). This very simple purely bioinformatic path of exploration, which initially does not require laboratory work, has been pretty much exhausted. However, there are still many blank spots as far as the characterization of sequences is concerned. One of the reasons is that vague resemblances in sequences exist, leading to a high number of

---

Box 25.1 Sequence Similarities

New targets are fairly easy to define if there is a strong sequence similarity to known targets. In this case, a new member is added to a known gene family. Short amino acid signatures can provide clues about some functions of the unknown protein. Two cysteines and two histidines at defined distances, for example, define a C2H2 zinc finger domain within a protein, to which DNA or RNA can bind. More extensive similarities can characterize protein families. Five relatively short amino acid sequence motifs are sufficient to characterize the family of DEAD-box RNA helicases (Table 25.1).

Table 25.1 Diagram of the position of common sequence motifs in members of the DEAD-box RNA helicase family in the Blocks database.

Name of sequence	Length	Sequence
IF4A_CRYPV/O02494	405	
IF4A_LEIBR/Q25225	403	
IF43_NICPL/P41380	391	
IF4N_SCHPO/Q10055	394	
FAL1_YEAST/Q12099	399	
DB45_DROME/Q07886	521	
RM62_DROME/P19109	575	
DBP3_YEAST/P20447	523	

The name of these helicases is derived from the one-letter code of one of the motifs. Even if there is little similarity between sequences, a similar organization of domains may hint at functional similarities. This is referred to as superfamilies. The G-protein-coupled receptors (GPCRs; see Chapter 3.1.1), having seven transmembrane domains in common, make up such a superfamily (Fig. 25.2).

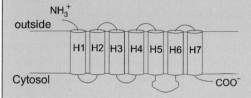

Fig. 25.2 Domain structure of GPCRs. H1–H7 are the seven α-helices, each of which represents a transmembrane domain.

false-positive results in bioinformatic comparisons. Proteins with high similarities in structure and function, by contrast, sometimes do not have any sequence similarities. Furthermore, bioinformatic methods for the correct prediction of the sequences of new genes are not very accurate yet (see Chapter 22). In the human genome, the coding sequences are hidden amongst 98% of noncoding material. Even without this extra hurdle, it is difficult enough to predict correctly all exons and transcription origins – a task further complicated by the presence of many pseudogenes. The prediction **of alternative splicing** is also patchy.

Being part of a certain **gene family** is therefore not a sufficient indicator for the function of a gene. In the pharmacologically relevant category of **nuclear receptors**, for example, it is relatively easy to identify proteins of the family through the **highly conserved sequence** of the DNA-binding domain, but these sequences do not help to identify the natural ligand or the target genes regulated by the receptors. Further experiments are needed to determine the function of the receptor in question.

In the favorable case that a sequence resembles a protein of which the **3D structure** has already been recognized, the structure of the new protein can be predicted with the help of a homology model. For homology modeling to be successful, the similarity between the sequences must be at least 30%. However, there are also many proteins lacking sequence similarities that can be functionally characterized by their similarity in structure. This is why great efforts are made to crystallize systematically all (bioinformatically predicted) proteins and investigate their structure in proteome projects. The available methods involving the expression, purification, crystallization, and structural analysis of proteins have, however, not yet reached the high-throughput rate achieved by DNA-based methods.

25.2.2
Comparative Genome Analysis

The **comparison of entire genomes** is a special case of sequence analysis. For example, a comparison of the genomes of the **two bacterial species** *Vibrio cholerae* and *Escherichia coli* has been carried out. *V. cholerae* is the infective agent causing cholera, while *E. coli* is a fairly innocuous intestinal bacterium, which only causes diarrhea in some isolated cases. The majority of the 4000 *V. cholerae* genes strongly resemble those of *E. coli*, but the comparison also led to the identification of 500 *V. cholerae* genes that have no equivalent in *E. coli*. These are the genes that are now undergoing further investigation because they probably hold the key to the **virulence and pathogenicity** of *V. cholerae*. It is also important, of course, to compare the **genetic inventory of pathogenic microorganisms** to that of their human host. This can give indications of metabolic pathways that can be utilized when developing active agents and will not harm the human host.

25.2.3
Experimental Target Identification: *In Vitro* Methods

Many experimental methods are used alongside pure sequence analysis to identify potential targets. Expression profiles can provide dues to the tissue specificity of targets, making it possible to group and identify genes according to their typical expression patterns (see Chapters 21, 22, and 24). **Differential expression profiles**, for example, can be used to identify genes that are expressed during a microbial invasion process. Comparisons between pathological and healthy tissue are also very important. If a protein interacts with other proteins and the functions of these proteins are known, it may be possible to draw conclusions from their interactions with the unknown protein. **Small interfering RNA (siRNA) screens** are used to inhibit the expression of a specific gene. The effects of

Box 25.2 Reporter Gene Assays

The agent that is being looked for (a low-molecular compound or a protein) nearly always has an impact on the strength of the expression of certain cellular genes. This can be harnessed to produce fairly straightforward assays known as reporter gene assays.

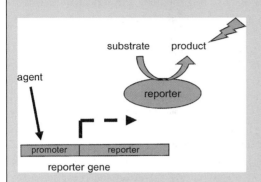

This involves replacing a coding region of a **regulated gene** by that of a gene that codes for a product that can be easily detected. This hybrid gene is called a **reporter gene** and the protein to be detected is the **reporter**. Apart from fluorescent proteins, suitable reporters can be found mainly in enzymes, which turn a suitable substrate into an easily detectable product. Examples for this are color reactions catalyzed by *β*-galactosidase or *β*-glucuronidase. In the widely used **luciferase reporters**, the light-producing enzymes from fireflies (*Photinus pyralis*) or cnidarians (*Renilla reniformis*) are quantified by measuring the light emitted during the consumption of the relevant substrates. In some reporter gene assays the reporter gene is directly modulated by the protein that is assayed (e.g., by nuclear receptors). Where that is not possible, the indirect effect of a protein can be measured. For example, if the effect of an extracellular compound on gene expression is mediated through an intracellular signaling process, it can also be determined by a reporter gene assay.

this inhibition on cellular processes can subsequently be detected with specific **reporter gene assays** (Box 25.2).

In some cases, the identification process is reversed – the **chemical modulators** are known, but not the **molecular target**. Here, again, the existing inventory of human cell components, built up by the Human Genome Project, is very useful. Comparing expression profiles that result from the application of certain compounds can also lead to the identification of targets. In many cases the compound can be immobilized and used for the affinity purification of binding proteins. Proteins that are enriched by this method may include the actual target of the compound. Their identities can subsequently be identified by mass spectrometry.

25.2.4
Experimental Identification of Targets: Model Organisms

Another way of identifying the role of genes involved in the development of disease is the use of **model organisms**. The **nematode (*Caenorhabditis elegans*), fruit fly (*Drosophila melanogaster*), zebrafish (*Danio rerio*)**, and **mouse (*Mus musculus*)** are the most widely used model organisms. Similar to siRNA screens, a systematic knock-out of all known genes can be used to study the phenotypic effect of each gene. An exhaustive phenotype analysis, particularly in mammals, is not feasible, however, as it would require a whole range of test procedures in the fields of biochemistry, histology, physiology, anatomy, and behavioral studies. If a relevant phenotype can be defined in a mammal this represents strong evidence for a role of the target protein. Then again, it is often found that knocking out genes thought to play a key role does not have the anticipated impact on the phenotype (see Chapter 28). As the most relevant models for genetic research do not always provide an adequate model for human disease, it is often necessary to choose a different species or to humanize the model organism by inserting a human gene.

25.2.5
Experimental Target Identification in Humans

Great hopes rest on the identification of **disease genes** through research in humans, which would eliminate the question of the transferability of results obtained from model organisms to humans. Although many methods cannot be applied to humans for ethical reasons, a range of procedures is open to researchers:

- **Linkage analysis** – observing how certain chromosome sections are linked in families to the onset of disease – has been successfully used to identify a number of genes responsible for **monogenic diseases**. The **CFTR gene** that is mutated in patients with cystic fibrosis, and the **BRCA1/2 genes that exist in several variants and are partly responsible for a hereditary risk of breast cancer, are well known examples.**
- **Genetic differences** are also found to be relevant in **nonfamilial diseases**. Most of these differences are deviations in single nucleotides in a DNA sequence. These are not rare mutations, but fairly common polymorphisms. These genetic variants in which the rare allele occurs with a frequency of at least 1% are known as **single nucleotide polymorphisms (SNPs)**. When comparing two homologous chromosomes, a SNP is found every 1–2 kb. It should be possible to accurately map disease genes by the statistical analysis of SNP data from whole genomes of a large number of individuals (**genome-wide association study**). The massive analysis of SNPs applying complex statistical methods could also lead to the identification of **polygenic causes of common diseases** if the effects of individual polymorphisms are not too small and if a high number of samples from stringently phenotyped subjects can be collected.
- **Genetic comparisons** between normal and diseased tissue samples can also show up correlations with symptoms of disease. In **cancer**, many of the genes relevant to the onset of the disease have been identified by detecting somatic mutations in cancerous tissue, which were absent in the surrounding normal tissue. The detection of the Abl oncogene as a cause for **chronic myeloid leukemia (CML)** is an example. A new efficient treatment was found by developing specific inhibitors of the **protein tyrosine kinase** activity in Abl (imatinib (Gleevec[rl5]® in the United States and Glivec® in Europe/Latin America/ Australia), a 2-phenylaminopyrimidine derivative). The progress of DNA sequencing technology allows the detection of somatic mutations by comparatively sequencing the genomes of affected tissues with reasonable operating expenses. The comparison of thousands of genomes of specific cancer forms will result in a data set of occurring mutations. The challenge consists in differentiating mutations with a causal effect (**driver mutations**) from those which are not relevant for the disease (**passenger mutations**). Again, this problem can only be tackled by analyzing a very high number of samples in these studies.
- Furthermore, it is possible to determine whether a **correlation** exists between the expression of the target gene in the affected tissue and the pathological state. For an initial approximation, it does not matter if this is done by specific repression or induction of the relevant gene, although induction is experimentally easier to assess.

25.2.6
Difference between Target Candidates and Genuine Targets

The procedures described above are very efficient for identifying target candidates that are definitely associated with the relevant physiological context or the disease. It is, however, much more difficult to establish a causal link between a gene and the onset of the disease it is thought to cause – perhaps via a crucial regulatory function. Once a gene has been causally validated, it is still possible that it is unsuitable as a target because the gene products cannot be modulated by small molecules or other therapeutic agents (see Box 25.3).

Box 25.3 The Low-Density Lipoprotein (LDL) Receptor: Promising at First Sight, yet Unsuitable as a Direct Target

It has been known for many years that the LDL receptor is responsible for the absorption and degradation of LDLs in liver cells, which are a major risk factor in **atherosclerosis** (Fig. 5.10). **Mutations** in the LDL receptor gene cause a hereditary pattern of pathologically **raised cholesterol levels** in humans. Although the LDL receptor has been characterized as a relevant limiting molecule in the cholesterol degradation pathway, it has not been possible so far to develop a medication that counteracts directly the production of LDL receptor. What is prescribed instead are **statins**, which inhibit one of the key enzymes in the cholesterol-synthesizing process in the liver, known as **3-hydroxy-3-methylglutaryl (HMG)-CoA reductase**. The resulting drop in intracellular cholesterol levels causes the liver cells to produce more LDL receptor, which, in turn, reduces the blood cholesterol level. This is an example of using metabolic pathways in order to indirectly achieve a higher concentration of this important receptor.

Since all target identification methods are based on correlations, all methods are associated with some uncertainty. The likelihood of detecting a genuine functional connection is greater if a target that has been identified by one method can be validated by independent experimental methods. In principle, all methods for the identification of targets described above can also be used for their validation. The methods available are summarized in Table 25.2. When it comes to describing the function of the target in the pathological process, the methods vary in their validity. The more clearly the role of the protein concerned can be shown by experimental methods, the more interesting it becomes for the pharmaceutical industry. However, high cost and low throughput can prove to be major obstacles to a wider use of methods that would deliver better results. This correlation has been visualized in Fig. 25.3 as a target validation pyramid. It should be remembered that apart from its medical relevance, a genuine target must have properties such as druggability and amenability to testing in robust assays. The most important milestone in validating a target before entry into the clinical phase is the trial of a drug candidate in an animal model. Eventually, the only conclusive form of target validation remains the clinical trial in humans.

25.2.7
Biologicals

Apart from conventional targets, modulated by low-molecular-weight compounds, there are biomolecules that can be applied directly as **therapeutic agents**. These are mostly proteins (biologicals), secreted by the cell into the body fluids, which act as messenger substances. They usually dock onto receptors on the cell surface. **Insulin** is a classical example – a secreted protein acting as im-

Table 25.2 Target validation.

Method	Underlying principle	Organism
Association analysis	Statistical correlation between genotypes and disease-relevant phenotypes	Humans (populations)
Linkage analysis	Correlation between segregation of disease-relevant genes and disease-relevant phenotypes	Humans (families)
Somatic mutations	Correlation between mutations in diseased tissue and disease	Humans, mammals, particularly in connection with cancer
Knock-out	Destruction of gene to generate functional loss	Mice, nematodes, yeast, bacteria
Mutants	Isolation of mutants in the target gene through random mutations	All
RNA interference	Loss of function through RNA interference	All
Overexpression	Enhanced function through overexpression of a gene	All
Expression of dominant-negative alleles	Loss of function through expression of alleles that inhibit the wild-type allele (e.g., by forming inactive multimers or through competition for binding partners)	All
Distribution of expression in tissue	Measuring expression in disease-relevant tissue, either at the mRNA level or preferably at the protein level via antibodies	Humans, disease models in animals (mice or rats)
Proteinprotein interaction	Binding a protein to other clearly disease-associated proteins	All
Pharmacological modulation of protein activity	Inhibition or activation of a target through low-molecular-weight compounds or antibodies	All

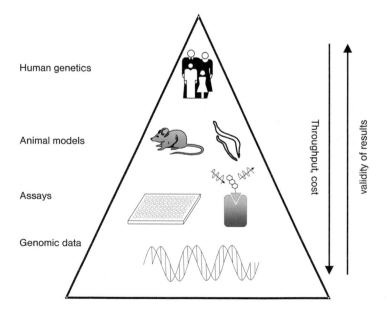

Fig. 25.3 Target validation pyramid. Genomic methods (sequence analysis, expression studies) are not very costly and can easily be automated. Purely bioinformatic methods are sometimes an option. Working with animal models or clinical trials on humans, by contrast, is very labor-intensive, but yields the most valuable information about *in vivo* conditions and the effects of individual gene products on (patho)physiological processes. Cellular and biochemical assays have their place in the middle of the pyramid.

portant messenger substance in the sugar and energy metabolism. The receptor is a cell surface protein mediating insulin activity on fat tissue, muscles, and the liver. Other examples of biologicals with therapeutic applications are **interferon-γ**, **growth hormone**, or a secreted form of the **tumor necrosis factor-α receptor** (etanercept (Enbrel®), from Amgen). **Antibodies** make also suitable active agents as they bind to their target molecules with high specificity and affinity (see Chapter 27). An antibody binding to **HER2 (human epidermal growth factor 2)** on the cell membrane of breast cancer cell is an example of an antibody used as an active agent. Overexpression of HER2 causes rampant growth in breast cancer cells. The antibody binding to HER2 inhibits its growth-promoting activity. It is produced by Genentech and sold under the name of Herceptin® (trastuzumab).

Biologicals have the disadvantage of not being able to pass cell membranes. This means they cannot be administered orally and must be injected. Unlike their conventionally synthesized chemical counterparts, biologicals can only be produced by genetic engineering (i.e., the production costs are considerably higher). They are therefore only available for the treatment of serious conditions, but not for lesser ailments.

25.2.8
DNA and RNA in New Therapeutic Approaches

It is only a matter of time until a new category alongside small organic molecules and therapeutic proteins will have proved themselves in the clinical trial stage – this approach uses **nucleic acids as active agents. Gene therapeutic methods** are currently being tested in which additional genetic material is introduced into somatic cells in order to cure hereditary gene defects or make tumor cells more easily identifiable for the immune system. The difficulty lies in developing suitable vectors that introduce the genetic material with high efficiency and cell type specificity (see Chapter 30).

There are methods in the preclinical stage that allow the **targeted degradation of mRNA**. Trials have been going on for a while, using RNA that is complementary to the transcript (**antisense RNA**) in order to prevent the translation of transcription. In another approach, **catalytic RNA molecules or ribozymes** are used to degrade transcripts. The recently discovered phenomenon of **RNA interference** also looks promising for therapeutic use. It involves introducing siRNAs to the cell. These elicit the sequence-specific degradation of transcripts (see Section 2.4 and Chapter 31). As far as vectors are concerned, the same requirements apply to these methods as described above.

25.2.9
Patent Protection for Targets

Every company has an interest in protecting its knowledge regarding the validation of a target. This is often done by patenting genes and their products. The patents usually include the composition of a gene and its protein to be used for the production of therapeutically active agents for certain indications (**composition of matter patent**). The patent can only be granted if the complete sequence of the relevant gene is not known at the time of patenting and a heretofore unknown description of its function is given. If a new disease association for a known gene is patented, a composition of matter cannot be given, as the **prerequisite of novelty** does not apply (see Chapter 35).

For biologicals where the gene product is also the therapeutic agent and for treatments based on nucleic acids (**gene therapy, RNA interference**, see Chapters 30, 31), the protective rights ensure that companies can economically exploit their research results. Where targets are to be modulated using chemical compounds or antibodies, only the use of the target is patented for the search of suitable active agents (**utility patent**). The therapeutic compound is protected by separate patents for a specific medical indication. The protection of therapeutic compounds, be they biologicals or low-molecular-weight compounds, is more important than the patent for using a target in order to develop such substances. To what extent patents on targets give their owner partial rights on the respective therapeutic compounds is unclear. Views vary between the many pharmaceutical and biotech companies. Only after a number of court cases have finally been settled can the value of these patents on targets target be assessed.

25.2.10
Compound Libraries as a Source of Drug Discovery

What use are the best targets, apart from biologicals, if you have no compounds to modulate them *in vivo*, which is why we look at compound libraries and screening procedures. In the early days of modern medicinal chemistry, the agents were mainly aromatic and aliphatic compounds that originated from the tar, coal, and dye industries. Over the last 100 years, the large pharmaceutical companies have collated extensive libraries of **synthetic compounds**. These usually contain several hundreds of thousands of compounds, individually synthesized and described by chemists. Libraries often contain natural products as well. Within the library, the compounds are often organized according to categories of molecules. The reasons for this are a historical focus on specific medical indications or on specific chemical reactions. **Sulfonamides, penicillins, steroids**, and **benzodiazepines** are good examples. The latter represent a class of chemical structures, the members of which have been involved in a wide range of actions on various target molecules (alongside γ-aminobutyric acid receptors, GPCRs, ligand-controlled ion channels, and kinases). The development of **combinatorial chemistry** (Box 25.4) has accelerated the synthesis of large compound libraries. The big pharma companies use libraries consisting of hundreds of thousands or even several millions of compounds.

Box 25.4 Combinatorial Chemistry

In combinatorial chemistry, large groups of components that share the same reactive principles are combined to produce all sorts of combinations. For example, the combination of three main components, each consisting of 20 individual units yields, in theory, 203 or 8000 different compounds.

In parallel with the widening range of chemical possibilities, new approaches to identifying targets (as described above) and new methods of producing large-scale assays have been developed. These screening procedures are explained in the following. An overview of the preclinical stages of drug development is given in Box 25.5.

Box 25.5 Preclinical Steps in Drug Development

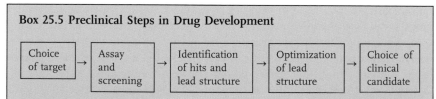

The first step comprises the choice and validation of a suitable target. Once a viable assay has been developed, compound libraries can be searched for active compounds and the hits tested in secondary assays. If the hits provide a suitable lead structure, medicinal chemists work on the optimization of pharmacological properties, developing various derivatives. When this has been successful and the target values have been met, appropriate candidates are chosen for further clinical development (see text for further details).

25.2.11
High-Throughput Screening

Historically, the pharmacological action of compounds was tested in **animal models**. In these time- and labor-intensive tests, the site of action for the active compounds had not been defined, and the results were evaluated in terms of impact of the compound on the development of the disease. Naturally, the throughput of these procedures was limited.

With the breathtaking speed of developments in molecular biology, genetics, genomics, and molecular medicine, the use of biochemical and cellular assays became increasingly popular. These have a defined molecular target gene or gene product on which the inhibiting or activating properties of compounds are tested in a **high-throughput screening** assay.

Since the late 1980s, high-throughput assays have been used that permit a systematic search of extensive compound libraries for suitable candidates. Depending on the system used, the throughput rate can exceed 400 000 individual tests per day. The past two decades have seen a marked trend towards further increases in throughput and miniaturization of assays, driven by the need to reduce the costs per data point and to use the limited resources of conventional compound libraries more efficiently. It is also a way of browsing large chemical libraries in order to test an increasing number of potentially interesting targets.

Table 25.3 Screening methods that can be accomplished in high throughput.

Type of screening	Target examples	Characteristics
Biochemical assays		
HTR-FRET (homogeneous time-resolved fluorescence resonance energy transfer)	Kinases, receptors, proteases, helicases, nuclear receptors	Fluorescence using lanthanides as fluorophores; through the long lifetime of the lanthanide emission, time-resolved measurements are possible
Fluorescence polarization	Kinases, receptors, proteases, nuclear receptors	A small fluorescent molecule will slow its tumbling motion when binding to a larger molecule; when excited by polarized light, the slower motion can be detected by changes in polarization of the emission
Alpha screening	Kinases, receptors, proteases, helicases, nuclear receptors	Luminescent proximity assay; donor bead, excited at 370 nm, generates reactive oxygen that causes light emission of 520 nm in a neighboring acceptor bead
SPA (scintillation proximity assay)	Binding assays (e.g., receptors, kinases), second messengers (e.g., cAMP)	Detection of radioisotopes in the neighborhood of a scintillator in a bead or on a plate
Filter-binding assays	Binding assays (e.g., kinases, polymerases, receptors)	A substrate is radioactively labeled (e.g., $[\gamma^{-32}P]ATP$ for a peptide); the label is held back in a filter when the substrate binds to the target
Precipitation/filtration assay	Binding assays (e.g., kinases, receptors)	Radioactive compound is bound to target and separated from unbound material through precipitation
ELISA (enzyme-linked immunosorbent assay)	Binding assays	Binding detected through antibodies
Cellular assays		
Reporter gene assay	GPCRs, nuclear receptors, transcription factors, kinases	Expression of a reporter gene (luciferase, alkaline phosphatase, β-galactosidase) as measuring parameter
Yeast two-hybrid or mammalian two-hybrid	Protein-protein interaction	Reporter expression as measuring parameter for interaction affected by compounds
High-content screening	GPCRs, kinases, proteases, transcription factors, etc.	Parallel measuring of intracellular target distribution or other cell biologically relevant marker molecules using confocal microscopy
FLIPR (fluorescent imaging plate reader)	GPCRs, ion channels, etc.	Measuring the uptake of Ca^{2+} using specific reporters (e.g., aequorin)
Phenotypic and physiological screening	Basically applicable to all targets	Measurable changes in phenotype (e.g., growth behavior of cells), similar to high-content screening

The main objective of screening for low-molecular-weight compounds is identifying structures that may be used as the basis for developing therapeutically useable compounds. Structures that feature the desired activity and can be chemically modified to produce derivatives are called **lead structures**. There are several screening methods in use, depending on the target category, medical indication, and chemical compound libraries used (Table 25.3). Two types of assays among the large variety of compound screening procedures have been chosen as examples and are depicted in Fig. 25.4.

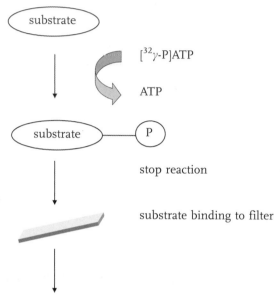

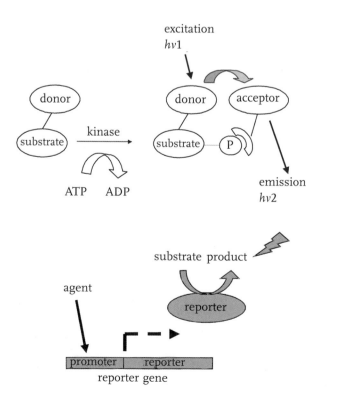

Fig. 25.4 Filter-binding and FRET assays. The top part shows a FRET assay for the indirect detection of kinase activity, while the bottom part shows a filter-binding assay for the direct detection of kinase activity.

25.2.12
High-Quality Paramounts in Screening Assays

Biochemical or cellular assays are used to identify inhibitors or activators of the analyzed target or to validate the targets of known low-molecular-weight compounds. The quality of the assay is crucial in order to distinguish between *bona fide* inhibitors (or activators) and false-positive results. The most important criteria for the evaluation of the quality of an assay, such as signal/noise ratio and signal/background ratio, enter the formula to calculate the **Z factor**:

$$Z = 1 - (3\sigma_{c+} + \sigma_{c-})/(\mu_{c+} - \mu_{c-}) \,, \tag{25.1}$$

where σ_{c+} stands for the standard deviation of the positive control in the assay, σ_{c-} stands for the standard deviation of the negative control, and $\mu_{c+} - \mu_{c-}$ stands for the mean value of positive and negative control. To be acceptable for high-throughput, assays must have a Z factor better than 0.5. Through the introduction of such quality criteria for assays it is possible to compare assay data that have been collected over a longer period of time (**stability of an assay**) or in different laboratories.

The choice of a suitable assay depends on many factors. **Cellular assays** often work well for receptors because a direct **binding assay** usually cannot distinguish between agonists and antagonists. Cellular assays can also identify activity-dependent modulators (e.g., for ion channels). **Biochemical assays**, however, have advantages when it comes to intracellular targets, often yielding a wider range of active chemical structures. The substances do not require a high potency or cell permeability and can often be tested at higher concentration levels. **Binding assays** provide detailed data for the chemical optimization of individual parameters such as binding affinities, which is particularly appreciated by medical chemists. When choosing a type of assay, the first consideration must be the nature of the target and its biological function, the available amount of target protein, and the possible substrates for an enzyme. Biochemical assays are either **separation assays** where the reaction product is measured after its separation from the starting material or **homogenous assays** that do not require a separation step. The **fluorescence resonance energy transfer (FRET)** assay (Fig. 25.4) is widely used in biochemical as well as cellular types of assays. In FRET-based assays, a fluorescent molecule (**donor fluorophore**) is excited by a certain wavelength, while a neighboring molecule acts as an acceptor fluorophore, picking up the emission of the excited donor fluorophore and, in turn, emitting at a different wavelength. The efficiency of the energy transfer is essentially determined by the distance between the two fluorophores. The protease assay is a simple example. A target peptide that is labeled by the acceptor and donor fluorophores, respectively, at the amino- and carboxy-terminals is used as an artificial substrate for the protease to be analyzed. The assay is able to identify compounds that have a positive (activators) or negative (inhibitors) effect on the action of the protease on the peptide substrate.

25.2.13
Virtual Ligand Screening

Virtual structure-based screening uses data of the **3D crystal structure** of target molecules (kinases, proteases, nuclear receptors, etc.) and their agonists or antagonists. The screening is carried out using a number of chemoinformatic algorithms. These include the **docking** of ligand–receptor association and the evaluation of the docked structures using scoring functions. The results of successful virtual screening have been published and the future will show if high-performance virtual screening methods can be developed that make them a viable alternative to more conventional methods.

In another type of **virtual screening**, the similarities between compound libraries are compared using algorithms. This requires the translation of the (physico)chemical properties and the spatial conformation of a compound into processable binary information. Known compounds with the desired properties are the starting point for the search for molecules in compound libraries that are expected to exhibit similar behavior. This computerized preselection process narrows down the number of candidates to undergo screening.

25.2.14
Activity of Drugs Described in Terms of Efficacy and Potency

The assays described above are used in a **primary screening** to identify compounds with the desired properties, called hits. Even very good assays yield a whole host of false-positive results. When screening 200 000 compounds with a proportion 0.5% of false-positive results, this means that 1000 hits will turn out to be useless in follow-up experiments. In order to select the genuine candidates, **a secondary screening** is carried out. It will be an advantage to use an independent assay system for the second round that generates a different type of readout.

Another important criterion for the further characterization of compounds lies in the dose–response relationship. The resulting **dose–response curve** of the substance is an indicator of the potency of a compound, measured as the concentration at which a concentration reaches half its maximum effect (also known as **effective concentration 50% (EC_{50})**; Fig. 25.5). Another way of describing a hit compound is the maximum strength of the desired effect upon compound binding. This permits a distinction between full and partial agonists and antagonists.

25.2.15
Chemical Optimization of Lead Structures

Having analyzed the secondary screening results, it is now possible to look at the chemical structure of the relevant molecules and select the lead structures for further development. The **definition of a lead structure** is defined by pharmaceutical or biotech companies according to their particular requirements. In most cases, the dose–response relationship of several active compounds sharing a basic chemical structure is the starting point. Medicinal chemists are also involved in the definition process of a lead structure, drawing from their wealth of experience in the **targeted modification** of chemical structures, and looking at the pharmacological, toxicological, and kinetic behavior of molecules. These factors limit the number of structures which can be selected for optimization. In the optimizing process of lead structures, chemical variations of the chosen structure are synthesized in iterative cycles and undergo testing in relevant biological and pharmacological assay systems. Most pharmaceutical companies use **multiparameter optimization**, running several tests (dose–response relationships, selectivity, solubility, cell permeation, toxicology, pharmacokinetics, in silico prediction) simultaneously in order to minimize delays and failure in the later development stages.

25.3
Preclinical Pharmacology and Toxicology

Before a new pharmaceutical compound can be studied in humans, it must undergo a whole series of pharmacological and toxicological tests. The efficacy and safety of the new compound must be experimentally proven. The data obtained must be carefully documented as part of the approval procedure, bearing in

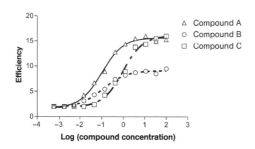

Fig. 25.5 Potency and efficiency. Compounds A and B are similar in potency, but the efficacy of compound B is lower. Compound C has the same efficacy as A, but is less potent.

mind the criteria of the regulatory agencies. **Preclinical trials** are intended to minimize the health risk for the participants of subsequent clinical studies. Pharmacological studies look at the pharmacodynamic and pharmacokinetic properties of a compound. The pharmacodynamic properties are defined as the effect a compound has on an organism. These are investigated by analyzing the dose–response and the structure–activity relationships. The **pharmacokinetic properties** are defined as changes in the concentration of the compound in the organism over a period of time. They depend on absorption, distribution, metabolization, and elimination in the organism. Toxicological studies look at the **side effects** of a medication. Pharmacological and toxicological studies are often referred to under the umbrella term "**ADME-T**" (absorption, distribution, metabolization, excretion, and toxicity).

As toxicological and pharmacological studies mostly investigate systemic effects, animal experiments are largely indispensable. Apart from pharmacological studies, it must be shown that the new compound can be produced to a high grade of purity as well as consistently high levels of **quality and stability**, which involves a whole host of chemical-analytical studies. Closely related to its pharmacokinetic properties is the pharmaceutical form of a compound, which is the subject of **galenics**. Many projects are terminated because an effective compound cannot be dissolved in a biocompatible way and no acceptable form of application can be found. Frequent injections or infusions, for example, are only acceptable in fairly serious medical conditions.

What looks here like a simple list of various tests is more complicated in real life because, usually, we are not talking about a single compound run through a preclinical testing program. In parallel, the stages of the further chemical optimization of candidate molecules in respect of complex biological and chemical parameters are is carried out. For example, the objective may be to enhance the bioavailability of a compound while retaining its good side-effect profile. If this requires several optimization cycles as well as testing in animal models, the preclinical development phase is likely to take years.

In order to avoid unpleasant surprises at a later stage of the development, there is a trend towards **early characterization** of pharmacological and toxicological properties in simple cell-based systems. This narrows animal model experiments down to the few most promising candidates. Tests on intestinal absorption, for example, are carried out using Caco-2 cells – a colon carcinoma cell line.

At the same time, bioinformatic and chemoinformatic methods are used to refine in silico prediction of pharmacotoxicological parameters. Often, data calculated in silico and experimental data are combined to make a prediction, as was the case in **Lipinski's famous "rule of five"** (1997), which sums up necessary characteristics of molecules concerning their bioavailability. According to this rule, a compound is likely to have poor absorption or permeation properties if more than one of the following criteria applies:

- The molecule contains more than five hydrogen bond donors.
- The molecular mass exceeds 500 Da.
- The $\log P$ value (octanol/water distribution) is larger than 5.
- The molecule carries more than 10 bound acceptors such as N or O.
- Compound classes that are substrates for biological transporters or endogenous compounds are exceptions to the rule.

The end of the preclinical phase is marked by the application for **investigational new drug (IND) status**.

Table 25.4 Overview of preclinical and clinical drug development.

	Preclinical	Clinical trials			Approval	
		Phase I	Phase II	Phase III	FDA/EMA	Phase IV
Years	3.5	1	2	3	2.5	
Tested on	Cells and laboratory animals	20–80 healthy volunteers	100–300 patients	1000–10000 patients	Approval procedure	
Objective	Safety and biological activity	Safety and dosage	Efficacy and common side effects	Efficacy, less-common side effects, long-term effects		Additional studies as specified by regulatory authority
Agents	5000 compounds	5 compounds (initially)		1 compound		

25.4
Clinical Development

The development from a candidate compound to an approved drug goes through a number of stages and takes on average 12 years. The longest and most expensive by far of these stages is the clinical phase. Table 25.4 gives an overview of the procedures involved.

25.5
Clinical Testing

Once a compound has successfully undergone preclinical trials, evidence of its **safety and efficacy** in humans must be obtained. Trials in humans must follow **Good Clinical Practice (GCP)** guidelines and must be authorized by an **ethics commission**. Clinical testing is carried out in four stages. In **phase I**, the medication is given to a small number of healthy volunteers. The dosage calculated on the basis of animal experiments is verified, pharmacokinetic data are collected, and side effects are monitored. As far as this is possible in healthy test persons, the desired effect (e.g., lowering of blood pressure) is also monitored. In **phase II**, the compound is given to a small group of patients for the first time in order to test its efficacy and innocuousness. Whether it is possible to conduct a classical placebo-controlled double-blind trial depends on the nature of the condition to be treated. For many diseases, treatments are available and the trials are intended to improve the existing treatments. It would be unethical in these cases to replace the treatment by a placebo, so comparative studies with an established therapeutic agent are carried out instead. Again, the documentation of possible side effects must be meticulous at this stage. **Side effects** and therapeutic use must be carefully weighed before a decision is taken to enter the next stage. **Phase III** is the **proper field trial** on several thousand patients where the efficacy of the compound in the majority of participants should be confirmed. The probability of picking up rarer side effects is also greater, due to the large number of participants. Phase III requires a large amount of logistics because patients in several centers are involved in order to ensure the comparability of random samples. All available data collected from preclinical trials up to and including phase III are submitted to the approval agency. The documentation submitted with an application for approval is between 40 000 and 100 000 pages long (Box 25.6).

Box 25.6 Regulatory Authorities

There are a number of regulatory authorities with different responsibilities (see Chapter 35). There are national authorities, such as the German Bundesinstitut für Arzneimittel und Medizinprodukte (BfArM; http://www.bfarm.de) in Bonn for the approval of drugs and the Paul-Ehrlich-Institut in Langen for the approval of vaccines and biologicals. In Switzerland, this is the task of Swissmedic in Bern and in Austria of the Bundesinstitut fuer Arzneimittel in Vienna. At the European level, there is the European Medicines Agency (EMA; http://ema.europa.eu) where applications for Europe-wide licenses can be submitted. The validity of national licenses can be extended to other member states of the European Union in a mutual-recognition procedure. Applications for approval in the United States are submitted to the Food and Drug Administration (FDA; http://www.fda.gov). Given the extensive documentation involved in each approval application, there is an interest in mutual recognition of approval and the reuse of the application documentation. The International Conference on Harmonization of Technical Requirements for the Registration of Pharmaceuticals for Human Use (ICH) was founded where representatives of the regulatory authorities in the United States, Japan, and Europe work in conjunction with the pharmaceutical industry on the harmonization of national approval criteria.

After the drugs have been approved, they are still subject to **pharmacovigilance** (i.e., experiences continue to be systematically collected) as rare **side effects** and **interaction** with other medication are not usually picked up during phase III studies. Furthermore, after the approval of a drug, targeted **phase IV** studies are carried out. The longer the development to market readiness takes, the less time there is for a patent holder to recoup and make a profit before competition from generics sets in. This is why recent years have seen a trend towards carrying out studies on approved drugs in order to document their effectiveness for new indications. If this is successful, an existing patent protection may be extended.

26
Drug Targeting and Prodrugs

Learning Objectives

For a highly effective drug to be truly successful, it is essential for it to be able to selectively target diseased cells while leaving healthy cells unharmed. This chapter describes the strategies used in drug targeting. We distinguish between passive targeting, physical targeting, active targeting, and targeting by the use of cellular carriers.

26.1 Drug Targeting

The concept of drug targeting is based on the observation that many drugs do not show any **selectivity** as to the place of their absorption or the place where they exercise their effect. Instead, they are distributed in an **untargeted way** through the body. As a result, some of their effects can be undesired or even toxic. The challenge of delivering a drug exclusively to the place where a disease process is occurring and achieving a selective pharmacological effect is considered one of the greatest problems facing modern drug therapy. This thought is one that has been pursued for some time, going back to Ehrlich's idea of the "magic bullet" as a treatment strategy in the early twentieth century. The successful implementation of this approach could be of particular value in the treatment of chronic diseases, where drugs are administered frequently or at high dose (e.g., the use of cytostatic drugs in cancer therapy).

The problem of drug targeting and the targeted release of active substances from a particular pharmaceutical form is relevant even for simple forms like tablets and capsules; many active substances are sensitive to the acid environment of the stomach or to enzymes in the upper digestive tract and simply cannot be swallowed. Special methods have to be developed to ensure that the drug is only released after it has passed through the stomach or when it reaches the distal parts of the intestines. To do this, tablets can be coated with **polymers** that dissolve in the intestine in response to pH (e.g., Eudragit® or cellulose acetate phthalate (CAP)) or with coatings that are not dissolved until they are degraded by special enzymes produced by the microbial flora of the large intestine. Technically, this type of targeted delivery is fairly easy to engineer using tank or fluidized bed coating equipment in which the coating material is applied to the tablets in dissolved form. It is much more difficult to deliver a drug to a particular organ or cell population in the body without letting it come in contact with other regions of the body. However, there are four possible ways that the problem can be approached:

- **Passive targeting** involving no modification of the active substance or drug carrier, but making use of special physiological characteristics of the target tissue.
- **Physical targeting**, based on nonphysiological pH values or temperatures in the target tissue, as well as magnetic targeting, in which drugs can be delivered to their place of action by paramagnetic carriers under the influence of an external magnetic field.

An Introduction to Molecular Biotechnology, 2nd Edition.
Edited by Michael Wink
Copyright © 2011 WILEY-VCH Verlag GmbH & Co. KGaA, Weinheim
ISBN: 978-3-527-32637-2

- **Active targeting** by means of modified active substances or carriers, to which target-seeking **vectors** are attached;
- **Targeting using cellular carriers**.

The various options are presented below by means of examples.

26.1.1
Passive Targeting by Exploiting Special Physiological Properties of the Target Tissue

Passive targeting is based on the fact that under certain conditions (e.g., in hypoxic regions following a heart attack or in rapidly proliferating solid tumors) blood vessels are more permeable than in healthy tissue. In such a **fenestrated endothelium**, drug carriers with a size of 10 500 nm (e.g., liposomes or nanoparticles) can permeate through the porous vessel wall and accumulate in the interstitial space. This effect is sometimes known as the **enhanced permeability and retention (EPR) effect** (Fig. 26.1). In order to be successful this type of drug targeting requires the chosen drug carrier to have a sufficient half-life. The longer a particle remains in the circulation, the more likely it is to enter the target tissue. The half-life of particle-based drug carrier systems in the circulation can be considerably reduced by the **reticuloendothelial system (RES)**, which includes circulating macrophages in the blood and cells in the liver and spleen. The uptake of drug carriers into RES cells is induced by the binding of serum proteins such as complement factors or antibodies (**opsonization**). There are various means of preventing excessive opsonization and these lead to an increase in half-life. Small particles up to a size of about 150 nm are less vulnerable to undesired interaction with the RES than larger ones. The incorporation of **polyethylene glycol (PEG)** chains on the exterior of the particles (e.g., Stealth liposomes) increases the hydrodynamic radius and has the effect of steric stabilization, resulting in reduced recognition and uptake by the RES. This concept is already being used for cytostatic-bearing Stealth liposomes in clinical studies and appears promising.

26.1.2
Physical Targeting

Temperature- and pH-sensitive liposomes can lead to increased accumulation of cytostatic drugs in tumor tissue. The use of these liposomes is based on the observation that neoplastic tissue can exhibit a lower pH value or a higher temperature (hyperthermia) than healthy tissue. It should therefore be the case that drug carriers that react to such stimuli will only release their contents in that type of environment (and therefore in proximity to a tumor), even if they are evenly distributed in the circulation (Fig. 26.2).

Another example of physical drug targeting is **magnetically controlled targeting**, which is also undergoing clinical trials. In this case, active substances such as cytostatic drugs are reversibly incorporated in magnetizable particles with a

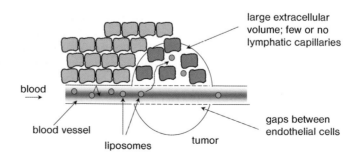

Fig. 26.1 EPR effect. Drug carriers permeate through the pathologically changed epithelium of a blood vessel and their size causes them to accumulate in the interstitial space.

region with changed pH value or
unphysiological temperature

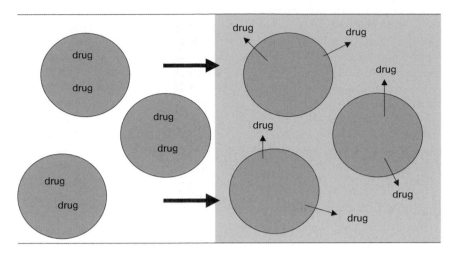

Fig. 26.2 Schematic diagram to illustrate physical targeting. The active substance or drug carrier is activated or released in the target region by a stimulus (pH, temperature, magnetic field, or ultrasound).

size of 50–500 nm. These can be held in the tumor tissue and made to accumulate there by applying an external magnetic field. A modified procedure, magnetic fluid hyperthermia, attempts to raise the temperature of tumor cells in a targeted way by magnetizing and demagnetizing them, so destroying them thermally.

26.1.3
Active Targeting

Significantly greater precision is possible with active targeting than with **passive targeting**. This arises from the fact that each cell type has its own characteristic properties that distinguish it from others, meaning that these properties can be utilized for **active cell recognition**. If these characteristics are located on the cell surface (e.g., adhesion proteins, receptors, or membrane transport systems), it is possible to use **ligands** or **antibodies** to these proteins, either for therapeutic targeting, or as a means of targeting drug delivery by coupling them to the active substances or drug carriers. If the potential target structures are in the cell interior, an active substance can be manipulated using a **prodrug strategy** in which the drug is only activated once it has been taken up into the interior of the cell (e.g., antiviral drugs such as acyclovir (Zovirax®), which is activated by viral protein kinases inside a virus-infected cell). The **cell surface recognition strategy** is somewhat limited by the variability of the target cells, especially when targeting tumor cells or viruses, when external factors can quite quickly lead to modifications of the surface structures.

Although the idea of **active targeting** is not new, a promising approach for clinical application has only come with the knowledge and experimental methods of modern molecular biology. **Recombinant methods** of production (e.g., of antibodies) have considerably extended the possibilities for applying this strategy, as the following examples indicate.

Each cell has proteins in its lipid membrane and these are expressed differently according to cell type. The **pattern of proteins** on the surface of a degenerate cell is often markedly different from that on healthy cells. **Suitable antibodies** can be used to target these surface epitopes selectively and this is why antibodies are used in cancer therapy, such as in the treatment of chronic lymphatic leukemia, even when they are not coupled to any other active substance. This disease is incurable by conventional means. Unlike erythrocytes and thrombocytes, almost all B and T lymphocytes, monocytes, thymocytes, and macrophages in the peripheral blood express the surface antigen CD52 at high density.

If antibodies (e.g., alemtuzumab (MabCampath®)) bind to this surface epitope, a complement is bound, and **antibody-dependent cytotoxicity** and **lymphocyte lysis** takes place. All the information available to date suggests that hematopoietic stem cells and precursor cells are not damaged. Today, antibodies for this sort of use can be produced by genetic technology. An example is that of **humanized monoclonal antibodies**, where certain regions of antibodies from the rat or mouse are incorporated into human immunoglobulins. Similarly, a monoclonal antibody (trastuzumab (Herceptin®)) is used in the treatment of **advanced breast carcinomas**. This antibody is used against cancer cells that overexpress the **HER2 protein** on the cell surface. The administration of the antibody in combination with chemotherapy has produced significant increases in survival times. Table 26.1 lists some antibodies that are currently either being used therapeutically in clinical trials or have already been introduced onto the market.

Table 26.1 Example antibodies in therapeutic use.

Antibody (drug name)	Antigen	Disease	Response rate	References
Trastuzumab (Herceptin)	Her-2/*neu*	Metastasizing breast cancer	1/43 CR 4/43 PR 2/43 SR 14/43 SD	Baselga *et al.*, 1999
MKC-454	Her-2/*neu*	Metastasizing breast cancer	2/18 OR	Tokuda *et al.*, 1999
Rituximab (Rituxan)	CD20	Non-Hodgkin's lymphoma	21/39 OR 14/39 SD or SR 15/26 CR	Hainsworth *et al.*, 2000
		Lymphoproliferative diseases after organ transplantation	2/26 PR	Milpied *et al.*, 2000
		Lymphoproliferative disease following bone marrow transplantation	5/6 CR	Milpied *et al.*, 2000
Infliximab (Remicade)	Tumor necrosis factor-*a*	Rheumatoid arthritis	428 patients, 20–50% improvement	Maini *et al.*, 1999
Pavilizumab (Synagis)	Respiratory syncytial virus F-glycoprotein	Respiratory syncytial virus infections in children	35 children <2 years; reduction in tracheal respiratory syncytial virus concentration, but not in nasal aspirate; no improvement in clinical picture compared to placebo	Malley *et al.*, 1998
Abciximab (ReoPro)	Platelet glycoprotein IIb/IIIa receptor	Ischemic complications during balloon angioplasty or atherectomy	Inhibition of platelet aggregation showed clinically relevant improvement after coronary intervention compared to placebo	Lincoff *et al.*, 1999
rhuMab HER2	Her-2/*neu*	Advanced breast cancer	9/37 PR	Pegram *et al.*, 1998
rhuMab VEGF	Vascular endothelial growth factor	Metastasizing renal cell carcinoma	9/37 SRSD 19/37 PD	http://www.clinicaltrials.gov
Humanized OKT3 (hOKT3$_{\lambda4}$)	CD3	Various malignant diseases	2/24 positive response	Richards *et al.*, 1999
Adalimumab (Humira)	Tumor necrosis factor-*a*	Rheumatoid arthritis, Psoriasis, arthritis, Bechterew's disease (ankylosing spondylitis), Crohn's disease	– – –	Scheinfeld, 2003
Tocilizumab (Actemra/RoActemra)	Interleukin 6-receptor	Rheumatoid arthritis	– – –	Paul-Pletzer, 2006
Bevacizumab (Avastin)	Vascular endothelial growth factor	Colon cancer, breast cancer, nonsmall cell bronchial carcinoma	– – –	Sirohi and Smith, 2008

rhuMab, recombinant humanized monoclonal antibody; CR, complete remission; PR, partial remission; SR, slight response; OR, objective response; SD, stable disease; PD, progressive disease.

In addition to the direct therapeutic application of antibodies, targeting structures can also be linked to an otherwise unselective drug. However, various factors may render this more difficult:

- **Lack of stability.** Binding between the vector and the active substance needs to remain stable for as long as the coupling product is still in the circulation. It should be cleaved to release the actual drug only after binding to or uptake into the target cell.
- **Lack of coupling efficiency.** A vector needs to be constituted in such a way that enough effector molecules (more than just one if possible) can be coupled to it.
- **Tolerability.** The coupling product must be well tolerated and as far as possible immunologically inert.

Various structures can be considered as possible vector molecules for drugs: antibodies or antibody fragments, lectins, saccharides, lipoproteins, or low-molecular-weight organic substances that are substrates for transporter proteins or enzymes (e.g., folate). In the simplest instance they can be coupled directly to a drug. The production of **immunotoxins**, consisting of a fragment of a natural toxin (e.g., a ricin A-chain or abrin A-chain) and an antibody, is a familiar example. The antibody serves to target a surface epitope of a cell, and the toxin enters the cell and produces its toxic effect by irreversibly blocking a metabolic pathway. Such immunotoxins are being tested for the treatment of leukemias, metastasizing melanomas, and colorectal carcinomas.

Another example of active targeting by **antibody coupling** is the production of a **chimeric peptide** that consists of a biotinylated vasoactive peptide (vasoactive intestinal peptide (VIP)) and an avidin-coupled antibody to the transferrin receptor. This coupling product can be used in targeting the bloodbrain barrier, where these receptors are expressed at high density. It was possible to demonstrate in animal testing that transcytosis of the active components via the transferrin receptor was brought about through the pharmacological effect.

The use of **bispecific antibodies** follows a similar principle (see Chapter 28). Antibodies are molecules with a symmetrical structure containing two identical binding sites able to recognize the same antigen. They are termed **bivalent**. Bispecific antibodies, in contrast, are asymmetrical in structure, and contain two different binding sites. As a result they are able to recognize both their target surface (e.g., that of a tumor cell) and an active substance (e.g., a cytostatic drug). Functionally they are monovalent. The production of this type of antibody can be done in several ways – by the chemical reassociation of monovalent fragments, by the heterogeneous aggregation of different monoclonal antibodies, or by biosynthetic production in **hybridoma cells**. This method involves codominant production of light and heavy fragments, by which up to 50% of the immunoglobulins formed represent the desired bispecific monoclonal antibody. Possible clinical applications for these bispecific antibodies are tumor imaging and therapy, and directed immunosuppression by the simultaneous recognition of immunosuppressive drugs and T lymphocytes.

The **low coupling efficiency**, already referred to above, is a great disadvantage of the systems mentioned here. Only a single effector molecule (or a small number) is coupled to a vector and so the number of effector molecules binding to a target cell is relatively low. It is therefore helpful to link supramolecular drug carriers with vectors. They need to be small enough to be administered intravenously and also fulfill the requirements given above for linking. Micelles, liposomes, and nanoparticles are potential vectors.

Liposome carrier systems have so far mainly been used in **tumor therapy**. The efficiency of chemotherapy for the treatment of tumors is greatly limited by the **low therapeutic index** of most cytostatic drugs. The causes – short half-life, lack of tumor selectivity, and associated side effects of the drugs – have given rise to the intensive search for drug carriers capable of avoiding these problems. Lipo-

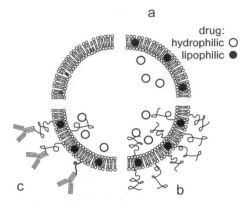

a

drug:
hydrophilic ○
lipophilic ●

c

b

Fig. 26.3 Structure of liposomes that can be used for drug targeting. (a) Conventional liposomes. (b) Stealth liposomes, in which polyethylene glycol chains are bound to the surface. This modification makes recognition by the RES more difficult. (c) Immunoliposomes in which targeting molecules (e.g., an antibody) are coupled to the polyethylene glycol chain. Use of these liposomes makes possible the active targeting of cell surface structures.

somes represent one possible means of delivering cytostatic drugs to their place of action in a targeted way. Tumor therapy is therefore the most important indication for active targeting with liposome formulations. One possibility for active liposome targeting is the encapsulation of a drug in **immunoliposomes**. These are liposomes with antibodies or antibody fragments bound to their surface to enable the liposomes to bind specifically to an antigen on tumor cells. For this to be successful the target must be easily attainable, so it can be useful not to target tumor cells directly, but instead to target a feature on tumor endothelial cells, such as the **KDR (kinase insert domain-containing receptor)** receptor. Tumors with a diameter greater than 1 mm secrete messenger substances to induce the formation of **new blood vessels**. If the formation of new blood vessels in the tumor can be prevented, or if these blood vessels can be destroyed, it would be possible to prevent further growth (**principle of antiangiogenesis**). Endothelial cells have receptors for these messenger substances on their surfaces and one of the receptors that endothelial cells express to an increased degree in the growing vascular bed is the KDR receptor. It can be directly blocked with the aid of the immunoliposomes. Other concepts pursue the coupling of antibodies against tumor cell surface antigens to Stealth liposomes loaded with adriamycin or daunomycin. Immunoliposomes of this type (Fig. 26.3) have been used successfully in animal trials against various metastasizing carcinomas.

A strategy for targeted drug delivery similar to that used in tumor therapy is currently being investigated to improve drug delivery to the central nervous system (CNS). In animal studies, much improved transfer of the cytostatic drug daunomycin into the brain was observed following the administration of daunomycin-loaded Stealth liposomes that had been coupled to antibodies against the transferrin receptor of brain endothelial cells. About 30 vector molecules were coupled to a liposome and over 30 000 molecules of the drug incorporated. Consequently, the transfer efficiency of the carrier exceeded several times that of a direct drug/vector construct. The high expression of low-density lipoprotein (LDL) and LDL-like receptors at the bloodbrain barrier makes the coupling of such apoproteins or fragments thereof interesting. For example, coupling of apolipoprotein E fragments (20 amino acids long) to liposomes yielded an increased uptake of these colloidal carriers into cerebral microvessels.

Apart from liposomes, **polymer nanoparticles** are of particular interest as carriers for drug targeting. Here, the type or polymer and the particle size can be used to direct the release and breakdown behavior. In this connection it has been shown that **poly(butylcyanoacrylate) nanoparticles** coated with polysorbate 80 have a 20-fold greater uptake into capillary endothelial cells of the bloodbrain barrier than uncoated particles. It is assumed that a endocytotic process is involved, possibly mediated by LDL receptors. By loading with fluorescent dyes it could be demonstrated that such nanoparticles indeed cross the bloodbrain barrier *in vivo*.

26.1.4
Cellular Carrier Systems

Cell envelopes or whole cells can also be loaded with drugs and used for drug targeting. Both bacteria and eukaryotic cells can be used in this way as carrier systems. These systems have the potential disadvantages of poor permeability of the cellular carriers through epithelial and endothelial barriers and immune reactions. So far cellular drug carriers have mainly been tested in animal studies on cancer therapy. An example of the systems tested is to take CD8-positive T cells that recognize a leukemia cell line and transfect them with a retroviral vector coding for a diphtheria toxin/interleukin-4 fusion protein. The intravenous injection of these transfected cells led to inhibition of tumor cell growth, and the hepatic and renal side effects were markedly reduced as compared with the side effects of the free toxin. In another approach, chimeric adhesion molecules were successfully incorporated into lymphocytes to achieve an antiangiogenic effect in the tumor endothelium.

26.2
Prodrugs

Combinatorial chemistry, **high-throughput screening**, and structure-based design are leading to the emergence of ever more specific drug molecules; however, such novel structures often have undesirable physicochemical, biopharmaceutical, or pharmacokinetic properties and a recourse to **prodrug strategies** becomes advisable. **Prodrugs** are inactive derivatives of drug molecules or carrier systems that undergo chemical or enzymatic **biotransformation** in the body to form the active parent substance or activated carrier system (Fig. 26.4). Prodrugs are used for the following:

- To improve the solubility of the drug in aqueous media.
- To increase chemical or enzymatic stability.
- To improve the ability of a drug to penetrate through biological membranes.
- To extend the duration of effect of a drug.
- To improve the targeted release of the drug.
- To minimize side effects.

Some examples to illustrate the various aims are described in the following sections.

26.2.1
Prodrugs to Improve Drug Solubility

Many modern drugs are poorly soluble in water, which means that they need to be treated with special solubilizing agents and are difficult to administer intravenously. **Phenytoin**, for example, which is used as an anticonvulsive drug, is mainly given in tablet form, since the poor solubility of phenytoin, of only about 0.08 mM in aqueous solvents, makes intravenous use difficult. The solubility in water of the prodrug fosphenytoin (Cerebyx®), on the other hand, is about 350 mM, making it suitable for intravenous use (Fig. 26.5).

L-dopa, which is used to treat Parkinson's disease, can be given in the form of water-soluble alkylester prodrugs, which are absorbed following nasal application with a bioavailability of about 90%.

26.2.2
Prodrugs to Increase Stability

Drugs for oral administration in particular have to fulfill higher demands of **drug stability**. Potential prodrugs, too, such as esters are sometimes cleaved so quickly that the prodrug principle breaks down. One way of countering this undesired effect can be to select a suitable salt. The choice of the right salt and the

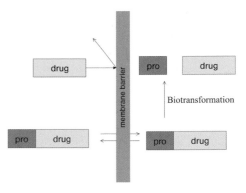

Fig. 26.4 Prodrug principle. The free drug cannot cross a membrane barrier. The prodrug is able to pass through the membrane and then undergoes a metabolic activation that releases the drug.

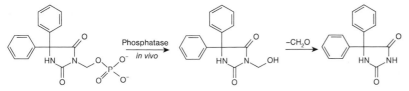

Fosphenytoin
solubility in water: 350 mM
produg suitable for i.v. administration

Phenytoin
solubility in water: 0.08 mM
poor suitability for i.v. administration
anticonvulsive drug

Fig. 26.5 Phenytoin and fosphenytoin. Fosphenytoin is around 40 times more water-soluble, which makes possible intravenous (i.v.) administration at relatively high doses.

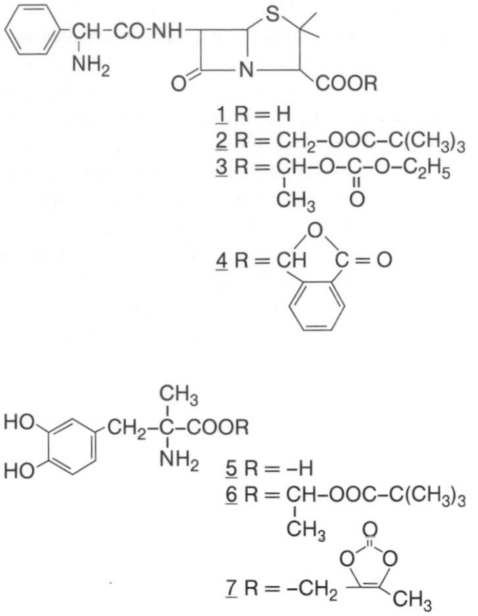

Fig. 26.6 Prodrugs of ampicillin. The functional acid group makes it difficult for the molecule to permeate through the membrane. It is masked by transformation it to an ester.

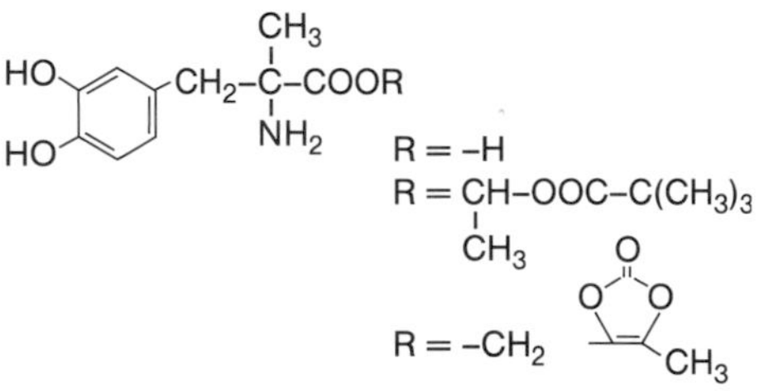

Fig. 26.7 Pivaloyloxyethyl ester of methyldopa. Esterification greatly improves the bioavailability of the drug.

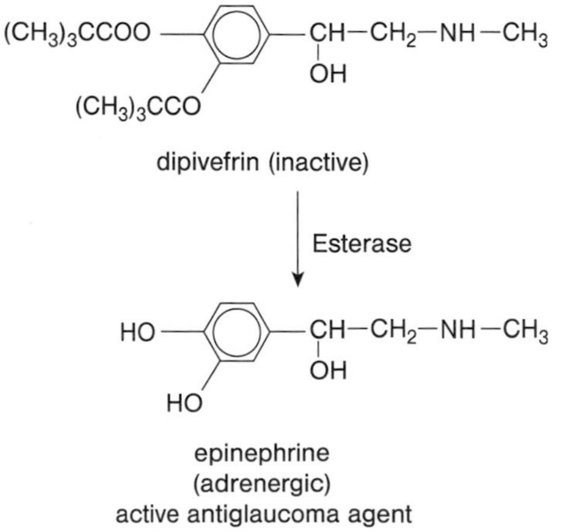

dipivefrin (inactive)

↓ Esterase

epinephrine
(adrenergic)
active antiglaucoma agent

Fig. 26.8 Dipivefrin, a dipivalyl ester of epinephrine, is used to treat glaucoma and administered by corneal application. The higher octanol/water distribution coefficient means that the dipivalyl ester crosses the cornea about 17 times better than the parent substance.

associated change in pH of the immediate environment of the drug can often bring about an improvement in stability, as was shown, for example, for glycoprotein IIb/IIIa receptor antagonists for oral use.

26.3
Penetration of Drugs through Biological Membranes

It is difficult for strongly polar or charged molecules to cross lipid membranes by passive diffusion (see Chapter 3). The preferred means of administering such compounds is therefore in the form of prodrugs, in which the charge is masked by **chemical modification** (Fig. 26.6). The transformation of the inactive prodrug into the active parent substance can be achieved by a variety of quite different chemical reactions. One of the most frequently used strategies is to transform the **inactive ester** into the active form. This type of transformation is an option because **esterases** can be found almost everywhere in the body. Drugs that have been used in the form of ester prodrugs include **acetylsalicylic acid**, **indomethacin**, and **β-lactam antibiotics**, which contain a carboxyl group in their active form, or **salicylic acid**, **chloramphenicol**, and **acyclovir**, which contain a hydroxyl group in their active form. One of the ways in which this principle has been used with success is to improve the bioavailability of ampicillin following oral administration. This led to the introduction onto the market of several **ester prodrugs** for this active substance: pivampicillin, bacampicillin, and talampicillin (Fig. 26.6). The bioavailability of the active substance following oral administration of the prodrug is around 6080% higher than after administration of the parent substance.

As well as simple prodrugs, it is possible to make **double prodrugs** (e.g., double esters such as acyloxyalkyl esters). These double esters are useful when simple ester prodrugs are not sufficiently reactive. **Methyldopa** is an example of this. The bioavailability of this drug following oral administration is found to vary considerably. Its pivaloyloxyethyl ester (Fig. 26.7), on the other hand, is almost completely absorbed from the gastrointestinal tract following oral administration.

A diester for corneal application is the dipivalyl ester of **epinephrine** (dipivefrin), which is used to treat **glaucoma** (Fig. 26.8). This diester has a higher octanol/water distribution coefficient and is some 17 times more efficient in crossing the cornea than the parent substance.

The use of redox systems to overcome the bloodbrain barrier is an interesting combination of prodrug strategy and a chemically directed targeting system. The **bloodbrain barrier** protects the CNS from xenobiotics and ensures the necessary ion homeostasis for brain function. It is permeable to lipophilic substances in both directions, but is to all practical purposes not crossed by charged or polar substances unless active carrier mechanisms are involved in their transport. The diffusion of a lipophilic substance back from the CNS into the blood can be prevented by coupling the drug to a carrier substance, which is sequentially changed by enzyme action, first into a charged form and then cleaved off. This can be carried out in a variety of ways, such as by coupling of (acyloxy)alkyl phosphonates that, once they have crossed the bloodbrain barrier, are first hydrolyzed, then changed into an anionic molecule, and finally cleaved. Another possibility is to use 1,4-dihydrotrigonellin derivatives, which are transformed into a hydrophilic quaternary molecular form after crossing the bloodbrain barrier, dependent on an NADH/NAD$^+$ redox system (Fig. 26.9). Following hydrolytic cleavage it is assumed that the vector molecule *N*-methyl-nicotinic acid is eliminated fairly quickly from the brain by active transport and the transported drug accumulates in the CNS. The efficiency of the **dihydrotrigonellin ↔ trigonellin vector** system has been demonstrated for a number of drugs, including steroid hor-

administration

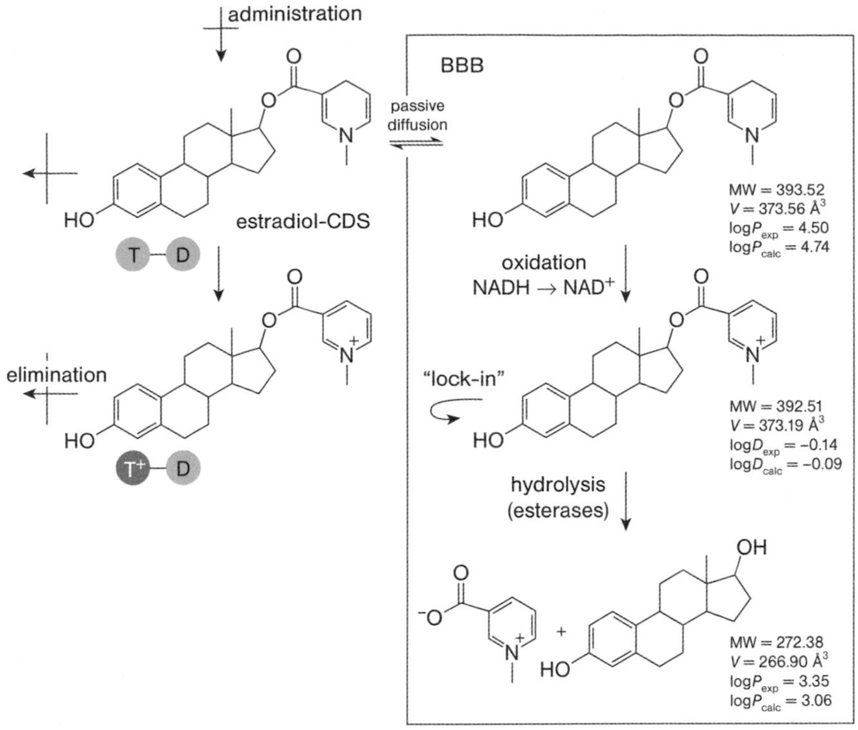

BBB

passive
diffusion

estradiol-CDS

T – D

MW = 393.52
$V = 373.56$ Å^3
$\log P_{exp} = 4.50$
$\log P_{calc} = 4.74$

oxidation
NADH → NAD$^+$

elimination

"lock-in"

T$^+$ – D

MW = 392.51
$V = 373.19$ Å^3
$\log D_{exp} = -0.14$
$\log D_{calc} = -0.09$

hydrolysis
(esterases)

OH

MW = 272.38
$V = 266.90$ Å^3
$\log P_{exp} = 3.35$
$\log P_{calc} = 3.06$

Fig. 26.9 Targeting of the CNS using a redox-based prodrug system. The drugs are coupled to 1,4-dihydrotrigonellin, which is converted into a hydrophilic quaternary form after passing through the bloodbrain barrier (BBB). It is then unable to diffuse back over the bloodbrain barrier. The drug is then cleaved off in the CNS. The diffusion coefficients $\log D$ and $\log P$ (octanol/water) are given in order to clarify the differences in distribution behavior of the substances. The lipophilic prodrug can easily cross the bloodbrain barrier, whereas the charged and much more hydrophilic intermediate can no longer do so. CDS, chemical delivery system.

mones, antiviral substances, antibiotics, neurotransmitters, cytostatic drugs, and even neuropeptides such as enkephalin or thyrotropin-releasing hormone.

For antiviral therapies this approach appears to be of particular interest to get drugs like azidothymidine into the CNS in order to treat hidden virus reservoirs.

26.4
Prodrugs to Extend Duration of Effect

In the case of drugs taken on a chronic basis, there are chemical modifications to extend the duration of the effect and so extend the intervals at which they need to be taken. Birth control steroids are a drug group that has been extensively researched. In addition to embedding in slow-release polymer carriers, compounds such as **norethisterone** were also covalently coupled to water-soluble polymers such as poly(N^5-hydroxypropyl-L-glutamine). The product is applied subcutaneously and the drug is then released over several months. Cytostatic drugs provide another example. After these have been covalently coupled to polymeric carriers such as dextrans, delayed release means that they show a more constant plasma concentrationtime profile and reduced side effects for a lower dose than the noncoupled drug. Similarly, **ranitidine (Zantac®)**, an inhibitor of acid secretion in the stomach, showed a considerably extended duration of effect after covalent linking with dextran than with the free drug given at the same dose.

26.5
Prodrugs for the Targeted Release of a Drug

Prodrug strategies have been used for some time to improve the availability of a drug after oral administration or so that it will only be released on reaching distal portions of the intestines. Prodrugs of **5-aminosalicylic acid** (sulfasalazine, olsalazine, and balsalazine), used to treat inflammatory bowel disease, have suc-

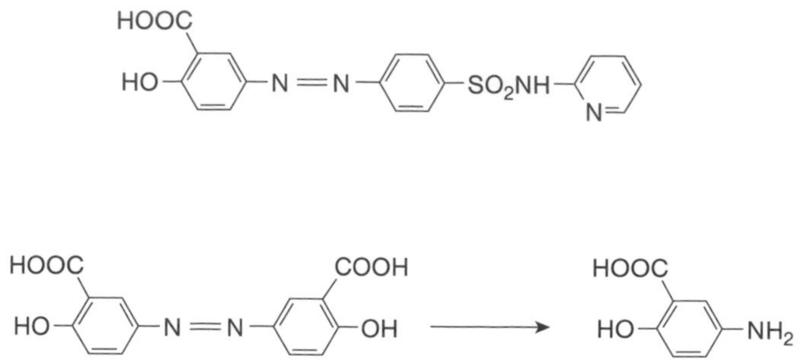

Fig. 26.10 Azo prodrugs of aminosalicylic acid, used to treat inflammatory bowel disease. Azoreductases in the large intestine cleave the prodrug and this breakdown releases the actual drug.

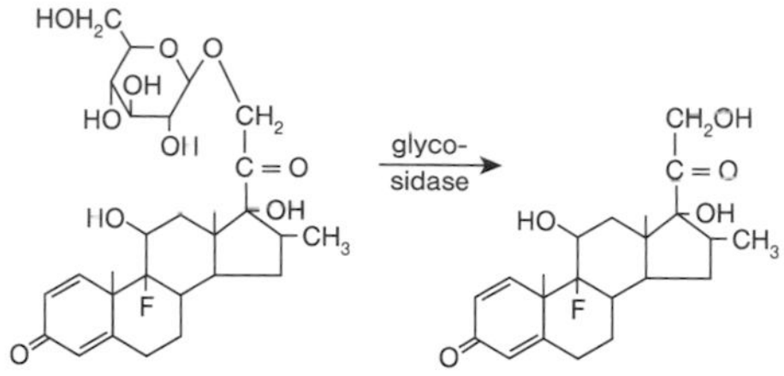

Fig. 26.11 Dexamethasone-21-β-D-glucoside. Following administration, up to 60% of the free steroid reaches the cecum. The sugar residue is cleaved off after absorption.

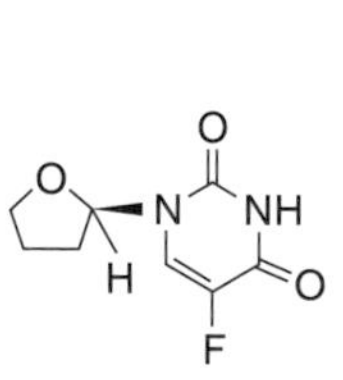

cessfully been introduced onto the market. It would be difficult for free 5-aminosalicylic acid to reach its target in the colon, since most of it is absorbed in the upper parts of the intestines. In the prodrug molecules, the drug itself is coupled to another molecule by an azo bond (Fig. 26.10). The prodrugs pass unchanged through the stomach and small intestine, and are then cleaved by **bacterial azoreductases** in the large intestine, when the drug is released. Prodrugs in which the drug is coupled to glucose or glucuronic acid work in a similar way. **Dexamethasone-21-β-D-glucoside** (Fig. 26.11) is mentioned as an example. It too is used to treat inflammatory bowel disease. Whereas the free steroid is almost completely absorbed from the small intestine, after administration of the prodrug almost 60% of the administered dose reaches the cecum in the form of the free steroid.

Polymeric coatings and coating materials have been developed in particular for peptide drugs, which are extremely labile in the gastrointestinal tract. These too are attached by azo bonds and only broken down on reaching the large intestine. Examples are styrols and hydroxyethylmethacrylates. The use of oligosaccharide or polysaccharide-coupled polymers, which are cleaved by glycosidases in the cecum, or that of dextran fatty acid esters, is also of interest.

26.6
Prodrugs to Minimize Side Effects

The targeted administration of drugs is often accompanied by a reduction in undesired effects. The therapeutic index of a prodrug can therefore be used as a measure to distinguish the desired properties and the toxic side effects of a drug. The cytostatic group of drugs includes some familiar examples of drugs with considerable side effects, where the prodrugs are less toxic. For example, treatment with **5-fluorouracil** can cause severe damage to the bone marrow or intestinal mucosa. The 1-(2-tetra-hydrofuranyl)-5-fluorouracil prodrug **ftorafur** (Fig. 26.12), which was discovered back in 1967, exhibits a similar antineoplastic effect along with significantly lower toxicity, but has a neurotoxic effect at very high dose. In recent years, a large number of other prodrugs have been synthesized starting from the basic structure of ftorafur. Some of these can be taken orally. They only release 5-fluorouracil after reaching the systemic circulation and are also less neurotoxic. The use of **doxorubicin** and **daunomycin** is limited, among other things, by their cardiotoxicity, the appearance of which may be acute, subacute, or chronic. Increased tumor selectivity and decreased toxicity have been demonstrated for various peptide prodrugs of cytostatic agents, both *in vitro* and *in vivo*. The observed effects of the prodrugs can be traced to differences in tissue distribution or to differences in the rate of uptake into target cells and distribution patterns in these cells.

Fig. 26.12 Ftorafur [1-(2-tetrahydrofuranyl)-5-fluorouracil]. This has an antineoplastic effect comparable to that of the free 5-fluorouracil, but is significantly less toxic.

27
Molecular Diagnostics in Medicine

Learning Objectives

Clinical chemical diagnostics play an important role in today's medicine. It can be assumed that the data acquired from medical laboratories contributes to 50 to 80% of diagnosis in modern clinical diagnostics. Alongside the many classical detection methods used in clinical chemistry (such as the determination of enzyme activity by means of a coupled enzyme assay, the quantitative detection of ions using flame photometry and ion-selective electrodes from metals through the use of atomic absorption spectrometry, the measurement of small molecules through chemical chromogenic methods, etc.), immunological detection methods and molecular genetic methods have gained in importance in the last 10–25 years. This is demonstrated in the rapid development of nucleic acid analysis, especially methods for efficient sequencing of genomes. A consequence is the substantial increase in our knowledge of genetic causes of diseases, which leads to a growing importance of molecular diagnostics in medicine. The term molecular diagnostics is used in various ways. As a broad definition, it encompasses all analytical methods used in the laboratory to analyze samples taken from patients, for the presence of single molecules or their function. As a narrow definition, the term is used simply as an abbreviation of molecular genetic diagnostics, and refers to methods that can be used for the detection of changes in the genetic information (DNA sequencing) and the interpretation of this information (gene expression) at the molecular level (DNA, mRNA).

27.1
Uses of Molecular Diagnostics

27.1.1
Introduction

The basis for molecular genetic diagnostics is the rapid increase and development of genome research and knowledge about the influence of genetic modification on the emergence and development of disease. The advancement and automation of DNA sequencing methods have made it possible to sequence not only the human genome, but also the genomes of many other organisms, such as viruses or pathogenic microorganisms (see Chapters 14 and 20). The sequence information and highly sensitive sequence-specific analytical methods now allow the rapid analysis of disease-specific changes in the DNA sequence, which can be used for diagnosis.

27.1.2
Monogenic and Polygenic Diseases

Diseases with a genetic basis are divided into two different groups. **Monogenic diseases** are those that are caused through the mutation or the loss of function

An Introduction to Molecular Biotechnology, 2nd Edition.
Edited by Michael Wink
Copyright © 2011 WILEY-VCH Verlag GmbH & Co. KGaA, Weinheim
ISBN: 978-3-527-32637-2

of a single gene. Diseases that stem from the co-occurrence of multiple disease promoting genetic modifications, which alone can be harmless, are referred to as **polygenic diseases**. Whereas monogenic diseases are easy to recognize due to their typical **inheritance pattern** (dominant or recessive), polygenic diseases are found merely to show a higher abundance in certain families without a clearly recognizable pattern of heredity.

A typical example of a **recessive monogenic disease** is the comparatively common (1:600) **cystic fibrosis**. This condition is caused by the loss of function in a chloride channel, which is coded for by the **cystic fibrosis transmembrane conductance regulator (CFTR) gene**. As a general rule, for every diploid cell, the loss of gene function occurs only if the corresponding gene is impaired through mutations on both homologous chromosomes. This means that the functional product can no longer be properly synthesized (see Chapter 4). **Heterozygous carriers** of this mutation occur with considerable frequency, although as the healthy gene can deliver an adequate amount of the functional protein they are mostly unaffected. Statistically only one in four of all offspring of parents who are heterozygous for the mutation is **homozygous for the gene defect**. Therefore the number of people affected by the disease is considerably lower than the number of heterozygous gene carriers. Due to their inconspicuous phenotype, the latter can only be identified by molecular genetics (see below). In populations where there is a high risk of specific, severe, recessive diseases, family planning already offers an analysis of the individual genotypes. However, the diagnostics are not straightforward; indeed, only a few mutations cause over 80% of the functional loss. There is, however, a large number of even rarer mutations that also impair the synthesis of a functional gene product. The diagnostics are so complex that it can only be indicated with a certain probability whether heterozygotes experience a loss of gene function.

Dominant diseases are considerably rarer than recessive diseases, due to the fact that only one affected allele is sufficient to cause the corresponding illness. This allele will also be passed on to every second child if one parent is affected. Therefore, due to strong selection pressures, only relatively mild dominant genetic diseases will survive for many generations in a population. Severe dominant diseases are normally caused by new mutations. Perhaps the most frequently occurring dominant disease (1 : 500) is **hypertrophic cardiomyopathy (HCM)**, which is caused by mutations in different proteins expressed in the cardiac muscle. The proteins most affected are the sarcomere proteins (e.g., b-myosin). A severe complication of this illness is **ventricular fibrillation**, which leads to cardiac death and mainly affects fit, healthy men. However, this particular complication is fortunately very rare. HCM strikes mainly in middle to old age through the occurrence of increased cardiac insufficiency (heart failure). Furthermore, it is known that for this well-researched disease pattern, the same mutation can be responsible for the different characteristics (from completely unaffected to severely ill) of the condition in different people (including individuals within the same family). This can be traced back to different genetic backgrounds or, to be more precise, to the co-occurrence of the mutation with other genetic variations in so-called modifier genes. Given that there are already a few hundred known mutations that cause HCM, none of which are particularly frequent, tailoring molecular diagnostics for this condition (as well as for many other mild dominant diseases) becomes markedly difficult and costly. With current methods, the screening of young people to determine the individual risk and to avoid sudden cardiac death is therefore far too expensive.

Polygenic diseases are caused by the interaction of several mutations in different genes. Individually, these mutations only have a minor effect, but together they can have catastrophic consequences. Due to the absence of a clear pattern of inheritance, the classic methods of family analysis to identify the risk genes fail. The association of genetic variants in functionally defined **risk genes** within a collection of nonrelated people or sibling pairs (sib-pair analyses) is useful for the identification of the responsible factors. Typical examples of diseases in this case are hypertonia and thrombophilia. Disease-associated alleles are often referred to as genetic risk factors. As each risk factor only contributes to a small extent to the eventual risk of disease, and as the presence of more disease-associated alleles does not inevitably mean that the corresponding disease will occur, the relative risk (calculated by cross-section studies) or the odds ratio (calculated from case-control studies) is given. Common mutations in clotting factor genes such as the Factor V Leiden mutation (R506Q) and a mutation in the 3′-untranslated region of Factor II mRNA (20210GA) are examples of thrombophilia risk alleles. As the allele frequencies for these mutations in central Europe are barely 3 and 1%, respectively, and hence heterozygous carriers occur with a frequency of just under 6 and 2%, respectively, these can also be referred to as **genetic polymorphisms**.

It is also worth mentioning that different genetic risks often combine. For example, many different inheritance patterns are known for **hypercholesterolemia** (the most important pathogenic factor for **myocardial infarction**): **familial hypercholesterolemia** is inherited dominantly and is caused either by mutations in the low-density lipoprotein receptor (LDL-R) or its ligand, apolipoprotein B100 (ApoB100), which leads to a wide loss of function with respect to LDL uptake. The inheritance of this disease is often also referred to as codominant. Whereas heterozygotes (frequency 1:500) already show increased plasma cholesterol concentrations, homozygotes (frequency 1: 1 000 000) are seriously affected. However, there is also a polygenic form whose frequency is given as 1:5 in the western industrialized world. This mild form of hypercholesterolemia is caused by the interaction of different polymorphisms (e.g., apolipoprotein E (ApoE) polymorphism) and mutations (which at the moment have not been identified) with external factors such as nutrition.

Mutations that have no direct effect on the gene product can also aid genetic diagnosis, particularly those mutations that are associated with a disease because they lie close to a functionally active mutation and are inherited along with it. The possibility of **genetic recombination** between two gene loci is inversely proportional to the distance between both loci. Therefore, it is possible to determine how far the mutation associated with the disease is removed from the functional mutation and, therefore, from the affected gene. In practice, this is used to identify the gene locus responsible for the disease; the association of many such genetic markers with a phenotype is investigated by **coupling or linkage analysis**. The genetic markers used should be spaced as evenly as possible throughout the entire genome. Markers associated with the disease can also be used for diagnostic purposes, but functionally this has nothing to do with the pathogenic mechanism.

As well as the previously described mutations, which are passed down through generations and therefore can be termed **germ line mutations**, a large number of new, so-called **somatic mutations** appear. These mutations occur in the many cell divisions during the development of an individual due to the intrinsic error rate of replication and the mutation rate. If these mutations lead to a loss of function of an important protein, the relevant cell will normally die off and will be replaced with a healthy cell. However, if the mutation leads to the activation of an oncogene or to the inactivation of a tumor suppressor gene, this can form the first step in a malignant transformation. Today it is assumed that a single mutation of this type can be quite harmless and that a minimum of three or more mutations of this type must accumulate in the cell before a malignant cell can develop. If and where this occurs is left purely to chance; admittedly the risk is directly dependent on the mutation rate, which in turn can be increased by external noxae (mutagenic substances, ionized radiation, etc.). Consequently, as well as being a polygenic disease, oncogenesis takes place at the somatic level. An exception to this is the classic inheritance of predisposed mutations in oncogenes and tumor suppressor genes, such as the retinoblastoma

(Rb) gene. In this case the risk of an individual being affected by a malignant tumor is increased, as the inherited mutation is already present in all somatic cells and therefore fewer additional, random events need to occur in order to trigger the disease.

27.1.3
Individual Variability in the Genome: Forensics

In addition to the pathogenically significant mutations described above, there are a considerable number of genetic variations in the human genome that have no apparent influence on the phenotype. Most of these regular polymorphisms lie in the noncoding regions of the DNA. This polymorphism is due to length differences in multiple repeats of short repetitive DNA sequence elements (**short tandem repeats**), of which most exist in the form of many alleles in the population. Due to their diversity it is possible to identify a person precisely or to determine family relations with high precision through the determination of a sufficient number of such polymorphisms. The typical polymorphism pattern of a person is also described as their **genetic fingerprint**, and is applied both to paternity analyses and in a large area of forensics (see Chapter 4).

27.1.4
Individual Variability in the Genome: HLA Typing

Individuals exhibit a high degree of polymorphism in the **HLA gene**, the gene products of which (i.e., major histocompatibility complex (MHC) I and II molecules) present antigens towards T cells. Since MHC I and II themselves function as immunogens, HLA incompatibility contributes heavily to the rejection of transplanted organs. For this reason organ donors and recipients have their HLA systems typed. A transplant is then only carried out if surface antigens such as the blood group characters, as well as the HLA types, are optimally suited to each other. In the past this was mostly done by analytical fluorescence-activated cell sorting (flow cytometry) using fluorescent labeled antibodies against specific HLA types; however, due to the accuracy of the results, this is now almost always done using genotyping.

27.1.5
Individual Variability in the Genome: Pharmacogenomics

Absorption, **metabolism**, and **elimination**, as well as the **specificity** for the target molecule, play fundamental roles in drug compatibility. In the normal approval procedure for a drug, these general terms are defined as **pharmacodynamics** and **pharmacokinetics** determined with a limited number of probands. Genetic differences between individuals, however, can lead to very large differences in these parameters. Decreased activity of the enzymes that participate in the metabolism or elimination of a drug leads to excessive concentrations and to undesired side effects, and in the worst case scenario can even result in death. A few of the important enzyme systems involved in the metabolism of drugs, such as the **cytochrome P450 oxidases**, are very polymorphic and lead to strong interindividual differences in the degradation of a large number of drugs. As these enzymes are mainly expressed in the liver, their characterization at the protein level as a general rule requires either specific indicator molecules capable of being metabolized and the quantification of the resulting metabolite or an *in vitro* study of liver biopsy samples. Since the most frequently occurring and also the rarer genetic variations are known, the variations that have an influence on the respective enzyme activity can be determined by genotyping. This can be done easily and reliably with a small amount of genomic DNA from **whole blood**. The known genetic variations can be determined before the start of therapy, especially when the drugs exhibit a small therapeutic window, display extremely toxic side effects, and are degraded and eliminated relatively slowly. Extrapolations indicate that many thousands of therapy-related deaths a year could be avoided.

27.1.6
Individual Variability in the Genome: Susceptibility to Infectious Diseases

Genetic variations of the patient also play an important role in **infectious diseases**: most microorganisms recognize specific cellular receptors in order to enter the host organism. Many pathogens also use the cellular systems of the host organism for replication and secretion. Both the receptors as well as intracellular susceptibility markers can be analyzed on the genomic level. Also, the innate and the adaptive immune systems can be genetically determined, as in severe cases of genetically caused **adenosine deaminase deficiency**, which is apparent in **severe immunodeficiency syndrome (SCID)**. An example of the influence of a genetic polymorphism on an infectious disease is the D32 deletion in the CCR5 gene. This gene codes for a **chemokine receptor** that serves as a **coreceptor** for the **human immunodeficiency virus (HIV)**. Individuals who are homozygous for the deletion are at a significantly reduced risk of being infected by HIV. The deletion leads to a displacement in the reading frame of the coding sequence (frameshift), and therefore to a shorter and altered amino acid sequence at the carboxy-terminus of the chemokine receptors. The physiological changes that occur as a result of these mutations are not known. In the future it should be expected that genetic predisposition analysis for infectious diseases will become more important and it is hoped that specifically tailored **individual therapies** will become available.

27.1.7
Viral Diagnosis

Classic virus diagnostics are based on the **detection of antibodies** produced by the patient in response to a viral infection. As the plasma concentration of antibodies following the immune response is considerably higher than that of the triggering virus, this diagnostic procedure is comparatively simple and inexpensive to carry out. However, considerable disadvantages do exist. On the one hand, the difference between a new infection and an immunity following a previous infection cannot be distinguished. On the other hand, a positive detection can only be seen after the production of antibodies, which normally occurs 12 weeks following the initial infection. In some case the diagnostic gap is considerably wider. An early improvement was the direct immunological detection of virus particles. However, this is only available for a few viral infections due to the high sensitivity required.

Molecular genetic diagnostics finally delivered a diagnostic breakthrough. This type of diagnostics not only makes extremely sensitive detection of viral **nucleic acids** possible (DNA or RNA by polymerase chain reaction (PCR) or reverse transcription (RT)-PCR, respectively), but also allows the typing of the virus with subsequent sequencing of the amplified genome segments. Furthermore, the number of viral genomes in the plasma can be ascertained by means of **quantitative PCR** (see Chapter 13). Perhaps the largest advantage of molecular genetic diagnostics is, however, the closing of the diagnostic gap: as soon as the first virus particles are circulating in the blood, they can be detected by a sensitive molecular genetic diagnostic technique. This advantage is particularly apparent in the testing of blood donors and blood products, as the diagnostic gap for HIV and **hepatitis C virus (HCV)** infections is particularly wide. In this gap the viremic phase declines with an extremely high viral load. In order to minimize severe viral infections that can often arise from blood transplants and from the therapeutic use of blood products such as clotting Factor VIII preparations, a molecular genetic test for HIV and HCV is a compulsory requirement. A further example of a recent advance in genetic diagnosis is the routinely carried out detection of **human papilloma virus (HPV)** for the prophylaxis of **cervical carcinoma**. These viruses are considered a risk factor for the development of cervical carcinoma, but only when the infection comes from a high-risk type. In contrast, infection by a low-risk HPV type presents little risk. The subtypes of these viruses are easily distinguished due to their different sequences.

27.1.8
Microbial Diagnosis and Resistance Diagnosis

Molecular genetic detection methods have also become an integral part of **microbial infection diagnosis**. The higher sensitivity, the possibility of differentiation between microbial subtypes, and above all the rapid detection of the pathogen constitute the greatest advantages over the classic methods based on incubation of the microbes, the typing and, when available, the sensitivity of the microorganism to antibiotic treatment. Whereas previously it took more than a week before mycobacteria isolated from the sputum of potential tuberculosis patients could be typed and an antibiogram produced, direct molecular genetic detection requires less time and also provides more informative genetic typing. From this, it is, for example, possible to deduce the potential resistance against established antibiotics.

This overview gives a very brief insight into the many difficult challenges relating to DNA analysis that occur daily in medicine. However, whether DNA analysis is applied depends on multiple factors; particularly important are the cost of the analysis and the relevance of the results for medical diagnosis. These two factors dictate that at present molecular genetic methods are only used where no other methods of analysis are possible, or alternatives are more expensive or too slow. It is expected, however, that rapid, cost-effective, reliable high-throughput tests will soon play a fundamental role in laboratory diagnostics and, therefore, the relative proportion of molecular genetic analyses will increase.

27.2
Which Molecular Variations Should be Detected

The goal of **molecular diagnostics** is the detection of **molecular genetic differences** that can lead to the development of a disease, or have an influence on the progression of the disease or therapy, respectively. Sequence variations can either influence the strength of expression of a relevant gene or lead to the production of an aberrant gene product. This product will either have lost certain properties due to a change in its amino acid sequence (recessive diseases, tumor suppressor genes) or gained a new property (oncogenes).

Genetic mutations can be divided into different groups according to their kind and function. The **functional classification** includes:
- Mutations/variations in the coding sequence.
- Mutations in regulatory elements (e.g., in promoters).
- Mutations in introns can affect RNA editing or lead to splicing variations.
- Changes in gene expression are also possible due to changes in the copy number of the gene (deletion, duplication, or amplification).
- Mutations in transactivating factors (e.g., transcription factors) that regulate the expression.
- Recombination between different genes, which, for example, can be caused by the translocation of chromosome fragments, leading to the formation of fusion proteins, which often display a different activity or function.

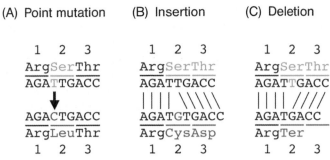

(A) Point mutation   (B) Insertion   (C) Deletion

Fig. 27.1 Mutation of a single nucleotide. A given nucleic acid sequence (top) can be selectively changed through a mutation. This leads to a modified sequence (bottom) and can also cause a change in the protein sequence itself (denoted by the three-letter genetic code). (A) An example of a point mutation, where a single base is replaced by another (in this case T is replaced with C). The mutation of the first base position in a codon always results in a change of the amino acid coded for (as in the example given), mutation at the second base position mostly results in a change of amino acid, whereas mutation at the third rarely leads to a change in the amino acid. The mutation is referred to as a significant mutation when the amino acid sequence of the coded protein is affected. If the protein sequence is not altered the mutation is deemed silent. The exchange of a pyrimidine for another pyrimidine (T→C or C→T) or a purine for a purine respectively (A→G or G→A) is known as a transition. The less frequently occurring exchange of a purine with a pyrimidine (A→T, A→C, G→T, or G→C) or that of a pyrimidine with a purine, respectively (T→A, T→G, C→A, or C→G), is referred to as a transversion. (B) Likewise, the insertion of a single nucleotide often leads to a change in the corresponding amino acid. However, as with the deletion of a single nucleotide (C) the largest problem is the resulting shift in the reading frame (frameshift mutation). This means that each frame following the deletion/insertion is now altered and results in a completely different amino acid sequence. This is also the case when two nucleotides are inserted/deleted simultaneously (not shown), whereas the insertion/deletion of three nucleotides results in the insertion/deletion of one amino acid in the resulting protein (not shown). (C) shows a single nucleotide deletion. This mutation leads to the appearance of a stop codon (Ter) and results in a premature abortion of the amino acid sequence.

In contrast, the **structural classification** describes the possible changes at the DNA level, such as point mutations, insertions, deletions, nucleotide repeats, deletion or duplication of entire genes, and recombination between genes on the same or different chromosomes (see Chapter 4). An example of which has been summarized in Fig. 27.1.

27.2.1
Point Mutations

Point mutations refer to exchanges of individual base pairs in genomic DNA (Fig. 27.1A) (see Chapter 4). For example, the desamination of cytosine leads to uracil that will base pair with adenosine in the following replication. Thus the new mutation can be fixed. Errors during DNA replication can also lead to point mutations. The mutations are known as **functional (significant)** or **silent mutations** depending on whether or not they lead to a change in the amino acid sequence and gene expression. If mutations occur in a regulatory sequence, they can also have an effect on the strength of expression and therefore have an effect on the phenotype. With the exception of mutations that display a distinct phenotype, the frequency with which point mutations occur in different chromosomal regions and the impact they have on the disease and a person's health is still extensively unclear. In order to improve this understanding, genetic variations that depend on point mutations (so-called **single nucleotide polymorphisms (SNPs)**) are being collected in great numbers as a supplement to the **Human Genome Project**.

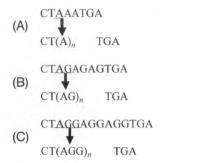

(A) CTAAATGA
→
CT(A)ₙ TGA

(B) CTAGAGAGTGA
→
CT(AG)ₙ TGA

(C) CTAGGAGGAGGTGA
→
CT(AGG)ₙ TGA

Fig. 27.2 Mutations through repeat expansion or reduction. The repeat of a single nucleotide (A), a dinucleotide motif (B), or a trinucleotide motif (C) can result in a change in the number of repeats in replication (although fortunately this rarely occurs), and leads either to an expansion or a reduction. In many cases, as with an insertion (see Fig. 27.1 C) this results not only in a change in the affected amino acid, but also in the case of mono- and dinucleotides, in a shift in the reading frame (frameshift) if the length of the repeat is changed by one or two repeats. Trinucleotide repeats simply lead to the insertion of further, identical amino acids in the resulting protein sequence. However, a modification of this type can also have severe consequences for the function of the coded protein.

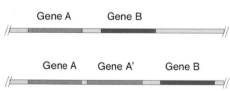

Fig. 27.3 Gene duplication. In a few cases the duplication of an entire gene occurs at the germ line (e.g., cytochrome P450 genes). An increase in the copy number of a gene mostly results in an increased expression and therefore also in an accumulation of the coded protein. Importantly, this type of mutation occurs frequently in somatic mutations in the course of cellular oncogenesis (e.g., c-myc amplification). The opposite of gene duplication is gene deletion, which can likewise be described for a range of genes. In the case of heterozygotes, this results in a reduced amount of the coded protein, whereas in the case of homozygotes the gene is completely lost (not shown). Partial gene duplications/deletions are also possible.

27.2.2
Insertions and Deletions

Insertion or deletion of one or more nucleotides can also lead to changes at the DNA level. Such mutations can, for example, occur due to mistakes during DNA replication, particularly in the area of short repeats (see Section 27.2.3). If such a mutation occurs within a coding sequence and does not correspond to a multiple of codon length (three nucleotides), this always leads to a change in the protein sequence through a shift in the reading frame (frameshift mutation).

27.2.3
Nucleotide Repeats

Repeats of a simple sequence motif often cause reading errors by DNA polymerase during replication. The consequence is an increase or decrease in the length of the DNA segment containing the sequence repeat (Fig. 27.2). Such nucleotide repeats are known as **dinucleotide repeats** (two base pairs) or **trinucleotide repeats** (three base pairs) according to the length of the repeated motif. Changes in the length of such nucleotide repeats also contribute to genetic diseases, such as Huntington's disease or Friedrich's ataxia.

In addition to the described short nucleotide motifs, longer repeated sequence motifs that occur repeatedly are also found in the genome. These comprise a significant part of the human genome and are present in different parts of the genome, often in strongly varying copy numbers. The frequency of variations can vary significantly for different motifs. Thus some particular motifs are well suited for the determination/differentiation of population groups and also individuals, if necessary. In forensics this variability is used for the generation of **genetic fingerprints**.

27.2.4
Deletion or Duplication of Genes

Changes in the genome can also include the complete deletion or the duplication of large segments of the genome (Fig. 27.3). These can occur both during replication and recombination. It is also common for partial deletions of a gene to lead to its inactivation. **Duplication of a gene** can lead to stronger gene expression and therefore disrupt the balance between cooperative or competitive genes. Multiple copies of a gene are described as **amplifications** and are especially found in the cytochrome P450 gene (see Section 27.1.5) in individuals with a rapid metabolism and also in tumors, particularly oncogenes such as c-*myc*.

27.2.5
Recombination between Chromosomes

A further possibility of genetic change is the exchange of gene segments between different chromosomes by recombination. These changes play a role in the emergence and development of particular diseases, and can also be detected through chromosome analysis. A known example is the formation of the **Philadelphia chromosome** in **chronic myeloid leukemia (CML)** from parts of chromosome 9 and 22. In this case, recombination often leads to the formation of a **fusion protein** (Bcr-Abl) that can no longer be regulated. This stimulates permanent cell growth, thus triggering the development of leukemia.

27.2.6
Epigenetic Changes

As well as changes to its sequence, the covalent modification of DNA via cytosine methylation (Fig. 27.4) also has an effect on the phenotype (see Chapters 2 and 4). **DNA methylation** leads to a decrease in the transcription in the concerned genomic regions and arbitrates epigenetic phenomena that are inherited maternally via the methylation pattern. At the somatic and cellular level, changes in methylation are involved in carcinogenesis and in this context are analyzed for diagnostic use.

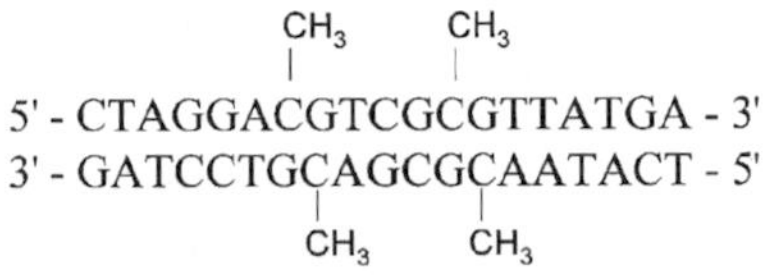

27.3
Molecular Diagnostic Methods

All of the methods used in molecular diagnostics are based on nucleic acid **sequence analysis**. Many techniques are available depending on the problem. The methods used often depend on the personal preference of the laboratory director, as up until now there have only been a few standard protocols for molecular diagnostics.

The following **goals of molecular diagnostics** are derived from the above-mentioned questions:

- Sensitive and, if required, quantitative analysis of genomic DNA or RNA in viral and microbial diagnostics.
- Genotyping of heterozygotes or homozygotes for particular known mutations.
- The search for mutations in long gene segments that are so far unknown.
- Measurement of gene expression by means of quantitative mRNA determination.

The realization of these goals first requires reproducible and sensitive methods of extraction of DNA and RNA from blood and other materials. Only then can the amplification and quantification of defined nucleic acid sequences be carried out. This enables genotyping of the DNA for known mutations and the identification of new mutations. Descriptions of the most important methods follow, illustrated by examples.

The methods described are all based on the following principles:

- Sequence-specific hybridization between reverse complementary strands of nucleic acids.
- New synthesis and formation of nucleotides by polymerases (sequencing, PCR).
- Sequence-specific cleavage of nucleic acids by restriction enzymes.
- Determination of the melting point and specific cleavage of mismatches.
- Determination of the molecular mass of nucleic acid fragments.

These methods are combined in various ways with detection and separation techniques such as gel electrophoresis, measurement of fluorescence, and enzyme reactions. The aim (where possible) is the detection of the changes described above both automatically and cost-effectively.

27.3.1
DNA/RNA Purification

All methods require the **purification of nucleic acids** from the sample material as the first step (see Chapter 9). For the analysis of DNA, a wide range of methods is available, most of which rely on the homogenization of the cells or tissue samples, lysis of the membrane, and removal of cellular debris. This is followed by separation of nucleic acids and proteins, and, if necessary, a further purifica-

$$CH_3 \qquad CH_3$$
$$|\qquad\qquad |$$
5' - CTAGGACGTCGCGTTATGA - 3'
3' - GATCCTGCAGCGCAATACT - 5'
$$|\qquad\qquad |$$
$$CH_3 \qquad CH_3$$

Fig. 27.4 Epigenetics: DNA methylation. The expression of a gene can also be fundamentally altered without a change in the DNA sequence. A change in the DNA methylation pattern causes imprinting (maternally inherited gene expression pattern) at the germ line and at the somatic level participates in a relevant part of molecular oncogenesis.

Fig. 27.5 PCR detection of a length polymorphism. Length insertions and deletions are easily detectable with PCR through the amplification of a corresponding gene segment. The resulting amplicons differ in length and can be separated by gel electrophoresis according to the speed at which they travel through the matrix.

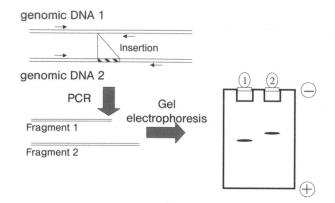

tion step. **mRNA** is largely unstable due its rapid degradation by **RNases** and chemical hydrolysis. Therefore, it is necessary to use methods that optimize the **purification** of intact RNA. For both problems, very reliable affinity chromatographic techniques are available from different manufacturers, nearly all of which can also be run completely automatically (see Chapter 9).

27.3.2
Determination of Known Sequence Variations

27.3.2.1 Length Polymorphism
The determination of a length polymorphism requires the gene region concerned to be amplified above the detection limits of the subsequent detection method. The most suitable way of doing this is via the **PCR** (see Chapter 13) using specifically flanking primers. The simplest way of separating the different lengths of amplification products is by gel electrophoresis (Fig. 27.5).

27.3.2.2 Restriction Fragment Length Polymorphism (RFLP)
Restriction enzymes recognize short, specific (mostly palindromic) nucleotide sequences where they cleave the DNA double strand (see Chapter 12). Naturally, it is possible that these sequences are modified by mutations. In this case, mutations are detected as changes in the resulting DNA fragment pattern (RFLP). First, the gene fragment concerned is amplified using PCR. The PCR product is digested using the corresponding restriction enzyme and subsequently analyzed using gel electrophoresis. If the fragment is cleaved by the restriction enzyme, the resulting fragments will be smaller than the original; if the restriction site is

Fig. 27.6 RFLP. If a mutation disrupts a given restriction enzyme recognition sequence or leads to the creation of a new one, point mutations can easily be detected through a restriction digest. The affected region is amplified using PCR and the PCR product is digested by the corresponding restriction enzyme. The mutation can easily be recognized after gel electrophoresis due to the differing band pattern (the splitting of a DNA fragment into small fragments). With heterozygotes both the intact fragments and the split products can be recognized simultaneously. Consequently it is also easily possible to distinguish between heterozygous and homozygous mutations. On the basis that restriction enzymes are extremely specific, RFLP analysis is considered as a very safe and robust detection method.

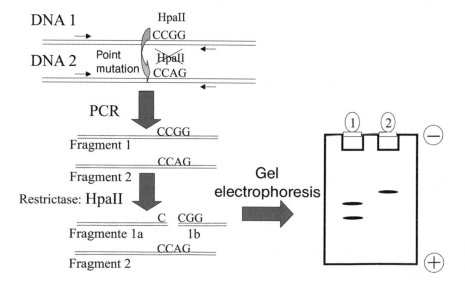

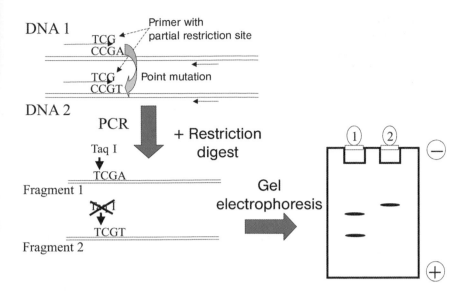

DNA 1

TCG ← Primer with
CCGA partial restriction site

TCG ← Point mutation
CCGT

DNA 2

PCR

Taq I

TCGA
Fragment 1

TCGT
Fragment 2

+ Restriction digest

Gel electrophoresis

Fig. 27.7 ARCS. If the mutation of interest does not alter a restriction enzyme recognition sequence (either by destroying it or creating a new one), site-directed mutagenesis can change the area of a point mutation so that the resulting PCR products contain a restriction enzyme recognition site that is dependent on the mutation. An RFLP is then carried out following a restriction digest and the mutation is consequently detected by gel electrophoresis.

mutated, these fragments cannot be detected (Fig. 27.6). Mutations can, however, also be detected as they lead to a new restriction site.

27.3.2.3 Amplification-Created Restriction Sites (ACRS)

If a mutation neither destroys a restriction site nor creates a new one, it is possible to utilize a variation of the RFLP analysis, where the mutation is detected through the **artificial generation of a recognition sequence**. With the help of a PCR primer that binds directly next to the mutated site, a restriction site can be introduced during PCR amplification. This new restriction site is either complemented or destroyed by the mutation. The subsequent detection occurs in analogy with the classic RFLP analysis (Fig. 27.7).

27.3.2.4 Amplification Refractory Mutation System (ARMS)

For the direct **detection of a mismatch without a restriction digest**, one of the PCR primers can be selected so that the 3′-end coincides with the mutation. A detectable PCR fragment arises only when the sequence of the target DNA matches that of the primer sequence. Normally, a further PCR analysis is performed as a control, in which the corresponding primer contains the mutated sequence and so leads to a complementary result. This method is based on allele-specific primers and is known as **ARMS**.

27.3.2.5 Mutationally Separated (MS)-PCR

MS-PCR is a variety of the ARMS analysis, where the wild-type primer and the mutated primer have different lengths. Both primers are combined in a reaction with the same reverse primer, and allow the wild-type and the mutated sequence to be distinguished from each other through the detection of the **length heterogeneity** by electrophoresis (Fig. 27.8).

27.3.2.6 Allele-Specific Hybridization

Different sequences can also be detected through **selective hybridization**. In **allele-specific hybridization**, different fluorescently labeled allele-specific primers are used in a PCR reaction and the product is subsequently detected by hybridization with probes for each allele (Fig. 27.9). This type of analysis can also be carried out using miniature probe arrays (microspots/DNA chips) (see Section 27.3.2.11).

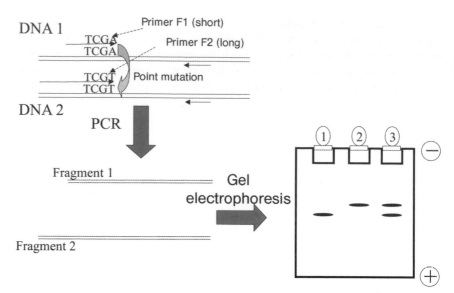

Fig. 27.8 ARMS. As with the LCR (see Section 27.3.2.7), PCR analysis of the negative influence of a mismatch on the efficiency of elongation can also be used in order to distinguish two alleles from each other. In this case two alternative primers are used. The 3′-end of these primers either corresponds to the wild-type sequence or to the mutated sequence, and the primers are either added to two independent PCR reactions or altered so they differ in length and used simultaneously in one PCR reaction with three primers. In the latter, which is depicted here, the mutation leads to the incorporation of the longer primer, which can be directly detected using gel electrophoresis based on the slower migration time of the longer PCR product.

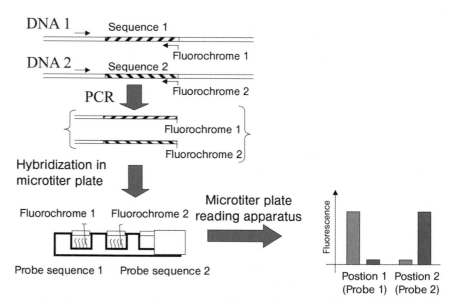

Fig. 27.9 Allele-specific hybridization. These detection methods also use the influence of mismatches on the stability of DNA hybrids in order to distinguish between the mutated DNA sequence and the wild-type sequence. This requires PCR amplification of the gene segment concerned and immobilization of the amplicon in a solid phase (e.g., microtiter plates). Two oligonucleotides that correspond to a wild-type and a mutated sequence are added. When the oligonucleotides hybridize with the immobilized amplicon under strict conditions the ideally suited oligonucleotide binds better than the mutated one. The detection of the binding of both nucleotides is carried out either in two different reaction vials or, if they are both labeled differently (e.g., with two different fluorophores), in the same reaction. The relative signal strength serves as an indicator for the wild-type sequence or the homozygous mutation. Heterozygotes bind to both oligonucleotides with equal strength.

27.3.2.7 Ligase Chain Reaction (LCR)

Ligation reactions can also be used to detect mutations. The ligation product is only formed when the two linkers match the linkage site exactly. Amplification is possible through the repetition of the reaction. The fragments can subsequently be detected using different methods (e.g., probe microarrays) (Fig. 27.10).

27.3.2.8 Minisequencing

Minisequencing describes a method that usually only adds one or a few nucleotides on a primer with the help of a sequence reaction. Through fluorescent labeling of a particular nucleotide, the bases that follow the primer sequence in the sample can immediately be read (Fig. 27.11).

27.3.2.9 Pyrosequencing

A variation of minisequencing is **pyrosequencing** (see Chapter 14). In this case, instead of fluorescently labeled nucleotides, the **amount of pyrophosphate** that is released during the formation of a nucleotide ester bond is measured. In each reaction only one nucleotide is used and a signal is only generated if the matching nucleotide was involved in the reaction. The intensity of the signal is proportional to the amount of incorporated nucleotide and enables the number of matching nucleotides that follow in the sequence to be determined.

27.3.2.10 Quantitative PCR

At present, perhaps the most important method used in molecular diagnostics is **quantitative PCR** (see also Chapter 13). One application is the **sequence-specific analysis of DNA fragments and their relative frequency**. Using this technique, mutations, allele-specific mutations, and also gene duplications and deletions can be detected. Quantitative PCR is also suitable for the sensitive detection of viruses, pathogenic microorganisms, and their antibiotic resistances. In the adaptation of this to quantitative RT-PCR for the detection of mRNAs, quantitative PCR is also used for the analysis of gene expression and for the analysis of changes at the DNA level, which are usually difficult to detect (such as the formation of splice variants and the expression of fusion proteins).

Quantitative PCR, just like standard PCR, uses two sequence-specific primers for the amplification of a defined target sequence. However, the detection of the amplification is not carried out through the analysis of the end product. Rather, fluorescent reagents are used, making it possible to determine the increase in amplified product in every cycle of PCR through online fluorescence measurement. Quantification is possible by comparing the onset of amplification at a specific reaction cycle with the onset of amplification in reference samples of known concentration. As the fragment size cannot be measured directly, different strategies are needed in order to confirm the specificity of the signal.

One possibility is the addition of double strand-specific fluorescent markers, such as **SYBR® Green** (Molecular Probes, Life Technologies). Although it binds all double-stranded DNA fragments (including those formed by unspecific amplification), determination of fragment size is necessary and is acquired through the final determination of the melting point (T_m) after the reaction is completed. This can be achieved using the same machine by generating melting curves that are characteristic for the length and the GC content of the double-stranded DNA fragment. The beginning of the amplification can only be used for the quantification if the melting point is the same for all samples.

Another possibility exists – the use of an internal sequence-specific probe. This additional third oligonucleotide enables the measurement of a signal only when the correct sequence is amplified between the two primers. There are

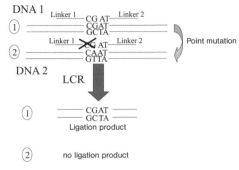

Fig. 27.10 LCR. In analogy with PCR, a point mutation can also be detected through a LCR. Four phosphorylated DNA oligonucleotides together with a thermostable ligase and ATP, can be incubated with the genomic DNA in a cyclic repeat of annealing/ligation and denaturing. Therefore the oligonucleotides (linkers 1–4) are selected so that the mutation is located precisely at the ligation site and a ligation is sterically hindered by the oligonucleotide attached to the genomic DNA matrix. Given that the prevention of the ligation is mostly incomplete, a complimentary reaction involving four oligonucleotides should be carried out. Ideally these four oligonucleotides can bind to the mutated DNA sequence and therefore act as an additional positive control for the mutation. This method is well suited for FRET detection methods.

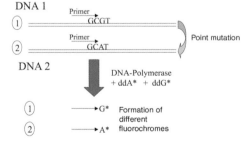

Fig. 27.11 Minisequencing and pyrosequencing. If, in a sequence reaction, instead of a mixture of dNTPs and fluorescently labeled ddNTPs, only ddNTPs are used, and the sequence primer lies directly in front of the position of the concerned point mutation, different ddNTPs that are distinguished by their fluorochromes will form in a sequence reaction that is independent of the mutation. Instead of separation of the sequence product by electrophoresis, the detection of the incorporated fluorochromes can be directly monitored. Pyrosequencing is a specialized form of minisequencing, where only one dNTP is present in each reaction and upon formation of a nucleotide ester bond the released pyrophosphates can be detected by a luminescence reaction. In contrast to minisequencing, the amplification of the template before the reaction is not required, by which the direct examination of genomic DNA is made possible.

many different possibilities for the design of this third oligonucleotide. One possibility is the **TaqMan® system** (Applied Biosystems). This involves a short oligonucleotide that has a fluorescent label covalently bound to one end and a quencher molecule bound to the other. No fluorescence can be detected as long as the TaqMan probe remains intact, as the absorbed energy of the fluorochrome is taken up by the quencher dye by **fluorescence resonance energy transfer (FRET)** (see Chapter 21). The oligonucleotide is further modified so that it cannot act as a primer itself, but so that it can be degraded by the $5' \rightarrow 3'$ endonuclease activity of Taq DNA polymerase. This only occurs if the polymerase binds a primer $5'$ from the TaqMan probe and synthesizes a new DNA strand from there. Through the degradation of the TaqMan probe the fluorescent dye and the quencher dye are displaced. The fluorescence transfer between the two can no longer occur and the fluorescence of the free fluorochrome can be detected. The sequence-specific amplification leads to an increase in the fluorescent signal as above. The increase in the signal can again be used for quantification. Unlike double strand-specific fluorescent markers, the signal from the TaqMan probe is sequence-dependent and a further determination of the specificity is normally not necessary. Through the use of more than one probe labeled with different fluorescent markers, it is possible to measure multiple amplifications in one reaction and therefore to quantitatively determine different sequences simultaneously (**multiplexing**).

27.3.2.11 Chip Technology

Microarrays, or gene chips as they are popularly called, are essentially miniature analysis systems in which a large number of probes are arranged in a pattern resembling that of a chess board in an **array format**. Normally only probes of the same type, such as nucleic acids (DNA chip) or antibodies (protein or antibody chip), are combined in an array. Today, arrays have a secure place in research laboratories and are used on a daily basis. In the future, a larger amount of molecular diagnostics will be carried out with the help of chip analyses.

Chip analysis is suited for the following applications:

- *Detection of the expression of a large number of genes (expression pattern).* Instead of analyzing the expression of a single, disease-relevant gene, the expression of a **multitude of genes** is simultaneously recorded and the pattern of relevant expression is used for the classification of diseases. This enables a better classification of tumors and the detection of individual differences. Therefore, the choice of therapy can be improved for each individual patient.
- *Detection of mutations, genomic deletions, and amplifications.* Microarray analyses are already available for the detection of mutations (SNPs) in different genes. They are used in order to predict the potential risk for the presentation of a particular disease or side effect from therapy. Microarrays are also in development for similar genomic hybridizations. These methods, known as **array comparative genome hybridization (CGH)** methods, promise a rapid and cost-effective alternative to the established genome analysis in diagnostics.
- *Detection of proteins and protein modifications.* In many cases it is important to detect proteins or their changes in patient samples. This is possible with the help of microarrays on which probes and antibodies are combined. The antibodies used recognize with a high specificity the proteins with or without their secondary modifications. A particular application is also the detection of antibodies in the serum of patients, which bind an immobilized antigen on a microarray (e.g., for the detection of antibodies that mediate allergic reactions against specific antigens).
- *Detection and subtyping of microorganisms and antibiotic resistance.* A very important application of microarrays for the analysis of nucleic acids is given by DNA chips that are used for the detection of microorganisms and viruses. The possibility of analyzing many DNA sequences simultaneously in the same

experiment enables not only the detection of microorganisms, but also the assignment of particular subtypes with different disease relevance. It is becoming more and more important to determine whether these microorganisms display **resistance against antibiotics**.

27.3.2.12 Production and Manufacture of Microarrays

As mentioned in Section 27.3.2.11, microarrays or biochips can be manufactured for the detection of nucleic acids or proteins. In the first case, nucleic acids, normally in the form of DNA, are used as immobilized probes. In the second case, the probes are either proteins or antibodies. Another difference concerns the way the chip is manufactured. In one case, the probes are **deposited in array format** (spotting); in the other case, nucleic acids or peptides are **synthesized by *in situ* synthesis** with different sequences in fixed positions of the array. In every case the position in the array (x, y) gives exact information as to the characteristics of the probe that exists at that position, which only analyzes a particular target molecule. The signal in a particular position of an array shows exactly which molecule is present in the analyzed probe and how much of that molecule there is. This can be used in relation to the other signals in the array for both qualitative and quantitative analyses. See Fig. 27.12.

As an example of microarrays, the Array-Tube® (Alere GmbH, Jena) is shown in Fig. 27.13, in which the array is positioned at the bottom of a reaction vial. In this system, binding and quantification of target molecules are detected through enzymatic deposition of a dye.

Most other microarray systems use the detection of a fluorescent dye. Through the use of different fluorescent dyes, different samples can be labeled with different fluorescence. Through competitive hybridization of two differently labeled samples, the relative relationship between molecules in both can be de-

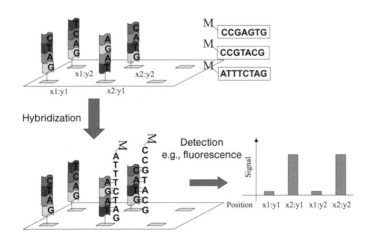

Fig. 27.12 DNA microarrays: the principle. DNA microarrays are a further development of sequence-specific hybridization. However, in this case it is not the DNA amplicon that is immobilized, but the sequence-specific probe. The target DNA is normally fluorescently labeled and hybridizes with the probes. Perfect match of the sequence causes strong binding and results in a strong fluorescent signal, whereas the binding of a mutated sequence is not as strong and therefore corresponds to a weaker fluorescent signal. The main advantage of microarrays is that a large number of different sequence segments can be analyzed parallel to each other with hybridization. For this reason, in principle even the presence of unknown mutations in large sequence sections can be detected. This application is based on the fact that the signal strength is also dependent on the amount of fluorescently labeled DNA. If, for example, mRNA from different cell preparations is copied into fluorescently labeled cDNA and hybridized on a DNA microarray, the ratio in signal strength of every dot represents the expression level of RNA. Thus, the induction or reduction of the expression of a particular gene can be observed.

Fig. 27.13 DNA microarrays: practical implementation.

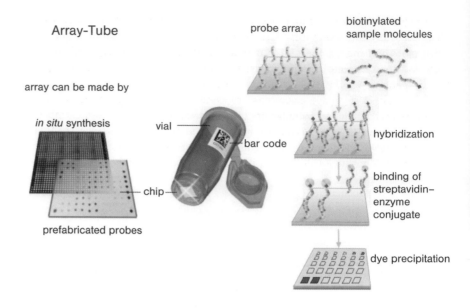

tected. This differential hybridization is preferentially used for arrays produced on the laboratory scale because it can be used with less precisely manufactured arrays.

27.3.2.13 Determination of Unknown Mutations

Although there are many known mutations in the genome that are medically significant, we have only just begun to investigate the molecular causes of disease onset and to find therapies for them. An important challenge for molecular diagnostics therefore is the **identification of previously unknown modifications** in the genome.

The development of cost-effective high-throughput sequencing technologies (next-generation sequencing, deep sequencing) made it possible to search for variations in the genome even without complex primary screens for mutations. With this approach it becomes possible to sequence the genome of one person almost completely at limited cost and in a short time. This is possible because the known human genome(s) can be used as reference for the assembly of an individual genome from many small DNA fragments (resequencing). In doing so all mutations in comparison to the known genome(s) are also immediately recorded. **How is it possible to sequence 3 billion base pairs in a short time?**

All **high-throughput sequencing** approaches are based on the rapid parallel sequencing of small DNA fragments. These are immobilized on a surface on which they are subsequently simultaneously sequenced in parallel either by incorporation of nucleotides following the basic Sanger sequencing reactions combined with fluorescence or luminescence (**pyrosequencing**) reading, or by ligation of short oligonucleotides (**sequencing by ligation**). All parallel sequencing reactions are recorded optically at each reaction step and the sequence of individual fragments is recorded by the specific reaction at one coordinate on the surface. This step is repeated until it is possible to assign a specific sequence of 30 or up to several hundred base pairs to the coordinate of all individual seed sequences. To obtain sufficient signal intensity in the sequencing reaction each immobilized seed sequence is amplified locally, so that each position on the sequencing chip contains a larger number of identical immobilized molecules that are sequenced together. Within one sequencing assay about 1 billion base pairs are sequenced in parallel. Thus with a few complementary assays the whole human genome of one individual can be sequenced. As only short sequences are obtained, computer programs use the known human reference genome for the assembly of the fragments to a new genome and to simultaneously detect differences between the sequences. The application of high-throughput sequencing will mainly depend on the costs for each analysis. (It can be expected that the price for one complete genome analysis by high-throughput re-sequencing of a human genome will be in the range of US $ 1000).

As only short sequence fragments are generated, not all genomic variations can be detected using high-throughput sequencing. Changes in copy number of repetitive sequences as well as point mutations in repetitive sequences and reorganization of chromosomes are not accessible or difficult to detect using this approach. These differences are still better accessible with other methods.

27.4
Outlook

The question concerning the methods that will dominate the future of laboratory diagnostics cannot be answered unambiguously. Present aims include the introduction of **larger integrated systems** that will enable all measurements in solution to be carried out rapidly and in parallel. The other path for laboratory diagnostics is that of continuous care for the patients. This takes the form of continuous **online monitoring** and the recording of various parameters, as is already done today with the continuous electrocardiogram or the measuring of oxygen saturation (spectroscopic measurement at the nail bed).

Developments in the field of molecular diagnostics are continuing to move forward at great speed in both directions integrated automated laboratory technology and near-patient monitoring (bedside diagnostics, personalized medicine). It is difficult to predict the long-term effects these developments will have in medicinal molecular diagnosis. It is certain, with the exception of answers to highly specific diagnostic questions, that the molecular diagnostic methods that will prevail will be those that are feasibly automated, cost-efficient, and highly reliable. These will be, in particular, the **PCR-based methods**, **miniature chips**, or other similar techniques. Applications in clinical diagnoses require highly reliable levels of analysis, which will certainly lead to the large use of microarray-based solutions in the long term. These appear to be the solutions of the future. With the miniaturized overhaul of all detection steps, a higher, laboratory-independent, quality of analysis will be possible. It is to be expected that optimized standard techniques will be adapted for an integrated, lab on a chip solution to answer every question in molecular diagnostics. Therefore, optimization of production should lead to an increased number of diagnostic applications and will actually result in a decrease in the cost of each individual analysis.

28
Recombinant Antibodies and Phage Display

Learning Objectives

Recombinant antibodies have become indispensable, particularly with the development of proteinogenic therapeutics. Through methods such as *in vitro* selection (i.e., the isolation of specific interactions outside of, and independent from, a living immune system) it is possible to manufacture a multitude of clinically useful antibodies. This is of particular value for making human antibodies. The most widely used method for this is phage display. This method is based on the selection of antibody genes by their encoded antigen-binding function. After obtaining a recombinant antibody clone, the fusion of other proteins/protein domains lends itself to the development of new characteristics in recombinant antibodies, which nature cannot provide. Furthermore, through the production of recombinant antibodies in different organisms, new possibilities for commercial mass production have arisen.

28.1
Introduction

The main function of antibodies in nature is the **specific labeling of pathogen determinants** for subsequent recognition by the immune system. A very large collection of antibodies with different binding specificities is present in our body for this purpose. However, it would be far too inefficient to store individual genes for each of the more than around 10^8 different antibodies present in the human organism and the amount of DNA required would exceed the size of the entire human genome itself. The solution to this problem depends instead on the principle of genetic combination. The final antibody gene is individually assembled in every B cell from a limited batch of different gene fragments. In addition to the recombination of gene fragments, further coincidental mechanisms introduce new, short sequences at one of the fusion points of the different gene fragments. The diversity of possible structures is further increased through the combination of two different protein chains (**light and heavy antibody chains**).

The generation of binding diversities, while maintaining constant functions of the effector mechanisms of the immune system, is made possible through the modular construction of the antibody. The largest part of the different antibody molecules is essentially identical (**constant domain**), and mediates effector mechanisms of the immune system upon contact with the antigen. Antigen binding is achieved by a small variable region of the molecule, which contains the **hypervariable regions**. Structurally, these constitute the **complementary determining regions (CDRs)** because they form a structure that is complementary to the antigen. Six of these hypervariable regions in total (three in the light chains and three in the heavy chains) are connected by conserved, structural regions (the so-called framework regions) so that they come together with the top end of the T- or Y-shaped structure of the molecule (Figs 28.1 and 28.2). This structure, made up of the CDRs and the framework regions, forms the **variable domain** of the antibody.

An Introduction to Molecular Biotechnology, 2nd Edition.
Edited by Michael Wink
Copyright © 2011 WILEY-VCH Verlag GmbH & Co. KGaA, Weinheim
ISBN: 978-3-527-32637-2

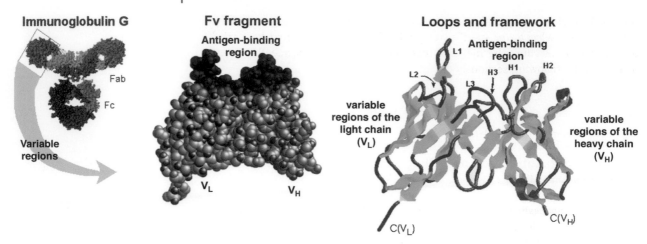

Fig. 28.1 (Top) Structures of immunoglobulin and the antigen-binding fragment (Fv). (Bottom) Structure of the immunoglobulin domains is defined by the β-sheet region (green) of the framework, whereas the antigen-binding sites are formed from six loops (H1H3, L1L3). The carboxyl-termini of both variable polypeptide chains of an Fv fragment (C) lie on the antigen-binding site at opposite ends of the molecules, therefore it is the preferred insertion site for protein fusion.

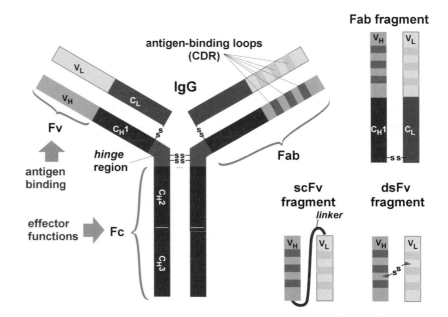

Fig. 28.2 Schematic diagram of IgG and the different antibody fragments derived from it.

The **constant domains** mediate the effector functions; different constant domains can therefore mediate a large range of different biological effects. The selection of the B cell that produces a corresponding antigen-binding antibody occurs after contact with the antigen through selective reproduction of this B cell (**clonal selection**). This selection was previously achieved using the immune system of animals (mainly mice or rabbits). This way either antisera (polyclonal antibodies) or B lymphocytes, which after immortalization produce monoclonal antibodies, were obtained. In each case, however, it was necessary to first immunize an animal. In the late 1980s, a third method of selecting antibody fragments, based on gene technology, became available. Using this method, the antibodies are no longer generated in research animals (or in humans), but instead they are generated *in vitro* in bacteria or cultured cells. This approach focuses on the antigen-binding part of the antibody. Following the initial production of mutants of available antibodies that differed in their molecular characteristics (e.g., the reduction of the immune response in humans – so-called **humaniza-**

tion) in **heterologous expression systems**, it was realized at the beginning of the 1990s that the selection of specific antigen-binding sequences could be carried out completely *in vitro*. The methods that were developed subsequently are explained in the following sections.

28.2
Why Recombinant Antibodies?

28.2.1
Recombinant Antibodies are Available *In Vitro* without Immunization

Recombinant antibodies can be obtained completely outside of a vertebrate organism. A few methods are especially interesting, as the biochemical properties for the binding can be precisely controlled during *in vitro* selection. For this reason antibodies are available that could never be produced in animals, such as antibodies against the transient conformation of a molecule after cofactor binding or against conformations that are normally masked *in vitro* by the presence of a competitor. Antibodies against antigens with which an immunization is not possible can also be obtained, such as antibodies against highly toxic substances, lethal pathogens, or evolutionary conserved molecules.

Natural human antibodies are rarely used, as an immunization for the production of antibodies in humans is only possible in exceptional cases. However, human antibodies for *in vivo* use (such as therapy and diagnosis) have been obtained, these antibodies are not recognized as foreign by the human immune system, resulting in a lower immune response by the patient against the antibody. This minimizes the risk of a patient's response that neutralizes the therapeutic antibody. Only recombinant antibody technology made the manufacture of **human antibody therapeutics** possible in a systematic manner (Fig. 28.3).

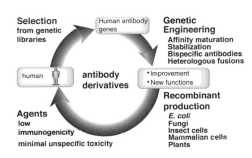

Fig. 28.3 From people, for the people: the production cycle of human recombinant antibodies.

28.2.2
Antibodies with New Characteristics Can Be Created

The antibody genes can be coupled to a gene of interest, so that both genes are fused together and expressed as proteins. The antigen-binding site and the carboxyl-terminal of the variable domain are on opposite sides of the molecule. Therefore, through the introduction of heterologous proteins to the carboxyl-terminus of a variable domain it is possible to provide the antigen-binding site of the antibody with new biochemical functions, which would not occur naturally. For example, an enzyme can be specifically directed to a tumor with the help of an antibody fragment – such **bifunctional antibodies** can lead to new cancer therapies.

The fusion of two different antibody-binding sites to form a **bispecific antibody** is also possible. Recombinant genetic fusions offer the advantage that the coupling site and the stoichiometry of both partners are exactly defined. An overview of the multitude of different constructs is given in Fig. 28.10.

28.3
Obtaining Specific Recombinant Antibodies

Antibodies require a complex molecular helper apparatus and an oxidative biochemical environment for this production. In bacteria, a functional production of antibody fragments is first of all acquired through a fusion to a bacterial signal sequence, which leads to the **secretion of the antibody fragment into the periplasmic space**. This periplasmic space lies between the two cell membranes of Gram-negative bacteria, and contains a biochemical environment that, unlike

Fig. 28.4 The three basic principles of humoral immune response, whose imitation *in vitro* enables a selection of recombinant antibodies.

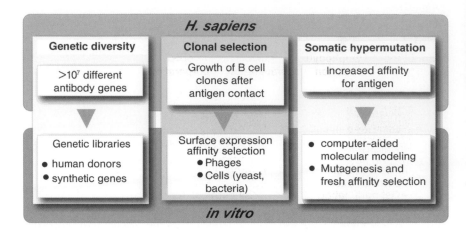

thc cytoplasm, allows the correct folding of the antibody and the correct formation of disulfide bonds. Therefore, it is possible to mimic in bacteria the decisive step in the generation of specific antibodies. This occurs through the imitation of three basic principles of the mammalian immune response (Fig. 28.4):

1. Construction of the gene collection encoding the multitude of antibody genes (in the form of recombinant gene libraries).
2. Effective selection of the correct gene from this initial repertoire (with the help of a surface expression vector).
3. Improvement of affinity and specificity of a selected antibody fragment (through *in vitro* mutagenesis and repeated selection).

28.3.1
Preparation of the Repertoire of Antibody Genes

It is estimated that in humans more than 10^8 **different antibody genes** are available through the random combination of gene fragments. This genetic diversity can be extracted from B lymphocyte mRNA through the **polymerase chain reaction (PCR)** (see Chapter 13) with the help of two oligonucleotide primers, which bind to conserved sequences at the ends of the antibody gene. The oligonucleotide primer contains restriction sites, so that the antibody genes can be assembled in *Escherichia coli* expression vectors (Fig. 28.5). If immunized, serum-positive donors are available (e.g., those that have survived an infectious disease), the chance of obtaining specific antibodies can be increased when the antibody genes for the library are prepared from circulating cells in the blood of the patient or lymphatic organs.

There are a number of methods for the generation of additional diversity through the introduction of synthetic sequences. These mostly include randomized sequences, which largely increase the diversity in the region of the antigen contact site. Completely synthetic gene libraries have also been successfully generated using randomized CDR regions. These libraries have provided the means for *in vitro* generation of antibodies, as well as the generation of diversity and selection of antibodies independently of an immune system. Antibody gene libraries of different origin and complexity are used according to the task at hand (Table 28.1).

28.3.2
Selection Systems for Recombinant Antibodies

28.3.2.1 **Transgenic Mice**
With the help of genetic methods individual genes can be specifically inactivated (**knock-out**) or foreign genes can be introduced in mice (**knock-in**) (see Chapter 28). Mice are bred whose immunoglobulin gene locus has been inactivated and replaced by homologous stretches of the human immunoglobulin gene. Human antibody genes rearrange in the same way as mouse genes; they can undergo

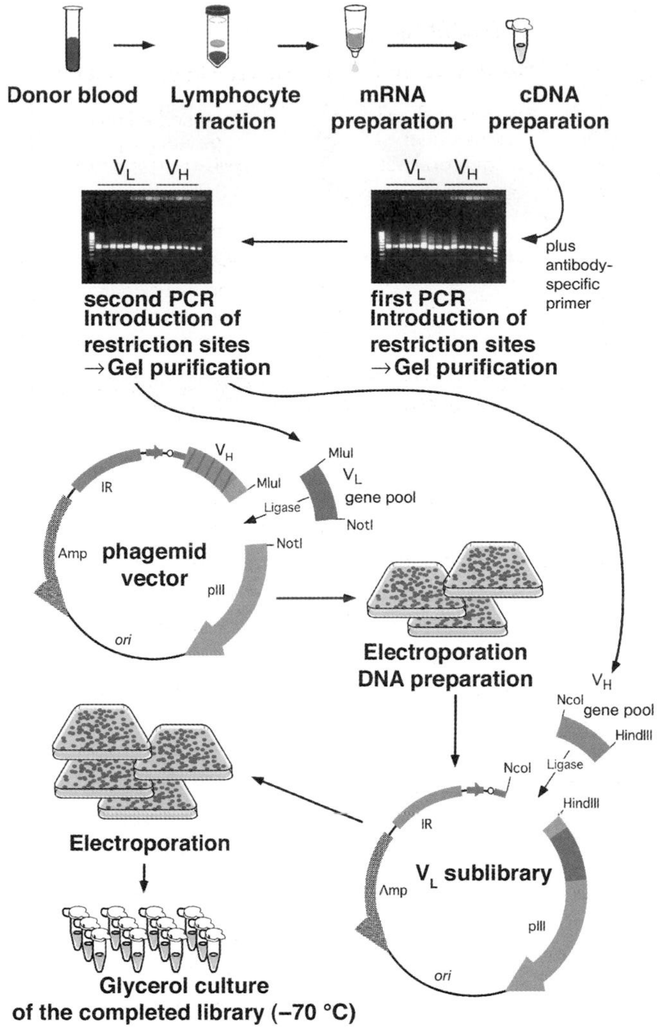

Fig. 28.5 Experimental flowchart for the production of an antibody gene library for phage display.

Donor blood → **Lymphocyte fraction** → **mRNA preparation** → **cDNA preparation**

V_L V_H

second PCR Introduction of restriction sites →Gel purification

V_L V_H

first PCR Introduction of restriction sites →Gel purification

plus antibody-specific primer

phagemid vector

V_H
MluI
MluI
Ligase
NotI

V_L gene pool

IR
Amp
pIII
ori

Electroporation DNA preparation

V_H
NcoI gene pool
HindIII

NcoI Ligase
HindIII

Electroporation

V_L sublibrary

IR
Amp
pIII
ori

Glycerol culture of the completed library (−70 °C)

Table 28.1 Diversity of different antibody gene libraries.

Type	Source	Complexity[a]
Hybridoma immortalization	Hybridomas	10^1 [b]
Immunized	Seropositive donors	10^6
Naïve repertoire (mixture of donors)	Mixture of donors	10^8
Genomic/germ line	Mixture of donors	10^{10}
Synthetic (random CDRs)	Sequence + oligonucleotide	10^{10}

a) Typical number of clones after primary transformation. This figure does not give the effective number of different antibodies, but the maximal upper limit. A direct determination is not possible and would reveal little about the functions of the library, which can be further restricted by other factors (number of complete inserts, inappropriate combination of heavy and light chains, folding efficiency).

b) Hybridomas often include the RNA of many different antibody genes/pseudogenes.

Fig. 28.6 Human antibodies can be generated by obtaining hybridoma cell lines from mice carrying the human immunoglobulin gene locus.

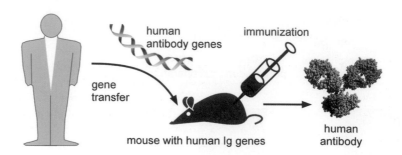

Fig. 28.6 Human antibodies can be generated by obtaining hybridoma cell lines from mice carrying the human immunoglobulin gene locus.

class switching and somatic hypermutation. These transgenic mice therefore produce human antibodies in mouse cells, which (as opposed to human hybridoma cells) can lead to stable hybridomas (Fig. 28.6). This serves as a system for the production of human antibodies, which methodically builts on the experience with mouse hybridoma technology to avoid the problems associated with the manufacture of human hybridomas. However, this system is based on the conventional method of immunization. This impairs the production of antibodies against small immunogens, toxins, or lethal pathogens. Affinity maturation occurs analogous to the human immune system, so that high-affinity antibodies can also be obtained. These methods have already been introduced successfully for the production of a few therapeutically relevant human antibodies.

28.3.2.2 *In Vitro* Selection Systems

In vitro selection systems (Fig. 28.7), in contrast to transgenic mice, do not require a complete immunoglobulin, but instead use their antigen-binding fragment, generally either in the form of a **single-chain Fv fragment (scFv)** or in the form of a **Fab fragment** (Fig. 28.2). The basic principle of these systems is to imitate our immune system by physically coupling function (antigen-binding domain) and genetic information (antibody-coding nucleic acids) into one particle. Such particles can be phage bacteria or yeast.

The most complex libraries to date use the expression of antibody fragments on the surface of filamentous phage (M13). **Phage display** has developed into the most robust *in vitro* selection system in the past two decades. Through the combination of the basic peptide display works of Smith (1985), with the possibility of production of antibody fragments in the periplasmic space of *E. coli*, a method was created for the *in vitro* selection of antibody fragments based on their binding characteristics (Figs 28.8 and 28.9).

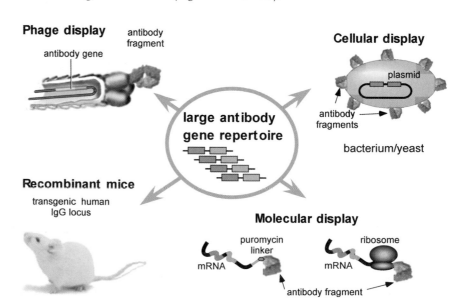

Fig. 28.7 Different *in vitro* selection systems for antibodies, which are based on the coupling of gene and function in a particle.

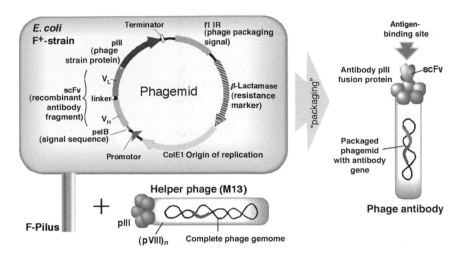

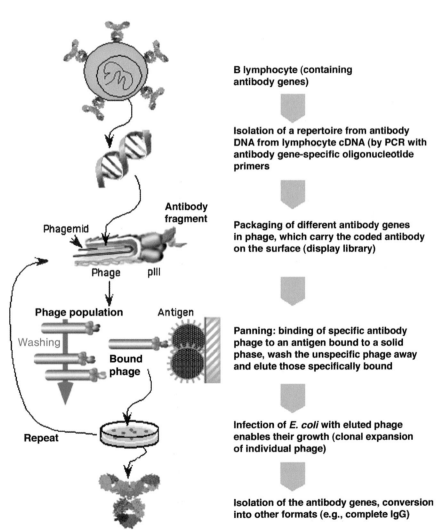

Fig. 28.9 Experimental flowchart of the selection of an antibody fragment via phage display.

B lymphocyte (containing antibody genes)

Isolation of a repertoire from antibody DNA from lymphocyte cDNA (by PCR with antibody gene-specific oligonucleotide primers

Packaging of different antibody genes in phage, which carry the coded antibody on the surface (display library)

Panning: binding of specific antibody phage to an antigen bound to a solid phase, wash the unspecific phage away and elute those specifically bound

Infection of *E. coli* with eluted phage enables their growth (clonal expansion of individual phage)

Isolation of the antibody genes, conversion into other formats (e.g., complete IgG)

Phage display is based on the fusion of scFv or Fab fragments with a surface protein, normally pIII (minor coat protein) of the **filamentous phage M13** (Fig. 28.8). Therefore, the antibody fragment is exposed on the virion (Fig. 28.8), which then can provide binding to a solid-phase immobilized antigen (Fig. 28.9). After reinfection with *E. coli* the selected clone is propagated and can be characterized further. It is important to the understanding of the potential, but also the possible experimental problems of phage display, to understand that the molecular interactions of only a few molecules are observed and used. Therefore, it is especially necessary to decrease the unspecific binding and to accurately characterize the obtained clones.

For the isolation of specific and high-affinity antibody fragments, the quality of the antibody library is crucial. In general, the affinity of the isolated antibody is proportional to the complexity of the initial library. Antibody libraries with more than 10^9 individual clones are currently being used and allow the selection of specific antibody fragments against different antigens. As a result of the development of this technology, antibody fragments against virtually any antigen can be obtained.

Furthermore, phage display allows to humanize monoclonal mouse antibodies. For this purpose the genes of an Fv or Fab fragment to be humanized are cloned in a phage selection vector and expressed as a fusion protein on the phage surface. The gene fragments of the light and heavy chains, respectively, are then substituted by a library of human antibody genes (**chain shuffling**). Through an affinity enrichment by for the antigen, a specifically binding antibody variant is then selected again. When these antibodies are assembled or if they are introduced consecutively for both chains (a process referred to as **guided selection**), then complete human antibody fragments can be obtained that recognize the same epitope as the original hybridoma antibody. Through the selective introduction of mutations specific to CDRs of the variable regions, the affinity of an antibody fragment for an antigen can be increased (**affinity maturation**).

Along this, direct coupling of antibody mRNA with that of its coded protein is also employed. With this method the coupling of the gene (mRNA) and the coded protein occurs directly on the translating ribosome or through a puromycin linker (Fig. 28.7), so that the selection of a specifically bound antibody fragment can be carried out completely in a cell-free *in vitro* system. The selection of a single clone from plates or membranes with tens of thousands of clones with the help of a robot is possible; however, it is not used often for the first selection round as the complexity of the library is still too large at this point.

28.4
Production of Recombinant Antibodies

28.4.1
Recombinant Production Systems

Recombinant antibodies in the form of antigen-binding fragments are initially selected and characterized in *E. coli* (see Chapter 15). *E. coli*, however, only has a limited suitability for the production of recombinant antibodies. Small functional fragments of immunoglobulins can easily be produced in *E. coli*, whereas production of complete IgG molecules has only been achieved in rare cases. The reason for this is the complexity of the molecule (four chains with many intra/intermolecular disulfide bridges) and the folding pathway involved, which is available in mammalian cells, but not in *E. coli*. A secretion into the periplasm is therefore essential for domain folding of the producible small fragments (scFv fragments, Fab fragments, scFab fragments, **diabodies**, among others) in *E. coli*. *In vitro* folding methods were developed, but they suffer from low yields of functional molecules. In *E. coli*, only a small fraction of the cytoplasmic synthesis capacity of bacteria can be used, as the cellular folding apparatus is already overloaded with less than 10% of the capacity for protein synthesis. Furthermore, due to the presence of an outer membrane, the production of Gram-negative bacteria requires a laborious processing of the bacteria in order to obtain the periplasmic extracts. All bacterial components have to be completely separated as they could act as an endotoxin.

Organism	Growth	Transformation	Yield	Glycosylation [a]
In vitro				
Reticulocyte lysate (rabbit)	Not necessary	Not necessary	Very low	No
Prokaryotic organisms				
E. coli				
Cytoplasm	Very fast	Simple	High/S–S refolding needed	No
Soluble fraction of the periplasm	Very fast	Simple	Low to medium	No
Periplasm inclusion body	Very fast	Simple	High/refolding necessary	No
Gram-positive bacteria				
Bacillus	Fast	Simple	High [b]	No
Streptomyces	Fast	Simple	High [b]	No
Proteus	Fast	Simple	High [b]	No
Eukaryotic organisms				
Yeast (*Pichia*) *Saccharomyces, Schizosaccharomyces*	Medium	Somewhat complicated	Variable [b]	Partially
Trichoderma	Medium	Complicated	High [b]	Partially
Aspergillus sp.	Medium	Complicated	High [b]	?
Baculovirus (insect cells)	Medium	Somewhat complicated	Variable to high	Partially
Mammalian cells (myeloma, CHO, COS)	Medium	Somewhat complicated	Variable to high	Yes
Transgenic Plants (tobacco)	Very slow	Very complicated	High [b]	Yes
Transgenic Animals	Very slow	Very complicated	High [b]	Yes

Table 28.2 Systems for the production of recombinant antibody fragments. (Modified from Breitling und Dübel, 1997 and Schirrmann *et al.*, 2008).

a) The type of glycosylation is very important for the biological function of the antibody. A completely correct glycosylation only occurs in mammalian cells with differences between the species and even cell lines.

b) With these systems a general estimation is not possible due to the small number of existing examples.

The optimal conditions for folding and **glycosylation** are provided by cells of the mammalian immune system, which are responsible for antibody production in our bodies. For example, in order for the antibody to bind the complementary component C1q or the cell surface receptor FcR, a correct glycosylation of Asn297 of the CH2 region is necessary. If, however, only the antigen binding is required, it is easier to produce the scFv, dsFv, or Fab fragments in yeast or bacteria. The production in transgenic animals is only advised if a large amount is required. Table 28.2 list a number of production systems for recombinant antibody fragments arranged according to specification and characteristics.

28.4.2
Purification of Recombinant Antibodies and their Fragments

In *E. coli* or eukaryotic secretion systems the supernatant is first obtained by centrifugation. The supernatant can then be separated from the lower molecular fraction by ultrafiltration and/or can be concentrated. In the case of intracellular expression or secretion in the *E. coli* periplasm, partial cell lysis (only of the outer membrane) is necessary. The further enrichment steps are normally carried out by **column chromatography** (see Chapter 10). Regularly used methods are ion exchange chromatography and molecular sieve chromatography. Better purification results are obtained by using specific binding as a separation principle.

These purification methods are named **affinity chromatography**. For most recombinantly produced antibody fragments, a two-step purification strategy has proved to be effective for the isolation of sufficiently clean material from cell extracts or cell culture supernatant. This is comprised of a combination of affinity chromatography with further column chromatography. In the case of *E. coli* expression of scFv fragments this second step is often molecular sieve chromatography, which serves to separate aggregates and dimers from monomers.

Therefore, affinity chromatographic methods have been implemented largely as the main purification step for recombinant antibodies. Two groups of purification methods exist. The first group, which can be characterized as antigen-specific methods, is based on the desired function of the recombinant antibody fragment itself (i.e., antigen recognition). In contrast, the second group applies the molecular characteristics of the antibody chain in order to achieve specific binding to the column material. These antibody-specific methods are not dependent on the antigen specificity. This is, however, only applicable when a corresponding binding region is still a part of the recombinant protein. Fc domains are necessary for effective chromatography with proteins A or G. If such interaction domains are not part of the antibody fragment, which is the case with scFc fragments, genetic fusions of the recombinant antibodies with small peptides (**tags**) are used instead. These tags specifically bind to the material in the column. The most common of these so-called tags is the His_6 tag, which binds to immobilized metals such as Ni, Co, Zn, and others (**immobilized metal affinity chromatography (IMAC)**) or tags that bind streptavidin or its variants (biotinylation tags or Strep tag) (see Chapter 16).

28.5
Formats for Recombinant Antibodies

Numerous changes to the constant domains of complete IgG molecules have been carried out. By changing the isotype (exchange of the C domains and retention of both V domains), the way of interaction with the immune system is affected. After identification of binding sites on the Fc fragment for cellular receptors, it was possible to modulate the binding characteristics of IgG antibodies via specific mutations of amino acids. This can lead to improved pharmacokinetic characteristics. Particular influence on therapeutic efficacy is attributed to the glycosylation pattern of the antibodies. Here, by using either selected cell lines with improved glycosylation or recombinant cells carrying a genetically changed repertoire of sugar processing enzymes, glycosylation can be optimized upon production.

Most studies, however, are concerned with the antigen-binding site itself and have developed a number of different formats with very different biochemical and pharmacokinetic properties, which are described in the following sections (Fig. 28.10).

28.5.1
Monospecific Antibody Fragments

Even if the antigen-binding characteristics of a monoclonal antibody in relation to its specificity and affinity appear ideal, its therapeutic use can be very restricted due to its isotype. The different isotypes are defined through the structure of the constant domain and affect the type of effector molecule that is activated after antigen binding. Many stem cells exhibit receptors for the Fc fraction of an immunoglobulin and therefore induce an antigen-independent background binding.

In applications that require the infiltration of an antibody in specific cells or the localization of an antigen (e.g., tumor imaging), it can be an advantage to re-

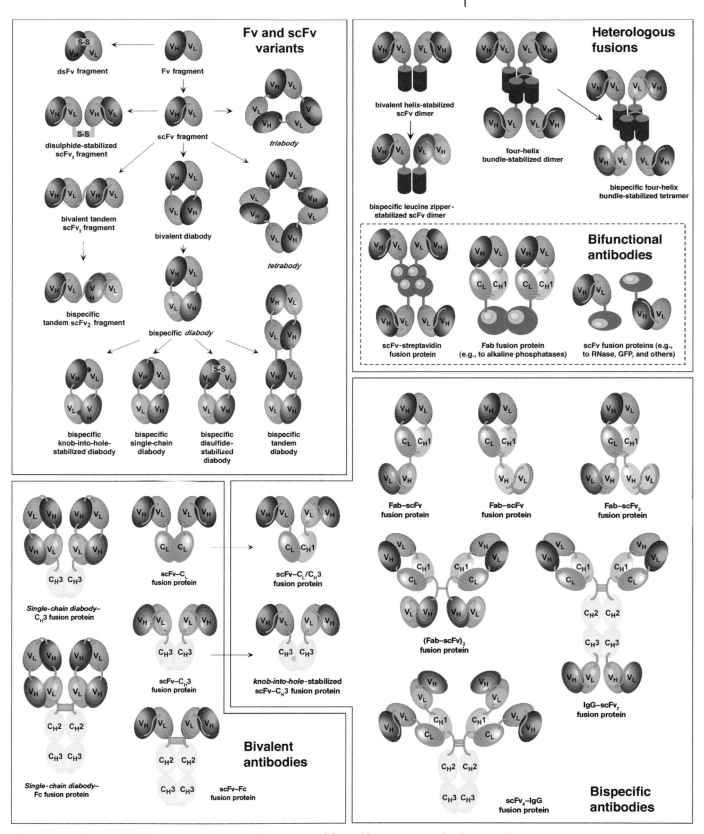

Fig. 28.10 A selection of different recombinant antibody variants and fusion proteins shows the enormous variety of constructs with new functions, which can be given to the antibody-binding site of immunoglobulin with the help of recombinant technology. Color scheme: red, V regions; blue, V regions with other specificity; turquoise, V regions tertiary specificity; yellow, regions from the Fc fragment; green, C$_H$1C$_L$ regions; orange and violet, heterologous fusion for di/oligomerization (orange) or with new functions (violet).

duce the size of an antibody through fragmentation or deleting the Fc fragment, respectively. Antibodies too small to be retained by the filtration limit of the kidney (approximately 60 kDa) suffer a very rapid clearance from the plasma. This only may be of advantage if rapid clearance is beneficial, such as in the case of some radionuclide-labeled antibodies.

28.5.1.1 Fab Fragments

In contrast to **bivalent IgGs**, **Fab fragments** are monovalent molecules. This can result in a reduction in their apparent binding affinity compared with the complete antibody if the antigen present is not monovalent. They no longer have the ability to form a precipitate with antigens. In solution they are comparatively stable like the complete IgG molecule, as their protein chains are linked via disulfide bridges. With production in *E. coli*, the necessity of synthesis of two different chains and their association in the periplasm means that the yield is decreased in comparison to single-chain production.

28.5.1.2 Fv Fragments

The Fv fragment of an antibody consists of the **variable domains of the heavy (V_H) and light (V_L) chain**. It constitutes the smallest entity of a human immunoglobulin that includes a complete antigen-binding site. Due to the often relatively weak cohesion of their variable domains, Fv fragments are rarely successfully produced in *E. coli*. Additional stabilization elements are necessary in order to prevent the dissociation of the subunits of the Fv fragment.

28.5.1.3 Single-Chain Antibody Fragments (scFv)

One possibility to apparently stabilize Fv fragments is to attach their V_H and V_L domains with the help of a short peptide to form so-called single-chain fragments (scFv, sFv) (Figs 28.2 and 28.10). The peptide linker must at the same time bridge the distance from 3.2 to 4.3 nm in order to guarantee the correct steric arrangement of both domains. Therefore, flexible, hydrophilic 15- to 20-amino-acid spacer peptides are normally used in order to link the carboxyl-terminal of the V_H with the amino-terminal of the V_L domain. With this arrangement of the variable regions the bridged distance is less than with the turned-around link. The $(G_4S)_3$ linker has proved particularly useful through its protease resistance and its large conformational flexibility.

The scFv fragments (26 kDa) usually have an antigen-binding activity comparable with the analogous Fab fragments (50 kDa) of the same antibody. Compared to Fv or Fab fragments, the construction of a single scFv gene greatly simplifies the production of recombinant antibodies. If the expressed protein domains exist on a polypeptide chain, the association of the V_H and V_L domains requires no further biomolecular reaction, but just an intramolecular rearrangement. Here, the concentration of both domains in the previous reaction space is high, their association to the Fv module is favored, and the scFv fragments are kinetically stabilized. A disadvantage of this format, however, is that some scFv molecules to form aggregates at high concentrations tend. In this case, a low affinity of both variable domains of a scFv module for each other leads to their temporary dissociation, after which the variable domains of a scFv fragment can associate with one of the others. This results in the formation of dimers, oligomers, and aggregates. The rate of dissociation depends on the length of the linkers as well as other external factors.

28.5.1.4 Single-Chain Fab Fragments (scFab)

Recently it has been shown that the introduction of a polypeptide linker between the Fd and LC chains of an antibody can have beneficial effects for the production of Fab fragments. This strategy combines the advantages of Fab (additional constant domains improve stability and provide detection by routine secondary antibody reagents in research) and scFv (better *E. coli* production). The elimination of the disulfide bond between Fd and LC proved particularly helpful to improve yields.

28.5.1.5 Disulfide-Stabilized Fv Fragments (dsFv)

An alternative approach to the stabilization of Fv fragments lies in the introduction of mutations in the scaffolding regions of both variable domains, which leads to the formation of a covalent linkage between the two contact surfaces. Structural analysis has identified a pair of amino acids in the complementary framework regions of the V_H and V_L domains, which, transformed in cysteine, are positioned at exactly the right distance from each other for the formation of disulfide bridges to occur. These dsFv fragments (Fig. 28.10) normally exhibit similar, sometimes even higher antigen-binding activity than their homologous scFv fragments. This is mainly a result of their improved stability against heat, denaturing agents, and proteases. Due to the direct covalent linkage between the contact surfaces of their variable regions they also tend not to aggregate at high concentrations. dsFv fragments can be more stable in blood serum than scFv fragments.

28.5.1.6 V_H and Camel Antibodies

Many experiments have been carried out to minimize further the antigen-binding region of an immunoglobulin. It has been attempted to separately express the V_H domains as the smallest functional entity of an antibody. However, this causes a lack of interaction with the corresponding V_L domain and therefore exposure of the hydrophobic contact surfaces, which in turn leads to reduced solubility and stability. A solution to this problem is the expression of V_H domains of immunoglobulins from camels and their relatives, since they have antibodies only formed from the heavy chains. The light chains as well as the C_H1 domains of the heavy chains are deleted in these antibodies. The improved solubility of these antibodies arises from substitution of amino acids on their surfaces, which would normally be masked by the variable regions of a light chain. Through the absence of the V_L domains, the complexity of this antibody repertoire is reduced. This is compensated with a CDR3 region longer and more heterologous than in human antibodies. So-called **single-domain VH antibodies** have been isolated from immunized camels and llamas. Single-domain V_H antibody gene libraries are also successfully used.

28.5.2
Multivalent Antibody Fragments

Natural immunoglobulins, especially the 10-armed IgM antibody, due to the fact that they are multivalent, display a significantly higher apparent affinity in comparison to the analogous monovalent Fab fragments. This is essentially a result of reduced dissociation of the antibody, where multiple interactions with two or more antigen molecules (e.g., from the cell surface) allow the rebinding of an initially dissociated Fv module without the entire molecule leaving the surface. An analogous increased avidity is achieved if Fab or scFv fragments are combined to form dimers, trimers, or large complexes.

The production of bivalent recombinant **F(ab')₂ fragments** was done by oxidation of the reduced cysteine of their **hinge** region. In addition, monovalent Fab fragments were expressed in *E. coli*, purified, and subsequently bound to homo-

with both constant domains of an IgG heavy chain. These strategies are admittedly not very suitable for producing bispecific heterodimers due to the symmetrical formation of the product. Nevertheless, the dimerization of bispecific single-chain diabodies through their fusion with the C_H3 domain or the entire Fc part of an antibody has been successful. Bispecific minibodies have been produced though the fusion of a scFv fragment with the constant domain of the light chain, as well as the fusion of other scFv fragments with the constant C_H1 domain of the heavy chain. The association of both of the different constant domains then resulted in the in vivo formation of bispecific heterodimers. In the previously described cases the antigen-bound scFv fragments, analogous to native IgGs, were always fused with the amino-terminal of the constant domains. Alternatively, scFv fragments can also be fused with the carboxyl-terminal of C_H1, C_L, or C_H3 domains, or with hinge regions, to generate such FabscFv or $FabscFv_2$ constructs with two or three antigen-binding sites. Fusion with bispecific $scFv_4IgG$ molecules, $(FabscFv)_2$, or $IgG(scFv)_2$ fusion proteins generates constructs with four antigen-binding sites.

A second method is the use of heterologous peptides or proteins as oligomerization domains. The short leucine zipper motif of the two transcription factors fos and jun (see Fig. 28.10) is useful here. These transcription factors form secondary structures, which tend towards dimerization, due to the formation of coiled-coils. This type of peptide could be used for the production of bivalent and bispecific antibodies.

Further, helical structures can serve as tetramerization domains for the formation of tetravalent antibody molecules. The short helix-loop-helix motifs function according to the same principle. Following formation of four-helix bundles, they dimerize and are used for the production of bivalent and bispecific scFv antibody constructs with two or four antibody-binding sites, respectively. Larger multivalent multimer scFv fusion proteins are obtained through the use of protein domains such as streptavidin, the tetramerization domain of the p53 protein, or the octamerization domain of the C4-binding protein.

28.6
Applications of Recombinant Antibodies

28.6.1
Clinical Applications

At present, about 24 antibodies are approved for clinical use, on a fast growing list. Over 500 products are currently being developed. As about a decade is needed to obtain market authorization for therapeutics, the current licensed antibodies reflect the state of **antibody engineering** from 10 years ago. From 1988 until the middle of the 1990s, recombinant humanization or chimerization (fusion of a mouse Fab' to human constant regions) developed as the methods of choice, while recombinantly produced complete human antibodies have since caught up as a main source of new developments of relevant antibodies. An overview of approved antibody products is given in Table 28.3.

The next generation of antibody therapeutics will include more examples using the possibilities of the bifunctionality/bispecificity approach, fusion proteins, and alternative production systems compared to the currently approved therapeutics, most of which consist of IgG molecules made in CHO cells.

28.6.2
Applications in Research and *In Vitro* Diagnostics

A large number of examples of the successful production of recombinant antibodies for research using phage display have already been described, including

Name (product/antibody)	Origin/type	Indication	Year of Approval
Orthoclone® (muromonab-CD3)	Mouse hybridoma	Graft rejection	1986
ReoPro® (abciximab)	Chimeric, Fab	Platelet aggregation	1994/97
Humaspect® (votumumab)	Human, radiolabeled	Colon cancer *in vivo* diagnostics	1996
Rituxan® (rituximab)	Chimeric	Non-Hodgkin's lymphoma	1997
Zenapax® (daclizumab)	Humanized	Graft rejection	1997
Simulect® (basiliximab)	Chimeric	Graft rejection	1998
Synagis® (palivizumab)	Humanized	Respiratory syncytial virus	1998
Remicade® (infliximab)	Chimeric	Rheumatoid arthritis, Crohn's disease	1998/99
Herceptin® (trastuzumab)	Humanized	Breast cancer	1998
Mylotarg (gemtuzumab ozogamicin)	Humanized immunotoxin	Leukemia	2000
Campath® (alemtuzumab)	Humanized	Leukemia	2001
Humira® (adalimumab)	Human	Rheumatoid arthritis	2002
Xolair® (omalizumab)	Humanized	Asthma	2003
Bexxar® (tositumomab-I-131)	Mouse, radioactive conjugate	Non-Hodgkin's lymphoma	2003
Raptiva® (efalizumab)	Humanized	Psoriasis	2003
Erbitux® (cetuximab)	Chimeric	Colon cancer	2004
Avastin® (bevacizumab)	Humanized	Colon cancer	2004
Tysabri® (natalizumab)	Humanized	Multiple sclerosis	2004
Lucentis® (ranibizumab)	Humanized, Fab	Macular degeneration	2006
Vectibix® (panitumumab)	Human	Colon cancer	2006
Soliris® (eculizumab)	Humanized	Paroxysmal nocturnal hemoglobinuria	2007
Cimzia® (certolizumab pegol)	PEGylated humanized Fab	Crohn's disease, rheumatoid arthritis	2008
Simponi® (golimumab)	Human	Rheumatoid arthritis	2009
Removab® (catumaxomab)	Mouse/rat bispecific	Ascites	2009

Table 27.3 Clinically approved antibody products. (Source: Rohrbach *et al.*, 2003; Reichert, 2009).

the first examples of antibodies that could not have been made in animals and that are specific for a particular conformation of the antigen. Despite the fact that commercial contractors of recombinant antibody production are usually focused on obtaining therapeutic antibodies, a growing number of academic initiatives now employ *in vitro* antibody selection to create the large numbers of antibodies needed for future proteomic and systems biology approaches. Three cases where these recombinant technologies are of particular advantage, or even offer the only way to succeed, are described below.

28.6.2.1 Recombinant Antibodies Selected to Avoid Cross-Reactivity

The advantage of recombinant production of antibodies is that it enables antibody-based testing for toxic substances or with specificities that cannot be achieved with immunization. For example, highly specific antibodies against the heroin metabolite 6-mono-acetyl-morphine (6-MAM) could be obtained from a phage display library, which in contrast to the previously available mouse antibodies can distinguish between this indicator of drug abuse and other morphinan alkaloids (e.g., those contained in cough syrup). This was achieved by adding the unwanted antigens as competitors during panning.

28.6.2.2 Intracellular Antibodies

In many cases, the deactivation of the corresponding antigens in the cell through the expression of antibody fragments in the cytoplasm has been attempted. This has frequently failed due to the inadequate folding of the antibody (**intrabody**) in the reduced environment of the cytoplasm. More success could be achieved if the natural intracellular IgG production pathway was used. Antibodies with signal sequences as well as retention signals for the endoplasmic reticulum (peptide sequence: KDEL) have been shown to functionally inactivate molecules from the secretory/membrane pathway. This offers a novel approach for functional genomics, in particular since potentially useful antibodies are provided by research initiatives in increasing numbers. Here, a knock-down phenotype of an unknown open reading frame may be observed requiring just one subcloning of the selected scFv into a respective eukaryotic vector and transfection into a suitable cell.

28.6.2.3 Recombinant Antibodies as Binding Molecules for Arrays

After sequencing of the human genome there is an enormous demand for binding molecules for the further functional analysis of the new genes. Here, antibodies are the tools of choice for the biochemical and cell biological characterization of the gene products and their functions (immunoblots, immunohistology, fluorescence-activated cell sorting (FACS), purification, pull-downs, etc.). The use of hundreds of different antibodies in a **protein microarray** (proteome chip) is already being intensively evaluated. Many proteins have splice variants or they are modified to form polymers or complexes. The detection of these changes is essential for the understanding of gene function and can only be assessed on the protein level. Therefore, the demand for antibodies is rapidly increasing beyond the number of 25 000 supposed human genes. Without recombinant technology the production of these antibodies would require around a million research animals! The effort of producing an antibody against every gene product and their relevant variants can therefore be best accomplished with a highly parallel *in vitro* selection approach, such as phage display.

28.7
Outlook

Recombinant antibodies are not only the most important, but also the most rapidly growing group of future protein therapeutics. Increasing numbers of recombinant antibodies will therefore be developed into high-affinity, protein-based therapeutics and diagnostics. In the next few years it is very possible that further solutions will be available for improved Fv stabilization with and without disulfide bridges, higher expression rates, as well as for the optimal design of fusion proteins. They will be used to activate the body's own T cells against tumors. It may also be possible to introduce recombinant viruses for cell-specific gene therapy. Furthermore, novel fusion proteins will open up a completely new field of application for recombinant antibodies. Alternative production systems will improve yield or activity, at lower cost. Nonmammalian systems (e.g., insect cells) as well as transgenic animals and plants should be able to establish themselves soon as worthwhile alternatives for industrial production (see Chapter 16).

Finally, the recombinant *in vitro* selection of antibodies offers an affordable way of producing antibodies in large numbers for functional genomics, proteomics, and interactomics projects, where tens of thousands of different antigens have to be analyzed. Here, their recombinant nature offers additional benefit by allowing applications (like intrabody studies) that cannot be attempted with animal-derived antibodies.

29
Transgenic and Gene-Targeted Mice and their Impact in Medical Research

Learning Objectives

The function of a single gene is revealed best when it is turned "off" or "over-expressed" in a living species. This chapter gives in overview of the methods employed for generating transgenic and gene-targeted mice. Selected examples describe the outstanding importance of genetically manipulated mice in biomedical research.

29.1
Overview

Most of our current knowledge on the function of individual genes is derived from experiments in the test tube or in cell culture. However, the controlled manipulation of genes in the mouse permits an analysis of gene function in living animals (**reverse genetics**). For example, additional cancer genes lead to enhanced tumor growth or additional growth hormone genes promote body size (Fig. 29.1). On the other hand, endogenous genes can be also inactivated. If a receptor gene for fast signal transmission in the brain is switched off, then the fast signal transmission in the brain is impaired, and consequently learning and memory of the animals are affected.

The sequencing of mammalian genomes has revealed more than 10 000 new genes and ongoing genetic screens of human patients provide a huge number of gene mutations that are correlated with human disease. The functional analysis of new genes and gene mutations will uncover their biological importance in live animals, and thus can contribute to novel strategies in biomedical research.

All global genetic modifications in mice are carried out in early embryos. Two fundamentally different genetic interventions can be performed (Fig. 29.2): a gene can be added to any genome position, in which case the generated mouse is called "**transgenic**", or a specific gene can be deleted or amended at its endogenous locus, in which case gene-modified mice are referred to as "**knock-out**" or "**knock-in**" mice, respectively. More up-to-date mouse models combine transgenic and knock-in animals to achieve cell-type-specific, conditional gene expression. These mouse models are called "**compound transgenic**", since they need to be generated by the breeding of several, independent transgenic and gene-targeted mice. In those mouse models a pharmacological controlled transgene expression determines the developmental time window and the cell type or tissue for the expression (or "shut down") of the manipulated gene(s).

29.2
Transgenic Mice

Transgenic mice are produced by infection of fertilized mouse oocytes with retroviral DNA or by injecting the pronucleus of oocytes with the DNA fragments to be inserted into the mouse genome (Fig. 29.2). After the viral or injected

Fig. 29.1 First example of a human gene (growth hormone gene) expressed in another animal (taken from the cover of *Science* 222 (4625), 1983). The mouse on the right contains a foreign gene for the human growth hormone. The sibling on the left has no human growth hormone gene and therefore is smaller. (Cover credited to R. L. Brinster and R. E. Hammer, School of Veterinary Medicine, University of Pennsylvania, Philadelphia, PA 19104.)

An Introduction to Molecular Biotechnology, 2nd Edition.
Edited by Michael Wink
Copyright © 2011 WILEY-VCH Verlag GmbH & Co. KGaA, Weinheim
ISBN: 978-3-527-32637-2

Fig. 29.2 Experimental flowchart. All gene manipulations are performed in early mouse embryos. The manipulated embryos are transferred to and delivered by foster mothers. In the offspring of the embryo-derived mice the introduced gene mutation or ES cell can be monitored by genomic PCR (of tissue samples) or simply by the coat color, respectively. Mice of the first-generation (F₁) that carry a heterozygous transgene (Tg) or targeting allele (+/–) are used to establish new mouse lines.

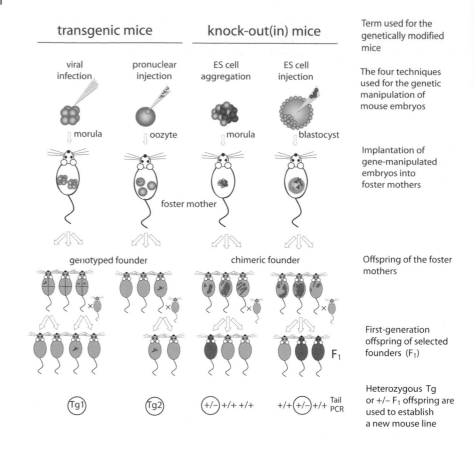

DNA has entered the nucleus it integrates itself at random into the genome of the cell.

29.2.1
Retroviral Infection

Retroviral vectors introduce transgenes in early-stage mouse embryos via retroviral vectors. These vectors use the characteristics of retroviruses to integrate themselves into a site in the genome of the infected cells. Therefore, the gene to be inserted is flanked by the integration elements of the retrovirus – the **long terminal repeats (LTRs)** (see Section 15.2.6.3). LTRs encode packaging and expression signals that promote, in so-called helper cell lines (see Section 29.5.1), the generation of infectious virus particles used to infect 2- to 4-day-old embryos (**morula**) (Fig. 29.2). The infected viral RNA is reverse transcribed to DNA (see Section 3.3) and integrates itself into the genome of the cell after the first or second cleavage division, so that many embryos consist of at least two cell populations (i.e., cells with and without the **retroviral transgene**). The infected embryos are then transferred into and delivered by foster mothers (Fig. 29.2). Six weeks after their birth the so-called transgenic **founder** mice are mated. The transgene is only passed on to offspring if the germ cells of the founder animals contain the retroviral transgene, whereby all offspring of a single founder represent one **transgenic line**. Mainly functionally important genes were identified with retroviral vectors. Thus, by random integration of the viral transgene, a functionally important gene can be destroyed, whose loss is strictly correlated with the appearance of an anomaly, illness, or abnormal behavior in the mouse concerned. In such cases, the destroyed gene must be identified in order to identify the underlying genetic components of the phenotype. The analysis of transgenic littermates that are **heterozygous** or **homozygous** for the viral insertion reveals if the virus-induced gene effect is **dominant** or **recessive**.

Recently, the development of lentiviral vectors, such as **HIV** (see Section 3.3), reactivated the application of virus infection for generating transgenic lines. The stable expression of virus-delivered transgenes can be obtained using lentiviral vectors.

29.2.2
Pronuclear Injection

Transgenic mice can simply be obtained by injecting minigenes into the pronucleus of fertilized oocytes (Figs 29.2 and 29.3). The injected minigenes are inserted in multiple copies into the genome of the oocyte randomly. For pronuclear injections there is no size limit for the injected DNA fragment. In retroviral vectors, however, huge transgenes cannot be packaged. Minigenes typically represent a eukaryotic transcription unit consisting of a promoter followed by a small intron, an open reading frame for the **gene of interest (GOI)**, and a transcriptional stop (see Section 15.2.6). Minigenes are constructed in *Escherichia coli* and released from plasmid backbones prior to their injection.

The embryos (in this case fertilized oocytes) for pronuclear injection are obtained from pregnant donor females. Two days before mating, the females are intraperitoneally injected with the serum of pregnant mice and on the day of mating ovulation is induced via injection with chorion gonadotropin, increasing the number of ovulated oocytes from five to eight through up to 40. Afterwards, the females immediately mate so that the fertilized oocytes can be collected the following morning via **oviduct flushing**. In the newly fertilized oocytes (**prezygotes**) the nucleus of the spermatozoon (the male pronucleus) remains separate from the nucleus of the oocyte until the latter has completed its meiosis with the formation of the female nucleus. During this phase, the male pronucleus is visible (Fig. 29.3). Immediately after the flushing, the pronuclear injection is carried out – using a holding capillary to keep the prezygote in place, 1–2 pl of a minigene solution is injected into the pronucleus using a fine injection pipette of a micromanipulator.

After microinjection, the **oocytes** are taken into culture. If the cleavage divisions are initiated, and only then, the embryos are reimplanted into foster mothers. The foster mother is prepared for the uptake of an embryo by mating with a vasectomized (sterile) male. In female mice the mating act is necessary for successful **nidation** (implantation). Around 100 microinjected oocytes are implanted into three or four pseudopregnant females. However, just a quarter of

Pronuclear injection

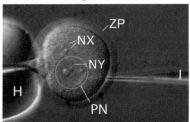

Cultured morulae

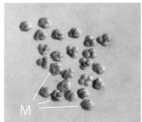

Blastocyst injection

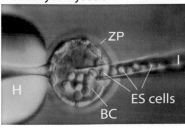

ES cell aggregation

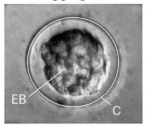

Fig. 29.3 Gene manipulations in early mouse embryos. Holding pipette (H), male pronucleus (PN), nucleoli male pronucleus (NY), nucleoli female pronucleus (NX), zona pellucida (ZP), injection needle (I), blastocyst (BC), four- to eight-cell morulae (M), embryonic body of ES and morula cells (EB), and Petri dish cavity (C).

the implanted embryos grow in the uterus and are delivered by the foster mother after 20 days.

Depending on the skills of the experimenter, 0–15% of the mice born carry the injected transgene. The presence of the transgene is revealed by **polymerase chain reaction (PCR)** (see Section 13). For this, genomic DNA is isolated from **tail biopsies** of potential transgenic mice. Tail-PCR-positive mice are called transgenic founders. A confirmation of the result by **Southern blotting** or through a second independent **PCR** is recommended in order to avoid false-positives (see Section 11.3.1). Each founder carries the transgene in its genome; however, copy number and transgene integration site differ from founder to founder. Both the gene environment and the copy number affect the expression pattern of the transgene; thus, its expression is highly variable among founders. PCRs on tail DNA from offspring demonstrate whether the transgene is stably inherited through the germ line of founder animals. Tissue expression and strength of transgene expression are determined in F_1 offspring (Fig. 29.?). Only such founders whose offspring show the desired characteristics are used to establish independent transgenic lines. F_1 offspring are heterozygous for the transgene. Brother and sister **inbreeding** produce offspring homozygous for the transgene. The failure to generate homozygous offspring is explained by the inactivation or impairment of a vital gene caused by the chromosonal insertion of the transgene.

The generation of a transgenic animal by pronuclear injection is routine, although success is haphazardly determined by the site of transgene insertion, copy number, and stability of the transgene. The integration affect is less pronounced when huge DNA fragments encoded by bacterial or yeast artificial chromosomes (**BACS and YACS**; see Section 15.2.1/Table 15.1), respectively, are use in pronuclear injections. The **GENSAT** program (www.gensat.org) used hundreds of **Green Fluorescent Protein (GFP)**-expressing BACS to visualize in transgenic mice the expression profiles of genes specifically expressed in the **central nervous system (CNS)**.

Nevertheless, the genetic manipulation of mice by random insertion of transgenes was complemented by a substantially more precise method – the exchange of genes via **homologous recombination**.

29.3
Homologous Recombination: knock-out (-in) mice

Homologous recombination (**gene targeting**) between identical DNA sequence segments can be used to modify an endogenous gene at its native gene locus. Homologous recombination is mainly used to delete and to modify specific genes in the mouse. These mice are described as gene "**knock-out**" and "**knock-in**" animals, respectively.

The gene targeting is performed in pluripotent **embryonic stem (ES) cells** growing in cell culture. Pluripotent ES cells are undifferentiated cells obtained from blastocyst embryos (see below). ES cells divide in cell culture and maintain the ability to differentiate into any cell type, including functional germ cells.

The DNA fragment that is used for ES cell manipulation (by simple cell electroporation and selection; see Section 15.2.7.3) is described as a **targeting vector** (see Section 15.2). Over a distance of 10–20 kb the targeting vector DNA sequence is identical to the gene to be altered. A **selection marker (neomycin resistance)** and the desired gene manipulation are inserted within this segment. This might be a missing exon, a point mutation, or also an additional indicator/reporter gene. Furthermore, the **targeting vector** also contains plasmid elements for its replication in *E. coli*. Upon homologous recombination with the endogenous gene of transfected ES cells, the plasmid elements are lost. The selection marker is exchanged together with the intended gene manipulation for the homologous region in the chromosome. This procedure takes place only at one of both alleles and the ES cell carries the targeted along with the wild-type allele (**heterozygote**).

After successful targeting the selected ES cells are injected into 3- to 4-day-old embryos (**blastocysts**) isolated again from superovulating females. Between 20 and 30 identically manipulated ES cells are injected per blastocyst (Fig. 29.3). The injected blastocysts are reimplanted in the uteruses of pseudopregnant foster mothers. As soon as the pups of the foster mothers show the first signs of fur, the efficiency of the ES cell integration can be evaluated (Fig. 29.4). If ES cells obtained from an embryo of a mouse line with brown fur color (**agouti**) are fused with blastocysts from a line exhibiting black fur (i.e., **C57Bl6**), the coat color of the new born mice is brown and black spotted (Fig. 29.4). Mice that are produced via ES cell injection are called **chimeric founders** (Fig. 29.4); chimeric because these mice are composed of two genetically different cell lines (Fig. 29.2). Offspring of the described brownblack chimeric founder have either brown or black fur when a mouse with the recessive black fur color (i.e., C57Bl6) is used for the mating. Since each offspring derives from a single germ cell (black or brown) of the chimeric founder the chimerism is not transmitted to the offspring. Thus, all cells in a single offspring are genetically identical. Either the genetic information of the dominant agouti (injected ES cells) or the recessive black coat color (injected blastocyst) is passed on to the offspring (Fig. 29.2). Since the injected "agouti" ES cell line is heterozygous for the targeted gene, the targeted gene is present in only 50% of the agouti colored offspring (Fig. 29.2).

Alternatively, ES cells can also be cocultured with four- to eight-cell embryos (morula). After 24 h the morula and ES cells form **embryonic bodies** (Fig. 29.3) that can be implanted in the uteruses of pseudopregnant mice, where they differentiate further and are finally born as so-called **aggregation chimeras**.

In contrast to pronuclear injection and viral infection, blastocyst-injected chimeric founder animals are all equivalent. They have been generated from cells of a single heterozygous ES cell clone. Fifty percent of the first-generation offspring (F_1) carry the inserted gene manipulation in one set of chromosomes (Fig. 29.2). The gene defect in the homozygous state can be examined following inbreeding. Mice that are homozygous for the deleted gene might be impaired in vital functions, so that many lines are bred as heterozygotes.

New experiments have been successful in producing chimeric mice almost exclusively from ES cells. For this, morulae were tetramerized prior to ES cell fusion. Since these the tetramerized cells contain two complete sets of chromosomes, they do not differentiate in embryonic tissue. However, they still participate in the formation of extraembryonic tissue, such as the **placenta**. Only cells from the injected ES cell line are recruited for the formation of the actual embryo. Therefore, it is possible to produce viable mice in a single step from ES cells.

Fig. 29.4 Chimeric founders. The efficiency of ES cell integration into blastocysts is monitored by bright spots in the fur of juvenile offspring. The injected R1-ES cells derived from SV129 mice with brown fur (left) and from E14-ES cells giving rise to cream-brown coat colors (right). Blastocysts for ES cell injection were isolated from black C57Bl6 mice (The picture on the right was kindly provided by F. Zimmermann, Uni-Heidelberg.)

29.4
Conditionally Regulated Gene Expression

In conventional gene knock-ins (-outs) the gene manipulation is global and not restricted to organs or cell types. For cell-type-specific and pharmacological controlled gene manipulations in the mouse, transgenic and targeted genes are combined (Fig. 29.5). **"Compound transgenic"** mice are generated by mating different independent mouse lines. They express a transgenic activator gene together with a targeted or transgenic responder gene. The activator gene determines the cell-type-specific expression, thus inducing the responder gene only in cell populations that contain the active responder protein. To achieve temporal control the activators are sensitive to antibiotics or hormones. The most commonly used drugs, **doxycycline** and **tamoxifen**, have no effect in wild-type mice;

400 | 29 Transgenic and Gene-Targeted Mice and their Impact in Medical Research

Cre/lox-induced gene inactivation

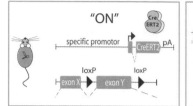
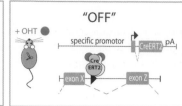

Tetracycline inhibited gene expression

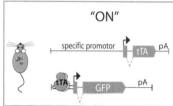
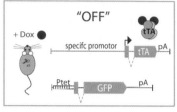

Fig. 29.5 Conditional gene expression in "compound transgenic" mice. (Upper part) The transgene for the 4-hydoxytamoxifen (OHT)-responding Cre recombinase (CreERT2) is expressed by a specific promoter in defined cell types of a gene-targeted mouse encoding a GOI with a functionally important exon (exon Y) flanked by two loxP sites. In OHT-treated mice the CreERT2 is released from cytoplasmic protein complexes; it can enter into the nucleus and removes exon 11 by inducing a recombination between the loxP sites. Thus, exon Y is deleted from both alleles of the floxed gene only in those cells that express CreERT2 (shown for one allele). (Lower part) The transgene of the tetracycline-sensitive transcription factor tTA (activator) expressed by a tissue-specific promoter and a second transgene encoding the tTA-dependent promoter (Ptet) driving GFP expression are encoded in the same mouse. In tTA-expressing cells tTA binds to Ptet, inducing GFP expression. Under doxycycline (DOX), a potent tetracycline derivative, the tTA/Ptet affinity is strongly reduced and GFP expression is stopped. pA, polyadenylation signal; dashed lines, introns. Activator and responder genes are combined by matings of independent transgenic and/or gene-targeted lines.

however, in the conditional compound mice they change the potency of the activator and as a consequence the responder gene is switched "on" or "off".

The **Cre/lox** and the reversible **tetracycline system** (see Section 15.2.6) are well-established technologies for conditional gene regulation in the mouse. Both systems permit gene function analysis in very well-defined cell populations in the mouse.

29.5
Impact of Genetically Modified Mice in Biomedicine

Genetic predispositions were identified in a large number of chronic diseases. The genes involved have been described to some extent. The respective candidate genes can be introduced into the mouse genome in such a way that they correspond to the human predisposition. If the genetically altered animals develop a similar disease as described for genetically predisposed humans, the mouse represents a useful animal model for this disease. The mouse model enables very detailed molecular, pharmacological, and behavioral analysis. Transgenic mouse models are used in preclinical trials to investigate the potency of drugs and to uncover unpleasant side effects. Mouse models for many genetically based diseases, such as arthritis, muscular dystrophy, cancer, hypertension, endocrine disorders, and coronary diseases, have already been described in the literature. Here, the focus is on neurodegenerative diseases and psychological disorders.

29.5.1
Alzheimer's Disease

Alzheimer's disease is a neurodegenerative disease characterized by the progressive loss of cognition and memory. Alzheimer's fibrils accumulate in neurons and thick, extracellular deposits are present in dendrites (referred to as **senile plaques**). The main component of senile plaques and **amyloid deposits** is β-amyloid (Aβ) of about 4 kDa – a proteolytic cleavage product of the βA4 amyloid precursor protein (APP). Molecular details for the formation of Aβ aggregates are still unsolved; however, gene mutations of APP and in presenilin-1 and -2 were identified in human patients with inherited Alzheimer's disease. Genetically manipulated mice carrying some of these mutations develop senile plaques in the brain with increasing age. Both the overexpression of the mutated APP gene in the brain of the mouse through pronuclear-injected transgenes as well as the gene-targeted mutations in the endogenous mouse APP gene were used for the generation of reliable Alzheimer mouse models. With these mouse models, biologicals or pharmacological substances that dissolve the Aβ deposits or prevent their formation can be developed. Other genes discussed in the onset of Alzheimer's disease (i.e., apolipoprotein E) are being intensively analyzed using similar mouse modes. There is justified hope that in the near future some therapeutic treatment of Alzheimer's disease can be realized.

29.5.2
Amyotrophic Lateral Sclerosis (ALS)

Similar to Alzheimer's disease, **ALS** is a progressive neurodegenerative disease with a late onset. Motor neurons degenerate, which leads to muscle weakness and atrophy, and most patients die from respiratory failure or pneumonia. In 5–10% of ALS patients an autosomal dominant inherited component is the cause of the disease. In those patient more then 50 gene variants of the ubiquitously expressed copper/zinc superoxide dismutase 1 (SOD1) were identified. Therefore, oxidative stress was linked to ALS. However, SOD1 knock-out mice show no ALS or ALS-like symptoms, indicating that the lack of SOD1 function is not causally linked to ALS. Similarly, in mice, transgenic overexpression of SOD1 variants induced accumulation of neurofilament (NF) comparable to NF aggregated in neurons of ALS patients, but the mice showed no ALS phenotype. This confirmed the dominance of the SOD1 mutations and showed that additional factors – possibly all participating in the same destructive pathways – contribute to ALS. For example, transgenic overexpression of an assembly-disrupting mutant version of NF-L led to selective degeneration of spinal motor neurons accompanied by the accumulation of NFs and denervation of skeletal muscle. This demonstrated that NF mutations can give rise to specific degeneration of motor neurons and muscle wasting. Using a conditionally controlled expression of the NF-L mutant would be perfect to examine whether the ALS symptoms are reversible and disappear when the NF-L mutation is switched off in compound transgenes. The clinical and therapeutic impact of such a finding would be tremendous.

The initial studies with SOD1 and NF-L mutants provided important new insights into molecular mechanisms underlying ALS. Many more additional studies with genetically manipulated mice are necessary to uncover the molecular details of this possibly multifactorial disease.

29.5.3
Psychological Disorders

In other nondegenerative diseases of the nervous system, such as **schizophrenia**, **depression**, **autism**, and **addiction**, the genetic components are far from being resolved even though the genetic disposition for these disorders is well established. Apart from genetic studies in humans, for example, in mutant mice more than 150 genes contributing to excessive **alcohol consumption** could be identified. Essential molecular players in addiction were described in functional detail using conditional **AMPA** (α-amino-3-hydroxyl-5-methyl-4-isoxazol-propionate) and **NMDA** (*N*-methyl-D-aspartic acid) receptor knock-out mice. It could be shown that drug-induced changes in the dopaminergic reward system in the brain are mediated by the cross-talk of the glutamate-responsive NMDA and AMPA receptor channels. In the first place the cross-talk is attenuated by one type of **metabotopic glutamate receptor** (mGluR-1). However, after repetitive drug exposure this "molecular brake" becomes nonfunctional and the drug-induced neuronal plasticity is distributed all over the entire reward system in the brain. Numerous other studies provide further evidence that pharmacological, physiological, and behavioral studies of genetically manipulated mice are solid experimental tools in research on mental disorders.

29.6
Outlook

Several new methods for the manipulation of genes in mammals that are currently being explored look very promising. Recently, ES cells were documented for rats and sequence-specific zinc finger nucleases have been used for targeted manipulation of rat genes during pronuclear injections. Induced pluripotent stem cells (iPSCs) open new gates for the generation of chimeric animals. The iPSCs can be obtained by reprogramming somatic cells. Several gene expression regulating factors (**Oct4**, **Sox2**, **myc**, and **Klf4**) need to be coexpressed in somatic cells to induce reprogramming. It is conceivable that the refinement of animal models for biomedical research will strongly benefit from the very advanced development of genetic manipulations.

30

Gene Therapy: Strategies and Vectors

Learning Objectives

Many hopes are pinned on the developing research area of gene therapy. There is the promise of momentous progress in healing many illnesses, including immunological problems, cardiovascular disease, metabolic dysfunction, and cancer. After a number of initial failures, the euphoria of the early years has now turned into cautious optimism. A prerequisite is multidisciplinary cooperation of scientists to introduce gene transfer medication as standard therapy. The genes that are involved have to be identified by molecular biologists, while virologists have to develop efficient and safe vectors that take the genetic cargo to its target – the diseased tissue. Cell biologists have to create methods for easier gene transfer and to find stem cells that can be used for the regeneration of failing organs. Physicians have to carry out clinical studies, using patient-friendly optimized vectors.

30.1
Introduction

Gene therapy is one of the key technologies of the twenty-first century. Hopes are high to find **new treatments** for cancer, AIDS, heart attack, stroke, and other common ailments. Everything seems feasible and the ideal of eternal youth seems within reach. However, success in gene therapy has been scarce so far and the setbacks have been numerous.

The major discoveries on which gene therapy is based go back only a few decades. In 1944, Avery showed that **DNA** is the substance that stores **genetic information**. In 1953, **Watson and Crick** proposed that DNA had a **double-helix** structure. The **triplet structure** of the **genetic code** was deciphered in 1961. In the same year, **mRNA** was discovered, revealing the basic mechanism for the **translation** of genes into proteins. By **cloning** eukaryotic genes into bacterial plasmids (for the first time in 1974), the *in vitro* production and analysis of genes was revolutionized. In 1977, methods for **sequencing DNA** were developed. In 1979, the first **cancer genes** (oncogenes) were identified. We know now that cancer can be due to genetic changes (mutations). In 2001, the **sequence of the human genome** was published and it came as a great surprise that less than 25 000 protein-coding sequences were found, of which more than 10% (over 4000 genes) are responsible in some way for the emergence of **monogenic hereditary diseases**.

In 1990, American scientists carried out the first gene therapy in a patient (Table 30.1). Ashanthi DeSilva, then 4 years old, had a serious **congenital immune deficiency**, which mostly results in premature death, that was caused by **adenosine deaminase (ADA) deficiency**. At the National Institutes of Health (NIH), she was injected with her own blood cells, transfected with an intact wild-type ADA gene. As the doctors did not have enough confidence in the new genetic treatment, she was also given ADA obtained from cattle blood. To this day, she is still given the traditionally produced enzyme. DeSilva is doing well, going to

An Introduction to Molecular Biotechnology, 2nd Edition.
Edited by Michael Wink
Copyright © 2011 WILEY-VCH Verlag GmbH & Co. KGaA, Weinheim
ISBN: 978-3-527-32637-2

Table 30.1 History of gene therapy.

1989	First gene labeling study (Rosenberg)
1990	First gene therapy study in patients with adenosine deaminase deficiency (Blaese, Culver, Anderson)
1994	First gene therapy in Germany
2000	First clinical evidence for the effectiveness of gene therapy (hemophilia B, immunodeficiency)
2001	More than 500 gene therapy studies involving more than 3400 patients carried out
2010	More than 2000 gene therapy studies worldwide

school, and exercising like a normal teenager. Whether this is a result of gene therapy or of conventional therapy remains an open question.

30.2
Principles of Somatic Gene Therapy

Gene therapy in general means the **introduction** of one or more foreign genes into an organism for the benefit of the individual concerned. The objective of somatic gene therapy is to permanently cure **hereditary** or **acquired genetic defects** by insertion of normal genes to target cells in the body. Whether a gene should be introduced into the germ line to cure hereditary diseases remains a controversial ethical issue. In most cases, gene transfer is carried out *ex vivo* (Fig. 30.1). The **target cells** are isolated from the organism and grown in culture where they are **transfected** with the therapeutic gene and then **reimplanted** into the body. This means that *ex vivo* gene therapy is only possible if it is relatively easy to take cells from the patient's body and propagate them in sufficient numbers outside the body. Often, the reimplantation of the cells transfected *in vitro* is only partially successful and the *in vivo* expression of the transgene is thus only effective for a short time. Two major problems are encountered in the development of an **effective gene therapy**. The first problem is that most diseases are not caused by a single deficient gene, but by **several**, and **environmental factors** such as nutrition, lifestyle, or infectious agents also nearly always play a role. The most **widespread diseases**, including hypertension, cardiac disease, apoplectic insult, and cancer, are therefore unlikely to be cured by sorting out a few genes. This is why many gene therapist are now focusing on **monogenic diseases**, which brings us to the second problem, technical in nature.

How can genetic material reach its target? In order to treat **cystic fibrosis** (Fig. 30.2) – the most widespread hereditary metabolic defect – it is a question of targeting a sufficient number of lung cells in order to introduce the normal gene. This has been attempted using **adenoviruses**. As these normally cause colds, many pathogenic genes were removed from the viral genome and replaced by the intact human genes. The manipulated adenoviruses reached the cells in the

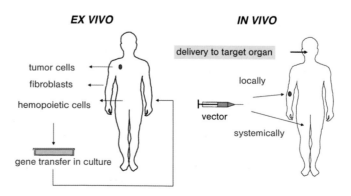

Fig. 30.1 Methods in somatic gene therapy.

Disease	Target cells	Therapeutic genes
ADA deficiency	Lymphocytes (blood), bone marrow stem cells	ADA
Cystic fibrosis	Lung epithelial cells	CFTR
Familial hypercholesterolemia	Liver cells	LDLR
Duchenne's muscular dystrophy	Muscle stem cells	Dystrophin
Hemophilia B	Liver cells, skin cells	Factor 9
Melanoma	Tumor-infiltrating lymphocytes	TNF, HLA-B7
Lung cancer	Lung cancer cells	p53
Brain tumors	Brain tumor cells	HSV-tk
Arthritis	Joints	Interleukin-1
AIDS	Lymphocytes	HSV-tk
Alzheimer's disease	Fibroblasts	NGF
Prostate cancer	Tumor cells	Interleukin-12
Hepatocellular cancer	Tumor cells	TK
Breast cancer	Tumor cells	MDR1

Table 30.2 Gene transfer studies (selection).

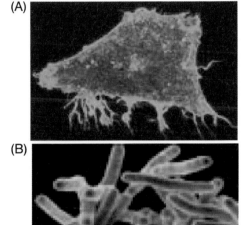

mucous membranes in the patients' lungs and expressed the **therapeutic gene**. However, the gene was only episomally present in the cytoplasm of the lung cells and was not integrated into the genetic material of the nucleus. After a few mitotic divisions, it had been diluted so far that its therapeutic effectiveness only lasted 4 weeks. Meanwhile, the adenoviruses had triggered an immune reaction in the body. The adenoviruses were destroyed and could no longer function as gene vehicles. Raising the dose was not the answer. As a result of allergic reaction to adenoviral vectors several patients developed pneumonia and the study had to be terminated. At present, there are several monogenetic diseases that are targets for the development of an effective gene therapy, including **Duchenne's muscular dystrophy**, **hemophilia**, and others (Table 30.2). Other approaches by gene therapy besides the replacement or the introduction of a missing gene include suicide therapy, such as with the herpes simplex virus thymidine kinase (HSV-tk). Cells containing HSV-tk may be selectively killed by infusion with ganciclovir, which is nontoxic unless metabolized by HSV-tk. The immune system is supported by the introduction of cytokine genes such as interleukin-12 or -6 and various other protocols are under investigation.

30.3
Germ Line Therapy

Under current **legislation**, it is illegal in Germany and the United Kingdom to introduce therapeutic genes into the germ line. UNESCO (United Nations Educational, Scientific, and Cultural Organization) has produced a *Universal Declaration on the Human Genome and Human Rights*, which was signed in November 1997, that also includes a ban on **germ line therapy**, and says there shall be no discrimination against individuals on the basis of genetic information and excludes so-called **reproductive cloning**. In the United States, there still is no legal ban in place. In 1998, 2001, 2004 and 2007, the United States House of Representatives voted whether to ban all human cloning, both reproductive and therapeutic. Each time, divisions in the Senate over therapeutic cloning prevented either competing proposal (a ban on both forms or reproductive cloning only) from passing. However, in 2009, the prohibition to use governmental funding for stem cell research has been revoked, removing barriers to human stem cell research. Germ line therapy would cure hereditary diseases by ensuring that only repaired genes are passed on. Increasing success in the creation of genetically modified mammals seems to sug-

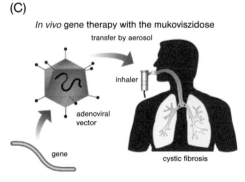

Fig. 30.2 Cystic fibrosis is the most common autosomal recessive metabolic disorder. Mutations in the cystic fibrosis transmembrane conductance regulator (CFTR) gene on the long arm of chromosome 7q31.2 were identified as the underlying mechanism in 1985. Up to now, around 900 pathogenic mutations of the CFTR gene have been found. The gene codes for a channel protein, shown here in the cell membrane (top left). The bottom left shows the *Pseudomonas* bacterium, which is harmless to normal individuals, but finds an ideal breeding ground in the viscous mucus of cystic fibrosis patients. The resulting bacterial colonies can be quite destructive to the lung. First therapeutic attempts were successful in introducing a normal CFTR gene into mucous membrane cells in the lung through inhalation.

gest that genetic modification of human germ cells is technically possible. However, there are **ethical and moral dilemmas** on which society has to find a consensus. An important point is to decide which diseases should be allowed to be treated by germ line therapy and what risks we are prepared to take in the process.

30.4
Setbacks in Gene Therapy

There has been at least one case in which **gene therapy** using adenoviruses ended with the death of the patient. Eighteen-year-old Jesse Gelsinger from Arizona took part in a clinical trial at the University of Pennsylvania in Philadelphia. He suffered from a hereditary metabolic condition with a deficiency in **ornithine transcarbamylase** – an enzyme involved in the urea cycle of the liver. Although the condition was manageable with conventional medication and an appropriate diet, Jesse wanted to take part in the study in order to help develop gene therapy treatment. On 17 September 1999, the teenager's liver artery was injected with a high dose of gene-packed adenoviruses and hours later he fell into a deep coma. He died 3 days later. The postmortem examination revealed that at some stages, his blood had contained more viruses than red blood cells. This caused a strong **immune reaction** that led to his death. In the case of Gelsinger, an NIH enquiry commission stated that the rules for medical trials had been disregarded. In preceding animal experiments, two monkeys had died of the same symptoms as Gelsinger. This known risk had not been disclosed to Gelsinger.

Playing down the risks of gene therapy when informing patients was by no means exclusive to this case. When, in the wake of this tragedy, the NIH toughened its rules, 652 belated reports of all sorts of serious complications were handed in. This is how a **second death** due to gene therapy came to light. In early May 1999, a man suffering from **coronary artery disease** had died in St Elizabeth's Hospital in Boston, 2 months after being treated with a viral vector. The vector carrying the blood vessel growth factor (vascular endothelial growth factor (VEGF)) had been applied through intramyocardial injection. Again, the cause of death had been the virus and not the gene. Such incidents highlight the risks involved in the development of new medical therapies, and led to a review and a tightening of regulations for future gene therapeutic studies.

This, however, could not prevent further incidents in the Necker Children's Hospital in Paris in late 2002. Alain Fisher was carrying out a gene therapy study to treat a condition named **X-SCID** – a severe combined immunodeficiency linked to an X chromosome. An *ex vivo* strategy was used to introduce the therapeutic gene through a **retroviral vector**. Fisher had already cured nine boys using this technique. However, two of his original 11 patients developed **leukemia**, caused by the retroviral vector. Retroviral vectors tend to insert at random into the host genome and thus may cause insertion mutagenesis. In both boys, the retroviral vector had inserted next to the LMO2 gene, which is involved in the onset of leukemia in children. These cases are a considerable setback for gene therapy and SCID patients. "The treatment was very effective", said Fisher, "but the risk is not acceptable".

30.5
Vectors for Gene Therapy

The transfer of therapeutic genes into patient cells depends critically on the further development of *in vivo* **gene transfer systems**. Such vectors should have certain properties in order to deliver the gene efficiently to the target cell (Table 30.3). They must be easy to produce and high-titer preparations of vector parti-

Easy and reproducible production in high concentrations ($>10^8$ viral particles mL^{-1})

Long-term gene expression through insertion into the genome or stable episomal persistence

Controllable gene expression

Tissue-specific expression

Low immunogenicity

Table 30.3 Properties of an ideal vector.

Table 30.4 Pros and cons of viral vectors in gene therapy.

Vector	Symptoms	Infection of nondividing cells	Insertion capacity (kb)	Stability of integration	Efficiency
Retrovirus	Tumors, AIDS	No (except for lenti-viruses)	<8	Stable, random	Medium
Adenovirus	Colds, conjunctivitis, gastroenteritis	Yes	8 (35 for gutless vectors)	Episomal	High
AAV	None known	Yes	<5	Stable, *in vivo* insertion unclear	Low
HSV-1	Cold sores, genital warts, meningitis	Yes	>25	Stable, episomal	High
Baculovirus	None in mammals, pathogenic to insects	Yes	>20	Unstable	High

cles should be reproducible. From a **safety aspect**, they should be nontoxic and should not elicit undesired effects such as immune reactions in the host. The **therapeutic gene** should be expressed persistently at a high level. **Viral vectors** should target defined types of cells or tissue and infect stationary as well as dividing cells because most cells in an adult patient are in a postmitotic state. The insertion to the host genome must be specific in order to avoid insertion mutagenesis. Specific insertion would render the repair of gene defects possible.

Over the past years, a wide range of **vector systems** of viral and nonviral origin has been developed. While methods such as the direct injection of naked plasmid DNA, gene transfer through a gene gun, or as liposome vesicles showed low transfection efficiency, experiments using viral vectors look more promising. Genomes of retroviruses, adenoviruses, and adeno-associated viruses (AAV) make up the bulk of the most frequently used viral vectors. Other, less commonly used viral vectors are derived from HSV-1, baculovirus, and others. Their properties are listed in Table 30.4. Viruses have evolved and adopted many properties of cells in the process, which enables them to identify target cells efficiently and penetrate them. They enter the cell and, depending on the virus, migrate from the cytoplasm to the nucleus to have their genes expressed by the host cell. This viral life cycle enables infectious virions to transfer genetic information with great success.

30.5.1
Retroviral Vectors

Retroviruses are a large versatile group of viruses with a genome consisting of single- or double-stranded RNA. They have a diameter of about 100 nm and they are covered by an **envelope**. The envelope contains a **viral glycoprotein** that binds to **cellular receptors**, thus defining the specificity of the host and cell type that is infected. The envelope protein furthers fusion with a cellular surface membrane or with endosomal compartments inside the cell. Depending on the arrangement of their genome, retroviruses are divided into two categories – **simple** and **complex retroviruses**. All retroviruses contain three essential genes: *gag* codes for structural proteins that constitute the matrix, the capsid, and the nucleoprotein complex, *pol* codes for reverse transcriptase and integrase, while *env* codes for the proteins of the envelope. There is also a **psi (Ψ)** packaging signal and two **long terminal repeats (LTRs)** with regulatory functions within the virus. The prototype for a **simple retrovirus** carrying only a small set of information is the **Moloney murine leukemia virus (M-MuLV)**. **Complex retroviruses** such as lentiviruses (e.g., HIV) contain additional regulatory and accessory genes. Initially, vectors for gene therapy were developed from simple retroviruses, very often M-MuLV. In order to develop retroviral vectors, knowledge of the **viral life cycle** became fundamental. After the infection of the host cell, the viral RNA is reversely transcribed into linear double-stranded DNA by **reverse transcriptase**. This process takes place in the cytoplasm and the viral DNA is then introduced into the nucleus followed by a stable insertion into the host genome.

The mechanism by which retroviruses are introduced into the nucleus of the host cell differs between simple and complex retroviruses. Whereas simple retroviruses can only enter the nucleus when the nuclear membrane is being dissolved during the mitotic process, lentiviruses have a preintegration complex that relies on an active cellular transport mechanism through the nuclear pores without destroying the nuclear membrane. Unlike M-MuLV, lentiviruses are therefore able to transduce stationary host cells. Once the virus has entered the nucleus, the viral enzyme **integrase** initiates the integration of the viral DNA into the host genome. The integrated viral DNA is called a **provirus**. It imitates a cellular gene and uses the host cell for gene expression. The transcriptional activity of the host is controlled by *cis*-**acting proviral LTR regions**. Complex retroviruses have additional *trans*-**acting factors** that activate RNA transcription (e.g., HIV-1, *tat*). After the

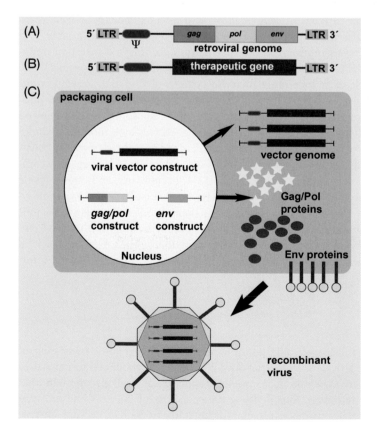

Fig. 30.3 M-MuLV-based retroviral vector. (A) The retroviral genome contains the genes *gag* (structural proteins), *pol* (reverse polymerase), and *env* (envelope proteins). Psi (Ψ) is the packaging signal distinguishing cellular RNA from viral packaging proteins. The viral genome is flanked by LTRs. (B) *gag*, *pol*, and *env* in the vector genome have been replaced by a therapeutic gene. (C) Gag, Pol, and Env are expressed by separate genes that are transfected into the packaging cell. If the viral vector construct is cotransfected with the transgene into the packaging cell, the protein products of the vector genome recombine with Gag/Pol to form infectious viruses that cannot replicate.

translation of the viral genes, the resulting protein products and the viral RNA form viral particles that are released from the cell via the cell membrane by budding. Most retroviral vectors currently used in **gene therapy studies** are based on M-MuLV – one of the first gene vehicles used in human gene therapy experiments. In order to produce viruses with a deficient replication mechanism that only replicate in the packaging cell and not in the host cell, the viral genes have been removed and replaced by a therapeutic gene. *gag*, *pol*, and *env* are expressed *in trans* in the packaging cell (Fig. 30.3). When the modified viral genome containing the therapeutic gene is transfected into the packaging cell, all required components are brought together to enable the formation of a **recombinant virus**. This virus can transfect target cells, but is unable to form infectious particles because the viral proteins are missing from its genome. **This is a safety measures often used in viral vectors. The viral genes responsible for the replication of the virus are physically separated from the rest of the genome, thus reducing the risk of a recombination of infectious particles.**

Lentiviruses are a subfamily of retroviruses with all the advantages of retroviral constructs plus the ability to also transduce **postmitotic cells and tissue**, including neurons, retinal, muscle, and hematopoietic cells. The most commonly used lentiviral vectors are based on the **HIV genome**. In order to avoid a recombination with infectious HIV particles, as many endogenous HIV proteins as possible have been deleted, without reducing the transduction and expression rate extensively. More recently developed vectors carry additional regulatory elements (Fig. 30.4). The **cPPT (central polypurine tract)** sequence facilitates the synthesis of the second strand and the transport of the preintegration complex into the nucleus, while the **WPRE (woodchuck hepatitis virus posttranscriptional regulatory element)** sequence enhances the expression of the transgene via a higher efficiency of the transduction and translation processes. An additional mutation of the 3′-LTRs results in **self-inactivation (SIN)**, thus reducing the risk of a recombination of infectious HIV particles. In order to further improve safety, efficient vectors have been developed from **lentiviruses** that are not pathogenic to humans, but also have the ability to transduce quiescent cells. The basic structure comes from the monkey-specific simian immunodeficiency virus (SIV), the cat-specific feline immunodeficiency virus (FIV), or the horse-specific equine infectious anemia virus (EIAV).

The potential risk for the application of lentiviral vectors lies in the possibility of **insertional mutagenesis** and a strong tendency in retroviruses to recombine with infectious foreign retroviruses, either within the transfected or the target

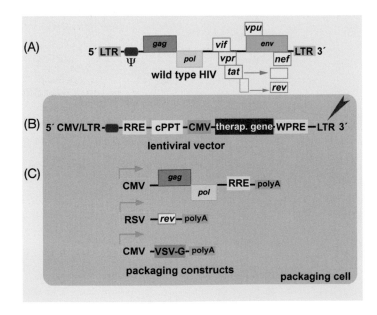

Fig. 30.4 Lentiviral vector. (A) Schematic representation of the wild-type HIV provirus. The HIV genome codes not only for Gag, Pol, and Env, but also for proteins such as Tat, Rev, Nef, Vif, Vpu, and Vpr. None of those, apart from Rev and Tat, is needed for the *in vitro* propagation of the virus. (B) The latest generation of the SIN lentiviral vector contains a cPPT to support the translocation of the vector into the nucleus. An additional WPRE sequence enhances the expression of the transgene. Nearly all viral elements have been deleted, apart from the LTRs (with a SIN deletion in the 3′-LTR, see arrow), Rev responsive element (RRE; essential for the nuclear export of viral RNA), and Ψ (which is needed for packaging). (C) The essential viral genes Gag, Pol, Tat (transactivates the HIV LTR promoter), Rev (enhances the export of unspliced genomic RNA from the nucleus after binding to RRE) and the envelope protein VSV-G are expressed by separated genes from different plasmids that are cotransfected into the packaging cell.

cells. It is also possible that new viruses may emerge through **recombination** with endogenous sequences. Thus, new infectious viruses with hitherto unknown properties could be spread by the use of retroviral vectors. These could not only affect other organs, but also the germ cells.

The latest developments in vector construction include nonintegrating episomal lentiviral vectors, reducing the risk of integrational mutagenesis. Furthermore, to improve tissue tropism, a variety of new and improved envelopes is available.

30.5.2
Adenoviral Vectors

Adenoviruses are primarily responsible for the infection of the respiratory and the gastrointestinal tracts. More than 50 adenoviral serotypes are known. Adenoviral vectors are based on **serotypes 2 and 5**, which do not cause serious disease in humans. Due to their wide tropism, adenoviruses can infect a wide range of host cells. Until recently, adenoviral viruses were very popular because they can be easily produced on an industrial scale, the virus titers are high, and they can transfect quiescent as well as dividing cells. The **linear double-stranded DNA** of adenoviruses codes for 11 proteins. The genome is packed into an **icosahedral protein capsule**, which is not surrounded by an envelope, but contains fiber envelope proteins. The fiber proteins combine with the surface receptors of the host cell to form a high-affinity complex. The endosomes are lysed by the adenoviral enzymes, but the genome is not integrated into the host genome and remains episomal. This results in a serial dilution of the adenoviral genome over several cell divisions. Unlike retroviruses, adenoviruses cannot be passed on via the germ line. Their high expression rate on a short-term basis makes them suitable for tumor treatment. Due to their wide host tropism, adenoviral viruses are not restricted to one compartment, but spread into surrounding tissue. This leads to toxic side effects, particularly on the liver. Furthermore, most patients have already been exposed to adenoviruses during their lifetime and have thus

Fig. 30.5 Adenoviral vectors. (A) Schematic representation of a serotype 5 adenovirus (Ad5) on which most of the adenoviral vectors described here are based. The vector genome is flanked by ITRs. Psi (Ψ) is the packaging signal. The adenoviral genes are highlighted in boxes. (B) First-generation adenoviral vector in which the genes E1 and E3 have been deleted. E1A plays a decisive part in viral replication as the initiator of the transcription of other viral transcription units. However, the gene is not needed for adenoviral replication within HEK-293 cells, which makes those cells ideal for virus production. The E3 gene product is not essential for viral reproduction, although its role in immunomodulation and suppression is important. The therapeutic gene is simultaneously transfected into the packaging cell, using a shuttle vector. It is than inserted into the adenoviral vector in exchange for the E1 gene. (C) Helper-dependent adenoviral vectors in which parts of the adenoviral genome (flanked by loxP recognition sites, triangles) have been excised in order to avoid immune reactions in the host. The expression of the therapeutic gene is driven by a promoter such as from cytomegalovirus (CMV). (D) In order to avoid potentially violent immune reactions of the host to adenoviral proteins, mini or gutless adenoviral vectors have been produced in which most of the adenoviral genes have been deleted.

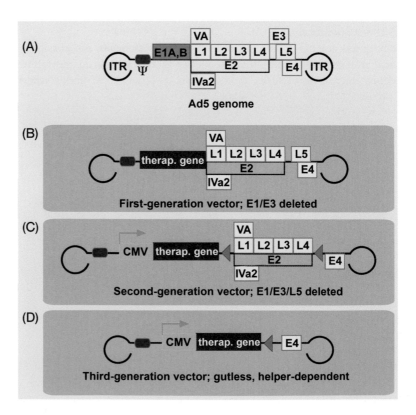

developed antibodies, which makes therapeutically relevant target tissues such as the epithelium of the respiratory tract as well as various tumors refractory to adenoviral infection. This could reduce the efficacy of adenoviral gene therapy. What is more, conventional adenoviral vectors could elicit a strong immune reaction in the host, mainly caused by the **adenoviral E2 protein**. While such an **inflammatory reaction** might well have an **antitumor effect**, there is also a high safety risk, as the death of a patient has demonstrated.

The replication defect in **first-generation adenoviral vectors** was the result of a deletion of the E1A and E1B genes (Fig. 30.5). In some of these vectors, the E3 gene was also deleted in order to improve their uptake capacity. However, they retain the other early and late viral genes that are expressed in small quantities after infection. In **second-generation adenoviral viruses** the E2 and E4 regions have also been deleted, and only the late genes are retained. Viral gene products induce an immune response against the transduced cells, resulting in a reduced expression of the transgene.

New strategies aim to completely avoid the immune response and to achieve a higher uptake capacity for foreign DNA in adenoviral viruses. This led to the development of adenoviral vectors in which all viral reading frames have been deleted. These are known as **gutless vectors** and contain only those viral DNA sequences that are active *in cis*, and are essential for the replication and packaging of viral DNA, such as **inverse terminal repeats (ITRs)**, which contain the polymerase binding sequence for the start of DNA replication and the DNA packaging signal psi. The original adenoviral gene region between the two ITRs has been replaced by foreign noncoding DNA. In recombinant vectors derived from gutless vectors, this space is partially taken by the transgene. Gutless vectors can only be produced with the assistance of a **helper virus**, which provides the proteins required for viral replication and packaging. However, the helper virus produces contaminations in the generated vector that must be removed. In Cre/loxP, a helper-dependent specific DNA excision system, the helper virus has lost its ability to replicate and can also be separated very efficiently, which makes it very safe. **Cre recombinase** from bacteriophage P1 cuts DNA sequences flanked by loxP recognition sequences. In HEK-303 packaging cells expressing Cre recombinase, the **Cre/loxP system** was able to remove 25 kb of the adenoviral vector genome containing loxP recognition sequences. However, the remaining vector has a size of only 9 kb and the DNA it contains has probably rearranged itself. This problem can be avoided by introducing gap-filling DNA in order to keep the vector size at 27 kb. Other problems that currently still plague third-generation vectors are associated with helper virus contamination and an insufficient titer concentration, making them unsuitable for clinical use. Latest developments concentrate on the modification of the viral envelope and the spikes, to improve their target range and to make them less or nonimmunogenic.

30.5.3
Adeno-associated Virus (AAV)

AAV, a member of the parvovirus family, is a promising candidate for the transfer of genes. AAV has an **icosahedral structure** and contains a **single-stranded DNA genome** of only 4.7 kb. It can only be replicated with the assistance of helper viruses such as adenoviruses or herpes viruses. Although a large proportion of the population is AAV-seropositive, so far no pathogenicity has been observed. In contrast to adenoviruses, AAVs are only weakly immunogenic. They can infect dividing as well as quiescent cells and integrate into the host genome, which is advantageous for long-term expression.

Wild-type AAV contains no more than two genes *rep* for replication and *cap* for encapsulation. The coding sequences are flanked by ITRs, which are needed for packaging DNA into capsids. In AAV vectors, the genes *rep* and *cap* have

Fig. 30.6 AAV vectors. The AAV genome contains sequences that are essential for the transduction process, such as ITRs and the genes *rep* and *cap*. (B) In the vector genome, *rep* and *cap* have been replaced by a therapeutic gene. If the therapeutic gene is larger than 4.5 kb, it is distributed over two concatemeric vector constructs. (C) The REP and CAP proteins are expressed by the packaging cells, and are needed for the production of single-stranded DNA genomes in a capsule consisting of proteins. A nonenveloped AAV virus collects in the nucleus. Helper proteins from adenoviruses, which are needed for replication, are also expressed in the packaging cell (not shown here). The AAVs are released from the packaging cell through the lytic adenoviral replication process.

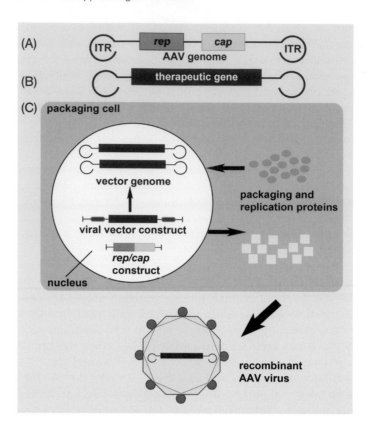

been replaced by a therapeutic gene (Fig. 30.6). In order to produce a recombinant virus, the AAV genes and adenoviral helper genes are expressed in a packaging cell *in trans*.

The **major advantage** of AAV-derived vectors is the ability to stably integrate into the target cell genome at a defined location in the chromosome. Location-specific insertion is mediated by a usually inactive 100-bp long region in the REP protein. However, since AAV vectors no longer contain the *rep* gene, targeted integration could only be detected using wild-type AAV. Furthermore, since AAV is widespread among the human population, the question arises if the AAV-specific insertion location has been occupied by other genetic material. This would have to be removed first before an AAV vector could be used efficiently. It is also unclear what would happen if the insertion location of an AAV-derived vector were not available – whether sequence-independent insertion or even chromosomal relocation would take place.

Another interesting property in AAV vectors derives from their **specific chromosomal insertion** – the capacity of **homologous recombination**. In specifically chosen reporter genes inserted into chromosome 14, it was possible to correct point mutations and deletions using an AAV vector, albeit at a very low frequency. This approach might also hold promising therapeutic possibilities.

At this stage, the production of **AAV vectors** is still a major problem, being very difficult and time-consuming. The *rep* gene and some of the adenoviral helper genes are cytotoxic to packaging cells, and there are no cell lines available for the large-scale production of pure recombinant viruses. AAV samples must be enriched in a cesium chloride gradient (using an ultracentrifuge) to remove contaminating helper viruses. Current purification methods achieve at least a 100 times concentration containing 70–80% of the original material. Another disadvantage is the low infection rate of AAV. Only 0.1–1% of viral particles are infectious. Second-strand synthesis is limited to specific areas. Finally, despite many limitations, AAV vectors are quite useful gene transfer systems, since they achieve excellent expression in muscle, brain, hematopoietic precursor cells, neurons, photoreceptor cells, and hepatocytes.

Another problem lies in the low packaging capacity of the vector of about 4.5 kb. However, this problem is currently being tackled with some success, making use of the observation that AAV genomes form **concatemers** after transduction. Thus, if two vectors, one carrying the first half of a transgene, the other the second, are transduced together into a cell, the viral genome is put together head to tail, resulting in the reconstitution of a functional gene. In this way, the maximum size of the gene to be transduced could be significantly enlarged. It remains to be seen if such concatemerizing vectors will be sufficiently stable to ensure long-term expression.

30.5.4
Other Viral Vectors

Herpes viruses infect the central nervous system and carry what is known as a **latency gene**. This enables them to evade immune defense mechanisms and survive in a latent state in certain cells and organs for the whole lifetime of the host. External factors may activate them, eliciting recurrent localized symptoms. Only in newborns are herpes infections able to invade the entire body, with fatal results if the brain is affected. Thus, herpes viruses may be used to transfer genes into neural tissue. **HSV-1** has a large double-stranded DNA genome containing over 80 genes, which gives it the capacity to carry larger transgenes. Furthermore, a HSV-1-based vector can carry several independently regulated transgenes simultaneously. As in adenoviral vectors, the viral genome remains episomally in the plasma of the nucleus instead of being inserted into the genome of the cell. It is therefore not replicated through cell division. However, due to the complex genome of the herpes virus, the production of efficient and safe vectors yielding high-titer concentrations proves to be difficult. As about 90% of humans have at least a latent infection with herpes virus, there is a **risk of a recombination** between recombinant and latent viruses, during which process transgenic sections could be inserted into wild-type herpes viruses and then spread.

Baculoviruses can also take up very large genes and express them highly efficiently. Until recently, it was thought that they were highly specific, infecting only insects and other arthropods, but they are now known to infect mammalian tissue as well. They have a high affinity to the human liver, which makes them ideal vectors for gene transfer into human liver tissue. So far, however, strong immune reactions from the host organism still prevent their therapeutic use *in vivo*. Currently, baculoviruses pose a safety risk due to their infectiousness to human tissue (see Chapter 16).

The quest for the ideal vehicle in gene therapy has led to the development of further viral vectors. Ongoing work focuses on vectors based on **smallpox viruses (vaccinia)**, **Sindbis viruses**, and many others.

30.6
Specific Expression

In addition to the efficacy of the gene transfer process, the cell type specificity of vectors is especially important for gene therapy. In *ex vivo* **gene therapy**, viral vehicles are selectively expressed in specific patient cells. Patient cells such as **hepatocytes** or **stem cells** are taken from the individual and grown in a culture, where they are transfected and then reintroduced into the patient (i.e., only the cells that have been transfected *in vitro* can express the transgene). In *in vivo* **gene therapy**, where the viral vehicle is directly introduced into the patient, it is more difficult to achieve the targeted expression of a transgene. The cell type targeted could be, for example, a certain type of cell in the hematopoietic system or a cell type within a specific tissue such as the brain. As the viral vectors mentioned above are mostly ubiquitous, their cell type specificity is insufficient and certain strategies had to be developed to ensure **cell type-specific expression**.

One of the strategies for modifying the tropism of a viral vector is based on **pseudotyping**. Various experiments have been carried out to try and utilize viral envelope proteins binding to particular receptors on certain cell surfaces. Viral envelope proteins can be swapped to produce chimeric vectors that bind to specific targets. Using HIV or certain SIV variants, for example, can result in CD4 cell specificity. Conversely, the envelope protein of the vesicular stomatitis virus (VSV-G) can be used to expand the tropism of viral particles to almost all cells. The Ebola Zaire (Ebola-Z) glycoprotein has been used as an alternative route to transfect lung epithelial cells efficiently.

Not only envelope proteins were tested, but also **adaptor proteins**. These are proteins that bind the virus to specific surface proteins of the target cell. Bispecific antibodies and other chimeric proteins (scFv fragments; see Chapter 27) were used, but strategies based on receptorligand binding might offer a viable alternative. These techniques are still being developed. Another promising approach could be the addition of tissue-specific promoters (e.g., liver, neuron, muscle, or CD4-specific promoters) to viral vectors. The problem lies in the transcriptional targeting of vectors. It is difficult to achieve a sufficiently high expression rate. Much work also remains to be done to ensure an optimum functioning of regulatory sequences in the context of recombinant viral genomes.

Another interesting approach is the use of regulatory systems that are tailor-made for the insertion of an inactive therapeutic gene into the target tissue. The therapeutic gene is only expressed when a cotransfected transactivator has been induced. Possible substances that could be used as **switches for therapeutic genes** include tetracycline, FK506, rapamycin, cyclosporin A, ganciclovir, RU486, and ecdysone. All these compounds can be taken orally and have a low toxicity for the patient. The limitations of the targeting strategies mentioned above lie in their often very low gene transfer rate, in the need for multiple infection, and in an insufficient expression period for the transgene.

31
RNA Interference, Modified DNA, Peptide Nucleic Acid, and Applications in Medicine and Biotechnology

Learning Objectives

This chapter describes possible ways of controlling gene expression through RNA interference (RNAi) or the use of modified DNA and its analogs by antisense technology. Biochemical mechanisms of RNAi are described in detail. The strategies for synthesis and placement of normal and chemically modified oligonucleotides are explained. The most important class of such molecules is peptide nucleic acid.

31.1
Introduction

DNA and RNA are the molecules used to preserve the genetic information in biological systems, and to provide a platform for translating this information into proteins. In this chapter, molecular biological possibilities for regulating and changing genetic information are explained. Genetic information can be altered by introducing foreign genetic information into the cell, in the form of plasmids and vectors. There are, however, cases in which the application of molecular biological methods is neither applicable nor practical to a given application. For example, in the treatment of human illnesses, the genes of the unhealthy cells causing an ailment cannot simply be altered. Therefore, in treating the ailment, other approaches, such as attempting to minimize the expression of an injurious protein, may be feasible. One approach, termed **antisense therapy** (Fig. 31.1), brings complementary RNA or DNA into the cell, which binds to the mRNA of the target protein (perhaps the translation start sequence), preventing the translation of the mRNA into protein. Alternatively, **antigene therapy** interferes in the transcription of DNA into RNA. However, as DNA is double-stranded in the cell nucleus, antigene (as opposed to antisense) therapy is significantly more difficult to carry out, and will not be discussed further.

A DNA oligomer, which should bind RNA through Watson Crick base-pairing, must meet several requirements to be effective in antisense therapy. First, the DNA oligomer must be stable under physiological conditions and should not be affected by nucleases and peptidases. In 1978, Zamecnik and Stephenson carried out the first **antisense experiments**, using normal DNA. Due to the rapid degradation of normal DNA in cells, clinical use could not be considered. To be useful, the DNA needed to be chemically stabilized under physiological conditions (i.e., a **DNA analog** was required). Secondly, the DNA analog must bind RNA sequence-specifically. Finally, the altered DNA must have good pharmacokinetic and pharmacodynamic properties, such as bioavailability, accessibility, and uptake into the cells.

DNA analogs need not only be used *in* vivo – analogs able to form strong sequence-specific bonds to complementary DNA (cDNA) offer possibilities of further analytical applications in the area of molecular analysis and clinical diagnostics, such as on **DNA chips** (see Chapter 26).

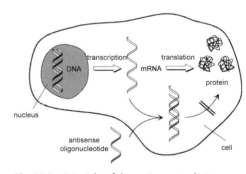

Fig. 31.1 Principle of the antisense technique. The mRNA is bound to a complementary antisense oligonucleotide preventing its translation into protein.

An Introduction to Molecular Biotechnology, 2nd Edition.
Edited by Michael Wink
Copyright © 2011 WILEY-VCH Verlag GmbH & Co. KGaA, Weinheim
ISBN: 978-3-527-32637-2

Interestingly, it was discovered that nature has long implemented antisense in the regulation of protein expression. This process is termed **RNA interference (RNAi)** (see Section 2.5). Short double-stranded RNA oligomers, roughly 21–23 bp long, are termed **small interfering RNA (siRNA)**. These siRNAs bind an enzyme complex (**RNA-induced silencing complex (RISC)**), which orients the siRNA so that the target RNA can be bound, sequence-specifically, to the siRNA. The endonuclease activity within the enzyme complex hydrolyzes the target RNA, at the location where the antisense strand is bound. This prevents the translation of RNA into protein, through the cleavage of the RNA. It is the same process as that exploited in antisense therapy (i.e., the disruption of mRNA through RNase due to the binding of an antisense oligonucleotide).

This chapter starts by describing antisense therapy with (modified) oligonucleotides. Subsequent parts detail the mechanisms of RNAi, followed by selected applications for both techniques. To sum up, the techniques of RNAi and antisense with modified oligonucleotides and analogs will be compared.

31.2
Modified Nucleic Acids

Figure 31.2 illustrates possible chemical modifications of RNA. Alteration in the place of the phosphate group is possible (e.g., the phosphate ester could be entirely replaced by other groups). Chemical modifications are also possible on the sugar moiety, as well as the bases. A large number of chemical modifications have been examined in the last decades; however, very few modifications have met the requirements described above. Figure 31.2 shows as an example of several successful modifications, two of which will be discussed below.

31.2.1
Phosphorothioate

Two well-investigated and successful DNA modifications are the **phosphorothioates** and the **methylphosphonates**. In both cases, a seemingly small chemical change has enormous consequences. In the case of phosphorothioate (Fig. 31.2 A), the only alteration is that an oxygen atom is replaced by a sulfur atom in the phosphate ester group. The original motivation to develop this class of substances was to insert a radioactive label into the DNA oligomer with [35]S. Sulfur is in the same group of the periodic table as oxygen, thus as expected the chemical properties of the oligomers do not differ significantly. However, phosphorothioates are more stable than normal DNA in the presence of nucleases.

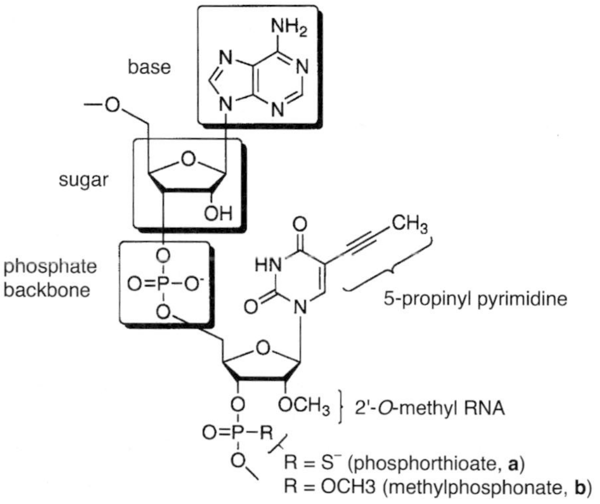

Fig. 31.2 Chemically altered RNA. In the upper monomer, the locations of possible chemical modifications are noted (base, sugar, and phosphate backbone). In the lower monomer, several successfully modifications are illustrated. However, contrary to what the illustration suggests, there is only one possible modification for each base. (a) Phosphorothioate and (b) methylphosphonate.

Box 31.1 Formivirsen (Vitravene™)

After becoming infected with **human immunodeficiency virus (HIV)** the full spectrum of **acquired immune deficiency syndrome (AIDS)** will, sooner or later, develop and the immune system will be systematically destroyed by the virus. As a result, pathogens that the healthy immune system hold in check will cause illnesses. These illnesses are termed opportunistic; their symptoms are very rarely found in the uninfected portion of the population. One such sickness is an eye infection caused by **cytomegalovirus (CMV retinitis)**. This infection is poorly treated with the usual virostatic drugs (e.g., ganciclovir). The first medication to work with an antisense mechanism was set to work against CMV retinitis. Formivirsen (ISIS 2922) is a 21mer oligo(phosphorothioate) developed by ISIS Pharmaceuticals. The medication, under the tradename Vitravene, was approved by the US Food and Drug Administration (FDA) in August 1998 and shortly thereafter also approved for use in Europe. Formivirsen has the sequence 5'-GCG TTT GCT CTT CTT CTT GCG-3' and all phosphates are racemic. The complementary mRNA sequence is specific for CMV and not present in humans. This region of viral mRNA codes for several proteins necessary for the replication of the virus. Through the binding of formivirsen on mRNA, RNase H will be activated, which destroys the mRNA. The synthesis of the protein is stopped and the replication of the virus is prevented. It is quite likely that unspecific, non-antisense effects contribute to the overall antiviral activity. Formivirsen is directly injected into the corpus vitreum of the eye. After an initial phase of three weekly injections, only a maintenance dose once every 2 weeks is necessary. The drug is normally well tolerated; the most common side effect was an increase of the inner eye pressure. Formivirsen also works with virus strains that are usually unresponsive to therapy. In preclinical trials, developing resistance was observed in only one virus strain in the laboratory and not seen with patients. The approval of formivirsen as the first antisense medication was an important milestone for the entire field and demonstrated that clinical applications of antisense therapy are feasible in principle. However, CMV retinitis, even in AIDS patients, is a very rare illness. Sales of Vitravene in Europe in 2001 were reported to be less than 100 cases per year. The approval of Vitravene for the European Market was voluntarily recalled in 2002 by Novartis Ophthalmics. In the explanation, it was emphasized that the recall was for economic reasons only; there were no grounds on which to question the safety and effectiveness of Vitravene.

Phosphorothioates are also internalized by cells more easily than normal DNA, as sulfur is bigger and more lipophilic than oxygen. Based on these advantageous pharmacokinetic properties, the first antisense medications were developed on the basis of phosphorothioates (formivirsen, Box 31.1).

31.2.2
Methylphosphonate

A methylphosphonate group results when a negatively charged oxygen from a phosphate ester group is replaced by a methyl group (Fig. 31.2 B). Methylphosphonates are stable in the presence of nucleases and, unlike phosphorothioates and normal DNA, are uncharged. Thus, an neutral oligo(methylphosphonate) binding to a complementary single-stranded DNA (which is a polyanion under physiological conditions) do not repel each other, in contrast to double-stranded DNA, where both strands are polyanions. This has consequences that will be discussed below

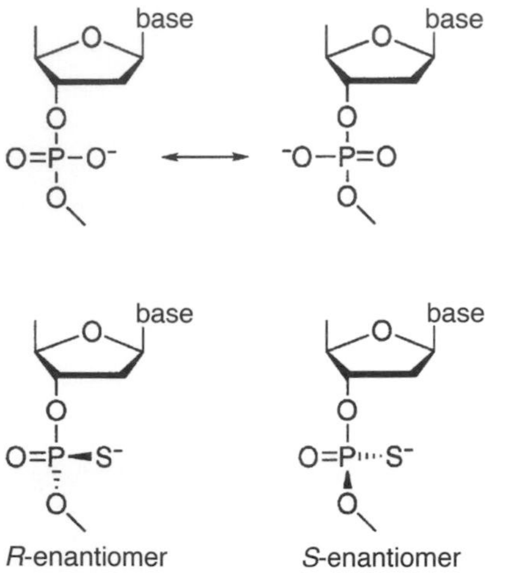

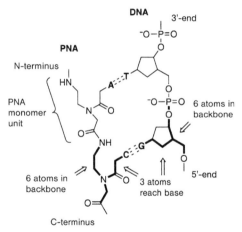

Fig. 31.3 (Top) Mesomer in the phosphate group of DNA. (Bottom) Both enantiomer forms of the phosphate atom in phosphorothioates.

Fig. 31.4 Section from a PNA·DNA double helix. A, C, T, and G symbolize the bases of DNA, behind these bases are the backbones; bonds between the bases and the backbones are in bold to illustrate the topographical similarities between DNA and PNA.

(see Section 31.3.1). However, a problem of methylphosphonates is that their uptake into cells is poor, essentially preventing medical use of this class of substances. Oligomers from phosphorothioates or methylphosphonates can be easily obtained through chemical synthesis. This synthesis uses suitable monomers on a solid support analogous to DNA solid-phase synthesis. All synthesis steps are well-prepared, and can be carried out automatically and in parallel, allowing the automated synthesis of large quantities of the substance.

Considering the arrangement of phosphorus atoms in phosphorothioates and methylphosphonates indicates that there is a stereogenic center. Note that the phosphorus atom in normal DNA is not chiral, as the mesomeric structures effectively make the two oxygen atoms identical (Fig. 31.3). However, every phosphorus atom in phosphorothioates can be either an *R* or *S* isomer. Thus, an *n*-oligomer with phosphorothioate bridges will be one of 2^n possible stereoisomers. These isomers are diastereomers, with all the related characteristics. The same situation applies to methylphosphonate. Thus, it can be expected that the stereochemistry of these molecules influences their physiological characteristics and also their binding to cDNA and RNA (see Section 31.3.1). It is currently possible to synthesize these molecules stereoselectively, yielding enantiomerically pure phosphorothioates or methylphosphonates (all-*R* or all-*S*, depending on the reagent).

31.2.3
Peptide Nucleic Acids (PNAs)

A further class of DNA analogs relevant to these issues are **peptide nucleic acids (PNAs)** (Fig. 31.4). In these molecules, the standard bases found in DNA molecules are bound via a methylcarbonyl group to an aminoethyl glycine backbone, rather than being bound to a sugar phosphate backbone as in DNA and RNA. Thus, individual monomers in these oligomers are linked by amide bonds. Strictly speaking, PNA oligomers are not peptides, as the monomers are not typical *a*-amino acids. For this reason, the alternative names **"peptoid nucleic acids"** or **"polyamide nucleic acids"** have been proposed for this class of molecule. However, the term **peptide nucleic acid** has become generally used for this class, despite not being chemically correct. Note that such verbal inaccuracies are not unusual in the realm of biochemistry – under normal conditions DNA is not the acid itself but is, more precisely, the polyanion of an acid. DNA and PNA appear at first glance to have very different structures; however, there are significant topographic similarities between the molecules. First, the repeat unit in the backbone is six atoms long. Secondly, there are two atoms between each base and the backbone. PNA monomers can be relatively easily synthesized out of simple building blocks, enabling the basic structural changes illustrated in Fig. 31.4. PNA oligomers are formed from these monomers using solid phase synthesis. For the synthesis of PNA oligomers, chemistry different from DNA synthesis needs to be used, with more similarity to peptide solid-phase synthesis. PNA oligomers up to a length of 15 bases are easily synthesized. There can, however, be serious problems in synthesizing longer oligomers, such as aggregation and adverse side reactions.

PNA oligomers are **uncharged**, similar to methylphosphonate. However, unlike methylphosphonate, PNA oligomers are **achiral**. They **bind well to cDNA and RNA**, and form WatsonCrick base pairs (Fig. 31.4).

Poor water solubility can be a problem with these compounds, depending on the sequence. To counter this problem, additional lysine residues are usually included at the carboxy-terminus of PNA oligomers. As these residues have a positive charge under physiological conditions, they increase the solubility of the oligomer. Since PNA oligomers, in a physiological sense, are neither true peptides nor nucleic acids, they are not affected by peptidases or nucleases and therefore are fully stable under physiological conditions. However, there remains the problem that they are poorly taken up by cells, such that they do not readily reach the cytoplasm. Solutions to this uptake problem have been investigated with the synthesis of PNA conjugated to peptides, sugar molecules, and DNA oligomers. No such conjugates can be produced using biomolecular methods, so they remain entirely in the realm of chemical synthesis.

31.3
Interactions of DNA Analogs with Complementary DNA and RNA

All previously discussed DNA analogs bind to complementary single-stranded DNA and RNA, and form WatsonCrick base pairs. This results in helical duplexes analogous to those of normal double-stranded DNA. Even though discussions of this subject tend to be predominantly DNA-oriented, similar considerations apply to RNA-related molecules. Two important parameters that need to be considered are the stability of the resulting double strands and the influence of individual base pair mismatches.

31.3.1
Melting Temperature

The stability of a **DNA·DNA duplex** can be assessed through the **melting temperature (T_m)**. To measure this, equimolar amounts of complementary strands are added to a suitable buffer solution. DNA helices form and the solution is then heated. Through the addition of thermal energy, the DNA double-helix strands separate and at higher temperatures this yields two cDNA single strands. During the heating, absorption of UV light by the bases is measured, conventionally at 260 nm. As the specific absorption of bases in the duplex is smaller than when the bases are in unpaired single strands, a decrease in absorption results at lower temperatures – an effect known as **hypochromicity**. A melting curve, obtained by plotting absorption A versus temperature T, has a sigmoidal shape due to a cooperative effect in duplex formation – the initial temperature increase breaks a few hydrogen bonds between the strands, making further separation of the bonds increasingly easy. The temperature at the turning point of the **melting curve** is defined as the T_m (Fig. 31.5). This signifies the temperature at which exactly half of the duplex has formed single strands. The higher T_m, the more stable the duplex. It should be noted that the T_m, at least with double-stranded DNA, is highly dependent on the concentration of the oligomer and the salt concentration. This point will be further discussed below.

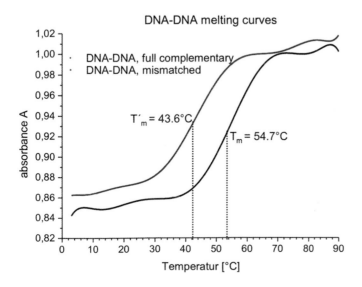

Fig. 31.5 Measuring the melting temperature T_m of DNA·DNA duplexes. The absorbance at 260 nm (A) is plotted against the temperature (T). Melting curves for a perfect base-matching duplex (black line, $T_m = 54.7\,°C$) and a mismatched duplex (red line, $T'_m = 43.6\,°C$). The decrease in melting temperature is $\Delta T_m = T'_m - T_m = 11.1\,°C$.

With slightly more effort, thermodynamic parameters such as standard enthalpy, ΔH^0, standard entropy ΔS^0, and the free energy ΔG^0 for the formation of the duplex can be determined. However, for a first comparison, measuring T_m under identical conditions is usually sufficient.

For many of the known DNA analogs, it is difficult to predict their affinity to the cDNA strand. Therefore, as a rule, experiments such as the one just discussed must be carried out to determine T_m. For example, a duplex made of adenosine phosphorothioates, all in S configurations, (All-S_P-d(A_{PS})$_{11}$A), with cDNA (dT$_{12}$) at 5 °C is more stable than the enantiomeric All-R_P-d(A_{PS})$_{11}$A·dT$_{12}$ duplex. If, however, both strands are exchanged so that the phosphorothioate strand is made up of T bases, the S enantiomer strand All-S_P-d(T_{PS})$_{11}$T·dA$_{12}$ becomes less stable than the R-configured double helix All-R_P-d(T_{PS})$_{11}$T·dA$_{12}$.

Clearly, in both cases (i.e., where the DNA analogs form helices with either poly(T) or poly(A) DNA), the number and type of hydrogen bonds, as well as the number and orientation of sulfur atoms, are identical. In both cases, the more stable duplex is also more stable than the unmodified dT$_{12}$·dA$_{12}$ duplex and therefore seems well suited for medical or biotechnological purposes. In addition, methylphosphonate have an affinity to cDNA comparable or slightly better than that of the natural oligodesoxyribonucleotide. An important difference is that oligo(methylphosphonate), as opposed to DNA oligomers, is uncharged. The T_m of natural double-stranded DNA is strongly dependent on the salt concentration and decreases with decreasing salt concentration as the repulsion of negatively charged DNA single strands is decreasingly compensated for through the interactions with the positively charged cations of the salt.

Generally, for T_m measurements a phosphate buffer with 0.1 M NaCl is used – under physiological conditions even bivalent cations such as Mg^{2+} and Zn^{2+} stabilize the DNA double helix. On the other hand, the T_m of (methylphosphonate DNA·DNA duplexes is to a great extent independent of the salt concentration at which it is measured.

The same argument applies to **PNA·DNA duplexes**, in that the T_m is also only slightly dependent on the salt concentration. For a PNA·DNA duplex, similar to double-stranded DNA, there are two arrangements of the strands with respect to one another. Of these two, the arrangement shown in Fig. 31.4 is more stable. The N-terminal end of the PNA is located where in double-stranded DNA, the 5′-end of a DNA strand, which replaces the PNA strand, would be expected. Such an arrangement, in analogy to double-stranded DNA, is described as an antiparallel arrangement. Generally, PNA·DNA duplexes are significantly more stable than the homologous double-stranded DNA. For oligomers of around 10 bases, an increase in T_m by approximately 1.5 °C per base pair is observed. PNA·DNA duplexes in the (more stable) antiparallel arrangement have been shown to form a right-handed helix, equivalent to that of naturally occurring DNA, although the PNA·DNA duplex is slightly compressed compared to naturally occurring double-stranded DNA. More specifically, B-DNA has 10 base pairs per turn while the PNA·DNA double helix has around 13 base pairs per turn. Also, in the PNA·DNA double helix the major groove spreads apart, and the minor groove is shallow and tight in comparison to normal double-stranded DNA. As PNA in this form is an achiral molecule, there are no stereochemical complications associated with such molecules. However, it should be noted, as already discussed above, that the stability of a PNA·DNA duplex is dependent in an unpredictable way upon which of the two strands is PNA and which is DNA. Even with the same sequence, T_m(PNA·DNA) $\neq T_m$(DNA·PNA), even if both combinations are more stable than DNA·DNA.

These observations demonstrate that many aspects of the structure and stability of DNA are not yet completely understood. Nevertheless, DNA analogs have found a wide variety of uses, several of which will be discussed below.

31.3.2
Mismatch Sensitivity

For biotechnological purposes, the recognition between cDNA strands should be as specific as possible. It is also important to be able to assess the effect of an individual mismatched base pair (e.g., G–A instead of G–C) on the stability of DNA duplexes with modified DNA oligomers. A change in the T_m is defined as:

$$\Delta T_m = T_m' \text{ (mismatch)} - T_m \text{ (perfect complementarity)} \qquad (31.1)$$

Usually $\Delta T_m < 0$, as every divergence from the WatsonCrick rules destabilizes the duplex (see Fig. 31.5). Obviously, it is impossible to give a specific number for every possible mismatch. If an incorrect base pair is found at the end of a strand, ΔT_m is generally smaller than if the same mismatch is found in the middle. The longer the oligomer, the greater T_m and therefore the smaller the effect of the mismatch on the general stability. For example, T_m is smaller in a 20mer than in a 10mer for the same mismatch on a comparable position. The ΔT_m values of PNA·DNA duplexes for the same mismatch are larger than with double-stranded DNA, often as much as twice as large. Better stated, PNA has a greater mismatch sensitivity than DNA. For DNA·PNA duplexes, ΔT_m values of up to − 15 °C have been measured. These extreme values open up a number of different applications for PNA that would be impossible using normal DNA and most modified DNAs. As PNA is bound much tighter to cDNA than DNA itself, and tighter than most of the DNA analogs, PNA can be reliably bound to shorter DNA sequences, under physiological conditions. The T_m of a PNA·DNA octamer, with mixed sequences, is typically around 40 °C. For comparison, the homologous DNA·DNA octamer has a T_m under 30 °C. As the PNA·DNA octamer has high mismatch sensitivity a single mistake in the DNA sequence leads to a decrease of T_m to under 35 °C. In practice, that means that this PNA sequence, under physiological conditions, already has the ability to bind short cDNA sequences; strong binding will, however, only be achieved with the perfect complementary sequence. Such exact DNA targeting of short sequences under physiological conditions is not attainable by any other method.

31.4
RNAi

RNAi is characterized as a cellular pathway in which the expression of genes is inhibited sequence specifically by double-stranded RNA (see Chapter 2.4). Inhibition of gene expression occurs posttranscriptional by the selective hindrance of the target mRNA and thus is also known as **posttranscriptional gene silencing (PTGS)** or **RNA silencing**.

RNAi is divided into three steps. First, either longer double-stranded RNA is expressed in the cell or is taken up into the cytoplasm from the extracellular matrix, followed by the processing into small double-stranded RNA by RNase III (Dicer and Drosha). Then, the duplex is unwound and only one RNA strand is incorporated into a protein complex (RISC). RNA strand incorporation turns the complex into its active state and enables the binding of complementary mRNA. One major component of the complex, an endonuclease from the argonaute protein family, cleaves the target RNA or constricts its translation. Thus, the sequence specificity of the RNAi pathway is determined by the incorporation of the RNA strand (template) and its activity is given by the argonaute protein.

31.4.1
Biogenesis of Small RNAs

Exogenous double-stranded RNA precursors from viral infections are cleaved into **short interfering RNAs (siRNAs)**, while **microRNAs (miRNAs)** are processed from genome-encoded hairpin sequences. The third class of small RNAs are **PIWI-interacting RNAs (piRNAs)**. In addition, endogenous siRNAs are known, which will not be considered here.

31.4.1.1 Biogenesis of siRNAs

Exogenous double-stranded RNA molecules are processed by RNase III Dicer into siRNAs (Fig. 31.6). siRNAs are double-stranded and consist of 21–25 nucleotides with 5′-phosphate groups and hydroxyl groups and two-nucleotide overhangs at both 3′ ends. The PAZ domain of Dicer binds siRNA and its two catalytically active RNase domains perform the cleavage of the siRNA.

31.4.1.2 Biogenesis of miRNAs

As shown in Fig. 31.7, two precursors appear in the course of the biogenesis of endogenous miRNAs.

The first precursor, **primary miRNA (pri-miRNA)**, is encoded in the genome and is transcribed by RNA polymerase II. pri-miRNA forms a stable hairpin structure, consisting of a 33-base-pair stem, a terminal loop, and single-stranded ends. pri-miRNA is processed by a large protein complex, a so called microprocessor. The microprocessor consists of the RNase III enzyme Drosha and DGCR8 – a double-stranded RNA-binding protein. Drosha cleaves pri-miRNA into the second precursor, **precursor miRNA (pre-miRNA)**. The cleavage site is given by DGCR8. pre-miRNA is 65–70 nucleotides long with two-nucleotide overhangs at 3′-ends. The 3′-overhang is identified by exportin-5 – an export factor from the nucleus into the cytoplasm. In the case that miRNA information is encoded within an intron, no cleavage by Drosha is necessary. The miRNA is processed directly by the spliceosome and finally cleaved to miRNA in the cytoplasm.

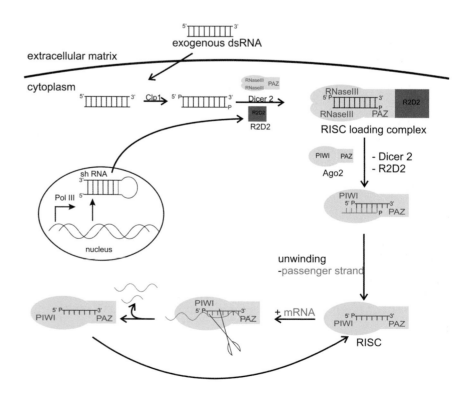

Fig. 31.6 RNAi on the basis of siRNA (red strand): siRNA biogenesis from exogenous double-stranded RNA or from shRNA in the cytoplasm by Dicer and its incorporation into the RISC loading complex; followed by the unwinding of the double strand and association of the guide strand with the argonaute protein to form the active RISC. ds, double stranded; Pol, polymerase; PIWI, P-element induced wimpy testis; PAZ, PIWI Argonaute Zwille; R2D2, Protein Kinase ARII-like.

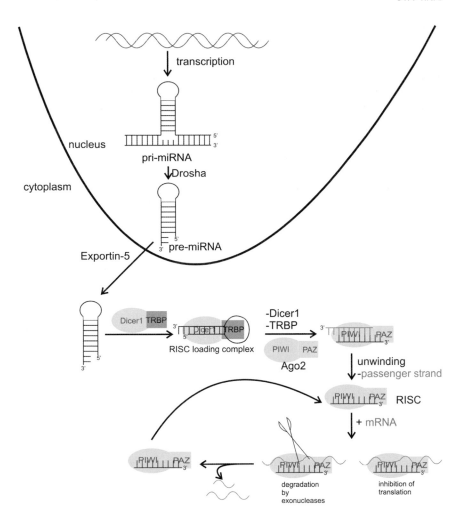

Fig. 31.7 RNAi via miRNA (red strand): genome-encoded pri-miRNA is processed in the nucleus into pre-miRNA, which is then exported into the cytoplasm. The loop of the pre-miRNA is cleaved and the duplex unwound. The guide strand associates with the argonaute protein. Protein synthesis is inhibited either by mRNA degradation or inhibition of translation (see Figure 31.6 for abbreviations).

31.4.2
Incorporation into RISC

The incorporation of small RNA into RISC requires the dissociation of the double-stranded RNA into its single strands. Only one strand, the guide strand, is loaded onto RISC. The complementary strand, called the passenger strand, is destroyed. In this step, the duplex must be unfolded and one of the strands must be selected as the guide strand. Strand selection is determined by thermodynamic asymmetry along the RNA duplex. The strand whose 5'-end is thermodynamically less stable (i.e. with more AT base pairs) is preferentially loaded onto RISC as the guide strand.

In human cells, pre-miRNA binds to the trimeric pre-RISC or RISC loading complex, which consists of an argonaute protein (Ago2), Dicer1, and TRBP (Fig. 31.7). The double-stranded RNA-binding domain (RBD) of TRBP arranges the binding of the pre-miRNA with the complex. Dicer performs the cleavage of the pre-miRNAs loop, which results in a 22-nucleotide long miRNA duplex with two-nucleotide overhangs at its 3'-ends. TRBP and Dicer dissociate from the complex, Double-stranded RNA is unwound, and the guide strand associates with the Argonaute protein and builds the active form of RISC (holo-RISC). For siRNAs, the incorporation process is similar, but the RISC loading complex consists of Dicer2 and its double-stranded RBD containing interaction partner R2D2 (Fig. 31.6).

31.4.3
Posttranscriptional Repression by miRNA und siRNA

Argonaut proteins feature a PAZ domain (PIWI Argonaute Zwille) and a PIWI (P-element induced wimpy testis) domain. The 3′ ends of siRNA or miRNA are identified and bound by the PAZ domain of the argonaute protein (Figs 31.6 and 31.7). The core of the PIWI domain belongs structurally to the RNase H protein family and catalyzes the cleavage of the target mRNA. In detail, the phosphodiester linkage between the target nucleotides that are base-paired to siRNA residues 10 and 11 (counting from the 5′-end) is cleaved. The mRNA cleavage products, which contain 5′-monophosphate and 3′-hydroxyl ends, dissociate from RISC, and are digested completely by cellular exonucleases. RISC is free again to cleave more target mRNA.

For posttranscriptional repression by miRNA, the target mRNA is recognized by its 3′-untranslated region sequence. Contrary to siRNA, the binding of the target mRNA to miRNA is not full complementary. Nevertheless, a fully complementary WatsonCrick base-pairing with the miRNA nucleotides 28 of the 5′-termini (seed region) is required for an effective repression. The prevalent view of posttranscriptional repression by miRNA was that the degree of complementary between miRNA and mRNA determines in which way the mRNA is repressed. Fully complementary mRNA is cleaved by the argonaute protein while mismatches inhibit the cleavage but also stop mRNA translation. From recent studies it has become clear that the degree of complementarity is not the critical factor. The inhibition of mRNA translation is the only preferred way in miRNA repression (Fig. 31.7). Degradation of mRNA is not due to the activity of the argonaute protein, but is caused by the deadenylation of the mRNA, which is then unstable and easily attacked by exonucleases (Fig. 31.7).

31.5
Applications

31.5.1
Antisense Technology with DNA Analogs

The **antisense technique**, as explained earlier, involves the disruption of mRNA and the blocking of locations important for translation through binding of complementary oligonucleotides. An antisense oligomer suitable for medicinal purposes is required to possess the following five characteristics:
- Easy uptake into the cell.
- Stability under physiological conditions (also against enzymes).
- Strong and sequence-specific binding to complementary RNA.
- Limited general toxicity.
- Activation of RNases.

The first three points have already been discussed in the context of DNA analogs. When considering the latter two points, it is clear that highly toxic reagents are unsuitable for application *in vivo*. Secondly, **activation of RNases**, in particular RNase H, is desirable, although not absolutely necessary. RNase degrades mRNA, thus permanently interrupting the production of the target protein. Unfortunately, very few of the known DNA analogs lead to activation of RNase H. Of the previously discussed molecules, only phosphorothioates have been found to have this property, but unfortunately neither methylphosphonate nor PNA invoke such activation. The structural parameters required to activate RNase H are not yet fully understood. As DNA·RNA duplexes also lead to the activation of RNase H, and therefore to the decomposition of RNA, conjugates were synthesized from PNA and DNA (Fig. 31.8). In these conjugates, a good recognition of

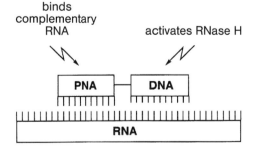

Fig. 31.8 Activation of RNase H through PNA·DNA conjugates. The PNA binds tightly to the complementary RNA. The DNA portion of the conjugate binds less strongly to the RNA than does the PNA portion, but is required to activate a DNA·RNA duplex RNase H.

RNA target sequences and strong RNA binding through PNA is achieved, as discussed in Section 31.3.1. The DNA component also binds to the associated RNA; however, the strength of the bond or the sequence specificity of this part is no longer important. The only factor of importance for the DNA part of the conjugate is the recognition of this region of the complexed RNA by RNase H, followed by the **degradation of the RNA**. An additional desirable effect of such complexes is the improved uptake of the conjugate into the cells, possibly by the recognition and active transport of the DNA component of the conjugate. These two positive effects are, unfortunately, offset by the considerable speed with which enzymatic reactions degrade the conjugated DNA. Analogous considerations can also be made for siRNA (see Section Section 31.5.2).

Additional medicinal applications of antisense oligonucleotides involve **treatment of tumors**. In this context, obvious goals for an antisense therapy are proteins that play a role in the regulation of cell growth and cell death. An example is the BcrAbl mRNA in **chronic myeloid leukemia (CML)**. These mRNAs are tumor-specific and are involved in the transformation of hematopoietic cells into tumor cells. For these reasons, they were the targets of several early antisense studies. Clinical studies were also carried out for the treatment of **non-Hodgkin's lymphoma (NHL)**. These tumor cells protect themselves from programmed cell death through the increased production of an apoptosis protection protein, Bcl-2 (B cell leukemia/lymphoma-2). It was shown that, using an anti-Bcl-2 mRNA oligonucleotide, the threshold for apoptosis of tumor cells was lowered. As many chemotherapeutics work through an induction of apoptosis, the combination of these chemotherapeutics with inhibition of the Bcl-2 expression is particularly promising. In cell culture experiments, a highly promising anti-Bcl-2 phosphorothioate oligonucleotide was discovered. It consisted of a racemic 18mer phosphorothioate against the first six codons of the Bcl-2 mRNA – this oligomer had the working title G3139. After successful *in vitro* experiments and mouse experiments, a clinical phase I study was carried out with a small group ($n=9$) of refractory NHL patients unresponsive to usual medications, through the additional administration of Bcl-2 mRNA oligonucleotides. One patient completely recovered and one patient had a significant decrease in the volume of the tumor. It was demonstrated that for these patients there was indeed a decrease in the Bcl-2 expression.

Additionally, the production of viral proteins can be decreased using antisense nucleotides. Successful studies were carried out with antisense nucleotides against HIV and CMV genes. However, the rapid adaptation of the virus through point mutations in the viral genome restricts the long-term usefulness of such antisense nucleotides for viruses. Many different sections of the virus life cycle are exploited in HIV therapy. Promising approaches with antisense oligonucleotides have been developed against reverse transcriptase, and against proteins of the viral mRNA transcription and translation machinery. Several clinical studies have examined these therapeutic approaches. In 1998, the first antisense medication was approved to treat a special viral conjunctivitis infection (CMV retinitis), which is an accompanying symptom of AIDS. The medication uses a 29mer phosphorothioate oligonucleotide directed against the mRNA of a CMV protein (see Box 31.1). Remarkably, there are neither further antisense pharmaceuticals nor any drugs based on RNAi that have reached clinical approval at present.

Interesting antisense effects can also be obtained with PNA in bacteria. Owing to the chemical structure and biophysical characteristics of PNA, especially the higher lipophilicity, PNA oligomers are especially suitable for uses in which the DNA analogs, with greater structural resemblance to DNA, fail. Biological membranes such as the bloodbrain barrier (BBB) or bacterial cell wall pose great challenges. An interesting experiment was carried out with antibiotic-resistant *Escherichia coli* bacteria. The bacteria produce an enzyme, β-lactamase, that inactivates many penicillins (including the penicillin derivative ampicillin) through the cleavage of β-lactam rings. Ampicillin-resistant bacteria were treated with a PNA 15mer, which was directed against the start codon of the bacterial β-lactamase mRNA. After subsequent treatment with ampicillin, these cells were killed by the antibiotic, even at low doses. Through suitable control experiments, a specific antisense effect of PNA was proven. Remarkably, in these experiments the true antisense effect presumably depends on permanent blockage of mRNA, as PNA does not activate RNase, at least not *in vitro*.

Furthermore, it was shown that PNA oligomers can cross biological membranes such as the BBB and bacterial cell walls, which is not possible for DNA oligomers or even many DNA analogs.

31.5.2
siRNA in Biotechnological Applications

Very early after the discovery of PTGS, small synthetic RNA molecules were developed for the treatment of diseases. These artificial systems differ in their mode of application. In two fundamentally different applications, either naked, synthetic (and chemically modified) double-stranded RNA penetrates cells directly (nonvectorial), or small double-stranded RNA is expressed from vectors in the nucleus of the cells.

31.5.2.1 Design of siRNAs
A lot of experiments were performed, and many algorithms were developed to construct efficient siRNAs and to predict their activity. It became clear, however, that the efficiency of the siRNA is influenced not only by the relative thermodynamic stability of both termini of the double strand, but also by the secondary structure of the guide strand. The 5′-end of the guide strand ought to be thermodynamically less stable and both ends should be unstructured (i.e., without helices). Furthermore, the accessibility of the 3′-termini of the target mRNA plays a central role in the efficiency of gene silencing. Nevertheless, a lot of factors are still not known. Notably, the effects that result from the interaction between mRNA and siRNA are not well understood yet.

In general, typical synthetic siRNAs are full complementary 19mer duplexes with a two-nucleotide overhang on their 3′-ends. It was observed that longer sequences are more efficient, if not longer than 30 nucleotides, which induces an unspecific interferon response. However, 27-nucleotide long duplexes are processed by Dicer to 21mer sequences once they penetrate the cell and enter much easier into the silencing pathway. Moreover, the siRNA sequences should contain up to 50% of the bases guanine and cytosine. The target mRNA should be accessible easily and should contain neither a protein-binding domain nor form intramolecular structures.

Chemical modification of siRNAs increases their stability towards nucleases and thus extends their half-life. In addition, fluorescent markers for intracellular localization studies or lipophilic groups for enhanced cellular penetration can be introduced into the sequence. Most chemical modifications were adopted from experience from antisense therapy (see Section 31.2). Chemical modifications, in particular in siRNA applications, benefit from the double-stranded nature of siRNAs and miRNAs. They can be introduced into the passenger strand, which will be mostly degraded without affecting the activity of the guide strand. However, some modified nucleotides are accepted at particular positions within the guide strand. In this regard, it must be considered that the 5′-hydroxy group of the guide strand should be unmodified and free for phosphorylation by kinase Clp1 to enter into the RNAi pathway. In contrast, the termini of the passenger strand can be provided with functional groups like fluorophors, cholesterols, or cell-penetrating peptides, without any loss of activity. The assembly of phosphorothioate RNA nucleotides, also known from antisense therapy, is accepted, but can induce toxic side effects.

Modifications at the 2′-position of the ribose, like 2′-*O*-methyl or 2′-fluoro substituents, increase the stability of the duplex proportional to their quantity within the duplex. Furthermore, 2′-fluoro-modified nucleotides do not reduce the efficiency of siRNA due to the small fluoro substituent. The more bulky methyl group inhibits RNAi if its frequency is too high within the duplex. However, 2′-*O*-methyl modified nucleotides have the advantages of decreasing an unspecified interferon response.

31.5.2.2 Nonvectorial Applications

For siRNAs as therapeutic agents, it must be considered how the target organ or tissue will be achieved. If it is easily accessible and facile, local injection is mostly sufficient. However, targets located deeply within the organism can only be reached by intravenous injections (**systemic application**). Here, especially pharmacokinetic effects must be considered – how does the therapeutic agent distribute in the organism, where does it accumulate, and how fast will it be eliminated?

In the case of lung infections, local treatments via intranasal or intratracheal delivery reach the lung epithelium directly and are already well established. Similarly, tumor growth in mouse was inhibited successfully via direct intratumoral injection of siRNApolyethyleneimine (PEI) complexes.

Systemic applications into tissues that are only reachable via the bloodstream make high demands on the therapeutic agents. Before the drugs reach the target, they should not be filtered out from the kidneys, be ingested by phagocytes, aggregate with proteins in the serum, or be digested enzymatically. Moreover, it must be kept in mind that molecules larger than 5 nm in diameter are not able to pass the capillary endothelium. They remain in the bloodstream until they are rinsed out from the organism. Only organs like the liver, spleen, and some tumors pick up molecules with a diameter up to 200 nm. Once the therapeutic agent reaches the target tissue, the cellular uptake should not be endosomal. In this case, the siRNA is not located freely in the cytoplasm and cannot be phosphorylated in order to step into the RNAi pathway. However, it is possible to wrap siRNAs into nanoparticles, which serve as vehicles and enhance the cellular uptake. Their properties are given by their surface composition. Positive charges improve the assembly with siRNA and increase the attachment on the negatively charged cellular membrane. **Polyethylene glycol (PEG)** and other hydrophilic conjugates on the surface inhibit aggregation with negatively charged proteins in the serum and with phagocytes, and decrease the activation of the immune system. Moreover, PEG determines the dimension of the nanoparticle and controls direct elimination along the kidneys, which is specific for molecules smaller than 50 kDa. Shimaoka *et al.* prevented successfully the overexpression of cyclin D1 in leukocytes, which cause gut inflammation (Peer *et al.*, 2008). Since leukocytes can only be attacked systemically, siRNA was enclosed in liposomal nanoparticles. In order to penetrate leukocytes specifically, the nanoparticles carried an antibody against a specific adhesion protein of the leukocytes on their surface. Additionally, the storage of siRNA in the nanoparticles was improved by a negatively charged protein, protamine, which shows high affinity to nucleic acids. Shimaoka *et al.* introduced not only a very specific way of application with low toxicity and without any immune response, but also a very efficient one due to the high packaging of siRNA into the nanoparticles.

31.5.2.3 Vectorial Applications

The biggest disadvantage of nonvectorial application is the transient and noncontinuous activity of the therapeutic agents due to degradation and dilution.

This can be avoided with expression systems that ensure a continuous synthesis of siRNA in the cells provided that the RNA is expressed stably within the cells.

For vectorial applications, siRNA is coded into the DNA sequence, cloned into a vector, and transcribed under the control of polymerase III promoters. The DNA sequence is composed of a sense and antisense strand, which are separated by a loop sequence. The transcript is then self complementary and also called **small/short hairpin RNA (shRNA)** (Fig. 31.6). In the stage of processing by Dicer, shRNA is cleaved to siRNA and enters into the RNAi pathway. Since the most commonly used vectors come from viruses like adenoviruses and lentivirus, they are referred to as viral vectors. They attach to the cells and infiltrate their genome into them very efficiently. However, inflammation reactions and increased immune responses must be taken into account. In order to produce stable transgenic animals, Westhusin *et al.* transfected lentiviral vectors, which contain the siRNA sequence complementary to the mRNA of the prion protein, into fibroblasts (Golding *et al.*, 2006). The fibroblasts then showed a permanent resistance against **bovine spongiform encephalopathy (BSE)**.

31.5.2.4 Other Applications for PNA

On the basis of their high affinity for cDNA, coupled with their increased mismatch sensitivity, PNA oligomers are suitable for many other applications in biotechnology. In the following, examples of **PCR clamping** through PNA and **fluorescent *in situ* hybridization (FISH)** with PNA probes will be discussed.

PCR clamping (also called PCR silencing) through PNA is a multifaceted and very sensitive technique, able to detect individual base mutations in DNA, even when these are present in a very small fraction of the total DNA. PNA oligomers cannot serve as a primer for PCR. Thus, if one runs a PCR reaction where one places a strand of a PNA oligomer complementary to a wild-type DNA, the primer can no longer bind to the DNA, as the stable PNA·DNA duplex blocks the primer binding site. As a result, duplication of the wild-type DNA does not happen.

However, should a point mutation exist within the PNA binding site, the PNA will bind significantly worse at this place and a passing primer will be able to bind, leading to amplification of the mutated DNA. PNA oligomers of 15 bp have proven to be successful in this task, with a T_m of 70 °C. A single mismatch in the middle of the sequence gives T_m reductions of around 15 °C. In this temperature window, the conditions of PCR could be optimized so that only the mutated, not the wild-type DNA, would be amplified. An important application of this technique is the detection of point mutations in proto-oncogenes, which are related to the onset of cancer. Traditional methods for identifying such mutations (e.g., DNA sequencing) are hindered because the mutations only exist in a small number of the examined cells, which later degenerate into cancer cells. One example is cancer caused by point mutations in the *ras* proto-oncogene, primarily in codons 12 and 13 of the gene. With a wild-type DNA, complementary to the PNA, its amplification can mostly be suppressed so that only the mutated *ras* DNA is amplified. Indeed, all known *ras* mutations in probes from different tumors could have been found with this technique, regardless of the type or location of the mutation. Under ideal circumstances, the sensitivity of the method can reach 1/20 000 (i.e., one single mutated cell in 20 000 normal cells).

Another interesting application of PNA is as probes for FISH – a technique that has many uses, such as in the identification of microorganisms. PNA oligomers, to which fluorescent dyes have been covalently bonded, can be used for this. The sequence of the PNA is chosen to be complementary to the rRNA of the species of interest. As rRNA is available in relatively high concentrations, individual cells can be visualized after incubation and hybridization with fluorescent PNA probes. As an example, the dangerous *Staphylococcus aureus* can be quickly and reliably identified using this technique. Today, RNA analysis is an established method for phylogenetic identification of microorganisms and is increasingly replacing the previously preferred biochemical or morphological analysis. As previously discussed, PNA probes, because of their preferred uptake into most cells and also because of their hybridization characteristics, offer a further improvement to this method.

31.5.3
Comparison of RNAi with DNA Analogs for Antisense Applications

Antisense therapy with (modified) oligonucleotides and RNAi feature a lot of similarities. The former findings on antisense therapy supported the progress in RNAi research, as some similar principles apply. This is especially true for the chemical synthesis and modification of RNA, as well as for methods to penetrate cells and stabilize RNA molecules. In both methods, mRNA is bound specifically, which leads to the inhibition of gene expression. Nevertheless, there are also important differences. In RNAi, the seed region (two to eight nucleotides) in siRNA is crucial for an effective binding of mRNA, while antisense oligonucleotides need longer sequences to bind mRNA. In addition, antisense oligonucleotides are single-stranded and also act in the nucleus. Under certain circumstances, and depending on their chemical structure, they are able to activate RNase H in the nucleus to induce the cleavage and degradation of the target mRNA. Oligonucleotides in RNAi are initially double-stranded and act only in the cytoplasm. Most notably, they enter into an endogenous intracellular pathway, which explains their high efficiency.

32
Plant Biotechnology

Learning Objectives

This chapter provides an introductory overview of the specific prerequisites and challenges of "green" biotechnology. The main focus is on techniques and method development for the production of genetically altered plants, including the requirements for optimal gene expression, the steps necessary for transferring gene constructs into a plant as well as the final quality control of the resulting transgenic event. References are made to selected aspects of the ongoing acceptance debate and research activities targeting the risk assessment of transgenic plants for humans, animals, or the environment.

32.1
Introduction

32.1.1
Green Genetic Engineering – A New Method Towards Traditional Goals

Biotechnological applications in **plant breeding** are based on either genome modifications achieved through cell culture techniques or on targeted modifications of a genome. Although in both cases the resulting plants have been genetically optimized, legislation on **genetically modified organisms (GMOs)** refers not to the presence of foreign genes *per se* in a plant variety, but only to the method used to insert genes into a plant (i.e., via genetic engineering).

For millennia, **crop plants** have been selected and optimized according to desired characteristics ("traits"). Well-known examples are the development of crop species of wheat in Mesopotamia and corn in Middle America. In both cases, advantageous characteristics were selected not only from the local genotypes, and not only through spontaneous mutation, but above all through crossing independent species with the consequence of mixing distinct sets of otherwise mutually foreign genes. This process was very successful, but slow and strictly restrained to the alleles in populations of the same or crossable genotypes. In the nineteenth century, these boundaries had already been circumvented in order to manipulate the plant genome, like in the case of Triticale (*Triticum aestivum* × *Secale cereale*) through forced crossing of wheat and rye with special cell culture methods. Three technologies, however, first enabled targeted changes of traits through the transfer of genes from the same or a completely different species:

- Recombinant DNA.
- Plant transformation.
- Plant regeneration *in vitro*.

Molecular biology techniques such as molecular markers and genome sequencing (see Chapter 21) became fundamental constituents of the progress in modern conventional breeding programmes. These methods are applied in order enable the efficient introgression of, for example, new resistance genes taken from

An Introduction to Molecular Biotechnology, 2nd Edition.
Edited by Michael Wink
Copyright © 2011 WILEY-VCH Verlag GmbH & Co. KGaA, Weinheim
ISBN: 978-3-527-32637-2

related species into high-yield varieties. However, it is important to note that legally speaking the resulting varieties do not qualify as genetically engineered.

Furthermore, the term **plant biotechnology** also encompasses areas lacking molecular methods. These include *in vitro* culture techniques of entire plants, plant parts, or individual plant cells and especially the regeneration of fertile autotrophic plants. These methods find wide applications in crop and horticultural plants, and in the production of secondary plant products.

This chapter concentrates on the method-based aspects of production, characterization, and utilization of genetically engineered plants according to Security Level 1 of the **Genetic Engineering Law** (known as GenTG for authorization of gene technical work and European Union Guidelines for the releasing of plants 2001/18/EG). Current information can be found at the *Official Journals of the European Communities* at http://eur-lex.europa.eu/JOIndex.do, where the importance of individual technologies for research and applications are discussed.

The products of plant biotechnology can be divided into two trait groups:

- **Input traits** refer to **agronomic characteristics** that serve to improve farming processes and are therefore also commonly referred to as **quantitative traits**. Most of the currently marketed genetically altered crops of the first generation belong to this group. The main traits are herbicide resistance and insect resistance, with the goal of increasing and securing yield while reducing the input of labor and classical agrochemicals. Since 2003, these two traits taken together represent more than 90% of the arable land used for the cultivation of genetically engineered soybeans, corn, cotton, and oilseed rape. Current information about cultivation development of these genetically engineered crops is supplied from the International Service for the Acquisition of Agribiotech Applications (http://www.isaaa.org).

- **Output traits** determine the **characteristics of plant products**, mostly seeds, and target either the improvement of the quality of agricultural products relating to nutrient content and nutrient composition (e.g., essential fatty acids for human nutrition or essential amino acids for animal feed) or the production of specific plant substances and proteins for large technical applications, such as industrial starches, technical enzymes, or pharmaceutically effective substances and proteins. Accordingly, these products are also classified as **qualitative traits**. It is expected that such products will dominate the second and third generation of genetically altered crop plants. A large number of projects in this area of second- and third-generation traits target **functional foods** (**nutraceuticals**), and the cultivation of plants that produce high-value products like antibodies and pharmaceuticals (**phytopharming**).

32.1.2
Challenges in Plant Biotechnology

Applications of plant biotechnology are at the center of so-called **biological safety research**; on the one hand, many products of genetically engineered plants are consumed by humans and animals; on the other hand, cultivation naturally requires the release of genetically engineered plants over large areas. The potential effects on humans or animals of ingesting genetically engineered products of altered (although substantially equivalent) composition are carefully monitored as is the safety of the agricultural ecosystem. These safety requirements, on the one hand, and the need for continuous growth in yield and efficiency, on the other, determine, for the most part, the applications, methods, and research goals of genetically engineered crops. In this regard, prominent trait projects include water use efficiency, drought tolerance, heat and cold resistance, salt tolerance, nitrogen use efficiency, yield increase, and many others.

In addition to the various trait projects, current research in the area of tool and technology development (**enabling technologies**) for genetic engineering focuses on the fine-tuning of gene expression, on optimizing multiple gene expression constructs for gene stacking for complex traits, on increasing gene transfer efficiencies on the molecular as well as cell culture level, and on high-throughput methodologies for plant analysis. Plant analysis targets the molecular as well as biochemical level and includes methods such as the **quantitative polymerase chain reaction (qPCR)** (see Chapter 13), and the application of next-generation sequencing (see Chapter 14) and metabolic profiling techniques. Many of the research areas listed, especially in the field of cell culture techniques and high-throughput genotyping, are the subject of automation and are worked on in close collaboration with engineers specializing in robotics. Modern bioinformatics and especially intelligent solutions for data mining (see Chapter 24) have became increasingly important for handling the vast amount of data produced during all of these steps. Innovation strategies extend to research on directional, site-specific integration of transgenes into the target genome as well as targeted DNA excision.

32.2
Gene Expression Control

Gene expression control is an area of intense research in basic as well as applied science. Since the addition of genetic elements for establishing a new or modifying an existent metabolic pathway usually requires the plant to redirect at least some of its energy resources, it is evident that the expression of such new or modified elements should be tightly controlled. This is even more important in view of the need to minimize potential interference of the newly established genetic element with existing features. Furthermore, specific metabolic attributes, like seed filling or oil deposition during seed formation and embryogenesis, require direct **spatial and temporal control of the expression** (Fig. 32.1) of the transgene added in tight correlation with the existing metabolic patterns of the targeted tissue. Therefore, research on gene expression control consists initially of the identification and characterization of plant-derived promoters and other elements exerting control, such as enhancers or introns of either the cognate or also very distantly related species. In view of the need for concerted expression control of several genes in a single construct for pathway engineering or for trait stacking in commercial applications, the compatibility of control elements and the potential for repeated use of a given element in a complex construct are further areas of in-depth investigation. Furthermore, **computational biology** is applied to generating tools for plant promoter prediction and the analysis of the modular composition from a variety of gene expression control motifs. Recently, attempts at recombining modular elements of natural promoter sequences into new, **artificial**, and partially **synthetic promoters** have gained momentum.

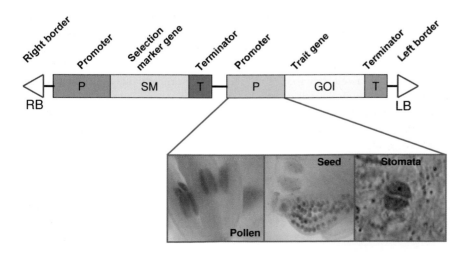

Fig. 32.1 Expression cassette for plant transformation and examples for plant promoters exerting temporal and spatial gene expression control. Between the left and right border the T-DNA exemplarily contains an expression cassette for the selectable marker gene and an expression cassette for the gene of interest (GOI). Both the selectable marker gene as well as the gene of interest are flanked by a promoter and a terminator element, respectively. The lower panel shows examples of tissue-specific gene expression monitored via reporter gene expression and GUS staining.

In addition to establishing additional expression, **down-regulation** up to a complete knock-out of existing gene expression patterns is a second area of intense research. While creating "antisense" constructs to down-regulate RNA translation into a functional protein used to be the method of choice in the first generation of transgenic research, the modern approaches predominantly involve the so-called **small RNAs**.

"**Gene switches**" (i.e., **chemically inducible promoters**) activate gene expression only while exposed to an activating chemical such as ethanol or steroid derivatives. While the most recent progress in research on chemically inducible promoters is described in the patent literature, the review by Jepson *et al.* (1999) summarizes the basics of the common systems and the associated challenges and benefits. So far, none of the known examples of inducible promoters has been used for establishing new trait characteristics in commercially important crop plants. However, inducible promoters as much as **transactivation systems** are valuable research tools when analyzing potentially lethal genotypes as well as for identifying the most suitable point in time and level of gene expression required for establishing a phenotype of interest.

32.3
Production of Transgenic Plants

The transformation of plants is defined as the **incorporation and expression of foreign genes in plants**. The first transformation was carried out through the insertion of a kanamycin resistance gene into the tobacco genome. Today, several hundred monocot and dicot plant species (as well as a few mosses and algae) are amenable to genetic transformation. In addition to a functional vector construct carrying the gene(s) of interest and suitable genetic control elements for the expression of foreign DNA, the prerequisites for successful genetic transformation are suitable transformation systems to transfer the DNA, selection systems to identify and select for the transformed cells, and regeneration systems to cultivate single transformed cells into fertile plants.

32.3.1
Transformation Systems

The transfer of DNA into plants can result in the **stable or transient** presence of foreign DNA. In the case of **stable transformation**, permanent integration of the foreign DNA into the target genome is achieved, so that the additional DNA will be passed to the following generations of plants (if there is complete regeneration) according to the general rules of Mendelian inheritance. In the case of **transient transformation**, the foreign DNA is introduced into the cell without integration into a genome. Due to a lack of integration into any autonomous genetic unit (a chromosome of the plant nucleus or the plastid or mitochondrial genome) the foreign DNA can be expressed but not passed on to daughter cells. Such introduced, but nonintegrated, DNA fragments can be preserved for up to a few weeks in a cell, but sooner or later they disappear, due to hydrolytic decomposition through the cell's own nucleases.

For the development of genetically engineered plants, stably transformed lines are indispensable, while transient transformation serves mainly for the analysis and testing of DNA constructs (e.g., the functionality of structural genes and especially their control elements, such as promoters or enhancers). Transient transformation occurs, except when using viral vectors (see Section 32.3.1.4), only with adequate efficiency for the nuclear genome. Otherwise, stable transformation can be applied on a regular basis to the nuclear as well as plastid genomes. The transformation of plant mitochondria is currently neither transiently nor stably successfully reproducible.

32.3.1.1 *Agrobacterium* as a Natural Transformation System

The first and so far most significant method for gene transfer in plant cells is provided by the natural characteristics of *Agrobacterium*. The genus of this soil bacterium includes two pathogenic types that generally infect dicot plants: *Agrobacterium tumefaciens* and *Agrobacterium rhizogenes*. Both can be used for transformation, but *A. tumefaciens* is used for most applications, because in its natural environment it infects above-ground plant parts and triggers the formation of so-called **crown galls**. *A. rhizogenes* infects only roots and causes aberrant proliferation – a phenotype known as **hairy roots**. These roots can be grown *in vitro*, and find applications for the production of secondary substances and recombinant proteins. Given the outstanding success of developing the *A. tumefaciens* transformation system, *A. rhizogenes* is of secondary importance to modern plant biotechnology.

Both *Agrobacterium* types contain plasmids of approximately 200 kb. These carry about 25 genes and replicate with autonomous replication origins. The *A. tumefaciens* plasmid is termed the **Ti-plasmid** (tumor inducing) and the *A. rhizogenes* plasmid is termed the **Ri-plasmid** (root inducing). They are constructed similarly and function by related mechanisms. The proteins encoded by the plasmid genes are responsible for the virulence of the bacterium (*vir*), the transfer of DNA from bacterium to plant (*tra*), and the induction of the above-mentioned local proliferation process (*onc*) of leaf, stem, or root tissue. In the end, only a DNA fragment of roughly 20 kb, the **transfer DNA (T-DNA)**, is actually conferred to the plant genome. The T-DNA is flanked by two 25-bp long, almost identical sequences, which are termed the **left border (LB)** and **right border (RB)**. They serve as recognition sites for the excision from the bacterial plasmid, and bind proteins for transport into the nucleus of the plant cell and for final integration into the nucleic genome. The transferred DNA contains (between the LB and RB) the so-called **oncogenes** (*onc*) that code for the enzymes of the synthesis of the plant hormones auxin and cytokinin. After integration of the T-DNA into the host genome, the expression of these oncogenes causes programmed tumor formation. The new tissue serves as a host site for the proliferating *Agrobacterium*. Apart from that, the T-DNA carries a gene for the formation of amino acid derivatives (**opines**). Depending on the *A. tumefaciens* strain, one can differentiate nopaline, octopine, and other types, according to the produced opines. The nopaline synthase (*nos*) of T-DNA forms, for example, nopaline from arginine and pyruvate. These nitrogen- and carbon-rich derivatives cannot be used as energy sources by the plant, but only by the *Agrobacterium*. The enzymes for the metabolization are again encoded on the Ti-plasmid, but outside of the T-DNA.

In conclusion, in natural infection processes for *Agrobacterium* the advantage of gene transfer consists of its own propagation through exploitation of energy-rich metabolites taken from plant cell metabolism and at the same time inducing increased local cell growth in the host plant. The biology of this bacterial plant interaction describes *Agrobacterium* clearly as a pathogen. The infection of plants by *Agrobacterium* is mediated by compounds like acetosyringone and other cell wall phenolics, which are secreted by the plants when injured. They are recognized as a signal by *Agrobacterium*, through a two-component regulation system, to locate the injury (for an overview, see Gelvin, 2003). In nature, through the infection of vegetative tissue only, inheritance of the integrated T-DNA to the offspring of the genetically altered crop does not occur. Furthermore, due to high metabolic cost the host plant would not be viable for an extended period of time when expressing any of the oncogenes.

For biotechnology applications the *Agrobacterium* Ti-plasmid has been developed into a tool for directed gene transfer under controlled conditions by deletion of the opine and *onc* genes (disarmed vector). After removal of these genes, *Agrobacterium* are apathogenic since the removal of the phytohormone genes and the opine synthesis genes prevents the uncontrolled growth of the infected cell as well as energetic support for the propagation of *Agrobacterium*.

(A) T-DNA with gene transfer cassette

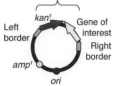
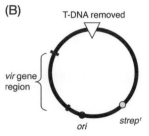

(B)

Fig. 32.2 Binary vector for plant transformation. The binary vector system consists of two plasmids: the binary plasmid (e.g. pBIN19; left) and the disarmed Ti-plasmid (e.g. pRK2013 helper plasmid; right) that coexist in the same *Agrobacterium* cell. Their properties with respect to infection and T-DNA are complementary to carry out the complete mechanism of gene transfer. The modified T-DNA usually carries a resistance marker gene for selection in eukaryotes and the gene of interest.

Today, the standard method is based on a binary system in which two plasmids are used: the so-called **binary vector** and **disarmed Ti-plasmid**, both of which exist in the same *Agrobacterium* cell (Fig. 32.2). This division into two components is possible because genes for virulence and integration are located outside of the T-DNA region of the Ti-plasmid and function independently of the T-DNA components. The binary vector carries the T-DNA and functions as such as the true transformation vehicle. It consists only of the flanking LB and RB sequences, a **multiple cloning site (MCS)** for the insertion of expression cassettes with genes of interest, and a selectable marker cassette for expression within the transformed plant cell, while all of the genes necessary for tumor growth and opine synthesis are eliminated. The capacity of binary vectors for foreign DNA between LB and RB is about 150 kb. The rest of the plasmid consists of a backbone of a classical *Escherichia coli*-compatible plasmid, like pBR322 with bacterial selection marker genes and compatible origins of replication for autonomous duplication in *E. coli* and *Agrobacterium* (average size without the transferred DNA, about 10 kb). Modern Ti-plasmids function in specially modified *Agrobacterium* strains as **helper plasmids** for the transformation. They carry their own bacterial selectable marker gene cassettes (e.g., spectinomycin resistance) to allow for their stable presence in *Agrobacterium* host strains, but as a separate entity they are no longer capable of inducing a tumor in plants.

Agrobacterium is used very successfully for the stable transformation of more than 100 commercially relevant dicot plant species, including oilseed rape, soybeans, sugar beets, potatoes, tomatoes, and tobacco. Monocots such as corn, rice, barley, and wheat can also be transformed on an increasing scale with the help of *Agrobacterium*, even though they are not natural hosts.

Usually, special organs and tissues are used for transformation: **callus tissue** (e.g., corn), **tissue pieces** (e.g., parts of leaves of potatoes or tobacco), or isolated immature **embryos** (wheat species) and **single organs** (e.g., buds of rapeseed). In addition to this, an *in planta* transformation protocol has been developed for the model plant *Arabidopsis thaliana* (thale cress), which uses *intact plants*. In this (floral dip) method, buds still on the plant are dipped into or infiltrated with suspensions of *Agrobacterium* that contain binary plasmid systems. With surprisingly high frequency, cells of the germ line are transformed before meiosis and therefore pass the integrated T-DNA directly on to the seeds of the following generation. Currently, this is the fastest and least costly method of *Arabidopsis* transformation and has already been successfully applied to other plant species (e.g., oilseed rape). Since this method avoids long-term regeneration protocols, it is especially attractive due to the reduced if not circumvented risk of somaclonal variations. The method depicted constitutes the basis of the T-DNA mutant collection of over 100 000 independent lines that are accessible for scientific purposes (The Arabidopsis Information Resource (TAIR); http://www.arabidopsis.org).

A further application consists of transient transformation in which plant parts (e.g., leaves) are infiltrated with suspensions of *Agrobacterium* carrying binary plasmid systems. Through the use of a target gene construct with **introns**, an influence on the bacterial expression can be completely ruled out, because only the plant itself can conduct the necessary splicing procedure.

On one hand, the limitations of the *Agrobacterium* system consist of its limited host range. On the other hand, T-DNA integrates solely into the nuclear genome. New biotechnology approaches, however, also involve the integration of the transformation of chloroplast genomes, which requires a completely different method.

32.3.1.2 Biolistic Method: Gene Gun

The limitations of biological gene transfer with *Agrobacterium* led to the development of alternative methods, of which particle-bound biolistic genetic transfer is the most successful and also the most widely employed. In the first experiments, DNA was packed onto particles that were used to load a 0.22-caliber shotgun and shot on leaves. This method was successful in the first transformation of a monocot (corn). Following significant optimizations, biolistic devices called "gene guns" are now commercially available. The biolistic method has several decisive advantages:

- It can be applied to every kind of plant or tissue.
- The cell wall – a major problem for most transformation methods – is physically overcome.
- In principle, all genomes in the cell can be reached.
- Stable and transient genetic transfers are optional.

For biolistic transformation plasmids reproduced in *E. coli* that carry corresponding bacterial elements as well as cassettes for the expression of plant genes are bound to tungsten or gold particles through precipitation with $CaCl_2$ and spermidine. The loaded particles are accelerated through air pressure and shot onto the plant. Apart from obvious tissue damage, a small fraction of the hit cells survive the impact of the particles. On the basis of mechanisms not yet understood in detail, the DNA is integrated into the chromosome and is therefore transferred to following generations once a complete plant has been regenerated from the transformed cell by *in vitro* culture methods. Most likely the mechanism is based on a partial solubilization of the DNA in the hit cells and the DNA is integrated through the cells' own recombination mechanisms, either into the nuclear genome where untargeted integration occurs or with significantly decreased frequency into the plastid genome, where homologous and therefore targeted recombination takes place. Physical parameters of the biolistic transformation process need to be optimized for each type of tissue subjected to transformation. The ability to reproduce the speed and power of the particle is guaranteed through a pressure chamber filled with helium and nitrogen. For most tissues, a particle speed of around $440 \, \text{m s}^{-1}$, a particle size of around 1–2 µm, and a particle density of $19 \, \text{g cm}^{-3}$ are suitable. In these machines, power and distance allow for properly adjusting the penetration of cell walls and the tissue depths to be reached.

An advantage of **biolistic transformation** is the comparatively small size of the binary vector carrying the gene constructs. It is around 3–4 kb, because basically only the DNA to be transformed is needed. Usually, expression cassettes for integration into the plant genome are amplified in *E. coli* by using simple standard cloning vectors. Expression cassettes are then removed from the vector backbone by restriction and then purified prior to direct transformation via particle bombardment. However, the whole size of the transformable fragments is limited by shear forces damaging large DNA fragments and therefore decreasing the efficiency. The experimental procedure in itself is simple and fast, but the yield of transformed cells with expression is only 1–5% of all cells hit. With this, a strong fluctuation of expression intensity in transient transformation experiments is observed. This is, however, largely irrelevant for **transient expression** experiments targeting subcellular localization of genetic products or the analysis of promoter functionality with the help of reporter gene fusions. The most prominent reporter genes in plant research are the *uidA* gene (*β*-**glucuronidase** from *E. coli*) and several varieties of the **Green Fluorescence Protein (GFP)** series (from the jellyfish *Aequorea victoria*).

For stable transformation, many transformants have to be regenerated from the targeted tissue in order to obtain lines showing stable expression. Due to not yet fully understood recombination events of vector DNA prior to integration or during the integration process the stably transformed plants often show complex integration patterns of the DNA of choice consisting of multiple copies integrated into one locus or even spread over several loci if independent integration events occurred in parallel. This can lead to stronger expression or to the silencing of the genes of interest. Therefore, each individual transgenic plant is always carefully checked for the expected integration pattern of a single copy at a single locus in order to be able to generate meaningful results. In addition to this, one can often observe changes in the placement of entire DNA segments in the target genome, which can cause phenotypical effects in the first-generation transformants. In such cases, several rounds of backcrossing in order to reconstitute the original genetic background must take place. However, several of the current market-certified genetically engineered species of corn, sugar cane, and soy have been produced on the basis of biolistic transformation.

Other physical transformation systems also transfer the "naked DNA" directly into cells and are usually based on the use of **protoplasts** (i.e. plant cells whose cell wall has been removed by **cellulases**). These systems are not commonly used. They are especially suited for transient transformation protocols, but can also allow the stable transformation of species provided they permit the regeneration of the cell wall.

32.3.1.3 Plastid Transformation

Bringing useful genes into the genome of plastids has several advantages compared to the transformation of the nuclear genome. So far, in high-throughput approaches, only heterologous recombination with unpredictable insertion locations in the nuclear DNA are possible. Consequently, T-DNA or DNA fragments from direct transformation can by chance hit and inactivate essential genes of the plant, so that the cell becomes impaired or even dies. An insertion into usually inactive chromosome areas, like heterochromatin, is also undesirable, because the gene of interest is then hardly expressed.

On the other hand, **plastid genomes** have the ability to perform homologous recombination supposedly because of their origin from prokaryotes according to the endosymbiont hypothesis (see Section 3.1.3). Accordingly, gene cassettes with flanked homologous areas can be produced for a plastid genome target location, just as with recombination in bacteria. Between sequences that are homologous to the target location lie genes for the selection of integration events and the gene of interest. **Selection markers** are usually antibiotic resistance genes. Both types of expression cassettes are controlled by plastid promoters. Strong plastid promoters can be derived from genes related to photosynthesis, like the light-regulated *psbA* gene of the D1 protein of photosystem II. The transfer is usually carried out by biolistic transformation.

The predominantly **maternal inheritance** of plastids is an important advantage for the expression of useful genes in these organelles. Basically, the plastids are passed on by the egg cell in many species, not by the pollen, so that the probability of **out-crossing of traits** of genetically engineered plants into related wild species through pollen transfer is close to zero.

The high number of copies of the transgene per cell with accordingly strong expression is another advantage. One leaf cell contains up to 100 chloroplasts and those have up to 100 copies of the plastid chromosome. Thus, up to 10 000 copies of the inserted gene per cell can be present. Through several rounds of selection, the plastid's own recombination system undertakes the complete transformation of all chromosomal copies of a chloroplast, resulting in a **homoplastomic** population of plastids in the cell. Supposedly because of the prokaryotic origin of the chloroplasts, no cases of gene silencing have been observed, despite the enormous number of genome copies. Therefore, plastid transformation is especially attractive for phytofarming approaches where large amounts of protein are to be produced from the transformed DNA. The prokaryotic nature of the plastids can be used to construct polycistronic transcription units for the simultaneous expression of several genes. While the coordinated expression of several transgenes in the nuclear genome requires several different promoters to prevent gene-silencing phenomena, the expression of such a polycistronic construction in the plastid genome can be controlled by a single suitable promoter.

The present limitations in the application of plastid transformation are:
- Low transformation frequencies.
- Intensive phase of tissue culture to regenerate intact homoplastomic plants.
- High risk of somaclonal variations due to long tissue culture passages.
- Plastid-expressed proteins cannot leave the chloroplast.
- General lack of posttranslational modifications (e.g., glycosylation) in plastids due to the prokaryotic origin of the organelle.

32.3.1.4 Viral Systems

Traits such as insect resistance need to be permanently conferred to a crop plant. If the goal is to express proteins in large amounts in order to purify the protein for industrial or pharmacological applications then **viral expression systems** can be used as an alternative to *Agrobacterium* and biolistic delivery. The autonomous multiplication of the viral genome in the infected plant is the main advantage of this still developing method, which potentially leads to a high expression rate, with simultaneous avoidance of **gene-silencing** effects. In addition, viral genomes are very small and therefore easy to manipulate. Disadvantages are the lack of heritability of the genetic changes, limitations regarding the sequence length of the gene of interest, and a limited applicability under field conditions. Of course, cultivation of plants infected with viral vectors for overexpression of proteins is carried out in greenhouses and not under field conditions.

The genomes of many plant viruses consist of one to three RNA molecules. These viruses often have a wide host spectrum and can reach a mass of $1–2 \text{ g kg}^{-1}$ host plant. Peptides of interest can be fused with a capsid protein in such a way that they are exposed at the surface of the virus. Such epitope presentations of suitable recognition sequences are present in multiple copies in the virus particles and can be used for **vaccination**. In the case of polypeptide expression systems, complete coding sequences of genes are cloned into the viral genome in such a way that the desired protein is efficiently translated, processed into a nonfused form, and finally purified with the help of protein chemical methods.

A successful example for the development of such an expression system is the **cowpea mosaic virus (CPMV)**, which has a split RNA genome. RNA1 carries genes for the replication in a host cell, and RNA2 carries genes for both capsid proteins and therefore the information necessary for the cell-to-cell spread in the host. Genes of interest can be translationally fused up to a size of several kilobases to the carboxyl-terminal end of the smaller of the two capsid proteins. All proteins on RNA2, including the foreign proteins, are translated by the plant cell as a single polypeptide; afterwards they are processed to their functional size by an RNA1-encoded sequence-specific endoprotease. The modifications of the viral genome are carried out at the cDNA level that is being cloned and multiplied in standard *E. coli* plasmids.

32.4
Selection of Transformed Plant Cells

The targeted genetic modification of plants is based on the transfer, integration, and expression of selected genes in plant cells, which can be regenerated into intact fertile transgenic plants. Since the efficiency of the stable gene transfer is as low as 10^{-3} to 10^{-4} even for the plant species with the best transformation rates, it is indispensable to apply systems for selection and identification of the transformed cells or tissue patches embedded in essentially nontransformed cells.

Selection marker systems in plants, which allow for the identification and selection of genetically altered cells after transformation, can be divided into two basic categories: **negative selection markers**, which allow transformed cells to detoxify a correspondingly toxic selection substance, while the untransformed cells die, and **positive selection markers**, which guarantee a physiological advantage for the transformed cells in comparison to the nontransformed cells, without which a regeneration would be slowed down or impossible.

Examples for negative selection are systems that rely on antibiotic or herbicide resistance. Examples for positive selection systems derive from, among others, plant sugar metabolism or hormone metabolism.

A further category of **selectable markers** are the so-called **counter-selectable markers**, which allow for conditional selective destruction of genetically engineered plants by transformation of a nontoxic or metabolically neutral substance into a phytotoxic compound through the enzymatic activity of the selection marker protein.

Furthermore, there are so-called **visual markers**, which rely on visual phenotypical traits to distinguish transformed from untransformed tissue. In this case, however, the number of untransformed plant regenerates is not reduced under selection pressure.

32.4.1
Requirements for an Optimal Selection Marker System

A selection marker system consists basically of three components: the **selective compound**, the **selection marker gene**, and the **material** (i.e., plant tissue) used for selection. Only the perfect interaction of these three modules permits successful selection. The material used for the selection plays an especially important role, particularly in the plant system. This is due to the versatility of the protocols, the variation in explants, and the use of different developmental stages and varying tissue culture conditions. The sensitivity of any given tissue culture system can vary greatly, because of the ever-differing genotypes, their sensitivity with regard to the selection substance, and the requirements for the expression of the selection marker gene. Generally, ubiquitous, constitutive promoters with different expression strengths are used for the expression of the selection marker gene in order to guarantee the optimal expression strength for the corresponding system. However, it is advantageous if the seeds as the final product and most valuable storage part of the plant remain excluded from expression of the marker protein. Promoters that induce expression in meristematic tissues are particularly desirable because these areas of growth are metabolically most active and therefore react especially sensitively to the toxic selection substances. As a general prerequisite for a useful selection marker system, it must not interfere with the plant's metabolic capacities in the long run in order to guarantee proper agronomic performance and, especially, to avoid any yield drag.

With respect to the choice of an **optimal selection marker system**, two molecular genetic parameters have to be especially monitored: the size of the expression cassette for the expression marker and the characteristics of the source organism. Concerning the size of selection marker genes, two points are crucial: the expression cassettes must be as small as possible in order to avoid an unnecessary size increase of the transformation construct and, furthermore, the selection marker should only consist of a single gene. Regarding the characteristics of the source organism from which the corresponding candidate gene is to be isolated, it is important that the organism concerned is traditionally a part of the food chain. If candidate genes are isolated from baker's yeast, food plants, or from symbiotic bacteria like *E. coli*, it has already been proven, through the tradition of nutrition, that the gene product and the respective metabolic products fulfill all criteria of common compatibility. It is therefore assumed to be unnecessary to evaluate a large number of new substances and their fate in the food chain.

From a biochemical point of view the mode of action of the selection marker and the characteristics of the selective agent are decisive. The detoxification of the selection substance should be irreversible and neither be based on an equilibrium reaction that in the cellular milieu would be reversible nor result in a circular process of detoxification and subsequent reconstitution of the toxic substance. Furthermore, it is important that the metabolism of the selective agent leads to intermediates that are already part of the cellular metabolism. Nevertheless, it is desirable that interference with primary metabolism can be ruled out as much as possible, which is most effectively achieved by using selection marker genes that encode for a reaction that is not present in the plant cell. This is especially important in view of avoiding marker gene-induced phenotypes that – to a limited extent – may be tolerable for research work in the laboratory, but could be detrimental to trait and yield stability under field conditions.

For **efficient and reproducible selection** procedures, a selection substance is needed that permits a fast and clear differentiation between genetically engineered and nongenetically engineered plants. Basically, there are two scenarios. Differentiation might be easier if the nontransformed plant cells die under selection pressure. However, it is also known that the regeneration ability of a cell in a cell colony is reduced if the concentration of metabolic products of dying cells increases and therefore becomes increasingly toxic for the surviving cells (in the described case, transgenic cells). As a result, the decision between pure retardation of nontransformed cells, on the one hand, and the efficient elimination of those cells, on the other hand, is a question of optimal balance for each tissue type.

32.4.2
Negative Selection Marker Systems

Most classical selection marker systems are based on **antibiotic resistance**. Among the most commonly used resistance genes are the ones encoding **neomycin phosphotransferase II** (*nptII*) and **hygromycin phosphotransferase** (*hpt*), both from *E. coli*. Through phosphorylation, the NPTII protein inactivates a number of aminoglycoside antibiotics, like **kanamycin, neomycin, geniticin (G418)**, or **paromomycin**. While geniticin is often used for the selection of transformed mammalian cells, the three others – with different efficacies for different species – are almost exclusively used in plant transformation systems. **Hygromycin** inactivation through the HPTII enzyme is a reliable selection marker system for a number of both animal and plant systems since hygromycin in general has a substantially more toxic effect on cellular metabolism than kanamycin. Other less widely used antibiotic resistance markers rely on **gentamycin, bleomycin**, or **phleomycin**.

During the regeneration of plants with low regeneration capacity, it has been shown that even in transformed cells the toxic effects of the antibiotic compounds have a negative influence on the regeneration success, so that for tissue culture work with transformation-recalcitrant plants like soybeans and sunflower the use of antibiotics is only of limited usefulness. An important aspect for the use of antibiotics for agricultural biotechnology is that the antibiotics and derivatives used in research today have almost no relevance as **medicinal therapeutics**.

The second large group of negative selection markers works by **herbicide resistance**. The selection marker genes of this category have plant and bacterial origins. In the scope of herbicide-based selection marker systems, genes for the production of selective tolerance or resistance against broadband herbicides like **glyphosate** (better known under the name Roundup) and **glufosinate** (better known under the names Basta®, L-phosphinotricin and bialaphos) are of special importance. Both herbicides are effective in dicot as well as monocot plants. Two different genes have been described for the generation of **glyphosate tolerance**: the *gox* gene encoding a glyphosate oxidoreductase from *Achromobacter* sp.

and the *epsps* or *aro*A gene for an enolpyruvate shikimate-3-phosphate synthase with alleles from *Agrobacterium*, corn, and petunia. While the *gox*-mediated resistance is based on metabolic detoxification of the selective agent, the latter class of genes allows for establishing tolerance by implementing a metabolic bypass for the sensitive reaction. For the production of **glufosinate resistance**, different alleles of phosphinotricin acetyltransferases from *Streptomyces hydroscopicus* and *S. viridochromogenes* have been isolated and established. The mode of action within this system is again the detoxification of the selective compound. The different alleles show varying degrees of selectivity in different plant systems. Further selection marker systems based on broadband herbicides include several alleles of acetolactate synthase genes from a number of organisms, among those model plants, crop plants, and also algae. The use of mutated alleles of acetolactate synthase grants resistance against the large group of sulfonylurea derivatives, imidazolinones, and thiazole pyrimidines. An example of a selection marker system on a herbicide basis, which can only be applied to dicot plants, is bromoxynil nitrilase from *Klebsiella ozaenae*. **Bromoxynil** is among the auxin analogs that exhibit a selective herbicide effect on dicot plants; they have no effect on monocots because of different anatomies of the sensitive shoot apical meristem.

32.4.3
Positive Selection Marker Systems

Positive selection is based on the use of nontoxic substances. As a result of enzymatic conversion through the marker gene product, such a substance allows for selectively compensating for a given auxotrophy of the tissue to be regenerated. Two scenarios dominate: either the catalytic conversion of a nontoxic precursor carried out by the marker gene product provides an essential compound exclusively only to the transformed cells, or the biochemical reaction catalyzed by the marker gene product yields a compound that provides a metabolic or physiologic advantage to the transformed cells when compared to the nontransformed cells.

An example for the positive selection is the **mannose phosphate isomerase** system that is based on mannose-6-phosphate as the only sugar source contained in the regeneration medium. Plant cells are only able to utilize this sugar after isomerization into glucose and fructose-6-phosphate. Thus, when providing mannose-6-phosphate as the only carbon source in the regeneration media only those cells that obtained the isomerase gene will be able to carry out the isomerization step, and will accordingly be able to use the sugar compound of the regeneration media and regenerate into whole plants. A similar example is the **xylose isomerase** system, the metabolic activity of which permits the use of xylose as an alternative carbohydrate source.

Further positive selection marker systems are based on the use of genes, the gene products of which catalyze the enzymatic release of plant hormones from specific precursors. Examples include the gene for an **indole acetamide hydrolase** (*iaaH*) from the Ti-plasmid of *A. tumefaciens*, the gene product of which catalyzes the hydrolytic release of **auxin (indole-3-acetic acid)** from indole acetamide, as well as genes encoding for **glucuronidases**, which release cytokinins required for shoot induction during plant regeneration from cytokinin glucuronids. These hormones are required to induce cell division and the regeneration of shoots and roots during tissue culture. In omitting hormones in the nutrient medium only cells with sufficient endogenous capacity of hormone formation are able to proliferate.

32.4.4
Counter-Selection using Bifunctional Marker Genes

Counter-selection is described as the selective elimination of plants that are transgenic for a selection marker gene, the product of which is capable of converting a metabolically neutral or nontoxic compound into a **phytotoxic** one. Accordingly, the previously described *iaaH* gene cannot only be used as a positive selection marker, but is also an example of a so-called conditional **counter-selectable marker** if using naphthylacetamide as selection substance. The hydrolysis of naphthylacetamide results in the release of the auxin analog naphthylacetic acid. During excessive release of naphthylacetic acid, the genetically engineered plants grow to death, while the nontransformed plants are not affected. This kind of selection marker is of special interest for transposon mutagenesis in order to allow for selection against the donor locus after successful transposition of the respective genetic element to a target locus had been achieved. Selection of the transposition event is usually carried out by using a classical negative or positive selection marker system. Counter-selection marker systems are also valuable tools in combination with recombination processes that require selective elimination of nonrecombined cell lines in order to enrich those cell lines that have undergone successful recombination. Applied examples can be found in the general literature discussing the production of marker gene-free plants through the use of recombination systems like Cre/lox or similar (see also Section 32.4.6).

32.4.5
Visual Markers

In addition to the marker systems with selective effects, there are the so-called **visual markers** with which genetically engineered tissue can be visually differentiated from nontransformed tissue. Basically, the use of any phenotypical characteristic is conceivable. The most well-known examples from the literature describe the use of classical reporter genes like *β*-D-**glucuronidase (GUS)** from *E. coli* or **GFP** from *A. victoria*. Furthermore, genes of the anthocyanine biosynthesis pathway or the ectopic expression of genes that control trichome formation have been proposed as candidates for visual markers, although the application in these cases was not targeted towards crop plants.

32.4.6
Selection Systems, Genetic Engineering Safety, and Marker-Free Plants

The success of products derived from agricultural biotechnology applications depends on the commercial registration of the specific plant species and plant product, and ultimately on **consumer acceptance** and demand. In particular, the use of marker genes for antibiotic resistance has led to public concern, especially if the use of antibiotic resistance genes may favor antibiotic resistance in medically relevant pathogens, and therefore pose a higher health risk to humans and animals. In response to these concerns, scientists worldwide have worked on the assessment of potential risks for the environment, for pest resistance management as well as nutritional safety with respect to crop plants carrying genetically engineered herbicide or insect resistance.

In conclusion, the scientific assessment clearly revealed that the use of **antibiotic resistance genes** does not pose a danger for humans, animals, or nature. Even though the originally isolated resistance genes are from bacteria, a reverse transfer from a genetically engineered plant into the original host organism has only a very faint probability since the modifications for successful expression of bacterial genes in a plant requires the complete exchange of regulatory sequences. Furthermore, the antibiotics used under laboratory conditions nowadays do not have any medical relevance.

In addition to considering the public concern regarding putative **horizontal gene transfer** of antibiotic resistance genes from genetically altered crop plants, the practical approaches towards agricultural plant biotechnology have revealed a technical need for the availability of a number of different selection marker systems:

- Production of a genetically altered variety with respect to several new traits may call for repeated transformation of an already genetically engineered line so that a second selection marker needs to be available for such a super-transformation. However, the latest technical progress has demonstrated the feasibility of successful crop plant transformation with large multigene constructs favoring direct gene stacking over super-transformation procedures.

- Development of tailor-made transformation protocols for each plant species has revealed that any given selection marker system does not work equally well in each plant species, each tissue type, or even each plant variety within a species. Therefore, it is advantageous to have a choice of different selection systems available. Privalle *et al.* (2000) reported that the *Agrobacterium*-mediated transformation of corn embryos using the mannose phosphate isomerase selection marker system allows for an average transformation efficiency of 30% (with maximum rates of up to 90%). In comparison, the same selection system results in transformation efficiencies of only 10–20% during the *Agrobacterium*-mediated transformation of rice embryos and only 1% efficiency could be reached with barley.

Further research focuses on strategies to eliminate selection marker genes after successful transformation in order to finally produce **marker-free plants**. Among the different technological approaches, two directions have proven to be the most promising:

- The technically more simple solution is **cotransformation**, which is based on separate localization of the trait gene and the selection marker gene on two different T-DNAs. This can be achieved by either placing the two expression cassettes into two different T-DNA cassettes on a single binary vector or by providing two different binary vectors, each of them carrying only one of the two expression cassettes. In the latter case the two binary vectors are either contained together within one *Agrobacterium* strain or each of the binary vectors is contained in a separate *Agrobacterium* strain and then both of these strains become mixed together for the actual transformation process. For the transformation of tobacco, it was shown that up to 70% of the individual transgenic plants integrated both T-DNA cassettes independently from one another. Accordingly, within the following generation, after **meiotic segregation**, one could obtain plants carrying the engineered trait gene cassette, but no longer carrying the selection marker gene. The only disadvantage of this otherwise elegant solution is the fact that after segregation only 25% of the original population of transgenic individuals carrying both expression cassettes will finally display – according to Mendel's laws – the desired genotype. Thus, cotransformation is only the method of choice if a protocol guarantees high transformation efficiencies anyway so that the loss of generally more than 75% of the transgenic plants can be easily compensated for.

- A technically more challenging alternative to producing marker-free plants consists of the use of **sequence-specific recombinases** that, following successful selection, catalyze the excision of the marker gene. All of the sequence-specific recombination systems successfully used in plants so far are of microbial origin and belong to the integrase family. They consist of a recombinase (Cre, Flp, or R) and a recognition sequence for the recombination (loxP, *frt*, or RS; Fig. 32.3). The process of recombination-mediated marker excision is based on the ability of these microbial recombinases to specifically cut DNA at the cognate recog-

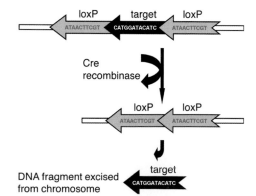

Fig. 32.3 Cre/lox-based DNA excision. The loxP (locus of excision) system of bacteriophage P1 is an established mechanism for targeted removal of DNA fragments in eukaryotic genomes. Cre recombinase recognizes the loxP sites and excises DNA fragments that are subsequently degraded.

nition sites, marked through so-called direct repeats, and to connect the two homologous ends. The latest developments provided systems that express the recombinase gene under the control of a chemically inducible promoter system.

Other approaches follow the use of specific promoters with strict tissue- and developmental-stage specificity in order to precisely determine the point in time of marker gene excision. Such promoters are chosen for being active only after the process of selective plant regeneration has been finalized. Another option for temporally controlled marker gene excision is to apply the recombinase activity to a later plant generation by crossing the original transgenic line carrying the trait gene cassette as well as the marker gene cassette with another transgenic line expressing the recombinase under a constitutive promoter. In all of these cases it is favorable to design the respective expression cassettes in such a way that the activity of the recombinase leads to a loss of both the marker gene cassette as well as the expression cassette of the recombinase itself. None of the technologies described above is sufficiently mature for routine application in an industrial setting, but remain under development.

32.5
Regeneration of Transgenic Plants

32.5.1
Regeneration Procedures

The transformation protocols using *Agrobacterium*, **gene guns**, or other methods are generally based on isolated plant parts like pieces of leaves, hypocotyls, or embryos. Often these explants are surface-sterilized or aseptically cultivated *in vitro* into callus tissue and prepared for transformation. The efficiency of most transformation protocols is still very low with all of the current methods, so that the few transformed cells have to be separated from the huge population of nontransformed cells by selection pressure and the use of a selection marker system.

This procedure is carried out in **aseptic tissue cultures**, in order to be able to cultivate the explants and to regenerate single individual transgenic cells into intact plants. Selection and regeneration are therefore tightly connected with one another, and have to be coordinated with one another. Usually, a given selection method as described in Section 32.3 is not necessarily compatible with any regeneration protocol and *vice versa*. The number of interacting factors, such as media composition, hormone regime, and light intensity, is already huge just for the optimization of a regeneration protocol for a single species. Accordingly, as soon as regeneration needs to be combined with selection, additional effects due to application of the selective compound have to be considered and included in the optimization program from the very beginning. Due to the multiple individual requirements of different plant species as well as plant tissues, it is not possible to refer to a single protocol within this chapter, but the reader is invited to refer to the specific literature on plant tissue culture. The complexity of tissue culture media as well as tissue culture protocols in combination with the limited transferability between the species renders regeneration work very demanding. The generally rather low transformation efficiencies, long duration, and thus high costs of regeneration account for **a critical process step in plant biotechnology** if applied to industrial-sized production procedures. Therefore, the recent intersection of robotics and modern cell culture techniques has attracted a great deal of attention.

Basically, there are two main paths of regeneration: somatic embryogenesis and adventitious shoot formation. In both cases, a precise hormone treatment is applied to either induce the development of whole embryos from somatic cells or to induce the development of shoots that afterwards become rooted by means of further hormone treatment. Each alternative has to be individually optimized for each plant species and selection method. Apart from systematic experimental optimization, these protocols are generally of empirical nature.

32.5.2
Composition of Regeneration Media

If an explant is produced for transformation, all inorganic and organic nutrients that otherwise would be made readily available by the intact plant have to be supplied in the tissue culture media to allow for growth. Accordingly, such media contain, first of all, **macro- and micronutrients** reflecting all **essential needs of the species**. Due to *in vitro* cultivation systems mostly for ornamental but also crop plant cultivation, such optimized mixtures are available for most commercially relevant species. Furthermore, isolated plant parts are often no longer totally photoautotrophic and therefore need to be supplied with carbon sources. **Sucrose** is usually the main transport form of sugar in plants. Many **vitamins**, such as nicotinic acid, thiamine, and pyridoxine, that plants otherwise synthesize themselves, are also not available in relevant concentrations and thus need to be added to the medium. As a rule, the vitamin cocktail has to be optimized more precisely as the explant becomes smaller.

The hormone composition within the media, especially their concentrations in relation to each other, is decisive for the type of tissue to be regenerated (embryo, shoot, or root). The main hormone groups are **auxins** and **cytokinins**, while gibberellins and other hormones are used only in special cases. Auxins and cytokinins are essential hormones (i.e., no null mutants are known in plants). Auxin (mostly indole-3-acetic acid) is mainly formed in the apical meristem and is distributed within the plant by a specific cell-to-cell transport process. It works as a general growth stimulant and enhances root formation. Cytokinins, such as zeatin, are purine derivatives and are mainly synthesized in the root meristem. From there, zeatin becomes conjugated as zeatin riboside and is then distributed within the plant over the phloem. Cytokinins stimulate cell division and work mainly in combination with auxins. Basically, the relative concentration of the two hormone classes determines if an explant develops into callus tissue, or forms adventitious roots or shoots. In many cases, further growth substances are necessary, such as polyamines during somatic embryogenesis to induce growth and differentiation of explant tissue. In some cases, the growth substances required for proper growth and development are not yet identified chemically and thus are replaced by complex additives. Classic sources for such unknown growth factors are coconut milk or extracts from maize kernels. Although challenging and labor-intensive, regeneration technologies for callus tissue, suspension or anther cultures, for tissue culture products subjected to cryo-conservation, for raising haploid plants, and for further reproducible methods towards the production of (not only genetically altered) augmentable individuals are one of the major core areas of plant biotechnology.

32.6
Plant Analysis: Identification and Characterization of Genetically Engineered Plants

32.6.1
DNA and RNA Verification

Transformation and regeneration of potentially transgenic plants is followed by the verification of the transformation event on the molecular level and the characterization of the expression of the introduced gene. The most important methods for this are **PCR**, DNADNA (**Southern hybridization**) and RNADNA (**Northern hybridization**) (see Chapters 11 and 13).

Plant analysis occurs during the whole phase of production and characterization of transgenic plants, but with different aims and targets at different points in time. The timing of the analysis steps in a high-throughput production pipeline needs to balance the need for, on the one hand, an early analysis to assure that virtually no resources are invested in less-favorable individual transformation events with unwanted integration patterns or even in false positives. On the other hand, it is important to harvest tissue for nucleic acid purification and analysis at a stage of plant regeneration where the tissue is already amenable to proper handling, where the process is fast, simple, and does possibly not interfere with the ongoing regeneration.

Generally, the various protocols for DNA and RNA isolation (see Chapter 9) from plant material are focused on the appropriate tissue. Yield and quality of clean nucleic acids from plants are nevertheless often less satisfactory as compared to bacterial, yeast, or mammalian cells. The reasons for this are the rigid cell walls of the plants, and the secondary metabolites derived from cell wall fragments, vacuoles (usually polyphenols), and chloroplasts (chlorophyll) often have oxidizing or contaminating characteristics. Furthermore, specialized organs, especially seeds, are challenging. The storage compounds contained therein, like starch and lipids, often interfere with classical extraction methods. For small amounts of materials or high-throughput applications, tailor-made isolation kits especially adapted to various plant tissues are commercially available.

Genomic PCR (see Chapter 13) is mainly used in the very early stages of the regeneration phases when only little plant material is available and a large number of samples have to be analyzed for the mere presence of the transgene, favorably as a single copy in a single location. Genomic PCR finds a further application during the sorting of offspring of the transgenic lines. While passing on the genotype, through selfing or crossing, the inserted genes split according to Mendelian law and depending on the number of copies into different parts: positive homozygote, heterozygote, or negative homozygote offspring. These segregating populations are most efficiently characterized through PCR. The identity of the PCR products is usually checked through specific controls, by sequencing or by the use of two nested specific primer pairs, or by hybridization with fragments of the transformed construct.

The exact determination of the **number of insertions** of foreign DNA via PCR methods is designed to circumvent laborious Southern hybridization of large sets of transgenic plants. It is generally preferred to limit Southern-based plant analysis to a smaller set of precharacterized candidate plants. Both, *Agrobacterium*-mediated as well as biolistic gene transfer may result in several insertions either at different genomic locations or as repeated copies within one location. For the reliable performance and inheritance of the genetically engineered trait, a single insertion carrying only one copy of the transgene is required. Such single-copy, single-insertion homozygous transgenic lines are easy to identify since they would no longer segregate for the transgenic trait locus and, as opposed to repeated copies, show more stable expression rates since they would more rarely (if at all) suffer from silencing effects. Hybridization with a choice of different probes reveals information about the intactness of the transformed expression cassette. Furthermore, for the interpretation of a hybridization experiment, it needs to be considered that many crop plants have very large genomes due to polyploidization of the chromosome sets. As an example, the genome of hexaploid wheat (*Triticum aestivum*) is about 68 times larger than that of humans. Thus, unambiguous molecular characterization is difficult due to the background signals of similar DNA sections within related genomes. Finally, the portion of plastid DNA of the entire DNA can be relatively high and disturbing. A green leaf cell can contain up to 10 000 copies of the chloroplast chromosome, thus influencing the signal intensity of the Southern analysis. On an industrial scale and based on the latest technological breakthroughs in the area of high-throughput sequencing hybridization technologies are being increasingly replaced by sequence determination of isolated PCR fragments.

The **expression of the trait gene** is usually examined after regeneration has been completed or in subsequent early generations. Expression strength constitutes a further criterion for event selection in addition to the need for single-copy, single-insertion lines. The expression rate of the trait gene on the RNA level can be determined through **Northern hybridization** using a specific probe or through **real-time PCR** on cDNA using reverse transcription protocols (see Chapter 13). Doing so usually only determines relative expression differences among the transgenic lines and between transgenic lines and control lines. However, more decisive data are derived from the analysis of content and functionality of the encoded protein. Recently, genome-wide expression analysis by means of **DNA arrays** has become increasingly popular in order to not only capture the expression behavior of the transgene but also to be able to examine the transgenic plants for the potential influence of the newly introduced gene on the expression of the plant's original genome. Thus, important conclusions on the functionality of an introduced gene and, at the same time, on the suitability of a given transgenic event *per se* can be drawn by identifying the overall changes of gene expression within a given transgenic plant.

32.6.2
Protein Analysis

The main targets for protein analysis in genetically modified plants are the products of the inserted trait genes, whereas the presence of the selection marker gene products today is rarely examined beyond functional selectivity. In a first step, the presence and relative concentration of the target protein are determined by using **specific antibodies**. The method of protein immunodetection through gel electrophoresis and transfer on filters (**Western blotting**) is a standard protocol. Methods of mass spectrometry are also useful to identify proteins (see Chapter 8). Knowledge of the relative expression frequency of the newly introduced proteins is an important criterion for the identification of candidate lines from earlier selected lines.

Highly abundant proteins, such as seed storage proteins, can also be detected visually without using antibodies by simply staining the protein pattern after size-based resolution on polyacrylamide gels. When the newly introduced protein displays specific **enzymatic activity**, the most decisive verification of functionality can be carried out by measuring the catalytic activity of the encoded protein. In order to confirm the results of immunological quantification experiments, enzyme assays are conducted to test for maximum reaction speed under substrate saturation, while kinetic or other specific characteristics like inhibition of a defined biological activity (e.g., *Bacillus thuringiensis* CRY proteins) have to be determined in specific tests, including laboratory tests, whole-plant greenhouse analysis or at a later stage even field trials.

32.6.3
Genetic and Molecular Maps

For research purposes the exact characterization of the insertion location of the transgene in the genome is only conducted if the transformation is carried out in order to create mutations or if the location of the gene or the insertion mechanism is the essential target of the research. In the area of applied plant biotechnology, however, knowledge of the exact location of an insertion is of practical relevance. The DNA construct could potentially hit an important gene and inactivate it. However, in a case when a gene for an essential primary function in metabolism is hit, the resulting mutation would usually be lethal. Potential hidden or conditional negative effects may become evident only later during the process of characterizing a candidate event if, for example, the locus of a gene involved in stress resistance responses would be affected by the transgenic insertion. A possible lower yield would not be recognized until the respective transgenic lines are exposed to natural stresses in the field. Knowledge of the insertion location is also important for further breeding steps. Initial transformation of the trait gene is often carried out using transformable varieties of a given crop plant species and only the promising events would be crossed into lead genotypes for further development of a final commercial variety. Furthermore, different elite genotypes adapted to farming in different climates and locations have been developed. In order to achieve the intended genetic improvement for a number of varieties of an agronomically relevant species the transgene would be introduced by crossing the initially transformed variety with the nontransgenic target varieties using a method called **marker-assisted (smart) breeding** that requires detailed knowledge of the genetic locus of transgene insertion. Overall, the deregulation of transgenic events with the respective authorities requires a detailed description of the insertion locus.

Genetic maps can be produced using phenotypic or molecular markers. Depending on the complexity of the genome as well as the availability of genomic information about the transformed plant species, classic genetic maps for the approximate localization of the inserted genes may be constructed first. Genetic maps are based on the recombination frequency between the gene of interest and the known phenotypic markers. Its relative distance to known neighboring markers describes the location of the newly introduced gene on a chromosome, after transgenic lines and marker lines have been crossed, and the resulting populations have been analyzed.

Molecular markers allow for considerably more precise localization in relation to already known markers. Molecular markers are based on variations of the DNA sequence at identical locations of different genotypes. Those variations mostly originate from base-pair exchange or deletion/insertion (**DNA polymorphisms**). In order to be able to make use of these DNA polymorphisms and to develop reliable molecular markers from these, the molecular differences between the different genotypes need to be simple and differentially detectable. Examples are restriction fragment length polymorphisms (RFLPs), random amplified polymorphic DNA (RAPD), and inter-simple sequence repeats (ISSR) (see Chapter 21).

The ultimate characterization of an insertion locus consists of sequence determination of the DNA flanking the inserted transgene. As described above, this can be of interest to basic research applications, whereas application of plant biotechnology in agriculture will routinely require the exact determination and characterization of the insertion locus of a transgene prior to deregulation for cultivation and commercialization.

32.6.4
Stability of Transgenic Plants

The most important aspect for genetic stability of the transgene first of all consists in the homozygous state of the newly inserted expression cassette. However, the mechanisms contributing to and controlling the stable inheritance of the genetically engineered trait are for the most part not yet fully understood. Thus, in today's practice, obtaining **stable inheritance patterns** is based on trial and error.

Only a very small number of transgenic lines fulfill the requirements of the thorough selection process for positive transgenic events, the last step of which consists of field experiments. An important reason is the inactivation or decrease of the expression of the transgene (gene silencing). Possible mechanisms are the **methylation of promoters** controlling transgene expression – especially if these are not of plant origin, but from plant viruses. A widespread example is the 35S promoter of the cauliflower mosaic virus (CaMV) that drives a strong and almost ubiquitous expression in plants. A further mechanism is the elimination of the RNA of the transgene via **small interfering RNAs (siRNA)** (see Chapters 2.4, 21, and 31). Production of siRNAs is a natural process in plants, supposedly serving as a virus defense mechanism, and occurs in transgenic plants during strong expression of the RNA of the newly introduced gene. A third mechanism comes into effect if integration of the transgene occurred in heterochromatic transcription-inactive areas. Therefore, the expression rate and pattern of a transgene has to be validated over a couple of generations until sufficient molecular evidence confirms a stable transgenic event.

33
Biocatalysis in the Chemical Industry

Learning Objectives
This chapter introduces industrial biotechnology and describes different industrial processes as well as their respective products. The most important aspects in research and development are the identification as well as the optimization of biocatalysts.

33.1
Introduction

Today, **biotechnology** is understood as the integrated application of engineering and natural sciences targeting the technical use of organisms, cells, or parts thereof. Biotechnological procedures are closely connected with the cultural history of mankind. In many societies, **fermentation processes** have been developed that serve the conservation of groceries or are used in the production of **alcoholic drinks**. Well-known examples in Europe are the production of sour milk products, sauerkraut, vinegar, the brewing of beer, or wine production. Enzymatic procedures like the use of chymosin for cheese production have been established for many centuries. In Asia, fermented foods have a long tradition. There are a number of food and drinks that are fermented before consumption. As examples, Indonesian *tempe* (fermented soy beans), Korean *kimchi* (fermented cabbage), and *saki* (Japanese rice wine) should be mentioned.

The corresponding production methods have been developed empirically; knowledge of the cellular and also molecular mechanisms is not necessary for the production of these products.

It was not before the seventeenth century that we were able to observe microorganisms through simple microscopes. In the nineteenth century, we began to understand the ability of microorganisms to conduct chemical syntheses. Important requirements for industrial biotechnology were the cultivation of microorganisms in pure culture and, connected with that, a sterile work technique. With the introduction of vaccinations, biotechnology was used for the first time in the pharmaceutical area.

In the twentieth century, biotechnology procedures were developed on an industrial scale, alongside food production. This is where enzymes were used, such as in leather tanning and also the use of fermentation processes for the production of chemicals. Before the heyday of petrochemistry, solvents like **acetone** and **butanol** were obtained by fermentation of the bacterium *Clostridium acetobutylicum*, as well as **citric acid**, through the surface cultures of the fungus *Aspergillus*. Interestingly, the biotechnological synthesis of *n*-butanol has undergone a renaissance in recent years (*cf.* Case Study 4 in Section 33.4.4).

An important milestone in the twentieth century was the discovery of penicillin and other **antibiotics**. More than 130 fermentative and around 50 semisynthetically produced antibiotics are used clinically to successfully fight infectious diseases.

An Introduction to Molecular Biotechnology, 2nd Edition.
Edited by Michael Wink
Copyright © 2011 WILEY-VCH Verlag GmbH & Co. KGaA, Weinheim
ISBN: 978-3-527-32637-2

New enzymatic and fermentative procedures were developed in the second half of the twentieth century including the production of insulin and other **therapeutic proteins**. Classical production processes have been revolutionized by modern genetic engineer methods. Genetic engineering and biochemistry are indispensable tools for the fast and systematic development of production organisms.

Products from biotechnology differ in volume and price. Comestible goods, such as beer, are produced worldwide in great amounts of 130 million tons per year. High-volume chemicals like glutamate (MSG) and citric acid as well as proteases are at least in the area of several hundred thousand tons. The production volumes of antibiotics or insulin are relatively small. However, higher prices can be reached. Table 33.1 lists production volumes and producers of important products. As mentioned above, traditional biotechnology is crucial in foodstuff production. **Starter cultures**, for example, are used for the controlled production of fermented products, according to today's quality requirements.

Further **biotechnologically produced products** like flavor enhancers, enzymes, aromas, and artificial sweeteners are food additives and adjuvants. Chemically, these are usually pure substances that are used for the refining or production of foodstuffs.

There are also many biotechnologically manufactured products in the area of agriculture. These range from feed additives in animal nutrition, like vitamins and amino acids, up to enzymes that are added to animal feed in order to increase the digestibility of the feed or genetically engineered plants as nutrition or food substances.

Enzymes are used in detergents because of their catalytic activity or are also used as **catalysts in the chemical industry**. Recombinant enzymes, antibodies, and protein hormones are widely used as pharmaceutically effective substances in medical applications. Table 33.2 shows the 15 top-selling recombinant proteins on the world market.

In the biotechnology, production processes can be differentiated between so-called **bioconversion** and the **fermentation processes**. Bioconversion (sometimes also called biotransformation) is an enzyme- or cell-catalyzed reaction of defined starting material(s) into defined products. Usually this is a one-step reaction; byproducts only appear in minor amounts. Often these reactions are not carried out by the corresponding biocatalysts in this manner *in vivo*. Bioconversions are often single reaction steps in chemical production processes (e.g., during the production of optically active products and intermediates).

The term fermentation derives from the Latin word *fermentare*, meaning to leaven or to brew. In biotechnology, the term fermentation is not limited to anaerobic fermentative metabolism, but is broader: the fermentative production of chemicals is the conversion of renewable raw materials (e.g., sugar) by living microorganisms. The product (e.g., an amino acid or vitamin) accumulates in the fermentation broth. Contrary to bioconversion, the substrates in fermentative processes go through the entire metabolic pathways and not just one single enzymatic step. Apart from the desired product, the fermentative procedure typically accumulates byproducts – waste substances and biomass. Generally, the synthesis sequence of a fermentative procedure is a naturally occurring biosynthetic pathway. Figure 33.1 schematically shows the differences between fermentative procedures and bioconversion.

Table 33.1 Biocatalytic processes.

Product group	Product	Amount (tons/annum)	Important producers	Technique[a]
Vitamins	keto-L-gulonic acid (KGA) (→ vitamin C)	>50000	Several Chinese companies	F
	B$_2$	>4000	BASF, Hoffmann-La Roche, several Chinese companies	F
	Pantolactone	>1000	Daiichi, several Chinese companies	B
	L-Carnitine	>100	Lonza	B
	B$_{12}$	15	Rhône-Poulenc/Aventis	F
	Q$_{10}$	>200	Kaneka, Mitsubishi	F
Amino acids	L-Glutamate	1900000	Ajinomoto, Vedan Enterprise, Daesang	F
	L-Lysine	1100	Ajinomoto, Changchun Dacheng Biochemical, ADM, Paik Kwang Industrial, Cheil Jedang, Evonik	F
	L-Threonine	150000	Ajinomoto, Evonik	F
	L-Tryptophan	>3400	Ajinomoto, ADM	
	L-Phenylalanine	>1000	Nutrasweet, Ajinomoto, Miwon	F
	L-Aspartate	1000	DSM, Evonik	B
	L-Methionine	<100	DSM, Evonik	B
	L-Valine	<100	DSM, Evonik	B
	L-*tert*-Leucine	<100	Evonik	B
Enzymes	Proteases	>300000	Novozymes, Genencor	F
	Amylases	>10000	Novozymes, Genencor	F
	Lipases	>4000	Novozymes, Genencor	F
	Phytase	<1	BASF, Novozymes	F
Optically active intermediates	D-Phenylglycine	>1000	DSM	B
	S-Methoxy-*iso*-propylamine	>1000	BASF	B
	Amines	>100	BASF	B
	L-DOPA	>100	Ajinomoto	B
	L-Malate	>100	Tanabe	B
	Alcohols	<10	BASF, Wacker, Kaneka, Evonik, and others	B
	Glycidylbutyrate	100	DSM	B
	R-Mandelic acid	100	BASF, Mitsubishi	B
	Thioisobutyrate	100	Tanabe	B
Intermediates/chemicals	Acrylamide	>10000	Nitto, DSM	B
	Citric acid	>5000	Several manufacturers	F
	6-Aminopenicillanic acid	>1000	DSM	F
	Lactic acid	>300000	Several manufacturers	F
	Hydroxynicotinic acid	>100	Lonza	B
	Nicotinic acid amide	>100	Lonza	B
	Steroids	>100	Bayer	F
	Ethanol	>50000000	Several manufacturers	F
	Fatty acid esters/ceramides	>100	Evonik	B
	Silicon acrylates	>10	Evonik	B
	Polyglycerol esters	>10	Evonik	B
	Glycidyl butyrate	>10	DSM	B
	1,3-Propoandiol		DuPont	
Polymers	Polylactide	>100000	Cargill	F
	Polysaccharides (xanthan)	>100	Several manufacturers	F
	Cyclodextrin	>5000	Cerestar, Wacker	
Active ingredients	Aspartame	>16000	Several manufacturers	B
	Antibiotics	>10000	Eli Lilly, DSM, and others	F
Others	Isosyrup	>1000000	ADM, Cargill, and others	B
	Cocoa butter	>10000	Several manufacturers	B

a) F: fermentation; B: bioconversion.

Table 33.2 The 15 top-selling recombinant proteins (million US $) in Aggarwal, 2007; 2008; 2009)

Insulins	Peptide hormone (diabetes)	>9500
Erythropoetins	Glycoprotein hormone (hematopoiesis)	>8400
Interferons	Immunostimulants (cancer therapy)	>5000
Octocog-a	Coagulation factor VIII	>1100
Thrombin	Coagulation factor IIa	>700
Aprotinin	Protease inhibitor	>400
Polymyxin B	Peptide antibiotic	>400
Trypsin	Serine protease	>100
Colistin	Peptide antibiotic	>100
Chymotrypsin	Serine protease	>88
Urokinase	Thrombolytic agent	>71
Chorion gonadotropin	Peptide hormone	>59
Streptokinase	Thrombolytic agent	>51
Ulinastatin	Trypsin inhibitor	>47
Streptodornase	Coagulation inhibitor	>39

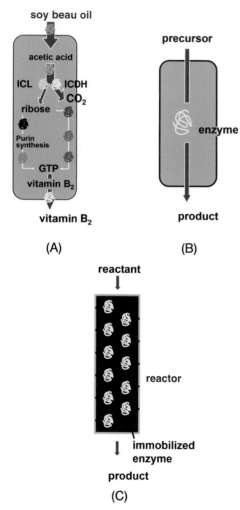

Fig. 33.1 Biotechnological processes can differentiate between fermentation and bioconversion. The industrial production of vitamin B_2 is a successful fermentative process (*cf.* Case Study 8 in Section 33.4.9). The biosynthetic pathway of *Ashbya gossypii* is used, shown in (A). ICL, isocitric lyase; ICDH, isocitric dehydrogenase. With bioconversion, only one (or a few) synthesis step(s) is carried out with a biocatalyst. Resting cells (B) or immobilized enzymes (C) can be used as catalysts.

33.2
Bioconversion/Enzymatic Procedures

In bioconversions, an enzyme as a highly active and selective catalyst is utilized in order to accelerate a chemical reaction. In doing so, enzymes can either be used as free or immobilized proteins, or contained in whole living or inactivated cells (Fig. 33.1). More than 120 technical bioconversions are documented in the literature. Industrial biotransformations are not at all new developments. As one can see from Table 33.3, the first industrial procedures were already established in the nineteenth century.

The most important requirements for a catalyst in technical processes are **selectivity**, **activity**, and **stability**. Enzymes are mainly used because of their high selectivity. Enzymes as chiral catalysts are often significantly superior to classical chemocatalysts with respect to their **stereoselectivity**. During the production of chiral compounds, an enantiomeric excess of 99% can be achieved. In the mid-1980s, enzymatic procedures saw a new upswing, especially in stereoselective synthesis, and the chemical industry is now unimaginable without them. The high substrate specificity of naturally occurring biocatalysts can, under certain circumstances, also be disadvantageous if only a limited number of substances are converted. The goal is a enzyme catalyst widely applicable in different technical processes.

Owing to their usually high specific activities, enzymes can be used in very small ratios relative to the substrate. In chemical catalysis the catalyst/substrate ratio is usually around 0.1–1 mol%; in enzyme-catalyzed reactions it is often only 0.0001–0.001%.

Chemical processes frequently run only under high pressures and under high temperatures. On the contrary, enzymes usually work under milder and less stressful conditions. In addition, bioconversions often allow an economical use of material. For the chemical industry, this means savings in terms of energy, raw materials, as well as the avoidance of waste, and therefore real financial advantages.

A frequent disadvantage of biochemical transformations is the lack of enzyme stability. Therefore, the costs of catalyst production can play an important role in the economy of a biocatalytic procedure. Hence, inexpensive and reproducible production of the corresponding enzymes is an important success factor of industrial bioconversion.

Finally, enzyme-catalyzed procedures are in constant competition with chemical processes. Only **economical advantages** will tip the balance in favor of biocatalysis in an industrial setting.

Product	Biocatalyst	Established
Vinegar	*Acetobacter aceti*	ca. 1820
R-Phenylacetylcarbinol (ephedrine precursor)	*Saccharomyces cerevisiae*	1932
Sorbitol/sorbose	*Gluconobacter suboxydans*	ca. 1930
Steroids	e.g., *Arthrobacter*	ca. 1950
High fructose corn syrup	Glucose isomerase	1965
6-Aminopenicillanic acid/ 7-aminodesacetoxycephalosporinic acid (precursors of semisynthetic antibiotics)	Penicillin amidase	ca. 1970
Aspartame	Thermolysin	1980
Acrylamide	*Rhodococcus* sp.	1985
L-*tert*-Leucine	Leucine dehydrogenase/ formate dehydrogenase	1981
L-Methionine	Aminoacylase	1979
R-Phenylethylamine	Lipase	1990

Table 33.3   Selected bioconversions.

Enzyme	Product	Amount (tons)
Glucose isomerase	Fructose	1000000
Nitrile hydratase	Acrylamide	10000
Lipase	Cocoa butter	10000
Penicillin amidase	6-Aminopenicillanic acid	1000
Aspartase	L-Aspartate	1000
Thermolysin	Aspartame	1000
Hydantoinase	D-Phenylglycine	1000
Hydantoinase/carbamoylase	D-Hydroxyphenylglycine	1000
Aldonolactonase	D-Pantothenic acid	1000
Fumarase	L-Malic acid	100
Aminoacylase	L-Methionine	100
Aminoacylase	L-Valine	100
β-Tyrosinase	L-Phenylalanine	100
Lipase	L-DOPA	100
Hydroxylase	L-Carnitine	100
Lipase	Glycidyl butyrate	10
Transglucosidase/lipase	Butylglucosides	10
Dextransucrase	Glucooligosaccharides	10

Table 33.4   Annual production volumes of different bioconversions.

The majority of industrially established bioconversions work exclusively with hydrolytic enzymes, to which **lipases**, **esterases**, and **proteases** belong. The use of enzymes, especially lipases and esterases in nonpolar organic solvents, has created new possibilities for biocatalysis.

Well-known and prominent examples of current bioconversions are the production of high fructose corn syrup, acryl amide, nicotinamide, optically active amines, R-pantolactone, and unnatural amino acids like d-*tert*-leucine. Most industrial procedures have shared characteristics: high product concentration and high productivity, no undesired byproducts, and robust, easily accessible enzymes that do not need expensive cofactors. Table 33.4 summarizes the most important bioconversion processes.

33.3
Development of an Enzyme for Industrial Biocatalysis

Biotechnologists who work in the field of biocatalysis see themselves confronted with two main challenges – the **identification of products** whose production by an enzymatic route is advantageous, and the **development of a process** in the shortest possible time and with the minimum of resources. The first challenge can only be solved in a team combining expertise from marketing, production, and engineering. If substrate and target molecules are known, the actual research and development work begins. The identification of a catalyst is obviously essential, but the synthesis of the starting material for the enzymatic step as well as downstream processing are also crucial issues. In doing so, the enzymatic step is often embedded in a complete procedure, in which classical chemical and enzymatic steps go hand in hand. Finally, it is decisive that the entire procedure is economical with respect to starting material, energy, and investment.

33.3.1
Identification of Novel Biocatalysts

The starting point for catalyst development can be **commercially available enzymes**. Knowledge of the catalyst's mechanism may be helpful during the selection of an enzyme, because often the field of application is broader than the name of an enzyme suggests. In this way, one can abuse known biocatalysts for unnatural reactions. Case Study 2 (Section 33.3.6) describes the successful application of this strategy during the development of a biocatalytic procedure for the production of optically active intermediates. If the desired enzyme is not found among commercially available enzymes, one can also test microorganisms from **strain collections** like the American Type Culture Collection or Deutsche Sammlung von Mikroorganismen und Zellkulturen (German Collection of Microorganisms and Cell Cultures). Figure 33.2 shows results from different screening experiments for the identification of novel biocatalysts.

Very often though, **microorganisms** that come **from nature** have to be enriched and screened according to enzyme activity. This is still a lengthy and laborious process. The common procedure for finding new enzyme activities consists of enriching microorganisms from soil samples and producing pure cultures (see also Case Study 1 in Section 33.3.5). The pure cultures are then examined for the presence of the desired new enzyme activity. During the **enrichment**, it attempted to link the desired reaction with the ability to grow.

Reaction		Number of tested strains	"Candidates"
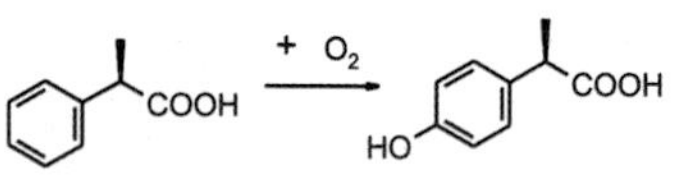 p-hydroxylation		7900	3
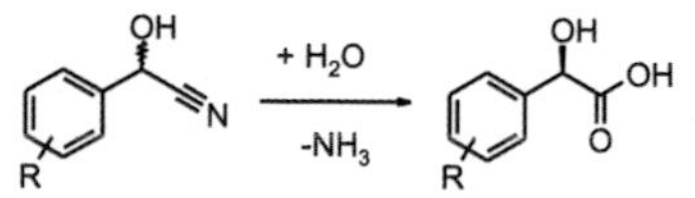 nitrile hydrolysis		1000	2
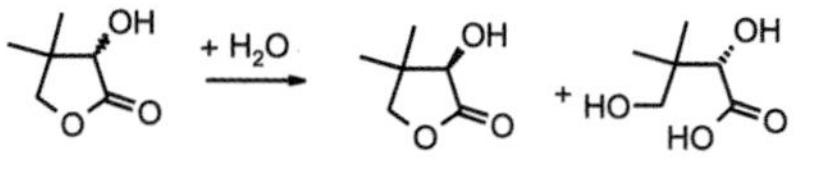 lactone hydrolysis		950	5
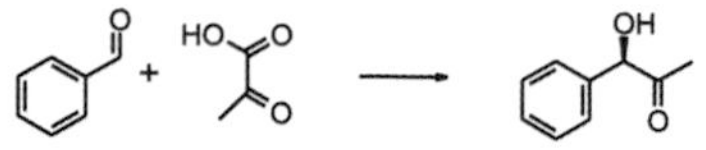 C–C bond formation		200	2

Fig. 33.2 Screening of strain collections can make new enzymes available. Selected examples show how many strains need to be tested to find promising hits. These candidates then present the starting material for catalyst development.

The desired substances for the bioconversion are made available as sole carbon or nitrogen sources in the enrichment cultures. Only microorganisms capable of converting these compounds will thrive, thus outgrowing other microbes present in the original sample. After cultivation of the isolates in pure culture, one must verify whether the growth of the microorganism can be traced back to an enzyme. This can be very lengthy, because often one has to characterize several hundred microorganisms in detail.

The biggest disadvantage of this procedure is that, admittedly, more than 90% of all microorganisms are not accessible in this way. The reason for this is mainly that a pure culture of many microorganisms is not possible because the exact growth conditions are unknown.

New methods have been developed that avoid a pure culture – DNA is directly isolated from the sample material and a recombinant expression library is established that is examined for new enzyme activities (Fig. 33.3).

The entire DNA that is to be isolated from a environmental sample is referred to as the **metagenome**. The metagenome contains the DNA of many different organisms that is now available for activity tests. To conduct this metagenomic approach successfully, a number of factors like isolation of DNA, normalization, expression, host strain, as well as fast and reliable test systems for the detection of the smallest amounts of an enzyme have to be established. The **screening of metagenome libraries** is a decisive improvement because it is now possible to screen nonculturable microorganisms for novel biocatalysts.

In the last decade the number of organisms with completely sequenced genomes has increased substantially (http://www.ncbi.nlm.nih.gov/sites/entrez?db=genome). A significant cost reduction and the technological progress in DNA sequencing have been the main drivers (Chan, 2005). Most of these data are publically available, thus accounting for another source of biocatalysts. The respective genes are either cloned by classical methods from the organisms or are available by *de novo* DNA synthesis. This approach is especially attractive when the original organism is either difficult to cultivate or poses a significant biohazard. In this context it often turns out to be beneficial to adapt the synthetic gene to the codon usage of the host organism that is ultimately used for the recombinant protein production.

So far there has been no significant contribution from bioinformatics to predict enzyme properties, such as stability, enantioselectivity, or substrate specifici-

Screening for New Biocatalysts

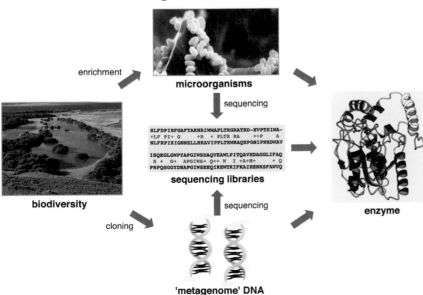

Fig. 33.3 In addition to classical screening of culturable microorganisms, an *in silico* search in genome databases or metagenome screening of environmental samples can also yield access to novel biocatalysts.

ty, from a DNA sequence alone (Sharan *et al.*, 2007). Major progress in this field is, however, expected in the mid-term future.

33.3.2
Improvement of Biocatalysts

Enzymes, as found in nature, are not necessarily suitable for use in biocatalytic processes. They can be too unstable or can be limited to their natural substrate. Therefore, it is sometimes necessary to adapt enzymes to the **requirements of industrial biocatalysis**. In essence, there are two different approaches.

During so-called **directed evolution**, the basic principles of evolution (i.e., **mutation, selection**, and **recombination**) are exploited in the laboratory in order to improve enzymes (Jäckel *et al.*, 2008; Johannes and Zhao, 2006). By untargeted mutagenesis, with error-prone **polymerase chain reaction (PCR)**, the biocatalyst gene is changed randomly. The modified genes are expressed for thousands of variants, which are then screened for a suitable biocatalyst. Often, this is done through highly automated testing procedures relying on robot-aided screening machines. Depending on the extent of the test, several thousand enzyme varieties can be tested daily. The best variants serve as a basis for further mutagenesis. Through **random mutagenesis** and **screening** (i.e., selection), the profile of an enzyme can be changed. In doing so, only the selection of variants decides the direction of the development after the mutagenesis steps. During sexual reproduction in nature, a mixture (i.e., recombination) of hereditary material occurs. This process accelerates the evolution. Also, the recombination of genes can be re-enacted in the laboratory and thus be used for the improvement of biocatalysts. In recent years, through directed evolution, enzymes have been obtained with improved thermostability, substrate specificity, enantiomer selectivity, and stability.

Rational design means the targeted change in the amino acid sequence for optimization of enzymes. Requirements for this procedure are not only the exact knowledge of the **structure-function relationship** in an enzyme, but also an understanding of the consequences of changes in the protein structure for the catalytic activity. Only with very few biocatalysts our understanding is deep enough to successfully employ this approach in a reasonable amount of time. Case Study 3 (Section 33.3.7) shows, using the example of pyruvate decarboxylase, how biocatalysts can be improved by rational design.

33.3.3
Production of Biocatalysts

In order to make the enzyme inexpensive and available in sufficient amounts, the next step is the development of a **recombinant production strain**. Wild-type strains, as isolated from nature, rarely produce enough of the enzyme. As a result, the gene is cloned and expressed in a suitable host strain (e.g., *Escherichia coli, Bacillus subtilis, Pichia pastoris*, or *Aspergillus*). After that, the fermentation conditions for the strain have to be determined – factors such as the optimal medium, aeration, and stirring speed, as well as the pH value and feeding profiles. The goal of all of this is to achieve the maximum amount of enzyme, specifically enzyme activity per liter of fermentation volume.

Further work is necessary to develop a technically applicable biocatalyst from an enzyme – in which formulation should the enzyme be used? The size of the production and, in particular, the cost play a role. For smaller productions up to 100 tons, the enzymes are mostly used in a stirred tank in an isolated form or in the form of whole cells of the production strain. If the production amount is very large and the enzymes are stable enough, it may be worth binding the enzymes on a carrier material. In this immobilized form, continuous processes are possible.

33.3.4
Outlook

A major challenge for the chemical industry is the **development of selective and sustainable production processes**. Enzymatic processes can contribute to the solution of this problem, when they not only provide a technical solution, but also offer an economic advantage over other alternatives. The fact that a certain process is done biocatalytically is rarely a decisive advantage on its own. The developments of recent years have opened new opportunities to improve weaknesses, such as lack of stability or substrate range. If the application of this technique is successful with technically relevant enzymes, one can expect further processes for the production of mass chemicals to be developed. Using **tailor-made biocatalysts**, enzymatic processes can further excel with **intermediates and specialty chemicals**. In particular, the high selectivity of enzyme catalysts can lead to tremendous simplifications and therefore cost savings. Currently, scientists are attempting to apply those advantages for product classes like e.g. polymers.

Development needs in the area of biocatalysis lie in the fast and effective access of new biocatalysts with desired characteristics. Closely connected to this are the establishment and automation of miniaturized high-throughput screening methods for the fast discovery and optimization of biocatalysts.

An important topic of recent work is the extension of the portfolio of technically suitable enzymes. In particular, interest is focused on the use of enzymes for **chemically difficult industrial reactions**, like carbon-carbon bonds or the sophisticated regioselective introduction of oxygen with the help of oxygenases. Enzymes that are active in **nonpolar organic solvents** are important in order to solve solubility problems between substrates and products by allowing a homogenous catalysis. With this, the potential of biocatalysis in the future can be exploited even further.

33.3.5
Case Study 1: Screening for New Nitrilases

Enantiomerically pure α-hydroxycarbonic acids are important building blocks for pharmacologically active ingredients. These compounds are, among others, accessible through a nitrilase-catalyzed reaction from the respective cyanohydrins (Fig. 33.4). Nitrilases hydrolyze α-hydroxynitriles into optically active carbonic acids and ammonium salts. In aqueous solution there is an equilibrium of cyanohydrin and aldehyde and prussic acid, respectively. Therefore, the enzymatic reaction can result in the quantitative conversion of the aldehyde/α-hydroxynitrile into optically pure α-hydroxycarbonic acids.

Microorganisms that have nitrilase activity can be enriched from soil samples by supplying nitriles as the sole nitrogen or carbon source in the growth medium. However, a problem with this approach is that false-positive strains may be isolated, which use the nitrogen or carbon source with the help of a different enzyme activity. Such organisms express, for example, nitrile hydratases – enzymes that convert nitriles into amides. Nitrogen fixation also circumvents nitrogen use from the nitrile. Different problems are the instability and toxicity of the nitrile. In order to solve these problems, one can use nontoxic model compounds, which are far more stable under enrichment conditions. However, under certain circumstances, this can lead to new hurdles because the reactivity and confirmation of the model compound are not always identical with the target molecules.

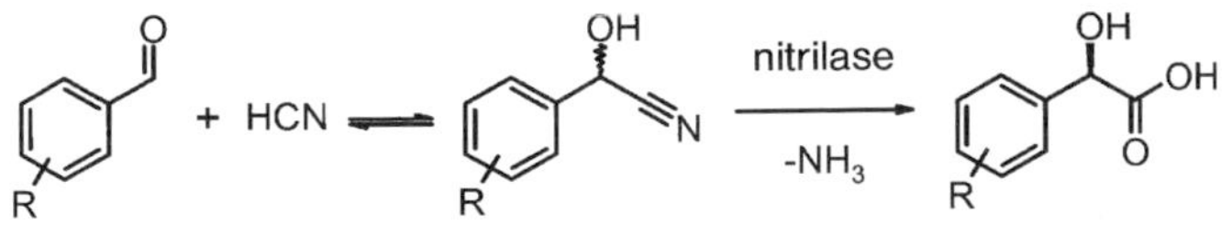

Fig. 33.4 Nitrilases are suitable biocatalysts for the production of optically active α-hydroxycarbonic acids from the corresponding nitriles.

In order to discover a highly selective nitrilase for the production of *R*-mandelic acid in classical screening, several hundred microorganisms had to be screened (Fig. 33.2).

33.3.6
Case Study 2: Use of Known Enzymes for New Reactions: Lipases for the Production of Optically Active Amines and Alcohols

The **understanding of the catalytic mechanism** of an enzyme can be very valuable. It may give new options in organic synthesis. A good example is the use of lipases in organic synthesis. **Lipases** are hydrolases that split ester bonds present in acylglycerides (e.g., fats). The understanding of the mechanism of catalysis was of decisive importance during the development of a completely different enzymatic process (Fig. 33.5). During the hydrolysis of an ester, an acyl enzyme complex is formed. The catalytic cycle starts by the nucleophile attack of a serine in the catalytic center of the enzyme on the carbonyl atom of the ester, which functions as an acyl donor. The serine residue is acylated, an acyl enzyme forms, and the alcohol is liberated. The enzyme acyl complex is then hydrolyzed by a nucleophile. *In vivo* water acts as the nucleophil. The fatty acid is set free and the enzyme is regenerated.

If the water can be substituted successfully with other nucleophiles, a whole number of interesting reactions are possible with the lipase catalysts. An essential requirement for this is, of course, that the lipases are also active in a water-free environment. In fact, lipases are active in certain organic solvents. Often possible solvents like alcohols or amines can act as nucleophiles. If chiral nucleophiles are used, usually only one enantiomer acts as a nucleophile (i.e., the enzyme catalyzes the transfer of the acyl function enantioselectively to one enantiomer).

Vinyl ester, anhydrides, or diketenes are technically used as acyl donors. With these acyl donors the reaction is practically irreversible. The formed esters as well as the alcohol can then be physically and/or chemically separated (e.g., by distillation). According to this principle, a whole number of optically active alcohols are technically produced, which serve as building blocks for active ingredient synthesis. In the lipase-catalyzed resolution of racemates it is crucial to completely avoid the presence of water. As a highly active nucleophile, water reacts much faster with the enzyme acyl complex than an alcohol or amine. In this case, only hydrolysis would occur and not enantioselective acyl transfer.

Amines can also be used as nucleophiles. Here, 2-methoxy acetic acid esters are suitable acylating substances for the lipase-catalyzing reaction. With 2-meth-

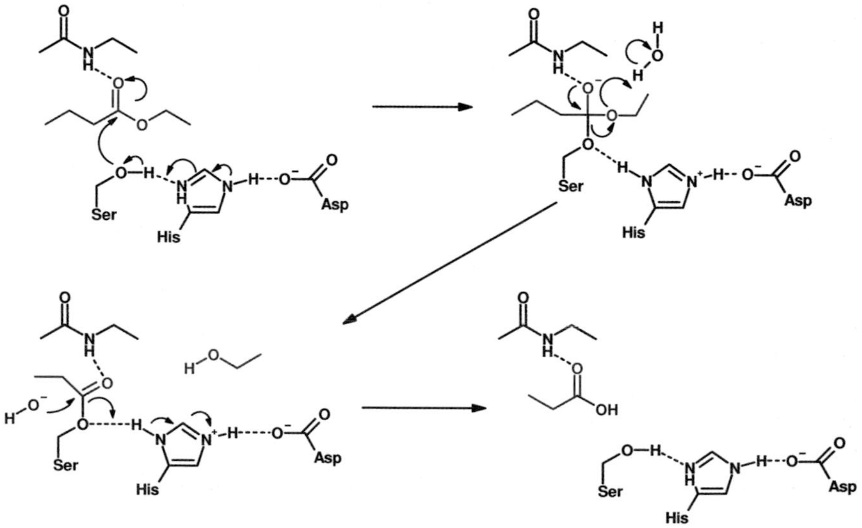

Fig. 33.5 Reaction mechanism of lipase.

oxy acetic acid ester, the initial velocity of the reaction is more than 100 times greater than with butyl acetate. The reason for this activating effect of the methoxy group is probably the higher carbonyl activity, induced through the electronegativity of the *a*-substituent. This procedure gives high selectivity and yield, in concert with high activity. The products, *R*-amide and *S*-amine, can again be separated by distillation. These processes can be conducted with a wide spectrum of amines (Fig. 33.6).

Also, from another point of view, lipases are a textbook example for the ideal biocatalyst – they are commercially available in sufficient quantities and qualities. In addition to this, the enzymes are very stable and active in organic solvents, and the substrate range is impressively wide.

33.3.7
Case Study 3: Enzyme Optimization with Rational and Evolutive Methods

Neuberg and Hirsch discovered in 1921 that yeast cells can catalyze the **formation of carbon-carbon bonds**. By feeding fermenting yeast with benzaldehyde, *R*-phenylacetylcarbinol is formed – an intermediate for the synthesis of the active ingredient ephedrine. This reaction was one of the very first industrial biotransformations. It is catalyzed by the enzyme pyruvate decarboxylase, which also has a carboligase side activity. Pyruvate, a product of glucose metabolism, is coupled with the benzaldehyde to *R*-phenylacetylcarbinol. *In vitro*, the yeast enzyme has not been successfully used in conducting this reaction because this enzyme is not sufficiently stable.

Therefore, the starting point for a cell-free enzymatic procedure was the respective enzyme from the bacterium *Zymomonas mobilis*, which is far more stable. The disadvantage of this enzyme is the weak carboligase activity. A comparison of the protein structures of *Zymomonas* and yeast enzymes shows an obvious relevant difference. The *Zymomonas* enzyme has a tryptophan residue in position 392 at the entrance of the catalytic site. When this tryptophan residue is substituted with alanine or methionine, one gets an enzyme with higher carboligase activity. A further advantage is that the bacterial enzyme not only works with pyruvate as a C-2 donor, but also with the far cheaper acetaldehyde.

Improving Biocatalysts by Directed Evolution

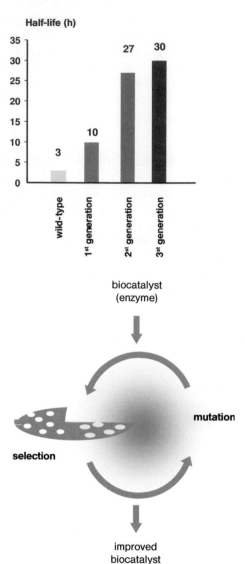

Fig. 33.7 Directed evolution increases the stability of pyruvate decarboxylase.

For a technical process, the stability in the presence of aldehydes was too low for an economical procedure. The protein structures did not provide any approachable points to improve the characteristics of this enzyme. Here, directed evolution of the enzyme using the methods described above was employed. More stable enzymes could be isolated by production of a number of enzyme variants through mutation followed by selection in the presence of acetaldehyde. After three mutation and selection cycles, an enzyme was evolved whose stability was higher than the starting material by a factor of 10 (Fig. 33.7).

33.4
Fermentative Procedures

Biosynthesic routes of microorganisms are used in fermentative processes in order to produce **chemically complex molecules**. It is an important goal of industrial research to develop economical processes and to further improve already existing procedures. Initially, the microorganism's growth, metabolism, and genetics are of central scientific interest. For a technical fermentation procedure, however, further factors are of relevance. The feed substances, preparation of the medium, as well as the operation and control of the fermentation process are also very important. Like other products, a fermented product also has to be purified, formulated, and packed. It is obvious that the development and operation of an industrial fermentation process is a complicated one, which requires the teamwork of experts from different areas. Nevertheless, the production organism is the first point approached in a process optimization.

33.4.1
Improvement of Fermentation Processes

It is the primary goal of strain optimization to maximize the amount and **concentration of the produced substance**, and to keep the fermentation time as short as possible. In addition, it is important to increase the **yield with respect to the raw materials**. This is the decisive measurement for the economic success of the process. Furthermore, it is necessary that the strains used for the production are genetically stable. With high growth rates, spontaneous mutations appear, like e.g. reversions, and the mutants can lose their desired characteristics again. Sensitivity against bacteriophages can also pose a problem. In such cases, one must try to make the production strain resistant to phage infection.

Only in a few exceptions (e.g., glutamate and lactic acid production) wild-type strains already show a satisfying performance for commercial use. Mostly, the endogenous syntheses are not sufficient in wilde-type strains of microorganisms. Therefore, these microorganisms have to be **optimized by strain development**.

Historically, from the high number of microorganisms, only a few suitable strains have turned out to be especially useful for fermentation (e.g., this is the case with *Penicillium*, *Corynebacterium*, and *Aspergillus*). Other production organisms have developed from common laboratory strains, because these were easy to handle and because an efficient genetic toolbox became available (e.g., yeast, *E. coli*, and *B. subtilis*).

As with bioconversion, there are basically two different strategies to improve the process of fermentation processes. On the one hand, the production organism is randomly mutated and among a great number of mutants those that have experienced an improvement in the desired characteristic are filtered out. This strategy re-enacts in the principles of evolution and is the basis of **classical strain optimization**. On the other hand, if the biosynthesis paths, their regulatory mechanisms, and the respective genes are exactly understood, one can conduct an optimization of the fermentation process through targeted intervention in the metabolism. This process is called as **metabolic engineering**.

33.4.2
Classical Strain Optimization

Microorganisms can be adapted to the requirements of industrial processes by **mutation and selection**. The potential of this strategy, which initially seems simple and scientifically boring, should not be underestimated. After all, these methods have successfully developed strains for important fermentation processes. In addition to the production of amino acids and vitamins, the production of antibiotics should especially be mentioned.

Classical strain development consists of two core elements – the generation of mutants and the selection of those strains that have the desired characteristics.

The mutants can either be developed by spontaneous mutations or by treatment of the cells with **mutation-inducing chemicals** or radiation. The number of spontaneous mutation events is in the range of 10^{-6} to 10^{-7}. A disadvantage is that many of these mutations are repaired, or revert functionally or genetically. In the process of strain development, one often chooses different mutagens, because they bring out different mutation types and it is believed that the resulting strains are more stable: UV radiation leads to thymine dimers, and nitrite deaminates adenine to hypoxanthine and cytosine to uracil (see Chapter 4). Point mutations also result from the use of alkylating substances acting on purines. Acridine orange intercalates and triggers frameshift mutations. The use of transposons allows the random inactivation of genes and thus their identification.

For the identification of improved strains, one needs a **screening system**. This consists of cultivation and a fast analysis. Usually this is done in Erlenmeyer flasks. The test system has to be extremely accurate and reproducible. Admittedly, the results from the shake-flask measurement are often not reproducible in the laboratory fermenter and even less so in the production fermenter. Therefore, by **downscaling** one tries to represent the physical conditions of the production fermenter on the laboratory scale as exactly as possible in order to avoid such difficulties.

A shake-flask screening is lengthy and labor-intensive. As a result, extensive work is being done to develop **automated test systems** (e.g., in microtiter plates). Some approaches have shown promising success; however, microplate systems are not yet routinely usable for all production strains.

The **goals of strain development** are to deregulate the desired metabolic pathway to the desired product, to abolish the flow of metabolites to byproducts, or to widen the substrate range of the organism. Initial enzymes of a biosynthetic pathway are often allosterically inhibited on the metabolic level by an intermediate or the final product of the biosynthesis. Threonine, for example, inhibits homoserine dehydrogenase – the first enzyme in the final threonine biosynthesis. The pathway is deregulated in mutants lacking feedback inhibition. Thus, the carbon flow towards threonine can occur undisturbed.

The enhancement of a metabolic pathway by **gene overexpression** can be a result of a point mutation in the promoter regions. Another well-characterized mechanism, on the level of DNA, is to increase the copy number of single genes or the whole gene cluster. This is the case with classically developed penicillin producers.

The prevention of byproducts is often achieved by searching for **auxotrophic mutants**. These carry one or more mutations in enzymes in the metabolic pathway, which lead to the undesired side product. The disadvantage of auxotrophic mutants is that it might become necessary to supplement the growth medium.

The ability of prokaryotes to adapt has been successfully exploited for the broadening of the substrate spectrum. Many bacteria can achieve the ability to utilize new substrates through spontaneous mutation events. This has been described for the utilization of rare sugars in a number of *Arthrobacter* and *Corynebacterium* strains.

Production strains that have been developed by classical strain development over many years and strain generations contain numerous mutations. By comparing the genome sequence of wild-types with classically produced production strains, it is known that up to 30% of all genes can contain mutations that lead to amino acid exchanges. However, not necessarily all mutants are responsible for the actual improved production characteristics. It is a known phenomenon that already highly developed strains do not improve further with methods of classical strain development.

33.4.3
Metabolic Engineering

With classical strain optimization, the genome of microorganisms is changed in a random fashion. From a great number of mutants, those mutants are chosen that display the improved characteristics. Only with hindsight the location and molecular effect of the genetic changes can be clarified. On the contrary, modern recombinant DNA methods allow targeted manipulation in the production of organisms. So-called metabolic engineering is therefore succeeding over the other methods. In essence, it means rational strain development with the help of genetic techniques. Apart from a good recombinant DNA toolbox, target identification plays a decisive role. The whole spectrum of modern biochemical and molecular biotechnology is used for this.

On the one hand, metabolic engineering can begin directly from a wild-type strain; on the other hand, classical production strains are also often further optimized with the help of metabolic engineering. The basis for all work is to have recombinant DNA tools available for the targeted changes on the specific production organism. If yeast or *E. coli* are involved, enough methods and experience are available. However, if the organism that is to be used belongs to an inadequately genetically investigated species, the tools for this have to be first established. Here, it is important to be able to transform the organism; vectors are needed together with a sufficient number of selection markers. It is often necessary to develop methods for the removal of selection markers (e.g., antibiotic resistance genes) available; the reasons for this lie in the product or plant licensing, or to improve the customer acceptance. For a number of production organisms, the genomes have been sequenced, which makes targeted operations in the metabolism much easier.

Products of classical fermentation processes are usually substances that also occur in the regular metabolism of the respective organism, albeit in much lower concentration. The metabolic pathways in production strains are similar or even identical to those in the wild-type strain. The regulatory controls of the wild-type strains have been lost or altered in the course of strain development. Additionally, the physiology of the cells during production differs drastically from natural conditions. For example, the organism's redox status might be completely unbalanced, ultimately reducing the overall viability of the strain. In order to make the fermentation process more efficient, the redox status of the production strain needs to be assessed and eventually tuned.

Starting material or the product of a fermentation process may have solvent characteristics, and can thus exert toxic effects on the production strain. This can be one reason for economically unattractive product concentrations with otherwise interesting strains.

Generally, the non-natural process conditions may lead to regulatory phenomena with a negative impact on productivity. These may be feedback inhibition on the synthesis of the pathways enzymes or adaptation on the genome level. In order to understand these effects the production organism needs to be understood as a system. It is investigated with regard to its metabolites (metabolome), proteins (proteome), and transcripts (transcriptome). Such a systems biological analysis augments genome-based models of metabolic networks and may even help to predict the metabolic behavior of a strain under certain conditions or after specific genetic modifications. Thus, **systems biology** helps to expand the possibilities of rational strain development.

An even greater challenge than the optimization of existing pathways is the realization of completely new metabolic routes, which do not exist in a given microorganism. There are many reasons to do so. It may be that similar routes are present in certain organisms, although these strains generate byproducts that are hard the separate from the desired compound or reduce carbon yield. When the deletion of the unwanted side activities is not successful, the design of a novel production organism may turn out to be the better solution. This seems to be the case for *n*-butanol fermentation (cf. Case Study 4 in Section 33.4.4).

In the near future we will certainly see more examples of fermentation processes leading to chemicals that are not the final product of natural biosynthetic pathways. Current examples from research and development are novel polyketide synthetases, which have a catalytic domain made of modules from different source organisms. The catalyst produces a metabolite that has not been found in nature (Wilkinson and Micklefield, 2007). The process for the biological production of 1,3-propanediol was developed by DuPont and Genencor according to this principle (Nakamura and Whited, 2003). The metabolic pathway to 1,3-propanediol was created *de novo* by combining catalyst genes from different organisms in one host.

Other research groups work on pathways tailor-made for the production of fuels or polymer precursors. For industrial applications, however, these novel production strains will need to comply with the rules of economics as well in order to outperform established routes.

33.4.4
Case Study 4: Fermentative Production of *n*-Butanol
(Jones and Woods, 1986; Lee *et al.*, 2008)

It was the French chemist Louis Pasteur who in the nineteenth century discovered that certain bacteria are able to produce *n*-butanol under anaerobic conditions. Chaim Weizman (1919) implemented the so-called ABE process into industrial practice during World War I in the UK. (ABE indicates the main fermentation products, i.e., acetone, butanol, and ethanol (Gabriel, 1928; Weizman (1919, GB patent 191504845A, 1919).) The microorganism used is *C. acetobutylicum*, which converts carbohydrates predominantly to acetone and *n*-butanol as well as smaller amounts of ethanol. This mixture of products is a major disadvantage of the ABE process, since the products ultimately have to be separated (e.g., by distillation). In addition, the production strain is susceptible to alcohol concentrations above 2%. These low titers make the work-up even more difficult.

Nevertheless, acetone and *n*-butanol were manufactured using this procedure until the middle of the twentieth century. However, raw materials made up to 60% of the production costs – a fact that in combination with technical weaknesses led to disappearance of the ABE process in the 1960s when more economic processes based on petrochemistry were established (such as the Reppe process, see Gabriel, 1928; Weissermel and Arpe, 1994). Only in the USSR, South Africa, and China could the ABE process maintain an industrial significance (Ni and Sun, 2009).

Increasing demand and costs for crude oil in combination with environmental concerns have led to a renaissance of research activity in the context of the ABE process in recent years. In particular, *n*-butanol is the focus of activities as this chemical is a favorable biofuel in comparison to ethanol. Although both are accessible from renewable resources, *n*-butanol outperforms ethanol (e.g., in terms of a lower water content, higher energy density, and lower vapor pressure). As

Fig. 33.8 *n*-butanol. Using metabolic engineering, pathways for the synthesis of *n*-butanol are realized in organisms that so far have not been able to produce this compound. Here, acetyl-CoA is the interface to glycolysis.

well as attempts to optimize the established production strains from *Clostridium*, it has also been attempted to tailor other genetically more amenable microorganisms.

The group of Liao at the University of California at Los Angeles developed a recombinant *E. coli* harboring the genes of the *n*-butanol pathway from *C. acetobutylicum* (Atsumi *et al.*, 2008). The respective enzymes catalyze the conversion of two molecules of acetyl-CoA and four of NADH to one *n*-butanol molecule (Fig. 33.8). A similar approach was chosen by Keasling (2008), who has transferred the *n*-butanol pathway from *Clostridium* to yeast (Steen *et al.*, 2008).

Even when yields with recombinant *E. coli* are lower than those of classical *Clostridium* fermentations, the potential of metabolic engineering is made clear in these examples. Further improvement may be the enhancement of the catalyst activity within *E. coli* and to funnel the overall metabolism more into the direction of acetyl-CoA generation (e.g., by knocking out unwanted genes). Other parameters such as tolerance for the product *n*-butanol will also be addressed. Eventually not only physiological aspects of the recombinant strain need to be optimized, but also technical aspects of the process itself can be improved. One approach might be continuous product removal in order to keep the concentration of *n*-butanol within the fermenter low and thus the microbes viable.

33.4.5
Case Study 5: Production of Glutamic Acid with *C. glutamicum*

Some microorganisms already have a naturally high potential for the synthesis of a desired substance. In these cases it is possible to use the isolated wild-type found in nature as a production strain. This is the case with *Corynebacterium*, used for the production of glutamic acid.

Glutamate, which is known under the product term MSG (monosodium glutamate), is used in Asia and also increasingly in America and Europe as a flavor enhancer. In Japan, brown algae are traditional groceries; so, it was attempted, at the beginning of the last century to identify the flavor components. In 1908, Ikeda succeeded in isolating glutamate as the main flavor component. It was first extracted from the algae and marketed. Ajinomoto Company has developed and executed the chemical synthesis of MSG. In the 1950s, Kinoshita, from the Kyowa Hakko Company, discovered a glutamate-producing bacterium. This organism was first called *Micrococcus glutamicus*, but is now known as *C. glutamicum*. These bacteria are Gram-positive, anaerobic, immotile, rod-shaped, with a high GC content. They are assigned to the so-called CNM group (*Corynebacterium, Nocardia, Mycobacterium*). Among these genera, there are many types with biotechnological significance. Together with *C. glutamicum* and related species, a number of other fermentation processes exist, such as for the production of lysine and nucleotides. In addition to this, one can also find pathogenic organ-

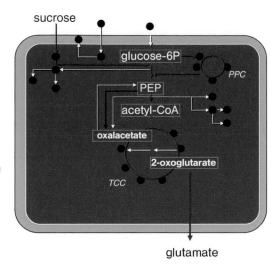

sucrose

**Important Factors in
Glutamate Production**

- biotin limitation

- detergents (Tween 40)

- detergent-hypersensitive
  mutants

- sublethal doses of penicillin

glucose-6P

PPC

PEP

acetyl-CoA

oxalacetate

2-oxoglutarate

TCC

glutamate

Fig. 33.9 Systematic representation of glutamate biosynthesis in *C. glutamicum*. PEP, phosphoenolpyruvate; TCC, tricarboxylic acid cycle; PPC, pentose phosphate cycle; glucose-6P, glucose-6-phosphate.

isms, like *Corynebacterium diphtheriae*, *Mycobacterium tuberculosis*, and *Mycobacterium leprae*.

Feedstocks for this **fermentative production of glutamate** are sugar and a nitrogen source. Typically, glucose, sucrose, or molasses are used as sugar. Common nitrogen sources are ammonia gas, ammonium salts, or uric acid. Under nonlimiting optimal growth circumstances, the *Corynebacterium* wild-type does not produce glutamate. Biotin limitation or addition of detergents are important for glutamate formation. In practice, polyoxyethylene sorbitan monopalmitate (Tween 40) is used. For the industrial production of glutamate, detergent-hypersensitive mutants have been selected. As a consequence, the amount of detergent can be kept small. Sublethal doses of penicillin also promote the formation of glutamate (Fig. 33.9).

Under these conditions, *C. glutamicum* is able to form up to 75 g·L^{-1} of glutamate per day. In the year 2005, the worldwide production of glutamate was estimated well above 1 billion tons.

33.4.5.1 Molecular Mechanism of Glutamate Overproduction
The **biosynthesis of glutamate** is conducted by the enzyme glutamate dehydrogenase. The substrate is 2-oxoglutarate – an intermediate of the tricarboxylic acid cycle. Glutamate dehydrogenase competes with 2-oxoglutarate dehydrogenase for the substrate. In *Corynebacterium* the 2-oxoglutarate dehydrogenase is very unstable. Owing to this, the enzyme could not be measured for a long time, as opposed to similar enzymes from other organisms. However, it has been shown that, with limited biotin and the addition of detergents or penicillin, the activity of 2-oxoglutarate dehydrogenase is lowered. The metabolites, therefore, flow preferably in the direction of glutamate (Fig. 33.10).

The **mechanism of glutamate overproduction** is not yet fully understood. Originally, it was thought that detergents and penicillin damage the cell membrane to such a great extent that an outflow of glutamate arises. In order to keep the intracellular glutamate pool constant, the cell would constantly resynthesize new glutamate. Meanwhile, there are biochemical data that point to a completely different molecular mechanism. The *dtsR1* gene seems to play an important role. Molecular analyses show that the *dtsR1* gene is involved in fatty acid synthesis. *dtsR1* deletion mutants are auxotrophic for certain fatty acids (i.e. these fatty acids cannot be synthesized any longer by the organism itself). These mutants are also especially sensitive to detergents. Interestingly they show a higher glutamate formation and a lower oxoglutarate dehydrogenase activity. Supposedly, the *dtsR1* protein functions as a b-chain of the biotin-dependent acyl-CoA carboxy-

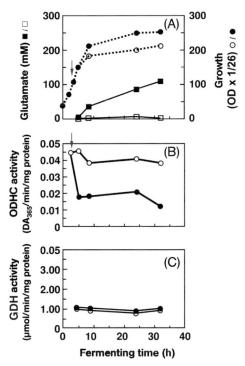

Fig. 33.10 Influence of penicillin on glutamate formation, and on the enzyme activity of 2-oxoglutarate dehydrogenase complex (ODHC) and glutamate dehydrogenase (GDH). A few hours after growth of the culture, sublethal doses of penicillin are given (arrow). Shortly thereafter, glutamate formation begins. At this point in time, the activity of the 2-oxoglutarate dehydrogenase complex declines, while the activity of glutamate dehydrogenase remains unchanged. 2-Oxoglutarate is channeled in the direction of glutamate. Open symbols, no addition of penicillin; closed symbols, penicillin addition. (Modified from Kawahara *et al.* (1997)).

lase, and is involved in the provision of building blocks for the synthesis of fatty acids and mycol acids. The second subunit of acyl-CoA carboxylase is dependent on biotin. This condition could be related to the above-described biotin limitation of the glutamate formation. An overexpression of *dtsR1* leads to decreased glutamate formation.

There are, however, initial clues of a specific active glutamate export, although the corresponding export protein has not yet been identified on the molecular level.

33.4.6
Case Study 6: Production of Lysine with *C. glutamicum*

Shortly after the discovery of *C. glutamicum* as a glutamate producer, new *Corynebacterium* strains were found that secrete the amino acid lysine into the medium. This discovery was used as an opportunity to systematically produce new mutants and to examine them for **lysine productivity**. Lysine has developed into the second largest biotechnologically produced amino acid, coming right after glutamate. While glutamate is sold as a product for human nutrition, lysine finds its applications mainly as an essential amino acid in animal nutrition. Small amounts are also marketed in human nutrition as well as for pharmaceutical applications. Various companies produce 500 000 tons of lysine per year.

33.4.6.1 Molecular Mechanism of Lysine Biosynthesis
The starting materials for the formation of lysine in *Corynebacterium* are oxalacetate and pyruvate – two metabolites of central metabolism. Oxalacetate is first converted into aspartate and then reduced by a transamination to aspartate semialdehyde. The corresponding enzymes, **aspartate kinase** and **aspartate semialdehyde dehydrogenase**, are encoded by the genes Ask (LysC) and *asd*, respectively. Both genes are organized in a single operon. The aspartate kinase is, as already explained above, allosterically regulated. Aspartate semialdehyde lies at a branching out point of metabolism. On the one hand, it can be channeled into the amino acids threonine, isoleucine, and methionine; on the other hand, it is a precursor of lysine and condenses as such, catalyzed by dihydrodipicolinate synthase (DapA) with pyruvate to dehydropicolinate. The **dihydrodipicolinate synthase** is, besides aspartate kinase, a further key enzyme in lysine biosynthesis. It has been shown that two copies of the *dapA* gene lead to lysine overproduction, as well as overexpression of the gene through a base exchange in the promoter region. The shared overexpression of *dapA* with Ask has a synergistic effect.

Catalyzed by a reductase (DapB), dihydropicolinate is transformed under NADPH consumption into tetrahydrodipicolinate. Starting from tetrahydrodipicolinate, two parallel biosynthesis routes exist. Both contain reduction by NADPH and the introduction of a second amino group, resulting in the first intermediate – *meso*-diaminopimelate. For this, the so-called succinylase pathway needs four single reactions, while in the dehydrogenase pathway this job is taken over by a single enzyme – **diaminopimelate dehydrogenase** (Ddh). Meso-diaminopimelate is also a building block for the cell wall and is converted in the last step of the biosynthesis by **diaminopimelate decarboxylase** (LysA) into lysine.

The fact that *C. glutamicum* contains **two parallel biosynthesis pathways** for the provision of the lysine precursor *meso*-diaminopimelate is uncommon. So far, this could only be shown for a few other bacteria. Flux analysis with ^{13}C-labeled substrates has shown that both metabolic paths contribute to lysine formation; depending on the ammonium concentration, they are used in different amounts/ratios. The succinyl path way uses glutamate for the incorporation of the amino group and receives the ammonium from glutamate dehydrogenase. The pathway is carried out preferentially at low ammonium concentrations. Diaminopimelate dehydrogenase has a small affinity for ammonium and therefore is rather used with high ammonium concentrations.

The LysA gene that codes for diaminopimelate decarboxylase, is, together with the gene of the arginyl aminoacyl tRNA synthetase (ArgS), organized in one operon and controlled by the same promoter. This is a hint that the lysine biosynthesis is coregulated with the metabolism of another, also nitrogen-rich amino acid – arginine. In addition, the recently identified lysine export protein (LysE) can recognize arginine as a substrate and transport it out of the cell. The transcription of LysE is activated from the regulator LysG during increasing lysine concentration. Overexpression of LysE leads to significantly improved lysine secretion.

33.4.6.2 Deregulation of the Key Enzyme Aspartate Kinase

Together with aspartate, methionine, threonine, and isoleucine, **lysine** belongs to the aspartate family of the amino acids. With *C. glutamicum*, the allosteric regulation on the enzymatic level plays a decisive role in the metabolic pathway. The initial enzyme, the aspartate kinase (Ask, respectively, LysC), is allosterically inhibited through the amino acids threonine and lysine in the wild-type. That means that with a physiologically satisfying amount of leucine and threonine, the enzyme activity of the aspartate kinase is reduced and therefore further amino acid synthesis is abolished. This makes sense in nature so that the microorganism does not unnecessarily consume resources and energy. An industrial production organism should, however, produce more lysine than needed for its own needs. If one succeeds in circumventing regulation, an important limitation has been overcome. Selection experiments with so-called antimetabolites have been successfully conducted to produce mutants whose biosynthesis regulation has been abolished,. In addition to the **natural feedback inhibitors**, lysine and threonine, the lysine analog amino ethyl cysteine (AEC) also has an inhibiting effect on aspartate kinase (Fig. 33.11). The inhibition is the strongest if AEC is used in combination cultures with threonine. If one cultivates mutant *Corynebacterium* on agar plates that contain AEC and threonine, most mutants cannot grow because the antimetabolite does not allow a synthesis of threonine and lysine. Among the few resistant mutants that grow in the presence of AEC, some mutants can be found that decompose AEC or do not transport it into the cell, so that the inhibitor cannot reach its active site. However, one can also always find such mutants that contain a modified aspartate kinase. Here the enzyme is altered in such a way that AEC and threonine cannot interact any more. Such

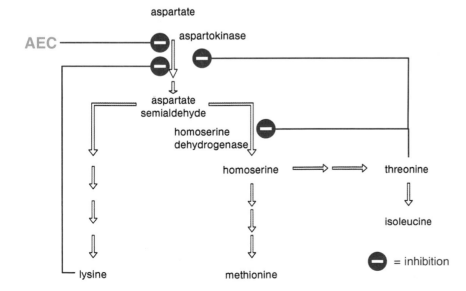

Fig. 33.11 Selection of feedback-deregulated mutants with antimetabolites.

mutations lead to lysine overproduction, because there is no longer any regulation. The organisms can synthesize more lysine than they need themselves; the amino acid is exported into the medium and accumulates there.

Meanwhile, the mutated Ask and LysC genes have been sequenced and analyzed. This has shown the different point mutations that can lead to the desired deregulating effect.

33.4.7
Genomic Research and Functional Genomics

For the targeted improvement of industrial production organisms, the exact knowledge of the specific genomes can be of a decisive advantage. Therefore, the genome of *C. glutamicum* has been decoded by several amino acid manufacturers. It contains 3.3 Mb and has approximately 3300 open reading frames. Two-thirds of those could be identified with the help of **bioinformatic annotation** methods. The knowledge of the genome accelerates genetic engineering significantly. In addition to this, all known metabolite pathways can be assigned to the corresponding enzymes and their genes. As a result, we have quite a comprehensive overview over the metabolism of *C. glutamicum*. The sequencing of the genome led to the discovery of new, so far unknown, metabolic pathways. It could be shown that *C. glutamicum* not only has one pathway to form the disaccharide trehalose, but actually three different ones.

In particular, the knowledge of the genome opens up new analytic possibilities for the identification of new target genes for strain optimization. Meanwhile, the first transcription analyses with **DNA arrays** or **DNA chips** have been conducted (see Chapter 22). By hybridization of small mRNA from fermentation samples with the immobilized *C. glutamicum* genes, it is possible to determine the activity of all genes for a certain point of time. In such experiments it has been discovered that the lysine biosynthesis genes are mainly constitutively expressed. Regulation on the level of transcription was only found with the gene for the oxalacetate glutamate amino transferase, the first step of the lysine biosynthesis, and with LysA. The LysA gene product, *meso*-diaminopimelate decarboxylase, catalyzes the last step of the lysine biosynthesis.

Apart from **transcription analysis**, **proteomics** play a more and more important role. In a 2D gel electrophoresis, up to 2000 proteins from one sample can be separated at the same time according to size and charge. The single proteins are then visualized, identified, and quantified. By a combination of transcription analysis, proteomics, and the already mentioned flux analysis, it is possible to achieve a far more exact image of the metabolism and its regulation than in the past, and possibly discover new approaches for rational strain development.

33.4.8
Case Study 7: Fermentative Penicillin Production

The antibiotic effect of the fungal metabolite penicillin was discovered in the 1930s by Fleming in the UK. The commercial production of penicillin started in 1941, with a *Penicillium notatum* strain in surface culture. As the productivity was not satisfactory, the search for better production strains in nature was initiated. This work paid off in 1943 when a *Penicillium chrysogenum* strain with better characteristics was introduced into production. Since that time, a whole number of classically optimized penicillin overproducers have been generated at many different pharmaceutical companies and universities. Over the years, productivity was increased by a factor of 100, compared to the starter strain. Only since the late 1980s have we been able to characterize **penicillin biosynthesis** and the corresponding genes. The *acvA*, *ipnA*, and *aat* genes code for the biosynthetic enzymes **D-(L-γ-aminoadipyl)-L-cysteinyl-D-valine synthase**, **isopenicillin-N synthase**, and **acyl-CoA:6-aminopenicillanic acid acyltransferase**. The first enzyme is a peptide synthase that catalyzes the formation of a tripeptide from the precursors aminoadipic acid, cysteine, and valine. In the next step, the β-lactam ring of isopenicillin-N is formed. Through the integration of additionally fed phenyl acetic acid, penicillin G forms in the third reaction. The three genes that code for the biosynthesis enzymes make up a 35-kb gene cluster.

P. chrysogenum strains, produced by classical strain development by the pharmaceutical company SmithKline Beecham, have been examined with molecular biological methods for the purpose of penicillin overproduction. In doing so, it has been discovered that the number of copies of the **penicillin gene cluster** has been increased with improved production strains. The penicillin titer achieved is in direct correlation with the number of copies. The best examined strains show up to 50 copies of the three biosynthesis genes. Not only the coded regions have been amplified, but also a 57.4-kb fragment on which the sequences are located that are responsible for the necessary recombination events. The analysis of the promoters of the three biosynthesis genes did not result in any clues as to the changes.

33.4.9
Case Study 8: Vitamin B$_2$ Production

Riboflavin is, as vitamin B$_2$, an essential component of the nutrition of humans and animals. After conversion into flavin adenine dinucleotide (FAD), or flavin mononucleotide (FMN), it takes part as a coenzyme in a number of redox reactions. In animal experiments, riboflavin deficiency leads to dermatitis, growth disturbances, and eye diseases.

For many decades, the vitamin has been produced by chemical synthesis in a multistep process. Since the end of the 1980s, biotechnological processes have superseded the chemical syntheses.

Three different production organisms are used for **riboflavin production**. The oldest process is based on the fungus *A. gossypii* – an ascomycete whose genes show great sequence similarities to the genome of *Saccharomyces cerevisiae*. A further process is being conducted with the yeast *Candida famata*. The third important production organism is *B. subtilis* that, as opposed to the first two organisms, is not a natural riboflavin overproducer. Furthermore, this production organism is a **genetically modified organism (GMO)** that has been developed into a riboflavin producer by targeted changes.

33.4.9.1 Riboflavin Biosynthesis
The biosynthesis of riboflavin starts out from guanosine triphosphate and ribulose-5-phosphate. Apart from one unspecific phosphatase, all enzymes for *Ashbya* and *Bacillus* have been described and their genes have been characterized. In bacteria, the deamination of diaminpyrimidinone seems to happen before the reduction of the ribityl side chain, while the sequence of these reactions is inverted in the fungus. Two bifunctional proteins are involved in the biosynthesis in *Bacillus*. RibA catalyzes the GTP cyclohydrolase II reaction. 3,4-Dehydroxybutane-2-onphosphate synthase is localized on the same peptide chain. RibG contains the deaminase and the reductase. In *Ashbya*, each functionality is localized on the same peptide chain.

The precursor **guanosine triphosphate** is provided via purine biosynthesis. This is a extremely long and complex metabolic pathway. An allosteric feedback inhibition of the first two biosynthesis steps by purines is known for several organisms.

33.4.9.2 **Classical Strain Development**

In the past, all three production organisms have been improved with the help of **classical mutation** and **selection**. The riboflavin synthesis of *C. famata* is inhibited by iron. Iron-resistant mutants showed improved riboflavin formation. With *Candida*, and also *Bacillus*, intensive work has been done on the deregulation of the purine biosynthesis, in such a way to produce mutants that show resistance against pure purine analogs. Improved *Candida* strains could be produced with the help of the antimetabolite **tubercidin** (7-deazaadenosine). For the optimization of *B. subtilis*, the purine compounds 8-azaguanine, decoyinine, and methionine sulfoxide, and the riboflavin analog roseoflavin have been used for the selection of mutants. **8-Azaguanine resistance** is being induced in connection with a strengthening of the expression of the biosynthesis genes. **Methionine sulfoxide**-resistant strains show a stronger conversion of inosine monophosphate to xanthosine monophosphate. The resistance against roseoflavin is induced by two different groups of mutants. RibC mutants have a significantly decreased enzyme activity of the riboflavin kinase, which transfers riboflavin in FMN. It has been shown that a single point mutation is enough to lower the activity of the riboflavin kinase over 90%. The second group of mutants shows point mutations in the noncoding leader region of the *rib* genes of *Bacillus*. The mechanism of the resulting strengthened riboflavin formation has not yet been clarified in detail.

Part IV
Biotechnology in Industry

34
Industrial Application:
Biotech Industry, Markets, and Opportunities

Learning Objectives

The chapter begins with a short historical look at the past, which takes us into traditional and molecular biotechnology. We then examine the various sectors of the industry, with a survey of possibilities for the application of molecular biotechnology in the fields of red, green, gray, or white biotechnology, as they are called, looking at the opportunities and markets. We close with a general view of the present state of the industry world-wide.

34.1
Historical Overview and Definitions of Concepts

The term **biotechnology** was coined by the Hungarian engineer Ereky in 1919. It is defined as the sum of all processes by which products are made with the aid of microorganisms or parts thereof. Aside from the invention of the term, with regard to its antiquity the original use of biotechnology dates back to well before the time of Christ. The main use then was in the area of foodstuffs, where it was applied to the **production of bread, cheese, beer, wine, and vinegar.** The agent responsible was unknown – the process simply exploited the effect of alcoholic fermentation and that of lactic and acetic acid fermentation. The technology was also used in tanning skins to produce leather. This stage of application, also referred to as **traditional biotechnology**, lasted into the eighteenth century when biotechnology for the first time was turned to industrial use.

The discoveries of Pasteur in 1864 laid the foundation for **applied microbiology**. The French chemist was the first to use a microscope to monitor the course of wine and vinegar production. He also developed pure cultures of microorganisms and the sterilization of their nutrient media (**pasteurization**). The period following Pasteur was initially characterized by the development of biotechnical procedures that did not absolutely exclude foreign microorganisms. Examples are the **fermentation** and surface culture of microorganisms for the industrial production of butanol, acetone, ethanol, and citric acid. Fermentation was also used for the biomass production of baker's yeast and feed-yeast. In the realm of public services, the introduction of aerobic and anaerobic purification of wastewater around 1900 was a milestone in the prevention of epidemics. The production of acetone and glycerin, which were used as raw materials to produce explosives during World War I, by fermentation methods gave the first impetus to the fermentation industry. During World War II, and following the chance discovery of the antibacterial effect of **penicillin** by Fleming in 1928–1929, the industrial production of antibiotics was set in motion. Over 1000 different antibiotics had been isolated by 1950 and many of these were used in large quantities in human medicine (a milestone in the treatment of infection), and increasingly also in animal production and plant protection. Another development that dates from 1950 is the industrialization of analytical biotechnology. This first used enzymes and later antibodies (based on the principles of immune analysis) for the highly selective detection of metabolites in body fluids.

An Introduction to Molecular Biotechnology, 2nd Edition.
Edited by Michael Wink
Copyright © 2011 WILEY-VCH Verlag GmbH & Co. KGaA, Weinheim
ISBN: 978-3-527-32637-2

The further developments occurring from the 1960s onward led to the application of biotechnological production methods that excluded foreign microorganisms and used selected strains. These were optimized in the traditional way (chemical and physical mutagenesis). Submersed processes, animal cell cultures, and microbial and enzymatic biotransformation made possible the production of virus vaccines, cortisone, vitamin B_{12}, and ovulation inhibitors. At the same time the integration and use of important research results from the fields of science and technology enabled the production of biopolymers by microbiological means and the immobilization of enzymes and cells. Examples of these products are protozoal proteins, enzymes (washing powders), polysaccharides (xanthan), and fructose syrup. The biotechnological developments that we have described, which are based on applied microbiology and biochemistry, are typical of the phase of traditional industrial biotechnology.

The foundations of **molecular biotechnology** and thus modern biotechnology were established in **1973** with the development of the *in vitro* **recombination of DNA** by Cohen and Boyer. In the United States, using this technology, the targeted transfer of a foreign gene into a host organism, where it was expressed, was achieved for the first time. This development was turned to commercial use with the founding of the company **Genentech** in San Francisco in 1976 (see Chapter 37). Over the course of 2 years the company succeeded in developing the first recombinant product – human insulin. This was later licensed out to the pharmaceutical company **Eli Lilly**, which brought it to market in 1982. Another pioneer of modern biotechnology is the U.S. company **Cetus**, founded in 1971, which was later subsumed into the company Chiron. Cetus developed **polymerase chain reaction (PCR)** technology and sold it to **Hoffmann-La Roche** in 1991. Cetus also developed interleukin-2, which has been used as a treatment for cancer since 1992, and β-interferon, which is used to treat multiple sclerosis. The founding of Genentech can be seen as the birth of the modern biotechnology industry, and consequently, in the United States, the building up of the industry has been happening for 34 years. As a fully integrated biopharmaceutical company, Genentech gained revenues of US $ 13.4 billion and a profit of US $ 3.6 billion in 2008. In spring 2009 the company was completely acquired by the Swiss pharma giant **Roche** in a deal valued at US $ 46.8 billion.

Other technologies that belong to the field of modern biotechnology (and this list makes no claim to be complete) can be cited as: modern cell and tissue technologies, metabolomics/system biology, RNA technologies, proteomics, combinatorial biology/chemistry, high-throughput screening, directed evolution, computer-aided drug development, nanobiotechnology, bioinformatics, and biochips or microarrays (see Chapters 21–23).

34.2
Areas of Industrial Application of Molecular Biotechnology

The industrial application of molecular biotechnology is often subdivided, so that we speak of **red, green, white, or gray biotechnology**. This distinction relates to the use of the technology in the medical field (in human and animal medicine), agriculture, industry, and the environment. Many companies also apply knowledge deriving from molecular biotechnology in areas that cut across these distinctions. Included in this category are companies that are exclusively or predominantly involved in providing services for the biotechnology industry or are suppliers for biotech firms. Companies that carry out contract-based production of biological molecules without conducting any development themselves are also included. According to a study conducted by biotechnologie.de on behalf of the German Ministry for Research and Education in 2009, 36% of all German biotech companies fall into this category. The highest percentage falls upon companies in the field of **red biotechnology (45%)** (e.g., developers of pharmaceuticals and diagnostics). **White biotechnology** is the focus of **10%** of all companies and **green biotechnology** refers to only **5%** of total industry. An additional **4%** of all companies is active in the field of **bioinformatics**.

34.2.1
Red Biotechnology

Within the field of red biotechnology, which deals with applications in human and animal medicine, there are various further distinctions that can be made: **biopharmaceutical drug development, drug delivery, cell and gene therapies, tissue engineering/regenerative medicine, pharmacogenomics** (personalized medicine), **molecular medical diagnostics**, and **system biology**.

34.2.1.1 Biopharmaceutical Drug Development
In the field of biopharmaceutical drug development, it is the development of **therapeutic human proteins** by recombinant methods for use as medicines that has the longest tradition. As mentioned in Section 34.1, recombinant human insulin was the first recombinant medicine in the world, developed by Genentech and brought to market in 1982. Today, recombinant human insulin has almost completely squeezed the other preparation of insulin (isolated from human or animal tissues) out of the market. Some examples of in Germany newly approved biopharmaceuticals in 2007 can be seen in Table 34.1.

Owing to the fact that proteins have a structure that is too complex to be synthesized chemically, the method used before the introduction of molecular biotechnology in medicine was to extract the active substance in question from human or animal blood or tissues. This caused various problems. Often the therapeutically effective substances were present in very low concentrations, making very large quantities of starting material necessary to obtain them. This called for extensive production procedures, which sometimes also had environmentally negative consequences. Proteins of animal origin, such as the insulin from pigs that was formerly used to treat diabetes, could produce severe intolerance reactions because the sequence differed from the human protein. Also, when a drug is being purified from blood from a human donor, there is a latent risk of contamination with pathogens. An example of this is coagulation factor VIII, which is not produced in males with hemophilia and leads to a life-threatening disorder of blood clotting. Factor VIII prepared from donor blood led in the past to many individuals becoming infected with HIV.

Table 34.1 Selected examples of in Germany/EU in 2007 newly registered biopharmaceuticals. (Source: vfa/Boston Consulting Group, Medizinische Biotechnologie in Deuschland, 2008).

Class	Agent	Indication
Antibody	Panitumumab	Treatment of colon cancer
	Ranibizumab	Age related macular degeneration
	Eculizumab	Paroxysmal nocturnal hemoglobinuria (PNH)
Other recombinant proteins	Abatacept	Rheumatoid arthritis
	Methoxypolyethylen-glycol-Epoetin beta	Anemia during chronic kidney diseases
	dursulfase	Hunter syndrom (mucopolysaccharidosis Type II)
	Mecasermin	Growth disorders at children/teens with a severe primary lack of IGF1-1
	Epoetin alfa	Anemia during renal failure, cancer or in advance of surgery
	Epoetin zeta	Anemia during renal failure, cancer or in advance of surgery
Vaccines	HPV vaccine	Prevention of cervical cancer

Table 34.2 Therapeutic antibodies by highest rank of sales (US $ million, 2006). (Source: according to Ernst & Young. Auf gutem Kurs, Deutscher Biotechnolgie-Report 2008.)

Drug	Brand name	Indication	Company	Sales
Infliximab	Remicade	Chron's disease	Johnson & Johnson	4.253
Rituximab	MabThera/Rituxan	Non-Hodgkins Lymphoma	Genentech/Roche	3.739
Trastuzumab	Herceptin	Breast cancer	Roche	3.136
Bevacizumab	Avastin	Breast, colon, lung, renal cancer	Genentech/Roche	2.365
Adalimumab	Humira	Rh. Arthritis	CAT/Knoll/Abbott	2.044
Cetuximab	Erbitux	Colon cancer	Imclone/BMS/Merck	1.100

The **advantages of using molecular biotechnology** in drug development are therefore clear:

- Lower risk of infection.
- Reduced side effects.
- Greater appropriateness to need.
- Extended therapeutic possibilities.
- More efficient and more environmentally friendly production.

Apart from the development of recombinant proteins (mainly hormones, growth factors, blood proteins, interleukins, and interferons), therapeutic antibodies are now becoming increasingly important (Table 34.2; see also Chapter 28). Therapeutic antibodies are already being used with success to treat cancer and rheumatoid arthritis. Additional indications that are being treated with antibodies include neurological and ophthalmic diseases. The success of therapeutic antibodies to date is mainly due to their selectivity, which means that the preparations are generally well tolerated. Therapeutic antibodies may be obtained as polyclonal or monoclonal antibodies and are also produced by recombinant methods. Recombinant production of antibodies is carried out in bacteria, yeasts, mammalian, and insect cell cultures, and in transgenic animals and plants.

The **first therapeutic antibodies**, especially monoclonal antibodies, came on the market in the late 1990s. In 2006, 20 approved therapeutic antibodies yielded revenues of nearly US $ 20 billion (Table 34.2). According to Datamonitor, revenues increased to US $ 26 billion in 2007 and are expected to reach US $ 49 billion in 2013. In 2008, revenues from therapeutic antibodies already accounted for more than 30% of the total pharmaceutical market. In comparison to other classes of compounds they are projected to have highest growth rates.

With an annual growth rate of 12.83% the **world-wide market for therapeutic proteins** will increase from US $ 57 billion in 2006 to more than US $ 90 billion US $ in 2010 according to the RNCOS report *Global Protein Therapeutics Market Analysis*. Amgen's blockbuster drug Enbrel® (a fusion protein for the treatment of rheumatoid arthritis) alone yielded revenues of US $ 3.6 billion in 2008.

Today, in addition to proteins, which currently play the most significant role in the biopharmaceutical field, new types of drugs based on RNA (antisense drugs, ribozymes, aptamers, spiegelmers (mirror-image oligonucleotides), and RNA interference) are also being developed on the basis of advances in knowledge in molecular biotechnology. These, however, are mainly at the stage of research or clinical development (Table 34.3; see Chapters 2.4, 21, and 31). The first antisense drug, ISIS Pharmaceuticals' Vitravene® (fomivirsen) for the treatment of retinitis, was approved by the US Food and Drug Administration in 1998. DNA itself is also thought to have therapeutic potential (see Chapter 31).

In 2008, the Pharmaceutical Research and Manufacturers of America, a pharmaceutical association, identified 633 biopharmaceuticals in clinical development or in registration for the US market. These are aimed at more than 100 diseases, with the focus on oncology and infections. In the development of ther-

Principle of action	Product name/ stage of development	Indication	Company
Antisense	Vitravene/Market	CMV retinitis	SIS Pharmaceuticals
Antisense	Mipomersen/Phase III	Cardiovascular	SIS Pharmaceuticals/ Genzyme
Antisense	Alicaforsen/Phase II	Crohn's disease	SIS Pharmaceuticals
Antisense	Trabedersen/Phase III	Brain tumor	Antisense Pharma
Spiegelmer	NOX-E36/Phase I (Jun 09)	Diabetic nephropathy	Noxxon Pharma
RNAi	ALN-RSV01/Phase II	RSV infection	Alnylam Pharmaceuticals
RNAi	RTP801i/Phase II	Macular degeneration	Quark Pharmaceuticals/ Pfizer

Table 34.3 Selected examples of therapeutic RNAs on the market or under development (Status: year end 2009).

apeutic agents, enormous opportunities are, however, balanced by great risks. Failures in these developments cost both time and money. Overall, the duration and cost of drug development are generally estimated at 8–12 years and US $ 500–1000 million for each successful registration (including failures).

Further potential for the application of molecular biotechnology, apart from its direct use in the development of biopharmaceuticals, is to apply it to the development of traditional drugs in the form of enabling technologies. These include the fields of genomics, proteomics, and bioinformatics (see Chapters 21–23). The use of these as enabling technologies should make traditional drug development quicker, cheaper, and better. However, there is at the same time a discussion as to whether the knowledge achieved in genomics, for example, will necessarily contribute to a simplification of drug development. In many cases all that has happened is that the number of targets has increased, without so far realizing any significant savings in time or expense.

34.2.1.2 Drug Delivery

Closely linked to the development of therapeutic agents are the means of achieving their **targeted delivery to their site of action**. These drug delivery systems are mainly used for drugs whose physical and chemical characteristics make them insufficiently stable in reaching their site of action intact. They can also be used to transport drugs in a targeted way to particular sites of action (tissue-specific targeting), or to overcome biological barriers such as the intestinal wall or the bloodbrain barrier (see Chapter 26).

Molecular biological approaches, unlike the traditional formulation of drugs, are primarily concerned with the targeted, **intelligent** transport of drug molecules. Usually, nucleic acids such as nucleotides, DNA, or RNA, which represent the actual drug, are **packaged** using other nucleic acids. The transport vehicle is usually a viral vector. This area is closely bound up with gene therapy.

Systems based on virus proteins, antibodies, receptors, or peptides can also assist the specific targeting of the drugs. Finally, lipids can be used, such as in the form of **liposomes** (Fig. 3.2).

34.2.1.3 Cell and Gene Therapy

Chronic diseases in particular, which are ultimately due to disturbances of cell function, are not capable of being cured by the drugs that have been available until now. Now, for the first time, **cell therapies** open up the possibility of treating the causes rather than just the symptoms of disease. Associations of cells are prepared outside the patient's body, and then used to replace diseased associations of cells and organs. The hope is not only to provide treatment in itself; for the first time there is also a great opportunity to save costs in the health system, since this method might make repeated medical treatment or hospital visits unnecessary. Treatments that have until now been highly cost-intensive and lifelong could be replaced by targeted cures. The market for cell therapies was estimated at nearly US $ 27 billion in 2005, and is forecast to reach US $ 56 billion in 2010 and US $ 96 billion in 2015.

The most recent discoveries concerning the potential for using human stem cells massively extend the spectrum of cell therapy. This holds out great possibilities, especially that of treating the root causes of common clinical situations such as organ failure (e.g., liver or heart), diseases of the joints and the intravertebral disks, mental disease (e.g., Parkinson's or Alzheimer's), and cardiovascular diseases. The role of embryonic stem cells as the basis for therapeutic products and procedures is only a subordinate one on account of the great technical difficulties of using them (i.e., in the developmental and production stages). The use of adult stem cells with the aim of regenerating diseased cell tissue is a more obvious route economically. This is the field of **regenerative medicine** (see also Section 34.2.1.4), in which **stem cells** are used as the raw material for the production of replacement tissue to be used to restore the function of destroyed tissue or organs in patients with chronic degenerative diseases, in a similar way to traditional organ transplantation.

Gene therapy is the targeted introduction of genetic material into the cells of sick individuals by suitable transfer methods, with the aim of achieving a cure or therapeutic improvement (see Chapter 30). The nucleic acids in this case serve as the active substance (the drug). In its wider sense, **gene therapy encompasses the replacement of defective genes by functionally intact copies** (gene addition), the **inactivation of pathogenic gene products** (antigene therapy, antisense therapy) and the **indirect curing of diseases by means of therapeutic genes**. This makes the use of genes as a drug in the general sense (treatment using genes) a conceivable option, going far beyond the correction of **inherited genetic defects (inherited diseases)**. The huge significance of gene therapy comes from the fact that this would truly be a treatment of the cause of disease. The first drug based on gene therapy is Gendicine – a genetically engineered adenovirus. It was approved 2003 in China for the treatment of patients with otorhinolaryngal tumors and developed by the Chinese company SiBiono GeneTech. Other gene therapies, however, remain at the research and development stage: more than 1500 clinical studies (mainly phase I) have been conducted so far in 28 countries (mainly the United States) with the focus on oncology (data from 2009). Leading companies in this field include AnGes MG, Cell Genesys, GenVec, Genzyme, Introgen Therapeutics, Oxford BioMedica, SiBiono GeneTech, Targeted Genetics, Transgene, Urigen Pharmaceuticals, and Vical. The biggest hurdle for gene therapy at the present time lies in the transfer systems or vectors, which are not yet fully developed. Theoretical risks of gene therapy are the development of tumors through the incorporation of the vectors at certain positions in the genome, the emergence of vectors with the capacity to reproduce, the establishment of new strains of viruses, and the excretion of the vectors into the environment. The use of viral vectors itself requires attention and careful consideration as to safety, because of the danger of recombination and the possible associated emergence of new, pathogenic viruses.

There has been news about increased cases of leukemia during a study in which retrovirally modified blood stem cells were used. Although the majority of all gene therapies were successful, these cases led to a setback with the consequence of stopping numerous other gene therapy projects world-wide. Nevertheless, according to Global Industry Analysts gene therapy will develop a global market of nearly US $ 500 million in 2015.

34.2.1.4 Tissue Engineering/Regenerative Medicine

A related technique is tissue engineering. Cell therapies are related to tissue engineering. This means the production of human cells, tissues, and whole organs from autologous cells. They are cultured and built into new tissue *ex vivo* (i.e., outside the body) using the patient's own cells and 3D structural scaffolds of cellular or synthetic origin. This requires knowledge of the biological interactions of tissue formation.

The purpose of **tissue engineering** today is not simply the construction of functioning tissue outside the body, but also to assist the body's capacity for regeneration.

The following reasons underlie the need for **tissue engineering products** in the long term:

- The age pyramid that is developing and the increase in chronic diseases (e.g., osteoporosis, diabetes, cardiovascular, and neurodegenerative diseases) associated with the rise in average age.
- The world-wide scarcity of donor organs for transplantation.
- The fact that implanted medical devices cannot fully replace the lost function of a tissue or organ and have only a limited life.

Regenerative medicine is closely linked to tissue engineering. It is concerned with products and technologies that protect or restore the normal function of tissues and organs.

This is carried out *in vivo* by stimulating or modulating the body's inborn capacity to regenerate damaged tissue. Growth factors and small molecules are examples of the means used. These stimulate the division of damaged cells and direct the restoration of the 3D structure of tissues, with the end result of renewed function, or assist the healing of damaged tissue by stimulating or inhibiting critical biological metabolic pathways.

Simpler tissues such as **skin, cartilage,** and **bone** are now routinely tissue engineered and marketed. Existing forms of treatment are being greatly improved by this means. The production of complete organs, which consist of several types of tissue in a complex 3D structure, is a more complicated matter. It is relatively difficult to produce larger, complex organs *ex vivo* by the methods currently available. The methods of regenerative medicine are therefore more suitable for the restoration of organs or organ systems. Life Science Intelligence estimated the world-wide market for tissue engineering and regenerative medicine in 2008 was US $ 1.5 billion and forecasts US $ 3.2 billion in 2013.

34.2.1.5 Pharmacogenomics and Personalized Medicine

Pharmacogenomics relates to the general investigation of all those genes that determine the body's reaction to a drug. **Pharmacogenetics** can be seen as being in effect a subdivision of pharmacogenomics. It relates to the investigation of inherited variations in genes for drug metabolism (e.g., variations in cytochrome oxidase genes). The terms are usually used synonymously in common speech.

From the point of view of the patient, the significance of pharmacogenomics lies in clarifying and ultimately avoiding **adverse drug reactions (ADRs)**, in which drugs have a negative effect on the body. The drugs that exist so far have in most cases been developed as a single solution for all, rather than on an individual basis. However, ADRs often bring about the need for further treatment or even the death of a patient. The promise of pharmacogenomics is that one day there will be individualized treatments, tailor-made for patients according to their genetic make-up. These should be more effective and associated with fewer side effects. Often this field is also known as **personalized medicine**. The method of choice to discover individual genetic variations is **single nucleotide polymorphism (SNP)** analysis (see Section 4.1.5). This uses DNA chips to identify the different gene sequences. SNPs occur every 100–300 base pairs in the human genome, which comprises 3 billion base pairs. The discovery of genetic variations associated with drug metabolism is, however, highly complex.

For the developing pharmaceutical and biotech companies themselves, the advantage of treatments tailor-made for certain subpopulations of patients or individuals is less obvious at first sight, as greater segmentation means a reduction in market sizes. Against this, in the field of clinical development, pharmacogenomics enables more precisely targeted clinical studies (i.e., tailored to drug responders), which are therefore smaller, quicker, and lower in cost. It is possible that, although the patient population is smaller, this is balanced by a larger market share because there are fewer side effects and increased efficiency, and that therefore a higher price can be set. Nearly US $ 5 billion were earned by the Swiss company Roche in 2008 with their drug Herceptin®. This drug can only be administered to 20–30% of all patients with breast cancer showing increased HER-2 receptor expression after a companion diagnostic test. There is the further possibility that drugs that had not originally been approved could be reactivated for a particular genetically defined subpopulation.

According to PricewaterhouseCoopers, the economic potential for personalized medicine on the US market is currently estimated at US $ 232 billion. Growing 11% per year, a market potential of US $ 452 billion could be reached in 2015.

34.2.1.6 Molecular Diagnostic Agents

The **diagnostic methods** of molecular biology have increasingly been used to complement the methods of determination used by traditional serological routine diagnostics and special diagnostics since the middle of the 1990s. They are based on PCR amplification of nucleic acids (see Chapter 13) from blood, urine, feces, sputum, or tissue from the patient and the use of genetic probes (biochips, gene chips) (see Chapter 11). Each complements the other, so providing a more complete picture in preventive and acute diagnosis as well as follow-up of the course of already established disease. Frequently, the presence of a disease can only be demonstrated with certainty by using molecular tests. The diagnosis of viral and bacterial infections by molecular biological methods is already part of everyday practice in modern laboratory diagnostics. The use of genetic cancer screening and tests for genetic predisposition to certain serious metabolic, endocrine, and cardiovascular disorders is also increasing. Molecular diagnostic methods are also applied today in paternity tests and forensic investigations (microsatellite PCR, see Section 4.1.1).

When molecular diagnostics are applied as part of indication-related disease management, the main aim is that of prevention, enabling future treatment costs to be avoided. The early discovery of life-threatening diseases is another benefit, as this makes it possible to provide earlier treatment and better monitoring of the course. Finally, molecular diagnostics is an important incitement for the above-mentioned personalized medicine. According to Datamonitor, revenues from molecular diagnostics reached US $ 2.6 billion in 2007. Global Industry Analysts expect revenues to increase to US $ 6 billion in 2015. The same source estimates revenues for biochips alone to reach US $ 3.4 billion in 2012. The highest portion is hold by DNA chips; the highest growth is seen by protein chips.

The Human Genome Project has had a lasting effect on the market for biochip products, both on its rapid growth and the high demand for these products. Since their introduction onto the market in the mid-1990s, biochips have revolutionized research within a short time by their capacity to collect and analyze enormous quantities of genomic data, largely automatically.

The field of **microarrays** has meanwhile developed beyond the original prototype of the DNA chip and today includes a variety of applications, such as protein arrays, antibody arrays, and even cell arrays. The typical DNA chip is nevertheless considered the most advanced from the point of view of market maturity and the use of microarrays in gene expression analysis has become established as the accepted standard method.

34.2.1.7 Systems Biology

Systems biology can also be seen as an important research direction in connection with the investigation of drug metabolism. Systems biology examines metabolic pathways that play a role in physiology and in disease (see Chapter 23). It is an interdisciplinary approach that aims for a comprehensive understanding of complex biological systems. It analyzes the complex interactions between genes, mRNAs, proteins, small molecules, and other elements in cells. It uses standardized data from "-omics" disciplines such as proteomics, genomics, etc., to develop predictive *in silico* (in computer) models, by means of mathematical and bioinformatic methods. In this way, system biology contributes to a better understanding of biological processes or regulatory networks, such as those that exist in cells.

An example of the commercial interest in systems biology is provided by Eli Lilly; the company invested over 5 years US $ 140 million in research in this field. The total world market for systems biology is estimated by Frost & Sullivan as being US $ 785 million in 2008.

34.2.2
Green Biotechnology

Green biotechnology is the application of biotechnology processes in **agriculture and food production** (see Chapter 32). The main dominant forces in green biotechnology today are agro giants with a world-wide area of operation such as BASF, Bayer CropScience, Monsanto, and Syngenta. They are concentrating considerable attention on molecular plant biotechnology, which is seen as a future growth factor in agroindustry. The traditional pesticide market, on the other hand, has been stagnating for years. A new field of application with a high growth potential is opening up for large companies through the use of new biological technologies, complementary to their previous activities. The substitution of traditional business segments is even a possibility as a result of modern green biotechnology.

34.2.2.1 Transgenic Plants

The main emphasis in modern plant biotechnology is the production of transgenic plants. The first use of gene technology to bring about changes in plants became possible at the beginning of the 1980s, around 10 years after the first experiment with bacteria. According to Global Industry Analysts, the market value of transgenic plants is estimated to be in excess of US $ 8 billion in 2010. According to the International Service for the Acquisition of Agri-biotech Applications (ISAAA), 125 million hectares are currently cultivated with transgenic plants world-wide, half of them in the United States.

A distinction is made in the genetic manipulation of plants between **input traits** and **output traits. Input traits** involve changing the agricultural characteristics of plants, offering the farmer technical advantages in cultivation. These include traits that affect the growth of the plant, such as herbicide or insect resistance, or tolerance to drought, cold, or lack of nutrients. In the United States, Canada, Argentina, and China, a large proportion of the harvest in cotton, corn, soybeans, and canola already depends on the use of transgenic plants, which demonstrate the corresponding cultivation advantages.

Output traits are the qualitative or quantitative improvement of characteristics relating to the condition of plants or the substances they contain. For example, attempts are being made to use gene technology to give plants and parts of plants a longer shelf life once they have been harvested (such as the famous genetically modified tomato). Other goals are to achieve a higher vitamin or protein content. Whereas input traits are of advantage only to the farmer, output traits aim to provide advantages that are of personal benefit to the end-consumer and offer improved processing quality to companies that carry out the further processing of the products. The aim in the latter case is to optimize the use of renewable raw materials.

Changing the agricultural and product qualities of plants is not the only goal in the production of transgenic plants. Another is **molecular pharming** (also sometimes called **gene farming** or **phytopharming**). Here, the plant is actually used as a **biofactory** for the production of biotherapeutics, diagnostic agents and other substances of interest.

34.2.2.2 Genomic Approaches in Green Biotechnology

Increasing use is being made of genomic approaches in green biotechnology. Knowledge about the function of the plant genome has a similar importance for the fields of seed breeding, agrochemicals, and food as for the pharmaceutical industry.

For seed breeding companies, the use of genomics leads, for example, to the much quicker development of varieties in comparison with conventional breeding. These companies were also the first to make use of genomic approaches in the nonpharmaceutical field. The information gained through the use of plant genomics enables not only an acceleration of plant breeding processes, but also the development of a greater range of seeds.

For the producers of agrochemicals, the possibilities offered by genomics are opening up new means of understanding the way plants function, or their metabolic processes, at the molecular level. This knowledge of molecular plant targets enables, among other things, the development of new kinds of herbicides. Completely new classes of products that work by means of entirely new mechanisms can be expected to come onto the market in this field. Here, too, a shortening of product development times and production costs can be expected.

34.2.2.3 Novel Food and Functional Food

New types of foodstuffs with novel properties are often called functional food. Another category that is often mentioned in this context is **nutraceuticals**. These are foods that (may) have a medicinal effect.

Functional foods are foods that have a higher vitamin content, for example, or that no longer contain certain undesired substances. The production of these foods stems mainly from the use of transgenic plants and is mainly carried on by large international groups of companies. According to Aroq, the market for functional food is estimated as US $ 167 billion world-wide in 2010.

34.2.2.4 Livestock Breeding

Modern biotechnology is being employed commercially to introduce novel performance features in **productive livestock**. The transgenic specimens then display, for example, different wool characteristics for sheep or improved milk characteristics for cattle. Intense efforts are put into the breeding of productive livestock races with accelerated growth by means of increased expression of growth hormones. The production of recombinant agents in animals and subsequent secretion in their milk is also being explored.

34.2.3
White and Gray Biotechnology

The term **white Biotechnology** has been coined for the application of biotechnological processes in industrial production contexts. The primary focus is the production of fine chemicals, in particular technical enzymes (see Chapter 33).

They can be found as proteases, lipases, cellulases, and amylases, such as in modern detergents, where they serve, amongst other purposes, as protein and fat solubilizers. The great demand for these enzymes is practically exclusively met by genetically modified producing strains, yielding resource savings and therewith cost savings that can reach dramatic proportions.

The availability of new enzymes can also improve chemical production. The chemical industry has developed numerous utilizing high concentrations, pressures, and temperatures that are carried out in organic solvents and that are afflicted with often severe environmental issues. In some cases, enzymes can be found that permit process steps under significantly milder conditions and in aqueous systems. Experts estimate revenues with white biotechnology to reach US $ 300 billion in 2015.

Gray biotechnology is considered as the application in environmental matters. Methods from molecular biology are primarily used for molecular **environmental diagnostics**. For instance, viruses and other pathogenic microorganisms are detected in water and the environment. In addition, harmful substances could be detected by transgenic plants. The public Danish company Aresa had developed RedDetect – a plant-based biosensor that changes color from green to red if located near land mines. Unfortunately the company went bancrupt. In contrast, direct applications for molecular biology in environmental protection (such as genetic modification of microorganisms for swifter degradation of pollutants) are to date commercially much less prominent. Molecular methods are also being used to determine whether genetically modified plants pass on their genes to natural species.

34.3
Status Quo of the Biotech Industry World-Wide

How are these commercial potentials actually implemented and turned into a (new) branch of industry? On the one hand, numerous small biotech companies have sprung up since the advent of industrial molecular biotechnology at the beginning of the 1970s; on the other hand, established medium-sized and large companies ("Big Pharma") are occupying themselves with this new technology. Appropriate statistics covering in particular the small and newly founded companies and their global distribution are listed below. All data applies to the situation in 2009 and has been exclusively extracted from the Biotech Reports of Ernst & Young.

34.3.1
Global Overview

In 2009, there were **over 4000 biotech companies** world-wide. Of these, **622 were listed stock companies**. Their turnover amounted to nearly US $ 80 billion, corresponding to an decrease of nearly 10% over 2008. Those companies registered on the stock market alone invested at least US $ 20 billion in research and development. The number of stock market registered companies as well as the number of employees declined. The people employed in the 622 companies numbered more than 175 000.

34.3.2
United States

With the successful development of pioneers Genentech, Amgen, and other companies, the United States has assumed a leading role in the world-wide biotech industry. This is not necessarily in terms of the number of biotech companies, since the number here, 1699, is smaller than that in Europe as a whole, which had 1790. However, the number of the 313 stock listed and thus financially strong companies, their number of employees, the turnover transacted, and the amount spent on research and development were all much higher in comparison to the adequate European companies; there were 109 100 employees, a turnover of nearly US $ 56.6 billion, and approaching US $ 17.2 billion of investment in research and development. The percentage of stock market registered companies at 18% (to Europe's 10%) means that the sector in the United States had greater maturity and better financing. The companies were more advanced and already had turnover-producing products on the market.

34.3.3
Europe

In Europe, the 171 biotech companies listed at a stock market had a total of nearly 50 000 employees, a turnover of nearly 12 billion Euros and an expenditure of roughly 3 billion Euros on research and development in 2008. This means that the US biotech industry far exceeds the equivalent European figures partly manyfold in terms of ratios such as employees, turnover, and research and development spending per company. The most mature biotech industry within Europe is that in the United Kingdom, which began to commercialize modern biotechnology by the foundation of new biotech companies back in the early to mid-1980s. France and Switzerland are also significant European players in the biotech industry. The biotech industry in Germany numbered according to Ernst & Young 387 companies with 9861 employees, including private companies. These companies achieved a turnover of 960 million Euros, and invested 746 million Euros in research and development. Germany had the greatest number of companies in Europe, but cannot compete with the United Kingdom in terms of other parameters, such as the number of companies listed on the stock exchange, and their number of employees, turnover, and investment in research and development.

35

Patents in the Molecular Biotechnology Industry: Legal and Ethical Issues

Learning Objectives

This chapter provides the reader with an overview of patenting in the pharmaceutical biotechnology industry, and summarizes some of the key legal and ethical issues related to the patenting of biomedical products and processes. It examines the legal aspects of patenting before considering ethical and policy issues. This chapter focuses primarily on US patent laws, which are very similar to European patent laws. Some differences between US and European laws are noted, and some relevant international intellectual property treaties are mentioned.

35.1 Patent Law

35.1.1
What is a Patent?

A patent is a type of **intellectual property**. All properties can be understood as collections rights to control a particular thing. Tangible properties give the property holder rights to control tangible things, such as cars or land. Intellectual properties, on the other hand, give the property holder rights to control intangible things, such as inventions, poems, or computer programs. **Tangible things** have a particular location in space and time; **intangible things** do not. The main types of intellectual property are **patents**, **copyrights**, **trademarks**, and **trade secrets**.

A **patent** is a private right granted by the government to someone who creates an invention. The patent gives the inventor the right to exclude others from making, using, or commercializing the invention. A patent may be awarded to more than one person. Today, most patents are awarded to groups of researchers, who are listed as coinventors. Once a patent is granted, the rights may be transferred, licensed, or assigned to other parties. Academic and industrial researchers usually assign their patent rights to their employer and receive a share of royalties. The employer then becomes the patent holder. Patent holders may also grant **licenses** to other parties in exchange for royalties or a fee. For example, a biotechnology company with a patent on a gene therapy technique could grant individuals or companies licenses to use the technique.

In the United States, a patent holder has the right to refrain from making, using, or licensing his/her invention. There, a patent confers rights to make, use, or commercialize a thing, but implies no corresponding obligations. As a result, some companies in the United States use patents to **block technological development** and gain an advantage over competitors. Some European countries, however, have **compulsory licensing**, which requires the patent holder to make, use, or commercialize his/her invention or license others to do so.

The **term of patent** in the United States and countries that belong to the European Union lasts 20 years from the time the inventor submits his application. A patent is not renewable. Once the patent expires, the invention becomes part of

An Introduction to Molecular Biotechnology, 2nd Edition.
Edited by Michael Wink
Copyright © 2011 WILEY-VCH Verlag GmbH & Co. KGaA, Weinheim
ISBN: 978-3-527-32637-2

the **public domain** and anyone can make, use, or commercialize the invention without permission from the inventor. In the pharmaceutical industry the average interval between discovery of a new drug and its final approval by the relevant regulatory agency is 10 years, which includes the time required to conduct clinical research, product development, as well as regulatory review. Thus, most pharmaceutical companies can expect that they will have about 10 years to recoup the money they have invested in a new drug before the patent expires. Once the patent expires, the name of the drug may still have trademark protection, but other companies can manufacture and market a generic version of the drug without obtaining permission from the company.

The main policy rationale for patent laws is that they promote the progress of science, technology, and industry by providing financial incentives for inventors, entrepreneurs, and investors. By granting **property rights** over inventions, the patent system gives inventors and research sponsors the opportunity to profit from their investments of time and money in research and development. Additionally, society benefits because the patent application becomes part of the public domain once the patent is granted, which gives other researchers the opportunity to learn from the invention and use the knowledge contained in the application. The agreement to grant patents rights in exchange for public disclosure is known as the **patent bargain**. The public benefits from this bargain because it encourages inventors share information instead of attempting to protect it through **trade secrecy**. A great deal of the world's scientific and technical information is disclosed in patent applications.

Although there is widespread agreement that patents benefit society, there is a dispute about whether some patenting practices and policies can have adverse effects. Some have argued that patents can actually inhibit innovation and discovery by discouraging researchers from sharing information and technology. (For further discussion, see Section 35.2.3). Since excessive private ownership of inventions can have negative consequences, patent laws, government agencies, and the courts attempt to strike an appropriate balance between public and private control of inventions. A good example of this balancing is the term of a patent – if the term is too short, companies and researchers will not have enough time to obtain a fair return on their investment; if the term is too long, the public will not have adequate access to technology.

35.1.2
How Does One Obtain a Patent?

To obtain a patent, one must submit a **patent application** to the patent office. In the United States, the **Patent and Trademark Office (PTO)** examines patent applications. The application must provide a description of the invention that would allow someone trained in the relevant practical art to make and use the invention. One or more individuals may be listed as **inventors** on the patent application. The application need not include a sample or model of the invention; a written description will suffice. The application will contain information about the invention, background references, data, as well as one or more claims pertaining to the invention. The claims stated on the patent application will determine the scope of the patent rights.

If the PTO rejects a patent application, the inventor may submit a revised application. The process of submission/revision/resubmission, otherwise known as **prosecuting** a patent, may continue for months or even years. If the PTO rejects the patent, the applicant may appeal the decision to a federal court. If the PTO accepts the patent, a **competitor** may still file a lawsuit challenging the PTO's decision. The PTO will award a patent to an inventor only if he/she provides evidence that his/her invention satisfies all of the following conditions (European Union countries have similar requirements):

1. **Originality.** The invention is new and original – it has not been previously disclosed in the prior art. The rationale for this condition is that the public does not benefit when the patent office grants a patent on something that has already been invented. Thus, if someone else has already patented the same invention, this would qualify as a prior disclosure. Also, disclosure could occur if a significant part of the invention has been published or used in public.

2. **Nonobviousness.** The invention is not obvious to someone who is trained in the relevant practical art. Prior disclosure of parts of the invention or similar inventions in the literature can undermine nonobviousness claims. The justification for this requirement is that the public does not benefit from granting obvious inventions.

3. **Usefulness.** The invention has some definite, practical utility. The utility of the invention should not be merely hypothetical, abstract, or contrived. A patent is not a fishing license. The rationale for this condition is self-explanatory: the public does not benefit from useless patents. In the late 1990s, the PTO raised the bar for proving the utility of patents on DNA in response to concerns that it was granting patents on DNA sequences when the inventors did not even know the biological functions of those sequences.

In addition to satisfying these three conditions, to obtain a patent in the United States, the inventor must exhibit due diligence in submitting an application and developing the invention. In the United States, the person who is the first to conceive of an invention will be awarded the patent unless he does not exhibit **due diligence**. If the first inventor does not exhibit due diligence, the PTO may award the patent to a second inventor, if that inventor reduces the invention to practice and submits an application before the first inventor. In European countries and many other nations, the patent goes to the first person to submit a complete and valid application, not to the first person to conceive of the invention.

35.1.3
What is the Proper Subject Matter for a Patent?

Under United States law, the PTO can award patents on articles of manufacture, compositions of matter, machines, or techniques or improvements thereof. EU countries allow patents on similar types of things. Although different patent laws use different terms to describe the subject matter of patents, there are three basic types of patents: patents on **products** (or materials), patents on **processes** (or methods), and patents on **improvements**. For example, one could patent a mousetrap (a product), a method for making a mousetrap (a process), or a more efficient and humane mousetrap (an improvement).

One of the most important doctrines in patent law is that patents only apply to inventions that result from **human ingenuity** (or inventiveness). Thus, US courts have held that one may not patent **laws of nature or natural phenomena**, since these would be patents on products of nature. Nearly three decades ago, a landmark US Supreme Court case, *Diamond v. Chakrabarty*, set the legal precedent in the United States for patents on life forms. Chakrabarty had used recombinant DNA techniques to create a type of bacteria that metabolizes crude oil. The PTO had rejected his patent application on the grounds that the bacteria did not result from human ingenuity, but the Supreme Court vacated this ruling and held that Chakrabarty could patent his genetically engineered life form. This decision helped to establish the legal precedent for other **patents on life forms**, such as patents on laboratory animals, livestock, and plants. EU countries have followed the United States in allowing patents on life forms that result from human ingenuity. Patents have also been granted on parts of living things, such as cell lines, tissues, bioengineered organs, **DNA**, and **proteins**, as well as biological and biochemical processes, such as **cloning** and **recombinant DNA techniques**.

In granting patents on organic compounds that occur in living organisms, such as animals or plants, patent agencies have distinguished between naturally occurring compounds and isolated and purified compounds. For example, DNA in its natural state occurs in virtually all organisms and is unpatentable in its natural state. However, scientists can use various chemical and biological techniques to create isolated and purified samples of DNA, which are patentable. Patents right apply to isolated and purified forms of DNA, not to DNA as it occurs naturally in animals, plants, or people. The reason why patent agencies allow patents on isolated and purified organic compounds is that they have determined that these products result from human ingenuity.

Another important doctrine in patent law is that patents apply to applications, not to ideas. Ideas are part of the public domain. For example, courts in the United States have ruled that mathematical algorithms are unpatentable ideas but that computer programs that use algorithms to perform practical functions are patentable. Courts have also held that scientific laws and formulas are not patentable, although applications of laws or formulas are patentable.

35.1.4
Types of Patents in Pharmaceutical and Molecular Biotechnology

There are many different types of patents that may be available to researchers and companies in the field pharmaceutical biotechnology. Following the distinction in Section 35.1.3 between products and processes, potential patents might include:

1. Patents on **pharmaceutical and biomedical products**, such as bioengineered drugs, proteins, receptors, neurotransmitters, oligonucleotides, hormones, genes, DNA, DNA microchips, RNA, cell lines, bioengineered tissues and organs, and genetically modified bacteria, viruses, animals, and plants.
2. Patents on **pharmaceutical and biotechnological processes**, such as methods for genetic testing, gene therapy procedures, DNA cloning techniques, methods for culturing cells and tissues, DNA and RNA sequencing methods, and xenotransplantation procedures.
3. Patents on **improvements** of pharmaceutical, biomedical, and biotechnological products and processes.

For any of these products, processes or improvements to be patentable, they would need to result from **human ingenuity**.

35.1.5
Patent Infringement

Patent infringement occurs when someone uses, makes, or commercializes an invention without **permission of the patent holder**. In the United States, the patent holder has the responsibility of bringing an infringement claim against a potential infringer and proving that infringement occurred. A court may issue an injunction to stop the infringement or award the patent holder damages for loss of income due to infringement. There are three types of infringement: **direct infringement**, **indirect infringement**, and **contributory infringement**. Patent holders may also settle infringement claims out of court. Researchers, corporations, and universities usually try to avoid any involvement in an infringement lawsuit, since patent infringement litigation is expensive and time-consuming.

Many EU countries have a defense to patent infringement known as the **research exemption**. The United States also has a research exemption (also known as the experimental use exemption), which has been used very infrequently. Under this exemption, someone who uses or makes a patented invention for pure research with no commercial intent can assert this defense in an infringement lawsuit to avoid an adverse legal decision. The research exemption is similar to the *fair use* exemption in copyright law in so far as it permits some unconsented uses of intellectual property. There are some problems with the exemption, however. First, the research exemption is not well publicized. Second, the research exemption is not well defined. Indeed, in the United States the research exemption has no statutory basis but is a creation of case law. Some commentators have argued that countries should clarify and strengthen the research exemption in order to promote research and innovation in biotechnology, and avoid excessive private control of inventions.

For many years academic researchers in the United States assumed that they were protected by the research exemption, but the landmark case in 2002 of *Madey v. Duke University* invalidated this assumption. The court ruled that Duke University had infringed Madey's patents when it continued to use a laser that he developed after he left Duke for another institution. Duke claimed that the research exemption applied to its use of the laser, but the court held that the laser helped to promote Duke's economic and commercial interests, such as the recruitment of students, the acquisition of grants and contracts, and the development of technology. Duke appealed the case to the US Supreme Court, but the Supreme Court declined to hear the case. Unless legislators amend the patent statutes in the United States to include a research exemption, the research exemption will have a very limited application there.

35.1.6
International Patent Law

Every country has the authority to make and enforce its own patent laws and to award its own patents. Thus, a patent holder must apply for a patent in every country in which he wants patent protection. For example, a corporation that patents a new drug in the United States must also apply for a patent in Germany, if it desires patent protection in Germany. Furthermore, complex matters relating to jurisdiction can arise when someone infringes a patent that is protected in one country but not another. For example, if someone infringes a US patent in Germany, but the invention is not protected by German patent laws, then the patent holder will need to bring a lawsuit in a court in the United States, which may or may not have jurisdiction.

To deal with international disputes about intellectual property and to harmonize intellectual property laws, many countries have signed **intellectual property treaties**. Most of these treaties define minimum standards for intellectual property protection and obligate signatories to cooperate in the international enforcement of property rights. The most important treaty related to patents is the **Trade Related Aspects of Intellectual Property agreement (TRIPS)**, which has been developed and negotiated by the **World Trade Organization (WTO)**. The TRIPS agreement defines minimum standards for patent rights. For example, it requires that patents last 20 years. Countries that have signed the agreement agree to adopt patent laws that provide at least the minimum level of protection under the agreement. Countries must also agree to cooperate in the enforcement of patent rights. TRIPS allows countries to override patents rights to deal with national emergencies, such as public health crises. TRIPS has been revised numerous times, most recently in 2008.

35.2
Ethical and Policy Issues in Biotechnology Patents

Having provided the reader with some background information on patenting in biotechnology, this section will briefly review some important ethical and policy issues.

35.2.1
No Patents on Nature

In the 1990s, a variety of writers, political activists, theologians, ethicists, and professional organizations opposed patents on biotechnological products and processes for a variety of reasons. Many of these critics argued that patents on living bodies, as well as patents on body parts, are **unethical** because they are patents on natural things. They argued that it is immoral and ought to be illegal to patent organisms, tissues, DNA, proteins, and other biological materials. Some of these critics based their opposition to biotechnology patents on religious convictions, while others based their opposition on a general distrust of biotechnology and the biotechnology industry. Some of the more thoughtful critics of biotechnology patents accepted some types of patents on biological materials, but objected to patents on other types of biological materials, such as patents on **genes** or **cell lines**, on the grounds that these types of patents attempt to patent nature.

A lawsuit filed in 2009 in the United States, *Association for Molecular Pathology et al. v. US Patent and Trademark Office et al.*, challenges Myriad Genetics' BRCA1 and BRCA2 gene patents claims on the grounds that these genes are products of nature and cannot be patented. The lawsuit also alleges that Myriad violated the defendants' **freedom of speech** by using its patent rights to restrict research by competitors. This lawsuit has not been decided as of the time of writing this chapter.

As noted in Section 35.1.3, patents on products of nature are not legally valid – a product or process must have resulted from human ingenuity to be patentable. However, how much **human ingenuity** should be required to transform something from an unpatentable product of nature to a patentable, human invention? Defining the boundaries between products of nature and human inventions is a fundamental issue in patent law and policy that parallels the tenuous distinction between natural and artificial. While most people can agree on paradigmatic cases of things that are natural, such as gold, and things that are artificial, such as gold jewelry, it is difficult to reach agreement on borderline cases, such as **DNA sequences**. On the one hand, DNA sequences exist in nature and can therefore be regarded as natural. On the other hand, isolated and purified DNA sequences do not exist in nature and are produced only under laboratory conditions. They are, in some sense, **human artifacts**. However, the nucleotide sequences in isolated and purified DNA are virtually identical to the sequences in naturally occurring DNA. There is probably no objective (i.e., scientific) basis for distinguishing between naturally occurring DNA and isolated and purified DNA. Likewise, there is probably no objective basis for distinctions between natural cell lines versus artificial cell lines, natural proteins versus artificial proteins, and natural organisms versus artificial organisms.

If the distinction between a product of nature and a human invention is not objective, then it depends, in large part, on human values and interests. It is like other controversial distinctions in biomedical law and ethics, such as human versus nonhuman and alive versus dead. The best way to deal with these **controversial distinctions** is to carefully consider, negotiate, and balance competing values and interests in light of the particular facts and circumstances. Laws and policies that define patentable subject matter should also attempt promote an optimal balance between competing interests and values and should carefully consider the facts and circumstances relating to each item of technology. Policies adopted by the United States and the European Union with respect to the patenting of DNA appear to strike an optimal balance between competing interests and values because these policies disallow the patenting of DNA in its natural state but allow the patenting of isolated and purified DNA.

35.2.2
Threats to Human Dignity

Critics of biotechnology patents also have claimed that patents on human body parts, such as genes, cell lines, and DNA, are **unethical** because they treat people as **marketable commodities**. Some have even compared patents on human genes to slavery. The issues concerning the commercialization of human body parts are complex and emotionally charged. They also have implications for many different social policies, including organ transplantation, surrogate parenting, and prenatal genetic testing. This section gives only give brief overview of this debate.

According to several different ethical theories, including Kantianism and the Judeo-Christian tradition, human beings have intrinsic moral value (or dignity) and should not be treated as if they have only extrinsic value. An entity (or thing) has intrinsic value if it is valuable for its own sake and not merely for the sake of some other thing. A commodity is a thing that has a value – a market value or price – which serves as a basis for exchanging it for some other thing. For example, one exchanges a barrel of oil for US$ 70 or exchanges a visit to the dentist for US$ 50. Treating an entity as a commodity is treating it as if it has only extrinsic value and not intrinsic value. Thus, it would be unethical to treat a **human being as a commodity** because this would be treating that person as if they have only extrinsic value and no intrinsic value. One reason why slavery is unethical is that it involves the buying and selling of whole human beings. People are not property.

Even though treating a whole human being as a commodity violates human dignity, one might argue that treating a human body part as a commodity does not violate human dignity. Human beings have billions of different body parts, ranging from DNA, RNA, proteins, and lipids, to membranes, organelles, cells, tissues, and organs. Properties that we ascribe to the parts of a thing do not necessarily transfer to the whole thing; inferences from parts to wholes are logically invalid. For example, the fact that a part of a automobile, such as the front tire, is made of rubber does not imply that the whole car is made of rubber. Likewise, treatment of a part of human being, such as blood or hair, as a commodity does not imply treatment of the whole human being as part. It is possible to commodify (or commercialize) a human body part without commodifying the whole human being.

This argument proves that buying and selling hair, blood, or even a kidney is not equivalent to slavery. Even so, one might argue that treating human body parts as commodities constitutes incomplete commodification of human beings, and that partial commodification of human beings can threaten human dignity even if it does not violate human dignity. Incomplete commodification can threaten human dignity because it can lead to exploitation, harm, and injustice, as well as complete commodification of human beings. For example, in the now famous case filed in 1990 of *Moore v. Regents of University of California*, the desire to patent a valuable cell line played an important role in the exploitation of a **cancer patient**. The researchers took cells from Moore's body that overexpress cytokines. The researchers did not tell Moore what they planned to do with tissue samples they took from him or that the samples could be worth millions of dollars. One might argue that treating human body parts as commodities inevitably leads to abuses of human rights and dignity as occurred in the Moore case. Although incomplete commodification of human beings is not intrinsically immoral, it can lead society down a slippery slope toward various types of immorality and injustices. In order to prevent this, society should forbid activities that constitute incomplete commodification of human beings, such as the patenting of cell lines and DNA, a market in human organs, surrogate pregnancy contracts, cloning for reproduction, and selling human gametes.

One could reply to this argument by acknowledging that the slippery slope poses a genuine threat to **human dignity** but maintain that it may be possible to prevent exploitation, injustice, and other abuses by developing clear and comprehensive regulations on practices that commodify human body parts. Regulations should require informed consent to tissue donation, gamete donation, and organ donation, as well as fair compensation for subjects that contribute biological materials to research and product development activities. Regulations should also protect the welfare and privacy human research subjects and patients. These regulations should also state that some human biological materials, such as embryos, should not be treated as commodities because treating these materials as commodities poses an especially worrisome threat to human dignity. Although an embryo is not a human being, it should be illegal to buy, sell, or patent a human embryo. However, one could argue that it should be legal to buy or patent **embryonic stem cells**, provided that society has appropriate regulations.

Patents on human embryonic stem cell lines have generated considerable controversy. The United States and United Kingdom have awarded patents on human embryonic stem cells, but not all countries have. Ireland, Italy, and Germany, for example, do not award patents on human embryonic stem cells. Canada and Denmark allow patents on human embryonic stem cells line derived from leftover embryos, but not from nuclear transfer. The patenting of human body parts has been controversial in Europe for many years. In 1998, the European Union adopted a Biotechnology Directive, which states that patents will not be allowed on inventions that are contrary to public morality. The **European Patent Office (EPO)** has recently ruled that patents cannot be granted on human embryonic stem cells lines whose derivation involves the destruction of embryos. Although the EPO does not have jurisdiction over any particular European nation, it develops uniform patent procedures for 38 European countries.

35.2.3
Problems with Access to Technology

One of the most important ethical and policy concerns raised by critics of biotechnology patenting is that patenting will have an adverse impact on access to information, materials, and methods that are vital to research and innovation in biotechnology as well as medical tests and treatments. In Section 35.1.1 we noted that the primary rational for the patent system is that it benefits society by encouraging progress in science, technology, and industry. However, this argument loses it force when patenting has the opposite effect. The issue of **access to data and materials** in biotechnology, like the issues discussed in Sections 35.2.1 and 35.2.2, is very complex and controversial. This chapter does not explore these issues in great depth, but attempts to provide the reader with an outline of the arguments on both sides.

Critics of patenting have argued that patents can interfere with innovation and discovery in biotechnology and biomedicine in a variety of ways. Although early criticisms of patenting were speculative, recent criticisms have been based on empirical studies, biomedical research, and intellectual property practices, such as surveys, interviews, and analyses of patenting trends. The major criticisms are as follows.

First, patenting can undermine the sharing of data and materials that is vital to academic research. Researchers may unwilling to share data and materials because they (and their employers) may want to protect their intellectual property rights. As noted earlier, disclosure of an invention prior to the filing of a patent application can invalidate a potential patent. Scientists conducting research on a patentable subject matter may refrain from sharing anything related to their research until they obtain patent protection. Even when scientists, companies, or universities are willing to share data and materials, they usually require recipients to sign **material transfer agreements (MTAs)**, which are legal documents that require time and resources to negotiate. MTAs may also contain restrictions on the use of data or materials.

Second, legal and administrative difficulties related to the licensing of patented inventions can interfere with research. If a researcher (or company) wants to develop a new product or process in biotechnology and biomedicine, he/she may need to negotiate and obtain dozens of different licenses from various patent holders in order to avoid patent infringement. The researcher or company might need to maneuver through a **patent thicket** in order to develop a new and useful invention. For example, DNA chip devices test for thousands of different genes in one assay. If dozens of companies hold patents on these different genes, then one may need to obtain dozens of different licenses to develop this new product. Although larger biotechnology and pharmaceutical companies are prepared to absorb the legal and administrative transaction costs associated with licensing, smaller companies and universities may find it difficult to maneuver through the patent thicket.

In some cases, companies may be unwilling to negotiate licenses because they want to use their patent rights to gain an advantage over competitors. For example, the lawsuit against Myriad Genetics (see Section 35.2.1) alleges that the company has used its patents to stifle competition.

In industries with many different interdependent products and processes, someone who holds a particular invention may be able to influence the development of subsequent inventions that depend on that prior invention. These prior inventions are also known as **upstream inventions** and the subsequent inventions are also known as **downstream inventions**. Some companies may obtain patents for the sole purpose of preventing competitors from developing useful inventions in biotechnology. In the United States, these companies would have no obligation to use, make, market, or licenses such inventions.

Third, **high licensing fees** could impose a heavy toll on research and innovation in biotechnology and biomedicine. Companies with patents on upstream inventions might issue **reach-through licenses** to capture a percentage of profits from downstream inventions. While downstream patent holders have no legal obligation to share their profits with upstream patent holders, downstream patent holders might negotiate with upstream patent holders to avoid costly patent litigation. Even companies that do not issue reach through licenses may still set high licensing fees. For example, many commentators have claimed that Myriad Genetics' high licensing fees for its tests for BRCA1 and BRCA2 mutations, which increase the risk of breast and ovarian cancer, have had a negative impact on research and innovation, and diagnostic and predictive testing.

These aforementioned **problems related to licensing** could undermine not only research and innovation, but could also have an adverse impact on health care by undermining access to new medical products and services, such as genetic tests. For example, if a company is unable to develop a genetic test, due to licensing problems, then patients will not benefit from that test. If a company develops a genetic test but charges a high fee to conduct the test or charges a high fee to license the test, then many patients may not be able to afford the test. In either case, problems related to the licensing of biotechnology products and processes could prevent the public from benefiting from new developments in biomedicine.

Many commentators and industry leaders have rebutted these criticisms of biotechnology patenting by arguing that problems are not as bad as critics suggest, and that there are mechanisms in place to overcome these problems. They have also conducted empirical studies to support their claims.

First, proponents of biotechnology patenting have argued that many of the problems with the sharing of data and materials methods have nothing to do with patenting *per se*, but are the result of practical difficulties. It takes time and money to share data and materials with other researchers. Researchers may not want to devote a large portion of their time and resources to sharing data and materials with other scientists. The administrative issues related to negotiating MTAs have more to do with the practicalities related to collaborative research than patenting. MTAs can be difficult for researchers to deal with even when no patent rights are involved.

Second, there a variety of ways of dealing with the licensing issues. Academic researchers often find ways of working around patented technologies or negotiating licenses with patent holders. In some case, academic researchers ignore patent protections and use the technology without a license. Although this practice may constitute patent infringement, so far patent holders have, for the most part, refrained from suing academic researchers. Private companies usually do not have any major difficulties negotiating and obtaining licenses because they understand the importance of cooperation in the biotechnology industry. Few companies use patents to block competitors because this strategy is usually prove unprofitable – company can make much more money from marketing or licensing a new invention than from keeping it on the shelf. Finally, high licensing costs are likely to decline in response to lower consumer demands, especially if competitors are able to enter the market by developing new inventions that work around existing ones (a **work-around invention** is an improvement on a patented invention or an alternative to a patented invention).

Industry leaders also point out that the potential licensing problems faced by the biotechnology industry are entirely new because many other industries have faced, and solved, similar problems. For example, many different companies in the semiconductor industry have worked together to develop licensing agreements. There are many interdependent products and processes in the semiconductor industry and many different patent holders, but companies have managed to avoid licensing problems and the industry has thrived. Indeed, the semiconductor industry is one of the most successful and innovative industries the world has ever known.

Commentators on both sides of this issue have argued that societies should **reform the patent system** to prevent licensing problems from occurring and to ensure that new biomedical technologies are affordable and accessible. These proposed reforms, some of which have been mentioned above, include the following:

1. Expanding and clarifying the research exemption in biotechnology.
2. Making it more difficult to obtain a patent by raising the bar on criteria, such as novelty, nonobviousness, and utility.
3. Restricting the scope of biotechnology patents in order to allow for work around inventions and to promote competition.
4. Applying antitrust laws to the biotechnology industry to promote fair competition.
5. Conducting an ethical review of patent applications to address ethical and policy issues before awarding patents.
6. Developing a patent pool in the biotechnology industry to promote efficient licensing.

Most of these proposed reforms would probably promote research and innovation in biotechnology and biomedicine without undermining financial incentives for researchers and companies. Many of these reforms could be enacted without any additional legislation, since patent offices and the courts already have a great deal of authority to shape patent law and policy through their interpretation and application of existing statutes.

35.2.4
Benefit Sharing

The final issue this chapter will consider involves the sharing of the **benefits** of research and innovation in biotechnology. Some critics of biotechnology patents have claimed that the distribution of the benefits of research and innovation is often unfair. According to these critics, pharmaceutical and biotechnology companies benefit greatly from research and innovation by earning large profits, but individual patients or research subjects, populations, or communities benefit very little. For example, to study a genetic disease, researchers need to take tissue samples from patients or subjects. Very often, researchers do not offer to pay subjects any money for their tissue samples or promise them any royalties from the commercialization of their research or its applications. If a company develops a profitable genetic test from free genetic samples, patients or subjects could argue that the company is not sharing benefits fairly. **Unequal distributions** of benefits could also occur between companies and entire communities or countries. For example, some pharmaceutical and biotechnology companies are now developing drugs based on knowledge obtained from indigenous populations concerning their medicinal plants. If a company develops a profitable medication from this indigenous knowledge and does not offer the population any compensation, the population could argue the company has not shared the benefits of research fairly. Unequal distributions of benefits could also take place between developed nations and developing nations. For example, if researchers, patients, and companies from the developed world benefit a great deal from biotechnology, but people in the developing world do not, one might argue the benefits of biotechnology have been distributed unfairly.

Several commentators and organizations have called for the **fair distribution** of the benefits of research in biotechnology. Some appeal directly to theories of justices, such as utilitarianism, egalitarianism, or social contract theory, to argue for a fair distribution of research benefits. Others appeal to the concept of a common heritage relating to human biological materials, such as DNA. Regardless of how one justifies a general principle of benefit sharing in biotechnology, the most important **practical problems** involve determining how benefits should be shared. What would be a fair sharing of benefits between researchers and companies and subjects/populations/communities? Should researchers and companies offer to give subjects/populations/communities financial compensation for providing research materials and methods, such as tissue samples of indigenous knowledge? Should researchers and companies offer to pay royalties for the commercialization of research to subjects/populations/communities? Although financial compensation might be useful and appropriate in some situations, such as giving communities royalties for indigenous knowledge or providing some subjects with compensation for their valuable tissues (as in the Moore case, discussed in Section 35.2.2), in other situations direct financial compensation may not be very useful or appropriate. For example, if a company collects thousands of tissue samples from subjects and uses knowledge gained from those samples to develop a commercial product, the financial benefit offered to any particular subject might be miniscule, since the benefits would need to be divided among thousands of subjects. Moreover, it may be impossible to estimate the potential benefits to subjects prior to the development of the product, since most new products are not profitable. Furthermore, subjects in some cultures might not be interested in financial rewards for participation. Perhaps the best way to share benefits in situations like these would be to offer to provide the population or community with nonfinancial benefits, such as improvements in health care, education, or infrastructure. In any case, these are complex questions that cannot be addressed in depth in this chapter. To answer questions about the fair distribution of research benefits in any particular case, one needs to apply **theories and concepts of distributive justice**.

Even though there is little consensus about the how to distribute the benefits of research and innovation in biotechnology, almost everyone with an interest in the issue agrees that subjects should be informed about plans for benefit sharing (if there are any). For example, the researchers in the Moore case should have told Moore that they planned to develop a cell line from his tissue and that they were not planning to offer him any financial compensation. If researchers conduct a study that involves an entire population or community, they should discuss benefit sharing plans with representatives of the community or population. Indeed, respect for human dignity requires nothing less than fully informing subjects of the material facts related to their research participation, including facts pertaining to the commercialization of research.

35.3
Conclusions

This chapter has provided the reader with an overview of the **legal**, **ethical**, **and policy issues** relating to the **patenting of products and processes** used in pharmaceutical and molecular biotechnology. Although it has attempted to provide the reader with up-to-date information, it is possible that some of this information may soon be out of date, due to changes in technology, case law, legislation, and international treaties. Since most of these issues are very complex and constantly changing, those who are interested in learning more about this topic should review the relevant documents, guidelines, and policies relating to their particular areas of research and development.

Acknowledgments   This research was supported, in part, by the intramural program of the National Institute of Environmental Health Sciences (NIEHS), National Institutes of Health (NIH). It does not represent the views of the NIEHS or NIH.

36
Drug Approval in the European Union and United States

Learning Objectives

This chapter provides the reader with an overview of drug approval procedures within the European Union and the United States, with particular focus upon products of biotechnology. Upon its completion, the reader will be familiar with: (1) the EU regulatory framework, as underpinned by EU regulations and directives, and as managed by the European Medicines Agency; (2) the regulatory approvals route for products of pharmaceutical biotechnology within the United States, as underpinned by the Food, Drug, and Cosmetics Act, and as enforced by the US Food and Drug Administration; and (3) the ongoing process of international pharmaceutical regulatory harmonization (i.e., the International Conference on Harmonization process).

36.1
Introduction

The **pharmaceutical sector is arguably the most highly regulated industry** in existence. Legislators in virtually all world regions continue to enact and update legislation controlling every aspect of pharmaceutical activity. Interpretation, implementation, and enforcement of these laws is generally delegated by the lawmakers to dedicated agencies. The relevant agencies within the European Union and the United States are the **European Medicines Agency (EMA)** and the **US Food and Drug Administration (FDA)**, respectively. Here, we focus upon the structure, remit, and operation of both these organizations, specifically in the context of biopharmaceutical products. A more comprehensive overview of the subject matter may be obtained from Tobin and Walsh (2008).

36.2
Regulation within the European Union

36.2.1
EU Regulatory Framework

The founding principles of what we now call the European Union are enshrined in the Treaty of Rome, initially adopted by six countries in 1957. While this treaty committed its signatories to a range of cooperation and harmonization measures, it largely deferred healthcare-related issues to individual member states. As a consequence, each member state drafted and adopted its own set of pharmaceutical laws, enforced by its own **national regulatory authority**. Although the main principles underpinning elements of national legislation were substantially similar throughout all European countries, details did differ from country to country. As a result, pharmaceutical companies seeking product marketing authorizations were forced to apply separately to each member state. Uniformity of regulatory response was not guaranteed and each country enforced its own

An Introduction to Molecular Biotechnology, 2nd Edition.
Edited by Michael Wink
Copyright © 2011 WILEY-VCH Verlag GmbH & Co. KGaA, Weinheim
ISBN: 978-3-527-32637-2

Table 36.1 Volumes comprising the rules governing medicinal products within the European Union.

Volume	Title
1	Pharmaceutical Legislation: Medicinal Products for Human Use
2	Notice to Applicants: Medicinal Products for Human Use
3	Guidelines: Medicinal Products for Human Use
4	Good Manufacturing Practices: Medicinal Products for Human and Veterinary Use
5	Pharmaceutical legislation: Veterinary Medicinal Products
6	Notice to Applicants: Veterinary Medicinal Products
7	Guidelines: Veterinary Medicinal Products
8	Maximum Residue Limits: Veterinary Medicinal Products
9	Pharmacovigilance: Medicinal Products for Human and Veterinary Use
10	Clinical Trials Guidelines

language requirements, scale of fees, processing times, etc. This approach created enormous duplication of effort, for companies and regulators alike.

In response, the **European Commission** (Brussels) began a determined effort to introduce European-wide pharmaceutical legislation in the mid-1980s. The commission represents the EU body with responsibility for drafting (and subsequently ensuring the implementation of) EU law, including **pharmaceutical law**. In pursuing this objective it has at its disposal two legal instruments: **regulations** and **directives**. Upon approval, a regulation must be enforced immediately and without alteration by all EU member states. A directive, in contrast, is a softer legal instrument, requiring member states only to introduce its essence or spirit into national law.

By the early 1990s some eight regulations and 18 directives had been introduced, which effectively **harmonized pharmaceutical law** throughout the European Union. In addition to making available the legislative text, the European commission has also facilitated the preparation and publication of several adjunct documents designed to assist industry and other interested parties to interpret and conform to the legislative requirements. Collectively these documents are known as the **rules governing medicinal products in the European Union** and they make compulsory reading for those involved in any aspect of pharmaceutical regulation. The 10-volume (Table 36.1) publication is regularly updated and hard copies may be purchased from the Commission's publication office (http://www.publications.eu.int) or may be consulted/downloaded (for free) from the relevant EU website (http://ec.europa.eu/enterprise/pharmaceuticals/eudralex/eudralex_en.htm).

36.2.2
EMA

Harmonization of pharmaceutical law made possible the implementation of an EU-wide system for the authorization and subsequent supervision of medicinal products. The **European Medicines Evaluation Agency** (EMA) (now renamed the **European Medicines Agency**, which is still abbreviated to EMA) was set up to coordinate and manage the new system (http://www.ema.europa.eu). Based in Canary Wharf, London, the agency became operational in 1995. The EMA's main responsibility is the protection and promotion of **human and animal health** within the European Union through the evaluation and supervision of medicines for human and veterinary use. It seeks to achieve this by:

- Providing high-quality evaluation of medicinal products.
- Advising on relevant research and development programs.
- Providing a source of drug and other relevant information to healthcare professionals/users.
- Controlling the safety of medicines for humans and animals.

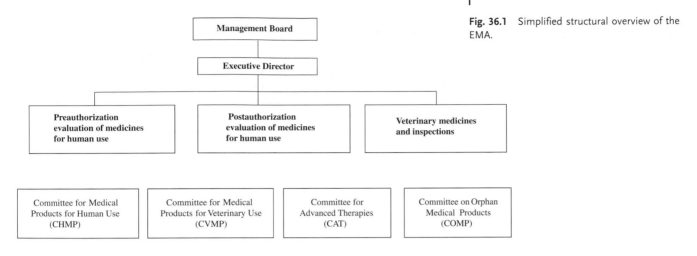

Fig. 36.1 Simplified structural overview of the EMA.

An outline structure of the EMA is provided in Fig. 36.1. From a technical standpoint, the most significant **organizational structures** are:
- The unit for preauthorization evaluation of medicines for human use.
- The unit for postauthorization evaluation of medicines for human use.
- The unit for veterinary medicines and inspections.

A more detailed description of these units and their responsibilities is available on the EMA homepage. Two additional structural units also exist: administration, and communications and networking. From a biotechnology drug approval perspective, four key scientific committees are at the core of the functioning of the EMA:
- Committee for Medicinal Products for Human Use (CHMP).
- Committee for Medicinal Products for Veterinary Use (CVMP).
- Committee for Advanced Therapies (CAT).
- Committee on Orphan Medicinal Products (COMP).

Each committee is composed of a number of (mainly technical) experts, the majority of whom are drawn from the national drug regulatory authorities of each EU member state. The function of the CHMP and CVMP in the context of new biotechnology drug approvals will be discussed in the next section. CAT serves mainly to assist the CHMP in the assessment of certain specific biotech product types, most notably gene and cell/tissue-based therapies, while the COMP's main role is to review applications seeking orphan status for products in development. Orphan products are those intended for diagnosis or treatment of rare diseases and as such can benefit from various technical and financial breaks within the regulatory system. In addition to these four committees, the EMA has at its disposal a bank of some 4500 **European technical experts** (the majority of whom, again, are drawn from the national regulatory and related authorities). The EMA draws upon this expert advice as required.

36.2.3
New Drug Approval Routes

The rules governing medicinal products in the European Union provide for two independent routes by which new potential medicines may be evaluated. These are termed the centralized and decentralized procedures, respectively, and the EMA plays a role in both (Walsh, 1999). The centralized procedure is compulsory for biotech medicines and as such is described in greatest detail below. This route may also be used to evaluate new chemical entities.

36.2.3.1 **Centralized Procedure**

Under the centralized route marketing authorization applications (dossiers) are submitted directly to the EMA. Before evaluation begins, EMA staff first validate the application, by scanning through it to ensure that all necessary information is present and presented in the correct format. This procedure usually takes 12 working weeks to complete. Biotech-based dossiers are termed **part A applications**, whereas new chemical entities are termed **part B applications**.

The validated application is then presented at the next meeting of the **CHMP (human medicine applications)** or **CVMP (veterinary medicines)**. This committee then appoints one of its members to act as **rapporteur** for the application. The rapporteur organizes **technical evaluation** of the application (product safety, quality, and efficacy) and this evaluation is often carried out in the rapporteur's home national regulatory agency. Another member of the committee (a **co-rapporteur**) is often also appointed to assist in this process. Upon completion of the evaluation phase the rapporteurs draw up a report, which they present, along with a recommendation, at the next CHMP (or CVMP) meeting. After discussion, the committee issues a scientific opinion on the product, either recommending **acceptance or rejection** of the marketing application. The EMA then transmits this scientific opinion to the European Commission (who represent the only body with the legal authority to actually grant marketing authorizations). The commission, in turn, issues a final decision on the product (Fig. 36.2).

Regulatory evaluation of marketing authorization applications must be completed within strict time limits. The EMA is given a **210-day window** to evaluate an application and provide a scientific opinion. However, during the application process, if the EMA officials seek further information/clarification on any aspect of the application this 210-day clock stops until the sponsoring company provides satisfactory answers. The average duration of active EMA evaluation of biotech-based product applications is in the region of 175 days – well within this 210-day timeframe. Duration of **clock stops** can vary widely, from 0 days to well over 300 days. Most applications, however, incur clock stops of the order of 30–80 days. Upon receipt of the EMA opinion, the commission is given a maximum of 90 days in which to translate this opinion into a **final decision**. Overall, therefore, the centralized process should take a maximum of 300 active evaluation days.

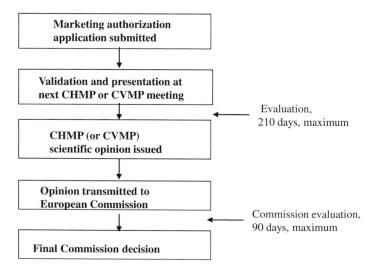

Fig. 36.2 Overview of the EU centralized procedure. Refer to text for details.

36.2.3.2 **Mutual Recognition**

The second route facilitating product authorization is termed mutual recognition or the **decentralized procedure**. This is open to nonbiotechnology products and the procedure entails the initial submission of an authorization application to a single national regulatory agency of an EU member state. This agency then assesses the application (within 210 days), formulates an opinion, and either grants or rejects the application. If authorization is granted the sponsoring company may then apply via mutual recognition to extend the market authorization to the remaining EU states. Theoretically, awarding of authorization in these remaining countries should follow almost automatically as the authorization requirements (dictated by pharmaceutical law) are harmonized throughout the European Union. Should disputes arise the EMA acts as arbitrator, itself forming a scientific opinion which it transmits to the European Commission who issue a final binding decision.

36.3
Regulation in the United States

The **FDA** is the US regulatory authority (http://www.fda.gov). Its primary mission is to protect public health. In addition to pharmaceuticals and cosmetics, food as well as medical and a range of other devices comes under its auspices (Table 36.2). Founded in 1930, it now forms part of the US Department of Health and Human Services and its commissioner is appointed directly by the US president.

The FDA derives its legal authority from the **Federal Food, Drug, and Cosmetic (FD&C) act**. Originally passed into law in 1930, the act has been amended several times since. The FDA interprets and enforces these laws. Although there are many parallels between the FDA and the EMA, its scope is far broader than that of the EMA and its organizational structure is significantly different. Overall the FDA now directly employs some 10 000 people, has an annual budget in the region of US$ 2 billion, and regulates over US$ 1 trillion worth of product annually. A partial organizational structure of the FDA is presented in Fig. 36.3. In the context of pharmaceutical biotechnology, the **Center for Drug Evaluation and Research (CDER)** and the **Center for Biologics Evaluation and Research (CBER)** are the most relevant FDA bodies.

Table 36.2 Product categories regulated by the FDA.

Foods, nutritional supplements
Drugs: chemical and biotech based
Blood supply and blood products
Cosmetics and toiletries
Medical devices
All radioactivity-emitting substances
Microwave ovens

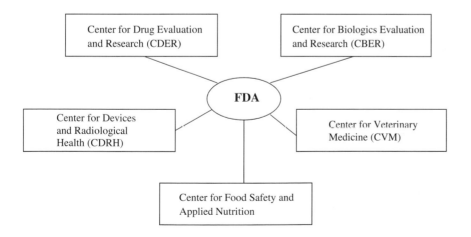

Fig. 36.3 Partial organizational structure of the FDA.

Table 36.3 Major biotechnology/biological-based drug types regulated by the CDER and CBER.

CDER regulated	CBER regulated
Monoclonal antibodies for *in vivo* use	Blood
Cytokines (e.g., interferons and interleukins)	Blood proteins (e.g., albumin and blood factors)
Therapeutic enzymes	Vaccines
Thrombolytic agents	Cell- and tissue-based products
Hormones	Gene therapy products
Growth factors	Antitoxins, venoms, and antivenins
Additional miscellaneous proteins	Allergic extracts

36.3.1
CDER and CBER

A major activity of the CDER is to evaluate new drugs and decide if market authorization should be granted or not. Additionally, the CDER also monitors the safety and efficacy of drugs already approved (i.e., postmarketing surveillance and related activities). Traditionally, the CDER predominantly regulated **chemically based drugs** (i.e., drugs that are usually of lower molecular weight and often manufactured by direct chemical synthesis). Included are prescription, generic, and over-the-counter drugs. The CDER has now also been assigned regulatory responsibility for the majority of products of **pharmaceutical biotechnology** (Table 36.3).

The CBER undertakes many activities similar to that of the CDER, but it focuses upon biologics and related products. Within **regulatory terminology**, the term biologic has a specific meaning, relating to a virus, therapeutic serum, toxin, antitoxin, vaccine, blood, blood components or derivatives, or allergenic products that are used in the prevention, treatment, or cure of diseases of human beings (http://www.fda.gov/BiologicsBloodVaccines/default.htm). The CBER therefore regulates products such as vaccines and blood factors, be they produced by traditional or modern biotechnological means (i.e., by nonrecombinant or recombinant means). Additional biological products, including cell and gene therapy and tissue-based products, also fall under the auspices of the CBER (Table 36.3).

36.3.2
Approval Procedure

The overall procedure by which biotechnology and other drugs are evaluated and approved by the CDER or CBER are, predictably, very similar, although some of the regulatory terminology used by these two centers differ. A summary overview of the main points along the drug development and approval road, in which the CDER/CBER play key regulatory roles, is provided in Fig. 36.4.

Once a sponsor (company, research institute, etc.) has completed preclinical evaluation of a proposed new drug, it must gain FDA approval before instituting clinical trials. The sponsor seeks this approval by submitting an **investigational new drug application (IND)** to either the CDER or CBER, as appropriate. The application, which is a multivolume work of several thousand pages, contains information detailing preclinical findings, methods of product manufacture, and proposed protocols for initial clinical trials. The regulatory officials then assess the data provided and may seek more information or clarification from the sponsor if necessary. Evaluation is followed by a decision to either permit or block clinical trials. Should clinical trials commence, the sponsor and regulatory officials hold regular meetings in order to keep the FDA appraised of trial findings. Upon successful completion of clinical trials the sponsor then usually **ap-**

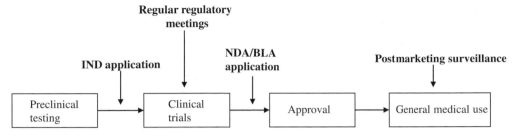

Fig. 36.4 Summary overview of the main points during a drug's lifetime at which the FDA plays a key regulatory role. Refer to text for further details.

plies for marketing authorization. In the CDER, this application is termed a **new drug application (NDA)**. NDAs usually consist of several hundred volumes containing over 100 000 pages in total. The NDA contains all the preclinical as well as clinical findings and other pertinent data and information. Upon receipt of an NDA, CDER officials check through the document ensuring completeness (a process similar to the EMA's validation phase). Once satisfied, they file the application and evaluation begins.

The NDA is reviewed by various **regulatory experts**, generally under topic headings such as medical, pharmacology, chemistry, biopharmaceutical, statistical, and microbiology reviews. Reviewers may seek additional information or clarification from the sponsor as they feel necessary. Upon review completion the application is either approved or rejected. If approved the product may go on sale but regulatory officials continue to monitor its performance (postmarketing surveillance). Should unexpected and/or adverse events be noted, the regulatory authority has the legal power (and responsibility) to suspend, revoke, or modify the approval, as appropriate.

The review process undertaken by CBER officials for biologic and related products is quite similar to that described above for CDER-regulated product. CBER-regulated investigational drugs may enter clinical trials subject to gaining IND status. The application process for marketing authorization undertaken by the sponsor subsequent to completion of successful clinical trials is termed the **licensure phase** in CBER terminology. The actual product application is known as a **biologics license application (BLA)**. Overall, the content and review process for a BLA is not dissimilar to that of the analogous CDER NDA process, as discussed above. The bottom line is that the application must support the thesis that the product is both safe and effective, and that it is manufactured and tested to the highest quality standards. Overall, the median time between submission and approval of product marketing application to the CBER/CDER stands at approximately 12 months.

While the majority of biotech-based drugs are regulated in the United States by either the CBER or CDER, it is worth noting that some such products fall outside their auspices. **Bone morphogenic proteins (BMPs)** function to stimulate bone formation. As such, several have been approved for the treatment of slow healing bone fractures. Product administration requires surgical implantation of the BMP in the immediate vicinity of the fracture, usually as part of a supporting device. In the United States, these products are regulated by the FDA's Center for Devices and Radiological Health (CDRH; http://www.fda.gov). Drugs (both biotech and nonbiotech) destined for veterinary use also fall outside the regulation of the CBER or CDER. Most such veterinary products are regulated by the FDA's Center for Veterinary Medicine (CVM), although veterinary vaccines (and related products) are regulated not by the FDA, but by the Center for Veterinary Biologics, which is part of the US Department of Agriculture (http://www.aphis.usda.gov/animal_health/vet_biologics/index.shtml).

36.4
Advent and Regulation of Biosimilars

It was recognized several decades ago that the high cost of originator, brand name pharmaceutical products limited access to their benefit on economic grounds. However, patent protection on such new drugs eventually runs out, opening up the possibility of competition from alternative manufacturers. A **legislative framework** for the development and approval of such generic pharmaceuticals was established in the United States when Congress passed the **Hatch Waxman Act** in the 1980s. Similar legislation was introduced in many other world regions, facilitating the advent of generic products. Products facilitated by generics legislation are invariably low-molecular-weight organic molecules manufactured by direct chemical synthesis via defined well-characterized chemical pathways, and are amenable to sensitive and exacting analytical characterization. The bottom line is that, in such cases, a product identical to the original one can be manufactured.

Biopharmaceuticals are amongst the most expensive of all pharmaceutical products. Once patent protection began to expire on earlier-approved biopharmaceuticals they became a target for generics manufacturers. However, biopharmaceuticals differ fundamentally from traditional pharmaceuticals. They are hundreds, usually thousands of times larger and are synthesized by biological processes, with all of the inherent variability that can entail. While genetic engineering can ensure the production of a recombinant protein with an amino acid sequence identical to any approved product, the exact details of manufacture (upstream and downstream processing) can and will influence the impurity profile of the product, as well as the exact details of any posttranslational modifications (e.g., glycosylation) present. Moreover, their complexity renders full analytical characterization of any such product extremely challenging. These complications make it highly improbable that a copy of a biopharmaceutical would be absolutely identical to the originator, hence the term "generic" (or "biogeneric") would be inappropriate in this context. What is achievable is the production of a product very substantially similar to the originator and hence the term "biosimilar" was coined.

Throughout the earlier part of previous decade the European Union developed legislative and regulatory provision for the approval of biosimilars. The EMA have put regulatory guidelines in place (http://www.ema.europa.eu/pdfs/human/biosimilar/043704en.pdf; Furlain, 2008) and thus far 13 such biosimilars have been approved within Europe. EU biosimilar regulations necessitate the generation of comparative data between the proposed new biosimilar product and the reference product, to which it claims biosimilarity. The reference product must already be approved for general medical use within the European Union. The company seeking biosimilar approval must submit the data generated in the form of a marketing application directly to the EMA for consideration via the centralized procedure (Section 36.2.3.1). The application dossier (relative to the one for the original reference product) will contain a full quality module (e.g., details of manufacture, analysis, etc.), as well as reduced clinical and nonclinical data modules.

The development of a biosimilar approval framework in the United States is moving much more slowly, although two bills relating to biosimilars are currently being considered by the US House of Representatives.

36.5
International Regulatory Harmonization

Europe, the United States, and Japan represent the three main **global pharmaceutical markets**. As such, pharmaceutical companies usually aim to register most new drugs in these three key regions. Although the underlining principles

Guideline number	Guideline title
Q5A	Viral safety evaluation of biotechnology products derived from cell lines of human or animal origin
Q5B	Quality of biotechnology products: analysis of the expression construct in cells used for the production of r-DNA derived products
Q5C	Quality of biotechnological products: stability testing of biotechnological/ biological products
Q5D	Quality of biotechnological products: derivation and characterization of cell substrates used for production of biotechnological/biological products
Q5E	Quality of biotechnological products: comparability of biotechnological/ biological products subject to changes in their manufacturing procedure
Q6B	Specifications: test procedures and acceptance criteria for biotechnological/ biological substances
S6	Preclinical safety evaluation of biotechnology-derived pharmaceuticals

Table 36.4 Finalized ICH guidelines that specifically focus upon products of pharmaceutical biotechnology.

are similar, detailed regulatory product authorization requirements differ in these different regions, making necessary some duplication of registration effort. The **International Conference on Harmonization of Technical Requirements for Registration of Pharmaceuticals for Human Use** (the ICH process) is an initiative aimed at harmonizing regulatory requirements for new drug approvals in these regions. The project was established in 1990, and brings together both regulatory and industry representatives from Europe (the EMA and the European Federation of Pharmaceutical Industries and Associations), the United States (the FDA and the Pharmaceutical Research and Manufacturer's of America), and Japan (The Japanese Ministry of Health, Labor, and Welfare and the Japanese Pharmaceutical Manufacturer's Association).

ICH is administered by a steering committee, consisting of representatives of the above-mentioned groupings. The steering committee determines ICH policies and procedures, selects topics for harmonization and monitors overall harmonization progress. This committee in turn is supported by an ICH secretariat, based in Geneva, Switzerland (http://www.ich.org). The main technical workings of ICH are undertaken by expert working groups charged with developing harmonizing guidelines. The guidelines are grouped under one of the following headings:

- Efficacy (clinical testing and safety monitoring related issues).
- Quality (pharmaceutical development and specifications).
- Safety (preclinical toxicity and related issues).
- Multidisciplinary (topics not fitting the above descriptions).

Thus far 63 guidelines aimed at both traditional and biotechnology based products have been produced and are being implemented (Table 36.4) and the process by which new guidelines are developed is outlined in Fig. 36.5.

One of the ICH's most ambitious initiatives to date has been the development of the common technical document. This provides a harmonized format and content for new product authorization applications within the European Union, United States, and Japan. When this and the other guidelines are fully implemented considerable streamlining of the drug development and, in particular, registration process will be evident. This will make more economical use of both company and regulatory authorities time, should reduce the cost of drug development, and should speed up the drug development procedure, ensuring faster public access to new drugs.

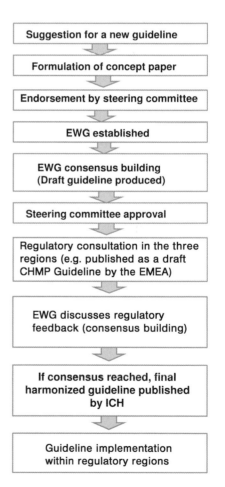

Fig. 36.5 Overview of the process of developing a new ICH guideline. EWG, Expert Working Group.

37

Emergence of a Biotechnology Industry

Learning Objectives
This chapter provides a background to the development of an industrialized pharmaceutical industry in the nineteenth and twentieth century. How a biotechnology industry could develop largely independently from this pharma industry and what the scientific foundations of this biotech industry are is illustrated using the example of Genentech Inc., the first and premier biotech company. Other nonscientific factors such as the availability of venture capital and an entrepreneurial culture willing to take risks are explained, which are at least as important as scientific talent to foster the emergence of a biotechnology industry.

If biotechnology already means the application of biological organisms, cells, or parts thereof for industrial or, more general, commercial purposes, what then is a **biotechnology industry**?

Biotechnology has had a long evolution from its ancient roots in **alcoholic fermentation** of beer invented by Sumerians or wine brought to an art by Greeks and Romans towards its modern variants in the form of genetic engineering and molecular biology. The **classical forms** of biotechnology such as beer brewing or vine fermentation were developed by ancient farmers. The knowledge of fermentation was then transmitted from the ancient people to medieval monks and from the monasteries back to farmers, and finally, with the advent of the industrial revolution, to breweries or distilleries. Some of these breweries have grown to huge industrial conglomerates, in particular in the United States and Japan, from their modest origins dating back to the nineteenth century.

A similar process of industrialization and emergence of huge globally operating companies took place in the development of the pharmaceutical industry. From their early start as local pharmacies, some companies developed industrial processes for the development and manufacture of drugs that were needed by a constantly growing population. The industrial revolution, which started as early as 1850 in England, and later in France and Germany, forced the increasing number of workers and their families to live together under uncomfortable and hygienically unacceptable conditions. This gave rise to huge infectious epidemics and therefore the need for vaccines or anti-infective treatments arose.

It was ingenious individuals such as Louis Pasteur and Robert Koch who discovered the **principles of vaccination** around the turn of the nineteenth to the twentieth century. However, the process of manufacturing a vaccine required attenuation of the virulence of the causative agent, infection of huge mammals with sufficient blood reserves, harvesting the antiserum, and effective delivery to the children and adults that required vaccination, and all this in a reproducible manner. The first companies that set up such processes, **Hoechst** and **Behringwerke** in Germany or **Institut Pasteur** in France, received the merits thereof in commercial terms. They expanded to global players over the course of the twentieth century. See Fig. 37.1.

The establishment of **effective chemotherapies** against infectious diseases was similarly a result of transmission of outstanding academic research results into

An Introduction to Molecular Biotechnology, 2nd Edition.
Edited by Michael Wink
Copyright © 2011 WILEY-VCH Verlag GmbH & Co. KGaA, Weinheim
ISBN: 978-3-527-32637-2

Fig. 37.1 From descriptive biology towards microbiology (adapted from Bayrhuber and Kull, 2003).

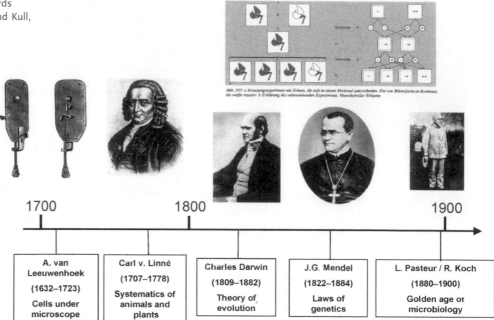

1700	1800		1900
A. van Leeuwenhoek (1632–1723) Cells under microscope	**Carl v. Linné** (1707–1778) Systematics of animals and plants	**Charles Darwin** (1809–1882) Theory of evolution	**J.G. Mendel** (1822–1884) Laws of genetics	**L. Pasteur / R. Koch** (1880–1900) Golden age of microbiology

an industrial process. Salvarsan, the first **magic bullet** against syphilis, a previously incurable disease, was invented by Paul Ehrlich, but commercialized by Hoechst AG of Frankfurt, Germany.

This vastly growing pharmaceutical industry yielded huge global player companies, first starting out in Europe and later, after World War II, this also happened in the United States. This growth was accompanied by further innovation: similar to the discovery and development of Salvarsan by Ehrlich in an academic environment, Alexander Fleming, a British physician, discovered the **antibacterial activity of** *Penicillium* **extracts** in 1928. However, it required the huge demand for anti-infective therapy brought about by the large number of wounded soldiers in World War II to establish appropriate techniques for the cultivation and extraction of penicillin from *Penicillium* cultures. After the war the further development of penicillins and cephalosporins as safe and effective antibiotics boosted the commercial triumph of the pharmaceutical giants. See Fig. 37.2.

However, this huge medical as well as business success prompted the pharmaceutical companies to still focus on classical **pharmacotherapy**. Their research and development departments became routine discoverers of improved drugs with better efficacy and side-effect profiles, but they neglected the roots of innovation – the serendipitous discovery of novel scientific concepts by talented individuals. Science made huge progress in the decades from the 1950s to the 1970s: Watson and Crick proposed a **structure for deoxyribonucleic acid (DNA)** as a physical substrate of heredity that encodes the genetic information in 1953. Then, in the early 1960s, Nirenberg and Matthei deciphered the alphabet of heredity: the translation map of the four DNA bases into the amino acid sequence of proteins, also known as the **genetic code**.

These scientific breakthroughs did not, however, catch the attention of the pharmaceutical senior management. Applications of these findings were too far thought, too much science fiction for people who were used to make double-digit revenue margins from selling small molecules made by **classical organic chemistry**.

The ultimate scientific breakthrough that turned the science of molecular biology into biotechnology, a bio or genetic engineering technique, came through an ingenious experiment performed by **Boyer** and **Cohen** in 1973: they took a short

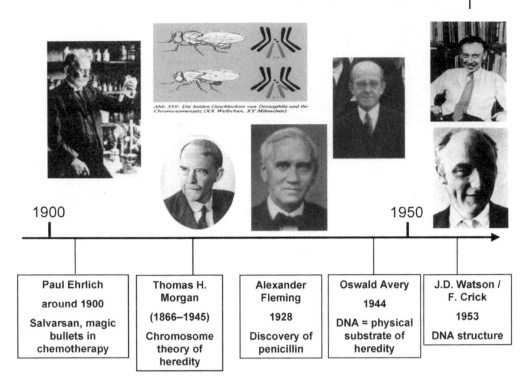

Fig. 37.2 Medical and genetic discoveries in the first half of the twentieth century (adapted from Bayrhuber and Kull, 2003).

1900   1950

Paul Ehrlich	Thomas H.	Alexander	Oswald Avery	J.D. Watson /
around 1900	Morgan	Fleming	1944	F. Crick
Salvarsan, magic	(1866–1945)	1928	DNA = physical	1953
bullets in	Chromosome	Discovery of	substrate of	DNA structure
chemotherapy	theory of	penicillin	heredity	
	heredity			

fragment of DNA that they obtained by digestion with a certain restriction enzyme and cloned it into a plasmid vector. In other words, they used the naturally occurring tools of restriction enzymes, DNA ligases, and bacterial plasmids to artificially combine or recombine fragments of unrelated DNA into a new alignment. This meant they not only described a new phenomenon in molecular biology, but they used these tools from nature in a construction kit fashion. Hence, the terms **recombinant DNA technology** or **genetic engineering** were coined to indicate the transition from descriptive science into engineering-type technology.

It took just a few years from this invention so that Boyer, one of the scientific pioneers of this biotechnology, could watch his findings put into a meaningful application – the recombinant expression of medicinally applicable insulin.

Boyer received a lot of academic attention for his findings, but it was hard for him to convince people in the pharmaceutical industry of the usefulness of this new form of bioengineering. Due to the huge success of the big pharmaceutical companies, they were not forced to foster real innovation, but rather they built on the perpetual success of their classical chemotherapy approach.

Boyer had good luck – he found someone who was as eager as himself to demonstrate the applicability of commercialization of this new technique. **Swanson** received a scientific education, but after finishing his doctoral degree he was more interested in generating commercial success than in an academic career. However, he was certainly not as simple-minded as many business people are, but rather he dreamt of a new scientific breakthrough that could be transformed into a real business success.

It was a just a matter of time before Boyer and Swanson found each other, like sticky ends of restriction enzyme-digested DNA: they shared the vision to create something big. Basically, together they invented a new industry and defined biotechnology as a business driven by startup companies. See Fig. 37.3.

Swanson came from Kleiner Perkins, a venture capital firm that was eager to invest into small but highly promising startup companies. This was exactly what they did with Swanson and Boyer's newly founded biotechnology enterprise, called **Genentech Inc.** Kleiner Perkins invested a sum of US $100 000 initially and took a certain number of Genentech shares in the form of an equity investment.

Fig. 37.3 1953–1976: from molecular genetics towards genetic engineering (adapted from Bayrhuber and Kull, 2003).

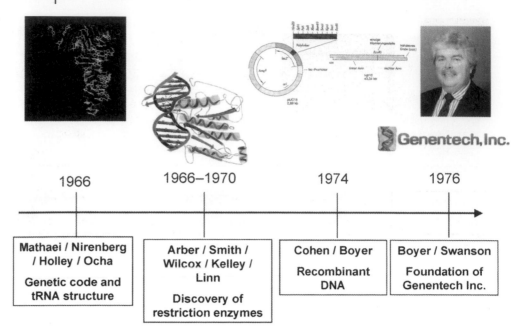

Genentech set up its labs in San Francisco in 1976, close enough to UCSF, Berkeley, and City of Hope from where they hired most of their initial scientific staff. Boyer had a strong enough reputation at that time that they got truly brilliant and talented young scientists such as **Ullrich**, **Seeburg**, and **Khorana**. These individuals made their fortune with Genentech, scientifically and money-wise. In fact, the scientific breakthroughs had been the prerequisites for their business success.

This became a general motif of the biotechnology ("biotech") industry: business success was mostly linked to scientific achievements. What have been the major achievements of Genentech and its scientists? Boyer and colleagues set themselves a very ambitious goal – they wanted to produce **insulin**, the polypeptide hormone famous for its blood glucose-lowering action, in bacteria using recombinant DNA technology. If they succeeded in this, all of the problems related to the limited supply of pancreas organs from slaughtered pigs, with the immune response against nonhuman insulin, and with impurities that came along with the purification process would have been overcome.

It was a long, long way and the task turned out to be more challenging than they had initially thought, but finally they succeeded. They even had to move part of their labs to France since this country had not ratified, at that time, the so-called **Asilomar moratorium**, which banned the use of recombinant DNA plasmid vectors and bacterial strains until they were proven to be safe. They were also competing against the group around **Walter Gilbert** in Cambridge, Massachusetts, which founded another biotech company, **Biogen**, that had similar aims to Genentech. Gilbert and colleagues faced even tougher problems in discussions with the educated, but in this respect, also anxious Boston area public opinion leaders in justifying their recombinant DNA "creator" work.

Once they had samples of recombinantly produced insulin at hand, the Genentech staff tested it successfully for its glucose-lowering potency in animal models. However, Genentech now faced a different problem. Further investment rounds had contributed enough capital to come to this point, but now the company had to invest into the very cost-intensive challenge of testing the recombinant insulin in humans. They recognized that in order to attract further investors and to convince the drug approval authorities with convincing clinical trials results, they needed a strong partner with a huge body of experience in insulin development.

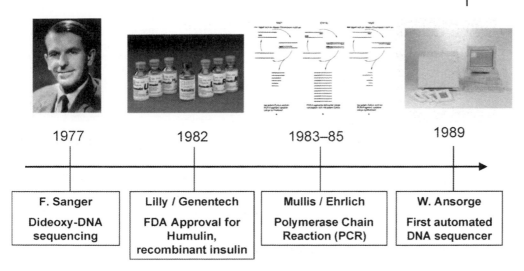

1977   **1982**   **1983–85**   **1989**

F. Sanger	Lilly / Genentech	Mullis / Ehrlich	W. Ansorge
Dideoxy-DNA sequencing	FDA Approval for Humulin, recombinant insulin	Polymerase Chain Reaction (PCR)	First automated DNA sequencer

Fig. 37.4 From genetic engineering towards biotechnology (adapted from Bayrhuber and Kull, 2003).

That partner was found in **Eli Lilly and Co.** Lilly is a typical example of a midwestern pharmacy that became a huge pharmaceutical giant over time. Lilly was also the company with the broadest experience of insulin development and marketing in the United States. So the two, Lilly and Genentech, ended up in a **strategic collaboration** in 1978 where Genentech contributed its patent and process of producing recombinant human insulin in bacteria and Lilly took over the role of the sponsor of clinical trials and the marketer of this new insulin. See Fig. 37.4.

Genentech used the good news as a warm wind from behind to raise additional capital for investments into further projects. However, this time Swanson decided to take the company public – to sell additional shares of Genentech in an **initial public offering (IPO)**, not only to institutional investors but basically to everyone.

Genentech's IPO on 14 October 1980 is still famous: during the first hour of public trading, the stock with the remarkable ticker symbol "DNA" ran up from US \$38 to US \$88. Genentech raised US \$35 million of fresh capital on that day. The stocks initially worth US \$100 000 that Kleiner Perkins had acquired in Genentech's first funding round were now worth US \$78.3 million – an increase by a factor of 783 within 5 years. This incredible value gain defined the role model for all biotech companies to come. Genentech had performed all of the major elements that defined a kind of **master plan for the evolution of a biotech company**:

1. The company was funded by **venture capital** (i.e., by a form of funding where the investor takes over a significant percentage of the company's equity against substantial funding). If the company succeeds, this will multiply the company's and, in parallel, the investor's value. However, the venture capital investor can only realize this value gain if he/she has the chance to **exit** the investment. An exit is either given through an IPO where the venture capitalist can freely sell their shares over the stock exchange or it is given by the acquisition of a company by a bigger one. After Genentech, many biotech investments were exited via both routes. It is essential to realize than many biotechs fail to return the investment. It is actually the minority of startups that yield a fruitful exit. Therefore, they have to multiply their initial funding to such a huge extent that this gain can cover all other failures in the venture capitalist's portfolio (Table 37.1).

2. Genentech had a very clear idea of what they finally wanted to deliver – a pharmaceutical product, an improved version of human insulin with clear benefits over other insulin forms. Over the forthcoming 25 years of the biotech industry, it turned out that only such pharma- or drug-oriented biotech

Table 37.1 Differences between biotech and big pharma companies.

Biotech start-up	Big biotech	Big pharma
• Age < five years	• Age > five years, mostly > ten years but < thirty years	• Age mostly > one hundred years
• < 100 employees	• 500 to 10 000 employees	• > 5000 employees
• Small sales volume, not profitable	• sales volume from 50 million US $ to 4 billion US $	• Sales volume from 1 billion to 25 billion US $
• Very intensive research: research and development cost per employee > 50 000 US $ per year	• intensive research: many collaborations with Big Pharma, academia and other biotechs	• Focus on development and marketing, majority of employees working in marketing and administration
• Focus on research and development of therapeutics, diagnostics, and technologies, not on marketing	• Products in late clinical trials or at the market	• Several products in clinical trials or at the market with a focus on blockbusters, i.e. with a sales volume of > 500 million US $ per year
• Frequently unconventional therapeutic strategies	• Frequently biologicals as products	• Frequently quoted on the stock exchange, financing from sales or from all financial instruments used by companies quoted on the stock exchange, e.g. bond issues
• Financing by venture capital	• Originally venture financed, later financing by stock exchange, i.e. public companies	
• Corporations or Ltd. not quoted on the stock exchange		

companies could survive independently. Other companies have focused on **agricultural biotechnology** ("agbio") or so-called **technology platforms**. There are only very few examples of technology platform companies (e.g., **Affymetrix**) that have turned into long-term sustainable businesses. For technology platforms as well as for agbio companies, the most likely exit route is the acquisition by a major pharma or agricultural company. One example of a successful agbio exit is the acquisition of **Plant Genetic Systems (PGS)** of Gent, Belgium by **Aventis** for US $ 750 million in 1996. See Fig. 37.5.

3. At the same time that Genentech decided to develop insulin, they also looked for a company with big financial and marketing muscle as a strategic partner. This also demonstrates the role of biotech companies as establishing pharmaceutical companies: biotechs represent the sources of innovation, new technologies, and new drug candidates for the pipelines of the big pharma companies. Biotechs and big pharmas therefore live together in a form of natural symbiosis.

4. Genentech is also a story of two individuals coming together that share a common goal, albeit with different approaches: **Boyer** was the scientist who wanted to demonstrate that his recombinant DNA technology had the potential to launch a new area of protein drugs; **Swanson**, the scientist-turned-businessman, had enough scientific understanding to capture the value of genetic engineering, but at the same time he coined the business model for a biotech company and paved the way for Genentech's financial and commercial success.

5. Finally, Genentech remained a source of innovation and new recombinant protein drugs because they managed to maintain the pace and perception as

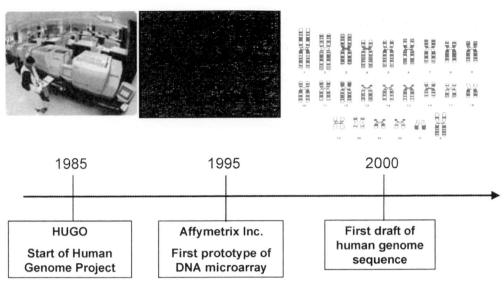

Fig. 37.5 From biotechnology towards genomics (adapted from Bayrhuber and Kull, 2003).

the place to be for young gifted scientists looking for a postdoctoral position in an industrial environment. Genentech and with it many other biotech companies established a culture of letting the best minds from all over the world compete and collaborate in striving for **scientific excellence** and **new innovations** with a clear benefit to the patient.

Very rapidly, other newly established biotech enterprises followed the footsteps of Genentech: **Amgen, Biogen, Genzyme, Immunex, Chiron**, and others all had their own recombinant DNA-borne success story. They and their further successors established an industry with about 2500 independent biotech companies and approximately 200 000 employees world-wide. Notably, the sum that this industry devotes to research and development expenditure per employee is still far higher on average than the same figure for the old pharma companies (approximately US $ 100 000 as opposed to US $ 30 000/year).

Finally, it should be noted that it was not coincidence that Genentech's success was born in California, a place unlike many others, where **academic excellence meets entrepreneurship** in an unmatched way. The third leg of the biotech industry's basis is contributed by the **availability of venture capital**. Other regions in the United States followed these guidelines, and also generated huge clusters of biotech companies and pharma research sites in conjunction with academic centers of excellence. European countries and other regions tried to mimic these success stories, but could hardly generate similarly sized biotech clusters. European biotech companies tend to be far smaller in size by numbers of employees and funding round sizes. Amongst many other factors this has to do with an unfavorable shift in the risk-to-reward perception of such high-tech/high-risk companies. Failure has a flavor of loosing out and is not very accepted in Europe, whereas ending a business and starting the next one is regarded as a new chance in the United States. These cultural differences are not only seen between Europe and the United States, but also between established big companies and young startup biotechs. This ultimately leads to the question: what differentiates a biotech company from a pharma company these days?

Today, the borders between biotech and pharma companies have tended to fade away. **Amgen**, by far the biggest so-called biotech company, is not really a biotech company any more. With its approximately US $ 6 billion in revenues and a market capitalization of around US $ 80 billion, it has superseded many established old pharma companies in numbers. It has also turned its attention to the development of small-molecule drugs and established a broad sales force for its own products. It has taken over other biotech companies such as Immunex and Tularik. On the other hand, Amgen now also faces the known problems of decreasing innovativeness with increasing research and development budget.

For a long time, **biotech drugs** have been considered to be **biologicals** only (e.g., recombinant hormones such as insulin, growth hormone, erythropoietin, or monoclonal antibodies). In fact, the first-generation of biotech companies has succeeded mainly with such types of recombinant proteins or antibodies.

Still, the majority of drugs that are developed by biotech are of protein nature, but the number of companies that develop classical small-molecule drugs is constantly increasing. However, whereas most of the pharma giants focus on huge **therapeutic applications** such as cardiovascular or metabolic diseases, osteoporosis, or central nervous system disorders, biotechs have a tendency towards diseases with huge medical needs, particularly cancer. The reason is that the hurdles for developing drugs that target life-threatening diseases are much lower than those for complex diseases that require chronic treatment. Therefore, biotech companies specializing in **oncology** or **rare diseases** need to invest less into clinical development and know earlier about the outcome of their efforts.

Looking at financial figures of publicly quoted biotech companies should surprise everyone who is not familiar with this industry. Only a handful of biotechs really generate profits – the vast majority of all roughly 2500 companies worldwide are substantially loss making. The reason is that the venture capitalists behind the company play a statistical game: a venture capitalist-funded company should not focus on generating early revenues but flat revenues and it is doomed to either launch a drug with skyrocketing potential profits or fail. The first and the last alternative would never pay off the equity investment into them. Thus, they burn one funding round volume after the other unless they can launch their product in the market.

This risk culture, again, is unknown to conservative European non-venture capital investors as well as pharmaceutical companies. They do not embark on high-risk projects where the return is not calculable with almost certain probability. This risk averseness has led to a situation where big pharma is generating fewer and fewer drugs on their own. The percentage of drugs in-licensed from biotech or other sources is getting higher, which has two main consequences:

6. The pharmaceutical giants will turn more and more into **marketing machineries**, whereas the small biotechs are doomed to remain research-driven and highly risky in nature.

7. As long as the financial supply is given, the biotech industry will have a bright future on a long-term perspective. There are current tendencies also for biotechs to shy away from research and in-license molecules, take them through clinical development or turn them towards other therapeutic applications. This approach embarks on the role of a biotech company as a high-risk shell for gambling than on the classical Genentech-type role of a biotech company.

Finally, the entrepreneurial environment of a biotech company, in particular of startups, attracts a different breed of people than the pharmaceutical industry. Whereas the former is more the playground of the ambitious and more quirky academics, the latter tends to attract more serious individuals. The biotech company is a place for people who can and have to take risks in an ambitious, sometimes close to maniac fashion, whereas the big pharma company is a refuge for persons who have families or want to start one.

These soft differences are certainly not as significant as the hard ones in terms of projects and figures. However, the more academic type of culture, at least for the younger and smaller biotechs, also establishes a transmission link between the basically nonregulated academic environment and the world of protocol-led meetings and reporting session embedded in a broad shell of **corporate guidelines** and personal politics that tend to discourage scientists in huge industrial environments over time.

38

The 101 of Founding a Biotech Company

Learning Objectives

Some of the readers will be soon faced with the question which way to go in their professional career: shall it be the academic route, a postdoctoral, or a different position in an industrial setting? A smaller startup? Or an established brand name company? If you are willing to take substantial risks, and you have skills and talents beyond science, then founding your own company is something that can be seriously considered. However, no one should think that he/she is poised to make the next Genentech success. This chapter presents a few rules that should be considered when starting your own biotech enterprise.

38.1
First Steps Towards Your Own Company

The establishment of a new company, in particular of a biotech startup, requires four main elements:
- A sound scientific rationale.
- A business plan how to turn the science into a product.
- A team of highly motivated, yet experienced and skilled individuals.
- By far the most difficult – money.

The first two points are not as easy to establish as it may seem. They present the first checkpoints whether the founders can really dedicate the skill set and the financial and personal resources, in particular the time, needed to set up a commercial biotech enterprise. Let us think of a standard situation that may become the starting point for the emergence of a **new biotech company**. One or more honored professors have discovered a new (family of) genes or proteins that seem to have a profound impact on animal or human health when altered in their physiological function (e.g., by pharmacological tools or by genetic means). The scientist(s) have published some exciting papers on this discovery in highly reputable journals. They file **patent applications** on their findings and they think that this may also become the starting point for a new therapeutic approach.

It is a far, far way from here to a successfully launched drug. However, it can be done. What is needed is an **appropriate environment** in which these first academic ideas can be brought into a meaningful therapeutic concept. Filing for drug approval might be the final goal, but for investors to build confidence they will certainly demand **milestones** to be set. Those milestones have to reflect different transition states from an academic idea into a drug-type treatment. Typical milestones would be:
- **Proof of principle** of therapeutic usefulness in an animal model or by human genetics or even in humans directly.
- **Development** of drug candidates for testing in animals;

An Introduction to Molecular Biotechnology, 2nd Edition.
Edited by Michael Wink
Copyright © 2011 WILEY-VCH Verlag GmbH & Co. KGaA, Weinheim
ISBN: 978-3-527-32637-2

- Teaming up with a partner from the pharmaceutical industry who takes care of further **preclinical** and **clinical development**.
- Alternatively, nomination of a clinical candidate for **investigational new drug (IND)** filing for human clinical trials.
- Different clinical milestones in **phases I and II**.

Each of these milestones requires a very broad set of skills and tremendous scientific efforts, which also means financial resources. A short written overview should note all the assets, be it technically, in business development, or in human resources, that are required for achieving those milestones and also their cost.

This is a step towards the second main criterion that has to be fulfilled before taking off as an independent company – the **business plan**.

The business plan should actually fulfill two main criteria:

- It should summarize the ideas of the founders and ensure themselves that their visions can be transformed into a marketable product. Thus, it should serve as a mirror reflecting the ambitions of the founders.
- It must be appealing and convincing to investors such that they will open their pockets and provide the appropriate level of funding to the new enterprise.

The writer of the business plan should dedicate his/her thoughts towards the second point. If he/she and others then read the final outcome of this business plan exercise, he/she will automatically also fulfill the purpose of the first point.

Investors and **venture capital** investors, in particular, have an entirely different way of thinking about science. For good reasons, because many startups did so, venture capitalists are afraid of pouring money into companies that define themselves as prolonged workbenches of their scientific founding godfathers. What they need is a clear separation between the scientific idea and the therapeutic rational. They do not need extensive descriptions of pathways of why and how a certain gene or protein exerts its effects. A precise definition why the approach is meaningful and what the key features of the sought after therapeutic looks like are of utmost importance.

The first and most important part (mostly the only one really read) of a business plan is the **executive summary**. This is the part that carries the idea, the flavor of the business model, and the potential financial gains to the investors – starting out from the scientific rationale. In other words, in these three to four pages, the writer has to transform a technical idea into business mechanics and wording. The reader, the potential investor, is solely and absolutely focused on the potential value that he/she can create while investing in this. In as few words as possible, he/she must be convinced that at this company, the right people are working on the right idea to make themselves and the investor rich.

It is a good idea to write the executive summary at the end as the last part when all of the other chapters are finished.

Then the more detailed part starts with the section that is usually entitled **Technology Platform, Market, and Product Development**. This is the place to broaden scientific explanations to the extent necessary. The reader shall learn where the uniqueness of the approach lies as opposed to the potential competition. The market description should contain figures for the therapeutic market to be addressed, who is currently serving it, how the market will develop, and who will serve it in future. Ideally, the company should address the currently emerging market potential and its underlying technology (and patent situation) must provide a readily visible competitive advantage.

Again, the shorter and more precise the explanation, the better. As Einstein said when asked how he explained his complicated theories, "as short as possible but not shorter".

The next section should concerns the **management team**. Some founders think that the people they trust will be the best choice for executive positions. However, what investors want to see is a **track record of success** for those people that should yield their return on investments. An ideal choice for investors is a combination of ingenious, respected scientists as scientific founders who remain in their academic position in conjunction with executives who have proven in at least two further positions that they can generate business success. Here, the degree of proven experience and reputation in the industry is it what makes the difference between a US $2 or 20 million first funding round.

Chapter 39 of this book, on **"Marketing"**, explains in more detail how the founders concluded that there is a huge nonserved market opportunity and how they are going to address it.

The **realization plan** section then focuses on defining appropriate milestones that will allow the investor to track the company's progress towards the development of the product. A thorough balance between a too ambitious goal and a conservative calculation is needed. Obviously, the consumption of financial resources must correlate with the achievement of the milestone goals set herein. This leads to one of the most important sections – the **financial plan**.

The financial plan is of utmost importance to the investor and therefore it deserves a dedicated section. The financial plan has to demonstrate:

- An overview of revenues, costs in detail, cash flow, and accumulated cash flow over the period up to successful product launch.
- How the money raised will be turned into milestone success of the realization plan.
- How the milestones correlate with an increase in value of the company.

The last point is hard to address in numbers. It aims to give the investors a feeling of how the value of their investment proportionally increases with the money burned by the company. They should be made confident that the achievement of a certain milestone in the realization plan really makes a quantum leap in the potential valuation of the company. The reason is that the investor has to know at every stage of the company how much his/her investment is worth. One reason for this is that the venture capitalist states this value as a certain asset in the balance sheet of his/her venture capital funds. He/she has to know how they can position themselves in negotiations with the management team and with other investors for the valuation of the next funding round.

All venture capital thinking, by definition, is **exit driven**. No equity investor is interested in making profits from the dividend that a company pays to him as a shareholder. This means that the venture capital investor will evaluate at each and every stage whether they either have to put more money into the company or whether they are able to realize the value gain by selling their shares. There are only two ways this can occur:

- An initial public offering (IPO).
- A trade sale to another company.

The **IPO** is by far the preferred exit route for venture capital investors. In the process of the IPO, a company offers newly issued shares to institutional investors under the guidance and supervision of so-called **underwriting banks**. The underwriters ensure that the IPO prospectus is set up properly, describing the assets of the company, the future goals, and the basis for the current valuation of the issuing company. **Institutional investors** are those that have dedicated industry expert groups for such biotech investments. Their analysts evaluate the issuing company by certain financial measures; however, since nearly all biotech candidates are loss making, the evaluation of the technology or product basis is much more important. They compare the candidate with similar companies in peer-group comparisons.

Institutional investors usually command huge funds that come from money that private people contribute from life insurance or other forms of investments. Classical pension funds are by far the biggest, but they only allocate a small percentage of their money in such **high-risk investments**.

The key management of the company accompanied by industry experts of the underwriting bank(s) undertakes a road show from funds manager to funds manager in order to convince them to sign up for the issued shares. The orders of these funds managers are collected in the order book, literally a handwritten compilation of the individual share orders. At the end of the road show, during the **book-building phase**, the underwriters analyze the sensitivity of the potential stock price against the order demand. What they think the optimum is then becomes the **issue price** (i.e., the price that the institutional investors have to pay for the stocks that they can buy in their allotment).

Provided there is a positive general financial climate, a company can raise in between US $20 and US $200 million in such an IPO. This is obviously dependent on the valuation of the company and the percentage of newly issued and then freely tradable shares – the so-called **free float**. This is the money that can be raised when there is already sound clinical data available and there is a clear route towards marketable products – for more than one product!

After all, it is mainly the actual market conditions, independent from the performance of the company, that will dictate what valuations can be realized in an IPO. If the shareholders want to achieve a too high valuation they might not get enough share orders signed up during the pre-IPO phase. This means the investors think the price is too high and they do not order enough shares to fill up the amount of shares issued. Alternatively, it might happen that the allotment of shares to institutional investors is reasonable, but then the public trading might bring the price down in a short time. These so-called post-IPO "underwater companies" then run into huge problems when they have to raise additional capital. The publicly quoted US biotech companies actually raised most of their capital not by the IPO, but by secondary issues. If a stock trades well, this means there is enough daily turnover of a stock and the price is constantly going up, and again additional shares can be issued. This is the main reason why companies want to see their stock price high even though they have already made their initial money during the IPO.

Another reason is that publicly traded stocks are a very good acquisition currency. A company with a complementing portfolio can be taken over by issuing new shares and giving them to the old owners of the acquisition target against changing ownership. These owners can then either convert this stock into cash directly or keep it.

There have been frenetic times such as the **genomic bubble** from the end of 1999 up to March 2000 when shares of each and every company that had a gene or bio in its name and was pretending to solve all of the problems of this world by any kind of -omics technology skyrocketed. Then, the so-called IPO window was wide open, which prompted many investors to fill up their funds with cash from NASDAQ or other stock exchanges. The mood of the public markets also has a dramatic impact onto the funding situation for early stage companies. If venture capitalists can make exits, they (a) have money to invest into startups and (b) their mood is optimistic in the sense that they strongly believe that the next wave of biotech companies can repeat their current success.

A bear time in the public markets obviously serves the opposite purpose: venture capitalists think they can never take their sheep to shearing! Unfortunately, this cannot be influenced by founders or company executives.

The only other form of an exit is the **trade sale** – the takeover of the company where the venture capital is invested in another company against cash or, usually, in a mixture with stocks of the other company (see above). Trade sales have been the classical exit route for smaller companies that could go as far as to move several programs into clinical development. Some technology platform companies have become famous for their trade sales. For example, Rosetta In-pharmatics, again at the peak time of the genomics bubble in 2000, was acquired by Merck and Co. for US $620 million – a sum that was regarded as far too high in the industry.

If you want to start a biotech business you need a **business plan** to convince yourself, but even more so investors, that this can be a meaningful undertaking. Investors do not want to know about scientific breakthroughs, only about a clear route towards the development of marketable products with attractive market share and revenues. The main purpose of the business plan is to demonstrate that scientific rationale, a realistic estimation of a development plan, and most importantly the people are there to make a success happen.

Biotech financing is a science in itself. In contrast to most other industry sectors, biotech companies are forced to burn money to prove as quickly as possible that they can deliver products or candidates at least with huge market potential. The enormous capital demands are supplied by consecutive financing rounds in which the equity of the biotech companies is stepwise increased among private investors (venture capital firms, typically). Venture capitalists want to sell the stock they took with a high return on investment through either an IPO or a trade sale of the portfolio company.

Box 38.1 Venture Capital and Biotech Funding

Biotech companies need venture capital to develop their concepts and products. Venture capital companies need promising biotech companies as targets of their **investments**. However, the relationship between biotechs and venture capitalists is mostly a tale of love and hate. In times of scarce funding money, biotechs strive for venture capital money and compete heavily against each other. In times of high biotech valuations, the situation turns into the opposite: biotechs can then invite venture capitalists to undergo a **beauty contest**. The nature of their interests is by definition contrary, although they both want to achieve success with the company: whereas the founders want to get as much money as possible for their startup against selling as few shares as possible, the venture capitalists want to obtain as many shares as possible. However, there is a range that should not be surpassed in one or the other direction in order to prevent problems in future negotiations. If the valuation of the first round is too high, the company runs into the problem that the **step-up** in valuation up to the next round has to be high enough in order to raise the additional sum needed for the next phases. If not, they will be faced with a down round in valuation (i.e., the pre-money valuation of the second financing round is lower than the post-money of the first round). The **general principle** should be that the investors that come in at later stages have to pay a higher price per share than the earlier investors, justified by the lower risk of their investment and by being potentially closer to an exit. This is the theory, but the practice looks much different. The **cyclical nature of biotech valuations** is more dominant to the actual numbers so that it outweighs the actual factors that contribute to the valuation (i.e., achievement of milestones). This may result in high step-ups over a few years when the general biotech sentiment is "bullish", meaning investors pay more for biotech in general. However, it can also result in substantial down rounds, as it happened between 2001 and 2004, which was detrimental to early-stage investments. Not only that, but pre-money valuations on rounds are below the earlier rounds, which results in dilution of the early investors. There are other mechanisms that provide a dominance of the later incoming investors (i.e., **liquidation preferences**). A liquidation preference is usually not only applicable to actual liquidation of the company but to all cases where ownership of the company stocks will change, in particular mergers or acquisitions. A one-fold liquidation preference guarantees the investors that in the case of an acquisition they first get the sum invested back. If there is surplus money from the transaction, the remainder will be given to all shareholders, including the founders, then, *pro rata* according to the relative percentage of their stock ownership. The instruments of liquidation preference and down rounds have been used to rescue biotechs that were overvalued in the boom time days but at the expense of the early-stage investors. This resulted in a negative mood for seed investments and let investors focus on this late-stage strategy to get in later and squeeze out early investors. In general, venture capital is by far not a mechanism that aims to contribute to the progress of science or medication. It is entirely motivated by yielding very high returns to the investors into venture funds. The venture capitalists themselves set the goal to outperform their competitors, which results in very harsh market conditions. Letting venture investors into your company certainly means a pact with the devil, but they are willing to invest money, and this is what is needed for growing a drug discovery and development enterprise.

38.2
Employees: Recruitment, Remuneration, Participation

This section deals with the most important aspect of building up your own company, the **entrepreneurs** and **employees**; in other words, the people who are supposed to be the driving engine of the company's success. Choosing the right individuals is crucial.

Investors often want to see **managers** in the panel with years of experience in the pharmaceutical industry and, above all, **business experience**. What exactly is meant by business experience? What are those highly valued qualities a founding team needs, apart from technological and scientific knowhow and common sense in order to lead a company from within and shape its corporate image?

These questions address the soft factors that help highlight significant differences in personality type. Let us first take a look at the top management level. The entrepreneurs need to be absolutely competent in the scientific and technological field they are doing business in, and in order to be able to raise funds from investors they must even be in the international premier league. Only then can they be successful against competition on their home turf. **Scientific and technological competence** could thus be described as the necessary condition for a successful business venture. The necessary conditions comprise a range of various abilities and personal qualities. In Table 38.1, we give examples of such qualities as opposite extremes on a spectrum. It is not only the competence of a person in a particular field, such as business management, that decides whether a person is a suitable project manager, sales director, or CEO. Table 38.1 gives an overview of relevant qualities in managers and employees.

Many of the qualities described as competences can only be developed on the basis of many years of experience in the biotechnological or pharmaceutical industry, which is why venture capital investors often want to see these experienced people in top management positions. However, a successful manager in a startup enterprise must be able to cope with its specific conditions, such as a lack of infrastructure or shortage of funding – conditions not always to the liking of managers coming from established companies. In any case, it is crucial that entrepreneurs and newly brought in managers act in complete agreement. Otherwise, conflicts of interest between the two groups tend to grow.

Table 38.1 Business attitude and experience.

Social skills

High	Low
• Actively approaches other people	• Retires into the lab
• Enjoys traveling	• Prefers to stay at home
• Cooperative, ready to compromise	• Confrontational
• Has years of experience working with customers	• Not used to accommodate customers' wishes
• Has the ability of turning theoretical approaches into affordable projects or products	• In spite of technical knowledge unable to come up with sellable products

Project management skills

High	Low
• Able to plan and structure the work	• Leaps before he looks as far as work is concerned
• Communicates with ease, open-mindedness and directness	• Does not like to talk, never approaches anybody directly
• Delegates tasks and responsibilities	• Only trusts himself and avoids involving other people

Leadership skills

High	Low
• Has the ability to inspire other people with visions, strategies and objectives	• Gets bogged down in detail instead of seeing the broader picture
• Has an open ear for the employees' concerns at a personal and technical level and is accepted by them	• Has no understanding of the science side and does not listen to employees
• Maintains contact with partners in all relevant fields	• Inexperienced, has no links to industry
• Balances the interests of scientists, investors, and partners	• Unilaterally supports the interest of a single group

You will not get anywhere without **qualified personnel**. A small startup company needs to cover far more areas than an academic research group. How do you find the people you are looking for?

A small company working in the field of conventional compound research, such as small-molecule drug discovery, has to cover various fields of competence in order to develop a project to the clinical approval stage (Table 38.2). Many functions need to be outsourced and in order to find competent partners in their fields, it makes sense to put a person in charge within your own company who can build on a broad experience. There are various ways of recruiting highly skilled lab assistants. The easiest and often the most effective way is by posting jobs in the relevant newsgroups on the internet. It can also be useful to publish appointment ads in the national newspapers, as this underlines the trustworthiness of the job offer. The same applies to ads in relevant journals and magazines (e.g., *Science* or *Nature*). In these days of job insecurity, many applicants tend to look for jobs with firms that look as if they might still be around a few years down the line. A professionally designed newspaper advertisement is more likely to create this image than an internet posting that has been hastily put together.

Another increasingly popular, although rather costly, recruiting method is the use of **recruiting agencies**, also known as **headhunters**. A recruiting agent that specializes in your field may be able to help you recruit a highly skilled person already holding a good position who would normally not answer job offers. In order to lure these very able people into your company you must make them an offer they cannot refuse.

This brings us to the **topic of remuneration**. The salaries paid by up-and-coming biotech companies vary greatly, not only between companies, but also within the hierarchy of companies. During the biotech euphoria of the late 1990s, experienced managers from the pharmaceutical industry were lured in droves

Table 38.2 Fields of competence and necessary qualifications in a startup company developing a project to the clinical approval stage.

Fields of competence/necessary qualifications	In-house vs. Outsourcing
Biologists, molecular biologists, biochemists, physiologists	In-house
Clinicians	Cooperation
Medicinal chemists, analytical chemists	In-house, partly outsourcing
Bioinformaticists, cheminformaticists	In-house
System administrator	Preferably in combination of function with bioinformaticists and cheminformaticists
Galenics	Outsourcing
Toxicologists, pharmacologists	In-house or outsourcing
Project management	In-house
Accounting, controlling	One person in-house, otherwise outsourcing
Business development	In-house
Marketing, public relations	Partly in-house, outsourcing
Finance	In-house
Legal matters	Partly in-house, outsourcing
Patents	Outsourcing

into small startup companies, with the help of **high salaries** paid from borrowed venture capital. Furthermore, on top of a good salary, stock options or direct shares may help sway long-serving managers or brilliant scientists to swap their good and secure jobs for the risky business of founding a new company.

As already mentioned, the prospect of a company multiplying its value within a few years is considered to be one of the main motivations for entrepreneurs as well as investors. Both usually hold direct shares. Why is there a need for additional stock options?

Buying stock is only worthwhile if they can be purchased at roughly nominal value and sold later at a much higher rate. If a company sells stock to employees or other people at a lower rate than their current market value, this could be considered a noncash benefit, as the buyer could make a risk-free profit by selling the stock at its market value, which would incur income tax. A new employee who buys stock at its market value, however, runs the same high risk as any other venture capital investor of losing everything if the company files for insolvency.

Stock options offer the advantages of stock without the drawback of having to pay cash upfront. A company offering an **employee stock option (ESO) scheme** reserves part of its capital for its employees. Depending on their position in the hierarchy, their entry date, and their individual performance, they are given an annual package of stock options (i.e., they acquire the right of converting the options into real stock). Since these options are not yet actual shares, the employee does not have to pay for them. Once the vesting period is over and the options can be exercised, their value may have risen and the employee paying the issuing value and obtaining stock at the current market value makes a profit. If, however, the options are not fungible (i.e., the company is not listed on the stock market) or if the value of the enterprise has decreased, the employee can either drop the options or perhaps exchange them against newly issued ones. Thus, an employee is protected from potential losses, but can have his/her share in profits made after he/she has joined the company.

ESOs can work as incentives for employees, especially in the management sector. In Europe, however, such schemes are still the exception and have so far failed to turn ordinary employees into rich people. This makes them less acceptable to the blue-collar section where many prefer a regular salary to a stock scheme. The setting of annual targets and individual performance-related annual premiums could be a viable alternative incentive.

Do not forget that apart from all those financial incentives, there are other factors that may have a more decisive impact on **employee's** motivation and their well-being, such as a good working atmosphere, short decision-making processes, clear-cut responsibilities, and self-determination.

"There is a tide in the affairs of men,
Which taken at the flood, leads on to fortune;
Omitted, all the voyage of their life is bound
in shadows and miseries.
On such a full sea are we now afloat,
And we must take the current when it serves,
or lose our ventures."

(Brutus from Shakespeare's Julius Caesar)

39
Marketing

Learning Objectives

In this chapter, we look at the marketing of up-and-coming biotech companies. Traditional marketing tools such as market research and product marketing are less important here than a business development strategy. This includes focusing on potential customers and partners as well as watching competitors and deal making. There are three kinds of biotech-biotech and biotech-pharmaceutical arrangements: fee-for-service contracts, where research is done for a paying customer; technology transfer or licensing agreements; and, somewhere in between those two, strategic alliances. Business communication comprises a set of tasks different from business development, and is often split into public relations and investor relations. Business communication is essential for new biotech firms who have not yet made a name for themselves, in order to develop a corporate image that attracts investors, partners in the pharmaceutical industry, the media, and employees. These are also known as stakeholders.

39.1
Introduction

In biotech companies, marketing involves much more than is usually meant by this term. In an established company selling a certain range of products, **marketing tasks** are pretty well defined:

- **Market research** in order to identify new product opportunities and to watch the moves of competitors.
- **Product marketing** starts in early development stages, liaising with customers, scientists, and technicians to define the specifications of the new product and supervising the developing process to market readiness. Product marketing also sets sales parameters.

While the positioning of the company, its performance, and its strategy are in the hands of **staff departments**, business results are a matter for the **public relations (PR)** and **investor relations (IR) departments**, which are not part of the marketing division.

In new biotech companies, all these functions amalgamate into what is labeled "marketing". In a large company selling consumer goods there are specially trained staff to cover each specific area of marketing or PR and these need not know anything about the technical side or the scientific applications of their products. In small biotech companies, by contrast, all these various tasks must be carried out by just a few employees, if not one individual. Sometimes, there is not even a job description for marketing. There are several reasons for this:

- Most biotech firms do not start out with a ready product, but are defined as **research and development (R&D) companies** that, if successful, will sell their

An Introduction to Molecular Biotechnology, 2nd Edition.
Edited by Michael Wink
Copyright © 2011 WILEY-VCH Verlag GmbH & Co. KGaA, Weinheim
ISBN: 978-3-527-32637-2

precursor product licenses to larger companies. These companies do not need a product marketing department of their own.

- Marketing science-based high-tech products is a demanding task and requires substantial knowledge in the relevant science. Not many scientists are prepared to put their efforts into marketing, which has a fairly low standing in the academic world. Traditional marketing experts, on the other hand, find it far more rewarding to work for an established company.

As a consequence, marketing has been replaced by **business development** in most biotech companies and each company has a different idea of what that involves. Business development means identifying potential cooperation **partners**, developing cooperation **concepts**, deal closing, and ongoing cooperation **management**. Business development describes a more limited range of activities in established pharmaceutical companies. The term has been adopted and its meaning modified by biotech companies where the individuals in charge must be prepared to take over the following tasks, since scientists are usually not inclined to familiarize themselves with them:

- **Watching competitors**. What are the strategies and product offers of other companies in the sector?
- **Market analysis**. What is the size of the target markets for which new products or new technologies are being developed?
- **Contacting potential partners** for cooperation, preparing cooperation agreements, and finalizing them; also known as deal making.

The first two tasks are traditional marketing functions, whereas the third activity is really a sales function. However, deal making is the main responsibility of the person in charge of business development. Why? Because successful cooperation is the only way biotech companies can create a turnover and this is the function in business development we are going to concentrate on in this chapter.

39.2
What Types of Deals are Possible?

Three basic types of cooperation between biotech firms and other companies can be distinguished:

- **Fee-for-service deals**. The biotech partner delivers a service that does not involve knowhow or technology transfer to a customer for a fee;
- **Technology transfer or licensing deals**. The biotech partner transfers or licenses its technology platform, its patented new active compounds, or a diagnostic to a partner company for further development or marketing.
- **Strategic alliances**. Two partners combine a large part of their resources. Such far-reaching deals are usually accompanied by an exchange or sale of company shares.

In a **fee-for-service deal** things are more or less clear-cut. The fee is determined by the kind and extent of the service provided. In a competitive world, the difficulty lies in agreeing on a fee that not only covers the **cost**, but also provides a **profit margin**. This may sound trivial, but is often ignored by the industry in a situation where many hundreds of biotech companies are jostling for contracts with only 20–30 pharmaceutical companies.

The nature of **licensing deals** can vary. For a therapeutic still in the preliminary preclinical stages (e.g., target candidate, lead series) the deal very often not only includes a license, but also cooperation in research. The **revenue flows** of the biotech firm may then comprise the following components:

1. An **upfront payment** when work is resumed. This should cover the preliminary investments of the biotech firm and is usually not refundable.

2. **Milestone payments** are due when the cooperation project reaches important development milestones.
3. **Royalties** (i.e., licensing fees) can only be charged where a patent exists. These must be distinguished from licenses on technologies where royalties must be paid during the research cooperation, and the licensing of, for example, new active compounds where royalties are only due after market approval of the medication. These are the royalties most biotech companies are most keen to receive. The following examples will illustrate the reasons behind this.

39.3
What Milestone or License Fees are Effectively Paid in a Biotech/Pharma Cooperation?

Let us assume that a biotech company called A-Gen wants to license a new therapeutic, say, a new low-molecular antitumor compound, to Pharma X. The **licensing deal** could look as follows. Pharma X agrees on a 2-year cooperation period with A-Gen to find out whether the compound would be suitable at all for clinical trials. Payments could be scheduled as in the following example of the licensing of a new antitumor compound for recurrent breast cancer:

- **Total market potential**: around US$8 billion in the seven major pharmaceutical markets, (United States, Japan, Germany, United Kingdom, France, Italy, and Spain), increasing by about 5% per annum.
- **Market penetration** (i.e., the market share of the product envisaged): around 20%, which would amount to peak sales of about $0.2 \times (\$8 \text{ billion} \times 1.05)^n$ (n = number of years until market approval, 5% market growth per annum.).

The claimable license fees depend on the value added by the biotech company at the time of licensing. Table 39.1 gives an overview of the sums in question.

It makes therefore perfect sense for biotech firms to develop all projects at least to **clinical phase II**, if only because the pharmaceutical industry is not interested in nonvalidated compounds. In the real world of scarce finances, however, many biotech companies are forced to sell their licenses much earlier.

Naturally, the amount of upfront and milestone payments also depends on **market potential**, and can range from several hundred thousand to a few millions of Euros in the early stages of development. What figures stand in the final contract depends on the negotiating skills of the **business development manager**. An experienced business development manager in the biotech industry will therefore carefully choose a negotiation partner who is most in need of the specific product and is prepared to pay the maximum. This is also a good strategy to drive up royalties.

A successful licensing deal not only provides the biotech company with a revenue, but substantially changes the basis for the valuation of the firm. This

Table 39.1 Potential revenue for the sale of developed compounds.

Development Status	Royalties in percent	Annual revenue for an anti-tumor compound in Million € (after approval in about 10 years time)
Validated target	0.5–3%	20–130
Lead Compounds	0.5–3%	20–130
Drug Candidate (i.e., optimized compounds including PK data, tox data, proof of principle in an animal model)	1–10%	40–400
Phase I/II Drugs	5–20%	200–800
Phase III NDA	10–50%	400–1700

PK, Pharmacokinetic; tox, Toxicology; NDA, New Drug Application

enables the company to obtain more capital on better terms the next time round, making deal making a pivotal marketing tool for a biotech company.

A **strategic alliance** differs from a licensing deal insofar as, in most cases, more than one component is licensed and the partner – a pharmaceutical or another biotech company – buys shares in the licensing company. As these are mostly sold at a premium rate, a strategic alliance could also be considered a **financing deal**, only this time the stock is not held by a venture capital company but by a pharmaceutical or biotech company. Two smaller companies that complement each other in value-added terms may also form a **strategic alliance**. They could either be companies owning complementary technologies in the same sector that want to improve their potential by **cross-licensing** or they could be, say, a company in the therapeutic sector and a chemical firm that want to join forces for the development of new compounds. Such alliances often herald a later **merger** or a **takeover**.

39.4
PR and IR in Biotech Companies

Although in most cases there is no direct link between a biotech company and the end user (e.g., consumers or patients), business communication (i.e., PR and IR) is extremely important, as **brand name recognition** is reflected in the value of a company. In particular, companies that prepare for listing on the **stock market** must pay great attention to their **corporate image**.

Biotechnology is considered high technology and hopes for future value-added are pinned on it. In times of economic crisis, the media are looking for the few promising examples of growth and new jobs, and biotech companies could use this image bonus to earn points with politicians and the wider public, bringing a glimmer of hope into the doom and gloom that besets the **traditional old economy industries**. The hype fired by wishful thinking could explain the high initial stock market valuation of some German biotech enterprises. However, when expectations proved to be unrealistic, their stocks began to plummet accordingly.

Extreme variation in the perception of the biotech industry over time shows that the companies concerned did not have a robust communication strategy in place. When in the spotlight of public attention, it is extremely important for a company to have worked out a communication strategy in order to sustain a positive corporate image.

This applies not only to firms quoted on the stock exchange, but also for small startups. The evaluation of a business by investors or potential cooperation partners in the pharmaceutical industry feeds on information obtained from several sources. Good PR not only reports on the current status of the enterprise, but endeavors to give an informed comment on major business decisions. All PR aims at generating a positive brand image shared by all stakeholders (i.e., employees, stock holders, investors, and cooperation partners). The name of the company must be associated with the positive values that form the basis of management decisions.

The **PR officer** in the company (often also acting as business development manager, general manager, or chairman) has to fulfill the following tasks:
- Define the objectives and values of the company.
- Convey these to the stakeholders of the business.

It is surprising to see how vaguely many startup biotech companies define their objectives. Often, the objectives are merely technical ones, such as **functionally validating targets** or to **find compounds to treat central nervous system diseases**. The objectives of a company must be independent of the success or failure of a project, or else they may shift indiscriminately. The company's mission statement should go hand in hand with an economic strategy. It may be a noble objective to cure untreatable diseases, but unless it is underpinned by a sound **economic strategy**, no investor will be prepared to part with their money. At the opposite end of the spectrum, a company defines economic success as its sole objective, neglecting the fact that the economic success of a biotech company can only be based on scientific and technological success. Only business objectives that have been well defined (e.g., the economic target of x billion turnover and y% profit for the year 20XX) can be presented to the public and attained.

What means of communication are available to the PR officer?

- **Press releases** sent out to various publications and journalists.
- **Company homepage** as a tool for shaping the corporate image.
- **Presentations** on the premises of customers or at conferences.
- **Addressing** potential partners personally.

All means of communication must be used in a concerted manner to create a coherent stream of information. Not only the contents of PR campaigns need to be coordinated, but also the visual image. Within the limited space of this textbook, we cannot even scratch the surface of topics like corporate identity. Suffice it to say that the choice of name, color scheme, and logo of a company can be used successfully to make it stand out from the crowd.

Appendix

An Introduction to Molecular Biotechnology, 2nd Edition.
Edited by Michael Wink
Copyright © 2011 WILEY-VCH Verlag GmbH & Co. KGaA, Weinheim
ISBN: 978-3-527-32637-2

Further Reading

Preface

Alberts, B., Johnson, A., Lewis, J., Raff, M., Roberts, K., Walter, P. (2008) Molecular Biology of the Cell, 5th edn, Garland Science, New York.

Ausubel, F.M., Brent, R., Kingston, R.E. et al. (eds.) (2009) Current Protocols in Molecular Biology. Wiley, Hoboken.

Ausubel, F.M., Brent, R., Kingston, R.E., et al. (eds.) (2010) Current Protocols in Molecular Biology, Wiley Interscience, Hoboken. [Updated every 3 months].

Campbell, N.A., Reece, J.B. (2008) Biology, 8th edn, Benjamin Cummings, San Francisco.

Katoh, S., Yoshida, F. (2009) Biochemical Engineering: A Textbook for Engineers, Chemists and Biologists. Wiley-VCH, Weinheim.

Kent, J.A. (2007) Kent and Riegel's Handbook of Industrial Chemistry: Handbook of Industrial Chemistry and Biotechnology, 11th edn. Springer, Heidelberg.

Krebs, J.E., Goldstein, E.S., Kilpatrick, S.T. (2011) Lewin's Genes X. Jones & Bartlett, Sudbury.

Renneberg, R. (2007) Biotechnology for Beginners. Springer, Heidelberg.

Sadava, D, Orians, G.H., Heller, H.C., Purves, W., Hillis, D. (2006) The Science of Biology. Freeman, San Francisco.

Soetard, W., Vandamme, E.J. (2010) Industrial Biotechnology: Sustainable Growth and Economic Success. Wiley-VCH, Weinheim.

Voet, D., Voet, J.G., Pratt, C.W. (2008) Fundamentals of Biochemistry. Wiley, Hoboken.

Walsh, G. (2007) Pharmaceutical Biotechnology. Concepts and Applications. Wiley, Hoboken.

Chapter 1–6

Alberts, B., Johnson, A., Lewis, J., Raff, M., Roberts, K., Walter, P. (2008) Molecular Biology of the Cell. 5th edn, Garland Science, New York.

Ambros, V. (2003) MicroRNA pathways in flies and worms: growth, death, fat, stress, and timing. Cell 113, 673–676.

Berk, A., Baltimore, D., Zipursky, S.L., Matsudaira, P., Darnell J. (1999) Molecular Cell Biology. 3rd edn, Scientific American Books, New York.

Carrington, J.C., Ambros, V. (2003) Role of microRNAs in plant and animal development. Science 301, 336–338.

Ciccarelli, F.D., Doerks, T., von Mering, C., Creevey, C.J., Snel, B., Bork, P. (2006) Toward automatic reconstruction of a highly resolved Tree of Life. Science 311, 1283–1287.

Dey, P.M., Harborne, J.B. (1997) Plant Biochemistry. Academic Press, San Diego.

Klug, W.S., Cummings, M.R. (1997) Concepts of Genetics. Prentice Hall, Upper Saddle River.

Krebs, J.E., Goldstein, E.S., Kilpatrick, S.T. (2011) Lewin's Genes X. Jones & Bartlett, Sudbury.

McManus, M.T., Sharp, P.A. (2002) Gene silencing in mammals by small interfering RNAs. Nat. Rev. Genet. 3, 737–747.

Voet, D., Voet, J.G., Pratt, C.W. (2002) Fundamentals of Biochemistry, upgrade edition. Wiley, New York.

Wink, M. (2010) Functions and Biotechnology of Plant Secondary Metabolites. Wiley-Blackwell Annual Plant Reviews vol. 39, Oxford.

Wink, M. (2010) Biochemistry of Plant Secondary Metabolism. Wiley-Blackwell Annual Plant Reviews vol. 40, Oxford.

Chapter 7

Deutscher M.P. (ed) (1990) Methods in Enzymology, Bd. 182, Guide to Protein Purification. Academic Press, San Diego.

Janson, J.C., Rydèn L. (1998) Protein Purification, Principles, High Resolution Methods and Applications. 2nd edn, Wiley VCH, Weinheim.

Scopes, R.K. (1994) Protein Purification, Principles and Practice. 3rd edn, Springer, Heidelberg, New York.

Sofer G., Hagel, L. (1997) Handbook of Process Chromatography. Academic Press, San Diego.

Chapter 8

Aebersold, R., Goodlett, D.R. (2001) Mass spectrometry in proteomics. Chem. Rev. 101, 269–295.

Kellner, R., Lottspeich, F., Meyer, H.E. (1999) Microcharacterization of Proteins, 2nd edn Wiley-VCH, Weinheim.

Kinter, M., Sherman, N.E. (2000) Protein Sequencing and Identification Using Tandem Mass Spectrometry. Wiley-Interscience, New York.

Makarov, A., E. Denisov, A. Kholomeev, W. Baischun, O. Lange, K. Strupat, S. Horning (2006) Performance evaluation of a hybrid linear ion trap/orbitrap mass spectrometer. Anal. Chem. 78:2113–2120.

Siuzdak, G. (1996) Mass Spectrometry for Biotechnology. San Diego, Academic Press.

Chapter 9

Ausubel, F.M., Brent, R., Kingston, R.E. et al. (eds.) (2010) Current Protocols in Molecular Biology. Wiley, Hoboken.

Chapter 10

Ausubel, F.M., Brent, R., Kingston, R.E. et al. (eds.) (2009) Current Protocols in Molecular Biology. Wiley, New York.

Sambrook, J., Russell, D. (2001) Molecular Cloning: A Laboratory Manual. 3rd edn, Cold Spring Harbor Laboratory, Cold Spring Harbor.

Chapter 11

Clark, M. (1996) In situ-Hybridization: Laboratory Companion. Wiley-VCH, Weinheim.

Grunstein, M., Hogness, D.S. (1975) Colony hybridization: a method for the isolation of cloned DNAs that contain a specific gene. Proc. Natl. Acad. Sci. USA 72(10), 3961–3965.

Jäger, R., Herzer, U., Schenkel, J., Weiher, H. (1997) Overexpression of Bcl-2 inhibits alveo-

An Introduction to Molecular Biotechnology, 2nd Edition.
Edited by Michael Wink
Copyright © 2011 WILEY-VCH Verlag GmbH & Co. KGaA, Weinheim
ISBN: 978-3-527-32637-2

lar cell apoptosis during involution and accelerates c-myc-induced tumorigenesis of the mammary gland in transgenic mice. Oncogene 15, 1787–1795.

Southern, E. M. (1975) Detection of specific sequences among DNA fragments separated by gel electrophoresis. J. Mol. Biol. 5, 98(3), 503–517.

Watson, J. D., Crick, F. H. C. (1953) Molecular structure of nucleic acids: a structure for deoxyribose nucleic acid. Nature 171 (4356), 737–738.

Chapter 12

Ausubel, F. M., Brent, R., Kingston, R. E. et al. (eds.) (2009) Current Protocols in Molecular Biology. Wiley, New York.

Sambrook, J., Russel, D. (2001) Molecular Cloning: A Laboratory Manual, 3rd edn Cold Spring Harbor Laboratory, Cold Spring Harbor.

Chapter 13

Mullis, K. B., Faloona, F. A. (1987) Specific synthesis of DNA in vitro via a polymerase-catalyzed chain reaction. Methods Enzymol. 155, 335–350.

Saiki, R. K., Gelfand, D. H., Stoffel, S., Scharf, S. J., Higuchi, R. et al. (1988) Primer directed enzymatic amplification of DNA with a thermostable DNA polymerase. Science 239, 487–491.

Chapter 14

Lander, E. S., Linton, L. M., Birren, B. et al. (2001) Initial sequencing and analysis of the human genome. Nature 409, 860–921.

Maxam, A., Gilbert, W. (1977) A new method for sequencing DNA. Proc. Natl. Acad. Sci. USA 74, 560–564.

Sanger, F., Nicklen, S., Coulsen, A. R. (1977) DNA sequencing with chain-terminating inhibitors. Proc. Natl. Acad. Sci. USA 74, 5463–5467.

Venter, J. C., Adams, M. D., Myers, E. W. et al. (2001) The sequence of the human genome. Science 291, 1304–1351.

Websites

http://www.celera.com
http://www.ncbi.nlm.nih.gov
http://www.appliedbiosystems.com
http://www.454.com

Chapter 15

Ausubel, F. M., Brent, R., Kingston, R. E. et al. (eds.) (2009) Current Protocols in Molecular Biology. Wiley, New York.

Ausubel, F. M., Berger, S. L., Kimmel, A. R. (1987) Guide to molecular cloning techniques. Methods Enzymol. vol. 152.

Ausubel, F. M., Goeddel, D. V. (ed.) (1990) Gene expression technology. Methods Enzymol. vol. 185.

Ausubel, F. M., Kaufman, R. J. (2000) Overview of vector design for mammalian gene expression. Mol. Biotechnol. 16, 151–160.

Ausubel, F. M., Sambrook, J., Russell, D. W. (2000) Molecular Cloning: A Laboratory Manual. 3rd edn, Cold Spring Harbor Press, Cold Spring Harbor.

Ausubel, F. M., Van Craenenbroeck, K., Vanhoenacker, P., Haegeman, G. et al. (2000) Episomal vectors for gene expression in mammalian cells. Eur. J. Biochem. 267, 5665–5678.

Chapter 16

Baneyx, F. (1999) Recombinant protein expression in Escherichia coli. Curr. Opin. Biotechnol. 10, 411–421.

Betton, J. M. (2003) Rapid translation system (RTS): a promising alternative for recombinant protein production. Curr. Protein Pept. Sci. 4, 73–80.

Buckholz, R. G., Gleesson, M. A. G. (1991) Yeast systems for the commercial production of heterologous proteins. Biotechnology 9, 1067–1072.

Giga-Hama, Y., Kumagai, H. (1999) Expression system for foreign genes using the fission yeast Schizosaccharomyces pombe. Biotechnol. Appl. Biochem. 30, 235–244.

Groß, G., Hauser, H. J. (1995) Heterologous expression as a tool for gene identification and analysis. Biotechnology 41, 91–110.

Higgins, D. R., Cregg, J. M. (1998) Introduction to Pichia pastoris. In: Methods in Molecular Biology, Bd. 103: Pichia Protocols (Higgins, D. R., Cregg, J. M. eds.). Humana Press, Totowa.

O'Reilly, D. L., Miller, K., Luckow, V. A. (1992) Baculovirus Expression Vectors: A Laboratory Manual. Freeman, New York.

Chapter 17

Hamill, O. P., Marty, A., Neher, E., Sakmann, B., Sigworth, F. J. (1991) Improved patch-clamp techniques for high-resolution current recording from cells and cell-free membrane patches. Pflügers Arch. 391, 85–100.

Neher, E., Sakman, B. Single channel currents recorded from membrane of dener-

vated frog muscle fibers, Nature 1976. 260, 799–801.

Walz, W. (eds.) (2007) Patch-Clamp Analysis: Advanced Techniques. Humana Press, Totowa.

Chapter 18

Doyle, A., Griffiths, J. B. (1998) Cell and Tissue Culture: Laboratory Procedures in Biotechnology. Wiley, New York.

Givan, A. L. (2001) Flow Cytometry: First Principles. 2nd edn, Wiley, New York.

Koepp, D. M., Harper, J. W., Elledge, S. J. (1999) How the cyclin became a cyclin: regulated proteolysis in the cell cycle, Cell 97, 431–434.

Murray, A., Hunt, T. (1993) The Cell Cycle. Oxford University Press, Oxford.

Nash, P., Tang, X., Orlicky, S., Chen, Q., Gertler, F. B., et al. (2001) Phosphorylation of a CDK inhibitor sets a threshold for the onset of DNA replication. Nature 414, 514–521.

Shapiro, H. M. (2003) Practical Flow Cytometry. 4th edn, Wiley, New York.

Stein, G., Baserga, R., Giordano, A., Denhardt, D. (1999) The Molecular Basis of Cell Cycle and Growth Control. Wiley-Liss, New York.

Chapter 19

Goldman R. D. and Spector D. L. (eds.) (2005) Live Cell Imaging, A Laboratory Manual. Cold Spring Harbour Laboratory Press, Cold Spring Harbor, New York.

Periasamy A. (ed) (2000) Methods in Cellular Imaging. Oxford University Press, New York.

Steinbrecht R. A. and Zierold K. (eds.) (1987) Cryotechniques in Biological Electron Microscoopy, Springer Verlag, Berlin.

Chapter 20

Niemz, M. H. (2007) Laser-Tissue Interaction. Springer, Berlin, Heidelberg.

Pawley, J. (2006) Handbook of Biologicial Confocal Microscopy. Springer, New York.

Prasad, P. N. (2003) Introduction to Biophotonics. Wiley, New York.

Siegman, A. E. (1986) Lasers. University Science Books, Sausalito.

Chapter 21

Adams, M. D., Dubnick, M., Kerlavage, A. R., Moreno, R., Kelley, J. M. et al. (1992) Sequence identification of 2375 human brain genes, Nature 355, 632–634.

Adams, M. D., Celniker, S. E., Holt, R. A., Evans, C. A., Gocayne, J. D. et al. (2000) The genome sequence of *Drosophila melanogaster*, Science 287, 2185–2195.

Altschul, S. F., Gish, W., Miller, W., Myers, E. W., Lipman, D. J. (1990) Basic local alignment search tool, J. Mol. Biol. 215, 403–410.

Ansorge, W., Sproat, B. S., Stegemann, J., Schwager, C. (1986) A non-radioactive automated method for DNA sequence determination, J. Biochem. Biophys. Methods 13, 315–323.

Balling, R. (2001) ENU mutagenesis: analyzing gene function in mice, Annu. Rev. Genomics Hum. Genet. 2, 463–492.

Baltimore, D. (1970) RNA-dependent DNA polymerase in virions of RNA tumour viruses, Nature 226, 1209–1211.

Benson, D. A., Karsch-Mizrachi, I., Lipman, D. J., Ostell, J., Wheeler, D. L. (2003) GenBank, Nucleic Acids Res. 31, 23–27.

Birren, B., Lai, E. (1996) Nonmammalian Genomic Analysis – A Practical Guide, Academic Press, San Diego.

Bonaldo, M. F., Lennon, G., Soares, M. B. (1996) Normalization and subtraction: two approaches to facilitate gene discovery, Genome Res. 6, 791–806.

Botstein, D., Risch, N. (2003) Discovering genotypes underlying human phenotypes: past successes for Mendelian disease, future approaches for complex disease, Nat. Genet. 33 (Suppl.), 228–237.

Brenner, S., Johnson, M., Bridgham, J., Golda, G., Lloyd, D. H. et al. (2000) Gene expression analysis by massively parallel signature sequencing (MPSS) on microbead arrays, Nat. Biotechnol. 18, 630–634.

Brent, R. (2000) Genomic biology, Cell 100, 169–183.

Caenorhabditis elegans Sequencing Consortium. (1998) Genome sequence of the nematode C. elegans: a platform for investigating biology, Science 282, 2012–2018.

Camargo, A. A., Samaia, H. P., Dias-Neto, E., Simao, D. F., Migotto, I. A. et al. (2001) The contribution of 700,000 ORF sequence tags to the definition of the human transcriptome, Proc. Natl. Acad. Sci. USA 98, 12103–12108.

Cantor, R. C., Smith, C. L. (1999) Genomics – The Science and Technology Behind the Human Genome Project, Wiley-Interscience, New York.

Carpenter, A. E., Sabatini, D. M. (2004) Systematic genome-wide screens of gene function, Nat. Rev. Genet. 5, 11–22.

Cecconi, F., Gruss, P. (2002) From ES cells to mice: the gene trap approach, Methods Mol. Biol. 185, 335–346.

Crick, F. H. (1968) The origin of the genetic code, J. Mol. Biol. 38, 367–379.

Crick, F. H. (1970) Central dogma of molecular biology, Nature 227, 561–563.

Das, M., Harvey, I., Chu, L. L., Sinha, M., Pelletier, J. (2001) Full-length cDNAs: more than just reaching the ends, Physiol. Genomics 6, 57–80.

Dear, P. H., Cook, P. R. (1989) HAPPY mapping, a proposal for linkage mapping the human genome, Nucleic Acids Res. 17, 6795–6807.

Diatchenko, L., Lau, Y. F., Campbell, A. P., Chenchik, A., Moqadam, F. et al. (1996) Suppression subtractive hybridization: a method for generating differentially regulated or tissue-specific cDNA probes and libraries, Proc. Natl. Acad. Sci. USA 93, 6025–6030.

Ding, C., Cantor, C. R. (2004) Quantitative analysis of nucleic acids – the last few years of progress, J. Biochem. Mol. Biol. 37, 1–10.

Donis-Keller, H., Green, P., Helms, C., Cartinhour, S., Weiffenbach, B. et al. (1987) A genetic linkage map of the human genome, Cell 51, 319–337.

Dunham, I. (ed.). (2003) Genomic Mapping and Sequencing, Horizon Scientific Press, Hethersett.

Dykxhoorn, D. M., Novina, C. D., Sharp, P. A. (2003) Killing the messenger: short RNAs that silence gene expression, Nat. Rev. Mol. Cell Biol. 4, 457–467.

Edwards, A., Voss, H., Rice, P., Civitello, A., Stegemann, J. et al. (1990) Automated DNA sequencing of the human HPRT locus, Genomics 6, 593–608.

Elahi, E., Kumm, J., Ronaghi, M. (2004) Global genetic analysis, J. Biochem. Mol. Biol. 37, 11–27.

Fauth, C., Speicher, M. R. (2001) Classifying by colors: FISH-based genome analysis, Cytogenet. Cell Genet. 93, 1–10.

Fleischmann, R. D., Adams, M. D., White, O., Clayton, R. A., Kirkness, E. F. et al. (1995) Whole-genome random sequencing and assembly of *Haemophilus influenzae* Rd, Science 269, 496–512.

Gardner, M. J., Hall, N., Fung, E., White, O., Berriman, M. et al. (2002) Genome sequence of the human malaria parasite Plasmodium falciparum, Nature 419, 498–511.

Goffeau, A., Aert, R., Agostini-Carbone, M. L., Ahmed, A., Aigle, M., et al. (1997) The Yeast Genome Directory, Nature 387 (Suppl.), 1–105.

Goldsmith, Z. G., Dhanasekaran, N. (2004) The microrevolution: applications and impacts of microarray technology on molecular biology and medicine, Int. J. Mol. Med. 13, 483–495.

Green, C. D., Simons, J. F., Taillon, B. E., Lewin, D. A. (2001) Open systems: panoramic views of gene expression, J. Immunol. Methods 250, 67–79.

Gubler, U., Hoffman, B. J. (1983) A simple and very efficient method for generating cDNA libraries, Gene 25, 263–269.

Hoheisel, J. D., Lehrach, H. (1993) Use of reference libraries and hybridisation fingerprinting for relational genome analysis, FEBS Lett. 325, 118–122.

Holt, R. A., Subramanian, G. M., Halpern, A., Sutton, G. G., Charlab, R. et al. (2002) The genome sequence of the malaria mosquito *Anopheles gambiae*, Science 298, 129–149.

Horak, C. E., Snyder, M. (2002) Global analysis of gene expression in yeast, Funct. Integr. Genomics 2, 171–180.

Hubank, M., Schatz, D. G. (1994) Identifying differences in mRNA expression by representational difference analysis of cDNA, Nucleic Acids Res. 22, 5640–5648.

Kurian, K. M., Watson, C. J., Wyllie, A. H. (1999) DNA chip technology, J. Pathol. 187, 267–271.

Lennon, G., Auffray, C., Polymeropoulos, M., Soares, M. B. (1996) The IMAGE Consortium: an integrated molecular analysis of genomes and their expression, Genomics 33, 151–152.

Lichter, P. (1997) Multicolor FISHing: what's the catch?, Trends Genet. 13, 475–479.

Lykke-Andersen, J. (2001) mRNA quality control: marking the message for life or death, Curr. Biol. 11, R88–R91.

Matz, M. V., Lukyanov, S. A. (1998) Different strategies of differential display: areas of application, Nucleic Acids Res. 26, 5537–5543.

Maxam, A. M., Gilbert, W. (1977) A new method for sequencing DNA, Proc. Natl. Acad. Sci. USA 74, 560–564.

McClelland, M., Mathieu-Daude, F., Welsh, J. (1995) RNA fingerprinting and differential display using arbitrarily primed PCR, Trends Genet. 11, 242–246.

McGall, G. H., Christians, F. C. (2002) High-density genechip oligonucleotide probe arrays, Adv. Biochem. Eng. Biotechnol. 77, 21–42.

Miyazaki, S., Sugawara, H., Gojobori, T., Tateno, Y. (2003) DNA Data Bank of Japan (DDBJ) in XML, Nucleic Acids Res. 31, 13–16.

Mullis, K. B., Faloona, F. A. (1987) Specific synthesis of DNA in vitro via a polymerase-catalyzed chain reaction, Methods Enzymol. 155, 335–350.

Myers, E. W., Sutton, G. G., Smith, H. O., Adams, M. D., Venter, J. C. (2002) On the sequencing and assembly of the human genome, Proc. Natl. Acad. Sci. USA 99, 4145–4146.

Orgel, L. E. (1968) Evolution of the genetic apparatus, J. Mol. Biol. 38, 381–393.

Pollok, B. A., Heim, R. (1999) Using GFP in FRET-based applications, Trends Cell Biol. 9, 57–60.

Primrose, S. B., Twyman, R. M. (2003) Principles of Genome Analysis and Genomics. Blackwell, Oxford.

Reits, E. A., Neefjes, J. J. (2001) From fixed to FRAP: measuring protein mobility and activity in living cells, Nat. Cell Biol. 3, E145–E147.

Salser, W. A. (1974) DNA sequencing techniques, Annu. Rev. Biochem. 43, 923–965.

Sanger, F., Coulson, A. R. (1975) A rapid method for determining sequences in DNA by primed synthesis with DNA polymerase, J. Mol. Biol. 94, 441–448.

Sanger, F., Nicklen, S., Coulson, A. R. (1977) DNA sequencing with chain-terminating inhibitors, Proc. Natl. Acad. Sci. USA 74, 5463–5467.

Sanger, F., Coulson, A. R., Friedmann, T., Air, G. M., Barrell, B. G. et al. (1978) The nucleotide sequence of bacteriophage phiX174, J. Mol. Biol. 125, 225–246.

Shi, H., Maier, S., Nimmrich, I., Yan, P. S., Caldwell, C. W. et al. (2003) Oligonucleotide-based microarray for DNA methylation analysis: principles and applications, J. Cell Biochem. 88, 138–143.

Simon, R., Mirlacher, M., Sauter, G. (2004) Tissue microarrays, Biotechniques 36, 98–105.

Smith, L. M., Sanders, J. Z., Kaiser, R. J., Hughes, P., Dodd, C. et al. (1986) Fluorescence detection in automated DNA sequence analysis, Nature 321, 674–679.

Soares, M. B., Bonaldo, M. F., Jelene, P., Su, L., Lawton, L., Efstratiadis, A. (1994) Construction and characterization of a normalized cDNA library, Proc. Natl. Acad. Sci. USA 91, 9228–9232.

Southern, E. M. (2000) Blotting at 25, Trends Biochem. Sci. 25, 585–588.

Stoesser, G., Baker, W., van den Broek, A., Garcia-Pastor, M., Kanz, C. et al. (2003) The EMBL Nucleotide Sequence Database: major new developments, Nucleic Acids Res. 31, 17–22.

Sung, Y. H., Song, J., Lee, H. W. (2004) Functional genomics approach using mice, J. Biochem. Mol. Biol. 37, 122–132.

Suzuki, Y., Sugano, S. (2001) Construction of full-length-enriched cDNA libraries. The oligo-capping method, Methods Mol. Biol. 175, 143–153.

Tabor, S., Richardson, C. C. (1987) DNA sequence analysis with a modified bacteriophage T7 DNA polymerase, Proc. Natl. Acad. Sci. USA 84, 4767–4771.

Tabor, S., Richardson, C. C. (1989) Effect of manganese ions on the incorporation of dideoxynucleotides by bacteriophage T7 DNA polymerase and *Escherichia coli* DNA polymerase I, Proc. Natl. Acad. Sci. USA 86, 4076–4080.

Tabor, S., Richardson, C. C. (1995) A single residue in DNA polymerases of the *Escherichia coli* DNA polymerase I family is critical for distinguishing between deoxy- and dideoxyribonucleotides, Proc. Natl. Acad. Sci. USA 92, 6339–6343.

Tucker, C. L. (2002) High-throughput cell-based assays in yeast, Drug Discov. Today 7, S125–S130.

Velculescu, V. E., Vogelstein, B., Kinzler, K. W. (2000) Analysing uncharted transcriptomes with SAGE, Trends Genet. 16, 423–425.

Waterston, R. H., Lander, E. S., Sulston, J. E. (2002) On the sequencing of the human genome, Proc. Natl. Acad. Sci. USA 99, 3712–3716.

Waterston, R. H., Lander, E. S., Sulston, J. E. (2003) More on the sequencing of the human genome, Proc. Natl. Acad. Sci. USA 100, 3022–3024; author reply 3025–3026.

Watson, J., Crick, F. (1953) A structure for deoxyribose nucleic acid, Nature 171, 737–738.

Weier, H. U. (2001) DNA fiber mapping techniques for the assembly of high-resolution physical maps, J. Histochem. Cytochem. 49, 939–948.

Chapter 22

Alon, U. (2006) An introduction to Systems Biology. Chapmann & Hall CRC Press, Virginia Beach.

Cormen, T. and Leiserson, C. (2003) Introduction into algorithms. McGraw Hill, New York.

Klipp, E. et al. (2009) Systems Biology – A Textbook. Wiley-Blackwell, Weinheim.

Sachs, L. (1984) Applied Statistics. Springer, Heidelberg.

Witten, I. and Frank, E. (2005) Data Mining. Morgan Kauffman, San Francisco.

Chapter 23

Aldridge, B. B., Burke, J. M., Lauffenburger, D. A., Sorger, P. K. (2006) Physicochemical modelling of cell signalling pathways, Nat. Cell Biol. 8, 1195–1203.

Alexa, A., Rahnenfuhrer, J., Lengauer, T. (2006) Improved scoring of functional groups from gene expression data by decorrelating GO graph structure, Bioinformatics 22, 1600–1607.

Ashburner, M., Ball, C. A., Blake, J. A., Botstein, D., Butler, H. et al. (2000) Gene Ontology: tool for the unification of biology, Nat. Genet. 25, 25–29.

Bandara, S., Schloder, J. P., Eils, R., Bock, H. G., Meyer, T. (2009) Optimal experimental design for parameter estimation of a cell signaling model, PLoS Comput. Biol. 5, e1000558.

Barabasi, A. L., Albert, R. (1999) Emergence of scaling in random networks, Science 286, 509–512.

Barkai, N., Shilo, B. Z. (2007) Variability and robustness in biomolecular systems. Mol. Cell 28, 755–760.

Becskei, A., Seraphin, B., Serrano, L. (2001) Positive feedback in eukaryotic gene networks: cell differentiation by graded to binary response conversion, EMBO J. 20, 2528–2535.

Beissbarth, T., Speed, T. P. (2004) GOstat: find statistically overrepresented Gene Ontologies within a group of genes, Bioinformatics 20, 1464–1465.

Bentele, M., Lavrik, I., Ulrich, M., Stosser, S., Heermann, D. W. et al. (2004) Mathematical modeling reveals threshold mechanism in CD95-induced apoptosis, J. Cell Biol. 166, 839–851.

Bonacich, P. (1972) Factoring and weighting approaches to status scores and clique identification, J. Math. Sociol. 2, 113–120.

Bruggeman, F. J., Westerhoff, H. V. (2007) The nature of systems biology, Trends Microbiol. 15, 45–50.

Cawley, S., Bekiranov, S., Ng, H. H., Kapranov, P., Sekinger, E. A. et al. (2004) Unbiased mapping of transcription factor binding sites along human chromosomes 21 and 22 points to widespread regulation of noncoding RNAs, Cell 116, 499–509.

Chuang, H. Y., Lee, E., Liu, Y. T., Lee, D., Ideker, T. (2007) Network-based classification of breast cancer metastasis, Mol. Syst. Biol. 3, 140.

Clodong, S., Duhring, U., Kronk, L., Wilde, A., Axmann, I. et al. (2007) Functioning and robustness of a bacterial circadian clock, Mol. Syst. Biol. 3, 90.

Dinu, I., Potter, J. D., Mueller, T., Liu, Q., Adewale, A. J. et al. (2007) Improving gene set analysis of microarray data by SAM-GS, BMC Bioinformatics 8, 242.

Edelstein-Keshet, L. (1988) Mathematical Models in Biology, Random House, New York.

Edwards, J. S., Palsson, B. O. (2000) Metabolic flux balance analysis and the in silico analysis of *Escherichia coli* K-12 gene deletions, BMC Bioinformatics 1, 1.

Edwards, J. S., Ibarra, R. U., Palsson, B. O. (2001) In silico predictions of Escherichia coli metabolic capabilities are consistent with experimental data, Nat. Biotechnol. 19, 125–130.

Edwards, J. S., Covert, M., Palsson, B. (2002) Metabolic modelling of microbes: the flux-balance approach, Environ. Microbiol. 4, 133–140.

Estrada, E. (2006) Virtual identification of essential proteins within the protein interaction network of yeast, Proteomics 6, 35–40.

Gardner, T. S., Cantor, C. R., Collins, J. J. (2000) Construction of a genetic toggle switch in Escherichia coli, Nature 403, 339–342.

Goeman, J. J., van de Geer, S. A., de Kort, F., van Houwelingen, H. C. (2004) A global test for groups of genes: testing association with a clinical outcome, Bioinformatics 20, 93–99.

Gustafson, A. M., Snitkin, E. S., Parker, S. C., DeLisi, C., Kasif, S. (2006) Towards the identification of essential genes using targeted genome sequencing and comparative analysis, BMC Genomics 7, 265.

Hartwell, L. H., Hopfield, J. J., Leibler, S., Murray, A. W. (1999) From molecular to modular cell biology, Nature 402, C47–C52.

Hoops, S., Sahle, S., Gauges, R., Lee, C., Pahle, J. et al. (2006) COPASI – a COmplex PAthway SImulator, Bioinformatics 22, 3067–3074.

Jeong, H., Tombor, B., Albert, R., Oltvai, Z. N., Barabasi, A. L. (2000) The large-scale organization of metabolic networks, Nature 407, 651–654.

Kanehisa, M., Araki, M., Goto, S., Hattori, M., Hirakawa, M. et al. (2008) KEGG for linking genomes to life and the environment, Nucleic Acids Res. 36, D480–484.

Kirsch, D. G., Doseff, A., Chau, B. N., Lim, D. S., de Souza-Pinto, N. C. et al. (1999) Caspase-3-dependent cleavage of Bcl-2 promotes release of cytochrome c, J. Biol. Chem. 274, 21155–21161.

Kitano, H. (2002) Computational systems biology, Nature 420, 206–210.

Kohn, K. W. (1999) Molecular interaction map of the mammalian cell cycle control and DNA repair systems, Mol. Biol. Cell 10, 2703–2734.

König, R., Eils, R. (2004) Gene expression analysis on biochemical networks using the Potts spin model, Bioinformatics 20, 1500–1505.

Koschützki, D., Schreiber, F. (2004) Comparison of centralities for biological networks, Proc. German Conf. Bioinformatics 53, 199–206

Kromer, J. O., Wittmann, C., Schroder, H., Heinzle, E. (2006) Metabolic pathway analysis for rational design of L-methionine production by *Escherichia coli* and *Corynebacterium glutamicum*, Metab. Eng. 8, 353–369.

Kuepfer, L., Peter, M., Sauer, U., Stelling, J. (2007) Ensemble modeling for analysis of cell signaling dynamics, Nat. Biotechnol. 25, 1001–1006.

Legewie, S., Bluthgen, N., Herzel, H. (2006) Mathematical modeling identifies inhibitors of apoptosis as mediators of positive feedback and bistability, PLoS Comput. Biol. 2, e120.

Luscombe, N. M., Babu, M. M., Yu, H., Snyder, M., Teichmann, S. A., Gerstein, M. (2004) Genomic analysis of regulatory network dynamics reveals large topological changes, Nature 431, 308–312.

Maiwald, T., Timmer, J. (2008) Dynamical modeling and multi-experiment fitting with PottersWheel, Bioinformatics 24, 2037–2043.

Milo, R., Shen-Orr, S., Itzkovitz, S., Kashtan, N., Chklovskii, D., Alon, U. (2002) Network motifs: simple building blocks of complex networks, Science 298, 824–827.

Mishra, G. R., Suresh, M., Kumaran, K., Kannabiran, N., Suresh, S. et al. (2006) Human Protein Reference Database – 2006 update, Nucleic Acids Res. 34, D411–D414.

Newman, M. E. J. (2008) The mathematics of networks, in The New Palgrave Encyclopedia of Economics, 2nd edn, Blume, L. E., Durlauf, S. N. (eds.), Palgrave Macmillan, Basingstoke.

Plaimas, K., Mallm, J. P., Oswald, M., Svara, F., Sourjik, V. et al. (2008) Machine learning based analyses on metabolic networks supports high-throughput knockout screens, BMC Syst. Biol. 2, 67.

Plaimas, K., Eils, R., König, R. (2010) Identifying essential genes in bacterial metabolic networks with machine learning methods, BMC Syst. Biol., 4, 56.

Price, N. D., Shmulevich, I. (2007) Biochemical and statistical network models for systems biology, Curr. Opin. Biotechnol. 18, 365–370.

Price, N. D., Reed, J. L., Palsson, B. O. (2004) Genome-scale models of microbial cells: evaluating the consequences of constraints, Nat. Rev. Microbiol. 2, 886–897.

Rahman, S. A., Schomburg, D. (2006) Observing local and global properties of metabolic pathways: "load points" and "choke points" in the metabolic networks, Bioinformatics 22, 1767–1774.

Raj, A., van Oudenaarden, A. (2008) Nature, nurture, or chance: stochastic gene expression and its consequences, Cell 135, 216–226.

Raue, A., Kreutz, C., Maiwald, T., Bachmann, J., Schilling, M. et al. (2009) Structural and practical identifiability analysis of partially observed dynamical models by exploiting the profile likelihood, Bioinformatics 25, 1923–1929.

Schramm, G., Wiesberg, S., Diessl, N., Kranz, A. L., Sagulenko, V. et al. (2010) PathWave: discovering patterns of differentially regulated enzymes in metabolic pathways, Bioinformatics 26, 1225–1231.

Schuster, S., Dandekar, T., Fell, D. A. (1999) Detection of elementary flux modes in biochemical networks: a promising tool for pathway analysis and metabolic engineering, Trends Biotechnol. 17, 53–60.

Schuster, S., Fell, D. A., Dandekar, T. (2000) A general definition of metabolic pathways useful for systematic organization and analysis of complex metabolic networks, Nat. Biotechnol. 18, 326–332.

Seringhaus, M., Paccanaro, A., Borneman, A., Snyder, M., Gerstein, M. (2006) Predicting essential genes in fungal genomes, Genome Res. 16, 1126–1135.

Slee, E. A., Keogh, S. A., Martin, S. J. (2000) Cleavage of BID during cytotoxic drug and UV radiation-induced apoptosis occurs downstream of the point of Bcl-2 action and is catalysed by caspase-3: a potential feedback loop for amplification of apoptosis-associated mitochondrial cytochrome c release, Cell Death Differ. 7, 556–565.

Stites, E. C., Trampont, P. C., Ma, Z., Ravichandran, K. S. (2007) Network analysis of oncogenic Ras activation in cancer, Science 318, 463–467.

Swameye, I., Muller, T. G., Timmer, J., Sandra, O., Klingmuller, U. (2003) Identification of nucleocytoplasmic cycling as a remote sensor in cellular signaling by databased modeling, Proc. Natl. Acad. Sci. USA 100, 1028–1033.

Yeh, I., Hanekamp, T., Tsoka, S., Karp, P. D., Altman, R. B. (2004) Computational analysis of *Plasmodium falciparum* metabolism: organizing genomic information to facilitate drug discovery, Genome Res. 14, 917–924.

Chapter 24

Baldi, P., Brunak, S. (2001) Bioinformatics – The Machine Learning Approach. 2nd edn, MIT Press, Cambridge.

Bard, J. (2003) Ontologies: Formalising biological knowledge for bioinformatics. Bioessays 25, 501–506.

Braun P., Tasan M., Dreze M., Barrios-Rodiles M., Lemmens I., Yu H., Sahalie J. M. et al. (2009) An experimentally derived confidence score for binary protein-protein interactions, Nat. Methods 6(1), 91–97.

Cramer, P., Bushnell, D. A., Cornberg, R. D. (2001) Structural basis of transcription: RNA polymerase II at 2.8 resolution, Science 292, 1863–1876.

Dayhoff, M. O., Schwartz, R. M., Orcutt, B. C. (1978) A model of evolutionary change in proteins, in (Dayhoff, M. O., eds.) Atlas of Protein Sequence and Structure, vol. 5, Suppl. 3, S. 345–352. Natl. Biomedical Res. Foundation, Washington.

Doolittle, R. F. (1990) Molecular Evolution: Computer Analysis of Protein and Nucleic Acid Sequences. Methods in Enzymology, vol. 183. Academic Press, San Diego.

Duda, R. O., Hart, P. E., Stork, D. G. (2000) Pattern Classification. 2nd edn, Wiley, New York.

Dudoit, S., Fridlyand, J., Speed, T. P. (2002) Comparison of discrimination methods for the classification of tumors using gene expression data. J. Am. Statist. Assoc. 97, 77–87.

Dudoit, S., Yang, Y. H., Speed, T. P. et al. (2002) Statistical methods for identifying differentially expressed genes in replicated cDNA microarray experiments. Statistica Sinica 12, 111–139.

Durbin, R., Eddy, S., Krogh, A., Mitchison, G. (1998) Biological Sequence Analysis. Cambridge University Press, Cambridge, UK.

Gentleman, R., Carey, V., Dudoit, S., Irizarry, R., Huber, W., (eds.) (2005) Bioinformatics and Computational Biology Solutions Using R and Bioconductor. Springer, Berlin.

Giegerich, R. (2000) A systematic approach to dynamic programming in bioinformatics. Bioinformatics 16, 665–677.

Gusfield, D. (1997) Algorithms on Strings, Trees, and Sequences. Cambridge University Press, New York.

Hastie, T., Tibshirani, R., Friedman, J. (2001) The Elements of Statistical Learning. Springer, New York.

Hillis, D. M., Moritz, C., Mable, B. K. (eds.) (1996) Molecular Systematics. 2nd edn, Sinauer Associates, Sunderland.

Kaufman, L., Rousseuw, P. J. (2005) Finding Groups in Data. An Introduction to Cluster Analysis. Wiley, New York.

Li, W. H. (1997) Molecular Evolution. Sinauer Publishers, Sunderland.

Lockhart, D. J., Winzeler, E. A. (2000) Genomics, gene expression and DNA arrays. Nature 405, 827–836.

Luscombe, N.M., Laskowski, R.A., Thornton, J.M. (2001) Amino acid-base interactions: a three-dimensional analysis of Protein–DNA interactions at an atomic level, Nucleic Acids Res. 29, 2860–2874.

Mandell, D.J., Kortemme, T. (2009) Computer-aided design of functional protein Interactions, Nature Chemical Biology 5(11), 797–807.

Moss, T. (eds.) (2001) DNA-Protein Interactions: Principles and Protocols (Methods in Molecular Biology), Humana Press, Totowa.

Mount, D.W. (2001) Bioinformatics: Sequence and Genome Analysis. Cold Spring Harbor Laboratory Press, Cold Spring Harbor.

Nature Genetics Supplement "The Chipping Forecast" (1999) Nat. Genet. 21 Suppl. 1.

Nature Genetics Supplement "The Chipping Forecast II" (2002) Nat. Genet. 32 Suppl. 2.

Nature Insight "Computational Systems Biology" (2002) Nature 420, 205–251.

Pohl, E., Holmes, R.K., Hol, W.G.J. (1999) A metal binding SH3-domain observed in the cobalt activated diphtheria toxin repressor-DNA complex, J. Mol. Biol. 292, 653–667.

Schena, M. (2002) Microarray Analysis. Wiley, New York.

Schulze, A., Downward, J. (2001) Navigating gene expression using microarrays – a technology review. Nat. Cell Biol. 3, E190–E195.

Schwikowski, B., Uetz, P., Fields, S. (2000) A network of interacting proteins in yeast, Nat. Biotechnol. 18, 1257–1261.

Science Special "Systems Biology" (2002) Science 295, 1662–1682.

Speed, T.P. (eds.) (2003) Statistical Analysis of Gene Expression Microarray Data. Chapman & Hall, Boca Raton.

Venter, C.J., Adams, M.D., Myers, E.W. et al. (2001) The sequence of the human genome, Science 291, 1343.

Zapatka, M., Koch, Y., Brors, B. (2008) Ontological Analysis and Pathway Modelling in Drug Discovery. Pharm. Med. 22, 99–105.

Zhang, M.Q. (2002) Computational prediction of eukaryotic protein-coding genes. Nat. Rev. Genet. 3, 698–710.

Websites

GenBank   http://www.ncbi.nlm.nih.gov/Genbank
RNAfam   http://rfam.sanger.ac.uk/
RNAdb   http://research.imb.uq.edu.au/rnadb/
ncRNA-DB   http://biobases.ibch.poznan.pl/ncRNA/
SwissProt   http://www.uniprot.org
PIR   http://pir.georgetown.edu/
Ensembl   http://www.ensembl.org
GenomeBrowser (Golden Path)   http://genome.ucsc.edu

BLOCKS   http://blocks.fhcrc.org
Prosite   http//www.expasy.ch/prosite
Pfam   http://pfam.sanger.ac.uk/
ProDom   http://prodom.prabi.fr
SMART   http://smart.embl-heidelberg.de
PDB   http://http://www.rcsb.org/pdb/home/home.do
SCOP   http://scop.mrc-lmb.cam.ac.uk/scop/
KEGG   http://www.genome.jp/kegg/
GeneOntology   http://www.geneontology.org
Biocarta   http://www.biocarta.com
EcoCyc   http://ecocyc.org/
Reactome   http://www.reactome.org/
SAGE   http://www.sagenet.org/
Arrayexpress   http://www.ebi.ac.uk/microarray-as/ae/
GEO   http://www.ncbi.nlm.nih.gov/geo
PubMed   http://www.ncbi.nlm.nih.gov/sites/entrez?db=pubmed
OMIM   http://www.ncbi.nlm.nih.gov/omim
GeneCards   http://www.genecards.org
Biomedcentral   http://www.biomedcentral.com
PLoS   http://www.plos.org
PAUP   http://paup.csit.fsu.edu
Phylip   http://evolution.genetics.washington.edu/phylip.html
EMBOSS   http://www.emboss.org
BLAST   http://blast.ncbi.nlm.nih.gov/
WU-BLAST   http://blast.wustl.edu/
BLAT   http://genome.ucsc.edu/cgi-bin/hgBlat?command=start
R   http://www.r-project.org
Bioconductor   http://www.bioconductor.org
MTEV   http://www.tm4.org/
dChip   https://sites.google.com/site/dchipsoft/

Chapter 25

Alberts, B., Johnson, A., Lewis, J., Raff, M., Roberts, K., Walter, P. (2008) Molecular Biology of the Cell. 5th edn, Garland Science, New York.

Babu, M.M., Balaji, S., Aravind, A. (2007) General Trends in the Evolution of Prokaryotic Transcriptional Regulatory Network. Gene and Protein Evolution, Volff J.N. (ed) Genome Dyn. 3, 66–80.

Cesareni et al. (eds.) (2004) Modular Protein Domains. Wiley-VCH, Weinheim.

Cho, B.K., Zengler, K., Qiu, Y., Park, Y.S., Knight, E.M., Barrett, C.L., Gao, Y., Palsson, B.Ø. (2009) The transcription unit architecture of the Escherichia coli genome. Nature Biotech. 27 (11), 1043–1149.

Collins, C.H., Yokobayashi, Y., Umeno, D., Arnold, F.H. (2003) Engineering proteins that bind, move, make and break DNA. Curr. Opin. Biotechnol. 14(4), 371–378.

Darnell, J.E. (2002) Transcription factors as targets for cancer therapy. Nat. Rev. Cancer 2, 740–749.

Gavin, A.-C., Superti-Furga, G. (2003) Protein complexes and proteome organization from yeast to man. Curr. Opin. Chem. Biol. 7, 21–27.

Ghosh, S., Karin, M. (2002) Missing pieces in the NF-kB puzzle. Cell 109 (Suppl.), S81–S91.

Gibson D.G., Benders G.A., Andrews-Pfannkoch C., Denisova E.A., Baden-Tillson H., Zaveri J., Stockwell T.B., et al. (2008) Complete chemical synthesis, assembly, and cloning of a Mycoplasma genitalium genome. Science 319, 1215–1220.

Golemis, E. (eds.) (2005) Protein-Protein Interactions – A Molecular Cloning Manual. 2nd edn, Cold Spring Harbor Laboratory Press, New York.

Janin, J. (2000) Kinetics and thermodynamics of Protein-protein interactions from a structural perspective, in: Kleanthous, C. (eds.) Protein-Protein Recognition, p. 1–31. Oxford University Press, Oxford.

Jen-Jacobson, L., Engler, L.E., Jacobson, L.A. (2000) Structural and thermodynamic strategies for site-specific DNA binding proteins. Structure 8, 1015–1023.

Jones, S., Thornton, J.M. (2000) Analysis and classification of Protein-protein interactions from a structural perspective, in: Kleanthous, C. (eds.) Protein-Protein Recognition, S. 33–59. Oxford University Press, Oxford.

Lodish, H., Berk, A., Kaiser, C., Krieger, M., Scott, M.P. Bretscher, A., Ploegh, H., Matsudaira, P. (2008) Molecular Cell Biology. 6th edn, Freeman, New York.

Luscombe, N.M., Austin, S.E., Berman, H.M., Thornton, J.M. (2000) An overview of the structures of Protein-DNA complexes. Genome Biol. 1(1), 1–10.

Pawson, T., Nash, P. (2003) Assembly of cell regulatory systems through protein interaction domains. Science 300, 445–452.

Salwinski, L., Eisenberg, D. (2003) Computational methods of analysis of Protein-protein interactions. Curr. Opin. Struct. Biol. 13, 377–382.

Seshasayee A.S.N., Bertone, P., Fraser, G.M., Luscombe, N.M. (2006) Transcriptional regulatory networks in networks in bacteria: from input signals to output response. Curr Opin Microbiol 9, 511–519.

Thorner, J. (eds.) (2000) Applications of Chimeric Genes and Hybrid Proteins, Part C: Protein-Protein Interactions and Genomics. Methods in Enzymology 328, Academic Press.

Urnov, F.D., Rebar, E.J. (2002) Designed transcription factors as tools for therapeutics and functional genomics. Biochem. Pharmacol. 64(5/6), 919–923.

Walsh, G. (2002) Proteins: Biochemistry and Biotechnology. Wiley, Chichester.

Chapter 26

Alyaudtin, R. N., Reichel, A., Löbenberg, R., Ramge, P., Kreuter, J., Begley, D. J. (2001) Interaction of poly(butylcyanoacrylate) nanoparticles with the blood-brain barrier in vivo and in vitro. Phase II study of weekly intravenous trastuzumab (Herceptin) in patients with HER2/neu-over-expressing metastatic breast cancer, J. Drug Target. 9, 209–221.

Baselga, J., Tripathy, D., Mendelsohn, J., Baughman, S., Benz, C. C., Dantis, L., Sklarin, N. T., Seidman, A. D., Hudis, C. A., Moore, J., Rosen, P. P., Twaddell, T., Henderson, I. C., Noron, L. (1999) Phase II study of weekly intravenous trastuzumab (Herceptin) in patients with HER2/neu-over-expressing metastatic breast cancer, Semin. Oncol. 26, 78–83.

Böhm, H.-J., Schneider, G. (2003) Protein-ligand interactions: from molecular recognition to drug design, in Methods and Principles in Medicinal Chemistry, vol. 19, Mannhold, R., Kubinyi, H., Folkers, G. (eds.), Wiley-VCH, Weinheim.

Brewster, M. E., Anderson, W. R., Webb, A. I., Pablo, L. M., Meinsma, D., Moreno, D., Derendorf, H., Bodor, N., Pop, E. (1997) Evaluation of a brain-targeting zidovudine chemical delivery system in dogs, Antimicrob. Agents Chemother. 41, 122–128.

Dembowsky, K., Stadler, P. (2000) Novel Therapeutic Proteins: Selected Case Studies, Wiley-VCH, Weinheim.

Demeunynck, M., Bailly, C., Wilson, W. D. (2002) DNA and RNA Binders: From Small Molecules to Drugs, vol. 2, Wiley-VCH, Weinheim.

Dolle, R. E. (2002) Comprehensive survey of combinatorial library synthesis, J. Combin. Chem. 4, 369–418.

Drews, J. (2000) Drug discovery: a historical perspective, Science 287, 1960–1964.

Drews, J. (2001) Pharmaceuticals: classes, therapeutic agents, areas of application, Drug Discov. Today 6, 1100.

European Commission (2010) Enterprise and Industry, Information on European Pharmaceuticals Legislation http://ec.europa.eu/ enterprise/sectors/pharmaceuticals/ documents/eudralex/index_en.htm.

Geysen, H. M., Schoenen, F., Wagner, D., Wagner, R. (2003) Combinatorial compound libraries for drug discovery: an ongoing challenge, Nat. Rev. Drug Discov. 2, 222–320.

Gosilia, D. N., Diamond, S. L. (2003) Printing chemical libraries on microarrays for fluid phase nanoliter reactions, Proc. Natl. Acad. Sci. USA 100, 8721–8726.

Hainsworth, J. D., Burris, H. A. R. D., Morrissey, L. H., Litchy, S., Scullin, D. C. J. R., Bearden, J. D. RD, Richards, P., Greco, F. A. (2000) Rituximab monoclonal antibody as initial systemic therapy for patients with low-grade non-Hodgkin lymphoma, Blood 95, 3052–3056.

Ho, R. J. Y., Gibaldi, M. (2003) Biotechnology and Biopharmaceuticals: Transforming Proteins and Genes into Drugs, Wiley, New York.

Hülsermann, U., Hoffmann, M. M., Massing, U., Fricker G. (2009) Uptake of apolipoprotein E fragment coupled liposomes by cultured brain microvessel endothelial cells and intact brain capillaries, J. Drug Target. 17, 610–618.

Huwyler, J., Wu, D., Pardridge, W. M. (1996) Brain drug delivery of small molecules using immunoliposomes, Proc. Natl. Acad. Sci. USA 93, 14164–14169.

Huwyler, J., Yang, J., Pardridge, W. M. (1997) Receptor mediated delivery of daunomycin using immunoliposomes: pharmacokinetics and tissue distribution in the rat, J. Pharmacol. Exp. Ther. 282, 1541–1546.

Imming, P., Sinning, C., Meyer, A. (2006) Drugs, their targets and the nature and number of drug targets, Nat. Rev. Drug Discov. 5, 821–834.

Kreuter, J. (2001) Nanoparticulate systems for brain delivery of drugs, Adv. Drug Deliv. Rev. 47, 6581.

Launhardt, H., Hinnen, A., Munder, T. (1998) Drug-induced phenotypes provide a tool for the functional analysis of yeast genes, Yeast 14, 935–942.

Licinio, J., Wong, M.-L. (2002) Pharmacogenomics: The Search for Individualized Therapies, Wiley-VCH, Weinheim.

Lincoff, A. M., Califf, R. M., Moliterno, D. J., Ellis, S. G., Ducas, J., Kramer, J. H., Kleiman, N. S., Cohen, E. A., Booth, J. E., Sapp, S. K., Cabot, C. F., Topol, E. J. (1999) Complementary clinical benefits of coronary-artery stenting and blockade of platelet glycoprotein IIb/IIIa receptors. Evaluation of Platelet IIb/IIIa Inhibition in Stenting Investigators, N. Engl. J. Med. 341, 319–327.

Lipinski, C. A., Lombardo, F., Dominy, B. W., Feeney, P. J. (1997) Experimental and computational approaches to estimate solubility and permeability in drug discovery and development settings, Adv. Drug Deliv. Rev. 23, 3–25.

Maini, R. S. T., Clair, E. W., Breedveld, F., Furst, D., Kalden, J., Weisman, M., Smolen, J., Emery, P., Harriman, G., Feldmann, M., Lipsky, P. (1999) Infliximab (chimeric anti-tumour necrosis factor alpha monoclonal antibody) versus placebo in rheumatoid arthritis patients receiving concomitant methotrexate: a randomised phase III trial. ATTRACT Study Group, Lancet. 354, 1932–1939.

Malley et al. (1998) Reduction of respiratory syncytial virus (RSV) in tracheal aspirates in intubated infants by use of humanized monoclonal antibody to RSV protein. J. Infect. Dis. 178, 1555–1561.

Mayer, T. U., Kapoor, T. M., Haggarty, S. S., King, R. W., Schreiber, S. L., Mitchison, T. J. (1999) Small molecule inhibitor of mitotic spindle bipolarity identified in a phenotype-based screen, Science 286, 971–974.

Milpied, N., Vasseur, B., Parquet, N., Garnier, J. L., Antoine, C., Quartier, P., Carret, A. S., Bouscary, D., Faye, A., Bourbigot, B., Reguerre, Y., Stoppa, A. M., Bourquard, P., Hurault De Ligny, B., Dubief, F., Mathieu-Boue, A., Leblond, V. (2000) Humanized anti-CD20 monoclonal antibody (Rituximab) in post transplant B-lymphoproliferative disorder: a retrospective analysis on 32 patients, Ann. Oncol. 11 Suppl 1, 113–116.

Overington, J. P., Al-Lazikani, B., Hopkins, A. L. (2006) How many drug targets are there? Nat. Rev. Drug Discov. 5, 993–996.

Paul-Pletzer, K. (2006) Tocilizumab: blockade of interleukin-6 signaling pathway as a therapeutic strategy for inflammatory disorders, Drugs Today, 42, 559–576.

Pegram, M. D., Lipton, A., Hayes, D. F., Weber, B. L., Baselga, J. M., Tripathy, D., Baly, D., Baughman, S. A., Twaddell, T., Glaspy, J. A., Slamon, D. J. (1998) Phase II study of receptor-enhanced chemosensitivity using recombinant humanized anti-p185HER2/neu monoclonal antibody plus cisplatin in patients with HER2/neu-overexpressing metastatic breast cancer refractory to chemotherapy treatment, J. Clin. Oncol. 16, 2659–2671.

Prendergast, G. C. (2004) Molecular Cancer Therapeutics: Strategies for Drug Discovery and Development, Wiley, New York.

Reimold, I., Domke, D., Bender, J., Seyfried, C. A., Radunz, H. F., Fricker, G. (2008) Delivery of nanoparticles to the brain detected by fluorescence microscopy, Eur. J. Pharm. Biopharm. 70, 627–632.

Richards, J., Auger, J., Peace, D., Gale, D., Michel, J., Koons, A., Haverty, T., Zivin, R., Jolliffe, L., Bluestone, J. A. (1999) Phase I evaluation of humanized OKT3: toxicity and immunomodulatory effects of hOKT3gamma4, Cancer Res. 59, 2096–2101.

Schapira, M., Abagyan, R., Totrov, M. (2003) Nuclear hormone receptor targeted virtual screening, J. Med. Chem. 46, 3045–3059.

Scheinfeld N. (2003) Adalimuman (HUMIRA): a review, J. Drugs Dermatol. 4, 375–377.

Seethala, R., Fernandes, P. (eds.) (2001) Handbook of Drug Screening (Drugs and Pharmaceutical Science Series 114). Dekker, New York.

Sibley, D. R. (2005) G protein coupled receptor-protein interactions, in Receptor Biochemistry and Methodology, George, S. R., O'Dowd, B.F (eds.), Wiley, New York.

Sipe, J. D. (2005) Amyloid Proteins: The Beta Sheet Conformation and Disease, vol. 2, Wiley-VCH, Weinheim.

Sirohi, B, Smith, K. (2008) Bevacizumab in the treatment of breast cancer, Expert Rev. Anticancer Ther. 8, 1559–1568.

Testa, B., Mayer, J. M. (2003) Hydrolysis in Drug and Prodrug Metabolism: Chemistry, Biochemistry, and Enzymology, Wiley-VCH, Weinheim.

Tokuda, Y., Watanabe, T., Omuro, Y., Ando, M., Katsumata, N., Okumura, A., Ohta, M., Fujii, H., Sasaki, Y., Niwa, T., Tajima, T. (1999) Dose escalation and pharmacokinetic study of a humanized anti-HER2 monoclonal antibody in patients with HER2/neu-overexpressing metastatic breast cancer, Br. J. Cancer 81, 1419–1425.

Torchilin, V. P. (2000) Drug targeting. Eur. J. Pharm. Sci., Suppl 2, 81–91.

Zhang, J. H., Chung, T. D., Oldenburg, K. R. (1999) A simple statistical parameter for use in the evaluation and validation of high-throughput screening assays, J. Biomol. Screen. 4, 27–32.

Chapter 27

Alyaudtin, R. N., Reichel, A., Lobenberg, R., Ramge, P., Kreuter, J., Begley, D. J. (2001) Interaction of poly(butylcyanoacrylate) nanoparticles with the blood-brain barrier in vivo and in vitro. J. Drug Target. 9, 209–221.

Baselga, J., Tripathy, D., Mendelsohn, J., Baughman, S., Benz, C. C. et al. (1999) Phase II study of weekly intravenous trastuzumab (Herceptin) in patients with HER2/neu-overexpressing metastatic breast cancer. Semin. Oncol. 26, 78–83.

Bickel, U., Yoshikawa, T., Landaw, E. M., Faull, K. F., Pardridge, W. M. (1993) Pharmacologic effects in vivo in brain by vector-mediated peptide drug delivery. Proc. Natl. Acad. Sci. USA 90, 2618–2622.

Bodor, N. (1987) Redox drug delivery systems for targeting drugs to the brain. Ann. NY Acad. Sci. 507, 289–306.

Goodsell, D. S. (2004) Bionanotechnology: Lessons from Nature. Wiley, New York.

Hainsworth, J. D., Burris, H. A. III, Morrissey, L. H., Litchy, S., Scullin, D. C. et al. (2000) Rituximab monoclonal antibody as initial systemic therapy for patients with low-grade non-Hodgkin lymphoma. Blood 95, 3052–3056.

Ho, R. J. Y., Gibaldi, M. (2003) Biotechnology and Biopharmaceuticals: Transforming Proteins and Genes into Drugs. Wiley, New York.

Hülsermann, U., Hoffmann, M. M., Massing, U., Fricker, G. (2009) Uptake of apolipoprotein E fragment coupled liposomes by cultured brain microvessel endothelial cells and intact brain capillaries. J. Drug Target. 17, 610–618.

Huwyler, J., Wu, D., Pardridge, W. M. (1996) Brain drug delivery of small molecules using immunoliposomes. Proc. Natl. Acad. Sci. USA 93, 14164–14169.

Huwyler, J., Yang, J., Pardridge, W. M. (1997) Receptor mediated delivery of daunomycin using immunoliposomes: pharmacokinetics and tissue distribution in the rat. J. Pharmacol. Exp. Ther. 282, 1541–1546.

Kayser, O., Müller, R. H. (2004) Pharmaceutical Biotechnology: Drug Discovery and Clinical Applications. Wiley-VCH, Weinheim.

Klefenz, H. (2002) Industrial Pharmaceutical Biotechnology. Wiley-VCH, Weinheim.

Knäblein, J. (2005) Modern Biopharmaceuticals: Design, Development and Optimization. Wiley-VCH, Weinheim.

Kreuter, J. (2001) Nanoparticulate systems for brain delivery of drugs. Adv. Drug Deliv. Rev. 47, 65–81.

Lincoff, A. M., Califf, R. M., Moliterno, D. J., Ellis, S. G., Ducas, J. et al. (1999) Complementary clinical benefits of coronary-artery stenting and blockade of platelet glycoprotein IIb/IIIa receptors. Evaluation of platelet IIb/IIIa inhibition in stenting investigators. N. Engl. J. Med. 341, 319–327.

Maini, R., St Clair, E. W., Breedveld, F. (1999) Infliximab (chimeric anti-tumour necrosis factor alpha monoclonal antibody) versus placebo in rheumatoid arthritis patients receiving concomitant methotrexate: a randomised phase III trial. Lancet 354, 1932–1939.

Malley, R., DeVincenzo, J., Ramilo, O., Dennehy, P. H., Meissner, H. C. et al. (1998) Reduction of respiratory syncytial virus (RSV) in tracheal aspirates in intubated infants by use of humanized monoclonal antibody to RSV F protein. J. Infect. Dis. 178, 1555–1561.

Massing, U. (1997) Cancer therapy with liposomal formulations of anticancer drugs. Lancet 354, 1932–1939.

Milpied, N., Vasseur, B., Parquet, N., Garnier, J. L., Antoine, C. et al. (2000) Humanized anti-CD20 monoclonal antibody (Rituximab) in post transplant B-lymphoproliferative disorder: a retrospective analysis on 32 patients. Ann. Oncol. 11, 113–116.

Molema, G., Meyer D. K. F. (eds.) (2001) Drug Targeting: Basic Concepts and Novel Advances. Wiley-VCH, Weinheim.

Niemeyer, C. M., Mirkin, C. A. (2004) Nanobiotechnology: Concepts, Applications and Perspectives. Wiley-VCH, Weinheim.

Paul-Pletzer, K. (2006) Tocilizumab: blockade of interleukin-6 signaling pathway as a therapeutic strategy for inflammatory disorders. Drugs Today 42, 559–576.

Pegram, M. D., Lipton, A., Hayes, D. F., Weber, B. L., Baselga, J. M. et al. (1998) Phase II study of receptor-enhanced chemosensitivity using recombinant humanized anti-p185HER2/neu monoclonal antibody plus cisplatin in patients with HER2/neu-overexpressing metastatic breast cancer refractory to chemotherapy treatment, J. Clin. Oncol. 16, 2659–2671.

Raffa, R. B. (2001) Drug Receptor Thermodynamics: Introduction and Applications. Wiley-VCH, Weinheim.

Reimold, I., Domke, D., Bender, J., Seyfried, C. A., Radunz, H. E., Fricker, G. (2008) Delivery of nanoparticles to the brain detected by fluorescence microscopy. Eur. J. Pharm. Biopharm. 70, 627–632.

Richards, J., Auger, J., Peace, D., Gale, D., Michel, J. et al. (1999) Phase I evaluation of humanized OKT3: toxicity and immunomodulatory effects of hOKT3gamma4. Cancer Res. 59, 2096–2101.

Saffran, M., Kumar, G. S., Savariar, C., Burnham, J. C., Williams, F., Neckers, D. C. (1986) A new approach to the oral administration of insulin and other peptide drugs. Science 233, 1081–1034.

Saffran, M., Kumar, G. S., Neckers, D. C., Pena, J., Jones, R. H., Field, J. B. (1990) Biodegradable azo polymer coating for oral delivery of peptide drugs. Biochem. Soc. Trans. 18, 752–754.

Scheinfeld, N. (2003) Adalimuman (HUMIRA): a review. J. Drugs Dermatol. 4, 375–377.

Schreier, H. (ed.) (2001) Drug Targeting Technology. Dekker, New York.

Sirohi, B, Smith, K. 2008, Bevacizumab in the treatment of breast cancer. Expert Rev. Anticancer Ther. 8, 1559–1568.

Testa, B., Mayer, J. M. (2003) Hydrolysis in Drug and Prodrug Metabolism: Chemistry, Biochemistry, and Enzymology. Wiley-VCH, Weinheim.

Tokuda, Y., Watanabe, T., Omuro, Y., Ando, M., Katsumata, N. et al. (1999) Dose escalation and pharmacokinetic study of a humanized anti-HER2 monoclonal antibody in patients with HER2/neu-overexpressing metastatic breast cancer. Br. J. Cancer 81, 1419–1425.

Walsh, G. (2003) Biopharmaceuticals: Biochemistry and Biotechnology. 2nd edn, Wiley, New York.

Chapter 28

Breitling, F. and Dübel, S. (1999) Recombinant Antibodies. John Wiley and Sons, New York.

Dübel, S. (ed.) (2007) Handbook of Therapeutic Anitbodies, Wiley VCH, Weinheim.

Grandi, G. (2004) Genomics, Proteomics and Vaccines. Wiley, New York.

Kambhampati, D. (2003) Protein Microarray Technology. Wiley-VCH, Weinheim.

Knudsen, S. (2004) Guide to Analysis of DNA Microarray Data. 2nd edn, Wiley, New York.

Kontermann, R., Dübel, S. (ed.) (2010) Antibody Engineering, Two vol., Springer, Heidelberg.

McLachlan, G. J., Do, K.-A., Ambroise, C. (2004) Analyzing Microarray Gene Expression Data: Wiley Series in Probability and Statistics. Wiley, New York.

Rapley, R., Harbron, S. (2004) Molecular Analysis and Genome Discovery. Wiley, New York.

Reece, R. J. (2004) Analysis of Genes and Genomes. Wiley, New York.

Reichert, J. M. (2009) Probabilities of success for antibody therapeutics, MAbs 1, 387–389.

Rohrbach, P., Broders, O., Toleikis, L., Dübel, S. (2003) Therapeutic antibodies and antibody fusion proteins, Biotech. Gen. Eng. 20, 137–163.

Smith, G. P. (1985) Filamentous fusion phage: novel expression vectors that display cloned antigens on the virion surface, Science 228, 1315–1317.

Wilson, G. N. (2000) Clinical Genetics: A Short Course. Wiley, New York.

Zhang, W., Shmulevich, I., Astola, J. (2004) Microarray Quality Control. Wiley, New York.

Chapter 29

Adams, G. P., McCartney, J. E., Tai, M.-S., Oppermann, H., Huston, J. S. et al. (1993) Highly specific in vivo tumor targeting by monovalent and divalent forms of 741F8 anti-c-erb B-2 single-chain Fv. Cancer Res. 53, 4026–4034.

Adams, G. P., Schier, R., McCall, A. M., Crawford, R. S., Wolf, E. J. et al. (1998) Prolonged in vivo tumour retention of a human diabody targeting the extracellular domain of human HER2/neu. Br. J. Cancer 77, 1405–1412.

Alt, M., Müller, R., Kontermann, R. E. (1999) Novel tetravalent and bispecific IgG-like antibody molecules combining single-chain diabodies with the immunoglobulin gamma1 Fc or C_H3 region. FEBS Lett. 454, 90–94.

Arndt, K. M., Müller, K. M., Plückthun, A. (1998) Factors influencing the dimer to monomer transition of an antibody single-chain Fv fragment. Biochemistry 37, 12918–12926.

Arndt, M. A., Krauss, J., Kipriyanov, S. M., Pfreundschuh, M., Little, M. (1999) A bispecific diabody that mediates natural killer cell cytotoxicity against xenotransplantated human Hodgkin's tumors. Blood 94, 2562–2568.

Atwell, J. L., Breheney, K. A., Lawrence, L. J., McCoy, A. J., Kortt, A. A., Hudson, P. J. (1999) ScFv multimers of the anti-neuraminidase antibody NC10: length of the linker between V_H and V_L domains dictates precisely the transition between diabodies and triabodies. Protein Eng. 12, 597–604.

Barbas, C. F. III, Kang, A. S., Lerner, R. A., Benkovic, S. J. (1991) Assembly of combinatorial antibody libraries on phage surfaces: The gene III site. Proc. Natl. Acad. Sci. USA 88, 7978–7992.

Bartlett, J. S., Kleinschmidt, J., Boucher, R. C., Samulski, R. J. (1999) Targeted adeno-associated virus vector transduction of nonpermissive cells mediated by a bispecific F(ab'gamma)2 antibody. Nat. Biotechnol. 17, 181–186.

Bera, T. K., Onda, M., Brinkmann, U., Pastan, I. (1998) A bivalent disulfide-stabilized Fv with improved antigen binding to erbB2. J. Mol. Biol. 281, 475–483.

Bera, T. K., Viner, J., Brinkmann, E., Pastan, I. (1999) Pharmacokinetics and antitumor activity of a bivalent disulfide-stabilized Fv immu-

notoxin with improved antigen binding to erbB2. Cancer Res. 59, 4018–4022.

Better, M., Chang, C. P., Robinson, R. R., Horowitz, A. H. (1988) *Escherichia coli* secretion of an active chimeric antibody fragment. Science 240, 1041–1043.

Better, M., Bernhard, S. L., Lei, S.-P., Fishwild, D. M., Lane, J. A. et al. (1993) Potent anti-CD5 ricin A chain immunoconjugates from bacterially produced Fab' and F(ab')2. Proc. Natl. Acad. Sci. USA 90, 457–461.

Biocca, S., Ruberti, F., Tafani, M., Pierandrei-Amaldi, P., Cattaneo, A. (1995) Redox state of single chain Fv fragments targeted to the endoplasmic reticulum, cytosol and mitochondria. Biotechnology 13, 1110–1115.

Bird, R. E., Hardman, K. D., Jacobson, J. W., Johnson, S., Kaufman, B. M. et al. (1988) Single-chain antigen-binding proteins. Science 242, 423–426.

Boder, E. T., Wittrup, K. D. (1997) Yeast surface display for screening combinatorial polypeptide libraries. Nat. Biotechnol. 15, 553–557.

Bornemann, K. D., Brewer, J. W., Beck-Engeser, G. B., Corley, R. B. et al. (1995) Roles of heavy and light chains in IgM polymerization. Proc. Natl. Acad. Sci. USA 92, 4912–4916.

Bradbury, A. (2003) scFvs and beyond. Drug Disc. Today 8, 737–739.

Breitling, F. and Dübel, S. (1999) Recombinant Antibodies. Wiley, New York.

Breitling, F., Dübel, S., Seehaus, T., Klewinghaus, I., Little, M. (1991) A surface expression vector for antibody screening. Gene 104, 147–153.

Brinkmann, U., Reiter, Y., Jung, S. H., Lee, B., Pastan, I. (1993) A recombinant immunotoxin containing a disulfide-stabilized Fv fragment. Proc. Natl. Acad. Sci. USA 90, 7538–7542.

Brocks, B., Rode, H. J., Klein, M., Gerlach, E., Dübel, S. et al. (1997) A TNF receptor antagonistic scFv, which is not secreted in mammalian cells, is expressed as a soluble mono- and bivalent scFv derivative in insect cells. Immunotechnology 3, 173–184.

Brüsselbach, S., Korn, T., Völkel, T., Müller, R., Kontermann, R. E. (1999) Enzyme recruitment and tumor cell killing in vitro by a secreted bispecific single-chain diabody. Tumor Targeting 4, 115–123.

Cai, X., Garen, A. (1996) A melanoma-specific VH antibody cloned from a fusion phage library of a vaccinated melanoma patient. Proc. Natl. Acad. Sci. USA 93, 6280.

Cai, X., Garen, A. (1997) Comparison of fusion phage libraries displaying VH or single-chain Fv antibody fragments derived from the antibody repertoire of a vaccinated melanoma patient as a source of melanoma-specific targeting molecules. Proc. Natl. Acad. Sci. USA 94, 9261.

Carter, P., Merchant, A. M. (1997) Engineering antibodies for imaging and therapy. Curr. Opin. Biotechnol. 8, 449–454.

Carter, P., Kelley, R. F., Rodrigues, M. L., Snedecor, B., Covarrubias, M. et al. (1992) High level Escherichia coli expression and production of a bivalent humanized antibody fragment. Biotech. 10, 163–167.

Casey, J. L., King, D. J., Chaplin, L. C., Haines, A. M., Pedley, R. B. et al. (1996) Preparation, characterisation and tumour targeting of cross-linked divalent and trivalent anti-tumour Fab' fragments. Br. J. Cancer 74, 1397–1405.

Clackson, T., Hoogenboom, H. R., Griffiths, A. D., Winter, G. (1991) Making antibody fragments using phage display libraries. Nature 352, 624–628.

Colcher, D., Pavlinkova, G., Beresford, G., Booth, B. J., Batra, S. K. (1999) Single-chain antibodies in pancreatic cancer. Ann. N. Y. Acad. Sci. 880, 263–280.

Coloma, M. J., Morrison, S. L. (1997) Design and production of novel tetravalent bispecific antibodies. Nat. Biotechnol. 15, 159–163.

Courtenay-Luck, N. S., Epenetos, A. A., Moore, R., Larche, M., Pectasides, D. et al. (1986) Development of primary and secondary immune responses to mouse monoclonal antibodies used in the diagnosis and therapy of malignant neoplasms. Cancer Res. 46, 6489–6493.

Cumber, A. J., Ward, E. S., Winter, G., Parnell, G., Wawrzynczak, E. J. (1992) Comparative stabilities in vitro and in vivo of a recombinant mouse antibody FvCys fragment and a bisFvCys conjugate. J. Immun. 149, 120–126.

Daugherty, P. S., Chen, G., Olsen, M. J., Iverson, B. L., Georgiou, G. (1998) Antibody affinity maturation using bacterial surface display. Protein Eng. 11, 825–832.

Daugherty, P. S., Olsen, M. J., Iverson, B. L., Georgiou, G. (1999) Development of an optimized expression system for the screening of antibody libraries displayed on the Escherichia coli surface. Protein Eng. 12, 613–621.

Denton, G., Brady, K., Lo, B. K., Murray, A., Graves, C. R. et al. (1999) Production and characterization of an anti-(MUC1 mucin) recombinant diabody. Cancer Immunol. Immunother. 48, 29–38.

Dolezal, O., Pearce, L. A., Lawrence, L. J., McCoy, A. J., Hudson, P. J., Kortt, A. A. (2000) ScFv multimers of the anti-neuraminidase antibody NC10: shortening of the linker in single-chain Fv fragment assembled in V(L) to V(H) orientation drives the formation of dimers, trimers, tetramers and higher molecular mass multimers. Protein Eng. 13, 565–574.

Dübel, S. (eds.) (2007) Handbook of Therapeutic Antibodies. Wiley-VCH, Weinheim

Dübel, S. (2007) Recombinant therapeutic antibodies. Applied Microbiology and Biotechnology, 74, 723–729

Dübel, S., Breitling, F., Klewinghaus, I., Little, M. (1992) Regulated secretion and purification of recombinant antibodies in *E. coli*. Cell Biophysics 21, 69–80.

Dübel, S., Breitling, F., Kontermann, R., Schmidt, T., Skerra, A., Little, M. (1995) Bifunctional and multimeric complexes of streptavidin fused to single chain antibodies (scFv). J. Immunol. Meth. 178, 201–209.

Eshhar, Z., Waks, T., Bendavid, A., Schindler, D.G. (2001) Functional expression of chimeric receptor genes in human T cells. J. Immunol. Meth. 248, 67–76.

Feldhaus, M.J., Siegel, R.W., Opresko, L.K., Coleman, J.R., Feldhaus, J.M. et al. (2003) Flow-cytometric isolation of human antibodies from a nonimmune Saccharomyces cerevisiae surface display library. Nat. Biotechnol. 21, 163–170.

Filpula, D., McGuire, J. (1999) Single-chain Fv designs for protein, cell and gene therapeutics. Exp. Opin. Ther. Patents 9, 231–245.

Fitz Gerald, K., Holliger, P., Winter, G. (1997) Improved tumour targeting by disulphide stabilized diabodies expressed in Pichia pastoris. Protein Eng. 10, 1221–1225.

Francisco, J.A., Campbell, R., Iverson, B.L., Georgiou, G. (1993) Production and fluorescence-activated cell sorting of *Escherichia coli* expressing a functional antibody fragment on the external surface. Proc. Natl. Acad. Sci. USA 90, 10444–10448.

Friedman, P.N., Chace, D.F., Trail, P.A., Siegall, C.B. (1993) Antitumor activity of the single-chain immunotoxin BR96 sFv-PE40 against established brest and lung tumor xenografts. J. Immunol. 150, 3054–3061.

Fuchs, P., Little, M., Breitling, F., Dübel, S. (1991) Recombinant antibodies at the surface of E. coli. Deutsches Patentamt Reg. Nr. P 41 22 5988.

Fuchs, P., Weichel, W., Dübel, S., Breitling, F., Little, M. (1996) Specific selection of *E. coli* expressing functional cell-wall bound antibody fragments by FACS. Immunotechnology 2, 97–102.

Gavilondo, J.V., Larrick, J.W. (2000) Antibody Engineering at the millennium. BioTechniques 29, 128–145.

Gilliland, L.K., Norris, N.A., Marquardt, H., Tsu, T.T., Hayden, M.S. et al. (1996) Rapid and reliable cloning of antibody variable regions and generation of recombinant single chain antibody fragments. Tissue Antigens 47, 1–20.

Glockshuber, R., Malia, M., Pfitzinger, I., Plückthun, A. (1990) A comparison of strategies to stabilize immunoglobulin Fv-fragments. Biochemistry 29, 1362–1367.

Hanes, J., Plückthun, A. (1997) In vitro selection and evolution of functional proteins by using ribosome display. Proc. Natl. Acad. Sci. USA 94, 4937–4942.

Helfrich, W., Kroesen, B.J., Roovers, R.C., Westers, L., Molema, G. et al. (1998) Construction and characterization of a bispecific diabody for retargeting T cells to human carcinomas. Int. J. Cancer 76, 232–239.

Hochman, J., Inbar, D., Givol, D. (1973) An active antibody fragment (Fv) composed of the variable portions of heavy and light chains. Biochemistry 12, 1130–1135.

Holliger, P., Winter, G. (1993) Engineering bispecific antibodies. Curr. Opin. Biotechnol. 4, 446–449.

Holliger, P., Prospero, T., Winter, G. (1993) "Diabodies: small bivalent and bispecific antibody fragments. Proc. Natl. Acad. Sci. USA 90, 6444–6448.

Holliger, P., Brissinck, J., Williams, R.L., Thielemans, K., Winter, G. (1996) Specific killing of lymphoma cells by cytotoxic T-cells mediated by a bispecific diabody. Protein Eng. 9, 299–305.

Holliger, P., Wing, M., Pound, J.D., Bohlen, H., Winter, G. (1997) Retargeting serum immunoglobulin with bispecific diabodies. Nat. Biotechnol. 15, 632–636.

Holliger, P., Manzke, O., Span, M., Hawkins, R., Fleischmann, B. et al. (1999) Carcinoembryonic antigen (CEA)-specific T-cell activation in colon carcinoma induced by anti-CD3 x anti-CEA bispecific diabodies and B7 x anti-CEA bispecific fusion proteins. Cancer Res. 59, 2909–2916.

Hoogenboom, H.R., Charmes, P. (2000) Natural and designer binding sites made by phage display technology. Immunol. Today 21, 371–378.

Hu, S., Shively, L., Raubitschek, A., Sherman, M., Williams, L.E. et al. (1996) Minibody: A novel engineered anti-carcinoembryonic antigen antibody fragment (single-chain Fv-CH3) which exhibits rapid, high-level targeting of xenografts. Cancer Res. 56, 3055–3061.

Hudson, P.J. (1999) Recombinant antibody constructs in cancer therapy. Curr. Opin. Immunol. 11, 548–557.

Hudson, P.J., Kortt, A.A. (1999) High avidity scFv multimers; diabodies and triabodies. J. Immunol. Meth. 231, 177–189.

Hust, M., Dübel, S. (2004) Mating antibody phage display to proteomics. Trends Biotechnol. 22, 8–14.

Hust, M., Thie, H., Schirrmann, T., Dübel, S. (2009) Antibody Phage Display. In: An, Z. & Strohl, W. (ed.) Handbook of Therapeutic Monoclonal Antibodies. Wiley-VCH, Weinheim.

Huston, J.S., Levinson, D., Mudgett-Hunter, M., Tai, M.-S., Novotny, J. et al. (1988) Protein engineering of antibody binding sites: recovery of specific activity in an anti-digoxin single-chain Fv analogue produced in Escherichia coli. Proc. Natl. Acad. Sci. USA 85, 5879–5882.

Iliades, P., Kortt, A.A., Hudson, P.J. (1997) Triabodies: single chain Fv fragments without a linker form trivalent trimers. FEBS Lett. 409, 437–441.

Jakobovits, A. (1995) Production of fully human antibodies by transgenic mice. Curr. Opin. Biotechnol. 6, 561–566.

Jordan, E., Al-Halabi, L., Schirrmann, T., Hust, M., Dübel, S. (2007) Production of single chain Fab (scFab) fragments in Bacillus megaterium. Microbial Cell Factories 6, 38

Karpovsky, B., Titus, J.A., Stephany, D.A., Segal, D.M. (1984) Production of target-specific effector cells using hetero-cross-linked aggregates containing anti-target cell and anti-Fc gamma receptor antibodies. J. Exp. Med. 160, 1686–1701.

Kipriyanov, S.M., Dübel, S., Breitling, F., Kontermann, R.E., Little, M. (1994) Recombinant single-chain Fv fragments carrying C-terminal cysteine residues: Production of bivalent and biotinylated miniantibodies. Mol. Immun. 31, 1047–1058.

Kipriyanov, S.M., Dübel, S., Breitling, F., Kontermann, R.E., Heymann, S., Little, M. (1995) Bacterial expression and refolding of single-chain Fv fragments with C-terminal cysteines. Cell. Biophysics 26, 187–204.

Kipriyanov, S.M., Moldenhauer, G., Strauss, G., Little, M. (1998) Bispecific CD3xCD19 diabody for T-cell mediated lysis of malignant human B cells. Int. J. Cancer 77, 763–772.

Kipriyanov, S.M., Moldenhauer, G., Schuhmacher, J., Cochlovius, B., von der Lieth, C.-W. et al. (1999) Bispecific tandem diabody for tumor therapy with improved antigen binding and pharmacokinetics. J. Mol. Biol. 293, 41–56.

Kirkpatrick, R.B., Ganguly, S., Angelichio, M., Griego, S., Shatzman, A. et al. (1995) Heavy chain dimers as well as complete antibodies are efficiently formed and secreted from Drosophila via a BiP mediated pathway. J. Biol. Chem. 270, 19800–19805.

Kontermann, R. (2000) Recombinant antibody fragments for cancer therapy. Mod. Asp. Immunobiol. 1, 88–91.

Kontermann, R., Dübel, S. (eds.) (2010) Antibody Engineering, Springer-Verlag, Heidelberg, New York.

Kontermann, R.E., Müller, R. (1999) Intracellular and cell surface displayed single-chain diabodies. J. Immunol. Meth. 226, 179–188.

Kontermann, R.E., Wing, M.G., Winter, G. (1997) Complement recruitment using bispecific diabodies. Nat. Biotechnol. 15, 629–631.

Kortt, A.A., Lah, M., Oddie, G.W., Gruen, C.L., Burns, J.E. et al. (1997) Single-chain Fv fragments of anti-neuraminidase antibody NC10 containing five- and ten-residue linkers form dimers and with zero-residue linker a trimer. Protein Eng. 10, 423–433.

Kostelny, S.A., Cole, M.S., Tso, J.Y. (1992) Formation of a bispecific antibody by the use of leucine zippers. J. Immunol. 148, 1547–1553.

Krauss, J., Arndt, M., Dübel, S., Rybak, S.M. (2008) Antibody-targeted RNase-fusion proteins (ImmunoRNases) for Cancer Therapy. Curr. Pharm. Biot. 9, 231–234

Kurucz, I., Titus, J.A., Jost, R., Jacobus, C.M., Segal, D.M. (1995) Retargeting of CTL by an efficiently refolded bispecific single-chain Fv dimer produced in bacteria. J. Immunol. 154, 4576–4582.

Le Gall, F., Kipriyanov, S.M., Moldenhauer, G., Little, M. (1999) Di-, tri- and tetrameric single chain Fv antibody fragments against human CD19: effect of valency on cell binding. FEBS Lett. 453, 164–168.

Li, E., Pedraza, A., Bestagno, M., Mancardi, S., Sanchez, R., Burrone, O. (1997) Mammalian cell expression of dimeric small immune proteins (SIP). Protein Eng. 10, 731–736.

Libyh, M.T., Goossens, D., Oudin, S., Gupta, N., Dervillez, X. et al. (1997) A recombinant human scFv anti-Rh(D) antibody with multiple valences using a C-terminal fragment of C4-binding protein. Blood 90, 3978–3983.

Lonberg, N., Huszar, D. (1995) Human antibodies from transgenic mice. Int. Rev. Immunol. 13, 65–93.

Luo, D., Geng, M., Noujaim, A.A., Madiyalakan, R. (1997) An engineered bivalent single-chain antibody fragment that increases antigen binding activity. J. Biochem. (Tokyo) 121, 831–834.

Mallender, W.D., Voss, E.W., Jr. (1994) Construction, expression, and activity of a bivalent bispecific single-chain antibody. Biol. Chem. 269, 199–206.

Mallender, W.D., Ferreira, S.T., Voss, E.W., Jr., Coelho-Sampaio, T. (1994) Inter-active-site distance and solution dynamics of a bivalent-bispecific single-chain antibody molecule. Biochemistry 33, 10100–10108.

Marasco, W.A., Haseltine, W.A., Chen, S.Y. (1993) Design, intracellular expression, and activity of a human anti-human immunodeficiency virus type 1 gp120 single-chain antibody. Proc. Natl. Acad. Sci. USA 90, 7889–3793.

Marks, J.D., Hoogenboom, H.R., Bonnert, T.P., McCafferty, J., Griffiths, A.D., Winter, G. (1991) By-passing immunization. Human antibodies from V-gene libraries displayed on phage. J. Mol. Biol. 222, 581–597.

Martsev, S.P., Dubnovitsky, A.P., Stremovsky, O.A., Chumanevich, A.A., Tsybovsky, Y.I. et al. (2002) Partially structured state of the functional VH domain of the mouse anti-ferritin antibody F11. FEBS Lett. 518, 177–182.

McCartney, J.E., Tai, M.S., Hudziak, R.M., Adams, G.P., Weiner, L.M. et al. (1995) Engineering disulfide-linked single-chain Fv dimers (sFv')2 with improved solution and targeting properties: anti-digoxin 26–10 (sFv')2 and anti-c-erbB-2 741F8 (sFv')2 made by protein folding and bonded through C-terminal cysteinyl peptides. Protein Eng. 8, 301–314.

McGregor, D.P., Molloy, P.E., Cunningham, C., Harris, W.J. (1994) Spontaneous assembly of bivalent single chain antibody fragments in Escherichia coli. Mol. Immunol. 31, 219–226.

Milenic, D.E., Yokota, T., Filpula, D.R., Finkelman, M.A.J., Dodd, S.W. et al (1991) Construction, binding properties, metabolism, and tumor targeting of a single-chain Fv derived from the pancarcinoma monoclonal antibody CC49. Cancer Res. 51, 6363–6371.

Milstein, C., Cuello, A.C. (1983) Hybrid hybridomas and their use in immunohistochemistry. Nature 305, 537–540.

Müller, K.M., Arndt, K.M., Plückthun, A. (1998a) A dimeric bispecific miniantibody combines two specificities with avidity. FEBS Lett. 432, 45–49.

Müller, K.M., Arndt, K.M., Strittmatter, W., Plückthun, A. (1998b) The first constant domain [C(H)1 and C(L)] of an antibody used as heterodimerization domain for bispecific miniantibodies. FEBS Lett. 422, 259–264.

Nakamura, M., Tsumoto, K., Ishimura, K., Kumagai, I. (2002) Phage library panning against cytosolic fraction of cells using quantitative dot blotting assay: application of selected VH to histochemistry. J. Immunol. Methods 261, 65–72.

Neri, D., Momo, M., Prospero, T., Winter, G. (1995) High-affinity antigen binding by chelating recombinant antibodies (CRAbs). J. Mol. Biol. 246, 367–373.

Nilsson, P. et al. (2005) Towards a human proteome atlas: high-throughput generation of mono-specific antibodies for tissue profiling. Proteomics 5, 4327–4337

Nizak, C., Monier, S., del Nery, E., Moutel, S., Goud, B., Perez, F. (2003) Recombinant antibodies to the small GTPase Rab6 as conformation sensors. Science 300, 984–987.

Nuttall, S.D., Irving, R.A., Hudson, P.J. (2000) Immunoglobulin VH domains and beyond: design and selection of single-domain binding and targeting reagents. Curr. Pharm. Biotechnol. 1, 253–263.

Pack, P., Plückthun, A. (1992) Miniantibodies: use of amphipathic helices to produce functional, flexibly linked dimeric FV fragments with high avidity in Escherichia coli. Biochemistry 31, 1579–1584.

Pack, P., Kujau, M., Schroeckh, V., Knüpfer, U., Wenderoth, R., Riesenberg, D., Plückthun, A. (1993) Improved bivalent miniantibodies, with identical avidity as whole antibodies, produced by high cell density fermentation of Escherichia coli. Bio/Technol. 11, 1271–1277.

Pack, P., Müller, K., Zahn, R., Plückthun, A. (1995) Tetravalent miniantibodies with high avidity assembling in Escherichia coli. J. Mol. Biol. 246, 28–34.

Pei, X.Y., Holliger, P., Murzin, A.G., Williams, R.L. (1997) The 2.0-A resolution crystal structure of a trimeric antibody fragment with noncognate V$_H$–V$_L$ domain pairs shows a rearrangement of V$_H$ CDR3. Proc. Natl. Acad. Sci. USA 94, 9637–9642.

Perisic, O., Webb, P.A., Holliger, P., Winter, G., Williams, R.L. (1994) Crystal structure of a diabody, a bivalent antibody fragment. Structure 2, 1217–1226.

Plückthun, A., Pack, P. (1997) New protein engineering approaches to multivalent and bispecific antibody fragments. Immunotechnology 3, 83–105.

Poljak, R.J. (1994) Production and structure of diabodies. Structure 2, 1121–1123.

Presta, L.G., Shields, R.L., Namenuk, A.K., Hong, K., Meng, Y.G. (2002) Engineering therapeutic antibodies for improved function. Biochem. Soc. Trans. 30, 487–490.

Reichert, J.M. (2001) Monoclonal antibodies in the clinic. Nat Biotechnol. 19, 819–822.

Reichert, J.M. (2009) Probabilities of success for antibody therapeutics. mAbs 1, 387–389.

Reiter, Y., Brinkmann, U., Webber, K.O., Jung, S.H., Lee, B., Pastan, I. (1994) Engineering interchain disulfide bonds into conserved framework regions of Fv fragments: improved biochemical characteristics of recombinant immuno-toxins containing disulfide-stabilized Fv. Protein Eng. 7, 697–704.

Reiter, Y., Brinkmann, U., Jung, S.-H., Lee, B., Kasprzyk, P.G. et al. (1994) Improved binding and antitumor activity of a recombinant anti-erbB2 immunotoxin by disulfide stabilization of the Fv fragment. J. Biol. Chem. 269, 18327–18331.

Reiter, Y., Kreitman, R.J., Brinkmann, U., Pastan, I. (1994) Cytotoxic and antitumor activity of a recombinant immunotoxin composed of disulfide-stabilized anti-Tac Fv fragment and truncated Pseudomonas exotoxin. Int. J. Cancer 58, 142–149.

Reiter, Y., Brinkmann, U., Kreitman, R.J., Jung, S.H., Lee, B., Pastan, I. (1994) Stabilization of the Fv fragments in recombinant immunotoxins by disulfide bonds engineered into conserved framework regions. Biochemistry 33, 5451–5459.

Reiter, Y., Brinkmann, U., Jung, S.H., Pastan, I., Lee, B. (1995) Disulfide stabilization of antibody Fv: computer predictions and experimental evaluation. Protein Eng. 8, 1323–1331.

Rheinnecker, M., Hardt, C., Ilag, L.L., Kufer, P., Gruber, R. et al. (1996) Multivalent antibody fragments with high functional affinity for a tumor-associated carbo-hydrate antigen. J. Immunol. 157, 2989–2997.

Richardson, J.H., Sodroski, J.G., Waldmann, T.A., Marasco, W.A. (1995) Phenotypic knockout of the high-affinity human interleukin 2 receptor by intracellular single-chain antibodies against the alpha subunit of the receptor. Proc. Natl. Acad. Sci. USA 92, 3137–3141.

Riechmann, L., Muyldermans, S. (1999) Single domain antibodies: comparison of camel VH and camelised human VH domains. J. Immunol. Methods 231, 25–38.

Robert, B., Dorvillius, M., Buchegger, F., Garambois, V., Mani, J.C. et al. (1999) Tumor targeting with newly designed biparatopic antibodies directed against two different epitopes of the carcinoembryonic antigen (CEA). Int. J. Cancer 81, 285–291.

Rodrigues, M.L., Snedecor, B., Chen, C., Wong, W.L.T., Garg, S. et al. (1993) Engineering Fab′ fragments for efficient F(ab′)2 formation in *Escherichia coli* and for improved in vitro stability. J. Immun. 151, 6954–6961.

Rohrbach, P., Broders, O., Toleikis, L., Dübel, S. (2003) Therapeutic antibodies and antibody fusion proteins. Biotech. Gen. Eng. 20, 137–163.

Schirrmann, T., Al-Halabi, L., Dübel, S., Hust, M. (2008) Production systems for recombinant antibodies. Front Biosci 13, 4576–4594.

Schmiedl, A., Breitling, F., Winter, C.H., Queitsch, I., Dübel, S. (2000a) Effects of unpaired cysteines on yield, solubility and activity of different recombinant antibody constructs expressed in *E. coli*. J. Immunol. Methods 242, 101–114.

Schmiedl, A., Breitling, F., Dübel, S. (2000b) Expression of a bispecific dsFv-dsFv′ antibody fragment in *Escherichia coli*. Protein Eng. 13, 725–734.

Schofield, D.J. et al. (2007) Application of phage display to high throughput antibody generation and characterization. Genome Biol 8, R254

Schoonjans, R., Willems, A., Grooten, J., Mertens, N. (2000) Efficient heterodimerization of recombinant bi- and trispecific antibodies. Bioseparations 9, 179–183.

Schoonjans, R., Willems, A., Schoonooghe, S., Fiers, W., Grooten, J., Mertens, N. (2000) Fab chains as an efficient heterodimerization scaffold for the production of recombinant bispecific and trispecific antibody derivatives. J. Immunol. 165, 7050–7057.

Schultz, J., Lin, Y., Sanderson, J., Zuo, Y., Stone, D. et al. (2000) A tetravalent single-chain antibody-streptavidin fusion protein for pretargeted lymphoma therapy. Cancer Res. 60, 6663–6669.

Segal, D.M., Weiner, G.J., Weiner, L.M. (2001) Introduction: bispecific antibodies. J. Immunol. Meth. 248, 1–6.

Sen, J., Beychok, S. (1986) Proteolytic dissection of a hapten binding site. Proteins 1, 256–262.

Sharon, J., Givol, D. (1976) Preparation of Fv fragment from mouse myeloma XRPC-25 immunoglobulin possessing antidinitrophenyl activity. Biochemistry 15, 1591–1594.

Sheets, M.D., Amersdorfer, P., Finnern, R., Sargent, P., Lindquist, E. et al. (1998) Efficient construction of a large nonimmune phage antibody library: the production of high-affinity human single-chain antibodies to protein antigens. Proc. Natl. Acad. Sci. USA 95, 6157–6162.

Skerra, A., Plückthun, A. (1988) Assembly of a functional immunglobulin Fv Fragment in *Escherichia coli*. Science 240, 1038–1041.

Smith, G.P. (1985) Filamentous fusion phage: novel expression vectors that display cloned

antigens on the virion surface. Science 228, 1315–1317.

Suresh, M.R., Cuello, A.C., Milstein, C. (1986) Bispecific monoclonal antibodies from hybrid hybridomas. Methods Enzymol. 121, 210–228.

Thie, H., Meyer, T., Schirrmann, T., Hust, M., Dübel, S. (2008) Phage display derived therapeutic antibodies. Curr. Pharm. Biot. 9, 439–446

Thie, H., Binius, S., Schirrmann, T., Hust, M., Dübel, S. (2009) Multimerization domains for antibody phage display and antibody production. New Biotec.

Todorovska, A., Roovers, R.C., Dolezal, O., Kortt, A.A., Hoogenboom, H.R., Hudson, P.J. (2001) Design and application of diabodies, triabodies and tetrabodies for cancer targeting. J. Immunol. Methods 248, 47–66.

Viti, F., Tarli, L., Giovannoni, L., Zardi, L., Neri, D. (1999) Increased binding affinity and valence of recombinant antibody fragments lead to improved targeting of tumoral angiogenesis. Cancer Res. 59, 347–352.

Winter, G., Milstein, C. (1991) Man-made antibodies. Nature 349, 293–299.

Wörn, A., Plückthun, A. (2001) Stability engineering of antibody single-chain Fv fragments. J. Mol. Biol. 305, 989–1010.

Wu, A.M., Chen, W., Raubitschek, A., Williams, L.E., Neumaier, M. et al. (1996) Tumor localization of anti-CEA single-chain Fvs: improved targeting by non-covalent dimers. Immunotechnology 2, 21–36.

Wu, A.M., Williams, L.E., Zieran, L., Padma, A., Sherman, M. et al. (1999) Anti-carcinoembryonic antigen (CEA) diabody for rapid tumor targeting and imaging. Tumor Targeting 4, 47–58.

Xu, L., Aha, P., Gu, K., Kuimelis, R.G., Kurz, M. et al. (2002) Directed evolution of high-affinity antibody mimics using mRNA display. Chem. Biol. 9, 933–942.

Yokota, T., Milenic, D.E., Whitlow, M., Schlom, J. (1992) Rapid tumor penetration of a single-chain Fv and comparison with other immunoglobulin forms. Cancer Res. 52, 3402–3408.

Zewe, M., Rybak, S., Dübel, S., Coy, J., Welschof, M., et al. (1997) Cloning and cytotoxicity of a human pancreatic RNase immunofusion. Immunotechnology 3, 127–136.

Zhu, Z., Ghose, T., Lee, S.H.S., Fernandez, L.A., Kerr, L.A. et al. (1994) Tumor localization and therapeutic potential of an antitumor-anti-CD3 heteroconjugate antibody in human renal cell carcinoma xenograft models. Cancer Lett. 86, 127–134.

Zhu, Z., Zapata, G., Shalaby, R. Snedecor, B., Chen, H., Carter, P. (1996) High level secretion of a humanized bispecific diabody from *Escherichia coli*. Biotechnology 14, 192–196.

Zhu, Z., Presta, L.G., Zapata, G., Carter, P. (1997) Remodeling domain interfaces to enhance heterodimer formation. Protein Sci. 6, 781–788.

Zuo, Z., Jimenez, X., Witte, L., Zhu, Z. (2000) An efficient route to the production of an IgG-like bispecific antibody. Prot. Eng. 13, 361–367.

Chapter 30

Engblom, D., Sanchis-Segura, C., Bilbao-Leis, A., Dahan, L., Perreau-Lenz, S., Balland, B., Mameli, M., Rodriguez Parkitna, J., Parlato, R., Sprengel, R., Lüscher, C., Schütz, G., Spanagel, R. (2008) Glutamate receptors on dopaminergic neurons control the persistence of drug-seeking. Neuron 59, 497–508.

Geurts, A.M. et al. (2009) Knockout rats via embryo microinjection of zinc-finger nucleases. Science 325, 433.

Heintz, N. (2004) Gene expression nervous system atlas (GENSAT). Nat Neurosci, 7, 483.

Inta, D., Monyer, H., Sprengel, R., Meyer-Lindenberg, A., Gass, P. (2009) Mice with genetically altered glutamate receptors as models of schizophrenia: A comprehensive review. Neuroscience Biohavioral Rev. 34:285–294

Joyner, A. (2000) Gene Targeting – A Practical Approach. 2nd edn, Oxford University Press, Oxford, New York.

Kühn, R., Wurst, W. (eds.) (2009) Gene Knockout Protocols, in: Methods in Molecular Biology, Springer Protocols, 2nd edn, Humana Press, Heidelberg, Berlin.

Li P, Tong C, Mehrian-Shai R, Jia L, Wu N, Yan Y, Maxson RE, Schulze EN, Song H, Hsieh C-L, Pera MF, Ying Q-L (2008) Germline competent embryonic stem cells derived from rat blastocysts. Cell 135, 1299–1310.

Nagy, A., Gerstenstein, M., Vintersten, K., Behringer, R. (2003) Manipulating the Mouse Embryo – A Laboratory Manual, 3rd edn (Inglis, J. ed.). Cold Spring Harbor Laboratory Press, Cold Spring Harbor.

Picciotto, M.R., Wickman, K. (1998) Using knockout and transgenic mice to study neurophysiology and behavior. Physiol Rev, 78, 1131–1163.

Schenkel, J. (2006) Transgene Tiere. 2nd edn, Springer, Berlin.

Spanagel R (2009) Alcoholism – a systems approach from molecular physiology to behavior. Physiol. Rev. 89, 649–705.

Sprengel, R., Hasan, M.T. (2007) Tetracycline-controlled genetic switches. Handb. Exp. Pharmacol. 49–72.

Sprengel, R., Eshkind, L., Hengstler, J., Bockamp, E., Conn, P.M. (2008) Improved models for animal research. In: Sourcebook of models for biomedical research, pp 17–24.

Yusa, K., Rad, R., Takeda, J., Bradley, A. (2009) Generation of transgene-free induced pluripotent mouse stem cells by the piggyBac transposon. Nat. Meth. 6, 363–369.

Chapter 31

Burns, J. C., Friedmann, T., Driever, W., Burrascano, M., Yee, J. K. (1993) Vesicular stomatitis virus G glycoprotein pseudotyped retroviral vectors: concentration to very high titer and efficient gene transfer into mammalian and nonmammalian cells. Proc. Natl. Acad. Sci. USA 90, 8033–8037.

Carthew, R. W., Sontheimer, E. J. (2009) Origins and Mechanisms of miRNAs and siRNAs. Cell 136, 642–655.

Castanotto, D., Rossi, J. J. (2009) The promises and pitfalls of RNA-interference-based therapeutics. Nature 457, 426–433.

Check, E. (2003) Second cancer case halts gene-therapy trials. Nature 421, 305.

Collins, S. A., Guinn, B., Harrison, P. T., Scallan, M. F., O'Sullivan, G. C. Tangney, M. (2008) Viral vectors in cancer immunotherapy: Which vector for which strategy? Current Gene Therapy, 8, 66–78.

De Mesmaeker, A., Häner, R., Martin, P., Moser, H. E. (1995) Antisense oligonucleotides. Accounts of Chemical Research 28, 366–374.

Ehrhardt, A., Haase, R., Schepers, A., Deutsch, M. J., Lipps, H. J., Baiker, A. (2008) Episomal vectors for gene therapy, Current Gene Therapy, 8, 147–161.

Follenzi, A., Ailles, L. E., Bakovic, S., Geuna, M., Naldini, L. (2000) Gene transfer by lentiviral vectors is limited by nuclear translocation and rescued by HIV-1 pol sequences. Nature Genetics 25, 217–222.

Follenzi, A., Sabatino, G., Lombardo, A., Boccacio, C., Naldini, L. (2002) Efficient gene delivery and targeted expression to hepatocytes in vivo by improved lentiviral vectors. Human Gene Therapy 13, 243–260.

Gewirtz, A. M., Sokol, D. L., Ratajczak, M. Z. (1998) Nucleic acid therapeutics: state of the art and future prospects. Blood 92, 712–736.

Golding, M. C., Long, C. R., Carmell, M. A., Hannon, G. J., Westhusin, M. E. (2006) Suppression of prion protein in livestock by RNA interference. Proceedings of the National Academy of Sciences of the USA 103, 5285–5290.

Krisky, D. M., Marconi, P. C., Oligino, T. J., Rouse, R. J., Fink, D. J. et al. (1998) Development of herpes simplex virus replication-defective multigene vectors for combination gene therapy applications. Gene Therapy 5, 1517–1530.

Kurreck, J., (2009) RNA Interference: From Basic Research to Therapeutic Applications. Angewandte Chemie Interantional Edition 48, 1378–1398.

Lotze, M. T., Kost, T. A. (2002) Viruses as gene delivery vectors: application to gene function, target validation, and assay development. Cancer Gene Therapy 9, 692–699.

Nielsen, P. E., Egholm, M. (eds.) (2002) Peptide Nucleic Acids – Methods and Protocols (vol. 208 of the Series Methods in Molecular Biology). Humana Press, Totowa.

Nielsen, P. E., Egholm, M. (eds.) (1999) Peptide Nucleic Acids. Horizon Scientific Press, Wymondham.

Nielsen, P. E., Haaima, G. (1997) Peptide Nucleic Acid (PNA). A DNA mimic with a pseudopeptide backbone. Chemical Society Reviews, 73–78.

Peer, D., Park, E. J., Morishita, Y., Carman, C. V., Shimaoka, M. (2008) Systemic leukocyte-directed siRNA delivery revealing cyclin D1 as an anti-inflammatory target. Science 319, 627–630.

Pfeifer, A., Verma, I. M. (2001) Gene therapy: promises and problems. Annual Reviews in Genomics and Human Genetic 2, 177–211.

Sanlioglu, S., Monick, M. M., Luleci, G., Hunninghake, G. W., Engelhardt, J. F. (2001) Rate limiting steps of AAV transduction and implications for human gene therapy. Current Gene Therapy 1, 137–147.

Siomi, H., Siomi, M. C. (2009) On the road to reading the RNA-interference code. Nature 457, 396–404.

Somia, N., Verma, I. M. (2000) Gene therapy: trials and tribulations. Nat. Rev. Genet. 1, 91–99.

Stein, C. A., Krieg, A. M. (eds.) (1998) Applied Antisense Oligonucleotide Technology. Wiley-Liss, New York.

Uhlmann, F., Peyman, A. (1990) Antisense oligonucleotides: a new therapeutic principle. Chemical Reviews 90, 543–584.

Walther, W., Stein, U. (2000) Viral vectors for gene transfer: a review of their use in the treatment of human diseases. Drugs 60, 249–271.

Whitehead, K. A., Langer, R., Anderson, D. G. (2009) Knocking down barriers: advances in siRNA delivery. Nature Reviews Drug Discovery 8, 129–138.

Wu, Z., Asokan, A., Jude Samulski, R. (2006) Adeno-associated virus serotypes: Vector toolkit for human gene therapy. Molecular Therapy 14, 316–327.

Chapter 32

Abbott, A. (2002) Techniques for gene marking, transferring, and tagging, in: Transgenic Plants and Crops (Khachatourians, G. C., McHughen, A., Nip, W.-K., Hui, Y. H., eds.) M. Dekker, New York, pp. 85–98.

Ansorge, W. J. (2009) Next-generation DNA sequencing techniques, New Biotechnology 25, 195–203.

Auer, C. and Frederick, R. (2009) Crop improvement using small RNAs: applications and predictive ecological risk assessments, Trends in Biotechnology 27, 644–651.

Bailey, M. J., Timms-Wilson, T. M., Lilley, A. K., Godfrey, H. C. J. (2001) The risks and consequences of gene transfer from genetically-manipulated microorganisms in the environment. Genetically-modified Organisms, Research Report No. 17, Department for Environment, Food and Rural Affairs, UK, 38 pp.

Carthew, R. W., Sontheimer, E. J. (2009) Origins and Mechanisms of miRNAs and siRNAs. Cell 136, 642–655.

Castanotto, D., Rossi, J. J. (2009) The promises and pitfalls of RNA-interference-based therapeutics. Nature 457, 426–433.

De Mesmaeker, A., Häner, R., Martin, P., Moser, H. E. (1995) Antisense oligonucleotides. Accounts of Chemical Research 28, 366–374.

Duggan, P. S., Chambers, P. A., Heritage, J., Forbes, J. M. (2000) Survival of free DNA encoding antibiotic resistance from transgenic maize and the transformation activity of DNA in ovine saliva, ovine rumen fluid and silage effluent. University of Leeds, Leeds, UK.

Gelvin, S. B. (2003) Agrobacterium-Mediated Plant Transformation: the Biology behind the "Gene-Jockeying" Tool, Microbiology and Molecular Biology Reviews 67, 16–37.

Gewirtz, A. M., Sokol, D. L., Ratajczak, M. Z. (1998) Nucleic acid therapeutics: state of the art and future prospects. Blood 92, 712–736.

Hare, P. D., Chua, N. H. (2002) Excision of selectable marker genes from transgenic plants, Nat. Biotechnol. 20, 575–580.

Hellens, R. P., Mullineaux, P. (2000) A guide to Agrobacterium binary Ti vectors, Trends in Plant Science 5, 446–451.

Hohn, B., Levy, A. A., Puchta, H. (2001) Elimination of selection markers from transgenic plants, Curr. Opin. Biotechnol. 12, 139–143.

Jepson, I., Martinez, A., Sweetman, J. P. (1999) Chemical-inducible gene expression systems for plants – a review, Pesticide Science 54, 360–367.

Khachatourians, G. C., Mchughen, A., Nip, W.-K., Hui, Y. H. (2002) Transgenic Plants and Crops, Dekker, New York.

Kurreck, J. (2009) RNA Interference: From Basic Research to Therapeutic Applications. Angewandte Chemie International Edition 48, 1378–1398.

Maliga, P. (2004) Plastid transformation in higher plants, Ann. Rev. Plant Biol. 55, 289–313.

Miki, B., McHugh, S. (2004) Selectable marker genes in transgenic plants: applications, alternatives and biosafety, J. Biotechnol. 107, 193–231.

Moore, I., Samalova, M., Kurup, S. (2006) Transactivated and chemically inducible gene expression in plants, The Plant Journal 45, 651–683.

Nielsen, P. E. (ed.) (2002) Peptide Nucleic Acids – Methods and Protocols (Bd. 208 der Serie Methods in Molecular Biology). Humana Press, Totowa.

Nielsen, P. E., Egholm, M. (eds.) (1999) Peptide Nucleic Acids. Horizon Scientific Press, Wymondham.

Nielsen, P. E., Haaima, G. (1997) Peptide Nucleic Acid (PNA). A DNA mimic with a pseudopeptide backbone. Chemical Society Reviews, 73–78.

Privalle, S., Wright, M., Reed, J., Hansen, G., Dawson, J., Dunder, E. M., Chang, Y.-F., Powell, M. L., Maghji, M. (2000) Proc. 6th Int. Sym. Biosafety of Genetical Modified Organisms. Extension Press, University of Saskatoon, pp. 171–178.

Razdan, M. K. (2003) Introduction to Plant Tissue Culture, Intercept, London.

Shahmuradov I. A., Solovyev V. V., Gammerman A. J. (2005) Plant promoter prediction with confidence estimation, Nucleic Acids Res. 33, 1069–1076.

Siomi, H., Siomi, M. C. (2009) On the road to reading the RNA-interference code. Nature 457, 396–404.

Slater, A., Scott, N., Fowler, M. (2003) Plant Biotechnology. The Genetic Manipulations of Plants. Oxford University Press, Oxford.

Stein, C. A., Krieg, A. M. (eds.) (1998) Applied Antisense Oligonucleotide Technology. Wiley-Liss, New York.

Taylor, N. J., Fauquet, C. M. (2002) Microparticle bombardment as a tool in plant science and agricultural biotechnology, DNA and Cell Biol. 21, 963–977.

Tzfira, T., Citovsky, V. (2006) Agrobacterium-mediated genetic transformation of plants: biology and biotechnology, Curr. Opin. Biotechnol. 17, 147–154.

Uhlmann, E., Peyman, A. (1990) Antisense oligonucleotides: a new therapeutic principle. Chemical Reviews 90, 543–584.

Whitehead, K. A., Langer, R., Anderson, D. G. (2009) Knocking down barriers: advances in siRNA delivery. Nature Reviews Drug Discovery 8, 129–138.

Yoshida K. and Shinmyo A. (2000) Transgene expression systems in plant, a natural bioreactor, Journal of Bioscience and Bioengineering 90, 353–362.

Zupan, J., Muth, T. R., Draper, O., Zambryski, P. (2000) The transfer of DNA from *Agrobacterium tumefaciens* into plants: a feast of fundamental insights, Plant J. 23, 11–28.

Chapter 33

Aggarwal, S. (2009) Nature Biotechnology 27, 987–993; ibd. (2008) 26, 1227–1233; ibd. (2007) 25, 1097–1104.

Atsumi, S., Cann, A. F., Connor, M. R., Shen, C. R., Smith, K. M., Brynildsen, M. P., Chou, K. J., Hanai, T.und Liao, J. C. (2008) Metab. Eng. 10, 305–311.

Chan, E. Y (2005) Mutat. Res , 573, 13–40.

Gabriel, C. L. (1928) Ind. Eng. Chem. 20, 1063–1067.

Jäckel, C., Kast, P. und Hilvert, D. (2008) Annu. Rev. Biophys. 37, 153–173.

Johannes, T. W., Zhao, H. (2006) Curr. Opin. Microbiol. 9, 261–267.

Jones, D. T., Woods, D. R. (1986) Microbiol. Rev. 50 484–524.

Kawahara, Y., Takahashi-Fuke, K., Shimizu, E., Nakamatsu, T., Nakamori, S. (1997)
Relationship between the glutamate production and the activity of 2-oxoglutarate dehydrogenase in Brevibacterium lactofermentum, Biosci. Biotech. Biochem. 61, 1109–1112.

Lee, S. Y., Park, J. H., Jang, S. H., Nielsen, L. K., Kim, J. und Jung, K. S. (2008) Biotechnol. Bioeng. 101 209–228.

Nakamura, C. E., Whited, G. M. (2007) Curr. Opin. Biotechnol. 14, 454–459.

Ni, Y., Sun, Z. 2009, Appl. Microbiol. Biotechnol. 83, 415–423.

Sharan, R., Ulitsky, I. und Shamir, R. (2007) Mol. Syst. Biol. 3, 88.

Keasling, J. D. 2008, Microb. Cell Fact. 7, 36.

Weizmann, C. (1919) GB patent 191504845A.

Wilkinson, B., Micklefield, J. (2007) Nat. Chem. Biol. 3, 379–386.

Chapter 36

Andrews, L., Paradise, J., Holbrook, T. et al. (2006) When patents threaten science. Science 314, 1395–1396.

California Supreme Court (1990) Moore v. Regents of the University of California, 793 P.2d 479.

Caulfield, T., Gold, R. (2000) Genetic testing, ethical concerns, and the role of patent law. Clin. Genet. 57, 370–375.

Caulfield, T., Cook-Degan, R., Kieff, F. et al. (2006) Evidence and anecdotes: an analysis of human gene patenting controversies. Nat. Biotechnol. 24, 1091–1094.

Caulfield, T., Zarzeczny, A., McCormick, J. et al. (2009) International stem cell environments: a world of difference. Nat. Rep. Stem. Cells doi:10.1038/stemcells.2009.61.

Chakrabarty, A. (1980) US Patent 4,259,444.

Cohen, S., Boyer, H. (1980) US Patent 4,237,224.

Cook-Degan, R., Chanddrasekharan, S., Angrist, M. (2009) The dangers of diagnostic monopolies. Nature 458, 405–406.

Crespi, R. (2000) An analysis of moral issues affecting patenting inventions in the life sciences: a European perspective. Sci. Eng. Ethics 6, 157–180.

Demaine, L., Fellmuth, A. (2003) Natural substances and patentable inventions. Science 300, 1375–1376.

Eisenberg, R. (1997) Structure and function in gene patenting. Nat. Genet. 15, 125–130.

European Commission (1998) Opinions of the Group of Advisors on the Ethical Implications of Biotechnology of the European Commission, European Commission, Brussels.

Fitt, R. (2009) New guidance on the patentability of embryonic stem cell patents in Europe. Nat. Biotechnol. 27, 338–339.

Hall, H., Ziedonis, R. (2001) The patent paradox revisited: an empirical study of patenting
in the US semiconductor industry. Rand J. Econ. 32, 101–128.

Hanson, M. (1997) Religious voices in biotechnology: the case of gene patenting. Hastings Center Rep. 27(6, Special Suppl.), 1–21.

Heller, M., Eisenberg, R. (1998) Can patents deter innovation? The anticommons in biomedical research. Science 280, 698–701.

Holman, C. (2008) Trends in human gene patent litigation. Science 322, 198–199.

Human Genome Organization (HUGO) (2000) Statement on Benefit Sharing. HUGO, Bethesda.

Jaffe, A., Lerner, J. (2004) Innovation and Its Discontents: How Our Broken Patent System is Endangering Innovation and Progress, and What to Do About It. Princeton University Press, Princeton.

Jensen, K., Murray, F. (2005) Intellectual property landscape of the human genome. Science 310, 239–240.

Joint Appeal Against Human and Animal Patenting (1995) Press Conference Text, May 17, 1995, Board of Church and Society of the United Methodist Church. Washington, DC.

Kevles, D., Berkowitz, A. (2001) The gene patenting controversy: a convergence of law, economic interests, and ethics. Brooklyn Law Rev. 67, 233–248.

Knoppers, M., Hirtle, M., Glass K. C. (1999) Commercialization of genetic research and public policy, Science 286, 2277–2278.

Koepsell, D. (2009) Who Owns You? The Corporate Gold Rush to Patent Your Genes. Wiley-Blackwell, Malden.

Lei, Z., Juneja, R., Wright, B. (2009) Patents versus patenting: implications of intellectual property protection for biological research. Nat. Biotechnol. 27, 36–40.

Marshall, E. (2009) Lawsuit challenges legal basis for patenting human genes, Science 324, 1000–1001.

Merz, J., Cho, M. K., Robertson, M. J. et al. (1997) Disease gene patenting is bad innovation. Mol. Diagn. 2, 299–304.

Miller, A., Davis, M. (2000) Intellectual Property. West Publishing, St Paul.

National Bioethics Advisory Commission (NBAC) (1998) Research Involving Human Biological Materials: Ethical Issues and Policy Guidance. NBAC, Washington.

Nuffield Council on Bioethics (2002) The Ethics of Patenting DNA. Nuffield Council, London.

Nielsen, P. E. (ed.) (2002) Peptide Nucleic Acids – Methods and Protocols (Bd. 208, Serie Methods in Molecular Biology). Humana Press, Totowa.

Nielsen, P. E., Egholm, M. (eds.) (1999) Peptide Nucleic Acids. Horizon Scientific Press, Wymondham.

Paradise, J., Andrews, L., Holbrook, T. (2005) Patents on human genes: an analysis of scope and claims. Science 307, 1566–1567.

Resnik, D. (2001a) DNA patents and scientific discovery and innovation:

assessing benefits and risks. Sci. Eng. Ethics 7, 29–62.

Resnik, D. (2001b) DNA patents and human dignity, J. Law Med. Ethics 29, 152–162.

Resnik, D. (2003a) A biotechnology patent pool: an idea whose time has come? J. Phil. Sci. Law 3.

Resnik, D. (2003b) Owning the Genome: A Moral Analysis of DNA Patenting. SUNY Press, Albany.

Resnik, D. (2003c) Are DNA patents bad for medicine? Health Policy 65, 181–197.

Resnik, D. (2007) Embryonic stem cell patents and human dignity. Health Care Analysis 15, 211–222.

Shapiro, C. (2000) Navigating the patent thicket: cross-licenses, patent pools, and standard setting, in Innovation Policy and the Economy. Jaffe, A., Lerner, J., and Stern, S. (eds.), MIT Press, Cambridge, pp. 119–150.

Shiva, V. (1996) Biopiracy: The Plunder of Nature and Knowledge. South End Press, Boston.

Stott, M. and Valentine, J. (2004) Gene patenting and medical research: a view from a pharmaceutical company. Nat. Rev. Drug Discov. 3, 364–368.

US Patent and Trademark Office (1999) Revised utility examination guidelines, Fed. Reg. 64(244), 71440–71442.

US Supreme Court (1980) Diamond v. Chakrabarty, 447 U.S. 303.

US Supreme Court (2002) Madey v. Duke University, 307 F.3d 1351.

Walsh, J., Arora, A., Cohen, W. (2003) Working through the patent problem. Science 299, 1021.

Walsh, J., Arora, A., Cohen, W. (2005) View from the bench: patents and material transfers. Science 309, 2002–2003.

Woolett, G., Hammond, O. (1999) An industry perspective on the gene patenting debate, in Perspectives on Gene Patenting, Chapman, A. (ed.) American Association for the Advancement of Science, Washington, DC, pp 43–50.

Yancey, A. Stewart, C., Jr (2007) Are university researchers at risk for patent infringement? Nat. Biotech. 25, 1225–1228.

Chapter 37

Furlain, P. (2008) Biosimilars, a regulatory review. Drug Inform. J. 42, 477–485.

Tobin, J. J. and Walsh, G. (2008) Medical Product Regulatory Affairs, Wiley-VCH, Weinheim.

Walsh, G. (1999) Drug approval in Europe. Nat. Biotechnol. 17, 237–240.

Chapter 38–39

Anonymous (1990) Drug treatment of stroke and ischemic brain: from acetylsalicylic acid to new drugs – 100 years of pharmacology at Bayer Wuppertal-Elberfeld. Satellite symposium of the XIth International Congress of Pharmacology, Scheveningen, The Netherlands, June 28–30, 1990. Proceedings, Stroke 21(12), IV1–175.

Borbye, L. et al. (2009) Industry Immersion Learning – Real-Life Industry Case-Studies in Biotechnology and Business, Wiley-VCH, Weinheim.

Cohen, S. N., Chang, A. C., Boyer, H. W., Helling, R. B. 1973, Construction of biologically functional bacterial plasmids in vitro. Proc. Natl. Acad. Sci. USA 70, 3240–3244.

Dhanda, R. K. (2002) Guiding Icarus: Merging Bioethics with Corporate Interests. Wiley, New York.

Drews, J. (1999) Research & development. Basic science and pharmaceutical innovation. Nat. Biotechnol. 17, 406.

Drews, J. (2000) Drug discovery: a historical perspective. Science 287 1960–1964.

Gruber, A. (2009) Biotech Funding Trends, Wiley-VCH, Weinheim.

Hall, S. S. (1987) Invisible Frontiers: The Race to Synthesize a Human Gene, Atlantic Monthly Press, New York.

Werth, B. (1995) The One Billion Dollar Molecule: One Company's Quest for the Perfect Drug. Touchstone Press, New York.

Websites

http://www.bio.org

http://www.ey.com/SearchResults?query=biotech-report&search_options=country_name;

Glossary

ABC transporter proteins: Membrane transport proteins using the energy of hydrolysis of ATP to transfer peptides and small molecules across membranes. Some ABC transporters mediate multidrug resistance, such as P-glycoprotein.

Absorption: Process whereby a cell, tissue, or organ takes up a substance.

Acetylcholine: Neurotransmitter that functions at cholinergic synapses. Occurs in the brain and in the peripheral nervous system at vertebrate neurotransmitter junctions.

Actin: Protein that forms actin filaments in all eukaryotic cells. The monomeric form is named globular or G-actin; the polymeric form is filamentous or F-actin. Many proteins are associated with actin, such as actin-binding proteins (myosin, actinin, and profilin).

Activation energy: Energy required by atoms or molecules in addition to their ground-state energy in order to undergo a particular chemical reaction.

Active site: Part of the surface of an enzyme to which a substrate molecule binds in order to undergo a catalyzed reaction.

Active transport: Energy-driven movement of a molecule across a membrane against an electrochemical or concentration gradient.

Adenoma: Benign swelling or tumor derived from gland tissue.

Adenovirus: Viruses that in humans cause upper respiratory infections or infectious pinkeye.

Adenylyl cyclase (adenylate cyclase): Enzyme that catalyzes the formation of cyclic AMP from ATP in membranes. Plays an important role in some intracellular signaling pathways.

ADME-T (absorption, distribution, metabolism, excretion, and toxicity): Summary of pharmacokinetic and toxicological parameters of a substance.

Adrenalin (epinephrine): Hormone released by chromaffin cells (in the adrenal gland) and by some neurons in stress responses, which binds to G-protein-coupled receptors. Adrenalin induces the flight-or-fight response, with enhanced heart rate and blood sugar levels. It also acts as a neurotransmitter.

Aerobic: Describes a process that requires, or occurs in the presence of, molecular oxygen (O_2).

Affinity chromatography: Chromatographic method in which the protein or oligonucleotide mixture to be purified is passed over a matrix to which specific ligands for the required protein or DNA are attached, so that the macromolecule is retained on the matrix.

Agonist: Drug or other chemical that can bind to a receptor to produce a physiologic reaction typical of a naturally occurring substance.

An Introduction to Molecular Biotechnology, 2nd Edition.
Edited by Michael Wink
Copyright © 2011 WILEY-VCH Verlag GmbH & Co. KGaA, Weinheim
ISBN: 978-3-527-32637-2

AIDS (acquired immunodeficiency syndrome): Advanced stage of a human immunodeficiency virus (HIV) infection.

Alignment: Determination of positional homology for molecular sequences, involving the juxtaposition of amino acids or nucleotides in homologous molecules.

Alkaloid: Chemically complex nitrogen-containing, small metabolite synthesized by plants as a defense against herbivores (e.g., caffeine, morphine, and colchicine). Alkaloids often affect neuronal signaling, and DNA and protein synthesis.

Alkylation: Organic reaction in which an alkyl group replaces a hydrogen atom in an organic compound, mostly a protein or nucleic acid.

Allele: One of a set of alternative forms of a gene. In a diploid cell each gene will have two alleles (one from the mother, the other from the father), each occupying the same position (locus) on homologous chromosomes.

Allergen: Substance inducing an allergic state or reaction.

Allergy: Acquired, abnormal immune response to a substance that can cause a broad range of inflammatory reactions.

Allosteric protein: Protein that changes from one conformation to another when it binds a regulatory ligand or when it is covalently modified. The change in conformation alters the activity of the protein and can form the basis of directed movements.

Alpha (*a*)-helix: Common folding pattern in proteins in which a linear sequence of amino acids folds into a right-handed helix stabilized by internal hydrogen bonding between backbone atoms.

Alternative RNA splicing: Production of different proteins from the same RNA transcript by splicing it in different ways (i.e., not all of the exons are used).

Amino-terminus (N-terminus): End of a polypeptide chain that carries a free *a*-amino group.

Aminoacyl-tRNA: Activated form of an amino acid, which is bound via a labile ester group of its carboxylic group with a hydroxyl group of a terminal ribose unit of a tRNA. Essential substrates for ribosomal translation, generally synthesized by aminoacyl-tRNA synthetases.

Aminoacyl-tRNA synthetase: Enzyme that attaches the correct amino acid to a tRNA molecule to form an aminoacyl-tRNA.

Amoebiasis: (Sub)tropical infectious disease induced by an protozoan, typically *Entamoeba histolytica*.

Amphipathic: Having both hydrophobic and hydrophilic regions, as in a phospholipid or a detergent molecule.

Anabolism: System of biosynthetic reactions in a cell by which large molecules are made from smaller precursors.

Anaerobic: Describes a cell, organism, or metabolic process that functions in the absence of air or, more precisely, in the absence of molecular oxygen.

Anaphase: Stage of mitosis during which the two sets of chromosomes separate and move away from each other. Composed of **anaphase A** (chromosomes move toward the two spindle poles) and **anaphase B** (spindle poles move apart).

Anesthetic: Substance that reduces or abolishes sensation, affecting either the whole body (general) or a particular area (local).

Angiogenesis: Formation of new blood vessels.

Antagonist: Drug that neutralizes or counteracts the effects of an endogenous ligand at a receptor.

Antibiotic: Active substance lethal to bacteria or that inhibits their growth. Most antibiotics are produced by bacteria or fungi.

Antibiotic resistance: Ability of microorganisms to cancel out the effect of antibiotics by synthesizing modified targets, through enzymatic modification of the antibiotic or via ABC transporters that export the antibiotics that had diffused into a cell.

Antibiotic resistance gene: Gene that confers upon a cell the ability to live and proliferate in the presence of an antibiotic.

Antibody (immunoglobulin): Protein formed by B cells in response to an antigen or invading microorganism.

Anticodon: Sequence of three nucleotides in a tRNA molecule that is complementary to a corresponding mRNA codon.

Antigen: Molecule or molecular structure that can induce an immune response.

Antigene therapy: Suppression of the formation of the proteins specific to a disease by inhibition of the transcription of the mRNA that codes for the protein.

Antigenic determinant (epitope): Specific region of an antigen molecule that binds to an antibody or T cell receptor.

Antisense oligonucleotide: Short, single-stranded oligomer made from modified DNA, which has a base sequence exactly complementary to that of the mRNA coding for the target protein and therefore blocks its function.

Antisense therapy: Suppression of the formation of a disease-specific protein through interception of the mRNA that codes for the protein.

Apex: Top of a cell, structure, or organ. In epithelial cells, the apical surface is that which faces outwards and is opposite to the basal side.

Apoptosis (programmed cell death): Cell suicide. Apoptosis is driven by several signaling cascades and special proteins (proteolytic caspases) that cause the cell to follow a precisely defined program leading to its death. This occurs through fragmenting of the DNA, atrophy of the cytoplasm, and changes to the cell membranes.

Archaea (*sing.:* **Archaeon**): Member of one of the two prokaryote kingdoms (see **Bacteria**), more similar to eukaryotes in genetic machinery.

Array: An ordered arrangement of nucleic acids, proteins, small molecules, or cells, which allows the parallel analysis of biological or chemical samples (usually on microscope slides). Also see **DNA chip**.

Arteriosclerosis: Lipid deposits in blood vessels that lead to narrowing and hardening of the arterial walls. It is a risk factor for heart disease, disturbances of the cerebral circulatory system, heart attack, and stroke.

Assay: Method to determine the effect of a (chemical) substance on the activity of a target. It is possible to isolate the target and measure its activity in either cell-free tests (e.g., measurement of the enzyme activity) or in cellular systems, in which the cellular reaction to the activity of the target is determined (e.g., expression of a fluorescent protein).

Atherosclerosis: Changes to arterial walls that cause arteriosclerosis.

ATP (adenosine-5'-triphosphate): Main carrier of chemical energy in cells. The terminal phosphate groups are highly reactive, in the sense that their hydrolysis or transfer to another molecule releases a large amount of free energy.

ATP synthase: Enzyme complex found in the inner membrane of mitochondria, in the plasma membrane bacteria, and in the thylakoid membrane of chloroplasts that catalyzes the formation of ATP from ADP and inorganic phosphate.

ATPase: Enzyme that catalyzes the hydrolysis of ATP.

Autoradiography: Radioactively labeled molecules darken x-ray film. If x-ray film is laid on an electrophoresis gel containing radioactively labeled proteins or nucleotides, an image is obtained. This image is known as an autoradiogram or an autoradiograph.

Autosome: Any chromosome that is not a sex chromosome.

Auxotrophy: When a cell line is dependent on one or more growth substances in the growth medium because, due to a mutation, it cannot (or is no longer able to) synthesize them itself.

Axon: Long protrusion from a nerve cell that facilitates the movement of nerve impulses quickly and over large distances. Synapses are found at the end of an axon.

Axoneme: Bundle of microtubules (9+2 pattern) and associated proteins that forms the core of a cilium or a flagellum in eukaryotic cells and is responsible for their movement.

B cell (B lymphocyte): Type of lymphocyte that forms antibodies.

B-DNA: Most frequent stable conformation in which DNA is found under physiological conditions. The DNA strand forms a right-handed double helix, in which the planar base pairs are perpendicular to the axis of the helix. The structure of B-DNA corresponds to that of the classic model suggested by Watson and Crick in 1953.

BAC: see **Bacterial artificial chromosome**.

Bacteria (*sing.:* **Bacterium**): Member of the Bacteria, one of the two large prokaryote kingdoms, the other being Archaea. Most bacteria are single celled and some cause diseases.

Bacterial artificial chromosome (BAC): Cloning vector that can accept DNA inserts of up to 1 million base pairs in length.

Bacteriophage (phage): Virus that infects bacteria. Bacteriophages have played an important role in the advance of molecular genetics; nowadays they are frequently used as cloning vectors.

Bacterium: All bacteria consist of a simple cell, surrounded by a cell wall. Their DNA is in the form of a single circular chromosome and they do not possess an endomembrane system.

Baculoviruses: Large, diverse group of DNA viruses that are only pathogenic to invertebrates and have, until now, primarily been isolated from insects. They have a double-stranded circular genome of between 100 and 180 kb in length.

Basal: Located at the base. The basal surface of a cell is situated opposite the apical surface.

Basal body: Short, cylindrical arrangement of microtubules and associated proteins located at the base of the cilia or flagella of a eukaryotic cell. The basal body supports the growth of the axoneme and is very similar in structure to a centriole.

Base: Component of nucleic acids. There are four different bases: the purines adenine (A) and guanine (G), and the pyrimidines cytosine (C) and thymine (T). In RNA, the pyrimidine uracil (U) replaces thymine.

Base pair: The four bases are always found as pairs in the DNA double helix. Due to their chemical structures, it is only possible for A to pair with T (or U in RNA) and for C to pair with G. A and T(U) and C and G are therefore described as being complementary.

Bead: Solid-phase carrier to which molecules are coupled during the implementation of an assay.

Benzodiazepine receptor: On the γ-amino butanoic acid (GABA) receptor, the binding site of benzodiazepine; point of attack of many sedatives.

Beta (β)-pleated sheet: Common structural motif of proteins, where two or more long polypeptide chains lie next to each other, linked by hydrogen bonds between the atoms of the polypeptide backbone.

Biologicals (protein drugs): Proteins, can be used as therapeutics.

Biomembrane: Barrier to permeation that surrounds every cell and cellular compartment. Cell membranes consist of phospholipids, cholesterol, and membrane proteins.

Blastomere: One of the cells that is formed when a fertilized egg cell divides.

Blastula: Early embryonic stage, consisting of a hollow ball of epithelial cells.

Bloodbrain barrier (BBB): The blood vessels of the brain have particularly tight endothelial cells lining their walls. This means that only selected substances can pass into the brain.

Blotting: Biochemical method whereby macromolecules separated on an agarose or polyacrylamide gel are transferred to a nylon membrane or a sheet of paper. This immobilizes them so that they can be analyzed further (see **Northern blotting**, **Southern blotting**, and **Western blotting**).

Bond energy: Strength of the chemical bond between two atoms, measured by the energy (in kilocalories or kilojoules) that is necessary to break it.

C-terminus: see **Carboxyl-terminus**.

Ca^{2+}-ATPase calcium pump: Transport protein that uses energy from the hydrolysis of ATP to pump Ca^{2+} ions out of the cytoplasm and into the endoplasmic reticulum.

Ca^{2+}/calmodulin-dependent protein kinase (CaM kinase): Protein kinase whose activity is driven by the binding of Ca^{2+}-bound calmodulin. In this way, Ca^{2+} indirectly drives the phosphorylation of other proteins.

Calmodulin: Calcium-binding protein, the activity of which is regulated by changes to the intracellular Ca^{2+} concentration. Ca^{2+}/calmodulin alters the activity of many enzymes and membrane transporters.

Calvin cycle: Main metabolic cycle used by plants to convert CO_2 and H_2O to carbohydrates during the second part of photosynthesis (carbon fixation).

cAMP: see **Cyclic AMP**.

cAMP-dependent protein kinase: Protein kinase that is activated by cAMP.

Cancer: Malignant tumor, usually with fast and uncontrolled cell division.

Capsid: Protein coat of a virus. Formed through self-assembly of one or more protein subunits to form a geometrically regular structure.

Carboxyl-terminus (C-terminus): End of a polypeptide chain that carries a free carboxyl group.

Carcinogenic: Property of any substance or type of radiation that can cause cancer.

Carcinoma: Cancer of epithelial cells. Carcinoma is the most frequent type of cancer in humans.

Carrier protein: Membrane transporter that binds a dissolved substance (solute) and channels it through the membrane, undergoing a series of conformational changes as it does so.

Caspases: Family of intracellular proteases involved at the start of the apoptosis cascade.

Catabolism: Metabolic process whereby organic material is broken down with the release of energy.

CD4: Coreceptor found on helper T cells. Binds to class II MHC molecules outside the antigen binding site.

CD8: Coreceptor found on cytotoxic T cells. Binds to class I MHC molecules outside the antigen binding site.

cDNA: see **Complementary DNA**.

cDNA library: Collected cDNAs of a type of cell, tissue, organ, or organism.

Cell cycle (cell division cycle): Well-ordered complex sequence of biochemical processes through which a cell copies its contents and then divides.

Cell division: Division of a cell to form two daughter cells. In eukaryotic cells this covers the division of the nucleus (mitosis) and the division of the cytoplasm (cytokinesis), which follows quickly afterwards.

Cell-free system: Fractionated homogenate of cells that contains a particular biological function of intact cells, and that can easily be used to investigate the biochemical reactions and processes of the cell *in vitro*.

Cell wall: Mechanically solid extracellular matrix that is secreted by a cell and is located outside the cytoplasmic membrane. It is of considerable thickness, and present in almost all plants, bacteria, algae, and fungi, but usually absent in animal cells.

Central nervous system (CNS): Main information processing organ of the nervous system. In vertebrates, it consists of the brain and spinal cord.

Centriole: Small, cylindrical arrangement of microtubules. One pair of centrioles is generally found in the center of the centrosome in animal cells.

Centromere: Region on the chromosome at which the kinetochore forms, which is also the site where microtubules of the mitotic spindle attach to the chromosome during mitosis. Sister chromatids are held together at the centromere.

Centrosome: Peripherally arranged multiprotein complex from which the microtubules (spindle poles) extend during mitosis. In most animal cells, the centrosome contains a pair of centrioles.

Channel protein: Membrane protein that forms a water-filled pore in the biomembrane, through which water-soluble substances, usually ions, can pass.

Chaperone: Proteins (e.g., heat shock protein 70) that help other proteins to avoid folding incorrectly. Incorrect folding might result in the formation of inactive or aggregated polypeptides.

Chemical lead: Common basic structure of a series of substances that show the desired activity in an assay and can be chemically modified for further optimization.

Chemotherapy: Cancer treatment involving the use of cytotoxic chemicals.

Chiasma (*pl.: Chiasmata*): χ (Greek=chi)-shaped connection observed to occur between paired chromosomes during metaphase I. The chiasma is where crossing-over occurs.

Chlorophyll: Light-absorbing green pigment (with Mg as the central atom of the porphyrin system) that is important for photosynthesis in bacteria, plants, and algae.

Chloroplast: Organelle found in green algae and plants that contains chlorophyll and carries out photosynthesis. It is a modified plastid that contains its own circular DNA and is capable of protein synthesis.

Cholesterol: A lipid, which is the most common steroid in the human body. It plays an important role in the fluidity of biomembranes and is important as a hormone precursor. A high cholesterol level is frequently a risk factor for heart disease.

Chromatid: Copy of a chromosome that is formed by DNA replication. The two identical chromatids, still joined at the centromere, are termed sister chromatids.

Chromatin: Complex of DNA, histones, and nonhistone proteins found in the cell nucleus; material of chromosomes.

Chromatography: Method of physically separating substances by allowing the mixture to be partitioned between a mobile phase and a stationary phase.

Chromosomal crossing-over: Exchange of DNA between paired homologous chromosomes (genetic recombination) during metaphase I of meiosis.

Chromosome: Specific linear arrangement of DNA with its associated proteins in a macromolecular complex. Chromosomes are visible as compact rod-like structures under a light microscope, particularly during mitosis or meiosis in plant and animal cells.

Cilium (*pl.: Cilia*): Threadlike structures found on the outer side of eukaryotic cells; they are made up of a bundle of microtubules and are capable of regular beating movements. Cilia are found in large number on the surface of many cells (e.g., bronchial epithelia) and are responsible for the swimming movements of many unicellular organisms.

Citric acid cycle (tricarboxylic acid (TCA) cycle, Krebs cycle): Central metabolic pathway found in aerobic organisms. Discovered by Hans Krebs. It involves the oxidation of acetyl groups obtained from food molecules to form CO_2 and H_2O. The reduction equivalent NADH is required for the oxidative phosphorylation in the respiratory chain.

Class I MHC molecule: Found on the surface of almost all types of cell; presents viral peptides on the surface of the cell when infected by a virus or microbe, which are then recognized by cytotoxic T cells. See **Major histocompatibility complex**.

Class II MHC molecule: One of the two classes of MHC molecule. Professional antigen-presenting cells (e.g., macrophages) have MHC II proteins on their cytoplasmic membranes and thus present foreign peptides to the helper T cells. See **Major histocompatibility complex**.

Clathrin: Protein that forms a polyhedral coating on endocytotic vesicles.

Clinical study: Development of new drugs takes place in four stages: (i) preclinical studies, (ii) clinical studies phase I, (iii) clinical studies phase II, and (iv) clinical studies phase III.

Clone: Population of identical cells or organisms that originate from a common ancestor through asexual reproduction.

Cloning: Creation of many copies of a required DNA fragment by means of recombinant DNA technology.

Cloning vector: Small molecule of DNA that usually originates from a bacteiophage or plasmid. Cloning vectors transport the DNA fragment that is to be cloned into the host cell, where it will be replicated.

CNS: see **Central nervous system**.

Codon: Sequence of three nucleotides in a DNA or RNA molecule that code for a specific amino acid in a growing peptide chain.

Coenzyme: Small molecule associated with an enzyme that takes part in the reaction that is being catalyzed (e.g., undergoes covalent bonding with the substrate). Examples include biotin, NAD^+, and coenzyme A.

Coiled-coil: Coil of usually a-helical regions of proteins.

Colony-stimulating factor (CSF): Generic term for the numerous signaling molecules that regulate the differentiation of blood cells.

Commercial application: Prerequisite for the granting of a patent; is demonstrated if the invention has an application in a commercial area including agriculture.

Commercial copyright: Preamble of a legal norm that serves to protect commercial-intellectual achievements and related issues (patents, trademarks, utility models, design patents).

Compartment: Membrane-bound region of the cell (cytoplasm, mitochondrion, nucleus, etc.).

Competence: Stage at which a cell can take up DNA (e.g., via transformation or transfection).

Complementary: Two nucleic acid sequences are described as being complementary when they can combine to form a double helix with perfect base pairing.

Complementary DNA (cDNA): DNA formed by reverse transcription of a mRNA molecule.

Complementary RNA (cRNA): RNA formed by *in vitro* transcription of cDNA. This is achieved through hybridization with oligonucleotide arrays, so a linear amplification of the molecule is possible. Several molecules of cRNA are formed from one cDNA template molecule.

Confocal microscope: Version of a light microscope that produces a clear image of a given plane within a probe. Laser light is used as a pinpoint source of illumination. A 2D "optical section" is produced by scanning with the laser beam across the plane.

Conformation: Three-dimensional arrangement of atoms in a macromolecule, such as a protein or nucleic acid.

Consensus sequence: Average or most typical form of a sequence that is found with only minor alterations in a group of related RNA, DNA, or protein sequences. At each position in the consensus sequence the nucleotide/amino acid given is the one that is found there most frequently.

Constitutive: Permanently formed in the same proportions; opposite of regulated.

Covalent bond: Stable chemical bond between two atoms formed through the sharing of one or several pairs of electrons.

cRNA: see **Complementary RNA**.

Crossing-over: see **Chromosomal crossing-over**.

CSF: see **Colony-stimulating factor**.

Curated database: Database, the contents of which may only be added to or altered by curators. Before an entry is made, checks on consistency and agreement are made against a defined system of concepts.

Cyanogenic glycoside: Secondary metabolite that is cleaved to form cyanide (HCN) and an aldehyde when a plant is injured.

Cyclic AMP (cAMP): Nucleotide formed from ATP by adenylate cyclase in response to the stimulation of cytoplasmic membrane receptors. cAMP is an intracellular signaling molecule (second messenger) and activates cAMP-dependent protein kinases (e.g., protein kinase A). It is hydrolyzed to AMP by phosphodiesterases. An analogous molecule is cGMP.

Cyclins: Proteins that control the cell cycle by activating to cyclin-dependent kinases (CDKs) and determining their activity and specificity.

Cyclooxygenase: Key enzyme in prostaglandin synthesis.

Cytokine: Extracellular signaling protein or peptide that acts as a local mediator in cellcell communication.

Cytoplasm: see **Cytosol**.

Cytoplasmic membrane: Membrane that surrounds living cells.

Cytoskeleton: System of protein filaments (actin filaments, microtubules, and intermediate filaments) found in the cytoplasm of eukaryotic cells that gives a cell its form and the ability to move in a specific direction.

Cytosol: Content of the main compartment of the cytoplasm, minus membrane-bound organelles such as the endoplasmic reticulum, mitochondria, and nucleus.

Cytostatic: Cytostatic substances inhibit the growth and proliferation of cells.

Cytotoxic: Cytotoxic chemicals are cell toxins that inhibit cell division or protein synthesis, block the generation of energy, or disturb the ionic balance. They lead to cell death (see **Apoptosis**). Cytostatic chemicals are cytotoxic in the long term.

Cytotoxic T cells: Type of T cell. Responsible for the death of infected cells.

Dalton (Da): Unit of molecular mass, approximately equal to the mass of one hydrogen atom (1.66×10^{-24} g).

Death receptor: Receptor at the biomembrane that induces apoptosis upon activation by extracellular ligands.

Deletion: Type of mutation in which a single nucleotide or a group of nucleotides is deleted from the DNA sequence.

Denaturation: Extreme changes to the conformation of a protein or a nucleic acid due to the effects of heat or chemicals. Usually leads to loss of biological function.

Dendrites: Nerve cells have hundreds of dendrites, which are short protrusions that communicate with the synapses of other nerve cells.

Dendritic cells: Cells of the immune system found in lymph and other tissues that are specialized for the uptake of particles through phagocytosis. They also act as professional antigen-presenting cells in the immune response.

Deoxyribonucleic acid (DNA): Polynucleotide made from covalently bound deoxyribonucleotides. It stores the cell's hereditary information and passes it from generation to generation.

Detergents: Type of surfactant that can dissolve membrane lipids and membrane proteins in an aqueous solution; consists of a polar (hydrophilic) region and a nonpolar (hydrophobic) region.

Diabetes mellitus: Illness characterized by raised blood sugar levels, either as a consequence of lack of insulin or reduced effectiveness of insulin.

Diacylglycerol (DAG): Lipid formed from the enzymatic cleavage of inositol phospholipids in response to an extracellular signal. Composed of two fatty acid chains linked to glycerol by an ester bond. Activates protein kinase C as a signal molecule.

Differentiation: Process by which an undifferentiated cell undergoes a change to become a specialized cell type.

Diffusion: Movement of molecules along a concentration gradient through statistical thermal movement (Brownian motion).

Diploid: Diploid organisms possess two sets of homologous chromosomes and, therefore, two copies of every gene or gene locus. See **Haploid**.

Distance: Measure of the displacement of objects. The mathematical requirements are: (i) A distance can only consist of 0 or a positive real number, (ii) the displacement of an object from itself must be 0, and (iii) The displacement between objects A and B must be equal to the displacement between B and A (commutative). For distances, it is requirement that they obey the triangle inequality: $d(a,c) \leq d(a,b) + d(b,c)$.

Disulfide bridge (–S–S–): Covalent bond between the sulfide groups of two cysteine residues. In extracellular proteins, an important method of linking two proteins or different parts of the same protein.

DNA: see **Deoxyribonucleic acid**.

DNA chip (microarray): Slide (made of glass or membrane) on which DNA fragments can be placed in a regular, rectangular order. These hybridize specifically with different mRNA species. The DNA fragments can be obtained from a cDNA library (cDNA chips) or can be synthetic oligonucleotides. They can be transferred by robots (spotted/printed chips) or synthesized directly on the chip (oligonucleotides only).

DNA footprinting: Technique used to determine the DNA sequence to which a DNA binding protein binds.

DNA library: Collection of cloned DNA molecules that represent either a complete genome (genomic library) or DNA copies of the mRNA made in a cell (see **cDNA library**).

DNA ligase: Enzyme that joins DNA fragments to each other; used in gene technology as a molecular glue.

DNA methylation: Addition of a methyl group to nucleotide bases (A and C). Extensive methylation of the cytosine residues (hypermethylation) in CG sequences is used in eukaryotes to permanently switch off genes (epigenetics).

DNA microarray: Method used to analyze the simultaneous expression of a large number of genes in cells. Isolated cellular RNA or cDNA is hybridized with short DNA probes that have been immobilized individually in large numbers on glass slides (see **DNA chip**).

DNA polymerase: Enzyme that synthesizes DNA through the condensation of nucleotides via phosphodiester bonds. DNA polymerase requires a complementary strand as template and a free 3'-primer end to start.

DNA primase: Enzyme that synthesizes a short RNA strand complementary to a DNA template.

DNA repair: Enzymatic correction of mutations or of mistakes in replication.

DNA topoisomerase: Enzyme that binds to DNA and reversibly breaks a phosphodiester bond on one (topoisomerase I) or both (topoisomerase II) strands, s o that the DNA at that point can uncoil. It prevents twisting during replication.

Domain: Structural and functional unit of proteins. They fold themselves independently of the other parts of the protein, are usually globular, and are generally between 40 and 150 amino acids in length.

Dominance: Refers to the inheritance of the half of the pair of alleles that is expressed in the phenotype of the organism when the other is not, regardless of whether both are present. Opposite of recessive.

Dorsal: Refers to the back of an animal, or to the upper side of a leaf or wing.

Dorsoventral: Describes the axis that runs from an animal's back to its front or from the upper side of a structure to its lower side.

Drug resistance: Cells or microorganisms can lose their sensitivity to an active substance, in that they inactivate it or pump it out of the cell.

Druggability: Property of a protein to bind small chemical molecules and thus alter its own activity.

Dynamic programming: Process by which all possible arrangements of two sequences are evaluated and the alignment found that optimizes the score value. For each subalignment, the score is recorded in a table. The cells of the table are filled via a recursion formula. The optimal alignment is obtained by following a path through the table that, at every step, takes the optimal score.

Dynein: Member of a family of large motor proteins that facilitate ATP-dependent movement along microtubules. In cilia, dynein forms the side arm of the axoneme that allows neighboring microtubule pairs to slide along each other.

EC_{50}: Effective concentration that results in 50% of the maximum possible effect being measured.

Efficacy: Term used to describe the maximum possible effect of a substance with regard to biochemical or cellular assays.

Elastin: Hydrophobic protein. Forms the extracellular elastic fibers that give tissues their elasticity and robustness.

Electrochemical gradient: Sum total of the effects of the difference in concentration of ions on either side of a membrane (concentration potential) and the difference in electrical charge across the membrane (membrane potential). It generates the power required for an ion to pass through a membrane.

Electrochemical proton gradient: Sum total of the H^+ (proton) gradient and the membrane potential.

Electron acceptor: Atom or molecule that takes up electrons with relative ease and is thus reduced (oxidizing agent).

Electron donor: Molecule that easily loses an electron and is thus oxidized (reducing agent).

Electrophoresis: Separation technique for proteins and nucleic acids, which migrate through a gel (agarose, polyacrylamide) when subjected to a strong electric field (also see **SDSpolyacrylamide gel electrophoresis**).

Electroporation: Use of electric pulses to induce the uptake of DNA by cells.

Embryogenesis: Development of an embryo from a fertilized egg or zygote.

Embryonic stem cells (ES cells): Cells obtained from the inner cell mass of an early mammalian embryo. They are still omnipotent and can, therefore, develop into any cell of the body. It is possible to cultivate them *in vitro*, modify them genetically, and then to insert them into a blastocyst.

Endocrine cells: Specialized animal cells that release hormones into the bloodstream.

Endocytosis: Uptake of molecules by a cell through the formation of vesicles by the cytoplasmic membrane (see **Pinocytosis** and **Phagocytosis**).

Endoplasmic reticulum (ER): System of inner membranes in which lipids, membranes, and proteins are synthesized. Many secretory proteins are modified postransationally.

Endoprotease: Enzyme that recognizes and cleaves peptide chains at specific recognition sequenes.

Endosome: Membrane-bound organelle found in animal cells that takes up endocytotic vesicles and passes on to the lysosome for digestion of their contents.

Endothelial cells: Flattened cells that form the endothelium (the layer of cells that lines all blood vessels).

Enhancer: Regulatory sequence of DNA to which gene regulatory proteins bind. Enhancers play an important role in the rate of transcription of a structural gene, which might be located many thousands of base pairs away.

Entropy (S): Thermodynamic measure of disorder within a system. The higher the entropy, the greater the disorder.

Enzyme: Protein that catalyzes specific chemical reactions (e.g., the hydrolysis of acetylcholine by acetylcholine esterase).

Enzyme-coupled receptor: Main type of membrane receptor, the cytoplasmic domain of which either possesses the ability to act as an enzyme itself or can bind to intracellular enzymes. The enzyme activity is triggered by the binding of a ligand to the receptor.

Epidermis: Layer of epithelial cells that covers the outer surface of the body. The outermost cell layer of plant tissues is also known as the epidermis.

Epinephrine: see **Adrenalin**.

Epithelium (*pl.*: **Epithelia**): Collection of one or multiple sheets of surface tissue that cover the outer surface of the body and line hollow organs.

Epitope: Region or sequence of a protein that possesses specific binding properties (e.g., is recognized by an antibody).

ER: see **Endoplasmic reticulum**.

ER retention signal: Short amino acid sequence within a protein that prevents it from leaving the endoplasmic reticulum. It is found in proteins located in the endoplasmic reticulum.

Erythrocyte (red blood cell): Small, hemoglobin-containing blood cell that transports oxygen to tissues and carbon dioxide away from them.

Erythropoietin: Growth factor that is formed in the kidney, and stimulates the red blood cell precursors in the bone marrow to differentiate and divide. Has been used as a performance enhancing drug in sports ("doping").

Escherichia coli: Bacterium found in the human gut. Variants of this coli bacterium (*E. coli* K12), which lack the specific properties of the wild-type necessary for survival in the wild; are frequently used in genetic engineering as so-called acceptor organisms in the cloning of recombinant DNA fragments.

Estrogen: Sex hormone of females.

Eukaryote (eucaryote): Uni- or multicellular organism, the cells of which possess a nucleus (protozoans, fungi, plants, and animals).

European Patent Convention (EPC): Convention for the acceptance of European patents, signed in Munich in 1973.

European Patent Office: Executive body of the European Patent Organization (EPO) – an intergovernmental institution formed on the basis of the European Patent Convention (EPC) whose members are the contractual states of the EPC. Operations are overseen by the Administrative Council of the organization, which is made up of delegates from the contractual states.

Exocytosis: Molecules (e.g., proteins) are packaged in small vesicles that fuse with the plasma membrane and release their contents.

Exon: Part of a gene that is transcribed into RNA and remains in the processed mRNA. Exons contain the protein-coding region of the gene. In general, an exon is located next to a noncoding region known as an intron.

Expressed sequence tag (EST): Transcripted DNA sequences obtained from cDNA libraries.

Expression vector: Cloning vector that contains the regulatory sequences necessary for efficient transcription and translation in an organism.

Extracellular matrix: Complex network of oligosaccharides (such as cellulose) and proteins (such as glucosaminoglucan or collagen) excreted from cells. It is a structural element of connective tissues.

Fas protein (Fas): Membrane-bound receptor (see **Death receptor**); the binding of a Fas ligand triggers apoptosis of the cell.

Fast protein liquid chromatography (fast performance liquid chromatography; FPLC): Form of low-pressure chromatography developed specifically for the purification of proteins.

Fat cell (adipocyte): Cell found in the connective tissues of animals that forms and stores fat.

Fc receptor: Member of a family of receptors that are able to recognize the nonvarying (Fc) region of immunoglobulins (with the exceptions of IgM and IgD). Different Fc receptors exist for IgG, IgA, IgE, and their subclasses.

Fermentation: Energy-producing anaerobic metabolic pathway by which, for example, glucose is converted to lactate or ethanol via pyruvate.

FGF: see **Fibroblast growth factor.**

Fibroblast: Most prevalent type of cell in connective tissue. Secretes an extracellular matrix that is rich in collagen and other extracellular matrix molecules. Fibrolasts move into wound tissue and multiply in tissue cultures.

Fibroblast growth factor (FGF): Protein growth factor that triggers cell division in fibroblasts and other types of cells.

FISH: see **Fluorescence *in situ* hybridization**.

Flagellum (*pl.:* **Flagella**): Long, whip-like cell modification that can propel a cell through a fluid medium by lashing. In eukaryotes, they are a form of long cilia. In bacteria, they are smaller and are completely different in both structure and behavior.

Fluid chromatography: In fluid chromatography, a mixture of solvents or buffers is used as a mobile matrix.

Fluid-phase endocytosis: Endocytosis, in which small vesicles of the cytoplasmic membrane are pinched off into the cell and bring with them extracellular fluid containing dissolved substances.

Fluorescein: Fluorescent dye that glows green under blue or UV light.

Fluorescence *in situ* hybridization (FISH): Method to stain DNA or RNA *in situ*, using fluorescently labeled specific probes.

Fluorescence resonance energy transfer (FRET): Method used to detect bonding between two fluorescently labeled molecules within a cell. Transfer of excitation energy from one fluorescent dye to another.

Fluorescent dye: Molecule that absorbs light of a particular wavelength and, as a result, emits light of a different wavelength (lower in energy).

Follicle cell: One of the types of cell that surrounds a developing egg or oocyte.

FPLC: see **Fast protein liquid chromatography**.

Free radical: Unstable oxygen species with an unpaired electron that can damage cells and cell components (DNA).

FRET: see **Fluorescence resonance energy transfer**.

Gel/2D gel: Method used to separate as many proteins (e.g., those resulting from the breakdown of a cell) from each other as possible. A combination of isoelectric focusing (separation of the proteins on the basis of their isoelectric point) and a denaturing SDSgel electrophoresis (which separates proteins by size) is used. The second gel is run at a right angle to the first, thus resulting in 2D separation.

G-protein: see **GTP-binding protein**.

G-protein-coupled receptor (GPCR): Receptor located in the cell surface membrane with seven transmembrane domains. After activation by specific extracellular ligands (signal molecules), it binds to GTP-binding proteins (G-proteins).

GABA receptor: γ-Amino butanoic acid receptor.

Galenics: Study of the pharmaceutical form of drugs.

Gamete: Haploid germ cell (oocyte, sperm) specialized for sexual reproduction.

Ganglion (*pl.:* **Ganglia**): Group of nerve cells and associated glial cells situated together outside the central nervous system.

GCP (Good Clinical Practice): Guidelines describing the ethical and scientific standards for clinical trials on humans.

Gel electrophoresis: Method of separating nucleic acid molecules or proteins embedded in a gel on the basis of their mobility in an electric field. The gels used are made from agarose or polyacrylamide.

Gene: Unit of hereditary information responsible for the expression of a characteristic trait. In this context it refers to a section of DNA that contains the genetic informaion required for the synthesis of a protein or functional RNA (e.g., rRNA, tRNA).

Gene cloning: Isolation and insertion of a gene into a cloning vector in order for DNA replication to take place.

Gene control element: General term for any protein that binds to a specific DNA sequence and in doing so alters the expression of a gene.

Gene control region: Sequence of DNA required to initiate transcription of a given gene and to control the rate of initiation.

Gene conversion: Process in which the DNA sequence from one DNA helix (which remains unaltered) is transferred to another DNA helix (the sequence of which changes). It happens occasionally during general recombination and through conversion different DNA sequences are made identical.

Gene expression: Transcription of a gene into mRNA and the ensuing translation of the mRNA into the corresponding protein.

Gene mapping: Analysis of an individual chromosome, in which the position of genes relative to each other is described by the frequency of genetic recombination between them, measured in centimorgans (cM).

Genetic code: Correspondence between nucleotide triplets (codons) in DNA or RNA to amino acids in proteins.

Genome: Sum of the genetic information that belongs to a cell or organism, in particular the DNA in which this information is stored.

Genomic DNA: DNA that constitutes the genome of a cell or an organism. Often used as opposed to cDNA (DNA obtained through reverse transcription of mRNA). Genomic DNA clones are made up of DNA cloned directly from chromosomal DNA. A collection of such clones from any given genome is known as a genomic DNA library or a genomic DNA bank.

Genomics: Subject area that deals with the investigation of DNA sequences and the properties of the genome as a whole.

Genotype: Genetic constitution of a single cell or an organism (in contrast to the phenotype).

Germ line: Line of descent of germ cells (which contribute to the formation of a new generation of organisms) in contrast to somatic cells (which form the body and do not leave any descendents).

GFP: see **Green Fluorescent Protein**.

Glial cells: Cells of the nervous system that provide support. These include oligodendrocytes and astrocytes in the central nervous system, as well as Schwann cells in the peripheral nervous system of vertebrates.

Glutathione-S-transferase (GST): Enzyme that transfers glutathione to different substrates. Frequently used in the purification of GST-binding protein by means of glutathione-coated carriers.

Glycolysis: Metabolic pathway found throughout the cytosol, through which sugar is broken down to pyruvate and ATP is generated; 2 mol of ATP and 2 mol of NADH are produced for every 2 mol of glucose.

Glycoside: Natural product that yields at least one simple sugar molecule when hydrolyzed.

Glycosylation: Addition of one or more sugar molecules to a protein or a lipid molecule.

Glycosylphosphatidylinositol anchor (GPI anchor): Possible anchoring of a protein in the biomembrane; coupled to a protein in the endoplasmic reticulum.

Golgi apparatus: Tubular compartment found in eukaryotic cells in which proteins and lipids originating from the endoplasmic reticulum are modified and sorted. It is the site of synthesis of many cell wall polysaccharides in plants and extracellular matrix glycosaminoglycans in animal cells.

Grana (*sing.:* **Granum**): Stacked tubes of membrane (thylakoid) of the inner membrane of chloroplasts. They contain chlorophyll, as well as proteins involved in electron transport, and are the site of the light-dependent reactions of photosynthesis.

Granulocyte: Type of white blood cell that is distinguished by the presence of clearly visible grains in the cytoplasm. There are three different types of granulocyte: neutrophils, basophils, and eosinophils.

GRAS (Generally Regarded As Safe): Classification of the US Food and Drug Administration for safe foods and drugs.

Gel filtration: Fractionation of proteins on the basis of differences in their sizes. Working under the assumption that proteins are a mixture of similar ball-shaped structures, the sequence of elution changes in proportion to the molecular weights.

Green Fluorescent Protein (GFP): Fluorescent protein (or gene, respectively) isolated from jellyfish (*Aequorea victoria*). Frequently used in cell biology as a marker and reporter protein.

Growth factor: Extracellular polypeptide signal molecule that can stimulate a cell to divide (e.g., epidermal growth factor and platelet-derived growth factor).

GTPase: Enzyme that hydrolyzes GTP to GDP.

GTPase-activating protein (GAP): Protein that binds to a GTP-binding protein, and inactivates it by triggering its GTPase activity and causing it to hydrolyze the bound GTP to GDP.

GTP-binding protein: Protein that is activated by binding GTP. Its GTPase activity eventually hydrolyzes the bound GTP to GDP, and thus inactivates the protein. G-proteins are important in intracellular signal transduction and consist of three different subunits: α-, β-, and γ-subunits. Members of the other very large families are monomers (small G-proteins or monomeric GTPases).

Haploid: Cell that possesses a single set of chromosomes (e.g., sperm cells or bacteria); in contrast to diploid, where cells possess two sets of chromosomes (as in somatic cells).

Heat shock protein: Protein formed in increased numbers in response to raised temperatures or other forms of stress. Important examples include HSP60 and HSP70, as well as HSP90. Can act as a chaperone.

Helix-loop-helix (HLH): Structural motif in many gene regulatory proteins, with which specific DNA sequences are recognized.

Helper T cell (T_h cell): Important type of T cells that helps B cells to form antibodies and activate macrophages in order to kill invading microorganisms.

Hemoglobin: Main protein in red blood cells, capable of transporting oxygen and CO_2.

Herpes simplex: Acute, primary, or secondary viral infection of the skin and mucus membranes (e.g., the lips and genitals).

Heterochromatin: Region of a chromosome with unusually condensed chromatin, which is transcriptionally inactive during interphase.

Heterodimer: Protein complex formed from two different subunits.

Heterozygote: Diploid cell or organism with two different alleles at one or more gene loci.

Heuristic: From Greek: to find, to advise. Describes algorithms in informatics that solve a problem almost optimally (i.e., find a solution that is almost optimal or is optimal in the majority of cases). Heuristics cannot guarantee an optimal solution. They are used in cases in which there are no algorithms that are capable of solving a problem optimally or in which such algorithms cannot be used because of complexity.

High-energy bond: Covalent bond, the hydrolysis of which causes the release of a large amount of energy. It is possible for any group bound to a molecule with such a bond to be transferred from one molecule to another. Examples include phosphodiester bonds in ATP and thioester bonds in acetyl-CoA.

High-performance liquid chromatography (high-pressure liquid chromatography; HPLC): Sensitive technique used to separate and analyze solutions or nonliquid substances in extract form. The grain size of the stationary phase is characteristic of HPLC at 3, 5, or 10 μm and this is what causes the high pressure of the mobile phase.

High-throughput screening (HTS): Search for a substance in a library of thousands of products.

Histone: Member of a group of small, common basic proteins that have a high arginine and lysine content. Histone binds to negatively charged DNA in eukarytes; four histones form a nucleosome.

Hit: Substance identified through a screening process.

HIV (human immunodeficiency virus): Virus that causes AIDS.

Hodgkin's lymphoma: Cancer of the lymphatic tissues, with tumors in the reticuloendothelial system and the formation of granuloma.

Homeobox: Short (180 bp) conserved DNA sequence that codes for a DNA-binding protein motif (homeodomain). It is found in many organisms in genes that control the early developmental processes.

Homolog: 1. *adj.* Term used for organs or molecules that are the same because they stem from a common precursor. 2. *noun* One of two or more genes that have the same DNA sequence, because they stem from the same ancestral gene. See **Homologous chromosome**.

Homologous chromosome (homolog): One of two copies of a particular chromosome found in a diploid cell; in every diploid cell one homologous chromosome is inherited from the mother and the other is inherited from the father.

Homozygote: Diploid cell or organism with two identical alleles at a specific gene locus.

Hormone: Chemical produced by an endocrine gland that is secreted into the bloodstream and controls another organ or tissue of the body.

HPLC: see **High-performance liquid chromatography**.

HTS: see **High-throughput screening**.

Hybridization: Formation of a duplex from two complementary, possibly modified single strands of nucleic acid. It is the basis of diagnostic and therapeutic procedures to find specific nucleotide sequences.

Hybridoma: Cell line used to obtain monoclonal antibodies. It is created by the fusion of antibody-forming B cells with lymphocyte tumor cells.

Hydrogen bond: Noncovalent bond that forms between an electropositive hydrogen and an electronegative atom.

Hydrophilic: Ability of a polar molecule to undergo interactions (e.g., hydrogen bonding) with water molecules; hydrophilic substances dissolve easily in water (from Greek: water loving).

Hydrophobic (lipophilic): Nonpolar molecules which cannot hydrogen bond with water molecules. They will not dissolve in water, but only in nonpolar lipids (from Greek: water hating (lipid loving)).

Hydrophobic interaction chromatography (HIC): Based on the interactions of hydrophobic protein regions with the hydrophobic ligands of the chromatography matrix. The proteins can usually be eluted using a linear salt gradient.

Hydrophobicity: Measure of the unwillingness of a substance to dissolve in water. By convention, the solvation enthalpy is the energy required for a substance to dissolve in water. For hydrophilic molecules, the value is negative (i.e., energy is released).

Hypertension: Raised blood pressure (above 140/90 mmHg).

Hypertrophy: Increase in size of a tissue or an organ due to an increase in the size or numbers of its cells.

Image processing: Computer editing of photographs gained from microscopy (e.g., for the reconstruction of 3D pictures).

Immune response: Reaction of the immune system to the entry of an antigen or a microorganism into the body.

Immunoprecipitation: Use of a specific antibody to isolate the corresponding protein antigen. Using this technique, complexes of interacting proteins in cell extracts can be identified through precipitation with a specific antibody against one of its protein components.

Immune system: Complex cellular and humoral system that protects against infection. Found in vertebrates.

***In situ* hybridization:** Technique whereby single-stranded DNA or RNA probes are used to localize a gene or mRNA molecule in a cell or tissue through hybridzation.

In vitro: From Latin: in glass (i.e., outside the organism).

***In vitro* transcription:** Transcription in the absence of a cell, usually by means of T7 RNA polymerase. Double-stranded DNA molecules containing the T7 promoters are required.

***In vitro* translation:** Translation in the absence of a cell by means of reticulocyte, wheat germ, or *E. coli* extracts. mRNA or *in vitro* transcribed RNA is required.

In vivo: From Latin: in life (i.e., in a living organism, animal, or human).

IND: see **Investigational new drug.**

Influenza: True flu; acute and highly contagious infectious disease caused by the influenza virus. The virus infects the mucosal cells of the respiratory tract.

Innovation: Prerequisite for the procurement of a patent; it is demonstrated when such an invention does not already exist in a publicly accessible form anywhere in the world.

Insert: DNA fragment that is inserted into a vector for the purpose of propagation or expression.

Insulin: Hormone synthesized by the pancreas (β cells of the islets of Langerhans) that regulates the blood glucose concentration. Insulin is synthesized in the form of a preprotein, from which a peptide is cut out.

Intellectual property: Basic principle of protection copyright and invention (intangible right).

Intercalation: Planar and lipophilic substances insert themselves between the bases of DNA. This can lead to frameshift mutations.

Intermediate filament: Fibrous strands of proteins (approximately 10 nm in diameter) that, when linked to each other, form networks in animal cells. One of the three important types of filaments that make up the cytoskeleton.

Intron: Genomic region of a gene that is initially transcribed into RNA, but is eliminated from the RNA during its processing by splicing. Introns generally do not contain any coding information. The 5'-end (GT) and the 3'-end (AG) are conserved in introns.

Invention: Device with a practical application whose claimed object or claimed job is feasible, repeatable, and of a technical nature, and which represents the solution to a problem through technical considerations.

Inversion: Mutation in which a DNA or chromosome segment has been inverted.

Investigational new drug (IND): Status of a new substance after successful authorization of clinical trials by the authorities.

Ion exchange chromatography (IEC): Ion exchanger with bound, loaded anions and cations that are exchangeable with other ions.

Ion channel: Transmembrane protein complex that forms a water-filled canal through the biomembrane. Specific inorganic ions can diffuse through it according to their electrochemical gradient.

Ionic bond: Noncovalent bond between two atoms, one with a positive charge and one with a negative charge.

Isoelectric focusing: Electrophoretic separation of molecules in a gel that contains a pH gradient. The proteins move through an electric field into the area of the gel where the pH corresponds to the isoelectric point of the protein.

Isoelectric point: pH value at which a molecule does not have a net charge, because the number of positive and negative charges are equal.

Karyotype: Complete set of chromosomes possessed by a cell, arranged by size, form, and number.

Kinesins: Class of motor proteins that utilize the energy released by the hydrolysis of ATP in order to move along microtubules.

Kinetochore: Protein complex (known as the centromere) of the mitotic chromosome to which microtubules bind. Important for the movement of sister chromatids into the newly forming cells.

Kinetochore microtubules: Microtubules that make up the mitotic and meiotic spindles, the ends of which bind to the kinetochore of a chromatid.

Knockin: Replacement of a gene in a model organism with a mutated gene.

Knockout: Destruction of a gene in a model organism.

LDL: see **Low-density lipoprotein.**

Lectins: Proteins that form strong bonds with a specific sugar. Lectins from plants (usually toxic, e.g., ricin, obtained from *Ricinus communis*) are often used as affinity reagents in order to purify glycoproteins or demonstrate their existence on the upper surface of cells.

Lethal mutation: Mutation that causes the death of the cell or organism in which it is found.

Leucine zipper: Structural motif found in many DNA-binding proteins; consists of two *a*-helices made up of individual proteins that together form a coiled coil similar to a zipper: a protein dimer.

Leukemia: Cancer of the white blood cells.

Leukocyte: General term for all blood cells that possess a nucleus and do not contain hemoglobin, including lymphocytes, neutrophils, eosinophils, basophils, and monocytes.

Ligand: Any molecule that binds to a specific site on a receptor or another molecule (from Latin: *ligare*=to bind).

Ligase: Enzyme that binds (ligates) one molecule to another in a process that requires energy. DNA ligase, for example, links two DNA molecules via phosphodiester bonds.

Ligation: Covalent linkage of the end of one DNA molecule to another, by means of a specific enzyme (DNA ligase).

Lipid: Substance that dissolves easily in a nonpolar solvent, but is insoluble in water.

Lipid raft: Localized area of the plasma membrane that is rich in sphingolipids and cholesterol.

Lipophilic: Lipid loving; see **Lipid.**

Liposome: Artificial vesicle with a phospholipid bilayer that forms when phospholipid molecules are suspended in a watery environment.

Locus (*pl.:* **Loci**): Location of a gene on a chromosome. Diploid organisms possess two copies of each loci and it is possible for the alleles at each loci either to be identical or slightly different. In a population, many loci demonstrate marked allele polymorphisms.

Long terminal repeat: Repetitive DNA-sequences, which flank certain genes and enable them to reintegrate into the genome (transposition).

Low-density lipoprotein (LDL): Complex formed from a single protein molecule and many molecules of cholesterol and other lipids. LDLs are responsible for the uptake of cholesterol from tissues and their transport in blood.

Lymphocyte: Class of white blood cells that is responsible for the specificity of the immune response. There are two types of lymphocyte: B cells and T cells. T cells are formed in the thymus gland and are the carriers of cell-mediated immunity. B cells are formed in the bone marrow of mammals and are responsible for the formation of antibodies that circulate in the blood.

Lysis: Destruction of the cytoplasmic membrane of a cell. Leads to escape of the cytoplasm and to cell death.

Lysosome: Compartment found in fungal and animal cells. Contains diverse digestive enzymes that are mainly active at low pH values. Proton ATPases pump protons into the lysosomes and thus ensure that the pH remains acidic.

Lysozyme: Enzyme that breaks down bacterial cell wall polysaccharides.

Macrophage: Phagocytic cells that develop from blood monocytes and are found in all tissues. Macrophages digest foreign organisms that enter the body and then present their peptides to the T cells.

Malaria: Parasitic disease triggered by the single-celled *Plasmodium*. The parasite is carried by the *Anopheles* mosquito.

Malignant: Describes tumors and tumor cells that grow invasively and/or are capable of metastasis. A malignant tumor is called cancer.

Mannose-6-phosphate: Modification of the oligosaccharide of some glycoproteins that are transported into the lysosomes.

MAP: see **Microtubule-associated protein.**

MAPK signaling pathway: Signaling pathway that starts with a signal being received by a cell membrane receptor and continues via mitogen-activated protein kinases (MAPKs) in the cell nucleus. Controls the regulation of genes.

Marker gene: 1. Gene that, when placed in a foreign organism, displays a property that is easily recognizable. 2. Gene that is investigated in place of another gene or genome (e.g., in phylogeny investigations).

Mass spectrometry (MS): Important method to identify small and large molecules on the basis of their exact mass-to-charge ratio and/or fragmentation patterns.

Materia medica: Various materials obtained from plants, animals, or minerals that have applications in medicine.

Matrix: 1. Central subcompartment of a mitochondrion, bordered by the inner membrane. 2. Corresponding compartment in a chloroplast (also known as the stroma).

Maximum likelihood (ML): Method by which a statistical model is chosen on the basis of its plausibility.

Maximum parsimony (MP): Parsimony implies that simpler hypotheses are preferable to more complicated ones. Maximum parsimony is a character-based method that infers a phylogenetic tree by minimizing the total number of evolutionary steps required to explain a given set of data, or in other words by minimizing the total tree length.

MDR protein: see **Multidrug-resistant protein.**

Medical indication: Criterion for the choice of patent category for a drug discovery, if the substance that forms the basis of the drug is already known but its application in medical terms or for the treatment of a particular illness is not.

Meiosis: Form of cell division by which egg and sperm cells are formed. Each round of cell division consists of two divisions of the genetic material following on immediately from each other, resulting in the production of four haploid daughter cells from each diploid (mother) cell.

Melanoma: Type of growth found on the skin and mucosal tissues that can be benign or malignant in nature. It is surrounded by pigmented tissue.

Melting temperature (T_m): Temperature at which the two halves of a nucleic acid double strand dissociate to form two single strands.

Membrane potential: Difference in voltage across a membrane caused by an abundance of positive ions on one side of the membrane and an abundance of negative ions on the other side of the membrane. Typically, the membrane potential of plasma membrane of an animal cell is 60 mV (the inside of the cell is negative with respect to the outside).

Membrane protein: Protein that is usually bound tightly to a cell membrane.

Membrane transport: Movement of molecules across a membrane, facilitated by a membrane transport protein (transporter, carrier).

Meristem: Organized group of dividing cells whose descendants make up the tissues and organs of a flowering plant. Key examples include the apical meristem at the tip of shoots and roots.

Mesoderm: Embryonic tissue that is the precursor of muscles, connective tissue, the skeleton, and many other internal organs.

Messenger RNA (mRNA): RNA polymerase copies a gene to form the corresponding mRNA, which specifies the amino acid sequence of a protein. In eukaryotes, mRNA is modified through RNA splicing to form a smaller RNA molecule.

Metabolism: Sum of all of the chemical processes that take place in a living cell.

Metaphase: Stage of mitosis in which the chromatids are firmly attached to the mitotic spindle in the region of its equator, but have not yet begun to move to the poles located at opposite ends of the cell.

Metastasis: Movement of cancer cells from their tissue of origin to another location in the body.

Methylphosphonate: Analog of DNA; one of the oxygen atoms normally found on the phosphate group of DNA is replaced with a methyl group. The nucleic acid is thus prevented from being broken down by enzymes.

MHC: see **Major histocompatibility complex.**

Microarray: Another term for a DNA chip (see **DNA chip**).

Microfilament: see **Actin filament.**

Microsome: Fragment of the membrane of the endoplasmic reticulum or Golgi apparatus that forms during the breakdown of a cell. Can be isolated as a vesicle fraction.

Microtubule: Linear tubular structure found in higher cells and made from tubulin dimers. Important for the formation of the spindle in cell division and for vesicle transport within the cell.

Microtubule-associated protein (MAP): Protein that binds to microtubules and alters their properties. There are many types of MAP protein, including structural proteins (e.g., MAP2) and motor proteins (e.g., dynein).

Mineral corticoid: Steroid hormone of the adrenal cortex (aldosterone) that regulates the salt content of the body.

Mismatch: In the context of the WatsonCrick Rule (G bonds with C and A with T or U, respectively), incorrect base pairing in double-stranded DNA, RNA, or their analogs.

Mismatch repair: DNA repair process, in which incorrectly paired nucleotides inserted during DNA replication are corrected.

Mitochondria: Important compartment in eukaryotic cells in which, for example, the citric acid cycle and respiratory chain (ATP synthesis) take place. Mitochondria contain their own DNA, replication and transcription enzymes, as well as their own ribosomes.

Mitosis: Division of the nucleus of a eukaryotic cell. The DNA (in the form of chromosomes) condenses so it is visible and the replicated chromosomes split to give two identical sets of chromosomes.

Mitotic chromosome: Highly condensed chromosome consisting of two sister chromatids. These chromatids will form the new chromosomes and are always joined together at the centromere.

Mitotic spindle: Arrangement of microtubules and associated proteins that is formed during mitosis and stretches between the poles, which lie opposite each other. It serves to pull the replicated chromosomes away from each other.

Module: Structural or functional unit of proteins (protein module) or nucleic acids.

Monoclonal antibodies: Antibodies secreted by a single hybridoma clone. As each clone descends from a single B cell, all antibody molecules formed are identical.

Monocyte: Class of white blood cell that leaves the blood circulatory system and matures to form macrophages in the tissues.

Monomer: Molecular building block that condenses with others of the same type to form a polymer.

Motif: Structural or functional element of a protein or a nucleic acid.

Motor protein: Protein that utilizes energy obtained from ATP to propel itself along a protein filament or another polymer molecule (e.g., actinmyosin system in muscle cells).

mRNA: see **Messenger RNA**.

Multidrug-resistant protein (MDR protein): Class of ABC transporters that can transport hydrophobic substances (drugs, e.g., some used to treat cancer) out of the cytoplasm of eukaryotic cells.

Multiple sclerosis: Disease of the central nervous system caused by the disintegration of the myelin sheaths surrounding the axons.

Multiple testing: Statistical test that checks for differences between two groups. Tests many variables independently. The p-value obtained from the test, which gives the probability that no difference exists between the two groups, decreases in validity as the number of tests increases and must therefore be corrected accordingly.

Mutagen: Substance that causes mutations.

Mutation: Change in the nucleotide sequence of a chromosome; heritable when occurring in the germ line.

Mutation rate: Rate at which detectable changes in a DNA sequence occur.

Myofibril: Long, extremely well-organized bundle found in muscle cells. Made up of actin, myosin, and other proteins.

N-terminus: see **Amino-terminus**.

Na$^+$/K$^+$ pump (Na$^+$/K$^+$-ATPase): Important ion pump found in animal cells. Utilizes energy obtained from the hydrolysis of ATP to pump Na$^+$ ions out of the cell and K$^+$ ions into the cell. Inhibited by cardiac glycosides.

Natural killer cells (NK cells): Cytotoxic cells of the innate immune system that can kill cells infected by viruses.

Necrosis: Death of cells and tissues.

Neuron (nerve cell): Type of cell from which a long axon and many dendrites extend. Specialized for the reception, conductance, and transfer of signals within the nervous system.

Neurotransmitter: Signaling substance found in neurons. Necessary for the transfer of an electrical signal from one nerve cell (presynapse) to the next (postsynapse). Important neurotransmitters include: acetylcholine, noradrenaline, adrenaline, dopamine, serotonin, histamine, glycine, γ-amino butanoic acid (GABA), glutamate, endorphins, and other peptides.

Neurovesicles: Small vesicles found in the presynapse that are filled with neurotransmitters.

Nitrogen monoxide (NO): Gaseous, signal molecule in animal and plant cells. In animal cells, it regulates the contraction of smooth muscle cells; in plant cells it is involved in the reaction to injury or infection.

NK cells: see **Natural killer cells**.

NMR: see **Nuclear magnetic resonance spectroscopy**.

NO: see **Nitrogen monoxide**.

Noncovalent bonds: Noncovalent bonds (hydrogen bonds, ionic bonds, hydrophobic interactions) are individually comparatively weak, but can, in large numbers, result in strong, highly specific interactions between molecules.

Northern blotting: Technique by which RNA fragments are separated by electrophoresis and then transferred to a nylon membrane. The desired RNA fragment is then located through hybridization with a labeled nucleic acid probe.

N-terminal: End of a polypeptide chain that has a free amino group.

Nuclear export signal: Sorting signal in molecules and complexes (e.g., RNA and ribosome subunits) that marks them for transport from the nucleus to the cytosol via the nuclear pore complexes.

Nuclear magnetic resonance spectroscopy (NMR): Type of spectroscopy used to determine molecular structures. It uses the resonance observed between individual atoms after they have been excited in a strong magnetic field.

Nuclear membrane: System of double membranes that surrounds the nucleus. It consists of outer and inner lipid bilayers (made from the endoplasmic reticulum), and is interrupted by nuclear pores.

Nuclear pore complex: Large multiprotein complexes that form the pores in the nuclear envelope. They facilitate the transport of selected molecules between the nuclear and cytoplasmic compartments.

Nucleolus: Structure in the nucleus, visible under a light microscope, in which rRNA is transcribed and ribosome subunits are assembled.

Nucleoporin: Protein that forms the nuclear pore complex.

Nucleosome: Rosary-shaped structure found in eukaryotic chromatin. It is made up of a short length of DNA wound round a core made of histone proteins.

Nucleus: In a eukaryotic cell, the nucleus contains the chromosomes. The nucleus is surrounded by a nuclear membrane derived from the endoplasmic reticulum (here with a double membrane). The nuclear pore complexes are important for the transport of substances in and out of the nucleus.

Okazaki fragments: Short pieces of DNA that form on the lagging strand during DNA synthesis. They are then joined to each other by DNA ligase to form a covalently bonded chain of DNA.

Oligomer: Short polymer made of amino acids (oligopeptide), sugars (oligosaccharide), or nucleotides (oligonucleotide) (from Greek: *oligos*=little, small).

Oncogene: Altered form of a gene (e.g., in retroviruses) whose product can cause a cell to divide uncontrollably. Typically, an oncogene is a mutated form of a normal gene (proto-oncogene) that acts to regulate cell growth or cell proliferation.

Ontogenesis: Sequence of differentiation and changes undergone in the development of a fertilized egg cell to a fully grown organism (plant or animal).

Open reading frame: Sequence without stop codons.

Operator: Describes a short, specific DNA sequence that is the binding site for transcription factors (positive or negative regulators).

Operon: Term used to describe several structural genes that are transcribed together and whose expression can be controlled positively or negatively.

ORF: see **Open reading frame**.

Organelle: Membrane-bound compartment found in eukaryotic cells. They have a marked structure, and macromolecular layout and function. Examples include the nucleus, chloroplasts, and the Golgi apparatus. Large multiprotein complexes are also sometimes described as organelles.

Origin of replication: Site on a DNA molecule at which DNA replication begins.

OTC (over-the-counter): Describes drugs that can be purchased without a prescription.

Oxidation: Loss of electrons by an atom through the substraction of hydrogen or the addition of oxygen to a molecule.

p53: Tumor suppressor gene that is mutated in many forms of cancer. It codes for a gene regulator protein that becomes active when the cell's DNA is damaged and inhibits the cell cycle.

Paracrine signal: Cellcell communication via secreted signaling molecules that have an effect on neighboring cells.

Parenteral: Method of administering an active substance that bypasses the digestive tract (e.g., intravenous (i.v.), intramuscular (i.m.), or subcutaneous injection (s.c.)).

Parkinson's disease: Neurological disease (shaking palsy) that is caused by the degeneration of the substantia nigra and a decrease in the concentration of dopamine.

Passive transport: Movement of a dissolved substance through a membrane across its concentration gradient or electrochemical potential.

Patent: State-issued and checked short-term trademark right for an innovative technical development.

Patent infringement: Professional manufacture, use, and tendering by a third party of an invention protected by a patent without authorization by previous usage, state arrangement, or legal act on the part of the patent holder (e.g., licensing).

Pathogen: Microorganism that causes disease.

PCR: see **Polymerase chain reaction**.

Peptide nucleic acid (PNA): Analog of DNA with very good biophysical properties (firm binding to complementary DNA and RNA sequences, and high mismatch sensitivity).

Peroxisome: Small membrane-bound organelle that uses molecular oxygen to oxidize organic molecules. Peroxisomes contain enzymes that form hydrogen peroxide (H_2O2) and others (e.g., catalase) that break it down.

pH value: General measure of the acidity of a solution; "p" refers to the negative power of 10 (from Latin: *pondus*=weight), "H" to hydrogen. It is defined as the negative logarithm of the hydrogen ion concentration, measured in moles per liter (M). On the pH scale, pH 7 (10^{-7} M H^+) is neutral, pH 3 (10^{-3} M H^+) acidic, and pH 9 (10^{-9} M H^+) alkaline.

Phage display: Method of identification of interacting proteins and peptides, based on the expression of peptides in phage particles that can be isolated using antibodies or other proteins and thus duplicated.

Phagemid: Vector that contains genetic elements of both plasmids and bacteiophages.

Phagocyte: General term for macrophages or granulocytic neutrophils that are specialized for the uptake of particles and microorganisms by phagocytosis.

Phagocytosis: Process by which bacteria and other particles are taken up by cells (see **Phagocyte**).

Phagosome: Large, intracellular, membrane-bound vesicle (endosome) that transports extracellular material taken up by the cell to, and then fuses with, the lysosome.

Pharmacodynamics: Area of pharmacology concerned with how drugs have an effect on the body (e.g., with which target they interact).

Pharmacokinetics: Area of pharmacology concerned with how drugs are taken up, distributed, metabolized, and excreted by the body.

Pharmacology: Study of the nature, properties, and uses of drugs; includes the study of endogenous active compounds.

Pharmacovigilance: Collection and reporting upon of undesirable side effects of medications and their scientific significance to the authorities.

Phenotype: Outwardly visible characteristics of a cell or organism.

Phosphodiesterase: Enzyme involved in signal transduction; inactivates cAMP or cGMP.

Phospholipase C: Enzyme involved in signal transduction. Causes the release of inositol phosphates, such as inositol 1,4,5-trisphosphate and diacylglycerol.

Phospholipid: Building block of plasma membranes. Linked to a phosphate group via an ester bond.

Phosphothioate: Analog of DNA; one of the oxygen atoms usually found on the phosphate group of the DNA is replaced with a sulfur atom. The nucleic acid is therefore protected from breakdown by enzymes.

Photosynthesis: Process used by plants, algae, and some bacteria to synthesize organic molecules from carbon dioxide and water using the energy of the sum.

Phylogeny: Evolutionary history of an organism or a group of organisms, often presented in the form of a phylogenetic tree or map of evolutionary relationships.

Pinocytosis: Form of endocytosis in which dissolved substances are taken up from the environment in vesicles (literally cell drinking, from Greek: *pinein* = to drink) (see **Fluid-phase endocytosis**).

PKA: see **cAMP-dependent protein kinase**.

PKC: see **Protein kinase C**.

Placebo: Dummy drug that, apart from being free of active substances, does not differ greatly from the original. Important in double-blind placebo-controlled clinical studies.

Plaque: Area of lysis or growth inhibition in a lawn of cells or bacteria caused by a virus (or bacteriophage, respectively).

Plasmid: Extrachromosomal, circular molecules of DNA that originate from bacteria and yeast, and can replicate independently of the main chromosomal DNA. Plasmids frequently carry genes for resistance factors (e.g., against antibiotics) that confer a selection advantage. Plasmids are important vectors in gene technology.

Plastid: Generic term for plant organelles that are bound by a double membrane, possess their own DNA, and are often pigmented (e.g., chloroplasts).

PNA: see **Peptide nucleic acid**.

Point mutation: Single-nucleotide exchange in a DNA sequence.

Polyhedrin: Protein of approximately 29 kDa in size, coded for by baculoviruses. Polyhedrin forms a stable storage matrix for baculoviruses in the environment.

Polylinker: Section of DNA placed in a vector with cutting sites for several restriction endonucleases (multiple cloning site).

Polymerase chain reaction (PCR): Method for the amplification of specific DNA sequences *in vitro*, through repeated synthesis cycles and the use of specific primers and thermostable DNA polymerase.

Polymorphism: Describes a characteristic present in many forms in a population (e.g., a gene locus with many different alleles).

Polymorphic: Occurrence of two or more alleles in a population, when the rarer allele is present with a frequency greater than or equal to 1% (see also **Single nucleotide polymorphism**).

Polyribosome (polysome): mRNA molecule to which a number of ribosomes simultaneously synthesizing a protein are bound.

Posttranslational modification: Processing reaction that occurs during or after translation. Examples include glycosylation, acylation, and phosphorylation.

Primary structure: Sequence of monomers in a linear polymer (e.g., the amino acid sequence in proteins).

Primosome: Complex made of DNA primase and DNA helicase that forms on the lagging strand during DNA replication.

Priority time: Period of time beginning of the day of patent registration, within which the patent holder is entitled to priority with regard to further applications for the same invention (i.e., further registrations have the priority of the first registration, provided that the object of invention is the same).

Probe: Defined section of RNA or DNA that has been marked radioactively or chemically and is used to localize specific nucleic acid sequences through hybridization.

Prodrug: Drug that is converted to an active substance in the body.

Prokaryote: Unicellular microorganism that does not possess a well-defined, membrane-bound nucleus. Prokaryotes make up two of the kingdoms of living things: Bacteria and Archaea. See **Eukaryote**.

Promoter: Short sequence of DNA to which RNA polymerase and transcription factors bind, and thus initiate transcription.

Prostaglandin: Chemical messenger found in the body that has many effects on tissues; is an important paracrine tissue messenger in the inflammation process.

Proteasome: Protein complex with built-in proteases whose main function is to break down defective proteins that are marked for breakdown by the protein ubiquitin.

Protein domain: Region of a protein that has its own tertiary structure and often its own function. Large proteins are often made up of several domains that are linked to each other by short, flexible polypeptide chains.

Protein drugs: Proteins that have therapeutic applications.

Protein glycosylation: Posttranslational addition of oligosaccharides to the side chains of proteins (*N*- and *O*-glycosylation).

Protein kinase: Enzyme that transfers the terminal phosphate group of an ATP molecule to a specific amino acid (serine/threonine or tyrosine) of a target protein. Important examples include protein kinase A and C.

Protein kinase C (PKC): Ca^{2+}-dependent protein kinase that phosphorylates specific serine or threonine residues on a target protein after it has been activated by diacylglycerol.

Proteome: Collective term for all the (currently existing) proteins present in a cell or an organism.

Proteomics: Analysis of the composition of the proteome as well as its dynamic development.

Proto-oncogene: Gene that controls cell proliferation and can, through mutation, be converted into a cancer-causing oncogene.

Protozoans: Free or parasitic, unicellular, mostly mobile eukaryotic organisms such as *Paramecium* and *Amoeba*. Free protozoans feed on bacteria or other microorganisms.

Provirus (prophage): Genome of a virus when it is integrated into the host DNA and replicated with it (it is usual for a viral genome to be integrated in this way).

Pseudogene: Gene that was once active but has, through evolution, undergone multiple mutations that have rendered it inactive and functionless.

Purine: Class of alkaloid; the bases adenine and guanine (found in DNA and RNA) are purines.

Pyrimidine: Class of alkaloid; the bases cytosine, uracil, and thymine (found in DNA and RNA) are pyrimidines.

Quaternary structure: Three-dimensional relationship and arrangement of different polypeptide chains in a protein complex.

Quinone (Q) (also ubiquinone, plastoquinone): Small, lipophilic electron-carrying molecules found in the respiratory and photosynthetic electron transport chains.

Rab protein: Representative of a large family of membrane-bound monomeric GTPases that confer vesicle-docking specificity.

Ran: Monomeric GTPase that is vital for the active transport of macromolecules both into and out of the nuclear membrane complex. It is presumed that the hydrolysis of GTP to GDP supplies the energy for this transport.

Ras protein: Best-known monomeric GTPase (or small G-protein) involved in signal transduction from the cytoplasmic membrane to the nucleus. Named after the *ras* gene, this was first identified in the retroviruses that trigger sarcomas in rats.

Reading frame: It is theoretically possible for an mRNA molecule to be read in each of the three reading frames, but only one reading frame allows for the formation of the functionally correct protein. The first codon for a protein is AUG, and codes for methionine in eukaryotes and formyl-methionine in prokaryotes.

Receptor: Protein (often a membrane protein) that possesses a binding site for another molecule (ligand); important in signal transduction within cells. Intracellular receptors, like steroid hormone receptors, bind their ligands intracellularly and then transport them into the cell nucleus.

Receptor-mediated endocytosis: Uptake of receptorligand complexes through the cytoplasmic membrane by means of endocytosis; aids the uptake of particular macromolecules (e.g., lipoproteins loaded with cholesterol.

Recessive: In genetics, refers to the member of a pair of alleles that is not visible in the phenotype if the dominant allele is also present. Also describes the phenotype of an organism that only possesses the recessive gene.

Recombinant DNA: DNA joined experimentally (e.g., plasmid DNA and newly expressed DNA obtained from another organism).

Recombination: Natural process of breaking and rejoining DNA strands to produce new combinations of genes and, thus, generate genetic variation.

Redox reaction (oxidation/reduction reaction): Reaction in which one component is oxidized and the other is reduced.

Reduction: Gain of electrons by an atom, occurring through the addition of hydrogen to a molecule or the loss of oxygen from a molecule. Opposite of oxidation.

Regulatory sequence: Sequence of DNA to which a gene regulatory protein (transcription factor) must bind before transcription can begin.

Repetitive sequence: Sequence of DNA that is frequently repeated.

Replication: Copying of the DNA double helix prior to cell division.

Repressor: Protein that binds to a specific region of a gene (located within the promoter) and prevents the transcription of the gene bordering it.

Respiration: Oxidation of sugars and other organic molecules within a cell. Whereas oxygen is used by the cell, CO_2 and H_2O are generated as waste products.

Respiratory chain: Electron transport chain situated in the inner membrane of the mitochondria NADH and $FADH_2$ are generated in the citric acid cycle. Electrons and protons are released in the electron transport chain, which generate a proton gradient across the membrane. This is then used to provide the energy for ATP synthesis.

Restriction endonuclease (restriction enzyme): Enzyme that recognizes palindromic sequences in DNA and cuts them. Examples include EcoRI, SmaI, and NaeI.

Reticulocyte: Highly specialized blood cells that lack a nucleus and are involved in the synthesis of hemoglobin. Reticulocyte lysate can be made from them and this can be used *in vitro* to synthesize proteins.

Retrotransposon: Type of transposable element (transposon) that moves by first being transcribed to form an RNA copy and then being changed back into DNA by reverse transcriptase. It then moves to another site in the chromosome and inserts itself into it.

Retrovirus: Virus that contains RNA and replicates in a cell. A double-stranded DNA intermediate is then formed through reverse transcription.

Reverse transcriptase: Enzyme found in retroviruses that copies single-stranded RNA and forms double-stranded DNA from them. Important for the formation of cDNA from mRNA.

Ribonucleic acid (RNA): Polymer formed from covalently-bound ribonucleotide monomers (see also **Messenger RNA**, **Ribosomal RNA**, and **Transfer RNA**).

Ribosomal RNA (rRNA): Specific RNA molecules that are involved in the structure of ribosomes and in protein synthesis. Often distinguished from each other by their sedimentation coefficients (28S rRNA or 5S rRNA, etc.). Transcribed as a single transcription unit.

Ribosome: Multiprotein complex that consists of rRNA and ribosomal proteins. Ribosomes bind RNA and catalyze the synthesis of proteins.

Ribozyme: RNA molecule with catalytic action that is involved in the sequence-specific degradation of mRNA.

RNA: see **Ribonucleic acid**.

RNA editing: Functional editing or trimming of an RNA molecule through the addition, deletion or exchange of single nucleotides after its synthesis.

RNA interference (RNAi): Selective intracellular degradation of RNA, through which foreign RNA (e.g., originating from viruses) is eliminated. Pieces of free double-stranded RNA bind to similar RNA sequences that are then destroyed. RNAi is frequently used to inhibit the expression of selected genes.

RNA polymerase: Enzyme that catalyzes the synthesis of an RNA molecule from nucleotide triphosphate precursors according to a DNA template.

RNA primer: Short sequence of RNA that is complementary to the corresponding DNA strand. Required by DNA polymerase in order to initiate DNA synthesis.

RNA splicing: Process that occurs during the processing of mRNA and other RNAs, in which the intron sequences are cut out of the primary RNA transcript.

RNase H: Enzyme that breaks down the RNA strand of an RNADNA double-stranded complex.

rRNA: see **Ribosomal RNA**.

Rough endoplasmic reticulum (rough ER): Endoplasmic reticulum covered with ribosomes on its cytosolic side. Involved in the synthesis of proteins that will be secreted and of membrane proteins.

Saponins: Glycoside of the triterpenes and steroids; while demonstrates lipophilic properties the aglycone, saponins are amphiphilic and water soluble;

there are two different types: monodesmosidic saponins with one sugar chain and bidesmosidic saponins with two sugar chains.

Sarcoma: Cancer of the connective tissues.

Sarcomere: Contractile, 2.4-μm long functional unit of muscles, which mainly consists of actin filaments and myosin, but also contains quite a few other proteins.

Sarcoplasmic reticulum: System of tubes in the cytoplasm of a muscle cell that contains high concentrations of Ca^{2+}. The Ca^{2+} is released during excitation of the muscle cells and pumped into the sarcoplasmic reticulum through the action of a Ca^{2+}-ATPase.

Satellite DNA: Area of highly repetitive DNA in a eukaryotic chromosome. Satellite DNA is not transcribed and its function is not known.

Saturated fatty acids: Fatty acids that do not contain any double bonds (found, for example, in the fat stores of animals or coconuts).

Score: Value used to choose between different statistical models (or different alignments).

Score matrix: Table used in the alignment of proteins that indicates how a pair of amino acids in an alignment should be valued. As the evolutionary pressure on amino acids varies due to their different physiochemical properties, amino acid exchange is observed with different frequencies. Such observed frequencies in alignments of protein families are used to calculate score matrices.

Screening: Systematic search through a library of substances in order to find substances with particular properties.

SDSpolyacrylamide gel electrophoresis (SDSPAGE): Form of electrophoresis in which the protein mixture that is to be separated is mixed with the detergent sodium dodecyl sulfate (SDS) and separated on a polyacrylamide gel.

Second messenger: Small molecule that forms in the cytosol in response to an extracellular signal or is released and, acting as a second messenger, helps to transfer the primary signal into the cell and amplify it. Examples include cAMP, inositol 1,4,5-trisphosphate, and Ca^{2+}.

Secondary metabolite: Mostly small molecular constituents with a large degree of structural variability that are used by plants as defense and signaling substances. Their origin is frequently restricted to a few plant groups. In contrast to secondary metabolites are the primary metabolites, which are vital to life for all plants and are therefore ubiquitously distributed.

Secondary structure: α-Helices, β-pleated sheets, and random coils form the secondary structure of a protein.

Secretion: Release of a protein out of a cell into its extracellular matrix. Usually mediated by specific signals.

Sensitivity: Measure in bioinformatics of how well a classifier can allocate two classes correctly. If TP and TN are the number of true positive and true negative cases, whereas FP and FN are the numbers of false positives and false negatives, respectively, sensitivity is defined as $Sens=TP/(TP+FN)$ (i.e., as the number of actual cases within a class). The specificity, on the other hand, is the number of actual positive cases of all those designated as being positive, therefore $Spec=TP/(TP+FP)$. Together, the sensitivity and the specificity give a measure of how well a classifier works, which is also visible on the ROC (receiver-operator characteristics) curve: the sensitivity plotted against (1 − specificity).

Sequence clustering: Grouping of a number of sequences by means of the similarities between them. Used to reduce redundancy in large banks of clones or in sequencing projects. The groupings are chosen so that each group contains sequences that contain a large number of fragments identical in sequence and thus yield redundant information. The clustering of expressed sequence tags (ESTs) is of particular importance; all EST clones that originate from the same mRNA are combined and the cluster is found in the UniGene databank.

SH2 domain: A protein domain found on many signal proteins. It binds a short amino acid sequence that contains a phosphotyrosine.

ShineDalgarno sequence: Bacterial ribosome binding site.

Shuttle vector: Cloning vector that can replicate in different organisms.

Signal recognition particle (SRP): Ribonucleoprotein particle that binds to endoplasmic reticulum signal sequences on a partly synthesized polypeptide chain and links it, along with its attached ribosomes, to the endoplasmic reticulum.

Signal peptidase: Enzyme that removes the signal sequence from the end of a protein after the sorting process has finished.

Signal peptide: Parts of an amino acid sequence that carry signal crucial to the localization of the proteins. Mitochondrial and plastid signal peptides, for example, are located at the N-terminal and are cleaved after their import into the organelle; nuclear localization signals are found at the C-terminal. A further example is the N-terminal sequence of approximately 20 amino acids that links growing secretory and transmembrane proteins to the endoplasmic reticulum.

Signal sequence: N-terminal signal sequence that directs proteins to the endoplasmic reticulum. They are then cleaved by signal peptidases.

Signal transduction: Process by which a cell converts an extracellular signal (a stimulus) into a usually intracellular answer.

Single nucleotide polymorphism (SNP): Differences between individuals at particular nucleotide positions on a segment of DNA. SNPs can serve as molecular markers for the recognition of individuals or of faulty genes.

siRNA: see **Small interfering RNA**.

Site-directed mutagenesis: Method by which a mutation can be inserted at a specific site in the DNA sequence.

Site-specific recombination: Form of recombination that does not require any great similarities between the two DNA sequences involved. Can occur between two different DNA molecules or within a single DNA molecule.

Small interfering RNA (siRNA): Naturally occurring small oligomers of RNA (21–23 bases in length) that bind sequence specifically to mRNA and initiate their destruction. This natural process, so-called RNA interference, is comparable in both mechanism and effect to antisense technology, which utilizes synthetic oligomers.

Small nuclear RNA (snRNA): RNA molecule that forms a complex with proteins in order to form the ribonucleoprotein particles required for RNA splicng.

Smooth endoplasmic reticulum: Area of the endoplasmic reticulum that is not covered with ribosomes. Important in lipid synthesis.

Smooth muscle cell: Type of muscle cell that possesses a single nucleus; is long and spindle-shaped, and does not have striated muscle fibers running through it. Such cells make up muscle tissues of arterial walls and the walls of the stomach, as well as other organs and tissues of vertebrates.

SNAREs: Large family of transmembrane proteins that occur in organelle membranes and the vesicles that form from them. They are involved in bringing the vesicle to the correct destination. They are found in pairs: a v-SNARE in the vesicle membrane that docks to a complementary t-SNARE on the target membrane.

SNP: see **Single nucleotide polymorphism.**

snRNA: see **Small nuclear RNA.**

Solid-phase synthesis: Sequential chemical synthesis on a solid carrier, primarily used for biopolymers such as DNA or RNA oligomers, peptides, and peptide nucleic acid oligomers.

Somatic cell: Every cell found in a plant or an animal that is not a germ cell or one of its precursors.

Southern blotting: Method by which DNA fragments that have been separated by electrophoresis are transferred to a nylon or nitrocellulose membrane. The immobilized DNA strands can then be detected using a labeled nucleic acid probe. Named after Edwin Mellor Southern, the inventor of the technique.

Spliceosome: RNA-processed protein complex that cuts the introns out of newly synthesized mRNA.

Statistics: Process that enables a decision to be made about acceptance or rejection of a hypothesis through use of a statistical test. The distribution of a statistic allows the probability of the accuracy of the hypothesis to be determined (the p value); if the probability is very small, the hypothesis can be rejected. The statistics can either be determined theoretically, which is necessary for most acceptances, or through permutation tests. Important tests used for statistical analysis include the t-test, F-test, χ^2-test, and Wilcoxon test.

Striated muscle: Skeletal and heart muscle; made from diagonally striped (striated) myofibrils.

Stroma: Large space found inside a chloroplast that contains the enzymes required for the CO_2 fixation to form sugar.

Structural gene: Section of DNA that codes for a protein or an mRNA molecule.

Symbiosis: Close relationship between two different organisms that has advantages for both of them.

Symporter: Protein that transports two different molecules through the membrane in the same direction along a concentration gradient.

Synapse: Neurons are connected to other neurons or to target organs by synapses, which are located at the end of axons. This is where an electrical impulse (action potential) is temporarily converted into a chemical signal (neurotransmitter) and transmitted from the presynapse to the postsynapse.

T cell (T lymphocyte): Lymphocytes that are responsible for cell-mediated natural immunity; includes both cytotoxic T cells and helper T (T_h) cells.

Tannin: Chemical containing many phenolic OH groups, which can undergo hydrogen and ionic bonding with proteins, and thus alter their conformations. There is a difference between gallotannins and catechin tannins, which are derived from epicatechin and catechin.

Target: Molecular site of attack for chemicals in the human body or in cells.

TATA box: Consensus sequence located in the promoter region of many eukarytic genes and to which general transcription factors bind.

Telomere: End portion of a chromosome; characterized by highly repetitive DNA. Telomeres prevent exonuclease action from damaging the chromosomes. When the telomeres have been broken down, cellular functions cease and cell death results.

Telomerase: Enzyme that extends telomere sequences in chromosomes; active in embryonic and cancer cells.

Template strand: Single strand of DNA or RNA, the nucleotide sequence of which is used as a template for the synthesis of the complementary strand.

Terminator: Transcriptional terminator in prokaryotes. Rho factor independent: GC-rich stem with loops and poly(U) tail. Rho factor independent: no specific motif.

Terpenes: Collective name for a very large group of plant secondary metabolites. Includes, among others: monoterpenes (with 10 carbon atoms), sesquiterpenes (15 carbon atoms), diterpenes (20 carbon atoms), triterpenes (30 carbon atoms), steroids (27 or fewer carbon atoms), tetraterpenes (40 carbon atoms), and polyterpenes.

Tertiary structure: Complex 3D structure of a folded polymer chain, particularly a protein or RNA molecule.

Testosterone: Male sex hormone.

Thylakoid: Flat membrane sack found in a chloroplast that contains chlorophyll and other pigments. Carries out the light-capturing reactions of photosynthesis. Stacks of thylakoids form the grana of the chloroplasts.

TIM complex: Protein translocation complex found in the inner membrane of the mitochondria. The TIM23 complex facilitates the transport of proteins into the matrix and the insertion of particular proteins in the inner membrane; the TIM22 complex facilitates the insertion of a subgroup of proteins into the inner membrane.

TOM complex: Protein translocase that transports proteins through the outer membrane of the mitochondria.

Toxicology: Scientific study of toxins and their effects on humans and animals.

Transcription: Copying of the nucleotide sequence of a gene into mRNA.

Transcription factor: General term for every protein necessary for the initiation or regulation of transcription in eukaryotes. It is used for both gene regulatory proteins and general transcription factors.

Transcriptome: Collection of all transcripts of a cell or organism present in it at a given time.

Transcriptomics: Study of the composition and dynamic changes of the transcriptome.

Transcytosis: Absorption of materials at a site on the cell through endocytosis, their vesicular transport through the cell and their excretion at another site on the cell through exocytosis.

Transfection: Introduction of DNA into a eukaryotic cell.

Transfer RNA (tRNA): Codon-specific tRNA molecules that are the mediators between mRNA and amino acid sequences in protein synthesis.

Transformation: Introduction of naked DNA into bacteria by means of specific reagents or an electric field.

Transgenic organism: Plant or animal that has successfully taken up one or more genes from another cell or organism.

Translation: Taking place in the ribosome, the translation of an mRNA sequence into the amino acid sequence of a protein.

Transmembrane protein: Membrane protein that extends right through the lipid bilayer.

Transporter: Membrane protein that specifically catalyzes the transport of a molecule across the biomembrane.

t-SNARE: see **SNAREs.**

Tuberculosis: Bacterial infection of the lungs and other organs with *Mycobacteium tuberculosis*; frequently chronic and usually fatal without treatment with antibiotics.

Tubulin: Protein subunit of microtubules.

Tumor: Visible swelling (growth) of the tissues of the body; can be benign or malignant.

Tumor necrosis factor (TNF): Signal protein formed by the cells of the immune system (e.g., macrophages) in response to infection and then released (e.g., in infections).

Tumor suppressor gene: Gene that appears to prevent the formation of a cancerous growth. Faulty genes increase susceptibility to cancer.

Two-hybrid system: Method to identify proteins that have a relationship with each other (cross-talk).

Ubiquitin: Small, highly conserved protein found in all eukaryotic cells. Binds enzymatically to the lysine residues of other proteins. The addition of a short chain of ubiquitin (ubiquitinization) marks a protein for proteolytic breakdown in a proteasome.

Uniporter: Membrane transporter responsible for the movement of a single dissolved substance from one side of the membrane to another.

Unsaturated fatty acid: Fat with one or more double bonds.

Vacuole: Very large compartment found in most plant and fungal cells that typically makes up more than a third of the cell's volume. Stores ions, primary metabolites, and secondary metabolites. Specific vacuoles store reserve proteins.

Van der Waals forces: Forces of attraction between atoms or molecules based on extremely short-lived inequalities in the distribution of charge within an atom or molecule, which lead to the formation of dipoles. Van der Waals forces are always present, but are relatively weak (20 kJ mol1 at the most).

Vector: DNA or agent (virus or plasmid) used to introduce genetic material into a cell or organism. Most vectors are derived from bacterial plasmids.

Ventral: Located on the underside (stomach down) of an animal or the underside of a wing or leaf.

Vesicle: Small, membrane-bound ball-shaped bubble found in the cytoplasm of eukaryotic cells (from Latin: *vesica*=ball).

Virion: Entire virus: nucleic acids surrounded by a protein shell.

Virostatic: Chemical that inhibits the proliferation of viruses.

Virulence gene: Gene that causes an organism to become pathogenic.

Virus: Infectious macromolecular complex that contains its hereditary information in the form of DNA or RNA; requires cells for its replication. Many viruses cause diseases (from Latin: *virus*=toxin).

Western blotting: Important method used in diagnostics in which proteins are separated by electrophoresis, immobilized on a cellulose, or nylon membrane and then detected and analyzed, usually immunochemically with the help of a labeled antibody.

Wild-type: Normal, nonmutated form of an organism; the form that is found in nature.

X-ray crystallography: Physical method used to establish the structure of proteins and other compounds that depends on the diffraction of x-rays by crystals. The protein that is to be investigated needs to be crystalline.

Zinc finger: DNA-binding structural motif found in many gene regulatory proteins; consists of a loop of the polypeptide chain, which is bent into a hairpin shape through the binding of a zinc atom.

Zygote: Diploid cell that results from the fusion of a male and female gamete.

Subject Index

An Introduction to Molecular Biotechnology, 2nd Edition.
Edited by Michael Wink
Copyright © 2011 WILEY-VCH Verlag GmbH & Co. KGaA, Weinheim
ISBN: 978-3-527-32637-2

- stability 355
- target 331 f.
- targeted delivery to the site of action 479
- targeting 349 ff.
druggability 332
Duchenne's muscular dystrophy 405
duplication 67
- gene 366
dye laser 213
dye primer sequencing 235
dynamic programming 281

e

early endosome 87
eccentricity 300
Ecdysozoa 96
economic strategy 531
effective concentration 50% (EC_{50}) 344
efficacy
- activity of drug 344
eicosanoids 13
eigenvector centrality 300
electron capture dissociation (ECD) 123
electron microscopy 197
electron tomography 200
electron transfer dissociation (ETD) 123
electrophoresis 130 f.
- nucleic acid 130 f.
- principle 103
electrophoretic mobility shift assay (EMSA or band shift) 326
electroporation 165 f.
electrospray ionization (ESI) 115 f.
- ESI-MS/MS 120
- mass spectrometry (ESI-MS) 115 ff.
- principle 116
electrospray tandem mass spectrometry
- peptide and protein analysis 115 ff.
elementary flux mode 301
Eli Lilly and Co. 513
elution volume 107
elutriation 190
EMBL (European Molecular Biology Laboratory) 276
EMBOSS 294
embryonic stem (ES) cell 53, 398, 494
emission 212
employee 522 ff.
- recruitment, remuneration, participation 522
- scientific and technological competence 522
employee stock option (ESO) scheme 524
enabling technology 432
EnbrelR 339
endocrine signal 33
endocytosis 30, 87, 166
endocytosis-exocytosis cycle 88
endomembrane system 38, 86
endonuclease 137
endoplasmic reticulum (ER) 30 ff., 81
- import signal peptide 85
- protein transport 85
- rough 85
- smooth 38
endosome 39, 87
- early 87
- late 87

endosymbiont hypothesis 42
endosymbiosis 44
energy carrier 21
energy-rich radiation 66
enhanced permeability and retention (EPR) effect 350
enhancer 72
Ensembl 276 f.
entrepreneur 522
entry vector 158
envelope 407
environmental diagnostics 485
enzymatic sequencing (Sanger-Coulson Method) 150
enzyme 20
- activity 454
- biocatalytic process 453
- class 21
- modification of nucleic acid 137 ff.
- selectivity 454
- stability 454
- stereoselectivity 454
enzyme-linked receptor 34 ff.
epidermal tissue 53
epifluorescence microscope 269
epigenetic change 75, 367
epigenetic inheritance 75
epinephrine 356
epithelia 53
equilibrium potential 181
ergosterol 13
erythropoietin 324
Escherichia coli 91
- extract 179
ester prodrug 356
esterase 89, 455
β-estradiol 13 f.
estrogens 14
etanercept (EnbrelR) 339
ethics commission 346
ethidium bromide (EtBr) 126, 130
N-ethyl-*N*-nitrosourea (ENU) mutagenesis 273
eubacteria 58
euchromatin 237
Euclidean distance 290
eucyte 42
Euglenozoa 93
Eukaryota
- genome sequencing 233
- molecular phylogeny 92
eukaryote 58
eukaryotic cell (Eukarya) 3 ff., 91
- biochemical and cell biological characters 5
- expression system 171
- structure 29
eukaryotic elongation factor EF2a 161
eukaryotic expression vector 159 ff.
- mammalian cell 161
- replication 162
- termination sequence 162
eukaryotic gene 19, 74
- exon 19
eukaryotic genome 60
eukaryotic nucleus 38
European Medicines Agency (EMA) 499 f.

European Medicines Evaluation Agency (EMA) 500
European Patent Office (EPO) 494
European Union (EU)
- drug approval 499
- medicinal product 500
- regulatory framework 499
evolution
- convergent 91
- organism 91 ff.
- statistical model 284
evolutive method 461
excisionase 157
exclusion chromatography 130
executive summary 518
exocytosis 30, 87 f.
exon 19, 73
- eukaryotic gene 19
exonuclease 66, 140
- exonuclease I 141
- exonuclease III 141
export 82 f.
export receptor 83
expressed sequence tag (EST) 57, 231, 243, 288
expression
- cell-free system 178
- *E. coli* 172
- gene, *see* gene expression
- heterologous 162, 169, 184, 379
- insect cell 176
- mammalian cell 177
- protein 177 ff.
- recombinant protein 169 ff.
- reticulocyte lysate 179
- specific 413
- stable 161
- transient 161, 177, 437
- yeast 174
expression screening 135
expression study 147
expression vector 163, 270
- eukaryotic 159 ff.
external calibration 116
extraction 126

f

F plasmid 156
F(ab')$_2$ fragment 389
Fab fragment 382 ff.
$FADH_2$ 46
FASTA 282
fatty acid 11
- unsaturated 12
FDA (Food and Drug Administration) 499 ff.
feature 258
- selection 290
fee-for-service deal 528
feed-forward loop (FFL) 327
fermentation 475
fermentation process 451 f., 462
- improvement 462
- penicillin production 470
ferric uptake regulator Fur 328
fiber FISH 232
financial plan 519
fingerprint 223
- genetic 362 ff.

rickets 14
right border (RB) 435
ring-shaped DNA (cpDNA) 44
RISC, *see* RNA-induced silencing complex
risk gene 360
RNA (ribonucleic acid) 23 ff., 247
– catalytically active 27, 339
– expression 144
– fragment 135
– interaction of DNA analog with
  complementary RNA 419
– isolation 125 ff., 367
– purification 367
– quantitative analysis 141
– structure 21 ff.
– therapeutic approach 339
– verification 446
RNA arbitrary primed (RAP)-PCR 257 f.
RNA binding domain 327
RNA fingerprinting 257
RNA *in situ* hybridization (RNA-ISH) 251 f.
RNA interference (RNAi) 27, 57, 184, 207,
  339, 415 ff.
T4 RNA ligase 139
RNA polymerase 25, 71 f., 141
– I 71
– II 71, 162
– III 71
RNA silencing 421
RNA-induced silencing complex (RISC) 27,
  416 ff.
– incorporation 423
RNAfam 276
RNase 25, 127, 240, 424
robustness 304
Rosetta Stone method 323
Roundup 441
royalties 529
rRNA (ribosomal RNA) 25 f., 76 ff., 125
ruby laser 212 f.

s
S phase 188
Saccharomyces cerevisiae 91, 174
– synchronization 190
SAGE, *see* serial analysis of gene expression
salicylic acid 356
salt bridge 319
Sanger-Coulson Method 150, 221
satellite DNA 61
satellite marker 227
scale-free network 299
scanning force microscope (SFM) 200
scanning transmission electron microscope
  (STEM) 198
Schizosaccharomyces pombe 174
SCOP (structural classification of
  proteins) 277
score matrices 285
screening 332, 458 ff.
– high-quality paramounts in screening
  assays 343
– high-throughput 341
– ligand 343
– metagenome library 457
– primary 344
– secondary 344
– strain collection 456

– system 463
– virtual 344
– virtual ligand 343
SDS-PAGE 104
second harmonic generation (SHG) 214
second messenger 36 f.
Seeburg, Peter 512
selection 176, 439, 458 ff., 472
selection marker 163, 172, 398, 438
– gene 440
– negative 440 f.
– positive 440 ff.
– system 439 f.
selection system 380 ff., 443
– *in vitro* selection system 382
selectivity 349
SELEX (systematic evolution of ligand by
  exponential enrichment) 326
self-inactivation (SIN) 409
senile plaques 401
sensory cell 54
separation
– general principle 107
separation assay 343
sequence analysis 279
sequence length polymorphism 231
sequence similarity 333
sequence tagged site (STS) 224 ff.
– marker 249
sequence-specific recombination 157
sequencing
– DNA 149 ff.
– ligation 374
– sequencing by synthesis 151
– ultraparallel 281
serial analysis of gene expression
  (SAGE) 253 f., 288
severe immunodeficiency syndrome
  (SCID) 363
sex chromosome 62
sexual hormone 14
SH1 domain 316
SH2 domain 316
SH3 domain 316
β-sheet structure 17
Shine-Dalgarno sequence 159
short/small hairpin RNA (shRNA) 428
short interspersed element (SINE) 61
short tandem repeat, *see* STR
shotgun approach 151
shotgun sublibrary 234
shrimp alkaline phosphatase (SAP) 142
shuttle principle 174
shuttle vector 164, 176
sialic acid 13
side effect 345 ff.
– prodrug 358
signal peptidase 82 ff.
signal recognition particle (SRP) 85
signal sequence analysis 279 f.
signal transduction 45
– biomembrane 33
signaling network 295
silencer 72
silent mutation 69, 365
silver stain method 106
similarity 280, 290
Sindbis virus 413

single nucleotide polymorphism, *see* SNP
single-cell PCR 184
single-chain Fab fragment (scFab) 389
single-chain Fv fragment (scFv) 382 ff.
single-input module (SIM) 327
siRNA (small interfering RNA) 25 ff., 270,
  416
– biogenesis 422
– biotechnological application 426
– chemical modification 426
– design 426
– mediated underexpression 251
– posttranscriptional repression 424
– screen 334
β-sitosterol 13
size exclusion chromatography 107
slot blot 261
small RNA 434
smallpox virus (vaccinia) 413
SMART 277
Smith-Waterman algorithm 281
SNAP25 87
SNARE protein 86 f.
snoRNA (small nucleolar RNA) 25, 76, 276
SNP (single nucleotide polymorphism) 69,
  152, 231, 250, 281, 336, 365, 481
snRNA (small nuclear RNA) 25, 73
somatic gene therapy
– principle 404
Southern blotting 131, 134, 251
SP6 promoter 157
specialization 54
specificity 265
spectral localization microscopy 205
spectral precision distance microscopy
  (SPDM) 205
spectrally assigned localization microscopy
  (SALM) 205
Spermatophyta 94
spermatozoon 64
spheroblast 166
sphingolipid 12
sphingomyelin 11 f.
sphingosine 12
spindle apparatus 64
spinning (Nipkow)-disk confocal microscope 204
splice acceptor sequence 251
splice site 287
splicing
– alternative 73, 239, 334
– differential 73
spliceosome 239
split pin 264
Spodoptera frugiperda 176
Sporophyta 94
sporozoa 59
spot 258
spotted chip 289
spotting 373
Src homology (SH) domain 316
SRP receptor 85
SSH (suppression subtractive hybridization)
  253 ff.
standard elution method 110
standard optical (bright-field) microscopy 202
Staphylococcus aureus 51
starch 9 f.